SUPERALLOYS 1988

SUPERALLOYS 1988

Proceedings of the Sixth International Symposium on Superalloys sponsored by the High Temperature Alloys Committee of The Metallurgical Society, held September 18-22, 1988, Seven Springs Mountain Resort, Champion, Pennsylvania.

Edited by
D.N. Duhl
G. Maurer
S. Antolovich
C. Lund
S. Reichman

A Publication of The Metallurgical Society, Inc.

A Publication of The Metallurgical Society, Inc.
420 Commonwealth Drive
Warrendale, Pennsylvania 15086
(412) 776-9024

The Metallurgical Society, Inc. is not responsible for statements or opinions and absolved of liability due to misuse of information contained in this publication.

Printed in the United States of America
Library of Congress Catalogue Number 88-62071
ISBN Number 0-87339-076-8

Authorization to photocopy items for internal or personal use, or the internal or personal use of specific clients, is granted by The Metallurgical Society, Inc. for users registered with the Copyright Clearance Center (CCC) Transactional Reporting Service, provided that the base fee of $3.00 per copy is paid directly to Copyright Clearance Center, 27 Congress Street, Salem, Massachusetts 01970. For those organizations that have been granted a photocopy license by Copyright Clearance Center, a separate system of payment has been arranged.

© 1988

Dedication

The Symposium and these proceedings are dedicated to honor the pioneering contribution of Mr. Herbert L. Eiselstein to the superalloy and gas turbine industry.

Preface

The purpose of this International Symposium is to bring together into one forum the most up-to-date technical knowledge on high strength, high temperature alloys commonly known as superalloys. Historically, this Symposium has been a "bench-mark" on what technical directions have been taken and what progress has been made during the four-year intervals between meetings.

The diversity of superalloy technology has greatly expanded during the twenty years this Symposium has been in existence. The initial symposium focused on the alloy problems associated with phase stability. Subsequent symposia highlighted conventional processing, minor element effects, strategic materials, and advanced processing such as single-crystal castings and powder metallurgy. This year's proceedings includes a wide spectrum of papers on superalloys which reflect the present diverse nature of the technology.

One unique aspect of this year's symposium is the attention paid to alternate materials. Awareness of the potential benefits of advanced metal matrix and ceramic composites is particularly useful in defining the full potential for superalloys in the future. There is no doubt that gas turbine engines of the future will require materials with capabilities different than those of current superalloys. Despite this trend, the uses and sophistication of superalloys are expected to grow, creating a more vital class of materials.

Each Seven Springs Symposium and the corresponding volume is dedicated to an individual as a means of honoring that individual for his or her contribution to the superalloy industry. The individual is selected by a long process of nomination, review, and final selection by the Seven Springs Committee and the High Temperature Alloy Committee of TMS. From a list of many individuals who have made major contributions, a dedicatee is selected whose accomplishments are clearly recognized by the entire industry. This Symposium is dedicated to **Herbert L. Eiselstein** for his contributions involving alloy design, development, and processing. During his 30 years at Huntington Alloys, Mr. Eiselstein authored and co-authored numerous publications dealing with alloying effects, heat treatment, and physical metallurgy of superalloys. Patents which bear his name include those for Inconel 718, Inconel 625, and Inconel 903. Inconel 718 is recognized as one of the most important superalloys used in gas turbine applications. The growing use of Inconel 718 in aerospace and other applications is further evidence of the importance of his internationally recognized accomplishments.

Each International Symposium on Superalloys transpires because of the efforts of those individuals who directly worked on this Symposium and those individuals who laid the groundwork at previous symposia. The following Seven Springs Committee represents the formal group which brought together a vast number of details to produce this international forum and to publish a proceedings that will always be a valuable reference to those in the industry.

Committee Members

General Chairman	Gernant Maurer
Secretary	Douglas Deye
Treasurer	Donald Muzyka
Program Chairman	David Duhl
	Carl Lund
	Stephen Antolovich
	Gernant Maurer
	Stephen Reichman
Arrangements Chairman	Robert Stusrud
	Robert Gasior
	Rebecca MacKay
Publication Chairman	Stephen Reichman
International Publicity	Michael Goulette
U.S. Publicity	Richard Menzies
Awards Committee Chairman	Charles Kortovich
	Stanley Wlodek
	Donald Muzyka
	Louis Lherbier
	William Boesch

Special recognition is also deserved by **Susan Floridio** of PWA for her special effort in putting together this book.

Table of Contents

Preface..v

Powder Metallurgical and Wrought Alloys

Development of a Damage Tolerant Microstructure for
Inconel 718 Turbine Disc Material..3
    A.K. Koul, P. Au, N. Bellinger, R. Thamburaj,
    W. Wallace and J-P. Immarigeon

Development of Damage Tolerant Microstructures in
Udimet 720...13
    K.R. Bain, M.L. Gambone, J.M. Hyzak and M.C. Thomas

Effect of Heat Treatment on Mechanical Properties
and Microstructure of Alloy 901..23
    R.B. Frank and R.K. Mahidhara

Metallurgical Stability of Inconel Alloy 718..33
    J.W. Brooks and P.J. Bridges

On Developing a Microstructurally and Thermally
Stable Iron-Nickel Base Superalloy...43
    J.P. Collier, A.O. Selius and J.K. Tien

Microstructure and Properties of Ni-Fe Base Ta-718....................................53
    S.A. Loewenkamp, J.P. Radavich and T. Kelly

N 18, a New Damage Tolerant PM Superalloy for
High Temperature Turbine Discs..63
    C. Ducrocq, A. Lasalmonie and Y. Honnorat

The Development of ODS Superalloys for Industrial
Gas Turbines..73
    R.C. Benn and G.M. McColvin

Abnormal Grain Growth of ODS Superalloys Enhanced
by Boron Doping or Torsional Strain..81
    Y.G. Nakagawa, H. Terashima and K. Mino

Causes and Effects of Center Segregation in Electro-Slag
Remelted Alloy 718 for Critical Rotating Part Applications........................91
    M.D. Evans and G.E. Kruzynski

Development of Gatorized® Merl 76 for Gas Turbine
Disk Applications..101
 R.H. Caless and D.F. Paulonis

HIP Modeling of Superalloy Powders...111
 J.C. Borofka, R.D. Kissinger and J.K. Tien

Dual Structure Turbine Disks Via Partial Immersion
Heat Treatment...121
 J.M. Hyzak, C.A. Macintyre and D.V. Sundberg

Development of Inconel Alloy MA 6000 Turbine Blades
for Advanced Gas Turbine Engine Designs...131
 B.A. Ewing and S.K. Jain

The Production of Advanced Turbine Blades from P/M
Superalloy Forging Stock..141
 I.C. Elliott, C. Cockburn and S.W.K. Shaw

The Physical Metallurgy of a Silicon-Containing Low
Expansion Superalloy..151
 K.A. Heck, D.F. Smith, J.S. Smith, D.A. Wells
 and M.A. Holderby

Determination of the Effects of Cooling Rate from
Solution Treatment on the Microstructure and Mechanical
Properties of a Precipitation Strengthened, Low Thermal
Expansion Alloy..161
 E.A. Wanner, D.A. DeAntonio and R.K. Mahidhara

Alternative Materials

Beyond Superalloys: The Goals, the Materials and
Some Reality...173
 C.T. Sims

Status and Prognosis for Alternative Engine Materials....................................183
 J.R. Stephens and M.V. Nathal

Reaction Kinetics Between Fiber and Matrix Components
in Metal Matrix Composites..193
 M.W. Kopp, J.K. Tien and D.W. Petrasek

Industrial Scale Processing and Elevated Temperature
Properties of Ni_3Al-Cr-Zr-B Alloys..203
 V.K. Sikka and E.A. Loria

Directionally Solidified/Single Crystal Alloys

Creep Deformation Anisotropy in Single Crystal
Superalloys 215
    P. Caron, Y. Ohta, Y.G. Nakagawa and T. Khan

High Performance Single Crystal Superalloys Developed
by the d-Electrons Concept 225
    N. Yukawa, M. Morinaga, Y. Murata, H. Ezaki
    and S. Inoue

Second-Generation Nickel-Base Single Crystal
Superalloy 235
    A.D. Cetel and D.N. Duhl

Enhanced Rupture Properties in Advanced Single
Crystal Alloys 245
    S.M. Foster, T.A. Nielsen and P. Nagy

Formation of Topologically Closed Packed Phases in
Nicle Base Single Crystal Superalloys 255
    R. Darolia, D.F. Lahrman, R.D. Field and R. Sisson

The Influence of High Thermal Gradient Casting, Hot
Isostatic Pressing and Alternate Heat Treatment on the
Structure and Properties of a Single Crystal Nickel
Base Superalloy 265
    L.G. Fritzemeier

The Effect of Temperature on the Deformation Structure
of Single Crystal Nickel Base Superalloys 275
    M. Dollar and I.M. Bernstein

Intermediate Temperature Creep Deformation in CMSX-3
Single Crystals 285
    T.M. Pollock and A.S. Argon

The Effect of Hydrogen on the Deformation and
Fracture of PWA 1480 295
    W.S. Walston, N.R. Moody, M. Dollar, I.M. Bernstein
    and J.C. Williams

An Atom-Probe Study of Some Fine-Scale Microstructural
Features in Ni-Based Single Crystal Superalloys 305
    D. Blavette, P. Caron and T. Khan

Solid Solution Strengthening of Ni$_3$Al Single Crystals
by Ternary Additions 315
    F.E. Heredia and D.P. Pope

High Speed Single Crystal Casting Technique..........325
    S. Morimoto, A. Yoshinari and E. Niyama

Effect of Chemistry Modifications and Heat Treatments
on the Mechanical Properties of DS MAR-M200 Superalloy..........335
    Z. Yunrong, W. Yuping, X. Jizhou, P. Caron and
    T. Khan

A Hafnium-Free Directionally Solidified Nickel-Base
Superalloy..........345
    D.L. Lin, S. Huang and C. Sun

The Processing and Testing of a Hollow DS Eutectic
High Pressure Turbine Blade..........355
    R.G. Menzies, C.A. Bruch, M.F. Gigliotti, J.A. Smith
    and R.C. Haubert

Advances in Processing

Superalloy Recycling 1976-1986..........367
    J.F. Papp

The Investigation of Minor Element Additions on Oxide
Filtering and Cleanliness of a Nickel Based Superalloy..........377
    S.O. Mancuso, F.E. Sczerzenie and G.E. Maurer

Evaluation of Electron-Beam Cold Hearth Refining (EBCHR)
of Virgin and Revert IN738LC..........387
    P.N. Quested, M. McLean and M.R. Winstone

Electron Beam Cold Hearth Refinement Processing of
Inconel Alloy 718 and Nimonic Alloy PK50..........397
    S. Patel, I.C. Elliott, H. Ranke and H. Stumpp

The Magnesium Problem in Superalloys..........407
    A. Mitchell, M. Hilborn, E. Samuelsson and
    A. Kanagawa

Liquid Metal Treatments to Reduce Microporosity in
Vacuum Cast Nickel Based Superalloys..........417
    R.E. Painter and J.M. Young

Compositional Control and Oxide Inclusion Level
Comparison of Pyromet® 718 and A-286 Ingots
Electroslag Remelted Under Air vs. Argon Atmosphere..........427
    D.D. Wegman

Determination of the Solidification Behaviour of Some
Selected Superalloys..........437
    U. Heubner, M. Köhler and B. Prinz

Effects of Chemistry on Vader Processing of Nickel
Base Superalloys..449
   P.W. Keefe, F.E. Sczerzenie and G.E. Maurer

Effect of HIP Parameters on Fine Grain Cast Alloy 718..................................459
   P. Siereveld, J.F. Radavich, T. Kelly, G. Cole
   and R. Widmer

History of Cast Inco 718...469
   O.W. Balloy and M.W. Coffey

Skin Effect of Hf-Rich Melts and Some Aspects in its
Usage for Hf-Containing Cast Nickel-Base Superalloys..................................475
   Z. Yunrong and L. Chenggong

Spray-Formed High-Strength Superalloys..485
   K.-M. Chang and H.C. Fiedler

Evaluation of the Potential of Low Pressure Plasma
Spraying and Simultaneous Spray Peening for Processing
of Superalloys...495
   J.V. Wright and J.E. Restall

The Deformation Behavior of P/M Rene '95 Under
Isothermal Forging Conditions...505
   J.M. Morra, R.R. Biederman and F.R. Tuler

Utilization of Computer Modeling in Superalloy
Forging Process Design..515
   T.E. Howson and H.E. Delgado

Property Optimization in Superalloys Through the Use
of Heat Treat Process Modelling...525
   R.A. Wallis and P.R. Bhowal

Processing of High Strength Superalloy Components
from Fine Grain Ingot..535
   P.D. Genereux and D.F. Paulonis

Isocon Manufacturing of Waspaloy Turbine Discs..545
   D.P. Stewart

Laser Drilling of a Superalloy Coated with Ceramic.......................................553
   P. Forget, M. Jeandin, P. Lechervy and D. Varela

Microstructure and Mechanical Behavior

The Effect of Microstructure on the Fatigue Crack
Growth Resistance of Nickel Base Superalloys..565
   R. Bowman and S.D. Antolovich

Isothermal and "Bithermal" Thermomechanical Fatigue
Behavior of a NiCoCrAlY-Coated Single Crystal Superalloy 575
   J. Gayda, T.P. Gabb and R.V. Miner

Effects of Aging on the LCF Behavior of Three Solid-
Solution-Strengthened Superalloys 585
   D.L. Klarstrom and G.Y. Lai

Oxide Dispersion Strengthened Superalloys: The Role
of Grain Structure and Dispersion During High
Temperature Low Cycle Fatigue 595
   D.M. Elzey and E. Arzt

Observations of Microstructural and Geometrical
Influences on Fatigue Crack Growth in Single Crystal
and Polycrystal Nickel-Base Alloys 605
   D.C. Wu, D.W. Cameron and D.W. Hoeppner

Accelerated Fatigue Crack Growth Behavior of PWA 1480
Single Crystal Alloy and its Dependence on the Deformation
Mode 615
   J. Telesman and L.J. Ghosn

Creep Behavior of Magnesium Microalloyed Wrought
Superalloys 625
   P. Ma, Y. Yuan and Z. Zhong

The Role of Mg on Structure and Mechanical Properties
in Alloy 718 635
   X. Xie, Z. Xu, B. Qu, G. Chen and J.F. Radavich

Torsional Creep of Alloy 617 Tubes at High Temperature 643
   H.J. Penkalla, F. Schubert and H. Nickel

Identification of Mechanisms Responsible for
Degradation in Thin-Wall Stress-Rupture Properties 653
   M. Doner and J.A. Heckler

Microstructural Development Under the Influence of
Elastic Energy in Ni-Base Alloys Containing γ' Precipitates 663
   M. Doi and T. Miyazaki

New Interpretation of Rupture Strength Using the
Potential Drop Technique 673
   I. Vasatis

A Model Based Computer Analysis of Creep Data (Crispen):
Applications to Nickel-Base Superalloys 683
   *A. Barbosa, N.G. Taylor, M.F. Ashby, D.F. Dyson
   and M. McLean*

Effect of Minor Elements on the Deformation Behavior of
Nickel-Base Superalloys..693
    D.M. Shah and D.N. Duhl

Superalloys with Low Segregation..703
    Z. Yaoxiao, Z. Shunnan, X. Leying, B. Jing, H. Zhuangqi
    and S. Changxu

Relation Between Chemistry, Solidification Behaviour,
Microstructure and Microporosity in Nickel-Base Superalloys.........................713
    J. Lecomte-Beckers

Phase Equilibria in Multicomponent Alloy Systems...723
    P. Willemin and M. Durand-Charre

Phase Calculation and its Use in Alloy Design Program
for Nickel-Base Superalloys..733
    H. Harada, K. Ohno, T. Yamagata, T. Yokokawa
    and M. Yamazaki

Repair, Post-Service Evaluation and Environmental Behavior

Aircraft Gas Turbine Blade and Vane Repair..745
    K.C. Antony and G.W. Goward

Rejuvenation of Service-Exposed in 738 Turbine Blades.................................755
    A.K. Koul, J-P. Immarigeon, R. Castillo, P. Lowden
    and J. Liburdi

Application of Melt-Spun Superalloy Ribbons to Solid Phase
Diffusion Welding for Ni-Base Superalloy...765
    K. Yasuda, M. Kobayashi, A. Okayama, H. Kodama,
    T. Funamoto and M. Suwa

Theoretical Research on Transient Liquid Insert Metal
Diffusion Bonding of Nickel Base Alloys..775
    Y. Nakao, K. Nichimoto, K. Shinozaki and
    C. Kang

Repair Weldability Studies of Alloy 718 Using Versatile
Varestraint Test..785
    C.P. Chou and C.H. Chao

Intergranular Sulfur Attack in Nickel and Nickel-Base Alloy...........................795
    J.P. Beckman and D.A. Woodford

The Effect of Service Exposure on the Creep Properties of
Cast IN-738LC Subjected to Low Stress High Temperature
Creep Conditions...805
    R. Castillo, A.K. Koul and J-P.A. Immarigeon

Degradation of Aluminide Coated Directionally Solidified
Superalloy Turbine Blades in an Aero Gas Turbine Engine...............................815
   P.C. Patnaik, J.E. Elder and R. Thamburaj

High Temperature Corrosion Fatigue and Grain Size Control
in Nickel-Base and Nickel-Iron-Base Superalloys......................................825
   M. Yoshiba and O. Miyagawa

Quantitative Microstructure Analysis to Determine
Overheating Temperatures in IN100 Turbine Blades..................................835
   H. Huff and H. Pillhöfer

Non-Destructive Analysis by Small Angle Neutron
Scattering..845
   P. Bianchi, F. Carsughi, D. D'Angelo, M. Magnani,
   A. Olchini, M. Stefanon and F. Rustichelli

Influence of Thermal Fatigue on Hot Corrosion of an
Intermetallic Ni-Aluminide Coating...855
   J.W. Holmes and F.A. McClintock

An Investigation on Magnetron Sputter Deposited Alloy-
Oxide Coating..865
   Y. Ruizeng, Z. Lang, C. Shouhua, G. Lian and L. Fanxiu

Subject Index...873

Alloy Index...879

Author Index...881

Powder Metallurgical and Wrought Alloys

DEVELOPMENT OF A DAMAGE TOLERANT MICROSTRUCTURE
FOR INCONEL 718 TURBINE DISC MATERIAL

A.K. Koul*, P. Au*, N. Bellinger**, R. Thamburaj***,
W. Wallace* and J-P. Immarigeon*

*Structures and Materials Laboratory
National Aeronautical Establishment
National Research Council of Canada
Ottawa, Canada K1A 0R6

**Dept. of Mechanical and Aeronautical Eng.
Carleton University, Ottawa, Ontario, Canada

***Orenda Division of Hawker Siddeley Canada Inc.
Toronto, Ontario, Canada

SUMMARY

A new modified heat treatment has been developed for Inconel 718. This heat treatment leads to substantial improvements in elevated temperature crack propagation resistance with apparently limited loss in resistance to LCF crack initiation as compared to the conventional heat treatment for this alloy. This is a result of tailoring the microstructure to obtain the optimum combination of grain size, grain boundary structure and matrix precipitate morphology.

In the modified heat treatment, the material is solution treated at 1032°C/1h (below the grain coarsening temperature) then furnace cooled to 843°C and held for 4h to produce profusely serrated grain boundaries by precipitating along the boundaries orthorhombic δ-Ni_3Nb needles, hereafter referred to as the Ni_3Nb phase. After this, the material is partially solution treated at 926°C/1h, to dissolve the coarse intragranular γ'' precipitates (also of the basic composition Ni_3Nb but having a body centred tetragonal structure) that previously formed during furnace cooling from 1032°C to 843°C/4h. Finally the material is subjected to the conventional double aging heat treatment.

Relative to the conventional heat treatment, the new heat treatment reduces the FCGRs and CCGR of Inconel 718 by a factor of 2 and 5 respectively at 650°C. The new heat treatment does not alter the LCF life as a function of total strain relative to the conventionally heat treated material at 650°C. This is a significant result if damage tolerance concepts are used in turbine disc design.

Introduction

Inconel 718 is used as disc alloy in a number of gas turbine engines. Disc forgings of Inconel 718 are generally solution treated (ST) at 955°C/1h then direct aged by cooling from this temperature to 718°C and held at 718°C for 8h, further cooled to 621°C and held at 621°C for 8h and finally air cooled (AC). This heat treatment has proved adequate from a safe life point of view, where disc life limits are established statistically on the basis of the number of low cycle fatigue (LCF) cycles required to form a detectable crack (~ 0.8 mm) in 1 in 1000 components. However, for those cases where damage tolerance design requirements must be satisfied, the conventional heat treatment may not be adequate since the crack growth rates may be too high to obtain a practical safe inspection interval (SII). In damage tolerance design, the fracture critical locations of discs are assumed to contain defects of a size corresponding to the detection limit of the nondestructive inspection (NDI) technique used to inspect the components. These defects are then assumed to act as propagating cracks and their rates of propagation are established on the basis of fracture mechanics principles, using experimental crack growth rate data.[1] The time or number of cycles to grow these inherent cracks to a predetermined dysfunction size are then used to establish a SII on the basis of which discs are repeatedly returned to service until a crack is eventually detected, Figure 1. If the crack propagation rates are excessively high, the SII may prove too short to be economically viable.

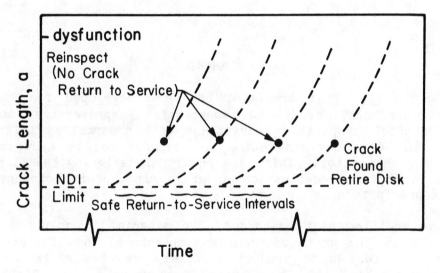

Fig. 1. Schematic representation of the damage tolerance based life prediction methodology.

Table 1. Chemical Compositions of two Inconel 718 heats in wt. %.

Stock Type	C	Si	Cr	Ni	Mo	Nb + Ta	Ti	Al	Fe
Bar	0.03	1.12	18.3	53.2	3.0	5.17	1.0	0.42	Bal
Plate	0.05	0.21	17.9	53.1	3.06	5.11	0.96	0.47	Bal

This paper describes work aimed at developing a heat treatment for Inconel 718 which reduces elevated temperature creep and fatigue crack growth rates (CCGR and FCGR) without substantial sacrifice in LCF crack initiation life relative to conventional microstructure.

A number of heat treatments have already been developed for improving the fracture toughness[2], notch rupture ductility[3], notch stress rupture properties [4,5] and CCGRs and FCGRs[6,7] of Inconel 718. However, none of these heat treatments are capable of producing a microstructure that improves CCGR and FCGR without sacrificing LCF life. It is also clear that microstructural variables such as the grain size, the grain boundary Ni_3Nb morphology and the size of γ'' precipitates have to be controlled for optimizing these properties.[7] The following strategy was therefore adopted for developing a damage tolerant microstructure for Inconel 718 discs. This included:

(i) selecting a solution treatment temperature for achieving full solutioning of the Ni_3Nb, γ' and γ'' precipitates without inducing excessive grain growth which can reduce LCF crack initiation life,

(ii) developing a post solutioning heat treatment sequence for reprecipitating the Ni_3Nb needles along the grain boundaries to form serrated grain boundaries which suppress grain boundary sliding and

(iii) developing an additional heat treatment sequence for precipitating the optimum amounts of γ' and γ'' precipitates to strengthen the grain interiors without altering the serrated grain boundary structures.

Experimental Materials and Methods

The commercially available hot rolled Inconel 718 was procured in the form of 22 mm diameter bars and 12.7 mm thick x 50.8 mm wide plates. The bar and the plate stock were from two different heats and their chemical compositions are given in Table 1.

Heat Treatments and Microscopy

The solution treatments were carried out at 1032°C, 1050°C, 1066°C, 1080°C and 1093°C for 1 to 16h. The grain boundary Ni_3Nb precipitation kinetics were monitored over a range of aging temperatures (818°C to 917°C) for starting solution treatment conditions of 1032°C/1h, 1066°C/1h and 1080°C/1h. In this case, the specimens were solution treated, furnace cooled (4 to 7°C/min) and direct aged for 1 to 6h. Transmission electron microscopy, using replica and thin foil techniques, was carried out on a selected number of solution-treated and direct aged and direct aged plus partial solution treated specimens.

A series of heat treatments was also conducted on a selected number of specimens that were direct aged in the Ni_3Nb precipitation range in order to solution the overaged γ' and γ'' precipitated during direct aging. These heat treatments were carried out at 917°C, 926°C, 955°C and 975°C for 1 to 10h.

Low Cycle Fatigue Testing

ASTM E606 axial fatigue specimens were used for conducting fully reversed, constant amplitude LCF tests in a closed loop electrohydraulic testing system under total axial strain control using a triangular wave form and a constant strain rate of 0.002/s. The test section of each specimen was polished manually in the axial direction with successively finer grit emery papers (grade 320, 400, 600) to remove circumferencial machining marks. The specimens were tested over a strain range ($\Delta\epsilon$) of 0.65 to 2% at 650°C in a laboratory air environment.

An x-y recorder was used to obtain cyclic total axial strain versus load plots, which were converted to engineering stress-strain hysteresis loops using the specimen cross sectional area. Specimen failure, N_f, was defined by the number of cycles for a 5% drop in the steady state tensile stress value.

Creep and Fatigue Crack Growth Rate (CCGR and FCGR) Testing

All CCGR and FCGR tests were conducted in an electrohydraulic testing system. Initially, a tapered double cantilever beam (DCB) fracture mechanics specimen, having a constant K region over 31.75 mm,[7] was used for CCGR and FCGR testing at a stress intensity factor and range (K and ΔK respectively) of 45 MPa $\sqrt{m}$ in laboratory air environment at 650°C. An R-value of 0.1 and a frequency of 0.1 Hz were selected for FCGR testing using a sine wave form. Fracture surfaces were studied by scanning electron microscopy (SEM) to determine the crack growth rates.[7]

Statistically significant FCGR data bases were further generated, using standard 50.8 mm wide and 12.7 mm thick compact tension (CT) specimens conforming to ASTM E647 specifications, at 650°C in a laboratory air environment. The CT specimens were precracked at room temperature and the FCGR tests were conducted at an R-value of 0.1 and a frequency of 1 Hz using a sawtooth wave form. A direct current potential drop (DC-PD) technique having an accuracy of 0.085 mm and a precision of 0.025 mm was used to monitor the crack lengths at 650°C. The FCGR data was generated over a ΔK range of 24 to 80 MPa $\sqrt{m}$.

Results and Discussion

Solution Treatment Selection

Figure 2 shows a plot of average grain size versus solutioning temperature for a range of solution treatment times. It is noted that excessive grain coarsening commences between 1040° and 1050°C and this is because primary NbC precipitates (solvus 1040-1093°C) begin to dissolve in this temperature range. The γ'' and Ni_3Nb solvus temperatures for Inconel 718 are 900°C and 982 to 1037°C respectively. A solutioning temperature of 1032°C will prove optimum in achieving full solutioning of the Ni_3Nb, γ' and γ'' precipitates without inducing excessive grain growth. A solution treatment time of 1h is considered adequate for homogenizing the microstructure. Longer solution treatment times could increase the grain size through Ostwald ripening of NbC precipitates even if the solution treatment temperature is kept below the NbC solvus temperature, Figure 2.

Selection of a Direct Aging Treatment to Form Serrated Grain Boundaries

In Inconel 718, the precipitation of the grain boundary Ni_3Nb needles may under certain optimum conditions create a serrated grain boundary structure.[7] In this case, the serrations arise from the cellular precipitation of Ni_3Nb following the Tu-Turnbull mechanism instead of the motion of heterogeneous γ' precipitates (a mechanism suggested to produce serrations in high γ' volume fraction alloys).[8] Serrations due to the latter mechanism exhibit well-rounded peaks and valleys whereas serrations arising from Ni_3Nb needles are angular in nature. It is therefore important to establish the kinetics of Ni_3Nb precipitation in Inconel 718.

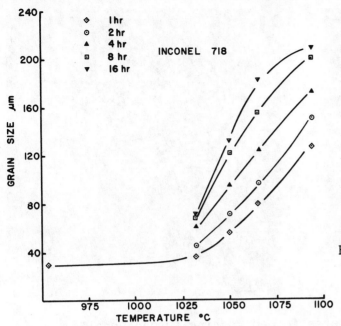

Fig. 2. Grain coarsening behaviour of Inconel 718

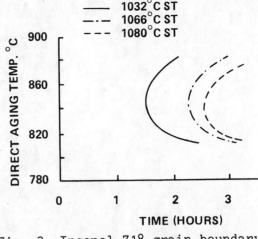

Fig. 3. Inconel 718 grain boundary Ni_3Nb time-temperature-precipitation (TTP) curves for samples solution treated (ST) at 1032°C, 1066°C and 1080°C.

A number of studies have documented the general kinetics of Ni_3Nb precipitation as a function of solution treatment and aging temperatures.[6] However, there is little or no information available for direct aging conditions where the material is slow cooled from the solutioning to the intermediate aging temperature and held at the aging temperature to produce grain boundary Ni_3Nb needles and serrations. For solution treatment temperatures of 1032°, 1066° and 1080°C, the time-temperature-precipitation (TTP) curves for a range of direct aging conditions are presented in Figure 3. It is somewhat surprising to note that lower solution temperatures enhance Ni_3Nb precipitation upon direct aging. Similar results have also been reported by other workers.[4]

It has been suggested that a lower solution treatment temperature produces a smaller grain size and a larger grain boundary area which accelerates the nucleation kinetics of Ni_3Nb needles.[4] Mechanistically, however, nucleation rate is sensitive to the solution treatment temperature only when quenching to the aging temperature is rapid and the equilibrium mole fraction of solute adsorbed at the grain boundaries varies markedly from one solution treatment condition to another.[9] Neither of these conditions are entirely satisfied in the present experiments because cooling to direct aging temperature is slow and the solution treatment temperatures lie close to or above the Ni_3Nb solvus thus minimizing the grain boundary solute concentration differences within the temperature range studied. It is possible that the differences in the Ni_3Nb TTP-curves in Figure 3 are instead related to the differences in the growth kinetics of Ni_3Nb during direct aging. From 1080°C, the cooling rate to the direct aging temperatures might be relatively faster than from 1032°C resulting in somewhat higher point defect densities in the 1080°C solution treated specimens. A higher defect density would precipitate a larger volume fraction of heterogeneous γ'' during direct aging thus relieving the matrix Nb super-saturation for grain boundary Ni_3Nb precipitation and vice versa, Figure 4. Therefore, the differences in the defect density and Nb supersaturation from one solution temperature to another may be responsible for delaying the growth of Ni_3Nb precipitates with increasing solutioning temperature during direct aging.

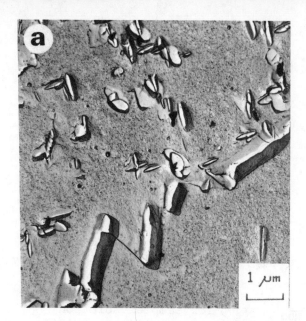 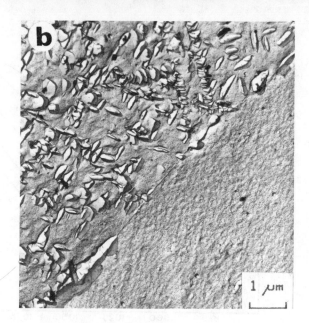

Fig. 4. Coarse γ'' precipitates formed during furnace cooling from the solution treatment temperature to the direct aging temperature of 843°C. (a) ST 1066°C (b) ST 1080°C.

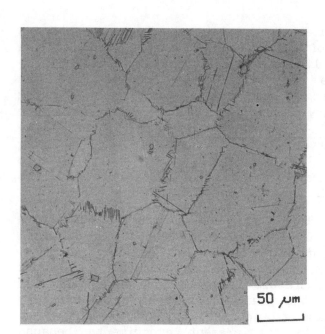 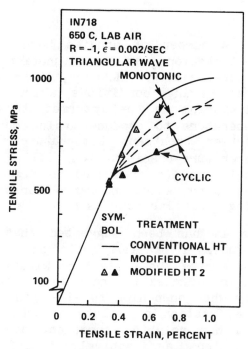

Fig. 5. Serrated grain boundary morphology in specimens solution treated at 1032°C/1h F.C.→ 843°C/4h/AC+926°C/1hr F.C.→ 718°C/8h F.C.→ 621°C/8h/AC.

Fig. 6. Monotonic and cyclic stress-strain data for specimens subjected to conventional and modified heat treatments.

The nose of the Ni_3Nb TTP-curves lies in the vicinity of 843°C in all cases, Figure 3. Direct aging at 818°C formed short Ni_3Nb needles with limited serrations at the grain boundaries whereas direct aging at 917°C led to Laves phase precipitation. A direct aging treatment of 843°C/4h produced profuse serrations, Figure 5, whereas longer aging times led to intragranular Ni_3Nb precipitation. Therefore, a direct aging treatment of 843°C/4h was selected for inducing a serrated grain boundary structure.

Selection of Aging Treatments for Precipitating Optimum Amounts of γ' and γ''

All direct aging treatments that produce serrated grain boundaries also induce coarsening of matrix γ' and γ'' precipitates, Figure 4, which may be harmful to LCF life. Therefore, it is necessary to introduce a partial solution treatment in order to dissolve the coarsened matrix γ' and γ'' precipitates and reprecipitate them in the optimum morphology without dissolving the grain boundary Ni_3Nb precipitates and straightening the serrated grain boundaries. After direct aging, partial solutioning at 917°C did not completely dissolve the coarse γ'' precipitates whereas at 955°C it led to excessive growth of Ni_3Nb needles while at 975°C it started to dissolve the Ni_3Nb needles. Excessive Ni_3Nb needle growth should be avoided because it removes Nb (the main γ'' forming constituent) from the grains thus decreasing the matrix strength. A partial solution treatment of 926°C/1h was thus deemed adequate for solutioning the coarse γ' and γ'' precipitates formed during direct aging. Following the suggestion of other workers,[4,7] it was further decided to reprecipitate γ' and γ'' through a standard double aging treatment, i.e. 718°C/8h F.C. → 621°C/8h/AC.

Two modified heat treatment schedules were selected to assess which would provide the best balance of mechanical properties relative to the conventional heat treatment, Table 2. A typical microstructure of the modified heat treated Inconel 718 is shown in Figure 5.

Mechanical properties

The modified heat treatments decrease the Inconel 718 yield strength relative to the conventional heat treatment, Figure 6 and Table 2. These trends are not unexpected because heavy grain boundary Ni_3Nb precipitation in modified heat treated materials removes some Nb (the element responsible for γ'' precipitation) from the matrix which leads to a decrease in the matrix strength.

Under LCF conditions, all materials hardened initially and then softened but cyclic softening was more pronounced in conventionally heat treated specimens. Upon plotting LCF life as a function of plastic strain range, Figure 7(a), it is evident that both modified heat treatments reduced the Inconel 718 LCF life by 40 to 50%. The superior LCF life (in terms of plastic strain range) of the conventionally heat treated specimens can be attributed to their finer grain sizes, Table 3, which promote homogeneous deformation and retard crack nucleation by reducing stress concentrations. Upon plotting the LCF data as a function of total strain range, Figure 7(b), all data fall within experimental scatter. In terms of disc LCF life, the transition fatigue life (N_t), i.e. LCF life where elastic and plastic strains are equal, is an important parameter.[10] For LCF life greater than N_t elastic strain predominates whereas for LCF life lower than N_t plastic strain predominates. The N_t values for the conventional and modified heat treated materials were of the order of 300 and 15 cycles respectively, Table 3. Typically turbine discs are designed to have a safe life of 10,000 cycles which is considerably greater than these N_t values. It is therefore likely that the total strain LCF data would be used to predict LCF lives of Inconel 718 turbine discs. It can thus be concluded that relative to the conventional heat treatment the modified heat treatments will not alter the LCF life of Inconel 718 discs at 650°C.

Table 2. Grain size and yield strength data for Inconel 718

Heat Treatment	Heat Treatment Schedule	Grain size μm	0.02% Proof Stress MPa
Conventional	955°C/h $\xrightarrow{F.C.}$ 718°C/8h $\xrightarrow{F.C.}$ 621°C/ 8h/AC	20-40	860
Modified H.T.1	1032°C/1h $\xrightarrow{F.C.}$ 843°C/4h+926°C/1h $\xrightarrow{F.C.}$ 718°C/8h $\xrightarrow{F.C.}$ 621°C/8h/AC	35-80	680
Modified H.T. 2	1032°C/1h $\xrightarrow{F.C.}$ 843°C/4h + 926°C/1h/AC + 718°C/8h $\xrightarrow{F.C.}$ 621°C/8h/AC	35-80	790

Table 3. LCF and constant K CCGR and Constant ΔK FCGR data at 650°C

Heat Treatment	Transition strain range in LCF	LCF life at transition strain in cycles	CCGR is mm/h at 45 MPa$\sqrt{m}$	FCGR in mm/cycle x 10^{-3} 45 MP$\sqrt{m}$
Conventional	0.83	300	2.67	3.0
Mod. H.T. 1	1.40	15	0.52	1.7
Mod. H.T. 2	1.20	15	0.58	1.8

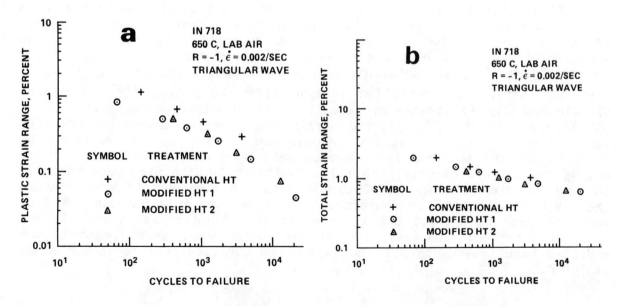

Fig. 7. LCF life as a function of (a) plastic strain range and (b) total strain range for Inconel 718 at 650°C.

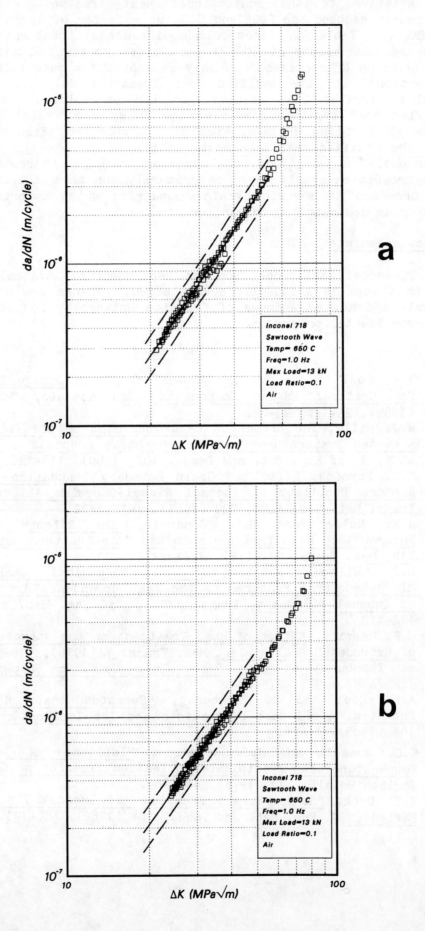

Fig. 8. FCGR data for Inconel 718 at 650°C, (a) conventional heat treatment and (b) modified heat treatment No. 2.

Relative to the conventional heat treatment both modified heat treatments reduced the FCGR and CCGR by a factor of 2 and 5 respectively at 45 MPa $\sqrt{m}$, Table 3. Since both modified heat treatments revealed similar CCGR and FCGR values at 45 MPa $\sqrt{m}$ under constant K and ΔK conditions, FCGR data on CT specimens was only generated for materials subjected to the conventional and the modified heat treatment No. 2. The CT specimen results revealed that relative to the conventional heat treatment the modified heat treatment improved the FCGR by a factor of 2 over a ΔK range of 30 to 80 MPa $\sqrt{m}$, Fig. 8. Furthermore, Figure 8 also indicates that the modified heat treatment might also improve the fracture toughness of Inconel 718. At 650°C, the superior crack growth resistance of modified heat treated materials could be primarily due to a coarser grain size and the presence of serrated grain boundaries which suppress grain boundary sliding as demonstrated in earlier studies.[7]

Acknowledgements

Financial assistance from the Department of National Defence, Canada, for this work is gratefully acknowledged. Thanks are also due to Mr. W. Doswell and Mr. R. Andrews of Carleton University for conducting parts of the experimental programme.

References

1. A.K. Koul et al, Proc. AGARD-SMP Conf. on Damage Tolerance Concepts for Critical Engine Components, San Antonio, Texas, AGARD-CP-393 (1985), 23-1 to 23-22.
2. W.J. Mills, The Effect of Heat Treatment on the Room Temperature and Elevated Temperature Fracture Toughness Response of Alloy 718, Trans ASME, J. of Eng. Mat. and Tech., 102 (1980), 118-126.
3. E.L. Raymond, Effect of Grain Boundary Denudation of γ' on Notch-Rupture Ductility of Inconel Nickel-Chromium Alloys X-750 and 718, Trans. Met. Soc. AIME, 239 (1967), 1415-1422.
4. J.F. Muller and M.F. Donachie, The Effects of Solution and Intermediate Heat Treatments on the Notch Rupture Behaviour of Inconel 718, Met. Trans, 6A(1975), 2221-2277.
5. D.J. Wilson, Relationship of Mechanical Characteristics and Microstructural Features to the Time Dependent Edge Notch Sensitivity of Inconel 718 Sheet, Trans ASME, J. of Eng. Mat. and Tech., (1973), 112-123
6. H.F. Merrick, Effect of Heat Treatment on the Structure and Properties of Extruded P/M Alloy 718, Met. Trans. 7A(1976), 505-514.
7. R. Thamburaj et al., Proc. Int. Conf. on Creep, Tokyo, Japan, JSME/ASME, 1986, 275-282.
8. A.K. Koul and R. Thamburaj, "Serrated Grain Boundary Formation Potential of Ni-Base Superalloys and Its Implication," Met. Trans., 16A (1985), 17-26.
9. K.C. Russell and H.S. Aaronson, "Influence of Solution Annealing Temperature Upon Precipitate Nucleation Kinetics at Grain Boundaries," Scripta Metall., 10 (1976), 463-469.
10. T.S. Cook, "Stress-Strain Behaviour of Inconel 718 during Low Cycle Fatigue," J. Eng. Mat. and Tech., 104 (1982), 186-191.

DEVELOPMENT OF DAMAGE TOLERANT MICROSTRUCTURES

IN UDIMET 720

K. R. Bain[*], M. L. Gambone[*], J. M. Hyzak[**], and M. C. Thomas[*]

[*]Allison Gas Turbine Division of G.M.C.
P.O. Box 420, Indianapolis, IN 46206

[**]Wyman-Gordon Company
Worcester Street
North Grafton, MA 01536

Abstract

Developments in the design of aerospace turbine engines are changing the material service requirements. In particular, the incorporation of damage tolerant design concepts has increased the need for high strength turbine disk alloys with good high temperature crack growth properties. A program was initiated to develop a crack growth resistant microstructure in the high strength nickel-base superalloy UDIMET[*] 720. This alloy has excellent tensile strength and fatigue strength when produced with a fine grained microstructure (ASTM 8-12). Several alternative structures with coarser grain sizes were produced via heat treatment and supersolvus forging which exhibited significant improvements in high temperature fatigue crack growth behavior. These alternative microstructures were evaluated for their effect on a variety of mechanical properties including tensile, stress-rupture, low cycle fatigue, and fatigue crack growth at several temperatures. The effects of gamma prime size and morphology along with grain boundary structure are also related to the observed behavior.

* UDIMET is a registered trade name of the Special Metals Corporation.

Superalloys 1988
Edited by S. Reichman, D.N. Duhl,
G. Maurer, S. Antolovich and C. Lund
The Metallurgical Society, 1988

Introduction

UDIMET 720 alloy is high strength corrosion resistant superalloy which is strengthened by gamma prime, $Ni_3Ti(Al)$. A review of the development of the UDIMET 720 alloy was given by Sczerzenie and Maurer(1). The alloy was developed for use in steam turbine blade applications but has recently been recognized for its outstanding strength and fatigue resistance when used with a fine grain size in gas turbine disk applications. With billet conversion and forging aimed at producing a fine grain (ASTM 8-12) microstructure, UDIMET 720 readily achieves average property levels of 220 ksi ultimate strength and 170 ksi yield strength at 1000°F. High temperature low cycle fatigue resistance is also very good in the fine grain condition with fatigue strength at or better than RENE-95 when corrected for the lower density of UDIMET 720 (.292 lb./in.^{3}). The measured fatigue crack growth rate of UDIMET 720 alloy in the fine grain condition at 1200°F with a 5 minute dwell at maximum load is 15x slower than PM/Extruded and Isothermally forged RENE-95(2,3). This improvement is expected when one considers the research by Whitlow(4) which indicates that the chemistry and grain boundary structure of UDIMET 720 inhibits diffusion of oxygen along the grain boundaries which accelerates intergranular crack growth at elevated temperatures.

In this program the benefits in high temperature fatigue crack growth observed in the fine grain version of UDIMET 720 are extended by producing microstructures which can further reduce the fatigue crack growth rates of the alloy at elevated temperatures. Considerable research has indicated that increasing the grain size of an alloy will dramatically reduce the high temperature fatigue crack growth rate of nickel-base superalloys (2,5-7). Increasing the grain size reduces the tensile and low cycle fatigue strength of the alloy. Therefore, the goal becomes to develop a process and structure which give the best compromise between the high strength of UDIMET 720 and improved fatigue crack growth resistance. Such a material will have considerable application in damage tolerant designs which must operate at temperatures at or above 1200°F, such as advanced gas turbine disks. The present paper reports on the progress made to date in developing the UDIMET 720 alloy for just such an application.

Experimental Results

The following sections will detail the processing, microstructure and experimental results obtained in this program.

Material

A total of five different microstructures were produced and evaluated in this program. The structures were obtained by controlled forging and heat treatment of the UDIMET 720 material. The chemistry is given below:

UDIMET 720 Chemistry

Ni	Cr	Co	Mo	W	Ti	Al	C	B	Zr
Bal.	17.8	14.5	2.98	1.21	4.94	2.45	.032	.032	.030

This material was vacuum induction melted and vacuum arc remelted, homogenized and forged to billet, to give a uniform fine grain size of ASTM 8. The fine grain billet was forged on hot dies at a subsolvus temperature. The gamma prime solvus for this alloy, as determined by differential thermal analysis, is generally 2090°F. The coarser grain microstructures were produced either by forging in the supersolvus regime or by supersolvus heat treatment of fine grain forgings. For this study, the supersolvus forgings were produced isothermally. One necklace structure was also produced. The forging stock was first supersolvus annealed to coarsen the grain size, then subsolvus forged on hot dies. The resultant microstructure showed large warm worked grains outlined by a necklace of fine recrystalized grains. The processing and heat treatment is summarized in Table 1. Representative photomicrographs of the structures produced are shown in Figure 1.

Table 1
UDIMET 720 Processing/Heat Treatment Combinations

	Forging Method/Temperature	Heat Treatment*	Grain Size
1.	Hot die /subsolvus	2080°F/2 hrs/Oil Quench	ASTM 8.5 (19 um)
2.	Isothermal/supersolvus (2.0 in./in./min. rate)	2040°F/2 hrs/Oil Quench	ASTM 4 (90 um)
3.	Isothermal/supersolvus (0.2 in./in./min. rate)	2040°F/2 hrs/Oil Quench	ASTM 3 (127 um)
4.	Hot die/subsolvus	2135°F/2 hrs/Oil Quench	ASTM 0 (359 um)
5.	Hot die/subsolvus (coarse grained billet)	2040°F/2 hrs/Oil Quench	Duplex ASTM 6-10

*Age: 1400°F/8 hrs/air cool + 1200°F/24 hrs/air cool

All of the above processes produced forgings with uniform microstructures. Supersolvus forging has demonstrated previously and in this instance, the ability to produce controlled grain structures. Special processing was also used to produce a uniform necklace structure throughout the majority of the pancake forging. UDIMET 720 alloy in these conditions is very forgeable and grain coarsens in a predictable manner with solution treatment temperature.

Tensile and Stress Rupture Strengths

Tensile test specimens were taken in the chordal direction from the forgings, and were tested at both room temperature and at 1200°F. Duplicate tests were performed and the average ultimate tensile strengths and 0.2% yield strengths are presented in Figure 2. The results show the fine grained (ASTM 8.5) structure exhibited the highest strengths while the ASTM 0 microstructure had the lowest strength. The duplex structure behaved like the uniform fine structure with high strength.

Smooth bar stress rupture testing was conducted in air at 1300°F with an applied stress of 100 ksi. The results of duplicate tests are shown in Figure 3. The coarser structures had the longest lives as expected. The ASTM 0 structure had approximately 2.5x longer lives than the uniform ASTM 8.5 microstructures. The duplex microstructure, which was selected because these structures generally maintain good stress rupture properties, lasted only half as long as the uniform fine grain size. The fine recrystalized grains in the duplex forging controlled the behavior.

Low Cycle Fatigue Results

Strain controlled low cycle fatigue testing was conducted to determine the relative effects of these microstructural variations on the number of cycles to initiate a crack. Five tests were performed at the same conditions on each of the five processes. All the tests were conducted at 1200°F on smooth, Kt=1, specimens, R-ratio = 0.0, with a frequency of 20 cycles per minute in strain control at a total strain range of 0.80%. The

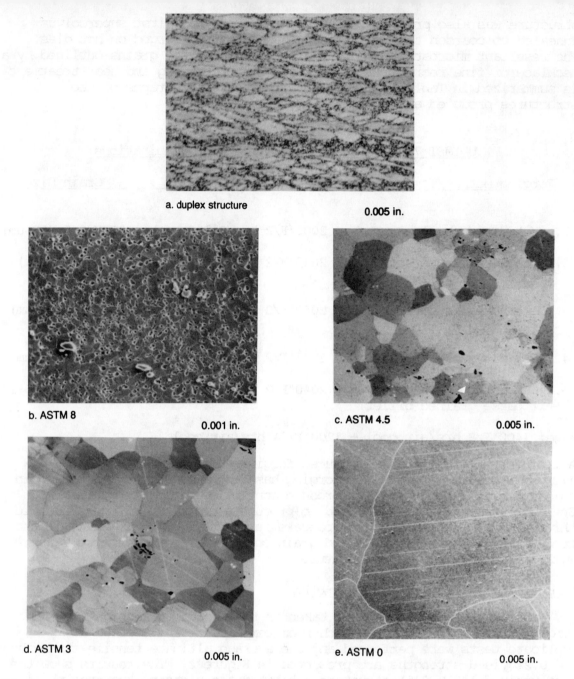

Figure 1. Microstructures of the five heat treatment/process variations of U-720 studied.

results indicated that as the grain size was decreased the fatigue lives increased. (see Figure 4) The ASTM 0 structure exhibited lives a factor of ten shorter than the fine ASTM 8.5 grain size structure. The duplex structure had low fatigue strength with the highest scatter in the results.

Fatigue Crack Growth Results

Duplicate fatigue crack growth rate tests were conducted in air on each of the five structures at two different test conditions. Duplicate tests at 800°F, R-ratio = 0.05, at 200 cpm were conducted to evaluate the lower temperature cyclic crack growth rate for each microstructure. Tests at 1200°F, R-ratio = 0.05, with a 5 minute dwell at the maximum load were conducted to evaluate the time dependent crack growth behavior of the five structures. Four of the five structures had the same 800°F crack growth rate which was similar to RENE-95 (2), but the ASTM 0 grain size process exhibited a factor of two reduction in the fatigue crack growth rate.

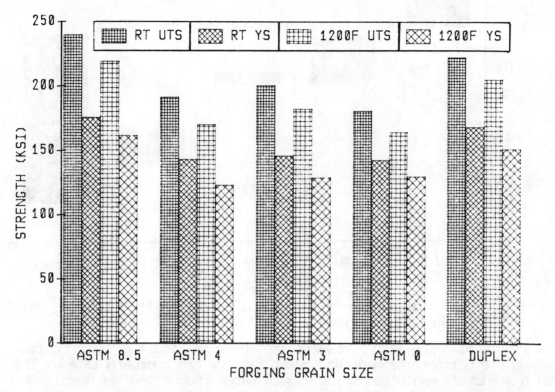

Figure 2. Tensile test results at room temperature and at 1200°f.

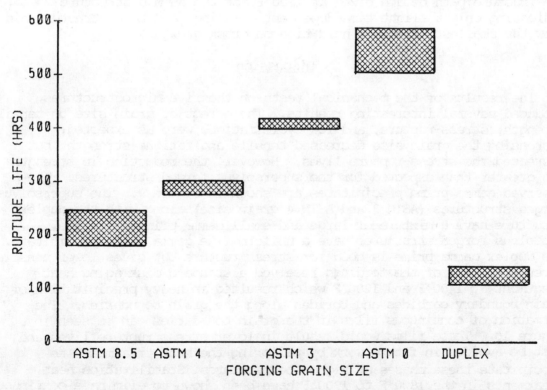

Figure 3. Range of stress-rupture results for the five microstructures tested at 1300°f/100 ksi.

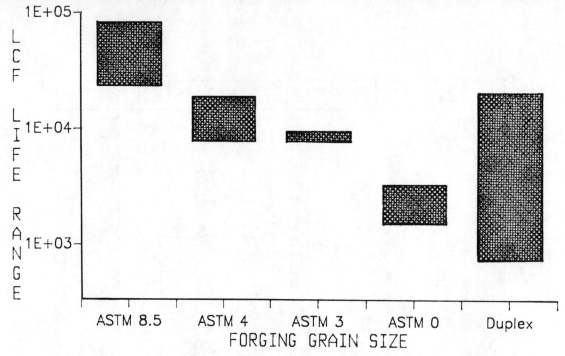

Figure 4. Low cycle fatigue test results. The results indicated that the finer structures are superior in LCF strength.

(see Figure 5). The 1200°F results are shown in Figure 6. The coarse grained microstructures had the lowest fatigue crack growth rates. The ASTM 0 structure exhibited crack growth rates four orders of magnitude lower than fine grained RENE-95 tested under similar conditions (3).

The fracture paths of each of the fatigue crack growth specimens was also examined. At 800°F Figure 7 shows that the fractures were transgranular with the fracture surface becoming very rough for the larger grain sizes. At 1200°F Figure 8 shows that the fracture paths were always intergranular as is common in time dependent crack growth tests. However, the fracture path became mixed at 1200°F for the ASTM 0 structure indicating only a slight time-dependent cracking, and this structure did show the smallest effect of hold time on crack growth.

Discussion

The results of the mechanical tests on the five microstructures produced several interesting results. The effect of grain size on tensile strength, stress-rupture, and low cycle fatigue were as expected. Increasing the grain size decreased tensile and fatigue strength, but increased the stress-rupture lives. However, the reduction in strength was greater than expected for the supersolvus forged structures. The observed gamma prime precipitates are shown in Figure 9. The supersolvus forged structures (ASTM 3 and ASTM 4 grain size) along with the duplex structure have a mixture of large and small gamma prime while the subsolvus forged structures have a uniform fine gamma prime precipitate. The duplex gamma prime is good for stress rupture but gives lower tensile strengths. All of the forgings received a standard disk aging heat treatment at 1400°F and 1200°F which resulted in heavy precipitation of grain boundary carbides and borides along the grain boundaries. The formation of continuous films in the grain boundaries can be seen in Figure 10. These films would result in lower stress rupture lives and must be avoided in future work by tailoring the heat treatments to precipitate these phases at higher temperatures. Stabilization heat treatments in the 1800°F to 1900°F have been shown to eliminate continuous grain boundary films in UDIMET 720 (1). The phases present in the grain boundary for the five microstructures studied is given in Table 2.

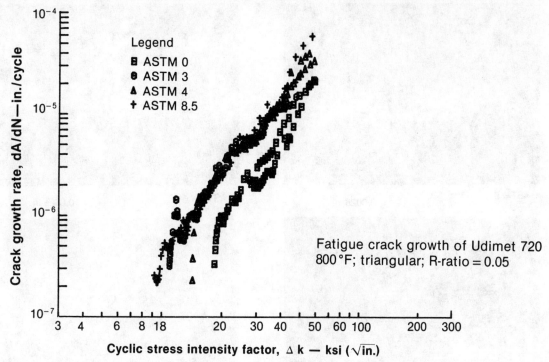

Figure 5. Fatigue crack growth at 800°f. The ASTM 0 grain size has an increased threshold and a reduced growth rate.

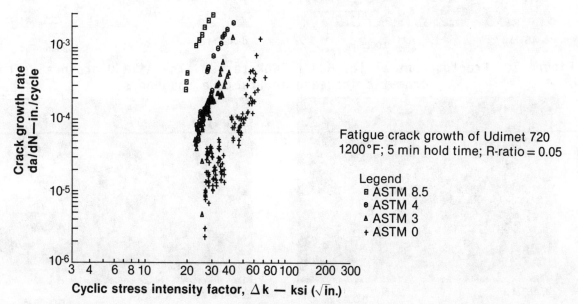

Figure 6. Fatigue crack growth at 1200°f with a 5 minute dwell at the maximum load. Fatigue crack growth rate decreased with increasing grain size.

Table 2
Grain Boundary Phases Observed in UDIMET 720

Grain Size	TiC	$CR_{23}C_6$	$(Cr,Mo,W)_3B_2$	M_xB_y
ASTM 0	X		X	
ASTM 3	X		X	X
ASTM 4.5	X	X	X	X
ASTM 8.5	X	X	X	
DUPLEX	X	X	X	X

The large ASTM 0 grain microstructure gave the best crack growth behavior of the five microstructures studied. At 800°F the large transgranular facets deflected the crack tip which would reduce the effective stress intensity at the crack tip and reduce the crack growth rate. The gamma prime morphology within the grains would have had an

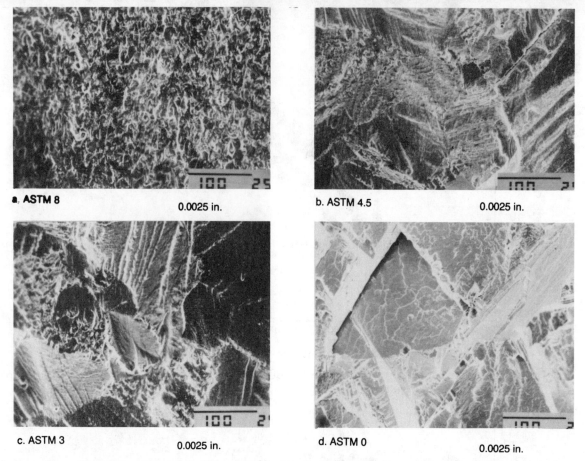

Figure 7. Fractographs of the 800°f FCGR tests. The ASTM 0 grain size has a dramatic increase in fracture roughness.

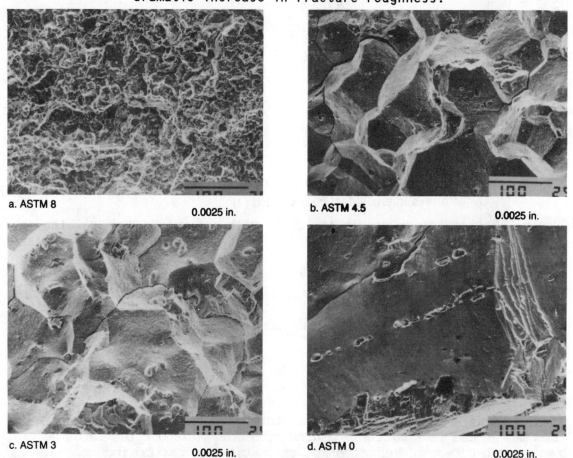

Figure 8. Fractographs of the 1200°f FCGR test specimens. The fracture path was intergranular as expected during time-dependent crack growth.

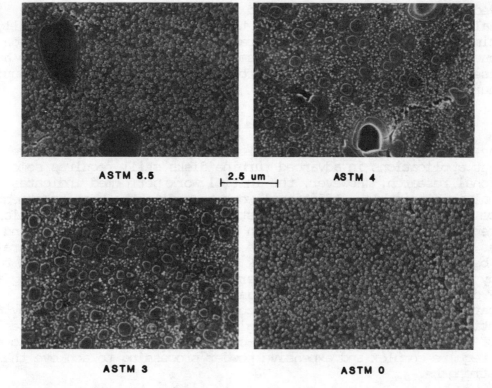

Figure 9. Gamma prime morphology and distribution in four of the microstructures tested. The supersolvus forged structures had a duplex large and small gamma prime distribution which is not optimum for high tensile strength.

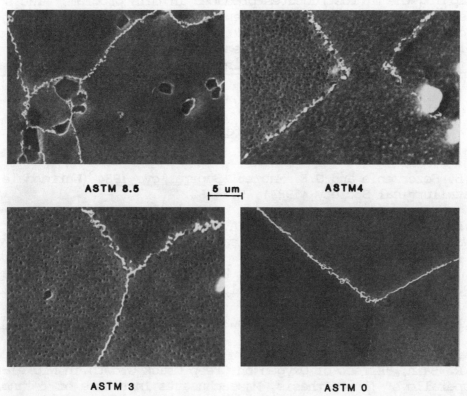

Figure 10. Heavy grain boundary decoration with carbides and borides was observed with all of the structures tested. Continuous grain boundary phases can reduce stress-rupture and low cycle fatigue strength.

effect on the crack growth rate at 800°F. At 1200°F the effect of grain size on the fatigue crack growth rate has been modelled (5). The effect of increasing the grain size from ASTM 8.5 to 0 should have reduced the crack growth rates by more than a factor of 100, and such a decrease was observed. The duplex structure acted more like a small grain size material, however research has shown that duplex structures generally have good time dependent crack growth characteristics. The precipitation of continuous carbide films around the small recrystalized grains may have increased the time dependent crack growth rates observed for the duplex structure.

Conclusion

The development of UDIMET 720 alloy for high temperature damage tolerant applications in advanced turbine disks still requires some additional research. However, the initial work performed indicates that the goal is attainable. Additional work to refine the heat treatments to enhance the low temperature tensile strength along with stabilization heat treatments to alter the precipitation of grain boundary carbides and borides to improve the stress rupture behavior of the microstructures must still be finished. The fatigue crack growth improvements obtained to date already are very significant and offer the potential for developing a 1300°F disk alloy for high strength damage tolerant turbine disk applications. In addition, the low cost of producing this alloy by the conventional VIM/VAR and hot-die forging route significantly reduces the cost of this alloy relative to other competitive high strength alloys which require complex and expensive powder processing to achieve their high strengths.

Acknowledgments

The authors would like to thank Prof. John Radavich for his work on photographing the microstructures produced in this program. The support of Allison Gas Turbine Division of General Motors Corporation and Wyman-Gordon Corporation are also acknowledged along with thanks for their permission to publish this work. Special thanks goes to Mr. Barry Holman and Mr. Paul Korinko for assistance in performing some of the mechanical tests in this program.

References

1. F.E. Sczerzenie and G.E. Maurer, Superalloys 1984 (Warrendale, PA; The Metallurgical Society, 1984), 573-582.

2. D.R. Chang, D.D. Krueger and R.A. Sprague, Superalloys 1984 (Warrendale, PA; The Metallurgical Society, 1984), 245-273.

3. S.J. Choe, S.V. Golwalker, D.J. Duquette and N.S. Stoloff, Superalloys 1984 (Warrendale, PA; The Metallurgical Society, 1984), 309-318.

4. G.A. Whitlow, C.G. Beck, R. Viswanathan and E.A. Crombie, Met. Trans. A, Vol. 15A, January 1984, 23.

5. K.R. Bain, "Effect of Oxygen on Creep Crack Growth in Nickel-Base Superalloys" (Ph.D. thesis, Massachusetts Institute of Technology, 1983), 102-112.

6. S. Floreen, Creep-Fatigue-Environment Interactions, (Warrendale, PA; The Metallurgical Society, 1980), 112-128.

7. R.M. Pelloux and J.S. Huang, Creep-Fatigue-Environment Interactions, (Warrendale, PA; The Metallurgical Society, 1980), 151.

EFFECT OF HEAT TREATMENT ON MECHANICAL PROPERTIES AND MICROSTRUCTURE OF ALLOY 901

R. B. Frank and R. K. Mahidhara

Carpenter Technology Corporation
Research and Development Center
P.O. Box 14662
Reading, PA 19612-4662
USA

Summary

Two types of heat treatments are commonly used for alloy 901, one to optimize resistance to creep and rupture and the other, to improve tensile and fatigue properties. Solution and aging-treatment parameters (temperature, time, quench) for both types of heat treatment were varied within the ranges permitted by common specifications to define heat treatments resulting in improved combinations of room-temperature strength and ductility along with good stress-rupture properties. Room-temperature tensile and 1200°F stress-rupture tests were performed and microstructures were evaluated using light and electron microscopy. Structure/property relationships were defined. The results showed that relatively small variations in heat-treatment parameters resulted in significant differences in mechanical properties and microstructure. Room-temperature tensile properties were optimized without degrading 1200°F stress-rupture properties. Mechanical properties were dependent on grain size and the characteristics of the grain boundary carbide and matrix hardening (γ') precipitates.

Introduction

Pyromet® alloy 901 is a nickel-iron-base superalloy used for aerospace and land-based turbine components requiring high tensile and stress-rupture strengths along with good hot-corrosion resistance at 1000-1200°F. Two different types of heat treatments are commonly used for the alloy, one to optimize resistance to creep and rupture at elevated temperatures (Type I) and the other, to improve tensile and fatigue properties (Type II). Commercial specifications for alloy 901 generally include solution and aging temperature ranges of 50-75°F and also allow some adjustment of treatment time and cooling rate from the solution temperature. Although the effects of heat treatment on alloy 901 have been summarized (1), a comprehensive study has not been published. The purpose of this study was to determine the effects of heat-treatment parameters, within the limits of commercial specifications, on mechanical properties and microstructure of alloy 901 with the goal of defining heat treatments resulting in improved combinations of room-temperature strength and ductility along with good stress-rupture properties at 1200°F.

Experimental Procedure

Material from a vacuum-induction melted/vacuum-arc remelted production heat of Pyromet alloy 901 was used for all experiments. The chemical composition of this material is listed below:

Element:	C	Cr	Ni	Mo	Ti	Al	B	Fe
Weight %:	0.03	11.5	42.1	5.3	3.0	0.3	0.012	Balance

Longitudinal sections of 6-inch diameter bar (2-inch square x 6-inch long) were heated to 2000°F and press forged to provide 3/4-inch square bars for testing.

Room-Temperature Tensile. Collective ranges for the two solution + double-age heat treatments commonly specified for alloy 901 are listed below:

Type I - 1950-2025°F ($\pm$25°F)/2h/AC or faster + 1400-1475°F ($\pm$15°F)/
2-4h/AC or faster + 1300-1375°F ($\pm$15°F)/24h/AC

Type II - 1800-1850°F ($\pm$25°F)/1-2h/AC or faster + 1300-1350°F ($\pm$15°F)/
6h min./AC or faster + 1175-1225°F ($\pm$15°F)/12h min./AC

Three-factor, two-level, full-factorial experimental designs were used to determine the effects of heat treatment on room-temperature tensile properties. For Type I heat treatments, the factors and levels chosen for initial evaluations were: solution temperature (1950,2000°F), mid-age temperature (1400,1450°F) and final-age temperature (1300,1350°F). For Type II heat treatments, the factors and levels were: solution temperature (1800,1850°F), mid-age temperature (1300,1350°F), and mid-age time (6, 18 hours). The final-age treatment was held constant at 1200°F/12h/AC for Type II treatments. Heat treatments representing the mid-points of each range were used to determine whether any non-linear relationships existed (curvature effects). Thus, a total of nine heat treatments of each type were evaluated initially.

Fully heat-treated 3/4-inch square blanks were machined to 0.252-inch gage-diameter threaded specimens for room-temperature tensile tests. Duplicate specimens were tested for each treatment. Sampling and testing procedures were designed to randomize any variations unrelated to heat treatment. Yield strength and elongation results were statistically analyzed (2) to identify factor and curvature effects that were significant with a 95%

® - registered trademark of Carpenter Technology Corporation

level of confidence. Following the initial experiments, additional tensile tests were performed to confirm optimum treatments, to define non-linear relationships, and to determine effects of cooling rate from the solution temperature (water quenching vs. air cooling).

<u>Stress-Rupture</u>. Testing was concentrated on heat treatments resulting in the best combinations of room-temperature strength and ductility although limited testing was done to define heat-treatment effects. Fully heat-treated blanks were machined to 0.178-inch gage-diameter combination smooth-notched (K_t=3.8) stress-rupture specimens and tested at 1200°F with a constant load of 90 ksi.

<u>Light and Electron Microscopy</u>. Longitudinal sections of heat-treated tensile specimens were metallographically prepared and examined using light microscopy to determine grain size and relative amounts of grain boundary precipitation. The fracture surfaces of several room-temperature tensile specimens were examined using scanning electron microscopy (SEM) to determine fracture modes. Thin foils, structural replicas (chromium-shadowed parlodion), and carbon extraction replicas were examined using transmission electron microscopy (TEM) to identify and characterize matrix and grain boundary phases. Phases were identified using convergent-beam electron diffraction analysis and energy dispersive spectroscopy (EDS). Precipitates were also identified using X-ray diffraction analysis of extracted residues. Larger matrix phases were analyzed in situ using the electron microprobe.

Gamma-prime (γ') size distributions were determined using thin foil electron micrographs representing several heat treatments. About 500 particle size measurements and about 50 measurements of interparticle spacing were made for each heat treatment. Because of the complexities involved in measuring fine γ' particles, the size distributions reported in this study should be used to compare heat treatments rather than to obtain absolute size data.

Results and Discussion

Type I Heat Treatments

<u>Room-Temperature Tensile Properties</u>. Room-temperature tensile and stress-rupture test results are listed in Table I. Ratings of room-temperature yield strength and elongation were calculated to indicate the degree to which the results exceeded minimum requirements of common specifications. One point was given for each increment of the pooled standard deviation (s) by which the result exceeded the minimum requirement plus one standard deviation e.g. 120 ksi + 3s = 2 points; 120 ksi + 1s = 0 points. The products of the yield strength and elongation ratings listed in Table I indicate the combination of strength and ductility offered by a particular heat treatment. It should be noted that the properties and ratings listed in Table I also depend on composition, forging procedure, and bar size but only heat-treatment effects will be discussed in this paper.

The minimum significant factor effects (95% confidence level) were 2 ksi for yield strength and 1% for elongation. Heat-treatment effects less than these values would not be separable from variation due to forging, sampling, specimen preparation and testing. Mid-age and final-age temperatures, and the interaction between these factors, had the most significant effects on yield strength. Solution and mid-age temperatures, and the interaction between the mid-age and final-age temperatures, had the most significant effects on tensile elongation. For Type I heat treatments, the lower solution, mid-age, and final-age temperatures resulted in the best combinations of room-temperature strength and ductility as evidenced by the ratings in Table I.

Table I. Mechanical Properties - Type I Heat Treatments

Heat Treatment(°F)*			Room-Temperature Tensile (avg.)					Stress Rupture (1200°F/90 ksi)		
Solution +	Mid Age +	Final Age	0.2% Y.S. (ksi)	U.T.S (ksi)	El. (%)	R.A. (%)	YS-El. Rating	Life (h)	El. (%)	R.A. (%)
1950/WQ	+ 1400	+ 1300	136	184	17	26	39	-	-	-
"	+ 1425	+ 1300	138	185	17	22	42	58	7.5	14
								294	10.5	18
								174	10.5	17
" /AC	+ 1425	+ 1300	135	186	16	25	24	-	-	-
" /WQ	+ 1450	+ 1300	140	186	16	19	32	-	-	-
"	+ 1400	+ 1350	135	183	16	20	30	-	-	-
"	+ 1450	+ 1350	133	184	16	21	21	-	-	-
1975/WQ	+ 1400	+ 1300	135	184	18	21	38	304	10.0	14
								314	9.0	14
" /AC	+ 1400	+ 1300	128	181	19	24	21	138	6.0	11
								201	6.5	9
" /WQ	+ 1425	+ 1300	138	183	14	19	6	289	11.5	15
								262	10.5	15
" /AC	+ 1425	+ 1300	131	183	17	25	22	286	9.0	11
								234	9.5	13
" /WQ	+ 1425	+ 1325	138	181	13	18	4	-	-	-
2000/WQ	+ 1400	+ 1300	134	175	15	18	19	-	-	-
"	+ 1425	+ 1300	-	-	-	-	-	271	10.0	17
								239	6.5	15
"	+ 1450	+ 1300	138	176	11	14	0	-	-	-
"	+ 1400	+ 1350	134	174	12	17	0	-	-	-
"	+ 1450	+ 1350	133	174	11	13	0	-	-	-
Minimum Requirements			120	165	12	15		23	5	-

* Solution treated 2h/WQ or AC + mid-aged 2h/AC + final-aged 24h/AC.

Results of the microstructural evaluation of heat-treated tensile specimens using light and electron microscopy are listed in Table II. X-ray diffraction and electron microscopy studies revealed that phases present in in the matrix were Ti and Mo-rich MC carbides (<1-25μm), Mo-rich M_3B_2 and the γ' [Ni_3(Ti,Al)] hardening phase (2-40 nm). Smaller amounts of Mo-rich M_6C carbide (<2μm) were also identified. Grain boundary precipitates were primarily Ti and Mo-rich MC carbides (50-250 nm). Similar phases have been found by others (1,3,4). The same types of phases were observed for all heat treatments although in varying amounts.

Figure 1 illustrates the detrimental effect of increasing the solution temperature from 1950°F to 2000°F on tensile ductility. It appears that some curvature exists in this relationship, depending on the aging treatment. Five of eight specimens solution treated at 2000°F before aging failed the minimum elongation requirement of 12%. The reduction in ductility is attributed to coarser grain size and more extensive precipitation of

Table II. Microstructural Observations - Type I Heat Treatments

Heat Treatment(°F)[1]			Tensile Fracture Mode[2]	ASTM Grain Size	Grain Boundary Precipitation		Gamma Prime Distribution			
Solution +	Mid Age +	Final Age			Rel. Amt.[3]	Contin-uity[4]	Avg. Diam. (nm)	Min. Diam. (nm)	Max. Diam. (nm)	Avg. Spacing (nm)
1950/WQ	+ 1400	+ 1300	50%DT/50%I	4/5	S	D	10	2	26	21
"	1425	+ 1300	--	4/5	M	S	14	3	35	19
"	1450	+ 1300	25%DT/75%I	3/5	M/L	C	15	3	30	16
"	1450	+ 1350	--	4/5	L	C	19	5	39	18
1975/WQ	+ 1400	+ 1300	--	3/5	S/M	D/S	12	3	30	19
" /AC	"	"	--	-	M/L	-	-	-	-	-
" /WQ	+ 1425	+ 1300	Mostly I	3/4	M	S	11	3	27	17
" /AC	"	"	Mostly I	3/5	M/L	S	10	2	22	18
" /WQ	+ 1425	+ 1325	Mostly I	3/4	M/L	C	15	4	29	15
2000/WQ	+ 1400	+ 1300	--	2/4	M	-	-	-	-	-
"	1450	+ 1300	Mostly I	2/4	L	C	16	6	29	15

1 - Solution treated 2h/WQ or AC + mid-aged 2h/AC + final-aged 24h/AC
2 - Fracture mode: DT = ductile transgranular; I = intergranular
3 - Relative amount: S = small; M = moderate; L = large
4 - Continuity: D = discontinuous; S = semi-continuous; C = continuous

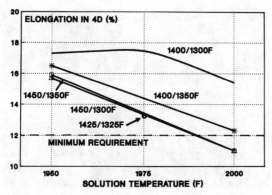

Figure 1. Effect of solution temperature on room-temperature tensile ductility - Type I treatments (mid-age/final-age temperatures are shown)

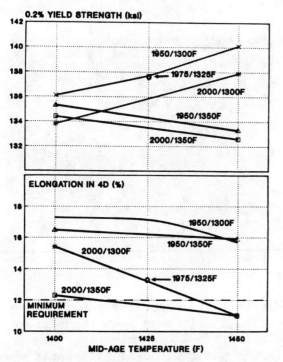

Figure 2. Effect of aging temperatures on room-temperature tensile properties - Type I treatments (sol'n/final-age temps. are shown).

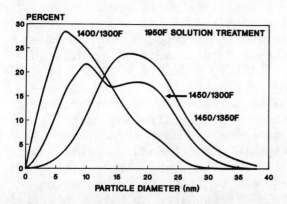

Figure 3. Effect of aging treatment on gamma-prime size distribution - Type I treatments.

semi-continuous or continuous grain boundary carbides (MC) which resulted in a predominantly intergranular tensile fracture mode. Higher solution temperatures were most detrimental when used in conjunction with higher mid-age temperatures (1425-1450°F) because grain boundary precipitation was further increased. Slower cooling from the solution temperature (air vs. water quench) decreased yield strength. Air cooling resulted in a slightly finer distribution of γ' compared to water quenching. Hammond and Ansell have reported that vacancies frozen in position during quenching from the solution temperature provide nucleation sites for γ' particles (5). It is postulated that water-quenching stresses may enhance the diffusion of solutes thereby accelerating the aging process.

Figure 2 illustrates the significant effect of mid-age temperature on tensile properties and the strong interaction between mid-age and final-age temperatures. Yield strength increased and elongation decreased with increasing mid-age temperature (1400-1450°F) when a 1300°F final age was used while yield strength and elongation both decreased slightly with increasing mid-age temperature when a 1350°F final age was used. Figure 3 contains size distribution plots showing the effects of mid-age and final-age temperatures on γ' particle size. Increasing mid-age temperature from 1400°F to 1450°F (1300°F final age) resulted in a coarser and more bimodal distribution of γ'. The higher strength of material aged at 1450°F + 1300°F appears to be a result of this duplex γ' size and reduced interparticle spacing. The lower ductility of samples solution treated at 1975-2000°F and mid-aged at the higher temperatures was associated with mostly intergranular tensile fracture modes (See Figure 4). Typical replica electron micrographs are shown in Figure 5. The relative amount, size and continuity of grain boundary precipitates increased as mid-age temperature increased leading to more intergranular fracture.

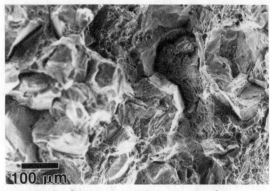 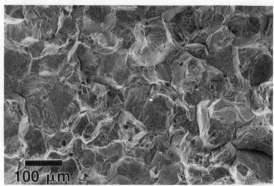

1950°F/WQ + 1400°F + 1300°F   1975°F/WQ + 1425°F + 1300°F
50% intergranular - 17.6% El   mostly intergranular - 13.9% El

Figure 4. SEM micrographs of fracture surfaces of room-temperature tensile specimens with different ductility values.

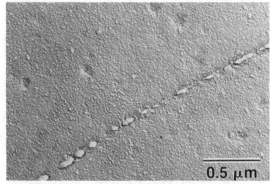 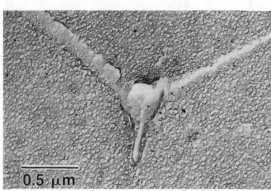

1950°F/WQ + 1400°F + 1300°F   1950°F/WQ + 1450°F + 1300°F

Figure 5. Replica electron micrographs showing the effect of mid-age temperature on gamma-prime and grain boundary carbide structure.

Results in Table I and Figure 2 indicate that increasing the final-age temperature above 1300°F provided no beneficial effect on room-temperature tensile properties, regardless of the solution/mid-age combination. The yield strength of material mid-aged at 1450°F was reduced significantly when final-age temperature was increased from 1300°F to 1350°F. Based on hardness measurements and γ' size distributions shown in Figure 3, it is concluded that the lower strength is a result of γ' coarsening (overaging).

Stress-Rupture Properties. All Type I heat treatments evaluated resulted in stress-rupture properties exceeding the requirements of common specifications (Table I). Ratings similar to those reported for tensile properties were not useful for stress-rupture data because of considerable variation between duplicate tests. The 1975°F/WQ + 1400°F + 1300°F treatment resulted in the best combination of room-temperature tensile and stress-rupture properties. Using the above treatment temperatures, air cooling from 1975°F resulted in significantly lower stress-rupture life and ductility than did water quenching. Air-cooled specimens contained more extensive precipitation of semi-continuous to continuous grain boundary carbides which resulted in larger amounts of intergranular fracture. Specimens with higher rupture lives generally displayed a mixed mode of fracture. Increasing solution temperature from 1950°F to 1975-2000°F appeared to have a beneficial effect on stress-rupture life.

Table III. Mechanical Properties - Type II Heat Treatments

Heat Treatment*		Room-Temperature Tensile (avg.)					Stress Rupture (1200°F/90 ksi)		
Solution (°F)	Mid-Age (°F) (h)	0.2% Y.S. (ksi)	U.T.S (ksi)	El. (%)	R.A. (%)	YS-El. Rating	Life (h)	El. (%)	R.A. (%)
1800/WQ	+ 1300/6	115	184	29	41	0	–	–	–
"	+ 1350/6	134	193	24	43	5/17	102	10.5	23
							42	8.5	27
"	+ 1300/18	135	194	22	43	2/11	–	–	–
"	+ 1350/18	148	198	20	35	0/6	–	–	–
1825/WQ	+ 1350/6	134	193	24	41	6/18	161	8.5	18
							72	6.5	14
"	+ 1325/12	135	191	23	40	5/16	112	9.5	16
							55	6.5	14
"/AC	+ 1325/12	127	190	24	38	0/8	87	Notch Break	
							20	Notch Break	
1850/WQ	+ 1300/6	107	176	34	45	0	–	–	–
"	+ 1350/6	129	188	26	42	3/15	186	9.5	21
							116	6.5	19
"/AC	+ 1350/6	122	190	22	40	0	–	–	–
"/WQ	+ 1325/12	133	190	23	39	2/12	186	10.5	19
							166	11.0	17
"	+ 1350/12	144	195	21	35	0/10	–	–	–
"	+ 1300/18	130	190	26	40	4/16	–	–	–
"	+ 1350/18	148	196	21	34	0/16	–	–	–
Minimum Requirements		120/125	170/175	18/20	25		23	5	

* Solution treated 1.5h/WQ or AC + mid-aged as shown/AC + final-aged 1200°F/12h/AC.

Type II Heat Treatments

Room-Temperature Tensile Properties. Room-temperature tensile and stress-rupture test results are listed in Table III. Yield strength/elongation ratings were calculated as described for Type I treatments except that typical minimum requirements of 120 and 125 ksi for yield strength and 18 and 20% for elongation were used for calculation. The minimum significant factor effects (95% confidence level) were 2.2 ksi for yield strength and 1.6% for elongation. Mid-age time and temperature had the largest effects on room-temperature tensile properties (Figure 6) although

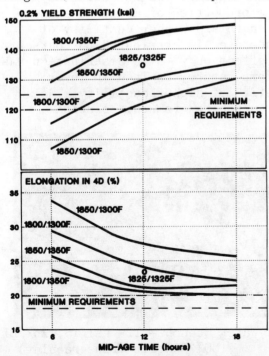

Figure 6. Effect of mid-age time on room-temperature tensile properties - Type II treatments (sol'n/mid-age temperatures are shown).

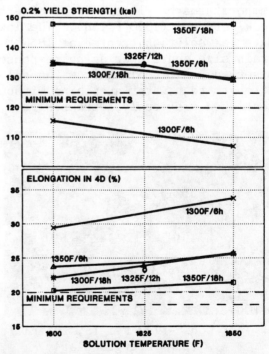

Figure 7. Effect of solution temperature on room-temperature tensile properties - Type II treatments (mid-age treatment is shown).

the solution temperature effect (Figure 7), and the interaction effect between mid-age time and temperature, were also significant with a 95% confidence level. The relationships between mid-age time and yield strength/elongation were not linear. For Type II treatments, a mid-age treatment at 1350°F for the minimum aging time of 6 hours resulted in the best combination of room-temperature strength and ductility.

Table IV. Microstructural Observations - Type II Heat Treatments

Heat Treatment[1]		Tensile Fracture Mode[2]	ASTM Grain Size	Grain Boundary Precipitation		Gamma Prime Distribution			
Solution (°F)	Mid-Age (°F)(h)			Rel. Amt.[3]	Continuity[4]	Avg. Diam. (nm)	Min. Diam. (nm)	Max. Diam. (nm)	Avg. Spacing (nm)
1800/WQ	+ 1300/6	DT	7/8	S	D	9	4	17	21
"	+ 1350/6	DT	7/8	M/L	S	12	4	22	17
"	+ 1300/18	DT	8	L	D/S	10	4	17	18
1825/WQ	+ 1350/6	-	6/7	S/M	-	-	-	-	-
"	+ 1325/12	Mostly DT	6/7	L	C	11	2	26	17
1825/AC	+ 1325/12	-	6/7	VL	S/C	10	2	22	15
1850/WQ	+ 1350/6	DT	5/7	S/M	D	12	4	22	14
1850/AC	+ 1350/6	75%DT/25%I	5/7	L	-	-	-	-	-
1850/WQ	+ 1325/12	-	5/7	S/M	-	-	-	-	-

1 - Solution treated 1.5/WQ or AC + mid-aged as shown/AC + final-aged 1200°F/12h/AC.
2 - Fracture mode: DT = ductile transgranular; I = intergranular
3 - Relative amount: S = small; M = moderate; L = large; VL = very large
4 - Continuity: D = discontinuous; S = semi-continuous; C = continuous

Results of the microstructural evaluation of heat-treated tensile specimens using light and electron microscopy are listed in Table IV. Phase identities were the same as those reported for Type I treatments except that some eta phase (Ni_3Ti) was observed in the matrix and grain boundaries of specimens solution treated at 1800°F. Matrix precipitates in specimens with Type II treatments were banded to a greater extent than those of Type I treatments, particularly when a solution temperature of 1800°F was used.

Increasing the solution temperature from 1800°F to 1850°F resulted in lower strength and higher ductility when lower aging temperatures and times were used (underage treatments) as shown in Figure 7. Higher solution temperatures had little or no effect on the properties of material aged to peak hardness (1350°F/18h). The lower strength and higher ductility of material solution treated at 1850°F and underaged is attributed to coarser grain size and slower aging response. No effect on γ' size distribution was observed. Air cooling from the solution temperature resulted in lower strength than water quenching without benefit of higher ductility. As mentioned previously, stresses from water quenching from the solution

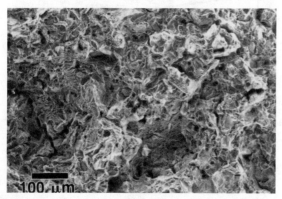

Air Cool - 122 ksi YS, 22% El
(25% Intergranular)

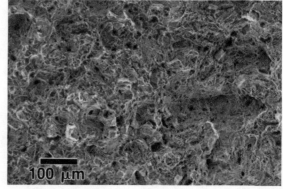

Water Quench - 129 ksi YS, 26% El
(100% Ductile Transgranular)

Figure 8. SEM micrographs of fracture surfaces of room-temperature tensile specimens with different ductility values.
Heat treatment: 1850°F/1.5h/AC or WQ + 1350°F/6h/AC + 1200°F/12h/AC

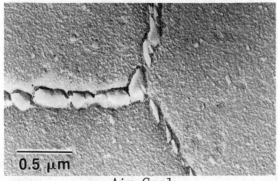 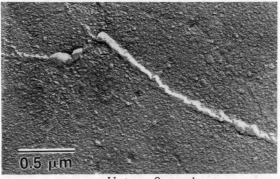

Air Cool   Water Quench

Figure 9. Replica electron micrographs showing the effect of cooling rate from the solution temperature on grain boundary carbide precipitation. Heat treatment: 1825°F/1.5h/AC or WQ + 1325°F/12h/AC + 1200°F/12h/AC.

temperature may accelerate the aging process, particularly for underage-type treatments. SEM fractographs and replica electron micrographs of air-cooled and water-quenched specimens are shown in Figures 8 and 9. The lower-strength air-cooled specimens did not have improved ductility because of more extensive precipitation of larger grain boundary carbides resulting in a partially intergranular tensile fracture mode.

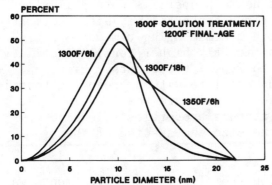

Figure 10. Effect of mid-age treatment on gamma-prime size distribution - Type II treatments.

The room-temperature tensile properties obtained using Type II treatments were primarily dependent on the size of the γ' particles precipitated during mid-aging and the resultant interparticle spacing (constant 1200°F/12h final age). The effect of mid-age treatment on γ' size distribution is shown in Figure 10. Yield strength increased and ductility decreased as mid-age time (6-18h) or temperature (1300-1350°F) increased because the size of the γ' particles increased and interparticle spacing decreased. Peak hardening was obtained only after about 18 hours at 1350°F, therefore most of the mid-age treatments evaluated resulted in underaging. Grain boundary precipitation had little effect on ductility as predominantly ductile transgranular tensile fractures were observed.

Type II treatments resulted in significantly higher room-temperature tensile ductility than Type I treatments at similar strength levels. The higher ductility was associated with ductile, microvoid-coalescence fracture modes (Figure 8) rather than the predominantly intergranular or mixed fracture modes characteristic of Type I heat treatments (Figure 4). The ductile transgranular fracture modes were associated with finer grain size and generally smaller amounts of finer and less continuous grain boundary precipitates.

Stress-Rupture Properties. Although there was considerable scatter in the data for Type II treatments, stress-rupture test results in Table III show that heat treatments which resulted in the best room-temperature tensile properties also resulted in 1200°F stress-rupture properties exceeding minimum requirements of common specifications. Type II heat treatments resulted in significantly lower stress-rupture lives than the Type I heat treatments previously discussed. The improvement in stress-rupture life obtained using a solution temperature of 1850°F is attributed to coarser grain size and smaller amounts of finer and less continuous grain boundary precipitates. Similar effects have been observed by others (3). The notch sensitivity of specimens air-cooled from the solution temperature was also related to an increase in the size and amount of grain boundary precipitates. The air-cooled specimens showed a primarily intergranular fracture mode in the fracture initiation zone while water-quenched specimens showed a mixed fracture mode.

Conclusions

Using heat treatments designed to optimize resistance of Pyromet alloy 901 to high-temperature creep and rupture (Type I), the best combinations of room-temperature strength and ductility were obtained using the minimum specified temperatures for solution treatment (1950-1975°F), mid-aging (1400-1425°F), and final-aging (1300°F). Excellent stress-rupture properties at 1200°F were also obtained using these combinations of solution and aging temperatures. Water quenching from the solution temperature generally resulted in better room-temperature tensile and stress-rupture properties than air cooling. Tensile properties were dependent on the size and spacing of γ' and grain boundary carbide precipitates.

Using heat treatments designed to optimize tensile and fatigue properties (Type II), room-temperature tensile properties were primarily dependent on mid-age treatment time and temperature. A mid-age treatment at 1350°F resulted in best properties after a minimum aging time of 6 hours. Type II heat treatments resulted in significantly higher tensile ductility but lower stress-rupture lives compared to Type I treatments. Higher solution temperatures improved stress-rupture life without reducing tensile properties. Air cooling from the solution temperature resulted in lower yield strengths and unacceptable stress-rupture properties. Tensile properties were primarily dependent on the size and spacing of the γ' hardening precipitates.

The lower tensile ductility resulting from Type I heat treatments was attributed to coarser grain size and partially or mostly intergranular tensile fracture modes. The amount of intergranular fracture increased with the size, amount and continuity of grain boundary precipitates. In contrast, Type II treatments resulted in finer grain size, finer and less continuous grain boundary precipitates and ductile, transgranular fracture modes.

Acknowledgments

The authors would like to thank D. A. Englehart for his assistance in conducting the experiments and in preparing figures for this manuscript, and G. F. Vander Voort for the development of γ' size distribution data.

References

1. N. A. Wilkinson, "Technological Considerations in the Forging of Superalloy Rotor Parts," Metals Technology, 4(7) (1977) 346-359.
2. E. I. duPont de Nemours & Co., Strategy of Experimentation (Wilmington, DE, 1974), 10-27.
3. J. A. Smith and J. F. Radavich, "A Structural Study of Super Incoloy 901," Microstructural Science, 10 (1982) 115-122.
4. F. Cosandey et al., "The Effect of Cerium on High-Temperature Tensile and Creep Behavior of a Superalloy," Met. Trans. A, 14A (1983) 611-621.
5. C. M. Hammond and G. S. Ansell, "Gamma-Prime Precipitation in an Fe-Ni-Base Alloy," Trans. ASM, 57 (1964) 727-738.

METALLURGICAL STABILITY OF INCONEL ALLOY 718

J.W. Brooks and P.J. Bridges

Inco Engineered Products Limited
Wiggin Street, Birmingham. B16 OAJ England

Summary

Extensive heat treatment and forging trials have been carried out on INCONEL* alloy 718 produced by modern vacuum melting practice in order to clarify the time-temperature transformation characteristics of the material. The effect of forging practice on microstructure and mechanical properties has also been determined together with the long term stability of the alloy.

The gamma star precipitation behaviour has been defined for annealing times up to 10,000 hours and the transformation kinetics have been compared and contrasted with those of conventional gamma prime strengthened superalloys. The work described provides an explanation for the relationship between grain boundary delta precipitation and creep ductility.

Introduction

The use of INCONEL alloy 718 as a disc material in gas turbine engines has increased significantly in recent years as it has good properties up to 650°C and is competitively priced due to the fact that the alloy contains no cobalt and has a relatively high iron content. However the material was originally developed as a weldable sheet alloy for service at intermediate temperatures and therefore much of the early work is not relevant to the current applications. Furthermore the data available on the precipitation reactions is ambiguous and is not sufficient to account for the behaviour of the alloy in service. Thus the aim of the work discussed in this paper was to define fully the microstructural effects of forging practice and heat treatment.

Experimental

The material used was commercially produced (INCO ALLOYS INTERNATIONAL LTD.) vacuum melted and refined alloy 718, homogenised and side forged down to 100mm diameter bar. The effects of solution treatment temperature (from 950 to 1200°C) and cooling rate (water quench, air cool or furnace cool) were investigated prior to isothermal annealing (0.1 to 10,000 hours from 500 to 1000°C) to determine the time, temperature, transformation behaviour. Forging trials were carried out using a 3:1 reduction at temperatures between 940 and 1120°C with up to three reheat temperatures and four interstage anneals. After conventional heat treatment (using either 955 or 980°C solution treatments) the forgings were given full mechanical property and metallographic assessments.

Selected fully heat treated forgings were subjected to long term exposure for 1000 and 10,000 hours at 650°C and 10,000 hours at 575°C and the structure and properties were then re-evaluated.

Metallographic analysis was carried out using conventional optical techniques combined with scanning and transmission electron microscopy. Mechanical testing was carried out on samples machined from the discs and consisted of tensile tests at room temperature or 650°C together with smooth and notched stress rupture at 650°C (730MPa) or 705°C (415MPa).

Results

(a) Effect of Annealing

The precipitating phases were all based on Ni_3Nb with varying amounts of titanium and aluminium substituting for the niobium. The principal strengthening phase was gamma star ($\gamma *$, sometimes called $\gamma"$ or gamma double prime) a body-centred tetragonal phase which formed as coherent platelets in three variants lying on the $\{100\}$ planes of the face-centre cubic matrix. The most stable Ni_3Nb precipitate was the orthohombic delta phase which nucleated at grain boundaries and grew into plates lying on the matrix $\{111\}$ planes. Gamma prime (γ') precipitates with the usual $L1_2$ structure were also observed, usually in matrix material which was niobium depleted due to heavy delta or $\gamma *$ precipitation. Similarly body-centred cubic alpha chromium (α-Cr) formed as globular particles at grain boundaries when the matrix became nickel depleted due to extensive precipitation of the intermetallic phases.

Solution treatment below 1010°C gave rise to delta phase while grain growth occurred above 1060°C although for fine grained material (<ASTM 7) grain growth started at 1020°C. The titanium and niobium primary carbides, which form during solidification, were stable throughout the temperature range examined and no evidence was found for the formation of Laves phase even at the highest temperature. It was found that variations in cooling rate from the solution treatment temperature had a marked influence on the kinetics of subsequent precipitation of the γ* phase during ageing. In particular rapid quenching enhanced the ageing response while slow cooling produced precipitation in the solution treated condition. Intermediate cooling rates suppressed precipitation but also reduced the nucleation kinetics. (Figure 1)

The long term exposure work showed that the γ* was stable, for 10,000 hours at 600°C, with very little particle coarsening. However, at higher temperatures and for times in excess of 3000 hours the γ* decomposed to form either γ' (650-850°C) or delta (750-1000°C). Also small quantities of globular α-chromium particles were observed, mainly at the grain boundaries, after 3000 hours at intermediate temperatures.

The t-t-t characteristics of these phases are shown in Figure 2. However, the formation of the α-chromium is very dependent on the stress state of the material and can form at much shorter times under creep conditions. Similarly M_6C does not form after annealing but was observed in samples deformed at temperature.

In an attempt to establish the precipitation kinetics the volume fractions of the γ* phase, formed during ageing at 700°C, were determined using both geometrical methods and microanalytical techniques. For comparison purposes, measurements were also carried out on NIMONIC* alloy 901, which has a similar composition to alloy 718 but is precipitation hardened with Ni_3(Al,Ti) γ' particles. The results are shown in Figure 3 and it is clear that the ageing responses are quite different such that the γ* precipitate volume fraction and size increase with time while the γ' coarsens with a constant volume fraction. In both cases particle coarsening follows the t α d^3 relationship observed in other nickel base materials (t is time and d is particle diameter).

(b) **Effect of Forging**

Typical forged microstructures are shown in Figure 4 and the micrographs are arranged so that the effects of the different forging parameters can be examined independently. It was clear that the final grain size became smaller as the forging temperature was reduced while an increase in the number of stages (reheats) gave some further refinement to the overall structure. However, this was accompanied by some sporadic grain growth. The resulting duplex grain structure was typical of all the multi-stage forgings and was most apparent in those forgings done at the higher temperatures. The greatest microstructural refinement was obtained by a multi-stage forging route carried out at successively lower temperatures and finished below 1000°C. The structures observed after this type of processing were characterised by a fairly uniform grain size (∼ASTM 10) with a fine distribution of the delta phase on the grain boundaries.

The 955°C solution treatment gave rise to fairly extensive grain boundary delta precipitation particularly in those samples which contained this phase in the as-forged condition. Annealing at 980°C, however, produced much smaller quantities of delta and tended to spheroidise that which was already present in the structure.

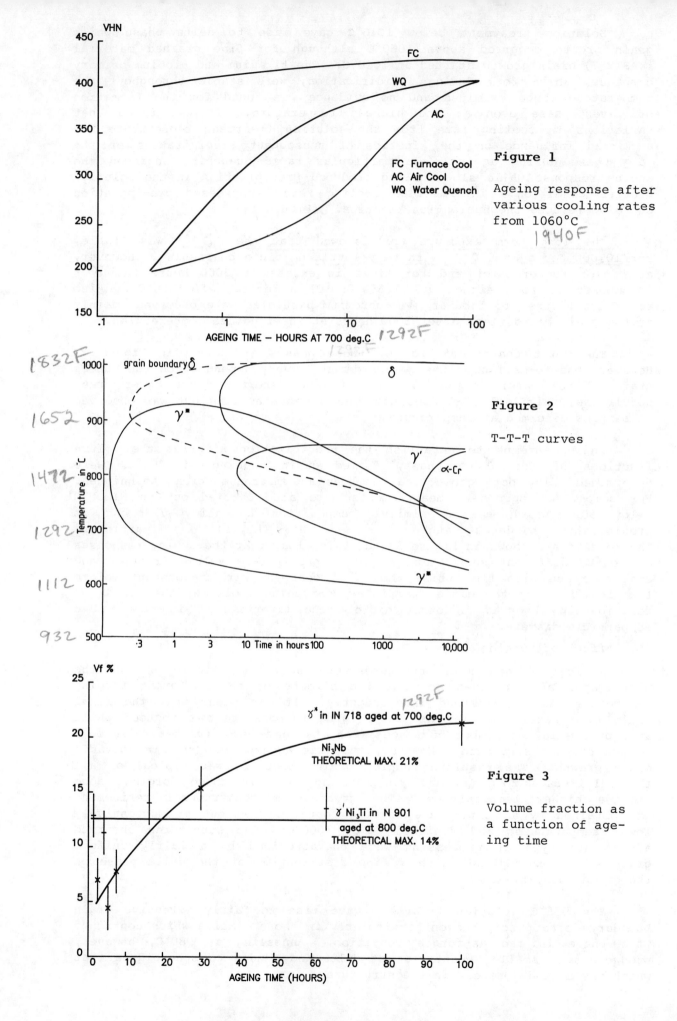

Figure 1

Ageing response after various cooling rates from 1060°C

Figure 2

T-T-T curves

Figure 3

Volume fraction as a function of ageing time

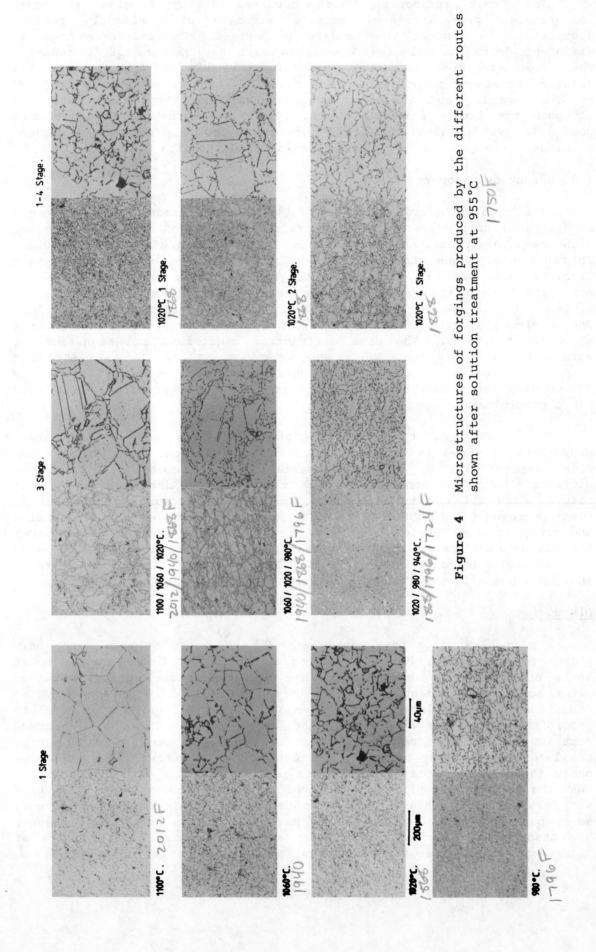

Figure 4 Microstructures of forgings produced by the different routes shown after solution treatment at 955°C

The grain refinement which occurred during forging at lower temperatures gave increased tensile strength with slightly reduced ductility. In general the tensile properties were unaffected by the different delta morphologies associated with the 955 and 980°C anneals. The stress rupture properties, however, varied with both forging and solution treatment temperature. The coarse delta-free structures produced by high working and annealing temperatures were very notch sensitive although the lives were quite high. Notch strengthening together with good lives was obtained by finish forging at low temperatures to produce a uniform fine grain size with some grain boundary delta precipitation.

(c) **Effect of exposure**

The optical microstructures of the forged discs before exposure consisted of uniform recrystallised grain structures of ASTM 8-11 together with even distributions of discrete grain boundary delta particles. This grain size was stable for all of the exposure conditions and this stability was also true of the intergranular delta phase although the precipitation in the samples exposed for 10,000 hours at 650°C showed evidence of coarsening with a tendency to form grain boundary films. The $\gamma *$ morphologies can be seen in the transmission electron micrographs shown in Figure 5. The same diffraction conditions and magnification were used to ensure that the images were comparable. It is clear that significant coarsening of the $\gamma *$ particles had taken place during the 10,000 hour exposure at 650°C and that this was concomitant with some γ' precipitation.

The effect of these microstructural changes on the mechanical properties is shown in Tables 1 and 2 which detail the tensile and creep data respectively. No signficant variations were observed between the different discs and consequently the values quoted are averages of the results from all the test pieces. The coarse structure present in the sample exposed for 10,000 hours at 650°C resulted in a loss in both tensile and creep strength while ductilities were relatively unaffected. However the slight particle coarsening evident after exposure at 575°C increased both strength and ductility while 1000 hours at 650°C had little effect, in accordance with the minor microstructural changes.

Discussion

It is clear from this work that $\gamma *$, γ' and delta are the only phases to precipitate during annealing at 500-1000°C for up to 100 hours while no significant differences were observed in the primary carbides after solution treatment at temperatures up to 1200°C. No evidence was found for grain boundary carbide films, of either M_6C or NbC, or metallic precipitates such as α-chromium or Laves phase. This is in marked contrast to some of the early literature (e.g. 1) but is supported by the later observations of Boesch and Canada (2) who showed that the Laves phase identified in previous work was probably delta. It is also likely that the greater susceptibility to carbide formation observed by Eiselstein was related to the higher carbon content of the air melted alloy used as compared to current vacuum melted material. It is not clear, however, how changes in NbC could take place anyway as these carbides form at high temperature, close to the solidus, and are very stable.

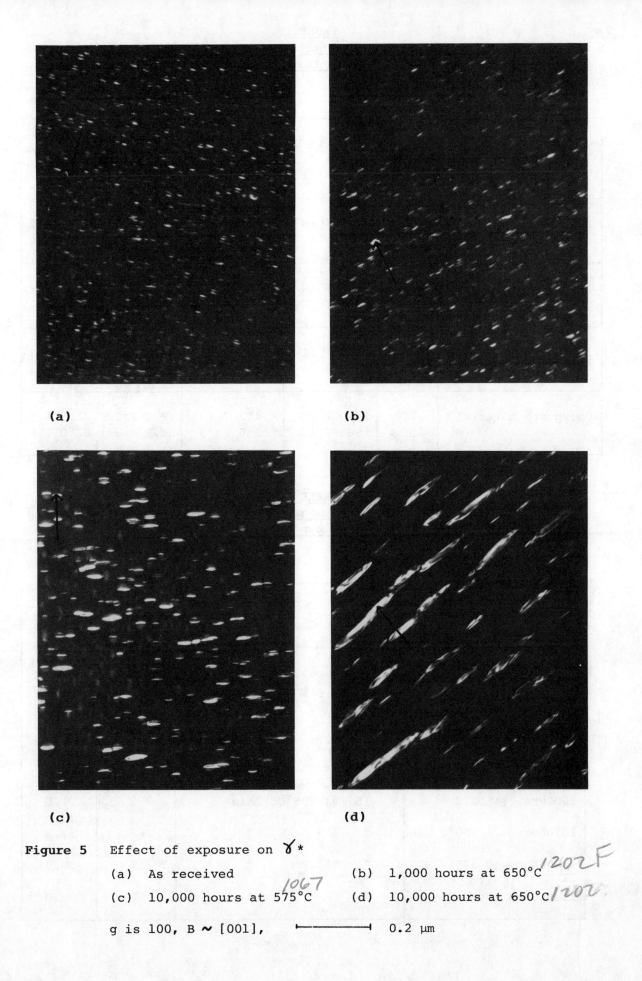

Figure 5 Effect of exposure on γ*

(a) As received
(b) 1,000 hours at 650°C 1202F
(c) 10,000 hours at 575°C 1067
(d) 10,000 hours at 650°C 1202

g is 100, B ~ [001], |⎯⎯⎯| 0.2 μm

TABLE 1
Average Tensile Properties of Forged Discs after long term exposure

Exposure Conditions	Test Temp. °C	0.2% PS MPa	UTS MPa	El.%	R.ofA. %
1000 hrs. @ 650°C	RT 650	1182 ∓ 18 987 ∓ 12	1450 ∓ 15 1188 ∓ 10	19.0 ∓ 1 27.1 ∓ 2	33.6 ∓ 2 60.7 ∓ 4
10,000hrs. @ 650°C	RT 650	1054 ∓ 19 847 ∓ 4	1382 ∓ 9 1109 ∓ 13	18.1 ∓ 2 28.3 ∓ 2	26.0 ∓ 2 61.4 ∓ 4
10,000hrs. @ 575°C	RT 650	1233 ∓ 11 1020 ∓ 8	1441 ∓ 19 1189 ∓ 16	19.9 ∓ 2 23.8 ∓ 3	34.6 ∓ 2 57.3 ∓ 4
Unexposed material	RT 650	1170 ∓ 20 965 ∓ 13	1412 ∓ 14 1156 ∓ 9	19.0 ∓ 2 25.3 ∓ 3	31.0 ∓ 5 60.0 ∓ 4

TABLE 2
Average creep rupture properties of forged discs after long term exposure

Exposure Conditions	Test Conditions	Life	Elong.
1000hrs. @ 650°C	730 MPa 650°C Notched	79	33.5
10,000hrs. @ 650°C	" " " "	11	31.0
10,000hrs. @ 575°C	" " " "	72	33.9
Unexposed material	" " " "	55	30.8
1000hrs. @ 650°C	415 MPa 705°C Plain	98	21.0
10,000hrs. @ 650°C	" " " "	69	32.8
10,000hrs. @ 575°C	" " " "	259	30.8
Unexposed material	" " " "	197	38.0

It can be seen that the three precipitation reactions observed, which are all based on Ni_3 (Nb,Ti,Al) compounds are a complex function of time and temperature. In particular the growth rate and volume fraction of the primary age hardening phase, the γ^*, is very dependent on cooling rate from the solution treatment temperature and this could certainly account for the variable mechanical properties observed in some large complex parts. The faster nucleation and growth rates observed in the rapidly cooled samples are probably associated with enhanced diffusion due to the quenched-in supersaturation of vacancies. The γ^* coarsening rates correlated well with the values quoted in the literature (e.g. 3 and 4) while the volume fractions obtained were slightly higher. The latter effect is probably due to the chemical extraction techniques used in the previous work as these can readily under estimate volume fractions as a result of partially dissolving the phase under examination. It is believed that the slow increase in volume fraction of the γ^* in alloy 718, as compared to the γ' in alloy 901 which reaches a maximum very rapidly, is due to the low diffusivity of niobium in nickel.

Various values have been quoted (e.g. 5 and 6) for the relative amounts of γ^* and γ' present in alloy 718 aged at 700°C which range from 0.2 to 0.4 for the volume fraction ratio of γ'/γ^*. This implies that half of the precipitation imaged in a typical <100> dark field micrograph should be γ' (as all the γ' but only one variant (i.e. 1/3) of the γ^* precipitates are in contrast) and this is clearly not true even of the images shown in the literature (e.g. 7). Generally γ' is differentiated from γ^* by imaging near the [001] zone axis (or crystallographically equivalent conditions) using 010, to show γ' and γ^*, and 1,½,0, to just show γ^*, so that contrast present in the former image but not the latter is attributed to γ'. However at this beam direction, and using the same notation, 010 is also close to ½,1,0 and consequently faint contrast can arise from another γ^* variant and it is likely that this has been interpreted erroneously as γ'. The analysis carried out in this work did not show any γ' to be present after ageing at 700°C for up to 100 hours but γ' was observed after ageing at 800°C such that the γ'/γ^* volume fraction ratio was 0.2 after 100 hours (long term exposure data indicates that γ' can form at 700°C after 1000 hours). Clearly the susceptibility of alloy 718 to γ' precipitation is highly temperature dependent and is probably affected by slight composition differences. Thus the precipitation characteristics of the Ni_3Nb γ^* phase present in alloy 718 are significantly different from those associated with the more commonly observed Ni_3 (Ti,Al) γ' precipitates present in the majority of superalloys since the kinetics are much slower. In particular the volume fraction of the γ^* precipitates produced on ageing between 500°C and 700°C increases substantially with annealing times up to 100 hours. An important consequence of this behaviour is that the conventional ageing treatment for alloy 718 does not produce the equilibrium structure, as the γ^* volume fraction is only $\sim$ 13% compared to a theoretical maximum of $\sim$21%. Thus it can be seen that the enhanced properties associated with long term exposure at 575°C are a result of increased precipitation without excessive particle growth while the degraded properties resulting from long times at 650°C are due to enhanced Oswald ripening at the higher temperature.

The principal failure mechanism in the test pieces was ductile and intergranular and the primary influence of thermal exposure was to soften further the grain boundary regions and so enhance the ductile appearance of the fracture. This grain boundary softening was the result of niobium depletion at the grain boundaries associated with the delta precipitation. Thermal exposure increases the depletion and eventually gives rise to the formation of γ' zones adjacent to the grain boundaries and these are weaker than the $\gamma*$ strengthened matrix.

Conclusions

(1) The primary strengthening phase in alloy 718 is body centred tetragonal, designated $\gamma* Ni_3Nb$.

(2) Creep ductility in alloy 718 is directly related to the amount of intergranular delta precipitation since this gives rise to niobium depletion and hence grain boundary softening.

(3) No loss of ductility occurs on thermal exposure at temperatures up to 650°C for times up to 10,000 hours.

References

(1) Eiselstein, H.L., 1965, ASTM STP 369, p62

(2) Boesch, W.J. and Canada, H.B., 1969, J. of Metals, 21, (10), p34

(3) Ya-Fang Han, Deb,P. and Chaturvedi, M.C., 1982, Met.Sci., 16, (12), p555.

(4) Paulonis, D.F., Oblak, J.M. and Duvall, D.S., 1969, Trans. ASM, 62, (3), p611.

(5) Chaturvedi, M.C. and Ya-fang Han, 1983, Met.Sci., 17, (3), p145.

(6) Cozar, R. and Pineau, A., 1973, Met. Trans., 4, (1), p47.

(7) Oblak, J.M., Paulonis, D.F. and Duvall, D.S., 1974, Met.Trans., 5, (1), p143

* INCONEL AND NIMONIC are Trademarks of the Inco Group of Companies

ON DEVELOPING A MICROSTRUCTURALLY AND THERMALLY STABLE

IRON-NICKEL BASE SUPERALLOY

J. P. Collier, A. O. Selius and J. K. Tien

Center for Strategic Materials
Henry Krumb School of Mines
Columbia University
New York, N. Y.

Abstract

Recent advances in the jet engine industry are placing new demands for an alloy which exhibits the excellent fabricability and heat resistant properties of IN718, but which is not hampered by the alloy's relatively low temperature ceiling. This ceiling is imposed by the microstructural instabilities that occur fairly rapidly at these higher temperatures. Seven new iron-nickel base superalloy compositions were designed to study the effect of systematically changing the Al, Ti, and Nb contents on the microstructural stability and long term elevated temperature mechanical properties. Rockwell hardness, tensile, and creep/stress rupture tests were performed on these alloys to determine their extended elevated temperature properties, while optical, scanning electron, and transmission electron microscopy, and x-ray diffractometry were used to analyze their phase stability. Increasing the Al + Nb content and the Al/Ti ratio in the alloy was found to (1) produce more of the stable γ' phase and less of the brittle δ phase, and (2) enhance the mechanical properties of the alloy. Preliminary results indicate that the alloy containing a greater Al + Nb content and Al/Ti ratio is more stable than Inconel 718.

Introduction

Inconel 718 (IN718), a precipitation strengthened iron-nickel base superalloy currently containing approximately 5.3 wt. pct. niobium, exhibits adequate strength, ductility and fatigue resistance up to 650°C. Since this temperature corresponds to the medium temperature range in jet engine rotating disk applications, and due to its good fabricability (as a consequence of the higher iron content), IN718 is one of the most used superalloys (1). At present, this alloy accounts for approximately 35 pct. of all wrought superalloy production. Recent advances in the jet engine industry are placing new requirements on turbine engines. These include a need for equal or greater strength at temperatures above the heat resistance of the otherwise highly fabricable IN718-type alloy.

Two precipitating phases are responsible for the high temperature mechanical properties of the matrix gamma (γ) phase. These two phases are gamma prime (γ') and gamma-double prime (γ'') (2,3). Gamma prime, the first of the two phases to precipitate during heat treatment, is a coherent, ordered $Ni_3(Al,Ti,Nb)$ face centered cubic structure. Gamma-double prime, which is reported to nucleate and coarsen on the γ' particles and in the matrix, is a coherent, but misfitting and ordered metastable $Ni_3(Nb,Al,Ti)$ body centered tetragonal structure (4).

Not welcome in this alloy is the brittle, needle-like delta (δ) phase. This phase is an incoherent, orthorhombic nickel and niobium rich phase that precipitates as a needle-like structure after extended times at elevated temperatures. Because of the crystal symmetry between the close packed (112) plane of the γ'' phase and the close packed (010) plane of the δ phase, it is believed that the γ'' particles act as nucleation sites for the δ phase (5). It is this δ transformation, accompanied by the overaging and partial dissolution of the precipitation strengthening phases that sets the temperature-time limits for the engine application of IN718 (6,7).

The goal of this study is to reduce the amount of overaging of the γ' and γ'' phases, and to increase the transformation time to form δ. The strategy involves systematically varying the chemistry of those elements responsible for forming these precipitating phases; i.e., aluminum, titanium, and niobium. The most promising modification appears to point at modifying the alloy in the direction of a higher (Al+Ti)/Nb atomic ratio. At present in Inconel 718 this ratio stands at approximately 0.7. Preliminary experimental results (8) suggest that allowing the atomic percent of aluminum and titanium to equal that of niobium may result in a more thermally stable γ'', due to the greater fraction of γ'' particles coarsening on γ'. Increasing the amount of aluminum and titanium should result in more γ' and especially in more γ' surface area for the γ'' to nucleate and grow. By increasing the number of γ'' nucleation sites, more γ'' particles should form. In doing so, this would produce finer sized γ'' particles upon reaching equilibrium, which could result in a reduction in the driving force to form δ.

Three sets of alloys were produced. The first alloy series examines the effect of increasing the (Al+Ti)/Nb ratio by increasing the Al+Ti content, while maintaining the niobium level at 5.25 wt. pct. The second alloy series compares the effects of aluminum and titanium on the stability of each phase by increasing the Al/Ti ratio from 0.86 to 1.67 while maintaining the (Al+Ti)/Nb ratio and titanium content. The third set, which was produced after an intensive investigation of the first two sets, follows the strategy of the second series by increasing the Al/Ti ratio to 1.78; however, this series also studies the effect of reducing the Ti content in the alloy. Work on this new series of alloys is still in progress, so not all comparative results have been acquired at this point.

Procedure

A master billet of SUPER IN718 was supplied courtesy of The Wyman-Gordon Company to be used as the base material to produce the standard and seven modified alloys used in this study. This billet was cut into equally sized pieces. Each piece was vacuum induction melted (VIM) courtesy of the Special Metals Corporation into 15 lb ingots and alloyed to the desired chemistry, see Table I. The alloy designated as IN718-1 is the standard "SUPER" IN718 alloy.

Table I. Alloy Chemistries (wt. %)

Alloy	Al	Ti	Nb	Ni	Fe	Mo	Cr	C	(Al+Ti)/Nb	Al/Ti
IN718-1	0.46	0.95	5.26	bal	18.25	3.05	18.00	0.031	0.65	0.86
4	0.63	1.34	5.28	bal	18.30	3.05	17.80	0.028	0.90	0.83
9	0.53	0.96	4.32	bal	18.30	3.02	18.15	0.31	0.85	0.98
10	0.68	0.97	4.91	bal	18.30	3.03	18.10	0.031	0.86	1.24
11	0.87	0.96	5.42	bal	18.00	3.02	17.80	0.031	0.90	1.61
11b	0.85	0.95	5.47	bal	17.97	3.05	17.70	0.035	0.87	1.59
12	0.94	0.96	5.72	bal	17.83	3.05	17.53	0.032	0.89	1.74
13	0.87	0.90	5.38	bal	18.05	3.04	17.80	0.035	0.88	1.72
14	0.86	0.86	5.66	bal	17.95	3.06	17.70	0.035	0.82	1.78

* Also contains approximately: 0.02 Mn, 0.11 Si, 0.02 Ta, 0.003 B, 0.002 S, 0.006 P, 0.006 Cu, 8 ppm Mg, 20 ppm Sn

A homogenization practice first had to be determined for the new experimental alloys. This was done by a solutionization and hot working step. The forging temperature window was determined for each alloy by heat treating cut pieces and optically analyzing them for any incipient melting (IMP) or δ phase (9).

Forty three pieces measuring 1.5 cm^2 x 0.15 cm were cut from each billet. Following this, they were solutionized at 1110°C for six hours, heat treated for various times and temperatures, and quenched. The precipitated phases were then extracted from each sample in a 1 pct. citric acid, 1 pct. ammonium sulfate, distilled water solution, and the extracted residue was filtered from the solution. X-ray diffraction patterns were produced from each extracted residue and the phases identified. In addition, diffraction patterns were produced for these alloys in their fully-aged condition to (1) determine whether any extraneous phases had formed due to these modifications, and (2) measure the changes in the relative amount and lattice parameters of each phase.

Microstructural information for this study was determined using optical, scanning electron microscopy (SEM), and transmission electron microscopy (TEM). The samples were electropolished in a 5 pct HCl, 5 pct $HClO_4$, ethanol solution, and etched with Kalling's waterless reagent. The TEM specimens were jet polished at -10°C in a 10 pct $HClO_4$, 30 pct n-butanol, ethanol solution.

Following commercial specification tests, ambient temperature, and 650°C

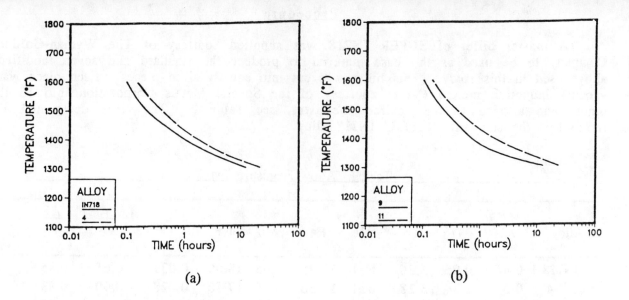

Figure 1 - Temperature-Time-Transformation Curves for Alloys (a) 1 and (4), and (b) 9 and 11

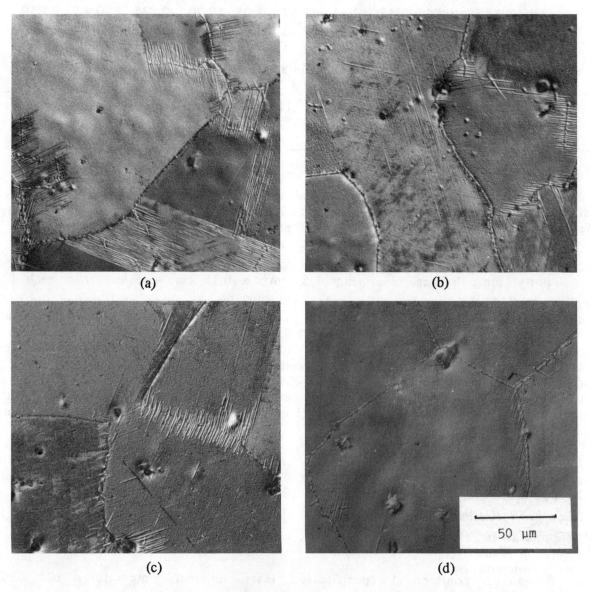

Figure 2 - Optical Micrographs of alloys (a) 1, (b) 4, (c) 9, and (d) 12 Heat Treated for 600 Hours at 760°C

tensile tests were performed on these alloys by the Wyman-Gordon Company.

Two types of tests were employed to examine the mechanical properties of these alloys after long exposure times at elevated temperature. The first of these, the Rockwell hardness test, is a simple and effective test used to indicate any changes in strength upon varying heat treatments which could be related to a change in microstructure. In this case the alloys were solutionized at 1110°C for six hours and heat treated at 760°C for times ranging from 1 to 600 hours. The second type of test consisted of comparing the results from two sets of 705°C creep-stress rupture tests. The first set was tested after the standard heat treatment practice, while the second set was tested after being heat treated at 732°C for 1000 hours following the standard heat treatment. Both sets were machined into creep bars after receiving their specific heat treatments.

Results

The diffraction patterns produced for the alloys in their fully-aged condition reveal that the alloys with the highest (Al+Ti)/Nb and Al/Ti ratios contain less δ and more γ' (9). Increasing the Al/Ti and (Al+Ti)/Nb ratios in the alloy causes an increase in the lattice parameters of the γ' and γ'' phases, most notably in the c-axis of the γ'' phase, see Table II. In addition, increasing these ratios delays the formational times of the δ phase, as seen in the δ Temperature-Time-Transformation curves presented in Figure 1.

Table II. X-Ray Lattice Parameters of γ' and γ'' Phases

Alloy	γ' a-axis (Å)	γ'' a-axis (Å)	γ'' c-axis (Å)
1	3.606	3.626	7.416
4	3.607	3.625	7.416
9	3.606	3.625	7.410
10	3.607	3.626	7.417
11	3.607	3.626	7.423
11b	3.607	3.626	7.422
12	3.608	3.627	7.429
13	3.606	3.625	7.422
14	3.607	3.629	7.418

Optical micrographs of selected alloys heat treated at 760°C for 600 hours are presented in Figure 2. It can be seen that the alloys with the highest (Al+Ti)/Nb and Al/Ti ratios, i.e., alloys 4 and 12, contain the least amount of δ.

Figure 3 reveals dark-field TEM micrographs of alloys 1, 4, 9, and 12 following a 100 hour heat treatment at 760°C. The average particle sizes of the γ' and γ'' phases are given in Table III. In addition, the volume fractions of the γ' and γ'' in these alloys were determined using the thin foil method, Table IV (10).

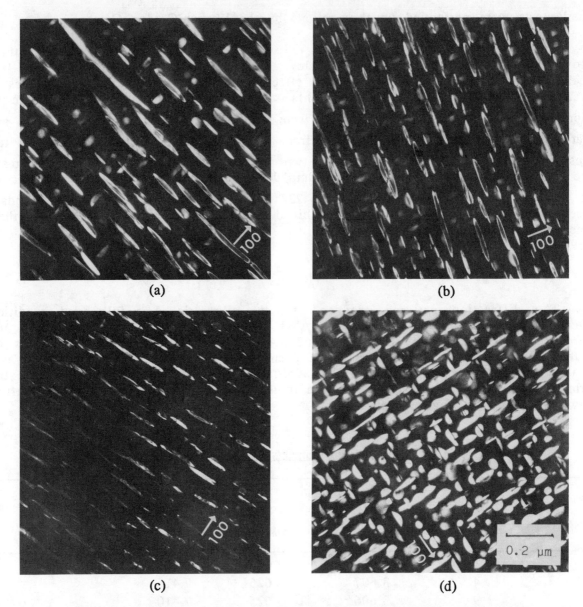

Figure 3 - Dark-field images of alloys (a) 1, (b) 4, (c) 9, and (d) 12 heat treated at 760°C for 100 hours. Dark-field micrographs were taken with the (100) ($\gamma' + \gamma''$) orientation.

Increasing the Al/Ti and/or the (Al+Ti)/Nb ratios in the alloy increases the size and volume fraction of the γ' phase, and decreases these values for the γ'' phase. The number of γ'' particles nucleating on γ' particles was found to increase by an order of magnitude with these increased ratios. In addition, Table III reveals that increasing the Al/Ti ratio causes a decrease in the γ'' particle length/thickness ratio.

Ambient temperature, and 650°C tensile tests are given in Table V. Figure 4 reveals the results from the Rockwell hardness tests, and Table VI gives the 705°C creep-stress rupture results from the alloys tested in their initial commercially heat treated condition, and also following a 1000 hour exposure at 732°C.

Increasing the Al/Ti and (Al+Ti)/Nb ratios enhanced the yield and ultimate tensile strengths of the alloy at both temperature ranges studied. The results from the Rockwell hardness and 732°C creep/stress rupture tests indicate a marked improvement in long time elevated temperature mechanical properties when the Al/Ti and/or the (Al+Ti)/Nb ratios are increased in the alloy.

Table III. Average γ' and γ" Precipitate Sizes After 100 Hour Heat Treatment at 760°C

Alloy	γ' Diameter (μm)	Length (μm)	γ" Thickness (μm)	Length/Thickness
1	0.045	0.090	0.016	5.6
4	0.048	0.064	0.011	5.8
9	0.032	0.084	0.011	7.6
10	0.044	0.065	0.008	8.1
11	0.051	0.052	0.006	8.7
11b	0.049	0.055	0.007	7.9
12	0.052	0.048	0.006	8.0
13	0.048	0.050	*	*
14	*	0.048	*	*

* - Data has not been acquired as of this time

Table IV. γ' and γ" Volume Fractions After 100 Hour Heat Treatment at 760°C

Alloy	γ' (%)	γ" (%)
1	3.5	10.1
4	5.3	6.5
9	*	*
10	4.9	3.8
11	8.1	3.0
12	9.1	2.6

Table V. Ambient Temperature and 650°C Tensile Results

Alloy	Ambient Temperature			650°C		
	0.2% Y.S. (KSI)	U.T.S. (KSI)	ε (%)	0.2% Y.S. (KSI)	U.T.S. (KSI)	ε (%)
1	172.6	204.2	27.8	143.2	165.6	29.9
4	173.0	214.4	19.3	144.4	173.6	22.1
11	168.0	211.0	22.8	143.2	173.6	17.2
11b	170.0	209.6	23.5	142.4	170.0	19.3
12	190.0	222.4	22.1	155.8	183.6	19.3
13	172.4	21102	25.7	141.2	170.4	19.3
14	185.6	218.8	21.4	149.8	174.8	16.5

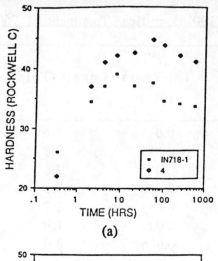
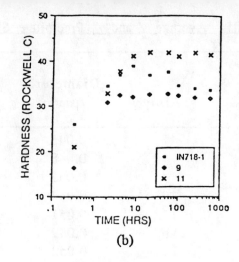
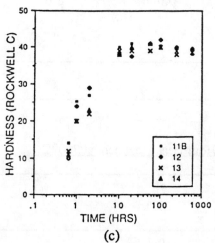

Figure 4 - Hardness Values of Alloys Heat Treated at 760°C

Table VI. 705°C Creep/Stress Rupture Results

(70 KSI)

	Initial Heat Treatment		+ 732°C - 1000 Hour Exposure	
Alloy	Rupture Life (hours)	ε (%)	Rupture Life (hours)	ε (%)
1	210.9	13.3	3.0	21.8
4	187.6	16.7	20.6	21.6
11	185.8	17.2	18.2	36.6

Discussion

One of the objectives in this study pertains to reducing the amount of overaging of the γ" particles in order to stabilize the mechanical properties of the alloy. By increasing the Al/Ti and/or the (Al+Ti)/Nb ratios in the alloy, the average length of the γ" particles was reduced by approximately 50 percent. The reason for this reduction in γ" size is twofold. First, the expansion of the γ" lattice parameters, especially the c-axis, increases the misfit of this phase within the γ matrix. As a result, it becomes more difficult for the γ" to nucleate within the γ at a given temperature. This difficulty is apparent in the observation of a greater fraction of γ" particles seen nucleating on γ' particles. The γ' particles have a larger lattice parameter than the matrix, and, therefore act as heterogeneous nucleation sites for the γ" particles. Second, the greater Al content increases the amount of γ' in the alloy. This γ' may be richer in Nb, which would deplete the amount of Nb available in the matrix to form γ".

The second objective of this study was to reduce the driving force to form δ at elevated temperatures. Figure 1 shows that increasing the Al/Ti and/or the (Al+Ti)/Nb ratios in the alloy increases the reactional times to form δ. As previously mentioned, it is believed that the close packed (112) planes of the γ" can act as the (010) habit plane for the δ phase. Increasing the Al/Ti and (Al+Ti)/Nb ratios not only decreases the volume fraction of γ", but also reduces the average size and thickness of these particles. As a result there is a decrease in γ" surface area for the δ to nucleate. Another explanation for the reduction in the amount of δ could be the result of the increased volume fraction of γ' in the alloy.

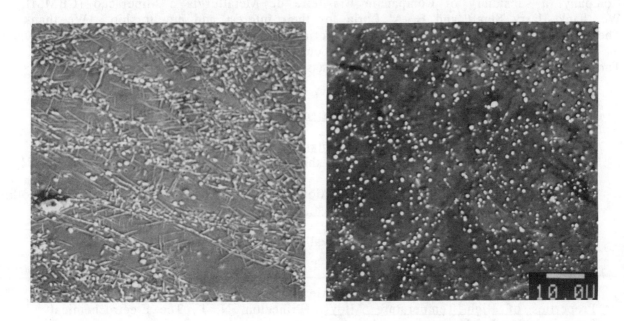

Figure 5 - Scanning Electron Micrographs of Alloys (a) 1, and (b) 11 after 1000 Hour exposure at 732°C

The mechanical results reveal that the properties of the alloys following the commercial heat treatment are somewhat enhanced with the increased Al/Ti and/or (Al+TI)/Nb ratios. These results are not surprising, since the loss in γ" volume fraction is accompanied by an increase in γ' fraction.

The results from the Rockwell hardness and the 705°C creep/stress rupture tests show a large improvement in the long term elevated temperature mechanical properties of the alloy when the Al/Ti and/or the (Al+Ti)/Nb ratios were increased. The microstructure of the standard IN718 and alloy 11 following a 1000 hour heat treatment at 732°C is presented in Figure 5. It can be seen that the standard alloy is filled with δ, while alloy 11 contains only a trace amount of this brittle phase. Considering the effect on the size of the γ" particles and the amount of δ in the alloy due to these modifications, these results are not surprising.

Concluding Remarks

These preliminary results indicate that increasing the Al/Ti ratio and/or the

Al+Ti content above that of the original IN718 results in a more thermally and mechanically stable alloy. This could be due to the higher volume fraction of γ', the smaller and more misfitting γ" particles, and/or the reduced amount of δ in the alloy. These results also indicate that more of the high temperature heat resistant niobium can be added into the alloy without any corresponding loss in stability.

The second phase of this alloy design program is currently under way, which will study the properties of a larger, commercial-sized ingot of the most stable alloy to date, alloy 12. The goal of this phase is to produce a more thermally stable to replace the existing "SUPER" IN718 alloy.

Acknowledgements

This study was made possible by a gift grant from The Niobium Products Company, a subsidiary of Companhie Brasileira de Metallurgia e Mineracao (CBMM). We thank Harry Stuart and E. A. Loria for their interest and monitorship. We thank The Wyman-Gordon Company for the material and mechanical working of the alloys, in particular, Steve Reichman and W. H. Couts for their technical advice. We thank The Special Metals Corporation for their assistance in casting the original alloys.

References

1. National Materials Advisory Board, <u>Tantalum and Columbium Supply and Demand</u>, Publication NMAB-391, (Washington D.C. National Academy Press, 1981).

2. I. Kirman, D.H. Warrington, "Precipitation in Nickel-Based Alloys Containing both Niobium and Titanium ,"<u>J. Iron Steel Inst.</u>, 99 (1967) 1264-5.

3. D.F. Paulonis, J.M. Oblak and D.S. Duvall,"Coherency Strengthening in Ni Base Alloys Hardened by DO_{22} γ" Precipitates ," <u>Trans. ASM</u>, 5 (1974) 143-53.

4. J.K. Tien and S. Purushothaman,"The Metallurgy of High Temperature Alloys," <u>Properties of High Temperature Alloys</u>, (Princeton, N. J., The Electrochemical Society, 1976) 3 - 41.

5. M. Sunararaman, P. Mukhopadhyay, and S. Banerjee, "Precipitation of the δ-Ni_3Nb Phase in Two Nickel Base Superalloys," <u>Met. Trans. A.</u>, 19A (1988) 453-65.

6. J.F. Barker, E.W. Ross, and J.F. Radavich," Long Time Stability of IN718," <u>Journal of Metals</u>, Jan 1970, 31-41.

7. D.F. Paulonis, J.M. Oblak, and D.S. Diuvall," Pecipitation in Nickel-Base Alloy 718," <u>Trans. ASM</u>, 62 (1969) 611-22.

8. R. Cozar, A. Pineau," Morphology of γ' and γ" Precipitates and Thermal Stability of Inconel 718 Type Alloys," <u>Met. Trans. A</u>, 4 (1973) 47-59.

9. J.P. Collier, S.H. Wong, and J.K.Tien, "The Effect of Varying Al, Ti, and Nb Content on the Phase Stability of INCONEL 718," <u>Met Trans A.</u>, in press.

10. J.W. Cahn and J.Nutting,"Transmission Quantitative Metallography," <u>Trans. Metall. Soc. AIME</u>, 215, (1959) 526.

MICROSTRUCTURE AND PROPERTIES OF Ni-Fe BASE Ta-718

Scott A. Loewenkamp and John F. Radavich
School of Materials Engineering
Purdue University, W. Lafayette, IN 47907

Tom Kelly
General Electric Co
Cincinnati, OH 45215

Abstract

The Nb was replaced by Ta on a one to one atom basis in alloy 718. Structural studies show that on solidification the MC, Laves, δ, and γ' phases which formed are similar to the corresponding phases in cast alloy 718. However, less Laves and δ phases form on solidification and homogenization occurs at a lower temperature in a shorter time than in cast alloy 718.

The γ' phase in TA 718 is stable up to temperatures of 1850°F which is higher than in alloy 718. The γ' to δ transition has not been observed in the normal heat treatments used in this study.

Cast TA 718 possesses tensile and stress-rupture properties as good as cast alloy 718. However, the heat treatments used in the test program are not necessarily optimum for this alloy.

INTRODUCTION

Cast alloy 718 has been used in many gas turbine applications because of its good castability, good weldability, and low cost. However, due to the high Nb content, as-cast components of alloy 718 become highly segregated during solidification requiring long homogenization cycles which affect weldability. Highly segregated areas tend to show excessive porosity which must be closed by HIP treatments. Current trends for greater Nb contents for higher strength means greater segregation will form and homogenization may not be achieved. Chemical segregation in alloy 718 produces variation in the precipitation of δ, γ' and γ phases which leads to differential overaging during engine exposure.

Recent studies of the effects of Ta in Ni base superalloys have shown that Ta stabilizes phases such as MC and γ' more than Nb. Little is known of the role of Ta as compared to Nb in 718 type alloys. To study the role of Ta on segregation and phase stability in 718 type alloys, Ta was substituted for Nb on an atom for atom basis. (Table 1) This alloy is called TA 718. While Ta and Nb are crystallographically similar and would be expected to form the same phases, it is believed that Ta might not produce the same amount of segregation as Nb. A lower tendency for segregation would then allow shorter homogenization cycles, produce high mechanical properties at higher operating temperatures and improve weldability.

MATERIALS AND PROCEDURE

The material for this study was conventionally cast into 1/4 inch thick plates. A number of the plates were cut into samples for a time-temperature study to determine the solvus behavior of the cast structures. Select samples were then given a 1600°F 1 hour heat treatment to determine residual segregation remaining after the various solution temperatures.

Other plates of the TA 718 were used for tensile and stress-rupture testing after being HIPped at 2050°F/14.7Ksi/3 hrs. Based on alloy 718 data, two post HIP heat treatments were given to the test bars: Heat treatment A was 1925°F/1 hr. + 1350°F/8 hrs. + 1150°F/8 hrs. while heat treatment B consisted of 2000°F/1 hr. + 1500°F/1 hr. + 1400°F/2 hrs.

Microstructural characterization was carried out using optical and scanning electron microscopy. Chemical analyses were carried out using EDAX type X-ray analyses on many structural features. Phase extractions were completed using a 15% HCl and 85% Methanol solution. A 5 volt potential was applied for 1 and 3/4 hours. Phase identification was then completed using X-ray diffraction patterns from the extracted residues.

Standard room temperature, 1200°F, and 1300°F tensile tests were run while stress-rupture tests were carried out at 1200°F and 1300°F at 90 Ksi.

RESULTS

Structural Study

Examination of the as-cast material reveals a prominent dendritic solidification pattern containing many carbides in the interdendritic regions as well as Laves and δ phases, Figure 1. Higher magnifications show the presence of γ' and γ in the interdendritic regions, Figure 2. Chemical analyses show the interdendritic regions to be high in Ta and Ti but have reduced levels of Fe, Cr, and Ni relative to the dendrites.

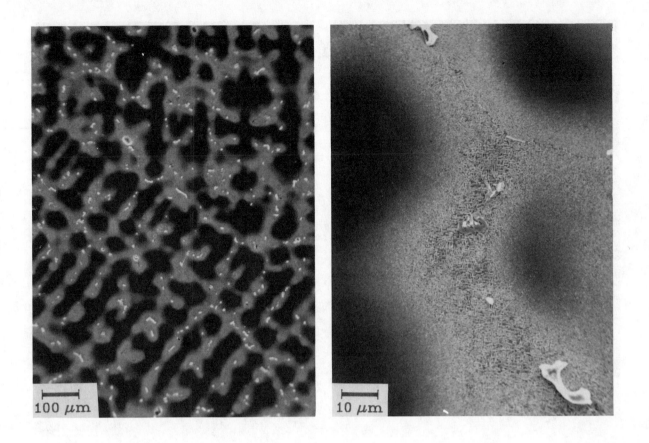

Fig. 1. TA-718 As Cast.

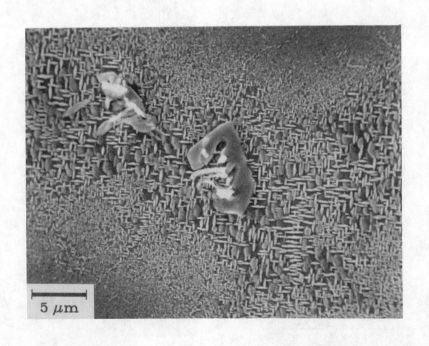

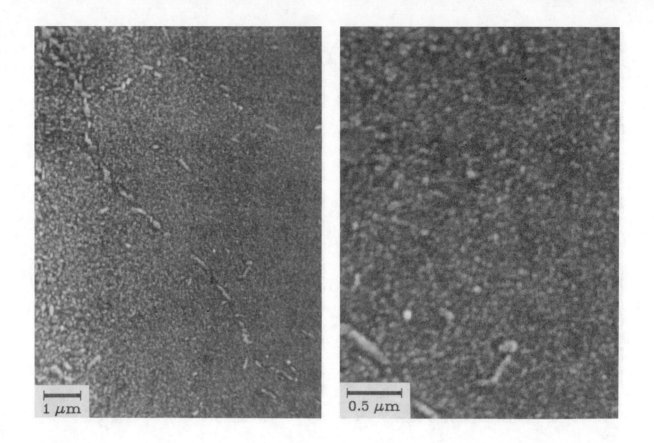

Fig. 2. TA-718 As Cast.

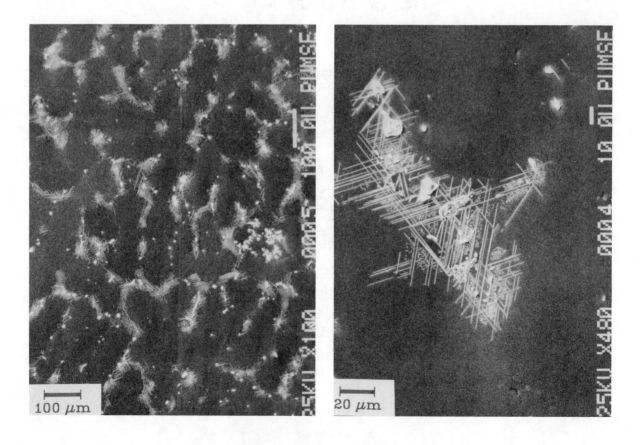

Fig. 3. Delta Phase 1900 °F/10 hrs.

X-ray diffraction patterns of residue extracted from the as-cast material confirm the presence of an MC phase, Laves phase, and γ'' phase. Lattice parameter measurements and EDAX analyses of the extracted residue show the MC phase to contain predominately Ta with lesser amounts of Ti. Ta, Ti, Fe, Cr, and Ni were detected by EDAX in the Laves phase while Ta, Ti, and Ni were detected in the γ'' residue. The γ'' phase was extracted using the 15% HCl and 85% Methanol solution.

Time-Temperature Study

When exposed 10 hours at 1900°F, plates of δ phase form as seen in Figure 3. Within the plate areas a fine disk shaped structure is present. After 2000°F for 1 hour, the MC and Laves distribution have undergone little change, but the amount of δ decreases and the disk shaped phase disappears, Figure 4.

The MC and Laves phases are increasingly solutioned by the application of higher temperatures and/or longer times. Figure 5 shows the amount of MC present after 2050°F for 1, 3, and 20 hours of heat treatment. As samples are heat treated at higher temperatures, a reduction is observed in MC carbies in both size and quantity.

Residual Segregation Study

The samples given the 4 hour heat treatments at 2000, 2050, and 2150°F were given a 1600°F/1 hour tag heat treatment to precipitate γ'' in areas of high Ta and Ti content. Figure 6 shows the effects of the 4 hour heat treatments on homogenization. Even after the 2150°F F/4 hour heat treatment, the original solidification pattern is visible indicating that residual segregation remains after this heat treatment.

Mechanical Property Results

The results of the mechanical properties are given in Table 2. Room temperature tensile values of over 155 Ksi show that the strength of TA 718 is as good as alloy 718 and far above the GE specification for tensile properties. Heat treatment B produces a smaller drop in YS than Heat Treatment A and an ultimate tensile strength of 130 Ksi at 1300°F and 1200°F indicates that TA 718 retains its strength at higher temperatures than that of alloy 718.

The stress-rupture life at 1300°F is only a few hours for both heat treatments, but the 1200°F life of TA 718 ranged from 30 to 140 hours which easily meets the GE specification for cast alloy 718. Such wide scatter is common in alloys like 718 and is expected in TA 718.

Structural studies were carried out on the broken test bars in order to understand the structural response of the TA 718 to post HIP heat treatments. Figure 8 shows the structures produced by A and B heat treatments. It is apparent that different thermal treatments can produce different responses in TA 718 and the heat treatments selected for the mechanical test program are not necessarily optimum for TA 718. Additional thermal treatments on solutioned TA 718 samples show that the γ'' phase precipitates as the major phase at a temperature of 1850°F without δ plate phase precipitation.

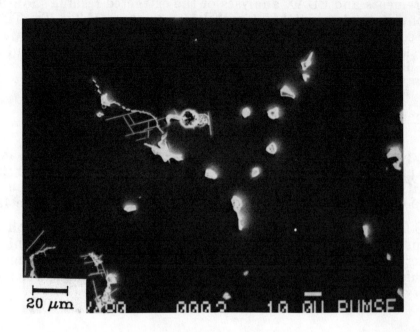

1 Hour

Fig. 4. Phase Stability @ 2000°F.

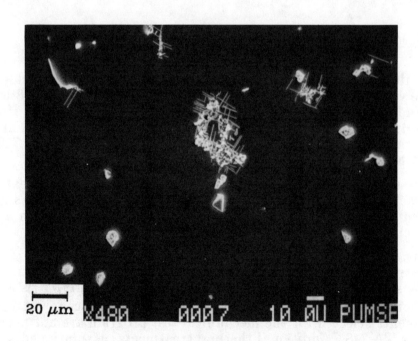

4 Hours

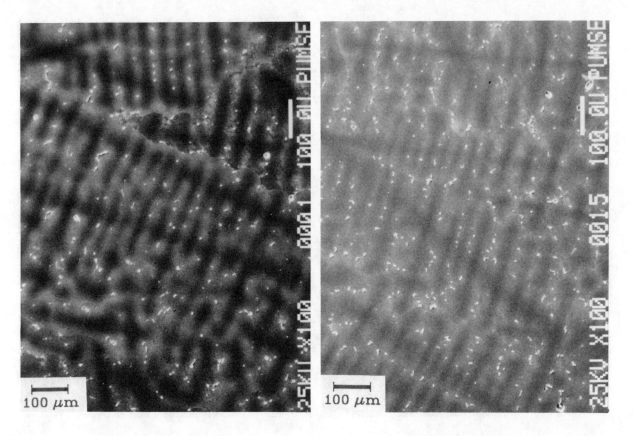

2000°F/4h+1600°F/1h   2050°F/4h+1600°F/1h

Fig. 5. Segregation Pattern after Heat Treatment.

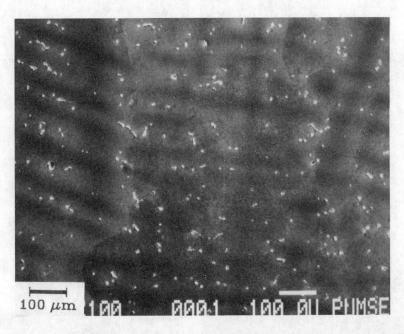

2150°F/4h +1600°F/1h

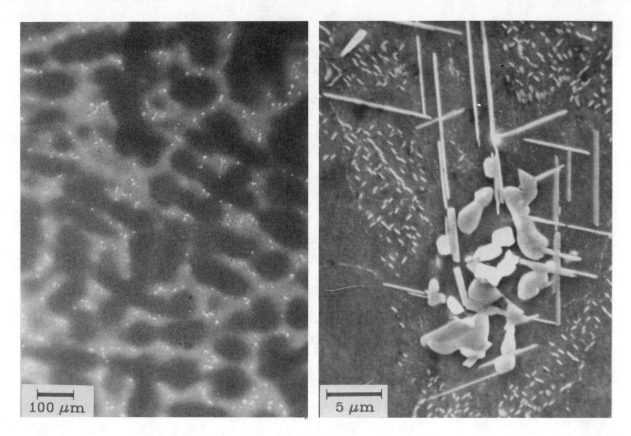

Heat Treatment A.

Fig. 6.

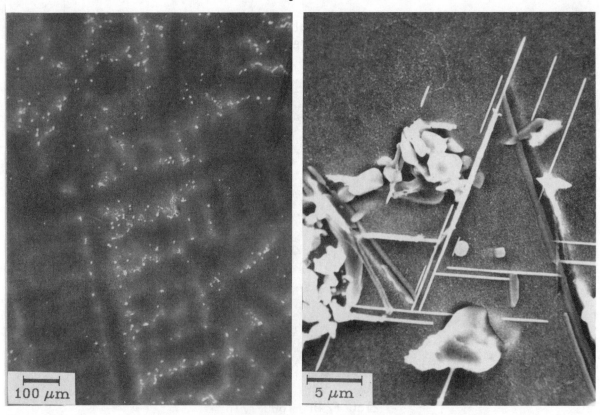

Heat Treatment B

TABLE I. TA 718 Composition.

TA-718		Typical IN-718	
wt%	at%	wt%	at%
48.6		Ni	52.5
19.2		Cr	18.5
18.0		Fe	18.5
0.02		Nb	5.1
9.1	3.1	Ta	---
3.0		Mo	3.0
1.04		Ti	0.9
0.47		Al	0.5
0.0043		B	---
0.044		C	0.04
0.02		Si	0.02
0.0018		S	---
0.0036		O	---

TABLE II. Mechanical Test Results.

TENSILE TEST

I.D. #	TEMP (°F)	ULT STRENGTH	.2% YIELD	% R.A.	% ELONG
A-1a	Room	155.2Ksi	114.9Ksi	24	20
A-1b	Room	155.4	121.2	34.2	17.9
B-1a	Room	154.5	131.3	22	9
B-1b	Room	160.7	136.7	21.6	10.8
Inconel 718*	70	149	122		9
A-3a	1200	122.4	105.2	25	11
B-3a	1200	129.5	113.6	37	8
Inconel 718*	1200	128	104		9
A-2a	1300	122.2	109.1	15	8
B-2a	1300	132.2	119.4	20	5
Inconel 718*	1300				

STRESS RUPTURE (90Ksi)

I.D. #	TEMP. (°F)	LIFE (Hrs)	% R.A.	% ELONG
A-5	1200	88.1	7.5	2.3
A-6	1200	66.2		
B-5	1200	30.0	2.0	1.8
B-6	1200	141.0	3.8	1.3
Inconel 718§	1200	20		
A-4	1300	1.2	13.7	3.3
B-4	1300	2.8	2.0	0.9

* Reference 2
§ Reference 1

61

DISCUSSION

Results of the structural studies of as-cast samples showed that Ta is similar to Nb in that Ta rich MC, Laves, δ, γ'', and γ' form during solidification. The as cast segregation contains high Ta content in the interdendritic regions while the Fe and Cr are enriched in the dendritic areas. The degree of segregation appears to be less in TA 718 as the Laves phase in Ta rich areas solutions in 4 hours at 2050°F while Laves does not solution in cast alloy 718 until temperatures of about 2150°F are reached.

The γ'' and γ' phases which form during cooling down of the as-cast material and then solutioned will re-precipitate during heat treatment in a higher temperature range than found in conventional 718. However, the γ' phase appears to be more difficult to show metallographically than that of alloy 718. The γ'' in TA 718 appears to be more chemically inert as it is extracted in solutions which dissolve regular γ'', and unlike the γ'' behavior in alloy 718, the γ'' precipitates out to temperatures of 1850°F without the presence of a transition δ phase.

A comparison of the Nb-Ni and Ta-Ni phase diagrams shows a higher degree of solubility of Ta in Ni than Nb in Ni. This may account for the reduced segregation effects observed in TA 718 and the ease with which TA 718 is homogenized.

CONCLUSIONS

Cast TA 718 possesses tensile and stress-rupture properties at least as good as cast IN 718. The MC, Laves, δ, and γ'' phases formed in TA 718 are similar to the corresponding phases in IN 718 with Ta taking the place of Nb. Less Laves and δ phases form on solidification and homogenization occurs at a lower temperature in a short time. The γ'' strengthening phase in TA 718 is stable at 1850°F which is much higher than γ'' in IN 718. The transition of γ'' to δ has not been seen in the normal heat treatments used in this study.

REFERENCES

1. Lecture Notes, MSE 540, Prof. J. F. Radavich, Purdue University, Fall 1987
2. C.T. Sims, N.S. Stoloff, and W.C. Hagel, <u>Superalloys II</u>, John Wiley & Sons, 1987
3. S. M. Jones, M.S. Thesis, Purdue University, 1987
4. F. A. Shunk, <u>Constitution of Binary Alloys 2nd Supplement</u>, 1969.

N 18, A NEW DAMAGE TOLERANT PM SUPERALLOY FOR HIGH TEMPERATURE

TURBINE DISCS

C. DUCROCQ, A. LASALMONIE, Y. HONNORAT

SNECMA - MATERIALS AND PROCESSES DPT - BP 81 - 91003 EVRY CEDEX

ABSTRACT

This paper describes the microstructure and the properties of N 18. This new P.M. alloy was designed for application in high temperature turbine discs. N 18 is strengthened by a high volume fraction of γ' (55 %). We describe how the N 18 composition was selected to fit the SNECMA requirements ; we show that it is possible in this alloy to control, the respective volume fraction of inter and intragranular γ', as well as the grain size by thermomechanical treatments ; as compared to other P.M. superalloys such as Astroloy, IN 100 and René 95, N 18 offers a unique combination of tensile strength, creep resistance and damage tolerance.

INTRODUCTION

The development by SNECMA of jet engines with improved efficiency and specific power requires, for the turbine discs, up to 750°C, the use of a limited density forgeable Ni based superalloy which has, a good compromise of mechanicals properties :

- high yield strength [$R\ 0,2 > 1000$ MPa for $T \leqslant 750°C$] so that the disc can tolerate overspeed without burst,
- high creep resistance,
- good damage tolerance : under cyclic deformation, the crack growth rate should be as low as possible even when environmental effects and hold time under stress are taken into account.

Several P.M. superalloys fullfil the strength specification such as René 95, IN 100, MERL 76 and Astroloy. Unfortunately the crack growth rate is an increasing function of yield strength (fig. 1) so that none of these alloys can satisfy the SNECMA goal, although the closest one is Astroloy.

As a consequence a research program was undertaken involving SNECMA, two laboratories : ONERA and EMP, and an alloy producer : IMPHY S.A., to define and develop, a new P.M. alloy, at the industrial scale.

EXPERIMENTAL TECHNIQUES

The conception and development of the new alloy progressed in three phases :

PHASE A : laboratory scale : a systematic study of the microstructural properties relationships based on physical considerations was made. Small batches of powders (10-30 kg) were atomized by REP and compacted by HIP. 47 alloys derived from Astroloy and René 95 were prepared to optimize the following parameters in three steps :
1 - the carbon content
2 - the nature and amount of γ and γ' forming elements
3 - the minor elements.

PHASE B : pilot scale : from the results of phase A, five compositions were selected for a more extensive characterization of mechanical properties and structural stability. The following fabrication route was used :

- atomization by rotating electrode process (50 kg)
- screening at 65 mesh (224 µm)
- compaction by HIP and isothermal forging.

The heat treatments were : a solutioning treatment (4h at 15°C below the γ' solvus) followed by a cooling at 100°C/min + 700°C/24 h air cooled + 800°C/24 h air cooled.

The comparison was made with the help of the following tests :

- tensile deformation ($T \leqslant 750°C$)
- creep on smooth and notched specimens ($650°C \leqslant T \leqslant 750°C$) ; fatigue crack growth rate measurements at 650°C with hold time 300 s.

PHASE C : industrial scale : the best composition (N 18) was argon atomized in the IMPHY S.A. plant. Several atomization batches of 100 kg were mixed to form large blends.

A blend of 300 kg of screened powders (diameter < 106 µm) was used for the N 18 characterization.

CHOICE OF THE OPTIMIZED COMPOSITION RANGE

The composition range of the N 18 family [1] is given in table 1. The choice of the elements obeyed the following objectives:

- the peak strength of the γ' phase must be shifted toward high temperatures,
- the intrinsic strength of the γ matrix must be high.

ALLOY	Cr	Co	Mo	Al	Ti	Nb	Hf	Other	C ppm	B ppm	Zr ppm
N 18 Patent	11-13	8-17	6-8	4-5	4-5	$\leq 1,5$	≤ 1		≤ 500	≤ 500	≤ 500
N 18 this study	11,5	15,7	6,5	4,35	4,35		0,5		150	150	300
IN 100	12,4	18,5	3,2	5,0	4,4			V:0,8	700	200	600
MERL 76	12,4	18,5	3,2	5,0	4,3	1,65	0,75		250	200	450
RENE 95	13	8	3,6	3,5	2,5	3,5		W:3,5	250	75	500
ASTROLOY	14,6	16,6	5,0	4	3,5				280	280	600

Table 1 : composition (Wt%) of N18 and other high strength P.M. superalloys

γ' forming elements :

The strenghtening capability of γ' can be improved by the addition of Ta or Nb (René 95). Because of its high density Ta was eliminated ; as for Nb, it has the drawback of increasing the notch sensitivity and the high temperature fatigue crack - growth rate [see René 95 on fig. 1] , so that the maximum content of Nb was limited to 1,5 weight %. As a consequence Al and Ti are the main γ' forming elements ; the required strength necessitates a volume fraction of γ' higher than 0,5. The sum Al + Ti was taken as 0,1 with a near unity ratio Al/Ti.

γ forming elements :

Mo and W are very efficient as high temperature solid solution strengheners of γ. A detailed study (2) evinced that Mo should be preferred to W :

- the partition ratio of Mo between γ and γ' was twice that of W
- for a given W + Mo content, the notch sensitivity during creep at 650°C increased drastically with the amount of W, whereas the rupture time for smooth specimen was only slightly improved (fig. 2) ; moreover the need of keeping a low density was in favour of Mo.

Other elements :

The choice of the chromium content (11-13%) gives to the N 18 family a good oxydation resistance, without excessive precipitation of intragranular carbides, detrimental for the ductility. For the same reason a low carbon content is prefered. Some Hf ($\leq 1\%$) is added to promote the formation of MC carbides without scavenging Mo, Ti and Cr ; thus Hf has an indirect effect on the strengthening ; some Hf remains in solid solution within γ and γ' and contributes directly to the strenghtening.

The composition range of the N 18 family is given on table 1 ; on the same table is given the composition which was extensively tested : on table 2 its properties are compared with those of the reference commercial alloys.

Fig.1 : Yield strength and fatigue crack growth rate at 650°C of PM alloys SNECMA results, except René 95 [9]

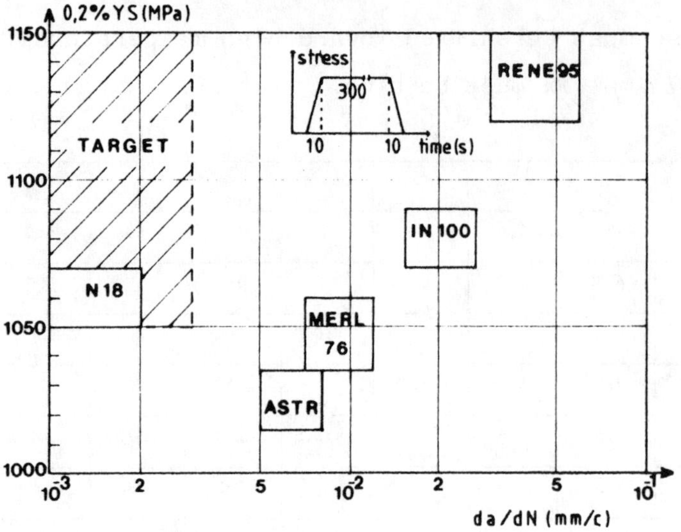

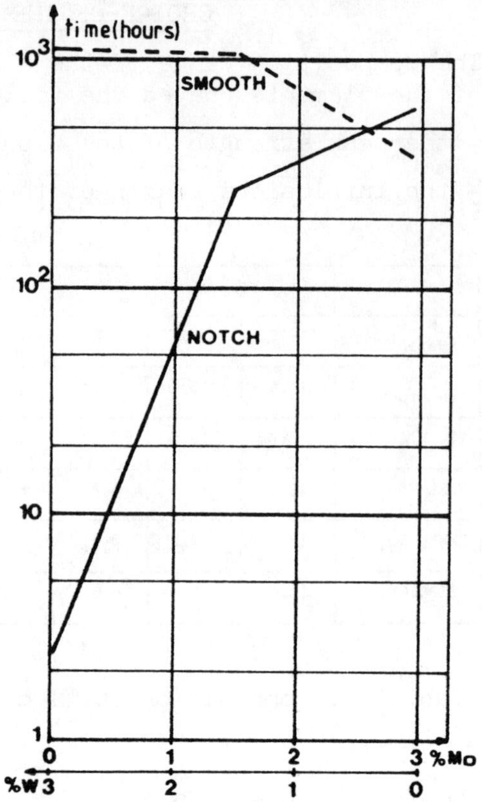

Fig.2 : Creep rupture time at 650°C 1000 MPa - Influence of W substitution (W+Mo=3%)

	Fraction γ'	solvus γ'	Density g/cm3	Nv	Md
IN 100	61 %	1185°C	7,9	2,471	0,928
MERL 76	64 %	1190°C	7,95	2,591	0,941
RENE 95	48 %	1155°C	8,27	2,217	0,915
ASTROLOY	45 %	1145°C	8,0	2,357	0,926
N 18	55 %	1190°C	8,0	2,361	0,932

Table 2 : Comparison between N 18 and commercial alloys

INDUSTRIAL PROCESSING

Powder production :

In view of mechanical characterization IMPHY S.A. produced a 300 kg blend of N 18 powder ; it was argon atomized, and then containerized in the clean rooms of IMPHY S.A. ; the transfer of powder in smaller extrusion containers was done in the clean rooms of SNECMA.

Powder control :

The present résults were obtained from powders smaller than 106 μm (150 mesh) 50 % of which was below 50 μm. The cleanliness was checked by the water elutriation procedure described in the SNECMA patent [3].

The cleanliness was equivalent to that of industrially produced Astroloy as shown in table 3. The gas content of the powder was 86 ppm oxygen and 34 ppm nitrogen (compared to 65 ppm O and 30 ppm N for Astroloy powder).

Size of inclusions	N 18*	ASTROLOY*	Specif. SNECMA
80 < size < 106 μm	6	3	< 20
63 < size < 80 μm	8	6	

* N 18   300 kg ∅ < 106 μm (150 mesh)
* ASTROLOY   2700 kg ∅ < 106 μm (150 mesh)

Table 3 : Water elutriation results on N 18 and Astroloy powders

Powder consolidation

Two routes were tested :
- compaction by HIP and conventional forging of 50 kg billets giving rise to a necklace structure,
- extrusion of 30 kg billets on the 4000 T. SNECMA press and isothermal forging of 5 kg billets on the 600 T. SNECMA press.

The second route proved to be better, because of an easier industrial practice and because of a better ultrasonic inspectability [4]. Small discs(∅ 170mm) were forged from bars (∅:90 mm). The heat treatments were similar to those given above for phase B ; several solutioning temperatures were tried, results are presented here for two cases :

4 h at 1110°C giving a small grain size (5 μm) (fig. 3a) or solutioning
4 h at 1165°C giving a medium grain size (12 μm) (fig. 3b).

In both cases the cooling rate after solutioning was 100°C/min and the precipitation treatments were 700°C/24 h A.C. + 800°C/4 h A.C.. For the medium grained alloy discs were forged at a reduced scale (∅ 180 mm) and bench tested (specially for overspeed resistance) at SNECMA (fig. 4).

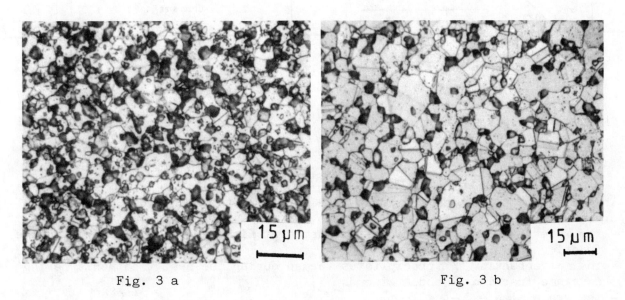

Fig. 3 a   Fig. 3 b

Fig. 3 : the two tested microstructures a) solutioning at 1110°C
                                      b) solutioning at 1165°C

Superplasticity

N 18 is superplastic at high temperature, for deformation rates smaller than 5.10^{-3} s^{-1}. The strain rate sensitivity exponent is about 0,7 as in Astroloy (fig. 5) ; for a 5 μm grain size, the flow stress is smaller than that of Astroloy and IN 100 ; this low flow stress allows the extrusion of larger bars for a given press.

Fig. 4 Isothermally forged disc preform

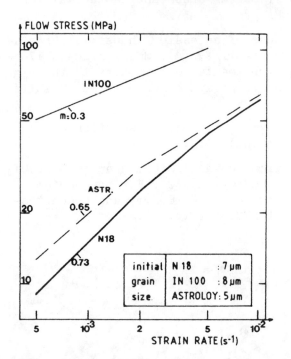

Fig.5 : flow stress of superplastic N18, IN 100 [5] and Astroloy at 1100°C

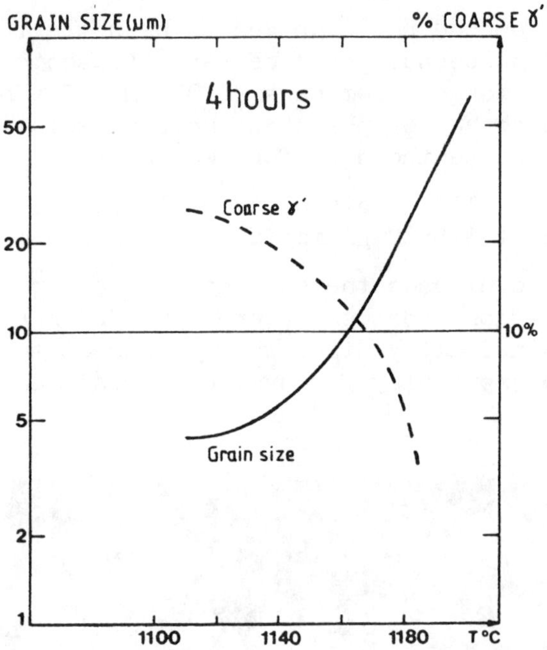

Fig. 6 : Influence of N18 solutioning temperature on fp and d

MICROSTRUCTURE

The main microstructural features which determine the mechanical properties are :
- the intergranular $\gamma'p$ precipitates formed at high temperature ; their volume fraction and size are fp and dp
- the intragranular $\gamma's$ precipitates formed during the heat treatments having a volume fraction fs and a size ds
- the grain size - d -.

The ability to control these parameters is critical when searching a balance between a high strength at low temperature and low creep rates above 600°C. For instance d must be small enough to induce a high Hall-Petch strengthening; on the contrary small d promote rapid grain boundary sliding during creep ; moreover it is well known that the crack growth rate at 650°C is a decreasing function of grain size [6].

Fig. 3 shows extruded - isoforged structure with the two different solutioning temperatures (1110°C and 1165°C). In both cases the grain size is homogeneous. In high strength P.M. superalloys, the grain growth kinetics is very low below 1100°C and drastically rapid near or above the γ' solvus ; as a consequence grain size control by heat treatment is difficult in Astroloy ; in N 18 it is easier to obtain a grain size in the range 15 μm - 30 μm due to the high γ' solvus (fig. 6). When the solutioning temperature increases from 1115°C to 1165°C, the fraction fp decreases continuously from 0,25 to about 0,15 ; since fs + fp $\simeq$ 0,55, the fraction fs of intragranular γ' is an increasing function of the temperature.

The γ' size resulting from the standard heat treatments is ds $\simeq$ 0,13 μm (fig. 7), whereas dp $\simeq$ 4 μm ; ds depends strongly on the cooling rate from solutioning but no systematic investigation of this parameter was made.

Fig. 7 : intragranular γ' after the standard precipitation treatment

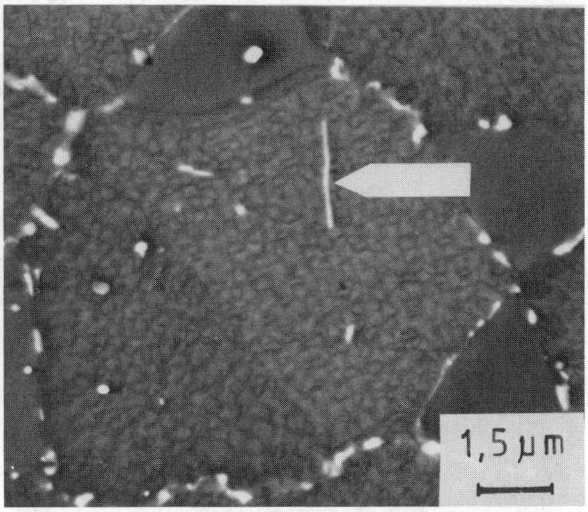

Fig. 8 : intergranular carbides and platelet precipitation (arrow)

THERMAL STABILITY

The Nv [7] and Md [8] parameters are given in table 2 for N 18 and other P.M. superalloys. The stability of N 18 as respect to T.C.P. phases (i.e. σ phase) is equivalent to that of Astroloy. This was confirmed by microscopic observations on specimens annealed between 600°C and 800°C. At 650°C a treatment of 3000 hours induced a precipitation of fine intergranular carbides but no σ phase was found, at 700°C some rare intragranular platelets were observed after 600 h ; these platelets disappear above 850°C ; their length was about 1 μm and thickness 0,1 μm (fig. 8), they were enriched in Mo and Cr and were found to contain carbon ; there is however some ambiguity about their nature : from the diffraction patterns they can be M_7C_3 or σ. In N 18 the precipitation of the platelets can be retarded by increasing the solutioning temperature. Up to 750°C the volume fraction of the platelets remains always small and does not seem to be detrimental for ductility.

No platelet but a much more abundant precipitation of carbides was observed in Astroloy annealed in the same conditions.

MECHANICAL PROPERTIES

Tensile properties

As explained before, the most extensive characterization was made on extruded iso forged structures, with grain sizes 5 μm and 12 μm.

The tensile properties are ploted on fig. 9 and compared to Astroloy and IN 100 (grain size $\simeq$ 5 μm). N 18 has the highest rupture stress ; its yield stress is intermediate between Astroloy and IN 100 up to 650°C but is less thermoactivated above 650°C.

The tensile properties below 650°C of N 18 are almost insensitive to the grain size for the range 5 μm - 15 μm. Indeed, when the grain size goes from 5 μm to 15 μm, the softening due to the Hall Petch mechanism is $\Delta\sigma_1 = k (d_2^{-1/2} - d_1^{-1/2}) = -140$ MPa for a Hall-Petch constant $k = 0.7$ MPa $\sqrt{m}$ [10]. It is balanced by an increase of the γ's volume fraction (fs goes from 0,3 to 0,4) giving a strengthening $\Delta\sigma_2/\sigma_2 = n \Delta f/f$ [for a constant ds]. $n = \frac{1}{2}$ if γ' is sheared and $n = \frac{1}{2}$ in the case of OROWAN bending so that 100 MPa $< \Delta\sigma_2 <$ 150 MPa. The total variation - 40 MPa $< \Delta\sigma_1 + \Delta\sigma_2 <$ 10 MPa, is inside experimental dispersion.

- The properties are given for a cooling rate 100°C/min from the solution treatment. The yield stress can be modified by changing the cooling rate as shown in other alloys such as Astroloy (fig. 10).
- The ductility of N 18 is always higher than 12 .

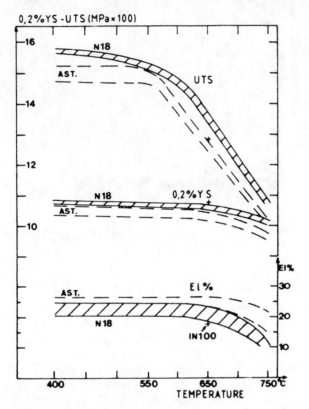

Fig. 9 : tensile properties of N 18 Astroloy and IN 10

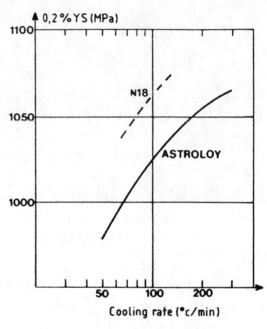

Fig. 10 : influence of cooling from the solutioning treatment on the yield stress at 650°C

Creep properties (600°C - 750°C)

In the usual Larson Miller plot, N 18 (d = 12 μm) is better than Astroloy and IN 100 (fig. 11). For a 5 μm grain size the creep properties are about the same as for Astroloy with a 5 μm grain size ; the creep ductility is about 7 % in the all range of temperatures.

Low cycle fatigue and crack growth rate

Fatigue life was measured both is stress and strain controlled tests at 550°C - 600°C and 650°C.
High stress (or deformation) levels were used so that the rupture initiated always on surface inclusions (mainly alumina) ; the fatigue life was found to be the same as in Astroloy tested in identical conditions : for instance at 600°C [f = 0,5 Hz, $\Delta\varepsilon$= 1% (Rε = 0)] the number of cycles ot failure is 2.10^3 for both alloys. The crack growth rate at 650°C with a dwell time of 300 s is given on fig. 12 ; da/dN follows the usual relatioship da/dN = C. $\Delta K^{2.5}$ [6] and is about half the rate observed in Astroloy which is one fo the most damage tolerant alloys. As shown on fig. 13 the rupture is in-

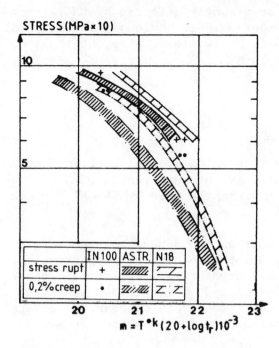

Fig. 11 : Larson Miller curve for creep deformation and creep rupture

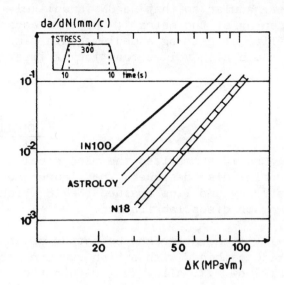

Fig. 12 : Crack growth rates at 650°C

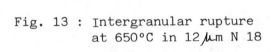

Fig. 13 : Intergranular rupture at 650°C in 12 μm N 18

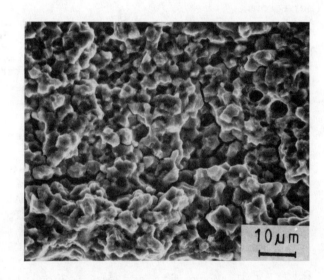

Fig. 14 : Minima design data at elevated temperature

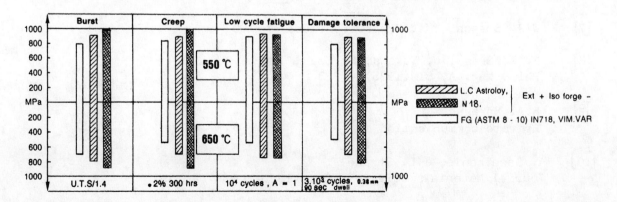

tergranular so that da/dN is unlikely to be strongly affected by the strengthening of the matrix ; the main advantage of N 18 is obviously in the possibility of getting a medium grain size (10-20 µm) while keeping, at the same time, a high volume fraction of intergranular γ'.

CONCLUSION

Figure 14 summarizes the mechanical properties which are critical for the turbine disc designer. The figure shows clearly the superiority of N 18 over Astroloy and fine grained IN 718 which are presently used by SNECMA for compressor discs fabrication.

N 18 is characterized by better creep properties and damage tolerance at 650°C combined with a high rupture strength ; the mechanical properties of this new P.M. alloy fall within the initial target (fig. 1) ; these properties are the result of easy grain size control and γ' morphology combined with optimum strengthening of the grain and of the grain boundaries. Several tons have now been extruded for bench testing of actual turbine discs.

ACKNOWLEDGEMENTS

We are grateful to DRET and STPA for financial support and to Dr J.L. STRUDEL (EMP), A. WALDER (ONERA) and J.H. DAVIDSON (IMPHY S.A.) for fruitfull collaboration.

REFERENCES

[1] C. Ducrocq, P. Lestrat, P. Paintendre, J.H. Davidson, M. Marty and A. Walder - French patent n° 86 01604 (1986)

[2] M. Marty, A. Walder and C. Diot
P.M. Aerospace materials - Luzern Nov. 87

[3] R. Morbioli - J. Ney - French patent n° 83 10717 (1983)

[4] D.R. Chang, D.D. Krueger and R.A. Sprague
Superalloy 1984 - Ed. by M. Coll, C.S. Kortovich, R.N. Bricknell W.H. Kent and J.F. Radavich
The metallurgical society of AIME (1984) - 245

[5] L.N. Moskovitz and al. : 2nd int. conf. on superalloys MCIP rep. 72-10, 1972, pp Z-1 to Z-25

[6] J. Gayda, R.V. Miner, Met. trans $\underline{14A}$ (1983) 2302

[7] W.J. Boesch - Met.Progr. 86 (1964) - 109

[8] M. Morinaga, N. Yukawa, H. Ezaki
Phil. Mag. A, 51 (1985) - 223

[9] R.A. Sprague -
Private communication

[10] A. Lasalmonie, J.L. Strudel
Journal Materials Science 21 (1986) 1837 - 1852

THE DEVELOPMENT OF ODS SUPERALLOYS FOR

INDUSTRIAL GAS TURBINES

R. C. Benn* and G. M. McColvin**

* Inco Alloys International, Inc.
Huntington, WV 25720 USA

** Inco Alloys, Limited
Hereford, HR4 9SL, England UK

Abstract

Oxide dispersion strengthened (ODS) alloys have the significant attribute of being able to retain useful strengths up to a relatively high fraction of their melting points. Moreover at the higher operating temperatures characteristic of advanced gas turbine engines, these alloys display long-term strength beyond the capabilities of conventional superalloys. The increasing use of ODS alloys, particularly in industrial applications, emphasizes the need not only to measure the creep-rupture performance and stability of these alloys but also to predict their behavior in long-term service. Characterization of the long-term properties and structural stability of commercial alloys such as INCONEL* alloy MA 6000 has identified some unique stress rupture behavior characteristics. Two distinct regions of the rupture stress vs. time curves have been identified which together show that higher design stresses for long-term service can be realized than predicted from only short term data. This behavior will be reviewed in relation to the development of a new class of ODS superalloys, typified by INCONEL* alloy MA 760, which combine the long-term strength attributes with the requirements for severe hot corrosion resistance typical of many industrial applications such as gas turbines and similar processes.

---

*INCONEL is a registered trademark of the Inco family of companies.

Introduction

The need for long-term data on new experimental alloys with only short-term results presents a familiar dichotomy. This problem of determining component performance from laboratory tests is fundamental in materials engineering. Frequently, high temperature strength data are needed for conditions where there is no experimental information. This is particularly true for long-time creep and stress rupture data where it is quite possible for the design engineer to require the creep strength to give 1% deformation in 100,000 h (~11 years) when the commercial alloy has been in existence only for about 2 years. The increasing use of ODS alloys, particularly in industrial applications, has created a need to measure the creep rupture performance of these alloys and to predict their behavior in long-term service. When other property requirements such as corrosion resistance, oxidation resistance, structural stability, etc. are progressively factored into the compositional development there are often problems of mutual exclusivity. For example, the extent to which strength is traded off against hot corrosion resistance must be balanced and preferably minimized.

From previous reported studies (1,2) on the long-term properties and related microstructures of ODS alloys it is known that INCONEL alloy MA 6000 (Table I) displays unusual behavior in that the rupture strength decreases slowly at high temperatures resulting in a leveling off (i.e. an upward departure from linearity) of the Larson-Miller curve. When the data are plotted as stress vs. log time, the upward inflection is clearly evident; as shown in Figure 1 for selected temperatures in the range 750-950°C. This unique behavior has important material design implications because the long-term stress capability of INCONEL alloy MA 6000 in the temperature range of, say, 750-1000°C is superior to initial estimates derived by extrapolation of relatively short-term data. Estimates of the long-term rupture stress capability in the 10^4-10^5 life range at a given temperature should be made strictly by using data beyond the inflection point of the stress rupture curve. As illustrated in Figure 1, σ_2 is the

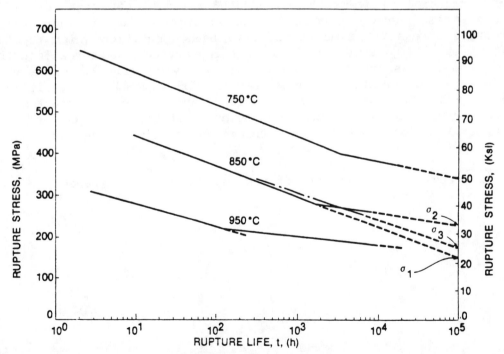

Figure 1. Implications of upward break on long term design stress estimates.

---

*INCONEL is a registered trademark of the Inco family of companies.

actual long-term stress capability rather than σ_1 or σ_3 which are extrapolations or interpolations respectively of short-term data where clearly $\sigma_2 \gg \sigma_3 > \sigma_1$.

While INCONEL alloy MA 6000 has excellent corrosion resistance in aerospace applications for which it was designed, new ODS superalloys are being developed by Inco Alloys International, Inc. for industrial applications that require extremely high corrosion resistance typical of IN-939. A wide range of experimental alloys, including the examples of alloys 1-3 given in Table I have been characterized with particular emphasis on long-term strength and corrosion/oxidation resistance. This work will be reviewed in relation to the development of a new class of ODS superalloys typified by INCONEL alloy MA 760.

Table I: Alloy Compositions (Wt. %)

Element	INCONEL alloy MA 6000	INCONEL alloy MA 754	IN-939	alloy 1	alloy 2	alloy 3	INCONEL alloy MA 760
Ni	Bal	Bal	Bal	Bal	Bal	Bal	Bal
Cr	15	20	22.5	19.7	19.5	20.8	20
Al	4.5	0.3	1.9	4.5	6.7	2.7	6
Ti	2.5	0.5	3.7	2.5	-	3.4	-
Ta	2.0	-	1.4	2.0	-	1.7	-
Nb	-	-	1.0	-	-	-	-
W	4.0	-	2.0	4.4	3.8	2.0	3.5
Mo	2.0	-	-	2.0	2.0	-	2.0
Co	-	-	19.0	-	-	9.7	-
C	0.05	0.05	0.15	0.05	0.044	0.057	0.05
B	0.01	-	0.009	0.012	0.011	0.014	0.01
Zr	0.15	-	0.10	0.075	0.15	0.20	0.15
Y_2O_3	1.1	0.6	-	0.6	0.6	0.6	0.95

Experimental Procedure

Experimental heats were mechanically alloyed initially in 10-S attritors (approx. 6 kg charge) used for compositional development. Selected alloys have been scaled up to commercial size 100-S (35 kg) attritors and a large-scale, commercial horizontal ball mill production route is now being developed. The mechanically alloyed powder was screened, canned and extruded followed by rolling as necessary. The bar products were directionally recrystallized, heat treated and evaluated for mechanical properties and structural characteristics. Cyclic and isothermal oxidation tests were run at 1100°C. Burner Rig hot corrosion tests were performed at 927° and 1093°C.

Results and Discussion

From consideration of the long-term rupture behavior of INCONEL alloy MA 6000 (1,2) it is reasonable to assume that the characteristics shown in Figure 1 may be advantageously used in the design of more sulfidation/oxidation resistant alloys. Three groups of ODS Ni-20 w/o Cr-based superalloys were developed through compositional variations nominally represented by experimental alloys 1-3 in Table I. The alloys contained, inter alia, Al contents in the range 3-7 w/o with various levels of other γ' phase and solid solution elements.

Table II shows that the sulfidation resistance of these alloys was considerably improved over INCONEL alloys MA 6000 and MA 754 being comparable with IN-939. Moreover, the oxidation resistance was considerably

Table II: **Oxidation and Burner Rig Hot Corrosion Resistance**

Alloy	Cyclic Oxidation* Descaled Mass Change (g/m^2)	Burner Rig Corrosion** Metal Loss (μm)	Metal Attack (μm)
alloy 1	-1.55	<2.54	55.88
alloy 2	-0.093	5.08	86.36
alloy 3	-16.521	2.54	91.44
INCONEL alloy MA 6000	-1.205	25.4	88.90
INCONEL alloy MA 754	-1.703	17.78	101.60
IN-939	-25.501	2.54	78.74
IN-738	-11.678	2.54	99.06
IN-100	-1.090	Destroyed in Test	

*504 h, 1100°C, Air + 5% H_2O, 24 h Cycle to Room Temperature
**504 h, 927°C, 1 cycle/h (58 min. in flame, 2 min. out in air)
30:1 ratio of air + 5 ppm sea water to fuel (0.3% S, JP-5)

improved over IN-939 with the high Al alloy group being outstanding. The mechanical properties of these three groups were generally similar to INCONEL alloy MA 6000 with slightly lower rupture strengths as shown in Figure 2.

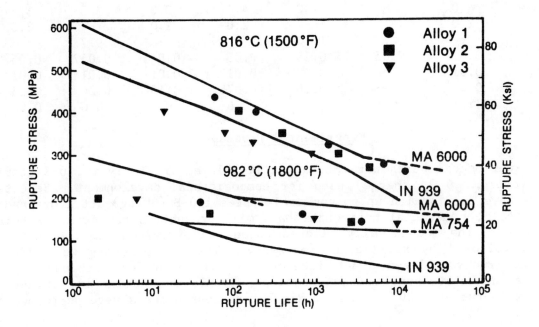

Figure 2. Comparison of stress rupture behavior of experimental alloys vs. commercial alloys.

Based on these results and some independent follow-up evaluations by an engine manufacturer, two alloys were selected from the high Al group of experimental alloys (Group 2) and one from Group 1 for scale-up production at Inco Alloys Ltd. using commercial-sized equipment. Evaluation of these alloys led to the selection of a Group 2 alloy designated INCONEL alloy MA 760 (see Table I for composition) for detailed commercial development. Similar practice to that used for other commercial ODS alloys was employed. In particular, a master alloy approach was adopted in the formulation of the final chemistry which has been found to simplify raw materials handling and enhance product reproducibility and, above all, cleanliness (3).

Based on previous studies (4) a Burner Rig hot corrosion test at 1093°C was selected as a more discerning method to compare the hot corrosion/oxidation performance of INCONEL alloy MA 760 against other materials. The results in Figure 3 and Table III clearly show the benefits of the compositional design of INCONEL alloy MA 760 in minimizing both metal loss and depth of attack.

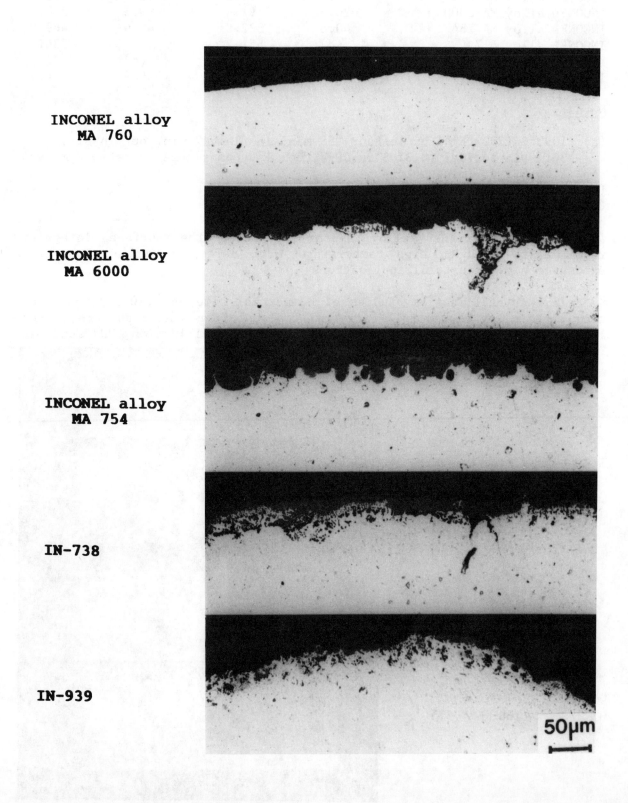

Figure 3. Micrographs showing surface integrity of tranverse sections of cylindrical specimens after 324 hours of 1093°C Burner Rig hot corrosion testing.

Table III: Burner Rig Corrosion Results[a]

Alloy	Metal Loss (μm) 324(h)	Metal Loss (μm) 504(h)	Depth of Attack (μm) 324(h)	Depth of Attack (μm) 504(h)
INCONEL alloy MA 760[b]	+7.6	+10.2	+10.2	172.7
INCONEL alloy MA 6000	20.3	91.4	81.3	358.1
INCONEL alloy MA 754	40.6	322.6	40.6	469.9
INCONEL alloy 617	(c)	551.2	(c)	736.6
IN-738	726.4	(c)	(c)	(c)
IN-939	1485.9	(c)	1546.9	(c)

Notes:

(a) 1093°C (2000°F) 1 cycle/hour (58 min. in flame 2 min. out in air) 30:1 air to fuel (0.3% S in JP-5 fuel + 5 ppm sea salt) ratio. Mass flow 5.5 ft^3/min.
(b) gain in diameter due to very adherent scale
(c) Not determined

Overall, the new alloy appears to meet the surface integrity requirements of both high temperature e.g. aerospace and/or long-term exposures e.g. industrial applications.

In recrystallization studies of commercial size bar (20 x 60 mm section), the alloy showed full recrystallization after static anneals in the 1200-1280°C temperature range. Directional recrystallization (DR) studies,

DR Temp. (°C)	Speed (mm/h)
1260	115
1260	95
1240	115
1260	75

Figure 4. Macrostructure of directionally recrystallized INCONEL alloy MA 760.

normally used to enhance the grain aspect ratio, gave excellent response, as shown in Figure 4, within this range of temperatures at speeds up to approximately 150 mm/h.

The room temperature tensile properties of the scale-up bar showed little variation between the DR conditions and were similar to results obtained from laboratory-produced material given in Table IV.

Table IV: **Preliminary Tensile Properties of INCONEL alloy MA 760***

Test Temperature		0.2 Y.S.		UTS		El	R.A.
°C	(°F)	MPa	(ksi)	MPa	(ksi)	(%)	(%)
21	(70)	1006	(146.0)	1107	(160.0)	3.6	4.0
404	(400)	1000	(145.0)	1091	(158.2)	4.0	4.0
427	(800)	960	(139.2)	1115	(161.7)	3.0	3.3
649	(1200)	969	(140.6)	1116	(161.9)	4.0	4.1
871	(1600)	615	(89.2)	652	(94.5)	4.0	13.7
1093	(2000)	141	(20.4)	141	(20.5)	14.7	41.4

*Directionally recrystallized and heat treated 1h/1240°C/FAC + 2h/955°C/AC + 24h/845°C/AC.

Similarly, stress rupture results to date on scale-up bar appear to be in general agreement with results on laboratory-produced bar which are shown in Figure 5. Some evidence of the upward inflection in rupture strength similar to INCONEL alloy MA 6000 is apparent from these results.

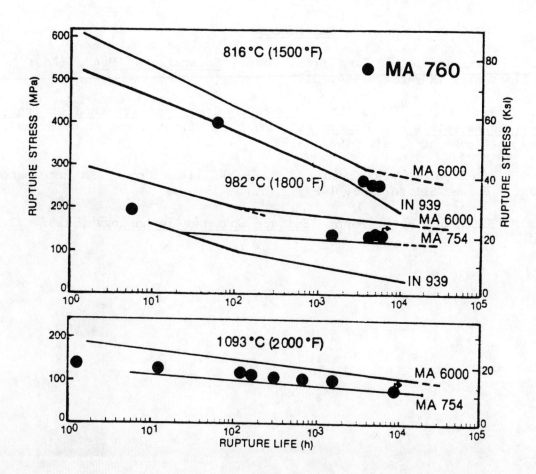

Figure 5. Preliminary stress rupture characteristics of INCONEL alloy MA 760.

A full program of mechanical property evaluations on scale-up bar is in progress. In addition, processing routes are being established to produce even larger production quantities of mechanically alloyed powder via horizontal ball mills and a range of larger section bar sizes.

Conclusions

1. Studies of the long-term stress and creep rupture characteristics of ODS alloys have identified unique stress-rupture behavior that implies significant advantages over conventional alloys particularly for long-term industrial applications.

2. A new class of ODS superalloys has been developed with strength characteristics approaching INCONEL alloy MA 6000 and hot corrosion resistance similar to IN-939 but with outstanding oxidation resistance for use, inter alia, in industrial applications.

3. An alloy composition designated INCONEL alloy MA 760 has been selected for commercial development and evaluation as, inter alia, industrial gas turbine airfoils.

Acknowledgements

The authors wish to thank Mr. G. D. Smith, with technical assistance from Mr. W. H. Wendler, for obtaining Burner Rig corrosion data and Mr. A. Zozom for technical assistance in the alloy development. Permission from Inco Alloys International to publish these results is gratefully acknowledged.

References

1. R. C. Benn and S. K. Kang Proc. Conf. "Superalloys 1984", (Eds: M. Gell et al.) TMS-AIME, 1984, 319.

2. R. C. Benn Proc. "Third Int'l. Conf. on Creep and Fracture of Engineering Materials and Structures", (Eds: B. Wilshire and R. W. Evans), The Institute of Metals, 1987, 319.

3. G.A.J. Hack, G. M. McColvin and M. P. Williams Proc. P/M Aerospace Conf., Berne, Switzerland, 1984, 2, Paper 19.

4. G. D. Smith and R. C. Benn, "Surface Coatings Technology", 1987, 32, 201.

ABNORMAL GRAIN GROWTH OF ODS SUPERALLOYS ENHANCED

BY BORON DOPING OR TORSIONAL STRAIN

Y.G.Nakagawa, H.Terashima, and K.Mino

Ishikawajima-Harima Heavy Industries Co., Ltd., TOKYO

Abstract

The abnormal grain growth of nickel base ODS alloys was promoted by various methods. The materials used in the study were as-extruded bars of MA6000 and Japanese experimental alloys, all of which showed no or very little grain growth by isothermal annealing. The samples were recrystallized into a large columnar grained structure using a directional zone annealing apparatus. Isothermal annealing could also induce the abnormal grain growth when:
1. a layer of nickel alloy containing boron was mounted on an end surface of a sample bar prior to the anneal, or
2. a large amount of plastic deformation was simultaneously exerted on a specimen during the anneal.

It was observed by TEM that there existed many grains with a high Cr content in the as-extruded MA6000. The recrystallized samples were free from dislocations and completely homogeneous in chemistry. The dynamical strain is considered to enhance solute diffusion for the homogenization, which is assumed to be one of the necessary conditions of triggering the abnormal grain growth.

Introduction

Oxide dispersion strengthened (ODS) superalloys are currently produced by the mechanical alloying process(1). Powders of oxides (yttria), elemental metals and alloys are mixed in a high energy ball mill to form composite powders with the dispersoid, from which bulk ODS ingots are obtained by hot extrusion. This dynamical consolidation process involving recrystallization during the extrusion is believed to give a fine grained and chemically homogeneous microstructure in the ingots. The transformation of the fine-grained structure to large elongated grains by abnormal grain growth is requisite for a superior high temperature strength of the ODS superalloys containing a large volume of γ' precipitates. The grain growth mechanism has not been well understood, but believed to be subjected to the initial (as-extruded) microstructure. The difference in the grain boundary energy is the primary driving force, but not sufficient for the coarsening in many cases where the reaction is considered to be triggered by additional factors, such as the γ' precipitate dissolution(2), oxide particle agglomeration(3), or release of the boundaries from impurity clouds (segregation). The effect of a plastic deformation on the grain growth has been also debated (4), but the strain energy stored by cold works in the form of high dislocation density is normally released before the transformation, and considered less important unless being induced during the heat treatment. In this paper the abnormal grain growth promoted by the zone annealing, boron doping, and high temperature deformation is described. Together with TEM microstructural studies for the as-extruded and the recrystallized ingots, some of the mechanistic discussion of the abnormal grain growth associated with the ingot production process will be introduced.

Experimental

The materials used in the present study were as-extruded bars of various nickel base ODS superalloys for which the nominal compositions are summarized in Table 1. The MA6000 is one of the commercial alloys, and an extruded sample having an 20 X 66 mm rectangular cross section was supplied by the courtesy of International Nickel Co.. The TMO-2 was developed by National Research Institute of Metals, and being used for Japanese alloy development excercises. The series of Alloy 4, 5, and 20 were also experimentally developed alloys in the authors' laboratory to study the processing parameters as a function of γ' quantity. These experimental alloys including the TMO-2 were extrusion-made at about 1050°C from attritted powders of the respective chemistries, and were supplied in the form of a 13 mm diameter bar. Some of the as-extruded samples showed no or very little abnormal grain growth by annealing at above γ' solvus temperatures, but were transformed into a large columnar grain structure by directional zone annealing. The zone annealing apparatus consisted of a 50 kHz induction heater, a pyrometer for temperature control, and a motor for travel of specimens through an induction coil. Zone anneal was done in air with typical temperature gradient of 25°C/cm across the specimen of 12 mm diameter. Iso-thermal annealing could also induce the abnormal grain growth when
1) a layer of nickel alloys containing boron was mounted on an end surface of specimen bars prior to the annealing(5), or
2) a large amount of plastic deformation was simultaneously exerted on the specimen during the annealing.

Slightly pressing amorphous tapes of Ni-15Cr-4B, or Ni-15Cr-1.7W-3.3W-1.7Ta-2.7B (wt%) against an end surface, specimens were heated in a vacuum furnace at 1200°C for 5h. The directional grain growth was observed in the materials of 5-6mm deep from the original tape interface. Small tensile bar samples with a gauge of 4 mm in diameter and 25 mm in length were machined from the as-extruded MA6000 billet, and heated at temperatures above 1100°C for 10 minutes while giving torsional deformations of various strain levels by gradually twisting the

Table 1 Chemical compositions of alloys investigated (wt%)

Alloy	C	Cr	Co	Mo	W	Ta	Al	Ti	B	Zr	Y_2O_3	Ni	γ' (vol%)
Alloy 20	0.09	14.30	-	1.75	3.66	1.77	4.15	1.82	0.025	0.09	1.1	Bal	47
Alloy 5	0.05	7.55	9.35	0.50	8.02	-	4.35	0.63	0.01	0.03	1.1	Bal	45
Alloy 4	0.05	7.20	8.25	0.51	7.79	2.14	3.24	0.60	0.01	0.03	1.0	Bal	41
TMO-2	0.05	5.9	9.7	2.0	4	2	4.2	0.8	0.01	0.05	1.1	Bal	55
MA6000	0.05	15	-	2.0	4	2	4.5	2.5	0.01	0.15	1.1	Bal	52

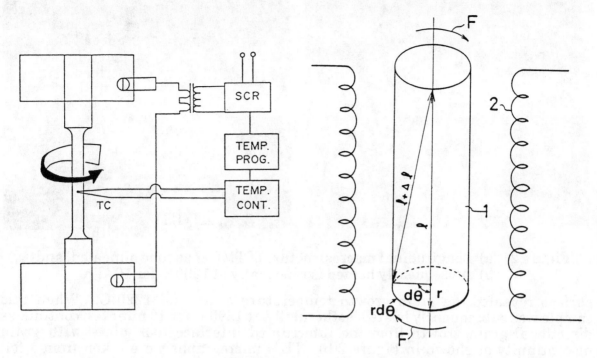

Figure 1 a) Set-up used for high temperature torsional straining
         b) Definition of torsional strain ($rd\theta/\ell$)

samples during heating. The set-up used for high temperature straining as schematically illustrated in Figure 1(a) were built in a vacuum bell jar. Heating was done by feeding current through a sample from the grips, while a strain was exerted by mechanical rotation of the grips. The precise control of the strain level was difficult but estimated in accordance to the definition described in Figure 1(b) by which a 90 - deg. axial rotation roughly corresponded to 20% deformation on the surface. Characterization of microstructures were mainly done by transmission electron microscopy (TEM) using Hitachi 700H with EDX for micro-chemical analyses.

Results and discussion

Isothermal and zone anneal of TMO-2

The as-extruded TMO-2 samples were isothermally annealed at temperatures between 1200°C and 1290°C for 30 minutes. Extensive grain growth with some directionality parallel to the extrusion direction, were observed at temperatures above 1230±10°C. It was noted that the lower temperature anneal brought the larger grain size. For instance very large grains of up to about 15 mm in length and 4 mm in diameter were observed in the samples annealed at 1240°C, close to the possible lowest for grain growth. The similar result have been reported for MA6000. Since the final grain size is thought to be determined by the nucleation rate of new grains, less nucleation for the abnormal grain growth has occured during the anneal at the lower temperature. If abnormal grain growth is triggered by either γ' phase dissolution or oxide coarsening, it is reasonable to assume that the lower temperature anneal will give the less chance of triggering. The γ' solvus of TMO-2 determined by a differential thermal analysis, and confirmed by TEM observations, is 1150 to 1160°C. From these observations it is concluded that the γ' dissolution might be a necessary but not sufficient for the abnormal grain growth.

A cylindrical specimen of 12 mm in diameter was zone annealed at travel speed of 100 mm/h. The maximum temperature in the furnace was about 1290°C. Figure 2(a) shows the longitudinal macrostructure near the final end of the zone-annealed specimen interrupted and cooled. The grain growth interface is seen in the middle of the figure. The temperature distribution measured by thermocouples at the

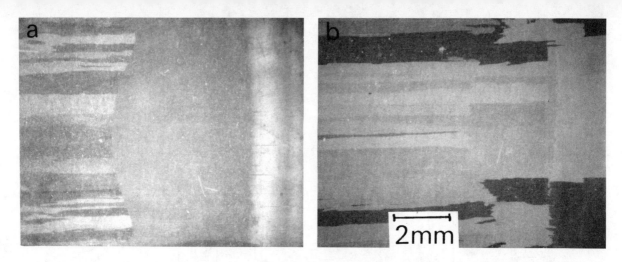

Figure 2 a) Longitudinal macrostructure of TMO-2 as zone annealed, and
b) subsequently heated isothermally at 1290°C for 10 min

surface revealed the grain growth temperature of around 1220°C. When this sample was subsequently heated isothermally at 1290°C for 10 minutes, continuous directional grain growth from the interrupted interface took place with grain discontinuity as shown in Figure 2(b). TEM micrographs were taken from a foil sliced parallel to the extrusion axis at the interface region of the zone-annealed specimen (Figure 2(a)). Figure 3 indicates the near interface structure, where the upper portion is a fine-grained area and the lower is the tip of abnormal grain growth of the columnar grains. The grain size of the fine grained region is about 0.5µm, and is not much different from the as-extruded grains. This means that the driving force for grain growth remains. Another finding was that the mean diameter of the dispersoids increased from 20 to 30 nm in the as-extruded condition to 30 to 50 nm near the interface, which would decrease back stress applied on grain growth, and thus, the dipersoid coarsening is considered as one of the secondary conditions for the grain growth triggering.

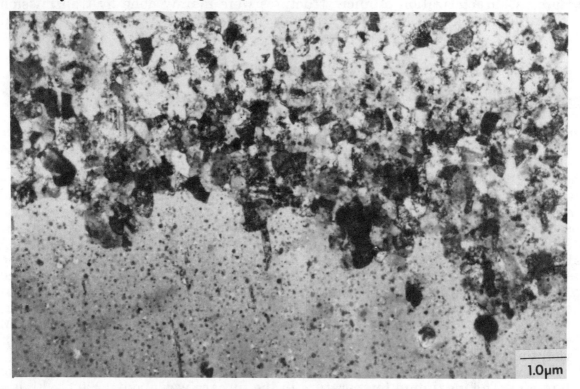

Figure 3 TEM microstructure of coarsely elongated grain and fine grain interface in the zone annealed TMO-2

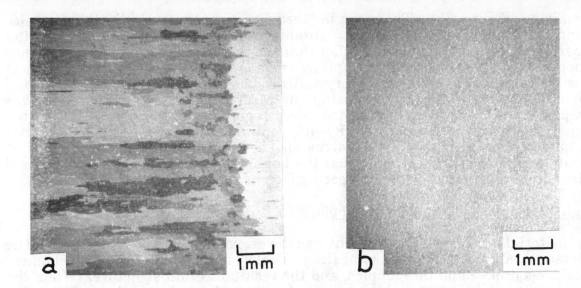

Figure 4 Longitudinal structure of Alloy 20 at 1200°C for 15h,
a) with mounting tape A, and
b) without tape

Boronizing for Alloy 4, 5, 20

The enhancement of the directional recrystallization by the boron deposition are seen in Figure 4(a) and (b), showing the longitudinal macrostructure of the Alloy 20 heated at 1200°C for 15 hours with and without boron deposition, respectively. In Figure 4(a) the tape was mounted in the left hand side of the picture, and the grain growth was developed to 6 mm in depth, which is about the distance expected for boron diffusion in nickel. No grain growth was found in the same sample without

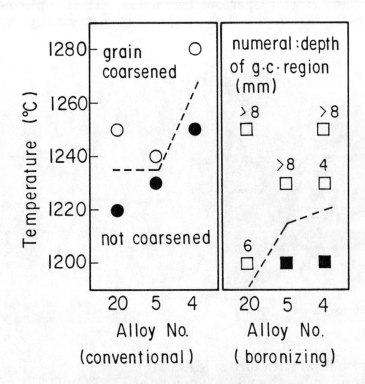

Figure 5 Effect of boronizing on abnormal grain growth temperature

boronizing (Figure 4(b)). The boron induced directional recrystallization was also found for Alloy 4 and 5. Figure 5 summarizes the results of the boronizing experiment for these alloy, suggesting that the recrystallization temperature has dropped by 20-40°C by boron doping from one end surface of the extruded bars. A possible cause of the enhanced recrystallization was initially thought due to γ' solvus temperature drop by increasing the boron content. A 300 mg slice was sampled from the boronized (near tape end) region of the Alloy 4, and from the region free from the boronizing. By differential thermal analysis for the two samples it was found that the γ' solvus for the both samples were same, and at about 1140°C. So it is concluded that the boronizing is nothing to do with the γ' solvus, and contributes to other triggering factors.

High temperature straining for MA6000

The material used for this experiment was the as-extruded MA6000 which could be recrystallized only by the zone annealing. Table 2 summarizes thermomechanical conditions imposed on the samples, and the resultant grain structure. When the total strain exceeded 30% and the temperature was above the γ' solvus of the alloy (1160 - 1170°C), the entire gauge length was transformed into the columnar grain structure parallel to the extrusion direction (TP13), as shown in Figure 6(a) and (b). The samples heated below the γ' solvus yielded no abnormal grain growth regardless of the strain imposition (TP08, 09), indicating importance of the γ' dissolution for the abnormal grain growth. Table 2 includes the result of microhardness measurements for the heat treated samples. Occurrence of the recrystallization was reflected in the local hardness value which dropped to near 500 Hv from 600 Hv level of the as-extruded or unrecrystallized materials. At 1180°C (above the γ' solvus), the strain level had a significant effect. When the strain was 30% or less the recrystallization took place only at a portion of the gauge length (TP12) as shown in Figure 7(a) and (b). The sample (TP10) to which 16% strain was given showed no grain growth in optical microstructures, but recrystallized grains (about 6μm) distributed in the fine grained (non-transformed) matrix were observed by TEM as illustrated in Figure 8. These recrystallized grains contained in some cases dispersoid free areas. TEM confirmed that the dislocation density in the strained but unrecrystallized specimens was low, and

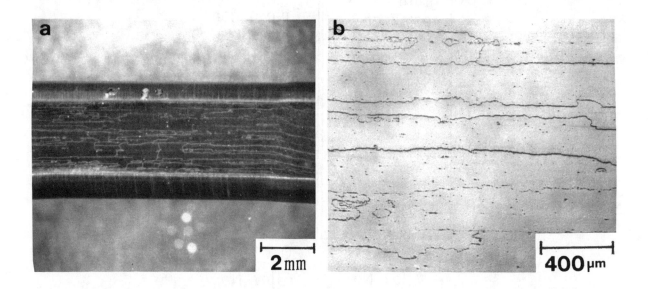

Figure 6 Longitudinal structure of MA6000
 a) strain (more than 30%) anealed at 1180°C, (TP13), and
 b) optical microstructure

Table 2 Thermomechanical conditions used for the high temperature straining tests and resultant macrostructure

× not recrystallized

Sample	Straining			Post-heat treatment		macro-structure	microhardness (Hv)	
	temp. (°C)	time (min.)	strain (%)	temp. (°C)	time (min.)			
08	1150	10	0			×	628	628
09	1150	5	7	1150	5	×	618	608
10	1180	5	16	1180	5	×	628	628
11	1180	10	0			×	618	628
12	1180	5	<30	1180	5	partial	505	608
13	1150	15	>30			full	513	

Figure 7 Longitudinal structure of MA6000
a) strain (less than 30%) annealed at 1180°C, (TP12), and
b) optical microstructure

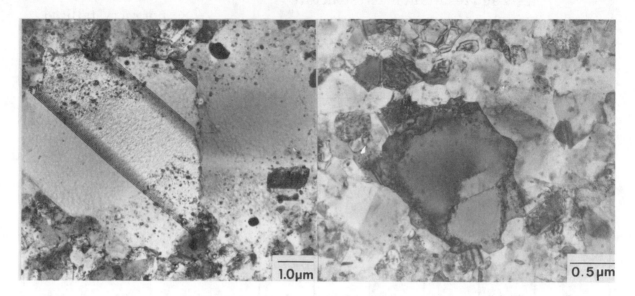

Figure 8 Recrystallized grains observed in MA6000 strain (16%) annealed at 1180°C, (TP11)

Figure 9 γ' grain observed in the as-extruded MA6000

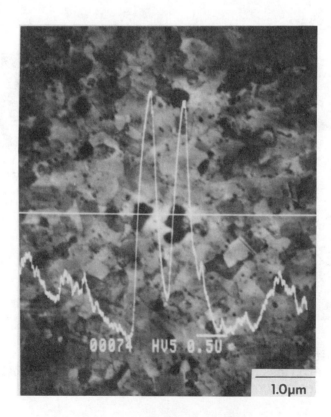

Figure 10 High Cr content grains in the as-extruded MA6000 (Cr line)

about equivalent to that of the fine grained structure of the as-extruded material. It was also observed by EDX and selected area diffraction that there existed many γ' grains (Figure 9) and grains with high Cr contents (Figure 10) in the as-extruded material. This implies that there is a strong chemical inhomogeneity in the microstructure. The γ' grains is a common feature of PM parts of high γ' nickel alloys. A typical composition of the high Cr grains measured by EDX was 75Cr-5Mo-15Ni (wt%). However, in the non-transformed area of the partially recrystallized samples (TP10, 12), non of the high Cr grains were found. The recrystallized samples were found about free from the dislocations and completely homogeneous in chemistry. Thus it is reasonable to assume that the abnormal grain growth must be preceded by the dissolution of the high Cr grains, which can be done by solute diffusion process. The roll of the strain is considered to enhance the diffusion by a number of means. The simplest model would be the pipe diffusion along the introduced dislocations, or excessive vacancies created by dislocation reactions. However the dislocation density observed in the strained sample was not much different from that in the unstrained, and not more than the density ($>10^{10}/cm^3$) normally required to see the effect. This suggests that the deformation was likely associated with the grain boundary, namely superplastic flows driven by boundary sliding and rotation where the solute transport through boundary diffusion is highly active. The recrystallization enhanced by boron doping may also related with the homogenization of microchemistry, since the amorphous tape was initially melted at the start of the heat treatment, which would reduce the chemistry difference at the near tape region.

Conclusions

1. The γ' dissolution is a necessary condition for the abnormal grain growth of nickel base ODS alloys containing a large amount of γ'.
2. It seems that the oxide coarsening triggers the grain growth.
3. The boron doping enhanced the directional recrystallization by isothermal annealing.
4. High temperature straining above the γ' solvus enhanced the directional grain growth.
5. In the as-extruded materials and the materials which did not show the grain growth, there were many grains with high Cr content. Non of them were found in the unrecrystallized region of the partially recrystallized materials.
6. The effect of the high temperature strain is considered to promote the chemical homogeneity prior to the recrystallization, which seems another triggering condition for the abnormal grain growth.

References

(1) J.S.Benjamin, "Dispersion Strengthened Superalloys by Mechanical Alloying", Met. Trans. 1(1970), 2943-51
(2) R.K.Hotzler and T.K.Glasgow,"The Influence of γ' on the Recrystallization of an Oxide Dispersion Strengthened Superalloy-MA6000E", Met. Trans., 13A(1982), 1665-74
(3) K.Mino, Y.G.Nakagawa, and A.Ohtomo, "Abnormal Grain Growth Behavior of an Oxide Dispersion Strengthened Superalloy", Met. Trans. 18A(1987), 777-784
(4) R.F.Singer and G.H.Gessinger,"The Influence of Hot Working on the Subsequent Recrystallization of a Dispersion Strengthened Superalloy-MA6000", Met. Trans., 13A(1982) 1463-70
(5) K.Mino, K.Asakawa, Y.G.Nakagawa, and A.Ohtomo, "Directional Recrystallization of an ODS MAR M247-1% Y_2O_3 Alloy", Proceeding of an International Conference on PM AEROSPACE MATERIALS, Paper No.18, (1984), Berne

CAUSES AND EFFECTS OF CENTER SEGREGATION IN ELECTRO-SLAG REMELTED ALLOY 718 FOR CRITICAL ROTATING PART APPLICATIONS

M.D. Evans, G.E. Kruzynski

Cytemp Specialty Steel Division
P.O. Box 247
Titusville, PA 16354

ABSTRACT

This paper discusses some of the possible causes and effects of center segregation in electro-slag (ESR) remelted alloy 718. This "defect" was investigated as part of a program to develop processing parameters that would result in consistently segregation free ESR 718 ingots suitable for gas turbine rotating part applications. Various remelt parameters; including melt rate, slag chemistry and volume, and electrodes immersion depth were evaluated. These remelt trials were also monitored by special equipment. In addition, ingot chemistry (Cb analysis) and billet conversion practices were investigated. Material from the processing trials was evaluated, or characterized, by means of macrostructure, microstructure, and mechanical properties. Results of the program to date reinforce the premise that electro-slag remelted 718 ingot product has a higher susceptibility to macro-segregation than the same product that is vacuum arc remelted. However, there is also substantial evidence to suggest that the center segregation can be affected positively by controlled remelt and conversion practices.

Introduction

Electro-slag remelting of superalloys has been an established process for several applications and alloys for many years, both in the US and abroad. In recent years, the ESR process has made strides towards acceptance in more critical applications and for the more segregation prone precipitation hardening superalloys. The quest to develop successful ESR practices is being motivated by the potential benefits of the ESR material relative to the vacuum arc remelted (VAR) material. Principle among these benefits are a significant improvement in cleanliness, especially as it relates to inclusion sizes; and the reduction, or possible elimination of "white spots." Reduction in size and number of inclusions or "defects" in the material has been shown in several testing programs to improve low cycle fatigue (LCF) life.

Alloy 718 is the largest volume precipitation hardening superalloy currently being pressed into service for critical rotating disc applications in gas turbine engines for the aerospace industry. While the ESR 718 material has demonstrated improved "cleanliness" relative to VAR 718, it has not been without other potentially serious problems. Alloy 718 has always shown the greatest tendency for macro- and micro-segregation of any of the high volume, wrought superalloys due to its large temperature differential between liquidus and solidus. The ESR process only adds to the potential problems relative to VAR because it is more complex and has more process variables which have an affect on the solidification of the ingot. (Figure 1.)

The driving force towards cleaner material and longer life for some highly stressed rotating discs has led one major engine company to somewhat of a compromise. So-called "Triple Melt" 718 has been in use for some years now with excellent results. This VIM + ESR + VAR process has given some of the virtues of each of the remelting practices to the quality of the final material. The ESR step has done a commendable job of cleaning up the VIM electrode, while the final VAR remelt assures a relatively consistent, segregation free ingot. However it is still a compromise since the VAR step may re-agglomerate some of the fine, widely dispersed inclusions resulting from the ESR step into larger "clusters." In addition, even though the "electrode" for the final VAR is relatively sound and homogenous, the probability of some "white spot" formation during the vacuum arc remelting exists. Of just as much concern, however, is the high cost of the "triple melt" material as a result of the extra process steps and very significant yield losses incurred.

Remelting and Processing Trials

All of the material involved in this program was melted, remelted and converted to billet at Cytemp Specialty Steel Division. The 718 electrodes were melted and cast in a 30 ton VIM facility in Titusville, Pennsylvania. The electro-slag remelting was accomplished in a newly refurbished ESR facility in Cytemp's Bridgeville, Pennsylvania plant. Billet conversion, ultrasonic testing, macro-etch inspection, mechanical property testing, and much of the microstructural evaluation was done in Titusville.

The ESR furnace used for this development program had just gone through an extensive upgrade, which was in fact an integral part of the overall program. Previous experience with ESR 718 made in this shop showed very inconsistent ingot quality with respect to segregation; some ingots were good, and others very poor. The worst problems were linked to poor

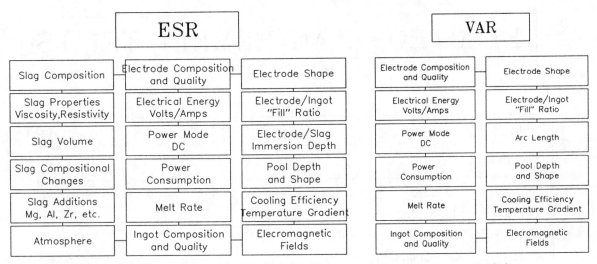

Figure 1. -- Process Characteristics of ESR and VAR (1)

electrode/slag immersion depth control resulting from an inadequate ram drive system and outdated voltage or "swing" control. In general the upgrade consisted of a new electro-mechanical ram drive system with solid state drive controls, a new load cell weighing system, and Consarc Corporation's Automatic Melt Control (AMC) system which included melt rate control and automatic voltage control.

Nine high columbium (~5.35%) ingots, 17" in diameter were remelted and evaluated in this specific program. Several other ingots of low columbium (~5.10%) in the same ingot size were melted before, and since, in order to test the newly refurbished furnaces, evaluate the effect of lower columbium on segregation, and on a production basis for other applications. Previous work to develop ESR 718 for critical applications[2] had established many of the remelt variables to relatively narrow ranges. For example, it has been shown through production experience and through determining local solidification times for ESR and VAR ingots that the melt rates required for relatively segregation free Alloy 718 are essentially the same. The slag composition has considerable effect on slag melting temperature and fluidity; and while some latitude exists for choosing the exact composition, the nominal slag composition used in these trials was held constant at the 60% CaF_2, 20% CaO, 20% Al_2O_3, with 5% TiO_2 arrived at in the earlier work. Another maxim that was arrived at earlier and adhered to in this program was that there would be no additions of the bulk slag composition during the remelt because of problems with any of the available slag addition systems. It would also be another process variable requiring close control and monitoring in a process already loaded with such variables. Also established in the earlier programs was the requirement for a very shallow electrode/slag immersion depth and carefully controlled and consistent start-up practices to assure thin slag "skins" and an adequate slag "cap" at the end of the remelt.

The remelting trials for the nine ingots are summarized in Figure 2. The first two ingots were remelted using the same parameters that were used to remelt two segregation free ingots prior to the furnace upgrades.[2] (The results of these two ingots were not reproducible, and for that reason the furnace upgrade was done.) The next two ingots were remelted using the same parameters except that a slag "conditioner" was added to lower the melting temperature of the slag and increase its fluidity. While the limits for melt rate had already been established to some extent, melt rate was varied within this range to determine the effect, if any, on the center

ELECTRO-SLAG REMELT TRIALS
Alloy 718

Heat Number	Practice	Ingot Surface	Slag "Cap" Height	Melt Rate Aim
L8742 R7	Standard	Top "necked"	4"	Standard
L8983 R1	Standard	Top "necked"	4"	Standard
L8983 R2	Add "Conditioner"	OK	4.5"	Standard
L8983 R3	Add "Conditioner"	OK	5.5"	Standard
L8983 R4	Standard	Bottom good	2.5"	Reduce 7%
L8983 R5	Add "Conditioner"	Good	6"	Reduce 7%
L8983 R6	Add "Conditioner"	Good	6"	Reduce 14%
L8983 R7	20% added Slag	Hvy "skin" btm 18"	6.5"	Reduce 14%
L8853 R3	High "fill ratio"	Top "necked"	3"	Reduce 14%

Figure 2. -- Summary of remelting trials.

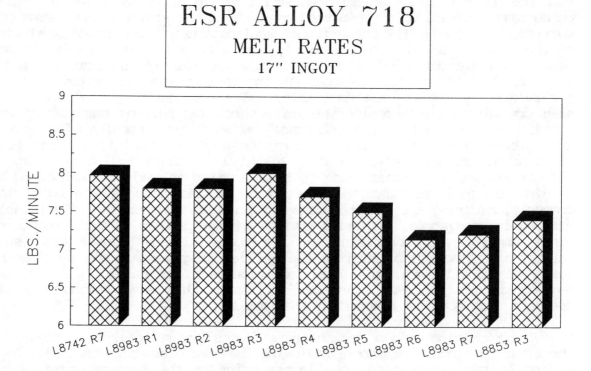

Melt rate defined by power-on to power-off.

Figure 3. -- ESR 718 melt rates.

segregation in the ingots. (See Figure 3.) The fifth ingot was remelted at a 7% lower melt rate, without the slag conditioner. The two ingots that followed were remelted at 7% and 14% lower melt rates than the first two, but both were remelted with the addition of the slag conditioner. Slag volume was increased ~20% for the eighth ingot while holding the lower melt rate and retaining the slag conditioner addition. The final ingot was remelted utilizing a higher electrode "fill ratio," the 14% lower melt rate, and the slag conditioner.

During the remelting of several of these ingots, additional monitoring devices were used to augment the standard furnace equipment. A six channel high speed recorder and a couple of oscilloscopes were used to confirm that the Consarc Automatic Melt Control, specifically the voltage "swing" control portion, was functioning adequately to maintain the electrode/slag immersion depth control that was considered necessary. Signals monitored included furnace voltage, voltage "swing" detector input, swing detector output, the automatic voltage control signal to the ram drive system, ram drive system response, and electrode speed. The additional monitoring confirmed that the furnace and controls were working to specification and supplying the control deemed imperative to the consistent remelting of ESR 718. Since it has been recognized that the successful electro-slag remelting of 718 will require a relatively small process "window," an IBM compatible PC was set up to collect and record the same signals so that they could later be analyzed in detail. The ability of the computer to handle these signals and statistically analyze them will enable the small process "window" to be precisely defined as well as monitor compliance afterwards. Examples of the "hard" copy output from this monitoring equipment is shown in Figures 4 through 6.

In addition to the supplemental monitoring of the ESR furnace controls and functions, another recorder and more measuring devices were set up to monitor the partitioning of current flow between the two possible paths. The two paths being directly down through the ingot to the base plate and out through the slag to the crucible wall and down to the base plate. The purpose of this experiment was to study the effect that short time "events" or longer term practices might have on the partitioning of the current. It has been surmised by many that some short time "events" may result in an immediate slag skin thickening and a corresponding change in current flow towards the center of the ingot which might cause a deepening of the molten ingot pool and more segregation. This scenario was not confirmed in these trials. Figure 7 shows the results obtained on three ingots. While long term conditions do affect the partitioning; such as the larger "fill ratio" and related thicker slag skin on ingot L8353 R3, short time "events" which were purposely induced during the remelting of L8779 R9 had no major effect. (While the ingot product of L8779 R9 was not evaluated as part of this program, it was a 718 electrode used to test the furnace controls and the various supplemental monitoring devices.)

Billet conversion parameters for eight of the nine ingots were kept constant. Ingot homogenization consisted of a nominal 2150° cycle for 48 to 60 hours. The ingots were then press forged to 8-½" billet using a straight draw fine grain practice. The fine grain practice was used to obtain the billet grain size required for many of the more highly stressed rotating discs and the low finishing temperatures result in a small amount of "delta" precipitation which accentuates any segregation that might be present in the macro-etch slices. One ingot, L8983 R5, was cut in half, with the bottom half converted by the same practice as the other eight ingots. However, the top half was converted by a "double-upset, double-homogenize" practice where the ingot is homogenized, upset and drawn back, re-homogenized, upset again, and finally drawn to final "hot" size.

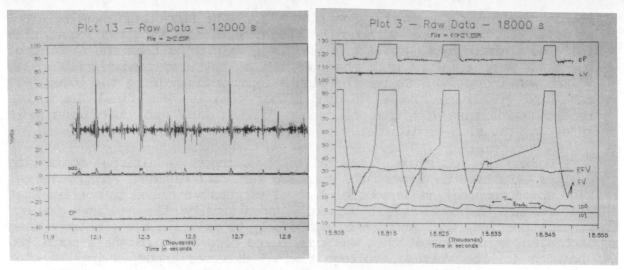

Figures 4 and 5. -- Examples of computer generated voltage signals.

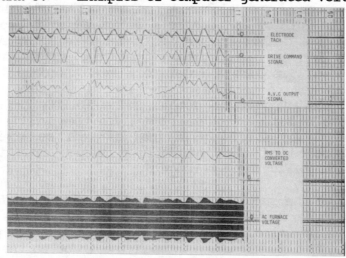

Figure 6. --Example of high speed recorder chart.

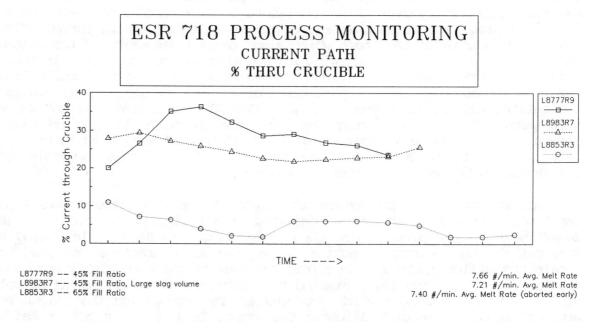

Figure 7. -- Current flow through crucible as precentage.

Discussion of Results

The effect of the various remelt trials on the ESR ingot surface quality, slag skin thickness, and remaining slag cap volume was significant. The first two ingots remelted utilizing what were considered the "baseline" parameters had fair surface appearances exhibiting a slight "necking" of the top ~12" of the ingots and 4" thick slag caps. The next two ingots remelted with the added slag conditioner had better slag caps at 4.5" and 5.5" respectively. Two methods of adding the slag conditioner were used for these two ingots. The one resulting in the larger slag cap was also the one allowing more consistency and control of the addition and was chosen to use for the remainder of the trials utilizing this slag addition. L8983 R4, remelted 7% slower without a slag addition had to be aborted with approximately 850 pounds of electrode remaining because a thicker slag skin used up too much of the slag covering. Only a 2.5" slag cap remained. The two ingots remelted at 7% and 14% lower melt rates with the slag conditioner exhibited to best ingot surfaces and finished with 6" slag caps. The ingot remelted at the low melt rate with 20% additional slag at the start showed a relatively heavy slag skin on the bottom 18" of the ingot and finished with 6.5" slag cap. The ingot with the larger electrode "fill ratio" also had to be aborted earlier due to a thick slag skin and running low on slag cover.

Billet slices were macro-etched in the "as-forged" condition in a solution of 90% hydrochloric and 10% nitric acid at 160°F. This procedure is excellent for revealing the slightest amount of center segregation that may be present in 718 wrought material. In general, the macro-etch results closely corresponded to the ingot surface quality and remaining slag cap heights. Overall, the best results were on ingots L8983 R5 an R6. However, the only segregation free product was from the top half of L8983 R5 which was converted to billet by the "double-upset, double-homogenize" practice. Other trends that become apparent were that the worst center segregation was found at the tops of ingots that had to be aborted early when the slag was running out. Ingots remelted at the higher melt rates or without addition of the slag conditioner exhibited more severe center segregation than those remelted at lower rates with the slag addition. As expected all the extreme bottom positions were free of center segregation. Representative photomacrographs of middle ingot positions are shown in figures 8 through 13.

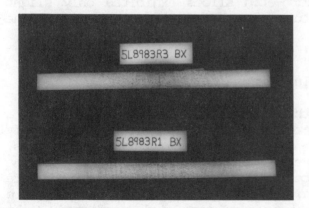

Figures 8 and 9. -- Longitudinal sections exhibiting various degrees of center segregation.

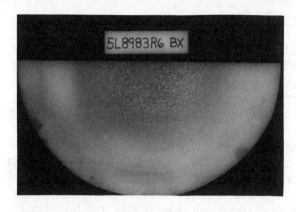 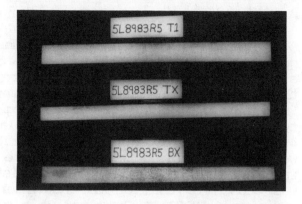

Figure 10. -- Transverse section.   Figure 11. -- Longitudinal sections of R5 ingot.

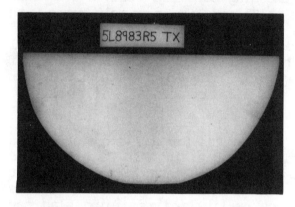 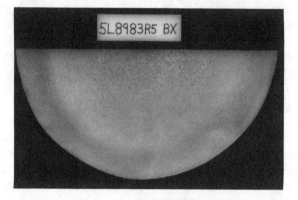

Figures 12 and 13. -- Adjacent transverse sections of R5 ingot.

Samples were cut from the center position of each macro-etch slice for metallographic examination. The various degrees of center segregation evident on the macro-etch slices show up microstructurally as "banding." This banding may show up as simply a difference in grain size, or in more severe cases, as bands of heavy "delta" precipitation. Examples are shown in figures 14 through 17. Samples were given various aging or combination of solution treat and age heat treatments to preferentially precipitate "delta" phase in the higher columbium areas of the material. Results of the metallographic examination very much reflected the macro-etch findings. However, in instances where the macro-etch slices exhibited only slight center segregation, the microstructure revealed little evidence of the segregation.

Conclusions

The ultimate objective of this program, that of producing a consistent, completely segregation free ESR 718 billet product to satisfy General Electric's requirements for critical rotating part applications, was not totally achieved. However, several conclusions as to the conditions which cause, or reduce, center segregation in the product were reached. The effects of slight macro center segregation on microstructure and some simple mechanical properties were seen and confirmed. In addition, ESR 718 material continues to function in many stator engine parts without problems or concern.

While the material results were not completely satisfactory, the ESR furnace refurbishment was successful. The supplementary monitoring devices did confirm that all furnace controls and drive systems were functioning

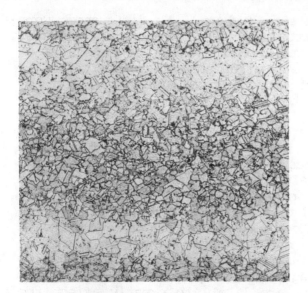 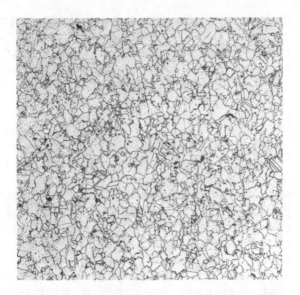

Figures 14 and 15. -- Longitudinal sections exhibiting little and no center segregation. (~100x)

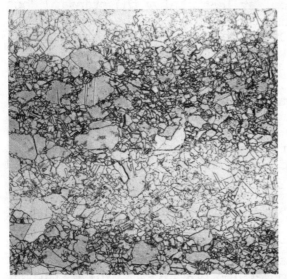 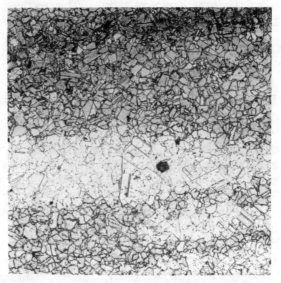

Figures 16 and 17. -- Longitudinal sections exhibiting more severe center segregation. (~100x)

properly and providing the control considered essential. The use of the computer to collect and record pertinent input and output signals from the furnace itself for "real time" monitoring and subsequent analysis of data was demonstrated. In the future this will be a valuable tool for defining the "process window" for electro-slag remelting of 718 more precisely and accurately than would be possible by other means. The computer will also provide more assurance of compliance to a specified standard procedure and is easily adaptable to the SPC techniques for quality assurance for customer satisfaction.

Clearly, center segregation of various degrees has been, and still is, the single hurdle remaining to get over in order to gain acceptance of ESR 718 product for critical rotating discs in gas turbine engines. Although the degree of center segregation varied with the various remelt and conversion practices utilized in the program, the amount was very slight on the ingot product remelted under the most favorable conditions. The microstructures examined on billet slices with only slight macro-segregation present revealed only slight grain size banding with a difference of only 1 to 2 ASTM grain size numbers. No differences were apparent in the basic mechanical properties, room temperature and elevated temperature tensiles and stress ruptures. Further work is scheduled to more precisely quantify the degree of segregation present in this material. Special "TAG"[3] heat treatments and subsequent SEM evaluation of the amounts of gamma' will be used along with chemical analysis via EDAX.

While there is no readily evident "cure" for the center segregation problem, several trends are apparent. Melt rate is still one of the most important parameters to control in order to affect pool depth and shape. Even the relatively small changes in melt rate investigated in this program were related to changes in the degree of segregation present in the ingot product. More effort will be expended to lower melt rate further, if that can be accomplished without causing the slag skin to thicken and increase resistance to heat flow out of the ingot and into the crucible cooling water. Controlling a very shallow electrode/slag immersion depth is also very critical to maintaining a thin slag skin; although, it does not appear that momentary "events" or disruptions are likely to greatly affect the slag skin thickness over the long term. As expected, the extreme bottom of the ingot, up to approximately one ingot diameter in height, does not exhibit any segregation. The extreme top, at least below the standard "hot top" crop, appeared to be no worse than the middle of the ingot over a total ingot height of approximately 100". The exception to this occurred when the slag volume was insufficient at the end of the remelt.

A minimum slag cap is required to maintain a consistent, desirable, molten pool profile. However, it appears that at the melt rates that are used for electro-slag remelting 718 that there is a practical limit on the volume of molten slag cover that can be maintained. The addition of a slag conditioner to the bulk slag composition proved to have a significant effect on the fluidity of the slag and resultant slag skin thickness and remaining slag cap volume.

The most dramatic effect, however, appears to have been the result of the ingot to billet conversion practices. The "double-upset, double-homogenized" fine-grain billet conversion practice showed a real reduction in the amount of center segregation present as compared to the single ingot homogenization, straight draw, billet conversion practice. This is further evidence that the segregation that is present is very slight and the remelting process is not far from possible complete success.

References

1. L.W. Lherbier and J.T. Cordy, "Superalloy Remelting Processes", <u>Refractory Alloying Elements in Superalloys</u>, Conference Proceedings, ASM 1984.

2. M.D. Evans, "ESR of Precipitation Hardened Superalloys for Critical Applications", <u>Proceedings - Vacuum Metallurgy Conference, 1984</u>, Pittsburgh, PA.

3. C.M. Lombard, "Effect of Conversion Practice on Segregation in ESR and VAR 718 Ingots" (M.S. thesis, Purdue University, 1987).

4. A. Mitchell, University of British Columbia, private communications, 1987-1988.

5. R.J. Roberts, Consarc Corporation, private communications, 1987-1988.

DEVELOPMENT OF GATORIZED® MERL 76

FOR GAS TURBINE DISK APPLICATIONS

R. H. Caless and D. F. Paulonis

Materials Engineering
Pratt & Whitney
400 Main Street
East Hartford, CT 06108

Abstract

The MERL 76 alloy was developed to produce components by hot isostatic pressing. To provide for enhanced shaping capability, improved material utilization, and to explore potential property benefits, Pratt & Whitney embarked on a development program to qualify Gatorized® MERL 76. The MERL 76 composition was fixed, and billets were consolidated by hot isostatic pressing (HIP), followed by Gatorizing® into the required component shape. Detailed studies were conducted to define both metal processing (e.g., HIP and forging) and heat treatment parameters. Comprehensive monotonic and fatigue testing of full-size components has demonstrated that Gatorized® MERL 76 met all program goals. This material has been used in commercial service since 1983, with excellent experience.

Introduction

The demand for higher performance aircraft gas turbine engines with longer operating lives and improved reliability is the driving force for the development of new material and processing techniques. Higher performance can frequently be achieved by operating gas turbine components under conditions which result in higher stresses and higher temperatures. To meet the need for a higher strength alloy with intermediate temperature capability (up to 1350°F) for use in disks, sideplates, etc., MERL 76 was developed in the mid-1970's for production of components by hot isostatic pressing (HIP). Alloy composition is shown in Table I. The direct HIP processing route was viewed as a method for manufacturing components very close to net shape for maximum material utilization and lowest cost (1,2).

Table I. MERL 76 Composition (Weight Percent)

Element	Ni	Cr	Co	Mo	Ti	Al	Hf	Cb	B	Zr	C
Max	Balance	12.9	19.0	3.6	4.50	5.15	0.9	1.6	0.024	0.08	0.035
Min	Balance	11.9	18.0	2.8	4.15	4.80	0.6	1.2	0.016	0.04	0.015

Intensive studies during the late 1970's demonstrated that for many larger components, typical of those required for commercial engines, the promise of near net shape manufacture by HIP was never realized (3). It became evident that for such components, Gatorizing® to achieve part shape provided greater net shape capability and improved material utilization. Furthermore, initial exploratory studies by Pratt & Whitney indicated potential property benefits for Gatorized® material. Therefore, a development program was initiated to qualify Gatorized® MERL 76. The basic approach was to utilize the existing MERL 76 composition, produce billet by HIP, and then Gatorize® (by isothermal forging) into component shapes. This paper traces the development of the process parameters for this product form and presents representative microstructures and mechanical property results generated during the development program.

Subscale Process Development

During the course of this program, forgings ranging in size from small 6-inch diameter subscale disks to full-size engine turbine disks were evaluated. To minimize the size of life limiting inclusions, -325 mesh powder, the same as used in the direct HIP product form, was selected as a product requirement. The forgings described in this paper were produced by several qualified production suppliers.

In the initial stages of this effort, HIP and forge process parameters were evaluated, and selections were made based on forgeability and resultant mechanical properties. In the evaluation of HIP parameters, time and pressure were kept constant at 3 hours and 15 ksi, respectively. These parameters represent common industry practice and also work effectively for direct HIP MERL 76. Temperature was the only variable. Small billets were hot isostatically pressed at 2050, 2100, and 2150°F (all subsolvus - the gamma prime solvus of MERL 76 is approximately 2170°F). Grain sizes of these billets increased slightly (from ASTM 10.8 to ASTM 10.2) as the HIP temperature increased; there were no major differences in gamma prime morphology as a function of HIP temperature, although the 2150°F HIP exposure did result in slightly coarsened primary gamma prime. Photomicrographs of this material are shown in Figure 1.

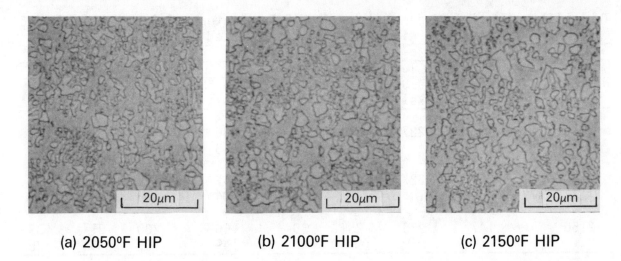

(a) 2050°F HIP   (b) 2100°F HIP   (c) 2150°F HIP

Figure 1 - Microstructure of MERL 76 processed at various HIP temperatures. Slight coarsening of gamma prime is evident for the 2150°F HIP condition.

These billets were then Gatorized® at 2050°F and a strain rate of 0.1 in/in/minute into 6-inch diameter subscale disk shapes (Figure 2). Forging proceeded smoothly with no evidence of cracking. The forge flow stresses (shown in Table II) increased from 6.6 to 12.0 ksi for the three HIP temperatures. Tensile testing of these subscale parts did not reveal any significant difference in mechanical properties among the three HIP temperatures. The tensile properties are listed in Table III. A HIP temperature in the middle of the range studied (2100°F) was selected based on the relative insensitivity of processing and property results to this temperature (thereby providing a wide tolerance band).

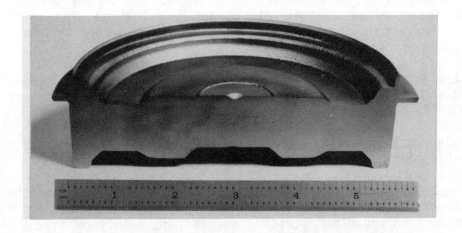

Figure 2 - Subscale Gatorized® MERL 76 disk.

Table II. Grain Size and Flow Stress of MERL 76 Versus HIP Temperature

HIP Temperature (°F)	Grain Size (ASTM)	Flow Stress (ksi)
2050	10.2	6.6
2100	10.5	9.5
2150	10.8	12.0

Table III. Tensile Properties Versus HIP Temperature of Gatorized® MERL 76 Subscale Forgings Solution Heat Treated at 2075°F

HIP Temp. (°F)	Room Temperature				1150°F			
	0.2% YS (ksi)	UTS (ksi)	%EL	%RA	0.2% YS (ksi)	UTS (ksi)	%EL	%RA
2050	150	227	31	34	146	198	31	33
					143	196	31	33
2100	149	227	29	32	147	199	31	35
					144	200	33	35
2150	147	226	25	30	150	200	31	33
					149	199	36	36

Having selected the HIP conditions, the next requirement was to select isothermal forging parameters. These were selected separately by each supplier based on trials conducted by the suppliers. The parameters selected were in the range of normal production practice: strain rate = 0.05 to 0.2 in/in/minute, and temperature 2050 to 2100°F. The mechanical properties were not affected over the range of parameters studied.

Significant development effort concentrated on the identification of a heat treatment cycle. For reference, the standard heat treatment for direct HIP MERL 76 (2) is:

o   Solution: 2125°F/2 hours/OQ
o   Stabilize: 1600°F/40 minutes/AC + 1800°F/45 minutes/AC
o   Age: 1200°F/24 hours/AC + 1400°F/16 hours/AC.

The solution cycle is conducted at a temperature below the gamma prime solvus and incorporates a rapid oil quench to solution most of the gamma prime. However, enough gamma prime is retained at temperature to pin grain boundaries and maintain a fine grain size.

The stabilize cycle is conducted at a sufficiently high temperature to relieve stresses created as a result of the prior oil quench and also to provide a uniform dispersion of intermediate size gamma prime.

The aging cycle creates a dispersion of very fine "aging" gamma prime which provides further increases in tensile strength.

Experience on the direct HIP product form demonstrated that these three heat treatment steps and an oil quench (OQ) from the solution temperature were necessary and would be retained; however, some refinement and simplification of the specific temperature/time parameters appeared desirable. The objective was to identify a sequence which: (1) was simple and low-cost, (2) minimized residual stress, and (3) resulted in the best balance of mechanical properties.

Separate studies were conducted for each step in the heat treat sequence. Most of this work was conducted on pancake forgings (produced by Wyman-Gordon) which were either 2 or 4 inches thick and weighed from 175 to 225 pounds. In addition, several 125-pound closed die disk-shaped forgings (produced and tested by Ladish Co.), which were 14-inch diameter x 4-inch thick, were used to evaluate alternative stabilize cycles.

The standard solution temperature for direct HIP MERL 76 is 2125°F. In this study, temperatures ranging from 2065 to 2125°F were examined. A 2-hour hold and oil quench (to maximize strength) were fixed at the outset, along with a standard 1200°F/24 hour + 1400°F/16 hour age. As shown in Table IV, solution temperatures in the higher part of the range produced a slight advantage. These results, coupled with some concern for quench cracking from the highest temperature (2125°F), resulted in selection of a 2090°F solution temperature.

Table IV. Tensile Properties of Gatorized® MERL 76 at Various Solution Temperatures

Solution Temp. (°F)	Room Temperature				1150°F			
	YS (ksi)	UTS (ksi)	%EL	%RA	YS (ksi)	UTS (ksi)	%EL	%RA
2065	163	236	23	25	151	206	27	28
2100	158	236	25	27	150	205	27	31
2125	165	241	24	25	153	208	27	33

Since a component in the solution and oil quenched condition contains very high residual stresses, adequate stress relief was considered a major criterion for a stabilize cycle. Two additional criteria were: (1) attainment of satisfactory mechanical properties, and (2) avoidance of embrittling grain boundary phases which can occur in this class of alloys at intermediate (1500 to 1700°F) temperatures. Temperatures in the range of 1400 to 1800°F were evaluated.

Initially, stress relief effects were evaluated by oil quenching a number of available direct HIP MERL 76 cylinders (8-inch diameter x 2-inch thick) from 2090°F, exposing each to a different stabilize heat treat cycle and then measuring the contraction which occurred on cutting the cylinder. An assessment of the effectiveness of stress relief was made by comparing these figures to the contraction measured on an as-quenched sample. Representative results from this study, as plotted in Figure 3, indicate that acceptable levels of stress relief can be accomplished at 1550°F and above. Limited property evaluation of selected conditions revealed a drop in room temperature (RT) ductility after 1550 or 1600°F exposure. Detailed metallographic, extraction replica, and energy dispersive spectroscopy (EDS) analyses (Figure 4) of samples with high and low ductility revealed the likely source of the property debit to be a semi-continuous film of sigma precipitation at grain boundaries in this intermediate temperature regime. Subsequent work has shown that by careful control of alloy chemistry, these microstructural effects (and associated property debits) can be minimized.

Similar property trends were observed on a series of closed die forgings Gatorized® and tested at Ladish Co. Using a solution treatment of 2090°F/2 hours/OQ and an age of 1350°F/8 hours/AC, tensile properties as a function of stabilize cycle are presented in Table V. Again, the RT ductility after the 1600°F stabilize dropped to slightly below goal level. All other properties exceed the goal by a substantial margin. Based on these property data and the residual stress results, a stabilize cycle of 1800°F/1 hour was selected.

A total of ten alternative aging cycles were examined. While there were no major effects on properties, a relatively simple cycle (1350°F/8 hours) provided a good property balance and was, therefore, selected as the final portion of the heat treatment sequence. The data are plotted in Figure 5.

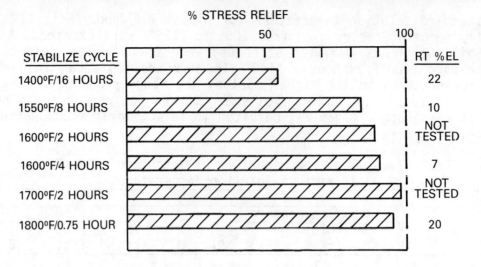

Figure 3 - Stress relief achieved with various stabilize cycles. Room temperature ductility values for selected conditions are also shown.

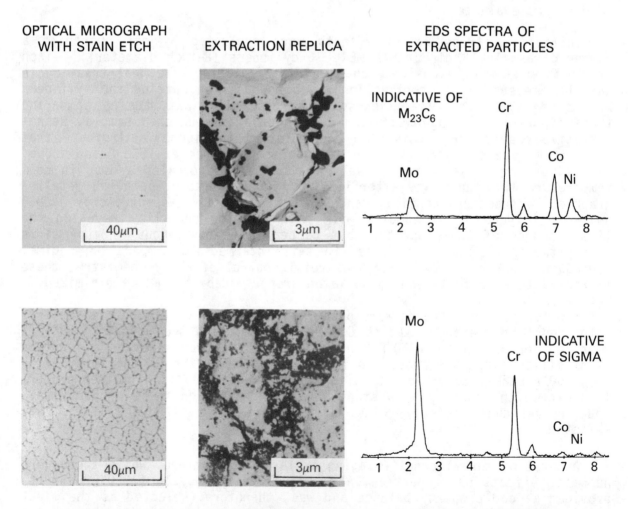

Figure 4 - Grain boundary precipitates of samples with high (top) and low (bottom) ductility.

Table V. Tensile Properties of Gatorized® MERL 76 Versus Stabilize Cycle
(Solution = 2090°F/2 Hours/OQ; Age = 1350°F/8 Hours/AC)

Stabilize Cycle	Average Tensile Properties							
	Room Temperature				1150°F			
	YS (ksi)	UTS (ksi)	%EL	%RA	YS (ksi)	UTS (ksi)	%EL	%RA
1500°F/1 Hour*	161	128	21	19	158	212	25	30
1600°F/1 Hour*	158	227	15	14	158	211	23	28
1750°F/1 Hour**	164	237	21	21	162	216	18	18
1800°F/2 Hours**	153	231	27	26	152	204	27	30
Goal	140	215	15	15	140	194	12	12

* Average of five tests
** Single test

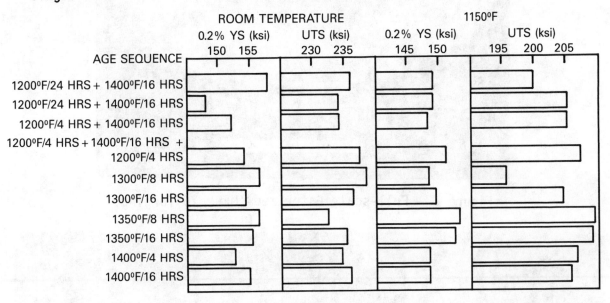

Figure 5 - Room and elevated temperature tensile properties versus age cycle.

Full-Scale Process Qualification

A comprehensive matrix of monotonic and fatigue testing was completed on full-size engine turbine disks (typical part shown in Figure 6), which were 20 to 35 inches in diameter and weighed 100 to 350 pounds. Test specimens were removed from the thickest portions of the bore, web and rim locations of these disks. A typical Gatorized® and heat treated MERL 76 microstructure is shown in Figure 7. This series of optical and replica micrographs reveals the three size populations of gamma prime (denoted as A, B, and C in Figure 7) which result from the three heat treat steps. The coarsest size (1 to 4μ) is retained during the solution cycle to prevent grain coarsening. The next larger gamma prime fraction is that which grows during the 1800°F stabilize step (size ≈ 0.2μ). The smallest size fraction (<0.05μ) is established during the 1350°F age. Grain size of this material, as shown in Figure 8, is ASTM 11-12. Tensile, stress-rupture, and creep properties (plotted in Figure 9) were similar to or slightly improved compared to those of the direct-HIP product form.

A large number of both smooth and notched low cycle fatigue tests were also conducted. As shown in Figure 10, the Gatorized® product has measurably improved properties compared to HIP MERL 76. Crack growth rate capability of the two product forms was also tested and found to be similar (Figure 11).

Since its introduction and qualification in 1983, Gatorized® MERL 76 has been specified for a variety of commercial gas turbine engine applications, most notably turbine disks, seals, and sideplates. To date, over 1000 parts have been placed in service, and performance of these has been excellent.

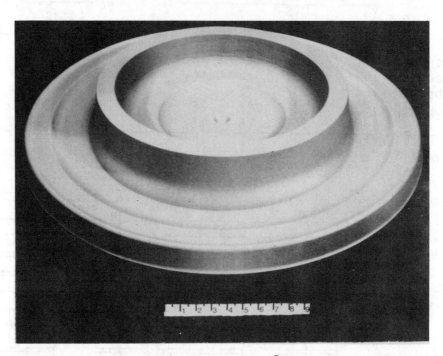

Figure 6 - Full-scale Gatorized® MERL 76 forging.

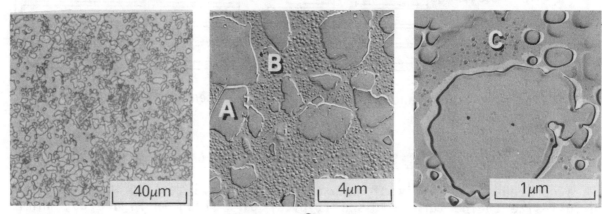

Figure 7 - Microstructure of Gatorized® and fully heat treated MERL 76 showing the three gamma prime sizes (A, B, C) as described in the text.

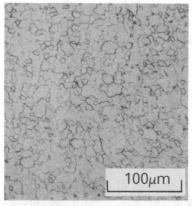

Figure 8 - Gatorized® MERL 76 sample etched to show grain structure. Grain size: ASTM 11-12.

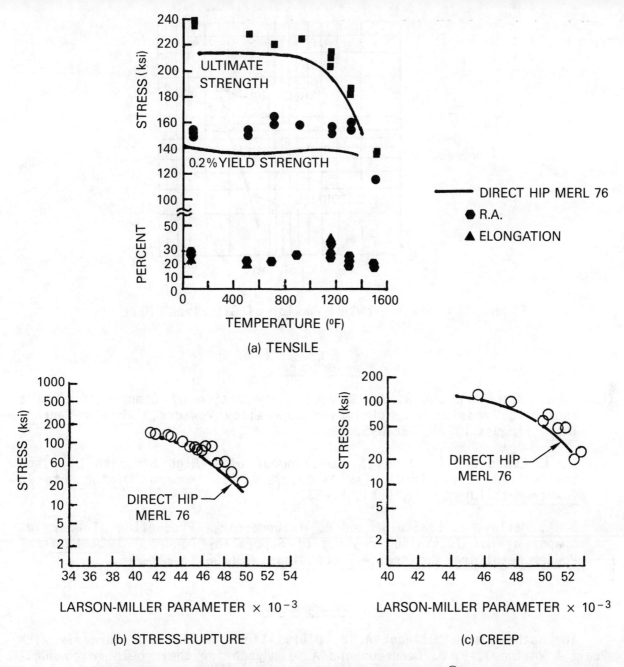

Figure 9 - Monotonic properties of Gatorized® MERL 76.

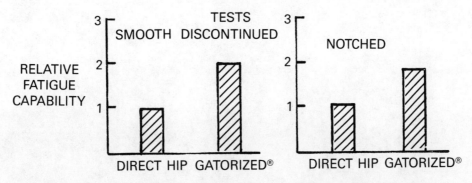

Figure 10 - Fatigue properties of Gatorized® MERL 76.

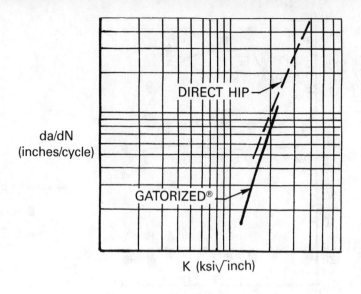

Figure 11 - Crack growth behavior of Gatorized® MERL 76.

References

1. M. J. Blackburn and R. A. Sprague, "Production of Components by Hot Isostatic Pressing of Nickel-Base Superalloy Powders," Metals Technology, (August 1977), 388-395.

2. D. J. Evans and R. D. Eng, "Development of a High Strength Hot Isostatically Pressed (HIP) Disk Alloy, MERL 76," Modern Developments in Powder Metallurgy, 14 (1981), 51-63.

3. D. R. Malley, J. E. Stulga and R. J. Ondercin, "Production of Near Net Shapes by Hot Isostatic Pressing of Superalloy Powder," 1982 National Powder Metallurgy Conference Proceedings, 38(1983), 229-246.

Acknowledgements

The authors wish to thank A. L. D'Orvilliers, R. D. Eng (formerly with Pratt & Whitney), P. D. Genereux and A. J. Nytch for their help in conducting the experimental work; M. W. Fox for his assistance in preparing the technical illustrations; and M. J. Blackburn for his consultation and continuing support.

HIP MODELING OF SUPERALLOY POWDERS

J.C. Borofka*, R.D. Kissinger**, and J.K. Tien*

*Columbia University, Henry Krumb School of Mines
Center for Strategic Materials
500 West 120th Street, 1124A S.W. Mudd Building
New York City, New York 10027-6699

**General Electric Company
Engineering Materials Technology Laboratories
1 Neumann Way, Box 156301, Mail Drop M85
Cincinnatti, Ohio 45215-6301

Abstract

Modeling of Hot Isostatic Pressing (HIP) allows for the prediction and optimization of densification kinetics and mechanisms. Monosized and bimodal powder distributions have been modeled. The results are best presented on HIP maps which show regimes of dominance for each mechanism in terms of HIP pressure versus density. In this work, the bimodal HIP model is extended to include kinetics of densification and both models are extended to consider densification by superplastic deformation. Kinetics from these two models are compared to experimental HIP densification data for René 95. Agreement between the models and experimental data is good. Through the use of HIP mechanism maps and with a detailed understanding of HIP kinetics, HIP cycles can be designed to control the dominant densification mechanism. Ramped pressure HIP runs were done in an effort to improve the as-HIP microstructure of René 95 powder, with promising results. The models can also be applied to other processing techniques as well as to HIP of new material systems and composites.

Introduction

Previously, the mechanisms and kinetics of HIP have been modeled for a single average powder particle radius (1,2) using the mechanisms of yield, power law creep, diffusional creep and diffusion. In a monosized powder distribution, the instantaneous stresses on all particles (at a given value of density) during HIP are the same and the deformation which leads to densification is uniform for all particles.

However, in a commercial distribution of powder (be it superalloy or any other material) a full spectrum of powder particle sizes is present. This leads to a non-uniform distribution of stresses and therefore non-uniform deformation of powder particles during HIP consolidation. Figure 1 shows an as-HIP superalloy, heat-treated to delineate the prior particle boundaries. The smaller particles have deformed to a greater extent than the large particles, in fact many large particles do not appear to have deformed at all, and have left behind smooth spherical (circular in cross-section) shells or crusts of surface segregation and/or contamination (3) in the densified material. These shells are known as prior particle boundaries (or ppb) and can degrade mechanical properties (4,5) by acting as easy propagation paths for fatigue cracks or by acting as low cycle fatigue initiation sites at extended reactive defects (6). Careful policing of the P/M process, which includes powder screening and container filling, improves the ppb problem in the as-HIP material, however some segregation or coarsening at the particle surfaces will remain. The stress and morphology differences that occur during HIP of commercial powder cannot be addressed by a model which only considers a monosized distribution of powder.

In order to more closely model reality, HIP of a bimodal distribution of powder sizes has been modeled (7,8,9). This model allows examination of the stress differences among powder particles, and can be applied to other material processing problems. In this paper, the monosized and bimodal HIP models have been extended to include densification by superplasticity, and densification kinetics have been added to the bimodal model. Experimental HIP data is compared to predicted results, and methods to improve the microstructure and properties of as-HIP superalloy are discussed.

HIP Modeling

The HIP models follow the technique used by Arzt et al. (1). Densification is modeled as the fictitious growth of spheres, which is mathematically equivalent to the movement of spheres closer together because volume is conserved over the powder bulk. The powder particles are considered to be spheres of radius R_1 and R_2 ($R_1 < R_2$, where $R_1 = R_2$ for monosized powder) present in weight fractions w_1 and w_2 (which can be related to v_i and n_i, volume and number fractions, respectively: $v_i = w_i D_o$, $n_i = (w_i/R_i)/\Sigma(w_i/R_i)$). The density is related to the change in radius (7):

$$D/D_o = \sum_i v_i (R_i/R_{oi})^3 \tag{1}$$

where D and D_o are the current and initial relative density respectively, and R_i and R_{oi} are the current and initial radius of particle i, where i takes the values 1 and 2 for the bimodal case. For the monosized case, no subscripts and sums are necessary, since all values are the same.

Contact Stresses

As densification proceeds and the spheres "grow", they approach one another and "overlap", as shown schematically in Figure 2. The overlap volume is mathematically distributed evenly over the exposed surface area of the spheres,

 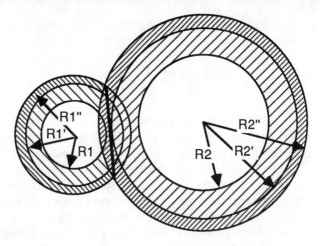

Figure 1. As-HIP René 95 (commercial powder distribution), heat-treated to emphasize prior particle boundary (ppb) morphology. The smaller particles in the distribution have suffered more deformation than the large particles.

Figure 2. The "growth" (coarse hatching) of powder particles leads to "overlap" volume (dark shading) which is distributed evenly over the exposed surface area (fine hatching). This leads to further overlap, and so on. The neck radius x_{ij} and density are calculated when R_i'' becomes invariant for each "growth step".

which in turn leads to a further increase in the radius and overlap volume, and so on. As the particles contact, they form necks of area a_{ij}. The contact stress (effective pressure) P_{ij}^* due to the applied HIP pressure P_{HIP} on these contacts can be calculated (7):

$$P_{ij}^* = (1/a_{ij})(\pi P_{HIP} / \sum_i v_i \sum_j N_{ij}/R_i^2) \tag{2a}$$

where N_{ij} is the number of j particles around an i particle, which is calculated from the radial distribution functions of a bimodal distribution of powder (10). The subscripts i and j take the values 1 and 2 in this and the following equations.

As this growth and formation of contacts continues, the contact stresses drop, and finally no exposed surface area remains on the spheres. In this stage, the material consists of isolated pores, and individual spheres cannot be readily identified. This stage occurs at about 90% relative density for the monosized case and 80% for the bimodal case. The final stage of densification is modeled as the closure of a distribution of isolated pores. The contact stress is taken as equal to the HIP pressure (2):

$$P_{ij}^* = P_{HIP}. \tag{2b}$$

The powder size dependence of HIP is seen explicitly in the stress equation for the initial stage of HIP from the contact area. The contact areas for small particles are smaller and lead to higher contact stresses than are experienced by the larger particles. Another effect of smaller particle radii is to change the initial density of the powder - a finer distribution of powder will have a higher initial density.

Densification Mechanisms and Densification Rates

The contact stress is used as input for constitutive equations for each mechanism under consideration and used to calculate a densification rate (Ḋ) for each mechanism. The rates are summed for all mechanisms (m) for each particle (i), then weighted by number fraction of each particle for a total densification rate:

$$\dot{D} = \sum_i n_i \sum_m \dot{D}_{im}. \tag{3}$$

Material parameters of René 95, a typical and commercially important P/M superalloy, are used as input for the modeling and are shown in Figures 3 and 4 and in the references (8,13). The mechanical properties of the powder particles most likely varies with particle size, because the grain size and crystalline morphology of the powder particles vary with particle size (14). As powder diameter decreases, the microstructure of the powder particles changes from coarse polycrystalline, to fine polycrystalline, to monocrystalline or nearly glassy. However, in this paper, the microstructure and therefore mechanical properties of all powder particles are taken to be equal.

Yielding. When the contact stress exceeds the yield strength of the powder particle material, yielding occurs at a very fast but finite rate, as with hot extrusion or hot forging. The yield strength of most materials is temperature dependent. However, the rate of yielding is not temperature dependent, which has lead in the past to the somewhat misleading labeling of this mechanism as athermal plastic flow. The temperature dependence of the yield strength of René 95 is shown in Figure 3. The densification rate for yielding is taken to be:

$$\dot{D}_{iy} = \begin{cases} 0 & P_{ij}^* < \sigma_y \\ 1 & P_{ij}^* > \sigma_y. \end{cases} \tag{4}$$

The particle size dependence affects the yield densification rate intrinsically through the effective stress on the particles, which is higher on the small particles.

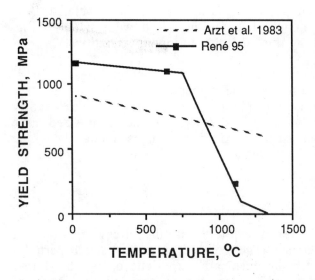

Figure 3. Temperature dependence of yield strength of superalloys at a strain rate of approximately 1.0 sec^{-1}. For this work, the data for René 95 is used (11,12).

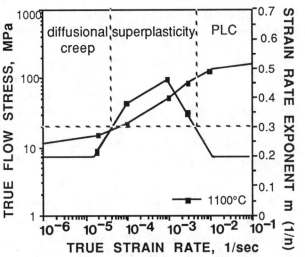

Figure 4. Flow stress-strain rate behavior of René 95. Strain rate exponents greater than 0.3 are taken to be indicative of superplasticity, and are used to determine which mechanism operates at the associated stress.

Power Law Creep and Superplasticity. These two mechanisms are considered to act exclusively of one another. The creep behavior of René 95 is used to determine at which stresses each mechanism operates, see Figure 4. The flow stress - strain rate behavior determines which mechanism operates at any particular stress through the usual criterion for superplasticity, that of a strain rate exponent m greater than 0.3 in the creep equation $\sigma = A\dot{\varepsilon}^m$, where $m = 1/n$ and n is the power law creep stress exponent. The appropriate creep parameters $(n, \dot{\varepsilon}_0, \sigma_0)$ for each regime are used in the densification rate equation (1,8).

Diffusional Creep (Nabarro-Herring/Coble Creep). This densification mechanism which involves grain boundary sliding and diffusion only operates when the grain size of the powder is small compared to the powder size. The grain size of argon-atomized superalloy powder is typically below 10 μm. The densification rate is given elsewhere (1,8).

Diffusion. Densification by long-range diffusion is similar to sintering, and is rather slow. The densification rate (2,14) is strongly affected by particle radius. Smaller particles have smaller radii of curvature at the interparticle necks and diffusion densification is enhanced.

HIP Maps

The densification rate equations are evaluated at a constant HIP temperature for a range of HIP pressures. The mechanism with the greatest densification rate is considered to be dominant at that HIP pressure and density. The densification rates are summed up and integrated to predict densification kinetics. The data is presented on a HIP map or diagram that shows fields of dominance for each mechanism in HIP pressure and density. Maps can also be created for a constant HIP pressure with a range of HIP temperatures. Figure 5 shows maps for distributions of powders with $R_2=R_1$, $R_2=2.3R_1$ and $R_2=5.0R_1$.

The HIP maps show fields of dominance for densification by yielding, power law creep, superplasticity, diffusional creep and diffusion. As is to be expected, at high HIP pressures, yielding dominates densification. At low HIP pressures, diffusional creep and diffusion dominate. Kinetics of densification will be discussed in the next section.

For the bimodal maps, each regime boundary has two values - one for the small particles and one for the large particles. As the difference in powder size increases, the split between the two values widens. For each mechanism, the boundary of the small particles is shifted upward compared to that of the large particles. This difference arises from the predicted and experimentally observed higher average stresses on the small particles. This difference means, for example at the yield/power law creep boundary, that the small particles deform by the very rapid yield mechanism longer (to a higher density) than the large particles. This effect leads to the difference in degree of deformation of particles which contributes to the ppb problem.

A way to minimize the difference in deformation between particles is to maintain HIP densification by a single dominating mechanism, especially by a mechanism such as superplasticity which is well-known for the uniformity of deformation over varying areas or gauge lengths. If the HIP pressure is increased during the HIP run, the interparticle contact stresses can be kept constant, rather than falling as the interparticle contact areas increase. In this way, a particular mechanism (in this case, superplasticity) can be allowed to dominate for the entire duration of the run, as shown by the dotted line in Figure 5a. The dotted line represents a density-pressure path for m=0.35, which maximizes the degree of superplastic deformation experienced by the powder particles. If the HIP pressure is held constant during a HIP run, any given mechanism will dominate for only a small range of the densification process.

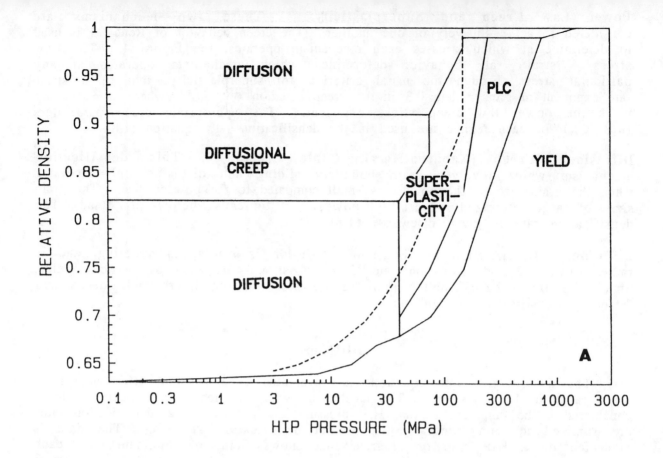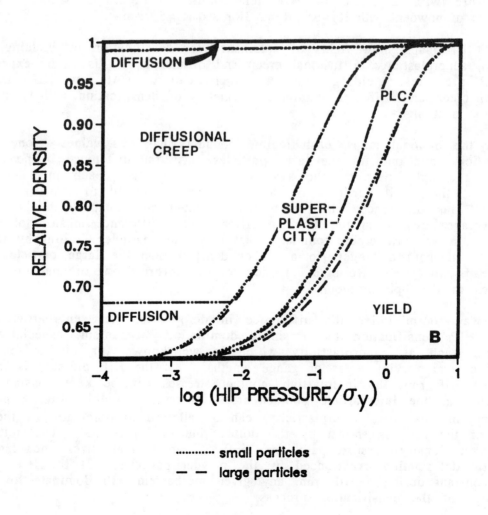

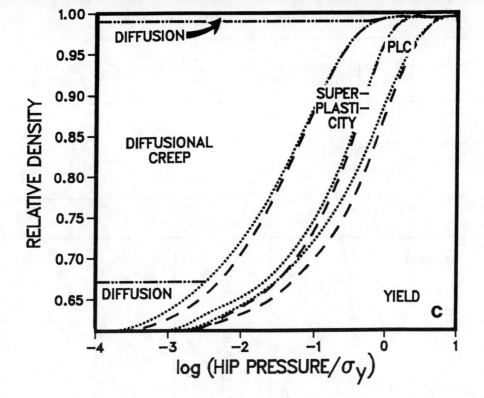

Figure 5. HIP maps for René 95 monosized and bimodal distributions (weight fractions of 10% and 90% for the small and large particles respectively). HIP temperature and R_2/R_1 are a) 1121°C and 1.0, b) 1100°C and 2.3, and c) 1100°C and 5.0. As the ratio of particle sizes increases, the split between the mechanism boundaries for the large and small particles increases. The dotted line in a) represents a density pressure path which maximizes superplasticity (m=0.35).

Experimental HIP Runs and Comparison to Modeling

Argon-atomized René 95 powder was sieved and classified into various distributions, as shown in Table I. The powder was transferred to stainless steel cans (either 4 cm diameter x 10 cm height or 15 cm x 18 cm). All handling was done in an inert atmosphere, and all cans were outgassed. The HIP conditions and initial and final heatup-cooldown schedules used are shown in Table II. The densities of the as-HIP material were then measured using a water displacement technique.

A comparison of densities of HIP runs from Table II and predicted densification kinetics is shown in Figure 6. Note that the model can not only predict the increase in density occuring during the hold time at maximum temperature and pressure, but can also be used to predict the increase in density at the start and end of the HIP cycle. Agreement between the experimental data and predicted kinetics is good.

Table I. Powder Size Distributions

Powder		Sieve Size	Average Diameter, μm	w_i, %	f_i, %	n_i, %
Monosized (M)		-170+200	81	100	100	100
Bimodal (B)	(small)	-400+500	28	10	6	24
	(large)	-170+200	81	90	55	76
Full (F)		-140	19.5	100	100	100

Table II. Experimental HIP Conditions.

HIP T,°C	Powder	HIP P, MPa	Pressure Up & Heatup, min	Hold Time, min	Pressure Down & Cooldown, min
900	B	103.0	60	5,15,60,180	60
1000	M,B,F	10.3	60	5,15,60,180	60
1121	M,B,F	103.0	60	5	60
1121	F	10.3	60	5,15,60	60

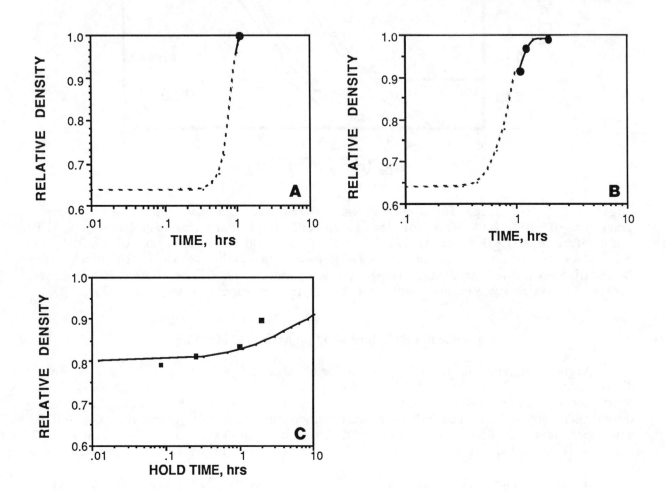

Figure 6. Comparison of densification kinetics predicted by the models (solid lines are densification during the hold time, dashed lines are densification during the ramp-up time) and experimental HIP data at a) 1121°C and 100 MPa, b) 1121°C and 10 MPa, and c) 900°C and 100 MPa.

Discussion and Conclusions

It must be said that currently the bimodal model is extrinsically bimodal only by consideration of differing powder particle radii. Even incomplete analysis of the HIP densification of a bimodal distribution of powder shows a split as shown on the HIP maps. Future work could include intrinsic differences in a distribution of powder, that is, the variation of material properties with powder particle radius.

The HIP model has been used to point the way to improved as-HIP superalloy material. Using HIP maps as guides, ramped pressure HIP runs have been done in an effort to promote uniform superplastic and creep deformation of superalloy powder particles, with promising results (8).

The model has also been extended to consider densification of superalloys by other techniques, such as CAP^{TM} (Consolidation at Atmospheric Pressure) (13) as well as HIP consolidation of other materials, such as superconducting oxide ceramics (15), as shown in Figure 7, and particulate composites (16). The model could also be used to predict closure of pores in castings by HIP.

Perhaps the major impact of the development of the bimodal HIP model is the ability to predict HIP of composites. The model easily treats HIP of a bimodal mixture of powder material of equal or different diameters. Each equation is summed for both particles, however for each powder, the material parameters are different. The geometry remains the same. For example, for densification by yielding, the densification rate becomes:

$$\dot{D}_{iY} = \begin{cases} 0 & P_{ij}^* < \sigma_{yi} \\ 1 & P_{ij}^* > \sigma_{yi} \end{cases} \tag{5}$$

Future work will explore various composite systems.

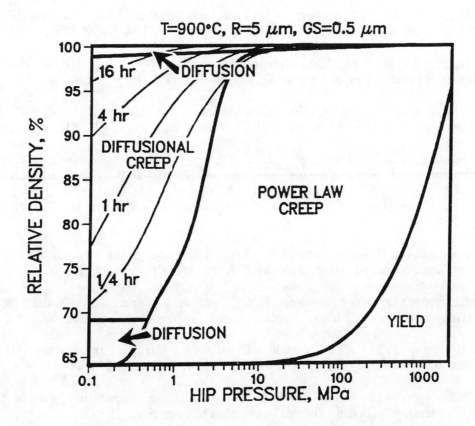

Figure 7. HIP map for monosized powder of the oxide superconductor $YBa_2Cu_3O_{7-x}$ at 900°C, with isochronal lines showing the density achieved at any HIP pressure in a given amount of time.

Acknowledgements

The authors thank the United States Air Force Office of Scientific Research for sponsoring this research under grant AFOSR-82-0352 monitored by Alan H. Rosenstein, and Special Metals Corporation and Wyman-Gordon Company for supplying and processing the materials used in this work.

References

1. E. Arzt, M.F. Ashby, and K.E. Easterling, "Practical Applications of HIP Diagrams: Four Case Studies," **Metall. Trans. A, 14A**(2)(1983), 211-221.

2. A.S. Helle, K.E. Easterling, and M.F. Ashby, "HIP Diagrams: New Developments," **Acta Metall., 33**(12)(1985), 2163-2174.

3. N.G. Ingesten, R. Warren, and L. Winberg, "The Nature and Origin of Previous Particle Boundary Precipitates in P/M Superalloys," **High Temperature Alloys for Gas Turbines 1982**, eds. R. Brunetaud et al. (Dordrecht, The Netherlands: D. Reidel Publ. Co.,1982).

4. R. Thamburaj, W. Wallace, Y. N. Chin and T.L. Prakash, "Influence of Processing Variables on Prior Particle Boundary Precipitation and Mechanical Behavior in PM Superalloy APK1," **Powder Metall., 27**(3)(1984), 169-180.

5. C. Aubin, J.H. Davidson, and J.P. Trottier, "The Influence of Powder Particle Surface Composition on the Properties of a Nickel-Based Superalloy Produced by HIP," **Superalloys 1980**, eds. J.K. Tien et al., Proc. 4th Intl. Symp. on Superalloys, Oct. 1980, Champion, PA (Metals Park, OH: ASM, 1980),345-354.

6. C.E. Shamblen and D.R. Chang, "Effect of Inclusions on LCF Life of HIP Plus Heat Treated Powder Metal René 95," **Metall. Trans. B, 16B**(12)(1985), 775-784.

7. S.V. Nair and J.K. Tien, "Densification Mechanism Maps for HIP of Unequal Sized Particles," **Metall. Trans. A, 18A**(1) (1987), 97-107.

8. J.K. Tien and J.C. Borofka, "Bimodal Modeling of HIP of Superalloys", First Intl. Conf. on HIP, Luleå, Sweden, June 1987 (Luleå, Sweden: CENTEK, in press).

9. R.D. Kissinger, S.V. Nair, and J.K. Tien, "Influence of Powder Particle Size Distribution and Pressure on the Kinetics of HIP Consolidation of P/M Superalloy René 95," **Superalloys 1984**, eds. M. Gell et al., Proc. 5th Intl. Symp. on Superalloys, Oct. 1984, Champion, PA (Warrendale, PA: TMS-AIME, 1984), 285-294.

10. S.V. Nair, B.C. Hendrix, and J.K. Tien, "Obtaining the Radial Distribution of Random Dense Packed Hard Spheres," **Acta Metall., 34**(8)(1986), 1599-1605.

11. D.R. Chang, D.D. Krueger, and R.A. Sprague, "Superalloy Powder Processing, Properties and Turbine Disk Applications," **Superalloys 1984**, op. cit., 245-273.

12. T.E. Howson, W.H. Couts, and J.E. Coyne, "High Temperature Deformation Behavior of P/M Rene 95," **Superalloys 1984**, op. cit., 275-284.

13. R.D. Kissinger, "The Densification of Nickel-Base Superalloy Powders by Hot Isostatic Pressing" (D.E.S. thesis, Columbia University, 1988).

14. F. Cosandey, R.D. Kissinger, and J.K. Tien, "Cooling Rates and Fine Microstructures of RSR and Argon Atomized Superalloy Powders," **Rapidly Solidified Amorphous and Crystalline Alloys**, eds. B.H. Kear et al. (New York: Elsevier Science Publ. Co., 1982), 173-178.

15. J.K. Tien, J.C. Borofka, B.C. Hendrix, T. Caulfield and S.H. Reichman, "Densification of Oxide Superconductors by Hot Isostatic Pressing," **Metall. Trans. A**, (to be published).

16. J.C. Borofka and J.K. Tien, "Modeling of Hot Isostatic Pressing of Powder Composites," (to be published).

DUAL STRUCTURE TURBINE DISKS

VIA PARTIAL IMMERSION HEAT TREATMENT[+]

J. M. HYZAK[*], C. A. MACINTYRE[**] AND D. V. SUNDBERG[**]

[*]WYMAN-GORDON COMPANY
NORTH GRAFTON, MASSACHUSETTS 01536-8001

[**]GARRETT TURBINE ENGINE COMPANY
PHOENIX, ARIZONA 85010-5217

ABSTRACT

A heat treat method has been developed to produce a dual structure turbine disk forging. The method is termed partial immersion treatment. It includes the partial immersion of a forging to a controlled depth in a high temperature (supersolvus) salt bath and total rotation of the forging so as to immerse the full rim section of the disk. The objective is to coarsen the grain structure in the rim of the disk while maintaining the as-forged, fine grain size in the bore. This is to improve the high temperature creep and stress rupture properties in the rim section without degrading strength and LCF properties of the bore.

Initial experiments on cast-wrought Astroloy forgings and powder-metallurgy AF2-1DA-6 forgings are described. The results show that uniform grain coarsening to a controlled depth can be achieved. Grain size in the Astroloy forgings coarsened from ASTM 8-10 to ASTM 1-3. The coarsened grain size in the AF2-1DA-6 forgings was ASTM 4 compared to the as-forged size of ASTM 10-12. Microstructural characterization of the coarse, fine and transition zones is included. Limited mechanical property data are presented. The results show a significant increase in creep and stress rupture properties in the coarse grain rim compared to the fine grain bore. There was also only a 10% decrease in tensile strength as a result of the coarsening heat treatment.

+ Patent applied for.

INTRODUCTION

The drive to improve gas turbine operating efficiencies generally results in increased turbine inlet temperatures with a corresponding increase in the thermal and mechanical stresses experienced by the turbine components. Within this aggressive environment, the design lives of the turbine disks are predicated on a fine balance between high temperature strength and creep properties. The rim section of a turbine disk experiences the highest temperature and can be life limited by creep and stress rupture properties. The bore section, however, operates at lower temperature and higher stresses, and is designed to meet tensile strength and low cycle fatigue requirements.

Cast-wrought (C/W) and powder metallurgy (P/M) nickel base superalloys are generally selected for high temperature/high speed turbine disk applications. These alloys, while possessing excellent strength properties up to 650°C, are somewhat limited in creep resistance. This, of course, is attributable to the thermo-mechanical processing of the part. In general, billet conversion and turbine forging operations have been designed to produce a finished part with a fine grain size. This is for maximum strength and low cycle fatigue properties. As a result, creep properties suffer.

In some turbine disk designs it would be beneficial to produce a dual structure forging with a fine grain size in the bore region to optimize strength and LCF properties and a coarser grain size toward the rim for improved creep strength. Various processing routes have been proposed for this. These have included redesigning of the forging process to minimize the deformation in the rim during the finish forge operation, and selective heat treatments to coarsen the rim. This paper describes a post forge heat treating technique, termed partial immersion treatment (P.I.T.), developed to produce a dual structure turbine disk.

The concept is shown schematically in Figure 1a. The forging is supported on a shaft through the center. The rim section of the forging is then immersed to a controlled depth in a high temperature (supersolvus) salt bath. The disk is rotated such that there is complete immersion (360° rotation) of the rim section. Grain coarsening occurs in the rim as a

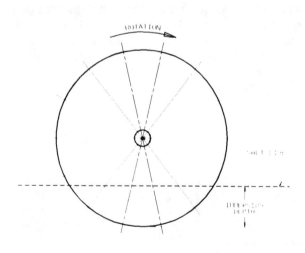

Figure 1. a) Schematic of Partial Immersion Treatment process; b) Spoke fixture for incremental rotation of forging.

result of the supersolvus heat treatment while the fine grain, as forged structure in the bore is maintained.

The partial immersion treatment is envisioned as a production process where several forgings can be supported on one shaft and heat treated at the same time. The salt bath temperature would be well controlled, ±2°C, as would the depth of immersion. The rotational speed of the forging in the salt bath would also be closely monitored so as to produce uniform, reproducible grain coarsening. This paper describes the initial experiments in the development of the partial immersion treatment to produce a dual microstructure disk. Fine grain forgings made from a P/M alloy (AF2-1DA-6) and a C/W alloy (Astroloy) were given the partial immersion treatment. The results include microstructural characterization of the coarsened zone, and mechanical property comparisons between coarse and fine grain microstructures.

EXPERIMENTAL DETAILS

PARTIAL IMMERSION TREATING

Heat treatments were performed in a commercial high temperature salt bath measuring 71cm X 94cm X 152cm deep. The temperature control and uniformity were within ±2°C. Disks were generally suspended on a steel shaft and lowered into the bath via an overhead crane; depth of immersion was controlled visually. Because this was the initial feasibility study for using P.I.T., the forgings were rotated manually rather than by a continuous drive motor (now in operation).

The forgings were heat treated in one or two revolutions over a 1-3 hour period. As a result, the forgings were incrementally advanced. A fixture with radial spokes was designed to aid in this manual advance (Figure 1b). Depending on the part diameter, forgings were rotated from 10° to 15° per advance. Forgings were held above the salt bath (≈550°C) before and after the P.I.T. immersion to minimize the thermal stresses.

MICROSTRUCTURAL CHARACTERIZATION

P.I.T. Astroloy Finish Forging

The initial experimental application of the partial immersion treatment was to coarsen the rim of a C/W Astroloy forging. The partial immersion treatment was performed after the finish forge operation. The geometry of the forging is shown in Figure 2a. This forging had an as-forged grain size of ASTM 8-10 with less than 10% unrecrystallized grains as large as (ALA) ASTM 5. This grain size results in good strength and LCF life but poor creep properties. The goal was to coarsen the grain size in the rim

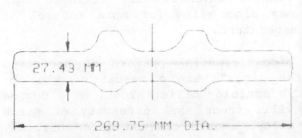 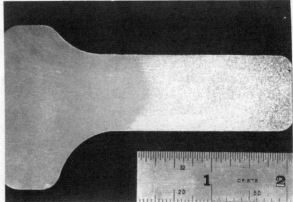

Figure 2. a) Geometry of Astroloy finish forging; b) Radial slice of forging after P.I.T.

to a depth of 50mm, as measured from the outside diameter.

The first forging was partial immersion treated in a 1163°C salt bath, 20°C above the nominal γ' solvus of the Astroloy stock (as determined by differential thermal analysis). The disk was immersed 50mm, as measured from the rim of the disk. The part was rotated 360° in 16 manual advances over 90 minutes. After removal from the salt bath, the disk was air cooled.

After the P.I.T., the disk was given a standard partial solution and age heat treatment for Astroloy: 1115°C/4 Hrs./Quench; 871°C/8 Hrs./AC; 982°C/4 Hrs./AC; 649°C/24 Hrs./AC; 760°C/8 Hrs./AC. The disk was sectioned radially. The etched macro slice is shown in Figure 2b. The depth of coarsening was 50mm as measured at mid-thickness. There was a 3mm difference in depth of coarsening between the mid-thickness and surface locations.

Figure 3 shows representative photomicrographs of the rim (coarsened), bore (uncoarsened) and transition regions. The coarsened grain size was ASTM 1-3. The transition from fine to coarse grain size occurred very abruptly with few grains of an intermediate size. Due to the chemical segregation (banding) in this cast-wrought alloy, the radial location of the transition from coarse to fine varied through the thickness depending on the local chemistry (γ' solvus). This nonuniformity in coarsening in the transition zone is shown in Figure 3c.

A second Astroloy disk was partial immersion treated. Again the immersion depth was 50mm in a 1163°C salt bath. The disk was rotated in increments of 10° with 2.5 minutes between advances. The disk was rotated a total of 450° in 112 minutes.

The forging was subsequently given a partial solution and age heat treatment, as before, and sectioned through mid-thickness to reveal the coarsening. The macro section, shown in Figure 4, had several interesting characteristics. First, the depth of coarsening (Loc. A) was uniform around the disk. The grains coarsened to ASTM 1-3 up to the depth of immersion (50 mm). There was also no significant difference in coarsening in the overlap region, that section of the disk immersed twice.

Secondly, there was a radial etch pattern (Loc. B) in the coarsened zone over three-fourths of the disk; in the remaining quarter, the etch response was uniform. The radial spokes were due to variations in gamma prime morphology produced during the P.I.T. from the incremental advance used in rotating the disks. That is, as each pie-shaped section was rotated out of the salt bath there was a gradient in cooling rate in that section and, as a result, a difference in γ' density. The quarter without spokes was that section of the forging in the salt bath at the conclusion of the P.I.T. that was uniformly cooled when the disk was lifted out of the salt bath. It is thought that continuous rotation of the disk in the salt bath to replace the manual incremental advance will eliminate this nonuniformity in microstructure. It may also allow for more control of the cooling rate from the supersolvus temperature.

Also apparent in Figure 4 is a circumferential pattern of preferred grain coarsening (Loc. C). These "tree rings" were a result of the inhomogeneity (chemical banding) in the C/W Astroloy billet stock. The degree of chemical homogeneity in an alloy will affect the uniformity of grain coarsening. In this regard, the P/M alloys should coarsen more uniformly than the C/W nickel-base superalloys.

Figure 3. Microstructure of Astroloy forging after P.I.T.; a) fine grain, b) coarse grain, and c) transition regions.

Figure 4. Macro etched Astroloy finish forging after P.I.T., A) depth of coarsening, B) radial etch pattern, and C) circumferential "tree rings"

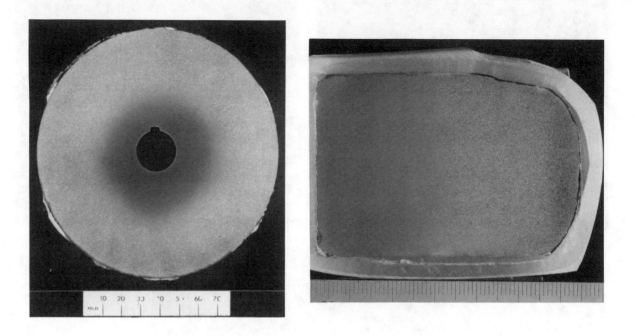

Figure 5. AF2-1DA disk after P.I.T.; a) full view and b) radial section.

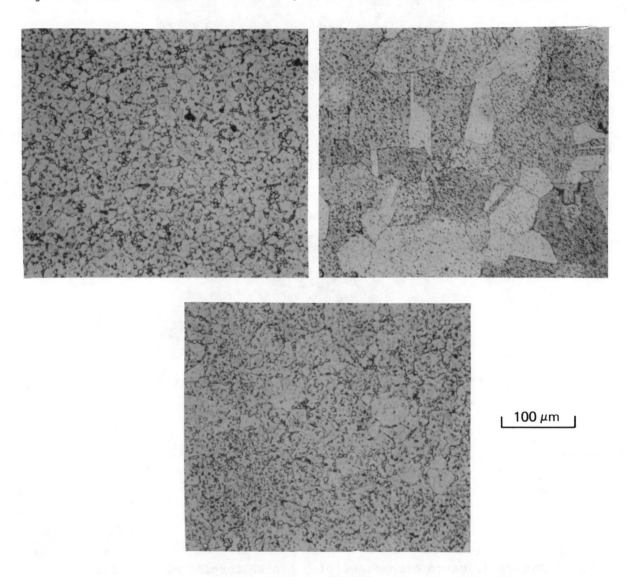

Figure 6. Microstructure of P/M AF2-1DA-6 forging after P.I.T.; a) fine grain, b) coarse grain, and c) transition zone.

P.I.T. AF2-1DA-6 Finish Forging

The partial immersion treatment was also used to develop a dual grain structure in fine grain P/M AF2-1DA-6 forgings. Small diameter (5.5" ⌀) turbine disk forgings were hot die forged from 4" ⌀ CAP plus extruded AF2-1DA-6 billet stock. The as-forged grain size was uniformly ASTM 10-12. The goal was to coarsen the grain size in the outer 33mm of the disk.

Forgings were immersed to a depth of 39mm, as measured from the outer diameter, in a 1232°C salt bath. The γ' solvus is nominally 1220°C for this alloy. The parts were rotated manually, as before, using a 24 spoke fixture (15° per increment). The disks were rotated a total of 540° in 216 minutes. A centerhole with a keyway to lock the forgings to the drive axle was introduced in these experiments. One AF2-1DA-6 forging is shown in Figure 5a. The orientation of the disk in the photograph (keyway up) is the same as when it was initially immersed in the salt bath. The uniformity of the coarsening was very good. Although not pronounced in this example, the P/M AF2-1DA-6 forgings exhibited similar etching characteristics after P.I.T. to those shown by the C/W Astroloy forgings. There were radial spokes over three-fourths of the disk and a uniform etching region in the last quarter.

A radial slice taken from the AF2-1DA-6 disk is shown in Figure 5b. The uniformity of coarsening through the thickness of the forging was good. The grain size coarsened to ASTM 4 to a depth of 31mm. Beyond the coarse grain zone there was a 9mm wide transition region over which the grain diameter decreased to the initial ASTM 10 size. The transition from fine grain to coarse grain was much more gradual in this P/M forging than in the C/W Astroloy disk due to the greater chemical homogeneity of the P/M billet stock. The forging was given a partial solution and age treatment (1149°C/2 hrs/FAC; 760°C/16 hrs/AC) after the P.I.T. Representative photomicrographs are presented in Figure 6 of the coarse grain, fine grain and transition regions.

P.I.T. Astroloy Blocker Forging

As already described, the initial P.I.T. experiments focused on the rim coarsening of finish (shape) forgings. An additional experiment was performed to determine the feasibility of grain coarsening the rim section of a blocker (preform) forging, and subsequently forging the part to the finish shape. The objective was to impart some work to the coarse grain-fine grain transition region without significantly affecting the coarsened microstructure in the rim. A blocker shape was designed so as to minimize the forging reduction in the rim section during the finish operation. The blocker shape is shown in Figure 7; the finish geometry is the same as shown in Figure 2a.

The blocker forging was given a partial immersion treatment in a 1163°C salt bath to a depth of 56 mm. It was rotated a total of 720° (two rotations) in increments of 10° (rotation). The total time for coarsening was 486 minutes.

Subsequent to the rim coarsening the blocker was hammer forged to its finish geometry. The resultant macrostructure is shown in Figure 8. In terms of final grain size, forging a grain coarsened blocker was successful. The coarse grain size in the rim section was maintained with no recrystallization (necklacing) noted. The coarsened grain size was ASTM 0-2 while the fine grain region was generally ASTM 10. The shape of the boundary between coarse and fine grain regions was also generally vertical, as desired. There was, however, a wide transition zone between

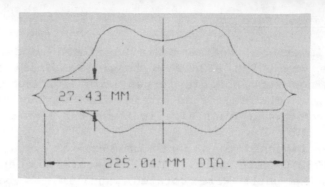

Figure 7. Geometry of Astroloy blocker forging.

the fully coarse and fully fine grain regions. This zone, measuring 125 mm, contained a mixture of fine (ASTM 10) and coarse (ASTM 0-2) grains. This nonuniform coarsening was due mostly to the chemical inhomogeneity (tree rings) in the Astroloy stock.

Although the grain size results were encouraging, there were some drawbacks to this approach. The blocker shape was much more difficult to coarsen to a controlled depth due to its thicker sections. Immersion times 5X greater than used for the finish forge geometry did not produce coarsening to the depth desired in finished shape. In addition, the poor forgeability of the coarse grain microstructure required special handling in the forge shop.

Mechanical Properties Comparison

The objective of the partial immersion treatment is to improve the high temperature creep and rupture properties in the rim of disk forgings compared to the original fine grain properties. Testing has been performed to determine the level of properties that can be attained. In general, the alloy's standard partial solution and age heat treatment has been used after the P.I.T. It is recognized that modifications to these procedures may further improve the properties. These changes may include better control of the cooling rate from the supersolvus heat treatment and the use of a subsolvus stabilization before the partial solution treatment so as to increase the fineness of the γ' precipitation.

C/W Astroloy Forging

The Astroloy finish forgings were given a partial immersion treatment

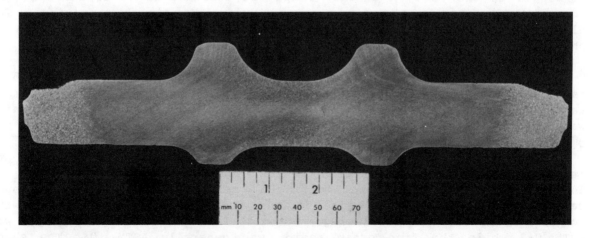

Figure 8. Macro slice of Astroloy blocker forging after P.I.T. and subsequent finish forge operation.

and subsequent full heat treatment as described earlier. The coarsening of the Astroloy forgings increased the 760°C/586 MPa stress rupture life 5X in the rim compared to the properties in the fine grain bore region. "Fir tree" testing was also performed. Dovetail slots were machined from material taken from the rim coarsened and the fine grain regions of the forging. These were tested at elevated temperature via dead weight loading of inserted dummy blades. The time to failure increased 10X for the rim coarsened material compared to the fine grain microstructure. In addition, there was only a 10% decrease in the room temperature and 760°C tensile properties in the coarse grain region.

Twenty high temperature LCF tests were performed in the grain size transition region. Specimens were oriented radially so as to contain coarse grain, fine grain and the transition zone microstructures in the gage section. All specimens failed in the coarse grain region well removed from the transition zone. Fracture initiated in the near-surface region, and the crack advanced initially in a Stage I crystallographic mode. The LCF lives were also typical of coarse grain Astroloy forgings. These results showed that the transition zone was not a site for preferred crack initiation.

P/M AF2-1DA-6 Forgings

Initial mechanical property comparisons between fine and coarse grain AF2-1DA-6 microstructures were made using pancake forgings given monolithic heat treatments. Properties generated from dual structure disks given the partial immersion treatment were to follow and were not yet available.

The fine grain properties were generated from pancake forgings given a partial solution and age heat treatment which retained the ASTM 10 grain size. Other pancake forgings were first grain coarsened at 1218°C/1 hr/AC and then given the same partial solution and age treatment. The resultant coarsened grain size was ASTM 4 ALA 3.

The smooth bar stress rupture properties showed a clear benefit to the coarse grain structure. At 704°C and 758 MPa, the average time to failure increased over 5X, compared to fine grain results. At 760°C and 586 MPa the advantage of the coarse grain structure increased 10X. Similar results have been reported for improvements in notch stress rupture times with grain coarsened microstructures (Ref. 1).

As expected, there was a decrease in room temperature and elevated temperature (593°C) tensile strengths with the coarsened grain size compared to fine grain properties. However, this reduction in yield and ultimate strengths at both temperatures was less than 10%. There was also a small increase in tensile ductility.

CONCLUSIONS

The partial immersion heat treatment has been developed as a means of producing a dual structure turbine disk forging with a coarse grain microstructure on the rim while maintaining the fine, as-forged microstructure in the bore. Partial immersion of a forging in a salt bath is a controllable process which can produce coarsening to a uniform depth over the full diameter of the part. In the C/W Astroloy forgings, the transition from coarse to fine grain was abrupt and was dependent on the chemical homogeneity of the billet. The P/M AF2-1DA-6 forgings had a larger transition zone with grain size coarsening from ASTM 10 to ASTM 4 occurring more gradually.

This early development study has relied on slow, manual rotation of the forgings in the salt bath. This has produced some nonuniformity in γ' distribution, and less than optimum mechanical properties. Even with this limitation, the increase in high temperature creep and stress rupture properties in the coarse grain region compared to the fine grain properties has been dramatic. The stress rupture lives increased 5X in Astroloy and up to 10X in AF2-1DA-6 with only a 10% decrease in tensile strengths.

Further development of the partial immersion treatment will include automated continuous rotation of the disk forgings. This should produce increased uniformity in the part and more flexibility in processing. Improved properties may be produced by controlling the cooling rate of the coarsened zone through the γ' solvus temperature. Hard fixturing of the disks during the P.I.T. will also improve the control over depth of coarsening.

ACKNOWLEDGEMENTS

The partial immersion heat treatment was conceived of by W.H. Couts, Jr. of the Wyman-Gordon Company. The partial solution heat treatments were performed at Sun Steel Treating, South Lyon, Michigan. Mr. Sergay Poborka of SST has contributed significantly to the development of P.I.T. The partial immersion treating and mechanical property testing of the AF2-1DA-6 forgings were performed as part of Air Force Contract F33615-81-C-5042, "Manufacturing Technology for P/M Superalloy Disks with Low Strategic Material Content". The principal engineer (prime contractor) was Mr. R. M. Gasior, of Cytemp Specialty Steel Division. The principal investigators (subcontractor) were Messrs. R. R. Paulson and C. C. Berger of Avco Lycoming/Textron.

REFERENCES

1. F. R. Dax, R. M. Gasior and D. J. Willebrand, "Manufacturing Technology for P/M Superalloy Disks with Low Strategic Material Content" (Report 1R-152-2 (XIX), Cytemp Specialty Steel Division, February, 1987).

Development of Inconel* Alloy MA 6000 Turbine Blades For Advanced Gas Turbine Engine Designs

B. A. Ewing and S. K. Jain

Allison Gas Turbine Division of General Motors Corporation
Indianapolis, Indiana

Abstract

This paper deals with Allison's work with the MA 6000 alloy aimed at near-net-shape forgings and thermal protection coatings for uncooled turbine blade applications. The work described is presented from the user's viewpoint and summarizes the ongoing efforts of a team that included leaders in oxide dispersion strengthening (ODS) alloy production and the forging of these alloys into near-net-shape airfoil configurations.

Results that were achieved in this program to date include the attainment of conventional forging schedules at both Textron Excello and Doncasters Monk Bridge for producing a near-net-shape shrouded second stage T406 turbine blade configuration. In addition, Doncasters was able to demonstrate their ability to EDM machine their forgings to required airfoil dimensions. The Excello and Doncasters stress rupture results were both competitive with single crystals in the high temperature/low stress regime, with the Doncasters blades showing the best performance. Also, results of oxidation testing performed at 2150°F on a new duplex coating system were very promising. The data showed that the new coating system was superior to that of contemporary coating systems and that the MA 6000 coating interface was free of porosity after 400 hours of exposure.

Overall, it was shown that near-net-shape forging to produce complicated MA 6000 turbine airfoils has the potential for significantly improved temperature capabilities relative to the CMSX-3 single crystal alloy in selected high taper ratio low stress turbine blade designs. For certain applications and configurations, the reduced input stock and attendant cost savings inherent with the conventional forging process appeared to justify this approach. For other applications that are less complicated from a configuration viewpoint the machining of the airfoils from extruded and hot rolled product may represent the best approach.

---

* Inconel is a registered trademark of the Inco family of companies.

Introduction

MA 6000 is an austenitic Ni, 15 Cr, 2.0 Mo, 2.0 Ta, 2.5 Ti, 4.5 Al, 4.0 W, 0.01B, 0.15Zr, 1.1 Y_2O_3 alloy that is produced by mechanical alloying techniques. The alloy is unique in that it combines gamma prime age hardening with Y_2O_3 dispersion strengthening providing it with a combination of intermediate and high temperature strength characteristics for turbine blading applications; it is generally produced as bar product and is used in the directionally recrystallized form.

The road to a major commitment in U.S. gas turbines has been relatively slow for MA 6000. This has in large part been due to competitive pressures from directionally solidified (DS) and single crystal (SC) castings and the relatively low intermediate temperature properties of MA 6000.

However, MA 6000's potential for superior high temperature capability at moderate stress levels continues to make it an attractive candidate for selected modern high taper ratio low stress designs where the alloy's dispersion strengthening mechanism can be used to advantage. Specifically, as is shown in Figure 1, the MA 6000 alloy has the potential to offer an approximate 125°F temperature advantage over a contemporary single crystal alloy such as CMSX-3 at a 15 ksi stress level. When this advantage is translated into either an elimination or a reduction in the amount of cooling air required for a turbine blade design, the resulting fuel savings can be significant.

During the early 1980s, the Garrett Turbine Engine Company, under NASA MATE (Materials For Advanced Turbine Engines) sponsorship, undertook a program to develop MA 6000 processing schedules for an uncooled HP turbine blade in the TFE 731 engine[1]. This effort, involved a dedicated activity to tailor a new blade design to the alloy's unique balance of moderate intermediate temperature capability and excellent high temperature properties. This work focused on a fabrication sequence for the new blade design that involved extrusion followed by hot rolling and directional recrystallization to develop barstock with the desired high aspect ratio (10:1 and greater) grain structure. The final blade configuration was then produced by a combination of airfoil electro chemical machining (ECM) and blade root grinding.

This program was very successful technically, however, because of its focus on a relatively simple unshrouded blade configuration and the design's relatively moderate blade operating temperature requirement, the program did not address two important issues. In particular, near-net-shape processing that could lead to improved materials utilization and subsequently reduced acquisition costs was not pursued. Also, the issue of Kirkendal porosity at coating/substrate interfaces following extended exposures in the 2100°F range remained as an impediment to the realization of maximum metal temperature operating capability. In the final analyses, the excellent Garrett work with MA 6000 firmly established the technical feasibility of utilizing MA 6000 in tailored engine designs. However, because of cost and maturity factors relative to DS and SC castings, the effort failed to transition to production implementation in their small commercial engine designs.

In recognition of the potential associated with the use of ODS blade alloys in large engine designs NASA funded, in the mid 1980s, a second, although, limited effort to dimension the capabilities of MA 6000 in a larger turbine blade configuration[2]. Emphasized for the first time was conventional forging activity aimed at reducing processing costs through improved materials utilization. The results of this effort, which was performed by Textron Excello, were encouraging in that the material was shown to be workable under carefully controlled conventional forging techniques. However, the limited scope of this program precluded a full scale optimization of the required thermo mechanical processes (TMP) and as a consequence work on the alloy by gas turbine manufacturers in the U.S. ceased.

This was the case in the U.S. until 1983 when Allison turned again to the potential that MA 6000 might have for utilization in several new turboprop engine programs that stood to benefit from the availability of a blade alloy with improved temperature capabilities relative to contemporary single crystal alloys. Of particular interest were T406 first- and second-stage turbine blade designs as shown in Figure 2. In the case of first-stage blade design, which is unshrouded and air cooled, it was desired to demonstrate that forged MA 6000 properties could be developed such that the cooled blade could be redesigned to be run uncooled. In the case of the uncooled second-stage shrouded blades, it was desired to demonstrate that cost advantages relative to cast single crystal designs could be realized with near-net-shape forging techniques. Consequently with a combination of U.S. Air Force and company support, a team effort was launched to develop the required process technologies. The vehicle for the work was a low stress, high taper ratio shrouded design for the second-stage T406 turbine blade. Included in the program were forging, machining and blade characterization efforts. Also included was coating development aimed at coating systems that would be diffusionally stable at temperatures to 2150°F. Team members included the following:

o Inco Alloys International (IAI)-Hereford, England and Huntington, West Virginia - Extruded and Hot Rolled MA 6000 Bar Supply
o Textron, Excello-Euclid, Ohio - Forging/Machining Optimization
o Doncasters Monk Bridge, Limited-Leeds, England - Forging/Machining Optimization
o Boone and Associates-Walnut Creek, California - Coating Consultants

Following is an overview of the work performed by the team and the status of the effort as it relates to Allison Gas Turbine engine design.

<u>Textron Excello Forging Development</u>

MA 6000 is forged from barstock which is in the fine equiaxed grain structure condition. Unless properly insulated the alloy can be prone to brittle fracture during conventional forging due to die chilling. In order

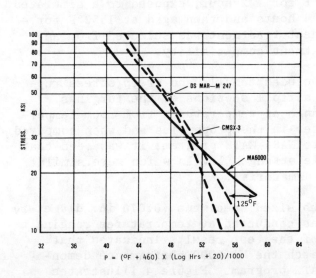

Figure 1. Stress-rupture data for MA 6000, SC CMSX-3, and DS Mar-M247, longitudinal orientation.

Figure 2. T406 first and second stage turbine blades.

to overcome die chilling problems, the Excello approach was to perform conventional mechanical press forging with the mild steel can left intact from the extrusion process.

The forging approach initially selected was based on a two step process involving preform and finish forge operations. Forge tool development involved an iterative approach to develop the finish forge blade design requirements. Forged blades representative of the last forging iteration were then to be directionally recrystallized, heat treated, and tested for mechanical properties.

To develop an optimum material input condition for the forgings, three approaches were pursued simultaneously as follows:

(1) Direct powder extrusion to a required one inch square cross section.
(2) Double extrusion which included an oversize extrusion followed by a second extrusion to the require cross section.
(3) Oversize extrusion followed by hot rolling to the required cross section.

The forging response of MA 6000 processed through the three approaches was more or less the same in that the material forged satisfactorily during the preform step. However, cracking was developed for each during the finish step due to die design difficulties.

The preliminary DR trials were conducted by IAI on preform as well as finish forgings. The results for preform forgings indicated that a complete recrystallization was achieved, however DR response for finish forging was only marginal. Further, the most favorable DR response was obtained with the Approach 3 material. Because the one-hit forgings (preforms) consistently exhibited a positive, coarser columnar grain structural response, a new set of dies was designed and built specifically for a one step, oversize forging approach. The new design was also aimed at minimizing metal flow patterns which could lead to cracking. Forgings representative of the new design were crack free, Figure 3, and exhibited a DR structure comparable to that observed previously with preform forgings.

Consequently, additional forgings, were made in order to generate tensile and creep/stress rupture data and to establish property performance levels for the MA 6000 forgings. They were directionally recrystallized at 2300°F, solution heat treated at 2250°F for 1/2 hour, exposed to a simulated coating diffusion cycle at 1975°F for 4 hours and then aged at 1550°F for 4 hours and 1400°F for 16 hours. This heat treatment was selected from an independent study on coating heat treatment compatibility.

Mechanical property evaluations included tensile and creep/stress rupture tests on specimens machined from airfoil sections of the forgings. Analysis of the tensile data over a range of temperatures from room temperature to 2150°F indicated little scatter in the properties and when compared with properties obtained in the Garrett NASA-MATE program, it was seen that the forged MA 6000 demonstrated tensile strength levels which were similar to those of the extruded and hot rolled material.

Stress rupture testing on small sub sized specimens (0.070 in. diameter) machined from airfoil stacking axes was conducted at temperatures ranging from 1400°F to 2150°F. The analysis of the test results indicated that, strength levels generally failed to match the MA 6000 performance demonstrated in the earlier Garrett NASA-MATE program. Figure 4 illustrates forged MA 6000 properties compared with MA 6000 properties obtained from large diameter test bars (0.250 in.) in the Garrett NASA-MATE program. Also shown are CMSX-3 single crystal properties for both large diameter and thin

 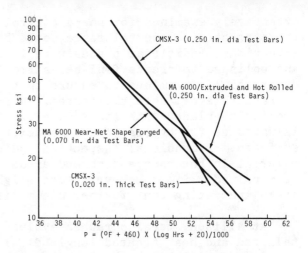

Figure 3. Second stage MA 6000 T406 forgings produced by Textron Excello.

Figure 4. Textron Excello forged MA 6000 stress rupture properties compared to Garrett-NASA MATE extruded and rolled MA 6000 bar and CMSX-3 single crystal test results.

wall (0.020 in.) test specimens. As shown, the small 0.070 in. diameter test bars machined from forged MA 6000 demonstrated that stress rupture strengths at the lowest test temperatures and the highest stress levels were similar to the strength behavior of the extruded material. However, as temperatures increased and stress levels decreased the forged MA 6000 properties degraded possibly as a consequence of environmental effects[3]. In addition, the forged MA 6000 results showed considerable scatter. By comparison to thin wall CMSX-3 data at 20 ksi stress levels and higher the MA 6000 results were generally lower. Below 20 ksi the MA 6000 results were competitive. Metallographic evaluations of several longitudinal sections indicated several microstructural deficiencies: (a) non-uniformity in grain aspect ratio within the airfoil regions as well as in shroud and root sections, (b) sporadic equiaxed grain structure throughout the blade, and (c) absence of a consistent coarse and high aspect ratio grain structure. These microstructural deficiencies may explain the scatter in the rupture strengths.

Doncasters Forging Development

Doncasters' MA 6000 near-net forging development effort was also aimed at the second-stage T406 turbine blade configuration. In this effort, Doncasters developed a two-step conventional forging route to produce crack-free near-net shape T406 blades with a high aspect ratio directionally recrystallized structure. The approach pursued involved the use of a screw press and a proprietary thermal barrier to minimize die chill. In this work, the general impact of forging and DR parameters on blade structures was identified. These experiments proved that similar recrystallized morphologies, in terms of grain size and aspect ratio, could be realized by either tailoring forging parameters for a fixed DR condition or independently selecting optimum DR parameters for a fixed forging practice. A wide degree of control over structures, therefore, was shown to exist by the variation of chosen parameters.

Subsequent efforts were aimed at improving the degree of structural control by "fine-tuning" the processing conditions as well as measuring creep rupture strength at elevated temperatures. This was followed by the forging of second-stage T406 turbine blades and EDM machining trials to produce airfoils for component testing. The blades from this forging campaign were extensively examined for shape control to ensure that finished blades of the correct size and tolerance could be produced. Following minor die modifications, additional blades were forged, heat treated and EDM machined and polished to create finished airfoils and platforms. Most of these were destructively examined to ensure that potentially deleterious effects, such as recast layer, were eliminated. A final forging run was then completed from which blades with finished airfoils and platforms were made. Figure 5 illustrates two typical forgings before and after machining of the airfoil surfaces. Dimensional inspection indicated that most blades had sufficient material in the root and shroud areas from which finished dimensions could be machined. Some, however, were slightly short of material in non-critical areas indicating that further die modifications would be desirable.

At Allison, metallographic evaluation of several directionally recrystallized and heat treated longitudinal sections showed that a desired coarse and high aspect ratio grain structure had been achieved in most of the airfoil cross sections. However, after careful review it was noted that there was a microstructural deficiency observed near the leading edge. These areas appeared not to have received the same TMP as the rest of the airfoil cross-section, possibly as a consequence of die chill developed during the forging operation. Stress rupture test results for specimens machined from airfoil stacking axes at temperatures ranging from 1400°F to 2150°F were then compared with 0.020 in. thin wall CMSX-3 single crystal alloy properties. With the exception of specimens that were machined from the leading edge area and which showed significantly reduced strength levels, the small 0.070 in. diameter test bars machined from the forged MA 6000 blades demonstrated stress rupture strength levels that were on the order of 50°F improved over thin section CMSX-3 properties at a 20 ksi stress level and 100°F at a 15 ksi stress level. A comparison of Excello and Doncasters thin section MA 6000 results to thin section CMSX-3 single crystal data is presented in Figure 6. Shown is the superiority of the Doncasters forgings over the Excello forgings. It is not clear as to the explanation for the different performance levels. However, several variables, ranging from starting input stock to DR heat treat practice could be significant factors.

The grain quality problems encountered at the leading edge of the Doncasters blades will require additional work, however, this is not expected to pose a significant roadblock to achieving the desired high aspect grain quality throughout the airfoil.

Figure 5. Typical Doncasters T406 second-stage turbine blade forgings before and after airfoil machining.

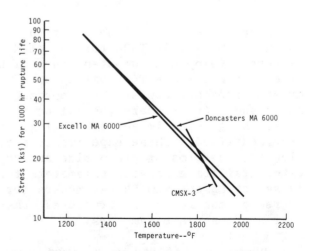

Figure 6. Thin section stress rupture capability for MA 6000 and CMSX-3.

In order to circumvent the grain quality problem at the leading edge, Doncasters has recently introduced a modified conventional forging procedure. This technique uses a multi-stage preform preparation cycle based on extrusion, prior to the standard finish forging operation. The preform contains a more uniform distribution of work, particularly in the airfoil section and is designed to eliminate the fine grained structure. Projected advantages for this approach include better overall structural control and reproducibility.

Trials are in progress to establish the correct working parameters to achieve coarse-grained structures in directionally recrystallized blades and improve the uniformity of structure across the airfoil. Additional blades are now being forged to verify and quantify the structural improvements in terms of stress rupture properties.

Coating Development

It has been shown in work conducted at Allison that the MCrAlY class of overlay coatings can provide excellent environmental protection on the MA 6000 alloy at temperatures up to and beyond those developed during second stage T406 blade operation. By contrast, the aluminide class of coating has much less temperature capability on MA 6000 and performs poorly under 1950°F test conditions due to the formation of Kirkendall porosity that occurs at the coating/substrate interface following extended exposure. The Kirkendall porosity is felt to develop as a consequence of outward diffusion of aluminum and chromium during the oxidation process, leaving behind vacancies or pores that ultimately lead to coating spallation and failure. As an example, Figure 7, shows Kirkendall porosity in an aluminide coated MA 6000 test specimen following 1000 cycles of oxidation testing between 1950°F and 600°F. Figure 8 shows a relatively porosity free MCrAlY coated MA 6000 test specimen following 1000 cycles to the same temperature and conditions as the aluminide.

Overall, conventional MCrAlY overlay coatings on MA 6000 have been determined to perform well in assorted hot corrosion tests to 1650°F and oxidation/thermal fatigue tests to 1950°F.

However, in order to utilize coatings on MA 6000 beyond 1950°F for extended periods of time, it was determined that it would be necessary to overcome diffusional stability problems that develop between the overlay

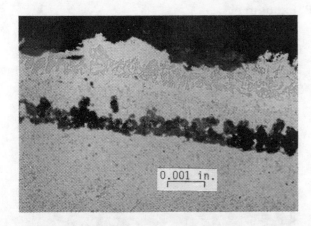

Figure 7. Microstructure of aluminide coating on MA 6000 after 1000 cycles, 1950°F/1 hr. - 600°F/10 min.

Figure 8. Microstructure of MCrAlY coating on MA 6000 after 1000 cycles, 1950°F/1 hr - 600°F/10 min.

coating and the MA 6000 substrate. Figure 9 shows an example of porosity developed on a MCrAlY/MA 6000 specimen following 400 hours of exposure at 2150°F. Diffusional stability of the coating with the substrate was felt to be particularly important for first-stage T406 blade applications as, in order to benefit from the full temperature potential of the alloy in advanced designs, a capability for using the alloy to metal temperatures in the 2150°F range was required. Consequently, an initiative was undertaken to develop a coating system with diffusional stability to 2150°F.

In this effort, a systematic screening approach was devised to evaluate a number of different coating systems. Included were aluminides, platinum aluminides and overlays. None, however, provided the diffusional stability sought for the MA 6000 alloy. Following a review of this work, it was conjectured that the presence of Kirkendall porosity could be eliminated by the application of a compositionally compatible interlayer to the MA 6000 over which a second high aluminum composition would be applied. The concept was to decrease the chemistry gradients between the outer coating and the substrate. In subsequent experimentation, the apparent validity of this concept was verified by extended static testing at 2150°F. Shown in Figure 10 is a MA 6000 specimen both with and without the interlayer but with an aluminide top coat following 400 hours exposure. Significantly, the duplex coating is free of porosity but the single layer coating is heavily decorated with pores.

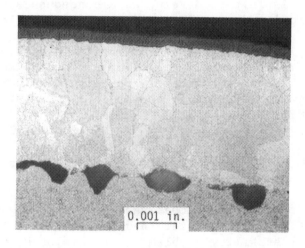

Figure 9. Photomicrographs of MCrAlY coated MA 6000 after 400 hours at 2150°F exposure.

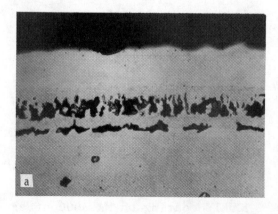 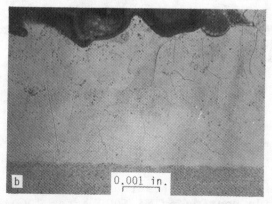

Figure 10. Photomicrographs of coatings on MA 6000 after 400 hours at 2150°F. (a) aluminide without interlayer and, (b) aluminide with interlayer.

Although more testing and evaluation of this new coating system is required, the results are felt to be very promising and offer the potential for fully exploiting the temperature capability of the MA 6000 in low stress/high temperature blading applications.

Concluding Remarks

Significant progress continues to be made in the development of MA 6000 for turbine blading applications. In particular, near-net-shape forging approaches have been identified that provide the potential for minimizing the amount of input stock and complimentary blade machining techniques have been demonstrated. Also, significant progress has been made on the problems of coating/substrate diffusional stability and as a consequence, the future appears promising for the protection of the alloy from oxidation attack in the 2100°F regime.

Although grain quality consistency issues have not been completely resolved, MA 6000 stress rupture results were developed that showed temperature capabilities in the 20 ksi and lower stress regime that ranged from competitive to superior relative to contemporary single crystal alloys. Therefore, for shrouded turbine blade designs of the type represented by the second-stage T406 turbine blade, the potential appears to exist for utilizing MA 6000 as a direct replacement for the single crystal CMSX-3 alloy. With forging mult weight on the order of 0.5 pounds, a materials utilization factor in the 3 to 4 range is projected for the near-net-shape forging. (The materials utilization factor is defined as input weight divided by finish machined weight). If the shrouded blades were to be machined from a solid piece of bar stock, a much larger forging mult would be required owing to the severe twist inherent in this design resulting in a materials utilization factor on the order of 20.0. As a consequence, the near-net-forging approach for this design saves approximately two pounds of the relatively expensive MA 6000 alloy. Even when finish machining costs are added to the cost of this configuration, the near-net-shape MA 6000 forging offers the potential for being cost competitive with cast single crystals.

In the case of the unshrouded first-stage T406 blade, the inherently simpler first-stage configuration lends itself far more readily to either forging or direct machining from extruded and hot rolled bar stock with a projected materials utilization factor of less than 10 for direct machining. With its inherently superior temperature capability at low stress levels and a suitable protective coating for high operating temperatures, the potential appears to exist for substituting an uncooled MA 6000 blade design for a cooled CMSX-3 blade design. Significantly, this would afford the opportunity for eliminating the need for cooling air with attendent fuel savings benefits. It would also provide the opportunity for reduced acquisition costs as an unshrouded uncooled MA 6000 turbine blade is projected as being lower cost than a cooled single crystal design.

It is anticipated that future work in both the shape processing and coating area should provide improved capabilities and lead to cost effective MA 6000 products that can be used in tailored turbine blade designs. These products will have the potential for being used at single crystal operating temperatures and beyond. However, to achieve eventual implementation into U.S. designs, MA 6000 products will need to demonstrate clear cut advantages over single crystals that have become firmly established in a number of U.S. applications. Further, they will need to demonstrate that process control is in place and that quality levels are reproducible and consistent with application requirements.

Acknowledgments

The authors wish to acknowledge the cooperative spirit and technical support afforded by a large cross section of the superalloy community. Particular attention is directed to Mr. Ken Kojola of the U.S. Air Force Manufacturing Technology Division, Mr. Ron Haeberle of IAI - Huntington, and Dr. Robert Dreshfield of NASA Lewis for their support, cooperation and advice. The authors also wish to direct attention for the technical contributions involved in the processing of MA 6000 to Dr. Eric Grundy of Doncasters Monk Bridge, Dr. Roch Shipley of Textron Excello, and Mr. Ray Benn of IAI - Huntington. Acknowledgment is also given to Dr. Sanford Baranow and Dr. Donald Boone for their efforts in the coating development area.

References

(1) P. P. Millan, Jr. and J. C. Mays, "Oxide-Dispersion-Strengthened Turbine Blades", Project Completion Report, NASA CR-179537, October 1986.

(2) D. J. Moracz, "Fabrication Development For ODS-Superalloy, Air Cooled Turbine Blades," Final Report, NASA CR-174650, January 1984.

(3) M. Doner and J. Heckler, "Identification of Mechanism Responsible for Degradation In Thin-Wall Stress Rupture Properties," Paper to be presented at the Sixth International Conference On Superalloys, Seven Springs, PA., October 1988

THE PRODUCTION OF ADVANCED TURBINE BLADES

FROM P/M SUPERALLOY FORGING STOCK

I.C. Elliott[+], C. Cockburn[+] and S.W.K. Shaw[++]

[+] Inco Alloys Limited, Holmer Road, Hereford, U.K.
[++] Inco Engineered Products Ltd., Wiggin Street, Birmingham, U.K.

Abstract

Powder metallurgy nickel based superalloys are frequently used in advanced aero engine gas turbines, but nevertheless are becoming of interest to more down to earth applications where high material integrity and enhanced properties can also be cost effective. Thus, for instance, increasing quantities of corrosion resistant tubing and tool steels are being produced by P/M routes. Another area of interest is that of blades for land based gas turbines. This paper describes the development of the nickel based superalloy APK6, a modified P/M version of the corrosion resistant casting alloy IN792, which possesses creep rupture strength superior to cast IN792 at 650 to 750°C together with fatigue resistance over two times greater than IN738 LC at 600°C and equivalent to wrought NIMONIC* alloy 90.

* NIMONIC is a trademark of the Inco family of Companies.

Introduction

The continual increase in the severity of gas turbine operating conditions has led to a progressive replacement of forged turbine blades from the first row of the high pressure turbine backwards to the lower temperature regions by cast blades in order to obtain increased creep and corrosion resistance and take advantage of the greater complexity of cooling passages which can be achieved in a casting. However, for the rear rows of blades, which in an industrial gas turbine can be very large, cooling is no longer necessary or even desirable and forged blades may then be preferred for their superior fatigue resistance, but forging such components from complex alloys presents certain difficulties.

Similar problems have been encountered in the development of alloys for gas turbine discs, particularly for aero-engine applications. Increasing thrust and efficiency requirements demanded greater tensile and creep strength which could be achieved, initially, by moving from steels to wrought nickel based superalloys. Further alloying of these materials resulted in increasing problems of segregation and forgeability, such that radical changes in manufacturing routes became necessary.

The solution to these problems has been found in inert gas atomised powders, which on compaction provide homogeneous billets for disc forgings and for the same reasons a powder version of a highly alloyed nickel based alloy previously developed for casting applications is being studied for large forged turbine blades in the rear rows of industrial gas turbines.

The P/M Approach

Before describing the current blade alloy development, the advantages of the powder metallurgy approach will be considered in more detail. The inert gas atomisation process uses the kinetic energy of a high velocity stream of a gas such as argon to fragment a stream of molten alloy into spherical droplets. Due to their size, which is typically less than 250 microns in diameter, and the quenching effects of the high volume of cold gas the droplets solidify rapidly to give, in effect, very small, spherical, segregation free ingots. The refinement of microstructure due to the high solidification rate can be translated to the bulk form by a suitable compaction process such as hot isostatic pressing or extrusion, resulting in homogeneous, albeit highly alloyed forging stock. Indeed, in some cases the grain size of the compacts is so fine that the material is superplastic and is ideal for isothermal forging.

The structure of a conventionally cast nickel based superalloy IN792 (nominal composition 12.4% Cr, 9% Co, 2% Mo, 3.9% W, 3.9% Ta, 4.5% Ti, 3.1% Al, balance Ni) is illustrated in Figure 1a and shows a coarse dendritic pattern typical of this type of alloy. In the solute rich interdendritic regions which are last to solidify, areas of degenerate gamma – gamma prime eutectic form together with concentrations of primary carbides, and such highly alloyed and segregated structures are very difficult to hot work. In contrast, the same alloy produced as inert gas atomised powder shows a considerably more refined microstructure in the powder particles (Fig. 1b) which when

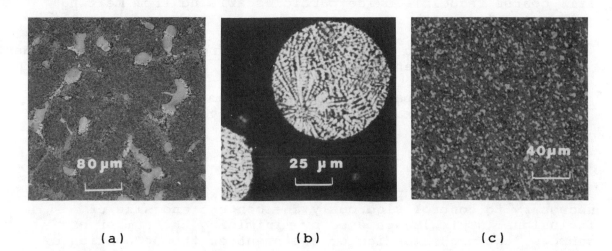

(a) (b) (c)

Figure 1 - Microstructural appearance of (a) as-cast
(b) inert gas atomised powder and (c) HIP plus extruded
powder alloy IN792.

HIPed and extruded results in an extremely homogeneous
microduplex gamma - gamma prime structure (Fig. 1c) with none of
the deleterious segregation features which are characteristic of
the cast version. It will be shown later that this microduplex
structure can be highly forgeable.

In order to take full advantage of refined microstructures
that can be generated by the atomisation process, a high level
of process development and quality assurance is required (1).
Principally, the Inco Alloys Limited atomisation unit consists
of a 500 kg capacity vacuum induction furnace mounted on top of
an atomisation tower some 6 metres high. A feature of the
powder handling equipment is the ability to prevent the powder
coming into contact with air throughout the sequence from raw
materials to finished compact. The atomised powder is cooled
and stored under argon until all quality control checks have
been performed. The powder is then sieved and blended before
being transferred to the degassing unit. Here the powder is
degassed under vacuum in order to remove any absorbed gases.
Still under vacuum the powder passes into cans which are sealed
when full, by means of a hydraulic crimp welder.

Perhaps the single most important factor which has emerged
from the development of Ni-based atomised powders is the great
importance that non-metallic contaminant inclusions have on the
ultimate performance of products made from pre-alloyed powder.
As most pre-alloyed powder development work has been directed
towards the production of gas turbine discs the effect of these
inclusions has been most apparent in reducing the fatigue life
of the component (1), but this is also true for large turbine
blades. The types of inclusions that may be found in
pre-alloyed powder are both numerous and varied; the following
is only a short list of the types and morphologies that can be
found:-

 i) fine (~ 20 um) stable oxide particles arising from
    deoxidation of melts.

 ii) coarse (~ 50 um) stable oxide particles arising from
     furnace linings and raw materials.

iii) coarse reducible oxide particles arising from melting practices.

iv) varied inorganic particles arising from pollution of powder handling equipment for example rust, cement, weld slag etc.

v) varied organic particles arising from powder handling equipment for example vacuum seals, vacuum pump oils, plastic tubing.

vi) coarse metallic contamination from other alloy powders.

For the optimum properties in P/M alloys it is therefore necessary to control rigorously the content and size of inclusions. This is achieved by minimising the number of sources, thorough cleaning of equipment or the dedication of equipment, the sealing of equipment from environmental pollution, careful control of raw materials, material handling equipment, melting practice and indeed, the very design of the equipment in order to facilitate cleaning etc. The effects of the latter two points on the non-metallic inclusion content of a Ni-based P/M disc alloy NIMONIC alloy AP-1 as measured by the water elutriation technique is illustrated in Figure 2, following the installation of the atomisation equipment in 1976.

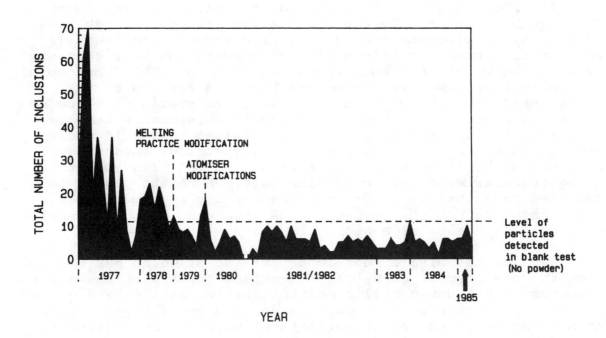

Figure 2 - Total number of inclusions found by water elutriation in the +106/150 micron fraction of a 0.5 kg sample of powder NIMONIC alloy AP-1 over the period 1977 to 1985, illustrating the effects of melt practice and atomiser modifications on powder contamination levels.

Similar reductions in inclusion content may be achieved by the introduction of such practices as melt filtration using reticular ceramic foam filters. This is particularly effective in minimising the number of fine inclusions arising from deoxidation (Fig. 3) as shown by the reduction in numbers of inclusions found metallographically in an extruded powder bar after filtration was introduced in 1982.

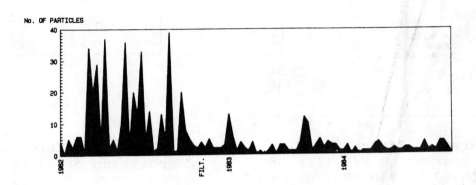

Figure 3 - Total number of particles less than 0.2 mm in length found metallographically in extruded power NIMONIC alloy AP-1 between the years 1982 and 1984, illustrating the effect of introducing melt filtration on powder contamination levels.

Ultimately therefore one has to rely on the highest standard of production housekeeping and quality control testing to ensure the integrity of the finished component. To put the problems in perspective, current quality standards equate to finding one inclusion in about a hundred million powder particles. The techniques available include visual and metallographic inspection, water elutriation, electrochemical machining and ultrasonic testing.

Choice of Alloy

Since the application in this case is to be industrial gas turbine blades, the choice of alloy was based largely on the balance of corrosion resistance and high temperature strength. The detailed reasoning for the current alloy choice is given elsewhere (2). However, as many cast blades for this type of application are in IN738LC, the target is similar corrosion resistance at 650-750°C but with a higher strength level than can be achieved by IN738LC (3). Alloy IN792, a Ni-Cr-Co casting alloy strengthened with W, Ta, Ti and Al, was identified as having the necessary balance of properties.

It has previously been demonstrated (4) that by moving to a P/M route for this alloy reducing carbon and correspondingly the carbide forming element Ta and Ti resulted in improved stress rupture properties. However, further enhancement was achieved by optimising the (Ti + Al) content (2) such that maximum intermediate temperature stress rupture strength occurs at a (Ti + Al) content of 8% (Fig. 4).

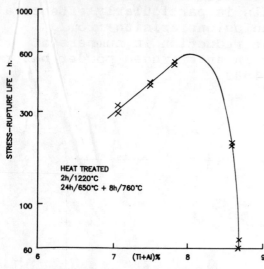

Figure 4 – Effect of (Ti + Al) content on S/R life of P/M IN792 at 740 MPa/700°C

Table 1 – Chemical Composition of alloy APK-6

0.03%C, 12.5%Cr, 9.0%Co,
2.0%Mo, 3.9%W, 3.0%Ta,
4.6%Ti, 3.4%Al, 0.10%Zr,
0.01%B, bal.Ni.

The resulting powder alloy, designated APK-6 therefore has the composition given in Table 1.

Production of Forging Stock and Blades

Laboratory extrusions of powder alloy APK-6 showed that high levels of ductility could be achieved with extrusion temperatures below about 1120°C (Fig. 5). The precise

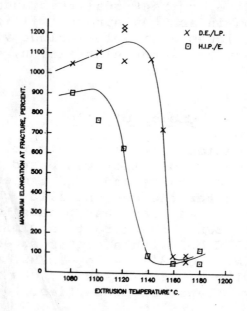

Figure 5 – Variation of maximum hot tensile elongation with extrusion temperature.

conditions under which this maximum ductility occurs for a given extrusion temperature may vary depending on extrusion ratio (2). However, it is clear that under certain circumstances alloy APK-6 exhibits superplastic behaviour and that higher maximum ductilities are achieved with direct extruded loose powder than for powder initially consolidated by hot isostatic pressing prior to extrusion. This may be explained by the finer microduplex structure which forms by loose powder extrusion compared to a somewhat coarser structure in the case of HIP/extrude generated from the HIP structure, in which initial recrystallisation of the cellular powder particles has already taken place during HIPing rather than on extrusion.

Full scale extrusion was therefore carried out using direct extrude, canned powder billets in a Loewy 5000 tonne horizontal press at 1120°C and an extrusion ratio in excess of 30 to 1. The resulting bar was decanned, machined and ground to 30 mm diameter!

As with the laboratory extruded material high hot tensile elongations were observed.

Tests were carried out to determine 'm-values' using the step-wise technique with a few check tests at critical strain rates. A typical curve for hot ductility versus temperature is shown in Figure 6. Figure 7 illustrates the 'm-values' obtained at the temperature of maximum ductility (T) and at T plus and minus 40°C. It is generally accepted that if the m-value approaches 0.5, then the material is superplastic. For these extruded bars of APK-6 values up to 0.64 have been recorded.

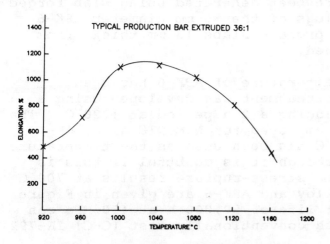
Figure 6 - Variation of hot tensile elongation of APK6 with temperature.

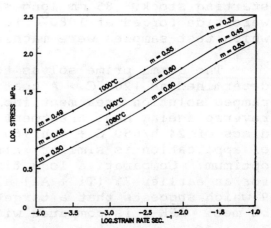
Figure 7 - Strain rate sensitivity of flow stress (m) for APK6.

While the ultimate aim of this development is to produce larger diameter forging stock for turbine blades 450-650 mm in length, the current intermediate phase was targeted on an existing rear row power turbine blade 180 mm long for a 5000 HP industrial gas turbine. Slug lengths of the 30 mm diameter APK-6 were therefore partially extruded to the preform for the aerofoil and root block of this blade. This is a critical step in the process, the main variables being forging temperature, speed and lubrication. The preforms were subsequently closed die forged to fully develop the aerofoil (Fig. 8).

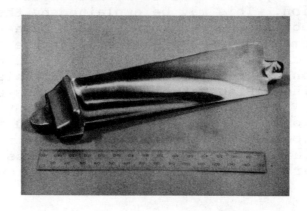

Figure 8 - 180 mm long forged APK-6 turbine blade for rear row of an industrial gas turbine.

Mechanical Properties of APK-6

Only a limited number of turbine blades were available for cut-up testing at this stage, as a considerable number are required for an engine test. Furthermore, it is only possible to machine a few small samples from each blade. The main property database has therefore been generated using slab forged starting stock. 85 mm long slugs of the 30 mm diameter APK-6 were side forged at $1080^{\circ}C$ to produce slabs 13 mm thick, from which test samples were machined.

The gamma prime solvus temperature of APK-6 has been determined as $1205^{\circ}C$. A heat treatment was developed using a ramped solution treatment including a 2 h period at $1220^{\circ}C$. The reverse ageing heat treatment employed for NIMONIC alloy AP-1 discs of $24\ h/650^{\circ}C + 8\ h/760^{\circ}C$ has been used as the temperature of application is similar, although it is doubtful if this is optimum. Comparative long time stress-rupture results at $700^{\circ}C$ for an earlier 7% (Ti + Al) alloy and APK-6 are given in Figure 9 which suggests that a target life of 50,000 h at 450 MPa can be met. APK-6 is compared with conventionally cast (C-C) IN-792

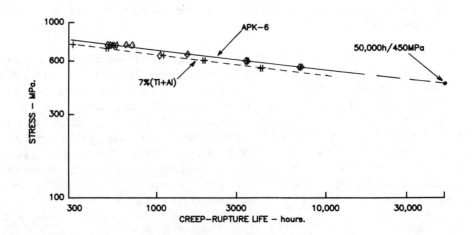

Figure 9 - Creep rupture properties of APK-6 at $700^{\circ}C$

and IN738LC in Figure 10 where it is seen that at high stresses and temperatures up to 750°C APK-6 is superior to C-C IN-792.

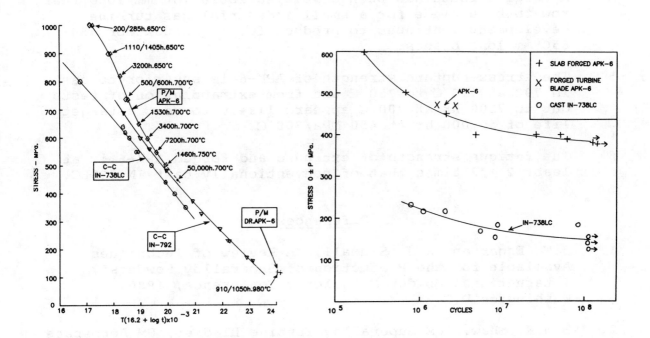

Figure 10 - Comparison of stress rupture properties of cast IN-738LC and APK-6

Figure 11 - Fatigue properties of APK-6 and IN738LC at 600°C, 0 +/- P

The main reason for considering P/M superalloys for forged rear row blades is for fatigue resistance, in combination with high stress-rupture strength and corrosion resistances. The latter 2 could nearly be provided by a C-C blade if satisfactory investment castings could be produced in so large a blade in the very difficult to cast, strongest alloys such as IN-792, but the fatigue resistance of C-C materials is always markedly inferior to those of forged materials. Figure 11 compares the fatigue resistance at 600°C, stress 0 +/- P of IN738 LC C-C in ideal carrot-shaped test bar blanks with those of APK-6 extruded and slab forged. Duplicate results are also shown for fatigue specimens cut from a 180 mm long forged blade of APK-6 showing that the forged blade possesses slightly superior fatigue resistance to the slab forgings which themselves equalled the fatigue resistance of the highly fatigue resistant forged alloy NIMONIC alloy 90.

Conclusions

1. The high quality standard production routes developed for inert gas atomised powder aero engine disc alloys can be applied to powder alloys for other components where high integrity is required, such as turbine blades.

2. A P/M superalloy, APK-6 has been developed from the casting alloy IN-792 which provides an improved combination of mechanical properties and hot corrosion resistance.

3. Extrusion compaction of powder APK-6 produces a forgeable alloy which under certain circumstances exhibits superplastic behaviour.

4. A forging route has been developed for a 180 mm long rear row turbine blade for a small industrial gas turbine. Development continues to produce forging stock for 450-650 mm long blades.

5. The stress-rupture strength of APK-6 is superior to cast IN 792 at 650°C to 750°C and from extrapolation of tests out to 7200 hr at 700°C appears likely to meet a target life of 50,000 hr at 450 MPa/700°C.

6. The fatigue strength of extruded and forged APK-6 is at least 2 1/2 times that of conventionally cast IN-738LC.

References

1. J.W. Eggar and R.J. Siddall, "A Review of Techniques Available for the Production of Superalloy Powders", International Powder Metallurgy Conference, 1980, Washington D.C.

2. S.W.K. Shaw, "PM Superalloy Turbine Blades", PM Aerospace Materials 87 Conference, Luzern, Switzerland, 1987.

3. W. Betteridge and S.W.K. Shaw, "Overview - Development of Superalloys", Materials Science and Technology, September 1987, 3, 682-693.

4. J.M. Larson, "Carbide Morphology in PM IN792", Metallurgical Trans., A, 7A, October 1976, pp 1497-1502.

THE PHYSICAL METALLURGY OF A SILICON-CONTAINING LOW EXPANSION SUPERALLOY

K. A. Heck, D. F. Smith, J. S. Smith, D. A. Wells, M. A. Holderby

Inco Alloys International, Inc.
Huntington, West Virginia 25720

INCOLOY* alloy 909 is an iron-nickel-cobalt-niobium-titanium controlled low thermal expansion superalloy with an intentional silicon addition. Recommended for use at temperatures to 650°C, the alloy is now specified for major new gas turbine engines. The pseudo-equilibrium time - temperature - transformation behavior of alloy 909 is presented as well as a brief description of the major phases: γ', ε'', ε, and Laves. Gamma prime (γ'), the major strengthening phase in the alloy, is shown to precipitate in the 538°C to 760°C range. The ε'' and ε phases, which are similar to γ'' and δ phase found in INCONEL* alloy 718, precipitate at intermediate temperatures (about 700°C to 950°C). Laves phase, which is used to control grain size, precipitates at the higher temperatures (800°C-1040°C) used during hot working and annealing. The alloy's Si addition significantly affects the formation of these phases during processing, thereby affecting mechanical properties. The interrelationship between the physical metallurgy and properties of the alloy are explored in a two level factorial study of processing (two levels of Laves precipitation) and age hardening heat treatments (the alloy's standard heat treat cycle and a short time cycle). Notched bar rupture tests and creep crack growth tests at 538°C demonstrate that resistance to stress accelerated grain boundary oxygen embrittlement (SAGBO) improves with precipitation of intergranular ε phase. Excess Laves precipitation lowers residual Nb content, consequently reducing ε abundance and reducing crack growth resistance. Combined with proper thermomechanical processing, the short aging cycle precipitates an ε grain boundary structure which significantly improves SAGBO resistance.

---

* INCOLOY and INCONEL are trademarks for products of the Inco family of companies.

Superalloys 1988
Edited by S. Reichman, D.N. Duhl,
G. Maurer, S. Antolovich and C. Lund
The Metallurgical Society, 1988

Introduction

INCOLOY alloy 909 is the latest development in a series of low expansion superalloys designed for use up to 650°C where age-hardened strength and close operating tolerances are required. Introduced at the Fifth International Symposium on Superalloys (1), this alloy is now specified in major new gas turbine engines and is being evaluated for many other applications.

Like other controlled expansion superalloys, alloy 909 derives its low expansion characteristics from electron spin interactions within its Fe-Ni-Co matrix. The high strength of these alloys is achieved by the precipitation of A_3B type phases, where B may be Al, Ti, or Nb, though INCOLOY alloys 907 and 909 rely only on Ti and Nb for strengthening. The necessity of a Cr-free matrix to achieve low controlled expansivity results in reduced oxidation resistance and a phenomenon known as stress accelerated grain boundary oxygen embrittlement (SAGBO), which occurs when material is stressed at intermediate temperatures in air.

In the earliest of these alloys, INCOLOY alloy 903, the threshold stress for oxygen embrittlement was extremely low except in the longitudinal direction of heavily warm-worked material (2). The restricted Al and increased Nb content of INCOLOY alloy 907, combined with an overaging heat treatment raised this threshold stress considerably (1,2). In alloy 909, the addition of 0.4% Si combined excellent SAGBO behavior with good tensile properties and eliminated the extreme processing and/or heat treatment measures of the earlier alloys. Substantial and complex differences in microstructure and physical metallurgy accompanied this small Si addition.

In this two part paper, Part I describes the alloy's time-temperature-transformation (TTT) behavior and its major precipitated phases. Part II applies this understanding to a two-level factorial study of thermomechanical processing and age hardening heat treatments.

I. Isothermal Time-Temperature-Transformation Study

Experimental Procedures

The TTT study was conducted on a commercially produced hot rolled flat containing (by weight %) 42.1% Fe, 38.3% Ni, 12.9% Co, 4.70% Nb, 1.58% Ti, 0.35% Si, 0.04% Al, 0.01% C, 0.01% Cr. Specimens were solution annealed at 1038°C for one hour, air cooled, then isothermally heat treated in 56°C increments from 538°C through 1038°C for 0.1, 1, 10, 100, and 1000 hours.

Bulk heats and extracted residues were chemically analyzed by X-ray spectrography and inductive-coupled plasma (ICP) analysis. Point probe analyses were performed on a microprobe and a SEM with EDX. Thin foils and extraction replicas were also examined by TEM at 100kV. X-ray diffraction (XRD) analyses of extracted phases were conducted on an automated diffractometer system using copper radiation. Previously unknown diffraction patterns were indexed by methods described by Cullity (3).

Conventional specimen preparation techniques are not suitable because the alloy is ferromagnetic, has poor oxidation resistance, and contains phases with modified structures. Because of space limitations, the new preparation and analysis techniques are not detailed here.

Review of Phase Analyses

Overview. Figure 1, the TTT diagram for solution annealed alloy 909, shows the precipitation C-curves for the intragranular γ′, ε″, ε phases and the grain boundary (noted as G.B.) phases of the Laves and ε types. G-phase silicides ($A_{16}B_6Si_7$) form in the Laves region after prolonged exposures (>100 hours) (8). Minor phases present, but not shown, include (Nb,Ti)(C,N), TiN and $Ti_2C_4S_2$.

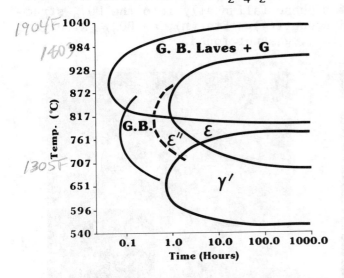

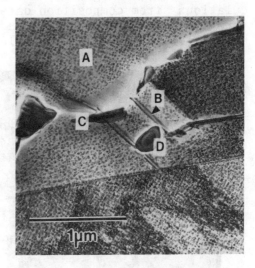

Figure 1: TTT diagram for INCOLOY alloy 909 (0.35% Si). Isothermal aging after 1038°C/1 hour anneal.

Figure 2: Thin foil TEM image after standard treatment (see Part II). A= γ+γ′, B= ε″, C= ε, D= Laves.

Table I (located at the end of Part I) gives X-ray diffraction data for extracted major precipitates which have the following approximate compositions (in atomic %):

Phase	Fe	Ni	Co	Nb	Ti	Other
γ′	8	51	12	13	16	---
ε″	12	46	12	14	16	---
ε	11	50	16	18	4	2 Si
Laves	18	38	14	18	3	5 Si, 2 C

Gamma prime (γ′). This major strengthening constituent precipitates between 538°C and 760°C, and is predominantly Ni_3(Ti,Nb) with an $L1_2$ structure. The fact that γ″ (a transition phase in other Nb strengthened alloys such as alloy 718) was not found is likely due to the effect of Co and Fe on stacking fault energy. The transmission electron photomicrograph in Figure 2 shows intragranular γ′ along with the ε″, ε and Laves phases precipitated during the alloy's standard heat treatment (see Part II).

Epsilon double prime (ε″). At higher temperatures (700°C-950°C) a $(Ni,Fe,Co)_3(Nb,Ti)$ transitional phase of very fine inter- and intragranular platelets precipitates. Together, the diffraction patterns for ε″ and Laves phases (Table I) can be mistaken as delta (δ) phase. The ε″ phase also occurs in commercially overaged alloy 907, but the Si content of alloy 909 accelerates its precipitation rate and broadens its temperature range.

The acicular morphology of ε″, shown in Figure 3, is similar to δ or γ″. (This accentuated precipitation is shown in warm-worked and aged material.) While its exact structure is unknown, it is expected to be distorted hexagonal or FCC compatible with the transition of γ′ (abcabc) to the equilibrium ε (abab) phase.

Epsilon (ε). This phase initially precipitates at grain boundaries, around nitrides and carbonitrides, and intragranularly from ε". The angular, blocky or acicular intergranular and Widsmanstatten intragranular precipitates appear optically similar to δ phase in alloy 718. As with ε", Si enhances the precipitation of this phase. Whether ε precipitation is always preceded by ε" is uncertain. Figure 4 shows the platelet morphology of ε, which has a DO_{19} hexagonal superlattice (Ni_3Sn-type) structure. The Co content of ε is greater and the Ti content lower than shown for ε". Calculations from composition data of ε phase fall neatly into the DO_{19} structure field as defined by Watson and Bennett(4). Eta (η), a DO_{24} structure of Ni_3Ti with a similar morphology to ε, was not found.

Figure 3: Thin foil TEM images of ε". Note similarity of pattern to <001> FCC γ', and heavy streaking causing displacement of superlattice spots.

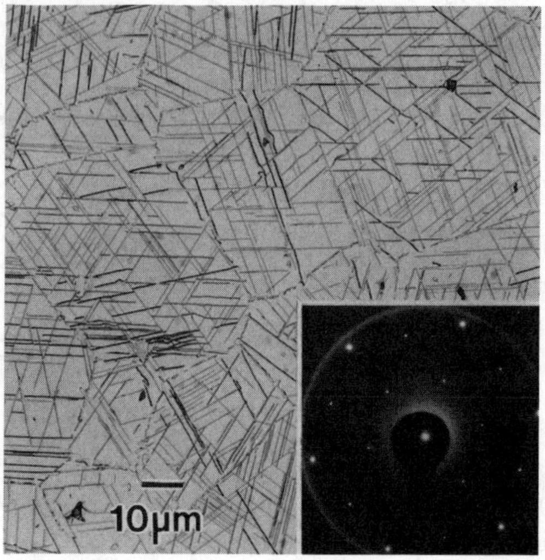

Figure 4: Optical micrograph of ε in an overaged sample. TEM diffraction pattern of extracted platelet is DO_{19} <001>. Ring is 0.204nm Cr.

Laves. Two and four-layered hexagonal Laves phases ($MgZn_2$ and $MgNi_2$ type crystal structures) form in grain boundaries at the high temperatures (800°C-1040°C) routinely encountered in hot working and annealing. The Si addition in alloy 909 induces the Laves precipitation by increasing the N_v of the matrix and stabilizing TCP phases (5). This effect is profound since Laves has not been observed in the lower Si alloy 907.

These Ni-rich phases have similar lattice parameters to Laves found in the Co-Nb binary system (6,7). Point-probe analyses of extracted Laves particles did not indicate any notable compositional differences. However, sample to sample variations in the d-spacings of the four-layered phase suggested slight differences could exist. The lattice parameters of the two-layered phase were equivalent from sample to sample. The two phases probably exist in varying ratios within a single, faulted particle.

Kinetic and Compositional Factors. In commercial practice, two effects modify the TTT behavior in Figure 1. First, the precipitation kinetics of the major phases are sensitive to internal energy levels. C-curves of material containing residual work or annealed at lower temperatures will be significantly displaced to shorter times with broadened temperature ranges. Second, the major phases can be significantly affected by variations in minor element content, especially Si, as well as the major elements.

Phase-property-SAGBO interrelationships. A primary concern in controlled expansion superalloys is oxygen embrittlement (SAGBO). Alloy 909 achieves its degree of SAGBO resistance by the formation of a fine grained structure and the precipitation of ε and perhaps ε" phases.

Grain refinement is achieved by controlling Laves quantity and distribution during thermomechanical processing. Sufficient Laves must be precipitated to permit grain boundary pinning, but very coarse Laves particles decrease the grain refining effect. A heavy Laves prior grain boundary network may contribute to reduced ductility. Excessive quantities are also undesirable as lower residual Nb may reduce ε and ε" phase precipitation during subsequent heat treatment. The resulting change in the Nb/Ti ratio may affect intragranular overaging rate and properties by altering coherency strains of strengthening precipitates.

Although the mechanism(s) is not fully understood, the presence of ε, and probably ε", in existing grain boundaries is required to offset oxygen embrittlement (as grain boundary δ benefits the rupture properties of alloy 718). The abundance and temperature regimes of these phases are sensitive to both composition and strain energy in the material before aging. Residual strain energy in the presence of Si promotes the precipitation of the beneficial intergranular phases and allows an excellent combination of SAGBO resistance and tensile properties to be achieved in commercially attractive aging cycles.

TABLE I - INCOLOY alloy 909: X-ray Diffraction Data - Phase Extractions

MAJOR PHASES							MINOR PHASE
Matrix	Matrix (Aged)	Gamma Prime	Epsilon Dbl.Prime	Epsilon	Laves Phases 2-layer	Laves Phases 4-layer	G
a=.3608	a=.3605	a=.3627	a=.518? c=.420?	a=.5177 c=.4196	a=.4770 c=.7759	a=.4810 c=1.589	a=1.125
d(nm) I	d(nm) I	d(nm) I	d(nm) I	d(nm) I	d(nm) I	d(nm) I	d(nm) I
						.405 2	
						.332 4	.325 16
				.306 3		.297* 21	
					.282 10		
			.2560 48				.256 6
					.238 40	.241 27	
						.229* 17	
			.223 17	.223 25			
.2083 100	.2081 100	.2084 100	.209 100	.2096 +80	.219 52		.216 100
					.206 6	.205 18	
					.203 100		
			.201* 70			.201 26	
					.1996 60	.1990 100	
			.196 90	.1972 100			.198 100
			.193* 35		.194 10	.1926 7	.190 9
.1804 20	.1802 20	.1806 50					.187 16
				.1527 15			
						.1401 8	
						.1356 4	
						.1327 5	.1326 12
.1276 21	.1275 21	.1278 48	.1292 23	.1293 15	.1298 10	.1234 2	.1303 6
					.1241 8		
				.1185 19			.1131 10
.1089 20	.1087 20	.1090 38	.1100 18	.1099 21			.1100 4
				.1080 12			.1088 5
.1042 7	.1041 7	.1043 6	.1051 3	.1047 3			
.0828 14	.0827 14	.0832 19					
		.0811 25					

+ High intensity due to texture.   * Peaks do not match calculated pattern.

II. Process - Aging Treatment Factorial Study

Procedure - Factorial Study

Experimental Design. To demonstrate the fundamental uses of the phases described in Part I and to illustrate their resulting effects on mechanical properties, it was decided to examine the interactive effects of processing and heat treatment using a two by two factorial design. Two hot working processes providing two levels of Laves phases were combined with two aging cycles.

Hot Working. The starting material for both experimental processes was 90 mm ϕ hot rolled and rough turned bar from a commercial melt with an average chemistry of 41.4% Fe, 38.4% Ni, 13.0% Co, 4.84% Nb, 1.65% Ti, 0.41% Si, 0.04% Al, 0.005% C, 0.14% Cr. Bars were rolled to a final size of 14 mm thick by 64 mm using two Processes: A and B.

In Process A, the bar was heated for one hour (1 h) at 1038°C and rolled to 41 mm square, reheated for 0.5 h at 1024°C, flat rolled to 20 mm thickness ($\geq$15% reduction below 927°C), reheated for 0.5 h at 996°C and rolled to final size ($\geq$15% reduction below 927°C).

In Process B, the bar was heated for 1 h at 996°C and rolled to 41 mm square, reheated for 2 h at 968°C, flat rolled to 32 mm thickness, reheated for 2 h at 968°C, rolled to 22 mm thickness, reheated for 2 h at 968°C and rolled to final size ($\geq$25% reduction below 927°C). The additional reheat, lower heating temperatures and extended soaking times for this process were selected to produce excessive Laves and reduce residual Nb content.

Heat Treatment. Test blanks were given a recrystallizing, fine grained anneal of 982°C for 1 h and air cooled (AC). Unlike the 1038°C treatment used for the TTT study, this anneal does not solution the Laves phase(s) but instead precipitates Laves in a prior grain boundary network. The retained energy from this lower annealing temperature accelerates the subsequent precipitation of intermediate temperature phases when aging with one of the following treatments (noted as STAND and SHORT):

STAND - 718°C/ 8 h furnace cooled (55°C/ h) to 621°C/ 8 h, AC.
The alloy's standard age gives an excellent combination of tensile and 649°C stress rupture strength and is widely used with alloy 718 (1).

SHORT - 746°C/ 4 h furnace cooled (55°C/ h) to 621°C/ 4 h, AC.
Earlier unreported work showed this treatment as an economical, yet effective age. Some sacrifice of tensile properties resulted, but rupture properties were quite good. Also, the TTT diagram suggested that a higher initial aging temperature (albeit shorter time) could promote more precipitation of ε and ε'', thought beneficial to SAGBO resistance.

Testing Procedures. The following mechanical property tests were conducted: room temperature tensile (20°C), 649°C high temperature tensile, 649°C, 510 MPa combination smooth and K_t 3.6 notched stress rupture bar and 538°C, 827 MPa K_t 2 notched stress rupture bar. (All smooth section gauge lengths were four times the gauge diameter. Smooth gauges and notches for high temperature tests were finished using a low stress grinding technique.)

Duplicate crack growth tests (9.5 mm thick compact tension specimens) were conducted in air at 538°C using both static and fatigue loading. Notches were oriented parallel to the rolling direction. Fatigue pre-cracking and crack growth testing were conducted in accordance with ASTM A647-85. Crack lengths were measured optically.

The K-increasing fatigue crack growth tests were conducted at Metcut Associates under constant peak load and amplitude using a simple linear ramp waveform at a frequency of 0.033 Hz and R ratio of 0.1.

Data Review

Mechanical Properties. Examination of processing and aging effects in factorial Table II shows:

1. Yield and tensile strengths were only slightly affected by processing. For Process B (excess Laves) strengths were lowered $\leq$ 1% at 20°C and $\leq$ 5% at 649°C. As expected, strengths were reduced by the short-time age. In this case, the decrease in yield and tensile strengths ranged from 7-10% and 4-8% respectively. Tensile ductilities were virtually unaffected by either variable.

2. All of the 649°C, 510 MPa combination smooth and notched bar stress rupture tests were notch ductile (i.e., broke in the smooth gauge) and showed excellent life and ductility. Given the normal variation in stress rupture testing there seemed to be little, if any, effect of processing on rupture life, and aging with the short cycle showed a mild negative effect at worst.

3. The 538°C notch results confirmed earlier findings (1) which demonstrated this test to be more sensitive to SAGBO than the 649°C test. Notch lives varied by an order of magnitude (i.e., from about 40 h to over 400 h). There was little effect of processing with the standard aging cycle but there was a strong, positive interaction with Process A (normal Laves) and the short-time age.

Crack Growth Results. The static crack growth (SCG) rates in Figure 5 were lowest in specimens given the combination of Process A and the short-time age. Rates were highest in Process B specimens given the standard age. The strong correlation between 538°C notch rupture and SCG behavior, coupled with their similar intergranular crack appearance (not shown), indicates the key to improving notch properties is decreasing SCG rate.

As discussed below, it is necessary to control processing to promote subsequent ε precipitation. Process control should include both grain size and Laves precipitation controls.

In contrast to SCG, fatigue crack growth (FCG) rates were relatively unaffected by process or heat treatment indicating that crack growth is dependent on the time under applied stress. One expects that fatigue profiles containing hold times at peak stress would show similar microstructural effects on FCG rate as observed on SCG behavior.

Microstructure Review. As expected, the two processes produced different amounts of Laves and age hardening precipitates. Extractions of the annealed Process A material contained 1.5 wgt% Laves compared to 3.5 wgt% for Process B, and EDX analysis of grain interiors showed a corresponding reduction of residual Nb with the latter process. Regardless of the aging cycle, extractions of fully heat treated materials showed Process A contained 6.1 wgt% A_3B age hardening phases vs. 4.1 wgt% for Process B.

Although it was intended that Processes A and B yield equal grain sizes, Process B material had an ASTM#9 grain size compared to Process A's ASTM#7.5 structure. Grain refinement has a strong positive effect on the SAGBO/ notch rupture strength of these alloys (2). Thus, on the basis of grain size alone, Process B should have had an edge over Process A.

TABLE II - INCOLOY alloy 909: Effect of Process and Aging Treatment on Mechanical Properties

		Age: STAND		Age: SHORT	
		Process:		Process:	
		A	B	A	B
20 C Tensile	YS (MPa):	1102	1089	998	996
	TS (MPa):	1366	1353	1309	1298
	El (%):	18	17	18	18
	RA (%):	40	42	42	44
649 C Tensile	YS (MPa):	931	888	832	826
	TS (MPa):	1093	1067	1009	984
	El (%):	16	22	21	23
	RA (%):	52	60	61	59
Comb. Rupture 649 C 510 MPa	Life (hr):	141	118	101	110
	El (%):	30	33	18	30
	RA (%):	48	57	30	41
Notch Rupture 538 C 827 MPa	Life (hr):	62*	80	417	100

* - Average of 41, 53, and 91 hrs.

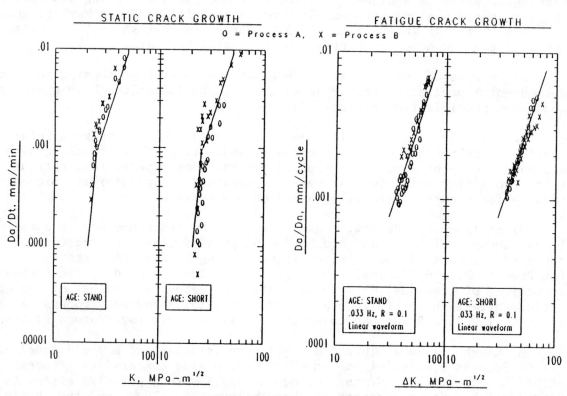

Figure 5. Effect of process and aging treatment on 538°C static and fatigue crack growth of INCOLOY alloy 909. The overall mean curves for both SCG and FCG are shown for comparison purposes.

Despite its coarser grain size, Process A material given the SHORT age showed substantial improvements in 538°C notch rupture and static crack growth behavior. This indicates that the location, quantity or morphology of age hardening precipitates are important factors in obtaining superior SAGBO resistance. Figures 6 and 7 show differences in grain boundary ε precipitation of Process A material given the two aging treatments. These optical and TEM photomicrographs demonstrate that the standard treatment has a less continuous, more globular precipitate compared to the more complete zipper-like acicular morphology found in the SHORT age material. The higher residual Nb content in Process A apparently shifts the ε C-curve to shorter times thus allowing more ε precipitation to occur. The 28°C higher intermediate aging temperature of the SHORT cycle more than compensates for its shorter time. Together, Process A and the SHORT cycle were very effective in accelerating ε precipitation, changing its distribution and morphology, and improving SAGBO performance. Unreported work showed a further increase in intermediate aging temperature would result in minimal improvement in SAGBO resistance compared to the loss in strength.

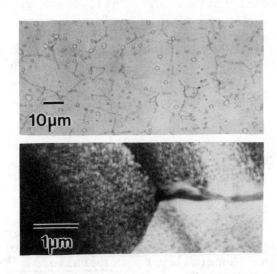

Figure 6: Optical (a) and TEM (b) micrographs of Process A material after STAND age.

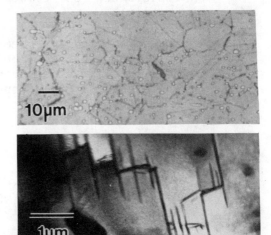

Figure 7: Optical (a) and TEM (b) micrographs of Process A material after SHORT age.

Summary

This investigation reveals the TTT behavior of INCOLOY alloy 909, an age hardenable Fe-Ni-Co material with a low coefficient of thermal expansion. These findings are applied in the design of a factorial experiment to examine the effects of processing and heat treatment on the mechanical properties and crack growth behavior of the alloy. Major findings include:

1. Gamma prime (γ'), the alloy's primary strengthening phase, precipitates at temperatures below about 760°C and consists of $Ni_3(Ti,Nb)$.

2. Epsilon double prime (ε'') and epsilon (ε) are transitional and equilibrium A_3B phases which precipitate during aging at intermediate temperatures. While ε'' and ε resemble the Ni_3Nb delta (δ) phase found in alloy 718, they are crystallographically distinct. Together, the grain boundary forms of these phases improve the alloy's resistance to stress accelerated grain boundary oxygen embrittlement (SAGBO).

3. Laves phases can precipitate in a prior grain boundary network at the higher temperatures used in commercial hot working and annealing. With proper process controls, Laves is essential to the grain refinement contributing to SAGBO resistance.

4. There are strong effects of matrix composition on TTT behavior:
   a.) Si raises the alloy's N_v which promotes the formation of Laves phases, and broadens the TTT curves for ε'' and ε.
   b.) Fe and Co affect the ordering of the A_3B phases and favor formation of ε-type structures (instead of γ'' and δ).
   c.) Excessive Laves precipitation reduces Nb in the matrix, affecting subsequent formation of beneficial ε-type phases.

5. The factorial study of processing and aging treatment showed tensile properties were little affected by processing. The short age cycle decreased yield and tensile strengths up to 10%, but combined with Process A (normal Laves), this age gave excellent 538°C SAGBO resistance (as evidenced by notch rupture and static crack growth results).

6. The dramatic improvement in SAGBO strength was accomplished by selecting a process/ heat treatment combination which created a beneficial microstructure containing both the grain-refining Laves phase and the grain boundary ε phase.

REFERENCES

1. D.Smith, J.Smith, S.Floreen, "A Silicon Containing, Low Expansion Alloy With Improved Properties", Superalloys 1984 (Conference Proceedings), TMS-AIME, 1984, p.591.

2. D.Smith, E.Clatworthy, D.Tipton, W.Mankins, "Improving the Notch Rupture Strength of Low Expansion Superalloys", Superalloys 1980 (Conference Proceedings), ASM, 1980, p.521.

3. B.Cullity, Elements of X-Ray Diffraction, (2nd Ed.), Addison-Wesley Co., Reading, MA, 1978, pp.324-348.

4. R.E.Watson, L.H.Bennett, "Transition Metals: d-Band Hybridization, Electronegativities and Structural Stability of Intermetallic Compounds", Physical Review B, Vol.18, No.12, Dec.,1978, p.6439.

5. R.F.Decker,"Strengthening Mechanisms in Nickel-Base Superalloys", International Nickel Co., New York, N.Y., c 1970.

6. J.K.Pargeter, W.Hume-Rothery,"Co-Nb Phase Diagram", J.Less Common Metals, Vol.12, 1967, pp.366-374.

7. B.Piearcey, R.Jackson, B.Argent,"The Mechanical Properties and Structure of Co-Ni-Nb Alloys", J.Inst.Metals, Vol.91, 1963, p.257.

8. F.X.Speigel, D.Bardos, P.Beck,"Ternary G and E Silicides of Transition Elements", Met.Trans., Vol.227, June, 1963, p.575.

DETERMINATION OF THE EFFECTS OF COOLING RATE FROM
SOLUTION TREATMENT ON THE MICROSTRUCTURE AND MECHANICAL
PROPERTIES OF A PRECIPITATION STRENGTHENED, LOW
THERMAL EXPANSION ALLOY

E. A. Wanner, D. A. DeAntonio, R. K. Mahidhara

Carpenter Technology Corp.
P.O. Box 14662
Reading, PA 19612-4662

Abstract

Segments of hot rolled Pyromet Alloy CTX-3 bar were solution treated at 1800°F and cooled to room temperature in various media, resulting in calculated cooling rates (from 1800°F to 1000°F) ranging from 0.1°F/sec. to ~66°F/sec. After subsequent overaging at 1425°F/1150°F, it was determined that both very slow and very rapid cooling after solution treatment resulted in extensive precipitation of the stable η phase and a decrease in γ' content. Since η contributes little to the strengthening of Alloy CTX-3, decreases in tensile and/or yield strength were associated with copious η precipitation. However, improvements in 1000°F notch stress rupture life were associated with abundant grain boundary η precipitation. SEM fractography of notch rupture specimens revealed increasing evidence of cleavage in specimens containing higher volume fractions of η phase, suggesting that η may reduce grain boundary sliding. It was also noted that after long term (>5000h) exposure to rupture test conditions, dissolution of coarse γ' and precipitation of a fine (<5nm), embryonic precipitate occurred. Results of this investigation indicate that incremental gains in notch rupture life are possible through control of the cooling rate after solution treatment in applications where slight decreases in tensile strength are tolerable.

Introduction

Pyromet® Alloy CTX-3 is a low thermal expansion superalloy based on a Ni-Fe-Co matrix of controlled composition. Titanium and columbium are present to promote formation of the metastable gamma prime (γ') phase (fcc $Ni_3(Ti,Al)$) in the austenitic matrix upon aging. To provide a preferred balance of tensile strength and ductility together with freedom from stress rupture notch sensitivity, extended aging treatments are typically specified for this alloy. Such "overaging" treatments result in the precipitation of both cuboidal γ' particles and the stable, plate-like eta (η) phase (hcp $Ni_3(Ti,Cb)$).

In the overaged condition, Pyromet Alloy CTX-3 typically exhibits coefficients of thermal expansion (α's) of approximately $4.0-4.5 \times 10^{-6}$ °F^{-1} to 780°F, values approximately 40-50% lower than exhibited by Alloy 718, an austenitic Ni-Fe-base alloy. Between 750 and 850°F, the austenitic Ni-Fe-Co matrix of Pyromet Alloy CTX-3 undergoes a ferromagnetic-to-paramagnetic transformation. The resultant decrease in magnetostriction causes mean α values from the inflection temperature upward to approach those of other austenitic superalloys. However, due to the substantially lower α in the ferromagnetic range, mean α values from room temperature to 1200°F remain approximately 30% lower than Alloy 718.

The typical microstructure of Alloy CTX-3 after solution treatment at 1800°F and overaging, as examined optically, contains patches of Widmanstatten η phase platelets in a partially recrystallized, austenitic matrix. Occasional regions of cellular η precipitation are sometimes observed along grain boundaries. In addition, coarse γ' is occasionally visible at the inner regions of grains.

In a previous, unpublished investigation(1), 1.45" x 1.38" cross section Pyromet Alloy CTX-3 bars which were solution treated, water quenched, then aged exhibited significantly lower room temperature and 1000°F tensile and yield strength levels than identically processed bars which were solution treated, air cooled, then aged. Optically, it was observed that the water quenched specimens contained a significantly higher volume fraction of η phase than the air cooled specimens. In addition, it was observed that the η platelets were substantially larger in the water quenched bars, frequently traversing entire grains. Examination by transmission electron microscopy (TEM) utilizing chromium shadowed parlodion structural replicas revealed large γ' denuded zones to be associated with the larger η platelets in the water quenched specimens. Additionally, an overall γ' deficiency was observed in the water quenched specimens. Based on these results, the present study was undertaken to attempt to quantify the relationship between cooling rate following solution treatment and the mechanical properties of Pyromet Alloy CTX-3. In addition, it was desired to determine the microstructural effects of post-solution cooling rate on the precipitation reactions accompanying overaging.

Procedure

Hot-rolled Pyromet Alloy CTX-3 bar stock (nominal composition, Table I), 1.45" x 1.38" in cross section was obtained from a production quantity. It was desired to vary cooling rate after solution treatment from a very slow cool (furnace cool @ 250°F/h) to a rapid quench (water quench of a 0.625" sq. bar). In addition, it was desired to investigate intermediate cooling rates simulating typical bar or rolled ring cross sections. Using a modification of Newton's law of heating and cooling(2), cooling rates were

---

® - registered trademark of Carpenter Technology Corporation

Table I: Nominal Composition of Pyromet CTX-3 Alloy

Element	Weight Percent
C	0.015
Si	0.15
Cr	0.2
Ni	38.25
Cb	4.85
Ti	1.6
Al	0.1
B	0.007
Co	13.6
Fe	Balance

approximated for various section sizes and quenching media.

Several bar segments were cut longitudinally to 0.625" sq., while the remaining segments were milled and ground to 1.25" sq. Specially fabricated fixtures were used to contain the 1.25" sq. ground bars during solution treatment and cooling, thereby simulating larger (2" sq. or 3" sq.) section sizes. The various cooling media used for the cooling simulations are shown in Table II, and for each simulation, the calculated cooling rate is shown in Table III, together with its equivalent in air cooled or water quenched bar/billet size.

Table II: Cooling Media Used for Pyromet CTX-3 Specimens

Specimen Identity	Cooling Method Employed
FC5	0.625" sq. Bar Furnace Cooled @ 250°F/h
VL3	3" sq.** Specimen Buried in Vermiculite
VL2	2" sq.** Specimen Buried in Vermiculite
AC1	1.25" sq. Specimen Cooled in Air
AC5	0.625" sq. Specimen Cooled in Air
OQ3	3" sq.** Specimen Quenched in Oil
OQ2	2" sq.** Specimen Quenched in Oil
WQ3	3" sq.** Specimen Quenched in Water
OQ1	1" sq. Specimen Quenched in Oil
OQ5	0.625" sq. Specimen Quenched in Oil
WQ1	1" sq. Specimen Quenched in Water
WQ5	0.625" Sq. Specimen Quenched in Water

*- Specimens Cooled after 1 h @ 1800°F
**- Indicates Simulated Bar Size (Fixtures Used)

Table III: Calculated Quench Rates* from Solution Temperature (1800°F) for Pyromet CTX-3 Specimens

Specimen Identity	Cooling Rate (°F/s)	Equivalent To:
FC5	0.1	8" rd./sq. Air Cool
VL3	0.2	5" rd./sq. Air Cool
VL2	0.3	3" rd./sq. Air Cool
AC1	0.9	1" rd./sq. Air Cool
AC5	1.5	---
OQ3	3.9	6" rd. Water Quench
OQ2	6.2	4" sq. Water Quench
WQ3	9.4	---
OQ1	13.0	3" sq. Water Quench
OQ5	20.8	1" sq. Water Quench
WQ1	39.0	---
WQ5	66.5	---

*- Rates Calculated for Cooling from 1800°F to 1000°F.

After solution treatment at 1800°F for 1 hour and cooling to room temperature at the desired rate, specimens for residual stress measurement (via x-ray diffraction) were removed from each of the 1.25" sq. bar segments. All bar segments were then overaged using a treatment of 1425°F/12h/furnace cool at 100°F/h to 1150°F/hold 8h/air cool. After overaging, the 1.25" sq. bar segments were quartered longitudinally. Room temperature tensile specimens with 0.252" gage diameters were single point machined from the overaged blanks. Notched stress rupture specimens, with 0.250" notch diameters, 0.354" shoulder diameters, and 0.0363" root radii with a 60° included notch angle ($K_t = \sim 2.0$) were also prepared from the overaged specimen blanks. Low stress cylindrical grinding was used to manufacture these specimens, since previous testing (3) had shown that single point turning produced specimens with inconsistent rupture lives. This effect was presumably due to non-uniform levels of residual stress in the notch.

Hardness measurements (BHN) were taken using samples representing each of the 1.25" sq. bars, as well as sample FC5 after solution treatment. Notch stress rupture testing was conducted at 1000°F with a constant load applied to generate an initial stress level of 120 ksi. Tests were continued to failure or to a maximum of 5000 h with no overloading. Tensile tests were conducted at room temperature at a free-running crosshead speed of 0.0625 in/min.

Longitudinal specimens for metallographic examination were removed from each fully heat treated bar prior to machining. In addition, thin foil specimens for TEM examination were prepared from material representative of each cooling rate. Foils were also prepared from selected broken mechanical test specimens. Chromium shadowed parlodion structural replicas were prepared to examine the γ' and η phase distribution over larger areas of the specimens.

Results and Discussion

Hardness Testing

Hardness test data for the solution treated specimens are presented in Table IV. These data show that the slowly cooled specimens (those cooled at rates less than ~1°F/sec.) possess significantly higher hardness than specimens cooled more rapidly. Based on computations performed using the calculated cooling rates of Table II for specimens VL-2, VL-3, and FC-5; the three specimens exhibiting higher solution treated hardness; a minimum of 20 minutes and a maximum of 1.3 hours elapsed during which these samples were within the favorable γ'/η precipitation temperature range as predicted by Muzyka, et.al.(4) for a similar alloy. Thus, it is likely that some precipitation of γ' and possibly η phase took place during the cooling of these specimens. The balance of the more rapidly cooled material passed through the favorable precipitation range in less than 6 minutes, an insufficient time period to allow for any substantial γ' or η precipitation.

Examination of selected transverse bar specimens using light optical microscopy showed no visible evidence of precipitation during cooling in specimens VL2 and VL3. However, it is possible that fine γ' or η phase, not resolvable using light microscopy, was responsible for the observed hardness increase.

Table IV: Hardness Results for Solution Treated Specimens Cooled at Various Rates

Specimen Identity	Hardness (BHN)
FC5	358
VL3	363
VL2	363
AC1	187
OQ3	179
OQ2	179
WQ3	174
OQ1	170
WQ1	174

Table V: Room Temp. Tensile Properties of CTX-3 Cooled at Various Rates after Solution Treatment
(Average of Duplicate Tests Reported)

Specimen Identity	Quench Rate (°F/sec)	0.2% Y.S. (ksi)	U.T.S. (ksi)	Elongation (%)	R.A. (%)
FC5	0.1	110.9	152.8	16.4	18.6
VL3	0.2	110.7	162.1	13.5	17.4
VL2	0.3	114.7	164.8	11.8	15.2
AC1	0.9	115.2	164.7	14.0	16.7
AC5	1.5	118.4	156.2	14.6	16.0
OQ3	3.9	111.9	155.5	11.8	16.0
OQ2	6.2	109.6	158.6	13.0	17.3
WQ3	9.4	113.2	161.0	13.2	17.0
OQ1	13.0	115.0	158.2	15.0	17.1
OQ5	20.8	108.2	152.8	16.6	20.3
WQ1	39.0	106.5	151.4	15.6	19.2
WQ5	66.5	92.8	140.4	16.6	19.2

All Specimens Solution Treated 1800°F/1h/Cooled then aged 1425°F/12h/Furnace Cool @ 100°F/h to 1150°F/1h/AC

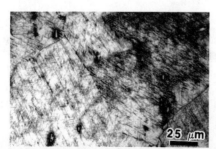

Figure 1: Optical Micrograph Showing Typical η Phase Distribution in Specimens VL2, AC1, and AC5. Etchant: Glyceregia

Tensile Testing

Results of room temperature tensile tests performed using overaged specimens representing each post-solution-treatment cooling rate are presented in Table V. These results indicate that an optimal combination of tensile and yield strengths in overaged Alloy CTX-3 are obtained when specimens are cooled at rates ranging from approximately 0.4 to 1.4°F/sec., as calculated using Newton's law of heating and cooling, from a solution treatment temperature of 1800°F.

Metallographic examination of specimens cooled within this range (specimens VL2, AC1, and AC5) revealed low to moderate η platelet concentrations, with mostly small platelet size (Figure 1). Lamellar grain boundary η precipitation was evident in each of these specimens, however the most extensive lamellar precipitation was observed in specimen AC1.

At the slower cooling rates, a decrease in yield strength was noted, with specimen FC5 (cooled at 0.07°F/sec.) also displaying a decrease in tensile strength. Specimens FC5 and VL3 were observed to possess markedly different η precipitation characteristics, as shown in Figure 2. While both specimens contained relatively small η platelets, a considerably higher volume fraction of η was present in specimen FC5. As shown in the TEM micrograph of Figure 3, localized regions of heavy Widmanstatten η platelet precipitation result in regions lean in γ', the primary strengthening precipitate in Alloy CTX-3. This is expected since both precipitating phases consume Ti. The effect of η platelet precipitation in lieu of or upon dissolution of γ' particles is reflected in the lower tensile strength of FC5 as compared to specimen VL3. Since specimens FC5 and VL3 possessed similar yield strengths with specimen VL3 displaying a significantly higher tensile strength, it can be surmised that, while the dislocation initiation was similar in both samples, dislocation motion was impeded to a greater extent in specimen VL3, further supporting the belief that a higher volume fraction of γ' existed in specimen VL3. This effect was also noted in that tensile ductility was slightly lower in specimen VL3 than in FC5.

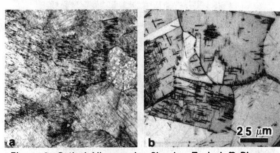
Figure 2: Optical Micrographs Showing Typical η Phase Distribution in Specimens FC5 (a) and VL3 (b). Etchant: Glyceregia

Figure 3: TEM Structural Replica Micrograph Showing γ' Denudation Associated with Locally Heavy Widmanstatten η Precipitation.

At cooling rates greater than 14°F/sec., tensile and yield strength levels were observed to decrease with increasing calculated cooling rate, with the exception of specimens OQ1 and WQ3. These specimens, cooled at 13.0 and 9.4°F/sec., respectively, possessed higher yield strengths than expected, as indicated by specimens cooled more slowly (OQ2) or more rapidly (OQ5). As can be seen in Figures 4 and 5, specimens OQ1 and WQ3 contained far fewer η platelets than did specimens OQ2 and OQ5, leading to a potentially higher γ' volume fraction than anticipated.

Specimens OQ3, OQ2, OQ5, WQ1, and WQ5; representing calculated cooling rates of 3.9, 6.2, 20.8, 39.0, and 66.5°F/sec., respectively; exhibited a clear trend of decreasing tensile and yield strength with increasing cooling rate. Metallographic examination of these specimens showed that both η platelet volume fraction and η platelet size increased with increasing cooling rate. This is especially evident when specimens OQ2, OQ5, WQ1, and WQ5 are compared (Figure 5). As stated previously, locally heavy η precipitation resulted in γ'-lean regions. In the specimens containing the elongated η platelet morphology, γ' denuded zones associated with the platelets have been observed to overlap, leading to relatively large intragranular regions essentially devoid of γ' precipitation. It is suspected that such large γ'-free zones were developed in specimens WQ1 and

WQ5, leading to sizeable decreases in tensile and yield strengths. A transmission electron micrograph showing such zones is presented in Figure 6. SEM fractography was also conducted using room temperature tensile specimens from each of the three previously mentioned conditions. This examination showed no difference in fracture mode based upon the extent of precipitation, indicating that within the range of η contents examined, η phase has little or no strengthening effect at room temperature.

Stress Rupture Testing

Data presented in Table VI show that rupture life at 1000°F/120 ksi for specimens heat treated as 1.25" sq. bar was

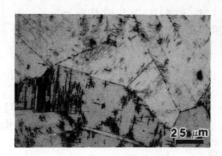

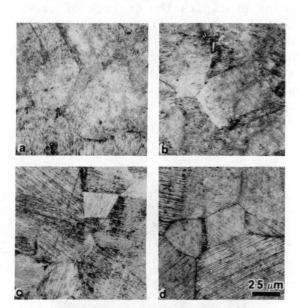

Figure 5: Optical Micrographs Showing Typical η Phase Distribution in Specimens OQ2 (a), OQ5 (b), WQ1 (c), and WQ5 (d). Note increasing η platelet length with increasing cooling rate. Etchant: Glyceregia

Figure 6: Structural Replica Micrograph Showing Absence of γ' in the Vicinity of Heavy η Precipitation

inversely related to tensile strength at room temperature. As shown in Figure 7, the exclusion of data generated using 0.625" sq. bar specimens results in a clear trend of decreasing notch rupture life with increasing room temperature tensile strength (to ~160 ksi).

Table VI: Notch Stress Rupture Properties of CTX-3 Cooled at Various Rates after Solution Treatment
(Average of Duplicate Tests Reported)

Specimen Identity	Quench Rate (°F/sec)	Rupture Life (hrs.)
FC5	0.1	>5000**
VL3	0.2	182
VL2	0.3	173
AC1	0.9	86
AC5	1.5	>5000*
OQ3	3.9	1431
OQ2	6.2	348
WQ3	9.4	164
OQ1	13.0	455
OQ5	20.8	>5000*
WQ1	39.0	2900√
WQ5	66.5	1009

* - Test Discontinued
** - One Test Discontinued, One Shank Failure
√ - Duplicate Shank Failures

All Specimens Solution Treated 1800°F/1h/Cooled then aged 1425°F/12h/Furnace Cool @ 100°F/h to 1150°F/8h/AC

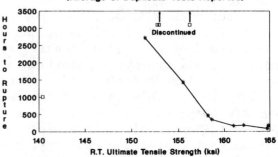

Figure 7: Notch Stress Rupture Life of CTX-3 Alloy Tested at 1000°F/120ksi
(Average of Duplicate Tests Reported)

— 1.25" sq. Specimens   □ 0.625" sq. Specimens

All Specimens Solution Treated 1800°F/1h/Cool plus Aged 1425°F/12h/FC @ 100°F/h to 1150°F/8h/AC.

The apparent correlation of increasing notch rupture life with decreasing tensile strength indicates a beneficial effect of η phase for notch strengthening. SEM fractographic analyses of three selected specimens representing short (AC1), intermediate (WQ5) and long (FC5) rupture lives, were conducted. It was observed in all conditions that fracture initiated in a region beneath the specimen surface, propagating toward the surface. In specimen AC1, where a rupture life of ~85 h was reported and minimal quantities of η precipitation were observed, only intergranular fracture was noted in the fracture initiation region (Figure 8a). Failure in the initiation region occurred by a grain boundary shear mechanism with extensive evidence of grain boundary sliding. In specimens FC5 and WQ5, where significantly longer rupture lives (~5000 h and ~1000 h, respectively) and higher volume fractions of η phase were observed, fracture initiation was mixed mode, with evidence of cleavage facets in many grains (Figure 8b and c). In all specimens, transgranular (cleavage) fracture was observed in the fast-fracture regions.

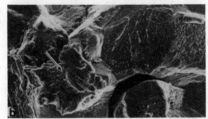
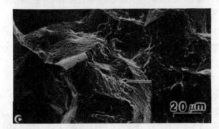

Figure 8: SEM Fractographs of Notch Rupture Specimens AC1 (a), 80 h life; WQ5 (b), 1000 h life; and FC5 (c), >5000 h life. Note presence of cleavage facets (arrows) in (b) and (c)

The formation of cleavage facets in the fracture initiation regions of these specimens suggests that some grain boundary strengthening mechanism, absent in specimens AC1, was present in specimens WQ5 and FC5. TEM examination revealed substantially thicker (>300nm) η platelets on the grain boundaries of specimens FC5 and WQ5 (Figure 9a) than in the matrix. While grain boundary η phase was also observed in specimen AC1, its thickness was similar to that of the matrix η platelets (~100 nm) (Figure 9b). The presence of fully developed η platelets on the grain boundaries of specimens FC5 and WQ5 suggests that this phase decreases stress rupture notch sensitivity by minimizing grain boundary sliding. The apparent propensity toward grain boundary sliding in specimen AC1 may be due to insufficient grain boundary η precipitation.

It is imperative when interpreting this notch stress rupture data that the effect of section size during heat treatment be considered. In the case of the 0.625" sq. bars, which exhibited very long rupture lives, the center axis of the test specimen coincided with the centerline of the bar during heat treatment. In the case of the 1.25" sq. bars, the specimen axes were not coincident with the bar centerlines. Because of this, the severity of thermal gradients established and the state of residual thermal stress at the specimen axis were not consistent for the two bar sizes evaluated. Similar effects were averted in the simulated 2" sq. and 3" sq. bar specimens, since thermal stresses are unlikely to have been transmitted across the interface between the fixture blocks and the 1.25" sq. bar surface. Since residual stresses have been shown to promote formation of the more stable η phase at the expense of the metastable γ' precipitate(5,6), and since increased η precipitation has been correlated with increased rupture life, it is likely that the anomolous stress rupture results of specimens FC5, AC5, and OQ5 can be explained by section size thermal stress effects. The shorter rupture

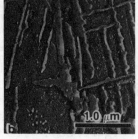

Figure 9: Structural Replica Micrographs Showing Thick Grain Boundary η Platelets in Specimens AC1 (a) and FC5 (b). Note substantially thicker platelets in FC5 (b).

life of specimen WQ5 may be explained by the fact that tensile strength was extremely low in this material. Since rupture testing was conducted at 1000°F/120 ksi and specimen WQ5 exhibited a room temperature UTS of only ~140 ksi, it is likely that the intrinsic strength of this specimen was lower than necessary to provide a long rupture life.

Additional Evaluations

It has been established that the tensile and notch stress rupture properties of Pyromet Alloy CTX-3 are related to microstructural features which are affected by cooling rate from the solution treatment temperature. As yet, the mechanism by which cooling rate affects the precipitation reactions during subsequent overaging has not been determined. Differences in residual stresses in the specimens solution treated and cooled as 1.25" sq. bars were measured by determining shifts in the 2θ angle for a (220) matrix reflection. No shifting of 2θ angles was observed for any of the eight specimens examined, indicating that any differences in residual stress which may exist were undetectable.

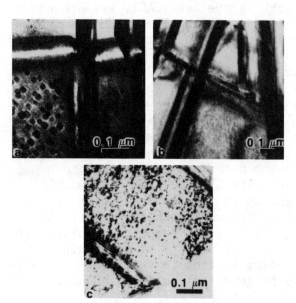

Figure 10: TEM Thin Foil Micrographs from Fracture Region of Notch Rupture Specimens AC1 (a), WQ5 (b), FC5 (c). Note coarse γ' in (a), absence of γ' in (b), and fine precipitate in (c).

Thin foil specimens from the notch areas of three selected notch stress rupture specimens were examined via TEM. Specimens AC1, WQ5, and FC5, exhibiting ~80, 1000, and 5000 h rupture lives, respectively, were selected for this study.

Examination of the specimen microstructures in the vicinity of the stress rupture fractures revealed distinct differences. Specimen AC1, showing a rupture life of ~80 h, was found to contain primarily γ' strengthening precipitates interspersed with η platelets (Figure 10a). Specimen WQ5, with a 1000 h rupture life, contained no detectable γ' precipitates (Figure 10b), with only η phase platelets observed. In specimen FC5, which exhibited a rupture life in excess of 5000 h, copious precipitation of fine (<5 nm) particles was observed, with little or no coarse (~40 nm) γ' (Figure 10c). Examination of FC5 specimens not exposed to the rupture test conditions did not reveal the presence of fine γ'; rather, only coarse, non-uniform γ' and η phase were observed. This indicates that during rupture testing, coarse γ' particles were dissolved and replaced by much finer particles of another crystal structure. This is supported by the fact that no particle ripening was observed. Since the fine precipitate was observed in its embryonic stage (diameter <5 nm), identification by electron diffraction analysis was not possible. It is also possible, due to the non-uniform precipitation pattern observed in these specimens, that the fine, embryonic particles observed were actually γ' particles precipitating in regions previously devoid of γ' precipitation. In either case, the fine particle size and relatively dense distribution would increase the resistance to dislocation bypass, thereby increasing rupture life at 1000°F.

Conclusions

Pyromet Alloy CTX-3 bar specimens were solution treated 1800°F/1h and cooled to room temperature at calculated cooling rates ranging from 0.07 to 66.5°F/sec. (rates calculated for cooling from 1800°F to 1000°F), followed by overaging 1425°F/12h/furnace cool @ 100°F/h to 1150°F/8h/air cool. Evaluations of the resultant room temperature tensile properties, 1000°F/120 ksi notch stress rupture properties, and microstructures led to the following conclusions:

1. Higher levels of yield strength were obtained at cooling rates ranging from 0.3-1.5°F/sec. These rates yielded microstructures containing low to moderate volume fractions of η platelets.

2. Both slower and more rapid cooling rates resulted in precipitation of larger quantities of η phase, leading to significantly lower levels of tensile and yield strengths.

3. Specimens cooled at extremely rapid rates after solution treatment were found to precipitate highly elongated, closely spaced η platelets, leading to pronounced decreases in tensile and yield strengths.

4. Notch stress rupture life at 1000°F/120 ksi was inversely related to tensile strength at room temperature.

5. Fractographic analysis showed that specimens exhibiting longer rupture lives contained mixed mode fracture in the initiation region, while those with shorter lives contained shear (intergranular) fracture only. In combination with the observations of optical and electron microscopy, this suggests a beneficial grain boundary strengthening effect of η phase.

6. After long term stress rupture testing, dissolution of coarse ($\sim$40 nm) γ' and formation of a fine (<5 nm) precipitate was observed.

7. Although air cooling resulted in optimized yield strength, relatively low notch rupture life was exhibited by the larger section (1.25" sq.) air cooled specimens. In applications where incremental gains in notch rupture life are required and slight decreases in yield strength are tolerable, cooling rates more rapid than air cooling may be worthy of investigation.

References

(1) Wanner, E. A., "Investigation of Low Tensile Strength in Hot Rolled Pyromet CTX-3 Bars," (Carpenter Technology Corp., 1987).

(2) Schumann, Jr., R. Metallurgical Engineering, Vol. 1, Engineering Principles (Reading, MA: Addison-Wesley, 1952), 259-283.

(3) Wanner, E. A., Unpublished Research (Carpenter Technology Corp., 1985).

(4) Muzyka, D. R., C. R. Whitney, and D. K. Schlosser, "Physical Metallurgy of a New Controlled-Expansion Superalloy," Journal of Metals, 27(7)(1975) 11-15.

(5) Mihalisin, J. R. and R. F. Decker, "Phase Transformations in Nickel-Rich, Nickel-Titanium-Aluminum Alloys," Trans. AIME, 218 (1960) 507-515.

(6) Wallwork, G. and J. Croll, "A Review of the Strengthening Mechanisms in Iron and Nickel Based Fe-Ni-Cr Alloys Used at High Temperatures," Reviews on High-Temperature Materials, 3(2)(1976) 106.

Alternative Materials

BEYOND SUPERALLOYS: The Goals, the Materials and Some
Reality

Chester T Sims
Rensselaer Polytechnic Institute, Troy, NY

Unusual national and international attention is now being focused on materials,- their science, their technology, their engineering. "Materials" are being challenged to give society great new benefits in areas such as superconductivity, information and computers, and the transportation field. It seems to materials engineers as if they had been suddenly "discovered", and investors are told of immense potential businesses in new materials, such as ceramics.

The field of high-temperature materials applies to one of the top two United States leading high technology businesses,- that of gas turbines for aircraft engines and industrial energy; it is in the forefront of this wave. The atmosphere is being charged by issue of unusually challenging goals for the performance of military aircraft gas turbine engines. A "quantum leap" forward is being sought. As a result, agencies which sponsor major research programs in materials identified revolutionary objectives for materials properties for building these engines in late 1984.

These objectives appear to add up to something like a doubling of both temperature and strength-to-weight capability in the forseeable future. This breaks into property requirements which require, at the very least, significantly increased absolute mechanical strength, materials densities which appear to eliminate materials more dense than alluminum and titanium, and surface stability to allow the materials to operate in atmospheres which could be oxidizing, neutral, or reducing. The goals also force intrinsic use of combinations of very dissimilar reactive materials in untried chemical combinations. All this is to be at temperatures which approach double current practice.

MATERIALS GOALS

The materials goals follow from the engineering objective, which is to increase significantly the thrust-to-weight ratio of military engines. These goals, as apparent by early 1985 were:

- Design and build engines with a minimal turbine inlet temperature of about 4000F (2200C)
- Design the engine(s) with very little cooling of hot-stage materials.
- Utilize materials with a maximum density of about 5g/cc.

o   Maximize energy from the fuel by combustion at
    stoichiometric conditions.

These goals are supported by extensive R&D programs which started in 1985 and in years thereafter at universities and other laboratories which invent materials for engines. The materials systems defined for study often are specified on the basis of very simple properties such as melting point, density, or modulus. Step-wise developments in conventional materials, such as superalloys, appear to be a thing of the past.

GOALS MODIFIED

After about three years the goals appear to have been somewhat modified. A temperature range of 3000-4000F is now being mentioned and a partial use of materials with greater density than 5g/cc is suggested as allowed, although this author has heard of no change in cooling or stoichiometric combustion goals. However, these objectives to a materials engineer still mean operation at 1.5 to 1.8 times the present materials temperature, replacement of much of the nickel, cobalt, and the iron-base alloys now used, and rejection of refractory metals (except, perhaps, for very minor portions of total engine weight through a trade-off with vey light-weight materials). Further, many more engine construction materials will approach actual combustion temperature due to restricted cooling, and if the stoichiometric combustion goal is retained, a now not-existing-in-nature class of materials,-resistant to both oxidizing and deducing conditions, will be needed.

The materials classes proposed for this job can be identified as follows:

   o   Ceramics and ceramic alloys
   o   Ceramic/ceramic composites
   o   Composites containing carbon (graphite fibers)
   o   Intermetallic compounds and IMC composites
   o   Metal-matrix composites

In addition, refractory metals are still mentioned, and, as implied above, superalloys exist in place as the standard, although both of these systems have densities far above 5g/cc.

WHAT DO THE GOALS MEAN?

It is appropriate to take a broad, engineering-like perspective of the stated goals, particularly since they are so far beyond current capability.

Temperature. Figure 1 places in perspective the regime of temperatures serviced to man. Importantly, it has taken 84 years to advance from about 500F to 2000F in turbine materials,- the last 20 years at about 15F per year. Now the objective is to reach specific bulk materials temperature of the 3000-4000F range in perhaps 10-15 years from 2000F,- a rate of about 100F per year.

Cooling. Figure 2 shows advances achieved in turbine engine performance from cooling alone in the last 20 years. It has exceeded advances through improved metallurgical invention.

Density. Figure 3 gives a simple listing of materials available based on a density consideration. It also lists the fracture toughness of the materials classes, showing that the elimination of metallic systems eliminates the toughness contribution common to date.

Stoichiometry. Stoichiometric combustion means complete burning of all the fuel hydrocarbons to a perfect neutralized balance of oxidizing and reducing components in the gas stream at 4000F. (4000F is about the stoichiometric combustion temperature for JP-5). If the gas stream becomes even slightly reducing, materials will be destroyed very rapidly. If reducing-resistant materials are used, even slight oxidation destroys them. Figure 4 shows the root section of a turbine bucket when a combuster went out of control allowing reducing conditions to exist. Over 100 blades like this one were burned off in 45 seconds. Immense energy was released,- a catastrophy. The slightest variation from stoichiometry in any system will produce this type of failure.

ARE THESE THE ONLY GOALS?

It is absolutely vital to remember that the above-stated "goals" are actually only a kind of ambient condition goal for materials, secondary to the design/engineering physical "requirements" to make the machine go:

- o *Strength.* The materials must be (at least) as strong as superalloys.
- o *Toughness.* The materials must be capable of sustained exposure to a wide variety of stress and temperature conditions, which absolutely require that the materials demonstrate a significant level of toughness,- at least a large portion of that available in superalloys.
- o *Surface Stability.* Surface recession rates allowing the small, thin, materials parts to retain their load-bearing capability for, at the very least, 100-200 hours are requisite,- with thousands desirable.
- o *Other Properties.* The above still does not face many other essentials,-HCF resistance, CTE (α) needs, retained structural stability, corrosion resistance, and the like.

GOALS IN RETROSPECT

Even this brief consideration of the materials property goals put forth in 1984, resulting in a number of now massive developmental programs, raises several questions.

- o Since the temperature and density goals eliminate metals and ductile IMC's, have the toughness requirements for future machines been very significantly reduced to compensate?

- Since stoichiometry is wanted (and oxidation resistance essential) is any class of materials with this combination of properties known?
- Why has development toward still more effective cooling stopped?

EVALUATION OF THE MATERIALS

While a number of major efforts were put in place to study or develop new materials based on empirical logic, several parallel programs were initiated to evaluate simply the basic physical/chemical properties of existing materials candidates. It is the interest here to cover, with high brevity, the salient features of several evaluations.

Existing Solid Materials. A broad survey comparing solid materials on the basis of density and (real or estimated) Youngs Modulus was compiled by Fleischer. A portion of his results are shown in Figure 5. It is quickly obvious that (on the basis of density and modulus only) carbon/graphite, carbides and oxides dominate the available materials. Metals are far behind. This is a shopping list. Fleischer did not pretend to consider toughness, chemical reactivity and the like.

Preliminary Evaluation. Hillig then attempted to apply a first level of reality to the (apparent) range of candidates. He considered not only density and modulus, but projected potential candidates for very-high temperature service on the basis of structural and surface stability (phase changes and vapor pressure), and mechanical behavior through estimates of creep performance. Directing the study into the composite area, he considered materials interaction as well. Hillig then attempted a balanced evaluation to identify possible composite systems. Examples of this approach as his recommendations are shown in Figure 6.

Creep Evaluation. This author attempted an engineering evaluation of creep by calculation intended to show whether *any* material exists capable of withstanding continued mechanical load of a usuable level at the original goal of 4000F (Figure 7).

Findings from this study are that the only material capable of usuable mechanical stress in turbines at 4000F is carbon/graphite. However, this is based on a single set of data of questionable character. Further, as discussed more below, graphite will burn explosively at 4000F in air, and toughness is almost non-existant.

Oxides exist at and have some strength at 4000F, but those found so far with calculable properties creep too rapidly. Further, the study suggests that diffusion is an underused technique for estimating creep.

Surface Stability. In Hillig's study, he evaluated the vaporization effects on surface recession rate of the now fairly obvious groups of "high-temperature" materials, with the results shown in Figure 8. The conclusion from this solid work also is clear,- that non-oxide systems either vaporize or oxidize too rapidly at the (original) goal of 4000F to be useful. Only oxide-type materials appear to have potential,- and that does not yet appear to have developed much depth. In any case, materials which do not satisfy the demands of this measurement should not be under study.

DISCUSSION AND SITUATIONS ANALYSIS

Materials Science and Engineering (MSE): The New Atmosphere
After years of being inventive followers, materials engineers and scientists are suddenly in the limelight. MSE is regarded as one of the most important high-tech industries of the future. This new atmosphere appears due to two major factors,-several spectacular developments in materials, and continued public emphasis, through both technical and non-technical news media of the (apparent) incredible future of "materials".

The technical developments are real. The public relations emphasis, however, is quite misleading. Growth of ceramics to a $300 *billion* business by the year 2000 has been trumpeted for about 5-6 years. The fact is, that no turbines, aircraft or industrial,- yet use ceramics, and the potential of other than trial demonstrations by the year 2000 is a myth.

Still, every materials magazine leads with articles on ceramics, composites and graphite, which, through exciting illustrations about *potential*, create an atmosphere that suggest such materials are just a half step away,- although prudent consideration of actual properties shows that they are many, years off, if at all. The situation is discussed in detail by TW Eagers, who tends to conclude that the real need is for common-sense attention toward quantities of economical advanced materials for clearly identified societial benefit, not curiosities. His article is needed reading.

Materials for Advanced Turbine Engines: The Cited Goals
As discussed previously, the goals presented to the technical community initially appear only in part to be "workable". Achnowledging this, recently the goals have been modified somewhat. For instance, the temperature goal has been stated now to be 3000-4000F instead of absolute 4000F. It is extremely important to realize that this relaxation may mean that stoichiometric combustion is not longer a hard requirement. The goals for combustion appear now to be "near-stoichiometric". If true, this is most fortunate, since it rescues design of advanced engines out of the realm of absolute impossibility, from the materials existance view, and suggests something might be available.

However, it still leaves a rather incredible hurdle,- that of changing the key hot-stage engine components from a superalloy technology to a ceramics/composites/graphitic technology. It seems assumed that the toughness problem will be eliminated.

Materials for Advanced Engines: The Candidates

Let us explore the potential materials on the assumption that they must be higher melting than superalloys. In order to be concise, this is done by a chart which generally categorizes the possible materials groups. There are no others. See Table I. The advanced engine goals are included but so are a few basic engineering requirements.

While there are eight items total on this chart, it is imperative to understand that, of the eight, two items are essential, regardless of goals. If the material does not have (1) strength and (2) *environmental resistance* or a rational, reasonable potential of obtaining both of these, further attention is a waste of time. The field considered here is very complex and the chart is heavily subjective, due to the cryptic nature of this presentation. However, this writer has some points to offer believed critically important.

1. **Graphite and Graphite Fibers** will be consumed in oxidation at a rate controlled only by ability to bring the reacting constituents together. Carbon is a fuel. Protection is conceivable, for a few seconds or minutes, but reaction thermodynamics has always shown very clearly that dynamic structural use at very high temperature is impossible.
2. **Monolithic Covalent Ceramics.** Despite millions expended, solving or bypassing the critical flaw size problem in these brittle materials has not been done, and the fracture toughness needs are not satisfied. Potentially, very tiny, high-cost parts may be possible to 2700F, the service limit of silica,- the generated oxide protective film.
3. **Composites.** The millions poured into ceramics before, are now pouring into composites,- because "fiber pull-out" demonstrates a slight increase in toughness over ceramics (still far below engine requirements) and also because of the tailored property potential, a real plus. Like ceramics, oxidation service is limited to 2700F.
4. **Unalloyed Oxides and Oxide/Oxide Composites.** The surface stability column in Table I shows that only these materials have potential to exist in the 3000-4000F range. The oxides are mechanically weak, but also largely unexplored.

From this, it is apparent that oxide ceramics, which have not been favored with any developmental attention, have the only real potential for 3000-4000F service. Further, the word *potential* used here is relative and subjective.

OXIDES FOR VERY HIGH TEMPERATURE SERVICE

To explore whether oxides have some real potential, a carefully plannedbody of work is needed. Some suggestions, and support comments follow.

1. At present, oxide ceramics of the alumina-chromia type are operating in the hot agressive atmosphere of military and all other engines to about 2100F; here they service in non-loading-bearing application as 1-5 mil protective coatings.
2. Oxides, and probably oxide/oxide composites, can be generally expected to have stability in oxidizing environment up to temperatures where they sublimate, disproportionate, or otherwise self-destruct.
3. Oxides are weak in creep and are generally brittle. Strengthening is a major objective.
4. Results of studies on oxides also will be applicable to other systems, all of which must have oxides to protect them.

ADVANCED AIR COOLING

The original goals specified "reduced cooling" at 4000F. With modified goals, it is now reduced cooling at 3000-4000F. It is instructive to observe the overall circumstance, and review this goal. Advances in gas turbine engine temperatures over the last 30 years have been due both to cooling and materials advances with cooling providing about 60% of the advance. At present, with materials real surfaces at about 2100F, gas stream temperature of above 3000F can be achieved for short periods when needed. This is a 1000F gas temperature capability above blade metal due to cooling. A cooling breakthrough, increasing cooling effectiveness by say 40% would put engine temperatures around 3700F,- probably as close to stoichiometry as practical.

SUPERALLOYS

Discarded as having no further potential by the original and revised advanced engine goals, it is important to understand that *significant advances in superalloy capabilities are still occuring* as industrial programs tail down. With their proven toughness, creep strength, and oxidation resistance, this is a vital signal. Five years later, superalloys are not done; they are not dead; they are very much alive.

FINAL OBSERVATIONS

The total situation, of course, is much more complex than can be shown in this short discussion. Recent changes in HPTE materials are laudable, -but this writer believes more changes are needed. A danger exists that development efforts will fall so far short (ceramics already) that the materials community will loose credibility. All participants want and expect challenging goals, sometimes impossible to fully attain,- but goals must be rational and supportable by sound engineering.

This writer recommends the following:

- Development programs on highly advanced superalloys should be strongly supported, and integrated with new advanced cooling programs.
- Very advanced work,- towards bulk material temperature in the 3000-3500F range should center on oxides and oxide systems.
- The very extensive developmental programs in composites containing ceramics, graphites or IMC´s should be faced with a specific goal in toughness at a level designers state is acceptable. If the goal is not met in five more years, composites should be abandoned.

REFERENCES

- Fleischer, R.L., High-Temperature, High-Strength Materials - An Overview. J. Met., 37, no. 12, Dec. 1985, 16-20.

- Hillig, W.B., Prospects for Ultra-High-Temperature Ceramic Composites. Tailoring Multiphase and Composite Ceramics, R.E. Tressler, et al., eds., Plenum, 1986, 697-712.

- Eagar, Thomas, W., The Promise of New Materials - Real or Imaginary? J. Metals 39, No. 4, April, 1987, p. 20-24.

- Hertzberg, R.W., Deformation and Fracture Mechanics of Engineering Materials, John Wiley & Sons, New York, 1976.

TABLE I

Potential of Advanced Materials Classes to Meet
Materials and Engineering Goals for Advanced Engines

Materials Class	Materials Goals				Materials/Engineering Goals			
	T_4= 3-4000F	Stoich. Combus.	Density <5g/cc	Less Cooling	Creep Strength	Toughness	Surface Stability	Available & Economic
SUPERALLOYS	Poss.	No	No	No	Yes	Yes	Yes	Yes
OXIDES & OX/OX	Yes	No	Poss.	Yes	Poss.	Poss.	Yes	Yes
GRAPHITE COMPOSITES	Yes	No	Yes	Yes	Yes	No	No	Poss.
COMPOSITES (Matrix)								
Metal & IMC	No	No	No	?	Poss.	No	?	No
Ceramic	Yes	No	Yes	?	Poss.	No	No	No
Oxide	Yes	No	Poss.	?	Poss.	?	Yes	No

YES - Has Potential   POSS. - Limited Potential   NO - No Potential

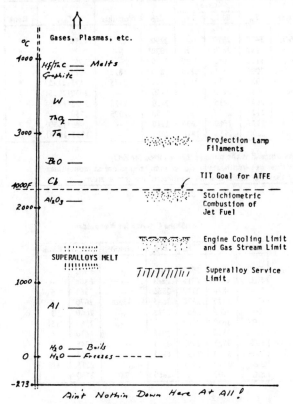

Fig. 1. Temperature and Its Relationship with Some Materials

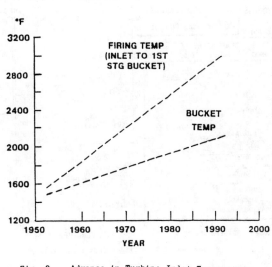

Fig. 2. Advance in Turbine Inlet Temperature

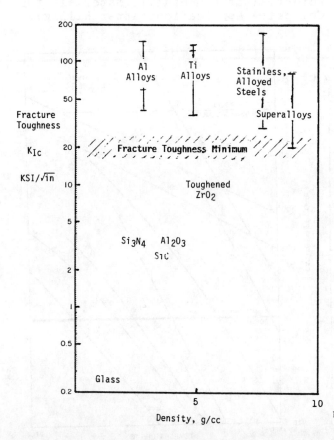

Fig. 3. Some Properties of Advanced Materials

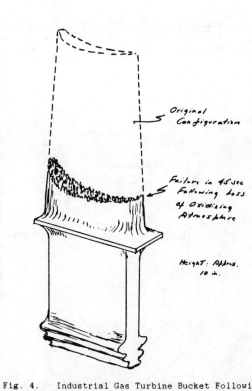

Fig. 4. Industrial Gas Turbine Bucket Following Exposure to Stoichiometric and Reducing Conditions

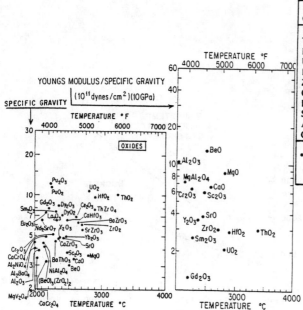

Fig. 5. Properties of Some Oxides (after Fleischer)

Combined Criteria for Oxides

Oxide	T_m	Creep .85T_m	.7T_m	E/ρ	Volatility	T_{min}	Rank
ThO$_2$	3493	2970	2445	2290	2375	2290	3
HfO$_2$	3117	2650	2180	2020*	--	2020	6
UO$_2$	3113	2645	2180	2050	--	2050	5
MgO$_2$	3098	2635	2170	2870	1825	1825	9
ZrO$_2$	3037	2580	2125	2195	2775	2195(2125)	4
CaO	2882	2450	2015	2430	--	2430(2015)	1
BeO	2843	2415	1990	2710	2325	2325(1990)	2
SrO	2727	2320	1910	1750	2075	1750	10
Al$_2$O$_3$	2327	1980	1630	2210	--	1980(1630)	7
Cr$_2$O$_3$	2300	1955	1610	2015	--	1955(1610)	8

* Assumes E is the same as that reported for ZrO$_2$
Values in parentheses in T_{min} column refer to use as monolithics if that use results in a lower estimated use temperature.

Combined Criteria for Nonoxides

Mat'l	T_m	.7T_m	E/ρ	Volatility	T_{min}	Rank
TaC	4258	2980	Fails	--	Fails	--
HfC	4163	2915	Fails	--	Fails	--
NbC	3886	2720	2365	--	2365	6
C	3825	2670	3685	~3500	2670	2
ZrC	3803	2660	2715	--	2660	3
HfN	3660	2560	--	--	2560	(5)
TiC	3530	2470	3035	--	2470	4
HfB$_2$	3523	2463	2345	--	2345	7
TaN	3360	2350	1855	--	1855	9
TaB$_2$	3310	2315	--	--	2315	(8)
SiC	3100	NA	2850	2780	2780	1

Fig. 6.

Potential of Nonmetallic Materials for Composite Application (after Hillig)

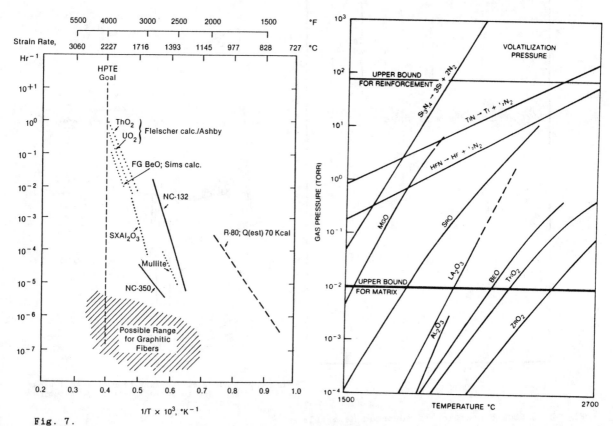

Fig. 7. Estimated Creep of Some Materials to 4000F

Fig. 8 Upper bound for service of non-metallic materials as defined by vaporization or dissociation (after Hillig).

STATUS AND PROGNOSIS FOR ALTERNATIVE ENGINE MATERIALS

Joseph R. Stephens and Michael V. Nathal

National Aeronautics and Space Administration
Lewis Research Center
Cleveland, Ohio 44135 USA

Abstract

The purpose of this paper is to review the current state of research and development of new materials for advanced aircraft engines. The advantages and disadvantages of intermetallic compounds and refractory metals as replacements for today's nickel-base alloys are discussed, along with some results of research directed at overcoming some of the problems which restrict their application. It is concluded that continuous fiber reinforced intermetallic matrix composites offer one of the best chances for success. However, major technical barriers still exist, especially in the development of suitable fibers, and introduction of these materials into aircraft engines is expected to take in excess of 5 to 10 years for most of these materials.

Introduction

Nickel-base and to some extent cobalt-base superalloys have been the primary materials in the hot zone of gas turbine engines since their commercialization. Although the properties of the superalloys continue to be improved via alloying, microstructural modifications, and innovative processing techniques, it is apparent that their maximum use temperature is rapidly being approached. Because of the limitation imposed by the melting temperatures of the superalloys and the need for more efficient, higher performance aircraft engines, the DOD and NASA have undertaken joint as well as independent programs to extend the use temperature and reduce the weight of future military and commercial engines. The potential payoffs for the use of new materials with higher temperature capabilities and/or lower densities include reduced cooling requirements, reduced weight throughout the engine, and increased thermodynamic efficiencies which will lead to reduced fuel consumption. For example, if the light weight aluminide intermetallics can be utilized in the turbine, the weight reduction can be translated throughout the engine and supporting structure with accompanying payoffs in both engine performance and cost savings. This paper will present some of the materials now under investigation, discussing their potential and the key technical issues that need to be addressed, and then briefly summarize our best estimate of when these materials may see application in engines for future aircraft.

Intermetallic Compounds

The potential of intermetallic compounds as high temperature structural materials has been recognized for some time now, although achievement of that potential has not been easy. Of the vast numbers of existing compounds, relatively few have been studied for use in structural applications. Table I lists some of these materials along with some of their advantages and disadvantages. In recent years the aluminides have received the most attention, as a result of their potential advantages of low density, high strength, and good environmental resistance. Although attempts to improve one property have frequently resulted in sacrifices of some of the benefits, significant progress has been made such that some aluminides should begin to see application within the next 5 years. For longer term application, a wide variety of other intermetallic compounds, including silicides, refractory metal aluminides, and beryllides, have also exhibited potential. The bulk of this discussion will be restricted to those aluminides which may be used within the next decade.

Table I. Some Potential Intermetallic Compounds

Compound	Advantages	Disadvantages
Near Term Application		
Ni_3Al	Ductility	Density, melting point
Ti_3Al	Density	Oxidation
$TiAl$	Density	Ductility
$FeAl$	Ductility, oxidation	Melting point, density
$NiAl$	Melting point, oxidation	Ductility
Far Term Application (Typical examples of the many potential compounds)		
Ti_5Si_3	Density	Ductility
Nb_5Si_3	Melting point	Ductility, oxidation
Nb_3Al	Melting point	Ductility, oxidation
$NbAl_3$	Melting point, density	Ductility, oxidation
$TiAl_3$	Density	Ductility
$MoAl_2$	Melting point	Ductility, oxidation
Nb_2Be_{17}	Melting point	Ductility, oxidation
$ZrBe_{13}$	Melting point	Ductility, oxidation

Figure 1 presents a summary of the density corrected yield strengths for various intermetallics and superalloys, which can serve as a starting point for rough comparisons. Because mechanical properties are sensitive to many microstructural and compositional variables, caution is required in interpreting this figure. Furthermore, short term monotonic strength is not sufficient for determining the suitability of a material for engine applications where life, environment, and cyclic conditions play critical roles.

NASA's program on intermetallics has focused primarily on the equiatomic aluminides NiAl and FeAl. These two compounds possess the B2 (ordered BCC) crystal structure, and have densities around 6 gm/cm^3. FeAl has a melting point of 1340 °C and therefore, has the most potential for intermediate temperature (i.e., below 1050 °C) applications. Iron rich

(Fe-40 at % Al) is one of the few aluminides with measurable low temperature ductility, about 3 to 5 percent elongation at room temperature (1). It can be processed by most conventional metal processing methods, such as induction melting and powder metallurgy techniques, extrusion, and hot and warm rolling. Alloying with Zr or Hf has been shown to improve strength up to about 700 °C while still maintaining ductility (2). Boron additions can have a small, but beneficial effect on the low temperature ductility of many FeAl alloys, where a change from intergranular to transgranular failure accompanies the increase in ductility (1,2). The FeAl alloys have excellent cyclic oxidation resistance up to at least 1000 °C (3), and are not susceptible to intermediate temperature oxygen-induced embrittlement (2). However, their strength and creep resistance decrease quickly above 700 °C, and it is not clear that further significant improvements in high temperature strength can be achieved without sacrifice of some other property. Although the combination of these properties and inexpensive raw material cost make FeAl attractive as a replacement for stainless steels in aircraft engines, its potential for aeropropulsion systems seems limited. However, it still has attractive density, ductility, and oxidation resistance for use as a matrix for a composite which could possibly see applications up to as high as 1100 °C.

NiAl has a much higher melting point, 1638 °C, and similar density to FeAl, and thus is more attractive for higher temperature service. NiAl has excellent cyclic oxidation resistance to at least 1300 °C, especially when alloyed with small additions of rare earth elements (4). Although some early work (5) reported room temperature ductility in polycrystalline NiAl, a more commonly observed behavior is a ductile to brittle transition temperature near 300 to 600 °C, depending on composition, grain size, and processing (5-8). Single crystals have exhibited ductility (7), but even single crystals suffer from a low cleavage strength. Grain refinement, which is achieved by thermomechanical processing (9) or by rapid solidification (10), appears to give some improvement in low temperature ductility (8). Additionally, compressive creep testing of powder metallurgy processed materials (9) has shown that fine grains are stable and do not harm creep resistance to at least 70 percent of its melting point. Studies have shown that it is relatively easy to improve the creep resistance of NiAl to equal or exceed that of superalloys (11,12), although achievement of ductility has remained more elusive. Strategies with the best chance for improving low temperature toughness appear to be macro-alloying in order to change the slip system from <100> to <110> type, or to reduce Al levels such that some Ni_3Al is formed. Although these alloying changes decrease the density, creep, oxidation and melting point advantages, an appropriate balance of properties may still be achieved for certain lower temperature applications such as turbine disks.

Ni_3Al has been studied extensively, primarily through the leadership of Oak Ridge National Laboratory (ORNL) (13). This compound has good low temperature ductility and good strength up to 800 °C, combined with good processability. However, sensitivity to an intermediate temperature oxygen-induced embrittlement has been observed. This problem has been reduced by alloying with about 8 at % Cr. However, the Ni_3Al based alloys with a balance of properties have only limited density and melting point advantages over superalloys, and do not appear to be able to offer large increases in engine performance.

Ti_3Al and TiAl have been developed under Air Force sponsorship (14-16) over the last 15 years, for use at temperatures up to 1000 °C. Ti_3Al has a low density, 4.2 gm/cm^3, which is sacrificed somewhat when Nb is added for ductility improvements. Oxidation resistance is lower than the Ni and Fe aluminides, and coatings will be necessary for applications above about

700 °C. TiAl has an even lower density, 3.9 gm/cm^3, and the potential for better environmental resistance and higher temperature capability, although ductility is lacking. Both Ti aluminides are also potential matrices for composites.

Intermetallic and Metal Matrix Composites

Intermetallic matrix composite (IMC) and metal matrix composite (MMC) materials are currently of extreme interest for future high temperature, high efficiency, high performance aircraft engines for both civil and military applications. NASA Lewis Research Center is currently investigating continuous fiber reinforced composites of FeAl and NiAl primarily under the NASA High Temperature Engine Materials Program (HITEMP) and Ti$_3$Al+Nb in the joint NASA-DOD program called NASP (National Aerospace Plane Program). In addition, many industrial organizations and universities under government contracts and grants are investigating the SiC reinforced Ti$_3$Al and TiAl systems. ORNL has some effort underway on SiC/Ni$_3$Al.

By going the composite route the low density of the intermetallic compounds can be utilized to good advantage and if low density, high strength fibers are available, the low strength of the intermetallic matrices becomes less of an issue. Thus, the matrix can be optimized for other properties, most importantly ductility, oxidation resistance, and density. The influence of the fiber on strength properties of a composite has been discussed (19) for the use of SiC in aluminide matrices where the predicted strength-to-density for SiC reinforced aluminide composites was shown to be essentially independent of matrix strength. Thus, the low matrix strength typical of many of the aluminides at the higher temperatures can be overcome by the use of high strength fibers present at a volume fraction ranging from 30 to 50 percent. As an example, it was shown (20) that in a 40 vol % W/Nb-1Zr composite (high strength fiber in a relative weak matrix), the matrix carries only about 3 percent of the load during high temperature creep testing. An example of the added strength properties achievable by a composite material is afforded by the comparison on a density corrected basis made in Fig. 2 where the properties of a 40 vol % SiC/Ti$_3$/Al + Nb composite are shown along with tensile properties of wrought nickel base and cobalt base alloys and the tensile strength of a single crystal superalloy, NASAIR 100. The strength advantage of the composite is evident in this comparison which provides the incentive for the research currently underway on these materials. Of course, there are many important problems that still must be solved, including transverse properties, creep, mechanical and thermal-mechanical fatigue, fabrication and joining, chemical compatibility between fiber and matrix, and differences in thermal expansion that would cause cracking during thermal cycling.

One of the major disadvantages of current composite systems under consideration is the chemical compatibility between fiber and matrix. This issue has been addressed (22) for the SiC/Ti$_3$Al+Nb system and the results to date are summarized in Fig. 3. The reaction is probably diffusion controlled and illustrates the extent of the fiber-matrix interaction that can take place upon exposing this composite to high temperatures. Based on these results it is evident that the properties of such a composite must be thoroughly explored as a function of use temperature and time of exposure.

A second major disadvantage is the thermal expansion mismatch between fiber and matrix. The commercially available fiber, SiC, that we have used for most investigations on the aluminide composites has a very low coefficient of thermal expansion (CTE), 5.5×10^{-6} °C^{-1} (23). In contrast, the aluminides have typically high CTE's ranging from 10×10^{-6} °C^{-1} for Ti$_3$Al+Nb (24), 14.5×10^{-6} °C^{-1} for NiAl (25), to as high as 20.9×10^{-6} °C^{-1} for FeAl

(25). The effect that this difference in CTE's can have during thermal cycling is illustrated for the SiC/Ti$_3$Al+Nb composite system (22) as shown in Fig. 4. Microcracks are noted to have developed after only three cycles of heating to 985 °C and cooling to room temperature. In contrast, the CTE of Al$_2$O$_3$ is 9.0×10^{-6} °C^{-1} (26) which is in excellent agreement with that of Ti$_3$Al+Nb. Thus, based on the criterion of matching CTE's, Al$_2$O$_3$ becomes a strong candidate as a reinforcement for this aluminide. Again, other factors such as chemical compatibility, lower strength, and availability compared to SiC must be weighed against this advantage of matching CTE which further illustrates the trade-offs that come into play when development of a composite system is being considered.

These initial results on the SiC/Ti$_3$Al+Nb composite help to illustrate why we consider the development of compatible fibers to be a key issue that limits the application of light weight composites for future aircraft engines. To address this critical technology a potpourri of properties are desirable for future high temperature fibers which include: matching coefficient of thermal expansion and chemical compatibility with the matrix, low density, high modulus of elasticity, high melting temperature, high temperature strength, good oxidation resistance, spoolable, and capable of mass production. To develop a fiber with these properties will take a considerable amount of funding, innovative research, and time. Still another approach to address these issues is to coat existing fibers such as SiC or graphite with a thin layer of a "functionally gradient material" that will be chemically compatible with both fiber and matrix and will have an intermediate graded CTE which will help to solve the thermal cycling problem. Such an approach is underway by Japanese investigators (27) as well as by several investigators within the U.S.

Composite fabrication also may be a limiting factor in the application of the aluminide matrix composites due to fiber-matrix reaction during the high-temperature fabrication processing and also because of the inherent attraction for oxygen by the aluminides which may lead to contamination and increased brittleness of the matrix. Our approach has been to fabricate monotapes by one of several techniques including powder cloth, arc spray, and plasma spray. Subsequent composite consolidation is normally achieved by vacuum hot pressing or by hot isostatic pressing. The powder techniques are particularly susceptible to oxygen contamination and thus the search for other more controlled techniques is necessary. The importance of final composite consolidation must also be considered. The processing parameters of time, temperature, and pressure must be optimized to obtain the best compromise of sometimes divergent property requirements. If processing time is too short and temperatures and pressures too low, insufficient bonding between fiber and matrix will occur. Conversely, if time is too long and temperature and pressure are too high, excessive reaction between the fiber and matrix can occur. Both of these extremes will lead to inferior composite properties. A final step in the fabrication of composites is the necessity to join them to other components. These techniques will have to be developed to make composites viable materials for future applications. Paralleling these developments will be the necessity of developing adequate NDE techniques to inspect composites during the initial fabrication steps to final component fabrication and use.

So far we have only discussed the use of continuous fibers. Another approach that is underway is the use of particulates in the light weight matrices. An example of this concept is a study on TiB$_2$ reinforced NiAl (28). The potential of strengthening NiAl with increasing volume percent of particulate is illustrated in Fig. 5 which provides a comparison with IN-601, TD-NiCr, and a single crystal superalloy, PWA-1480. The enhancement of strength was shown to be due to the effectiveness of the particles to interact with dislocations. For the matrix only a low dislocation density was observed after deformation. However, with the TiB_2 present particulate-dislocation interactions were evident and a much higher dislocation density was observed.

Refractory Metals

The aluminides under consideration have a peak melting temperature near 1750 °C. To utilize metallic materials rather than turn to ceramics or ceramic composites the refractory metals have once again come under consideration. A lot of resources were expended on the refractory metals, especially niobium base alloys for aircraft and space shuttle applications in the 1960's and 1970's. Today's interest once more is in niobium base alloys with some interest in molybdenum base alloys. Oxidation still remains the Achilles heel for these materials. Recent results on Nb alloying (29) with the sole purpose of improving the oxidation resistance have been reported which demonstrated the feasibility of forming protective alumina scales by the selective oxidation of Al. Even with this approach, achieving an oxidation resistant alloy for use at temperatures above 1350 °C for long periods of time remains a formidable task.

The Future for Aircraft Engine Materials

What is our prognosis for the incorporation of these experimental materials in flight hardware? To help us address this question, the developmental history of MA-6000 (30) is used as an example. The technology transfer from the initial focused research to the estimated time of actual flight hardware is seen in Fig. 6 to be in excess of 16 years and at a cost of 15 to 20 million dollars. It should be noted that some of the pitfalls in trying to make these types of predictions will be exemplified by a paper presented at this conference where one of the first intended uses of MA-6000 has been postponed due to unexpected technical shortfalls of the material (31). This time frame of 15 plus years has been shown to be reasonable for the introduction of other materials such as single crystal superalloys into gas turbine engines and as mentioned earlier in this paper the research on the titanium aluminides was initiated about 15 years ago and these materials still await introduction into engine use. In contrast, materials such as thermal barrier coatings were transferred from research to flight applications in about 7 years (30). Our estimate for the various materials presented herein are shown in Fig. 7. At best, the Ti base aluminides are probably 5 years away from engine flight hardware. In looking at Fig. 7, it should also be pointed out that new engines produce more specific thrust and thus low density materials can be very significant in terms of the overall engine even at small weight percents. The remaining intermetallics, composites, and refractory alloys are realistically in the 10 year plus time frame. Any prediction of course, depends on the funding that is sustained over a period of 5 to 15 years in order for the materials and structures technologists along with design engineers to be able to bring these materials to a point of readiness. In addition, the manufacturing technology will have to be developed and production facilities put in place.

In conclusion, we believe the payoffs of fuel efficiency, increased performance, and reduced operating costs make the current research on these advanced materials worthwhile, and will lead to exciting results which we and others will hopefully report at future Seven Springs Symposia.

References

1. M.A. Crimp, K.M. Vedula, and D.J. Gaydosh: in <u>High Temperature Ordered Intermetallic Alloys II</u>, MRS Proc. Vol. 81, N.S. Stoloff, C.C. Koch, C.T. Liu, and O. Izumi, eds., pp. 499-506, Materials Research Society, Pittsburgh, PA, 1986.

2. D.J. Gaydosh and M.V. Nathal: NASA TM-87290, 1986.

3. J.L. Smialek, J. Doychak, and D.J. Gaydosh: submitted to <u>Oxid. Met</u>. 1988.

4. C. Barrett: <u>Oxid. Met</u>., 1988, in press.

5. A.G. Rozner and R.J. Wasilewski: <u>J. Inst. Met.</u>, 1966, Vol. 94, pp. 169-175.

6. A. Ball and R.E. Smallman: <u>Acta Met.</u>, 1966, Vol. 14, pp. 1349-1355.

7. R.T. Pascoe and C.W.A. Newey: <u>Met. Sci. J.</u>, 1968, Vol. 2, pp. 138-143.

8. E.M. Schulson and D.R. Barker: <u>Scr. Metall.</u>, 1983, Vol. 17, pp. 519-522.

9. J.D. Whittenberger: <u>J. Mater. Sci.</u>, 1988, Vol. 23, pp. 235-240.

10. I.E. Locci and M.V. Nathal: <u>Proc. 45th Meeting of EMSA</u>, G.W. Bailey, ed., pp. 212-213, San Francisco Press, San Francisco, CA, 1987.

11. R.S. Polvani, W.S. Tzeng, and P.R. Strutt: <u>Metall. Trans. A</u>, 1976, Vol. 7, pp. 33-40.

12. J.L. Walter and H.E. Cline: <u>Met. Trans.</u>, 1970, Vol. 1, pp. 1221-1229.

13. C.T. Liu: <u>Micon 86: Optimization of Processing, Properties and Service Performance Through Microstructural Control</u>, ASTM STP-979, B.L. Bramfitt, R.C. Benn, C.R. Brinkman, and G.F. Vander Voort, eds., pp. 222-237, ASTM, Philadelphia, PA, 1988.

14. H.A. Lipsitt: in <u>High Temperature Ordered Intermetallic Alloys</u>, MRS Proc. Vol. 39, C.C. Koch, C.T. Liu, and N.S. Stoloff, eds., pp. 351-364, Materials Research Society, Pittsburgh, PA, 1984.

15. C.F. Yolton, T. Lizzi, V.K. Chandhok, and J.H. Moll: in <u>Titanium, Rapid Solidification Technology</u>, T.H. Froes and D. Eylon, eds., pp. 263-272, TMS-AIME, Warrendale, PA, 1986.

16. S.M.L. Sastry and H.A. Lipsitt: in <u>Titanium '80, Science and Technology</u>, H. Kimura and O. Ozumi, eds., pp. 1231-1243, TMS-AIME, Warrendale, PA, 1980.

17. <u>High Temperature, High Strength Nickel Base Alloys</u>, International Nickel Company, Saddle Brook, N.J., 1984.

18. T.E. Strangman, B. Heath, and M. Fujii: NASA CR-168218, 1983.

19. D.L. McDanels and J.R. Stephens: NASA TM-100844, 1988.

20. D.W. Petrasek and R.H. Titran: DOE/NASA/16310-5, NASA TM-100804, 1988.

21. P.K. Brindley: in <u>High Temperature Ordered Intermetallic Alloys II</u>, MRS Proc. Vol. 81, N.S. Stoloff, C.C. Koch, C.T. Liu, and O. Izumi, eds., pp. 419-424, Materials Research Society, Pittsburgh, PA, 1987.

22. P.K. Brindley: "High Temperature Behavior of a SiC/Ti_3Al+Nb Composite," Presented at AIME Annual Meeting, Jan. 1988, Phoenix, AZ.

23. J.A. DiCarlo: <u>J. Mater. Sci.</u>, 1986, Vol. 21, pp. 217-224.

24. P.W. Angel, R. Hann, P.K. Brindley: unpublished paper, NASA Lewis Research Center, Cleveland, OH.

25. R.W. Clark and J.D. Whittenberger: in <u>Thermal Expansion 8</u>, Thomas A. Hahn, ed., pp. 189-196, Plenum Press, NY, 1984.

26. P. Shaffer, <u>Handbook of High Temperature Materials</u>, Plenum Press, New York, 1963, p. 316.

27. <u>New Technology in Japan</u>, Vol. 15, No. 6, Sept. 1987. (Japan External Trade Organization (JETRO), New York, NY).

28. R.K. Viswanadham and J.D. Whittenberger: "Elevated-Temperature/Slow - Plastic Deformation of $NiAl/TiB_2$ Composites," Presented at the Materials Research Society 1988 Spring Meeting, Reno, NV, Apr. 1988.

29. R.A. Perkins: "Oxidation of Refractory Metals," Presented at the Materials Research Society 1988 Spring Meeting, Reno, NV, Apr. 1988.

30. J.R. Stephens and J.K. Tien: NASA TM-83395, 1983.

31. B.A. Ewing and S.K. Jain: "Development of Inconel Alloy MA6000 Turbine Blades for Advanced Gas Turbine Engine Designs," to be presented at the Sixth International Symposium on Superalloys, Champion, PA, Sept. 18-22, 1988.

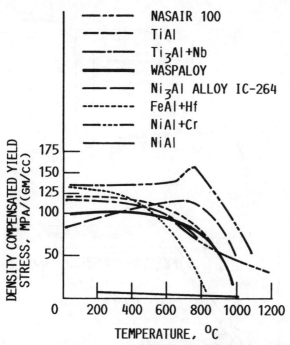

FIGURE 1. - TEMPERATURE DEPENDENCE OF DENSITY COMPENSATED YIELD STRESS FOR SEVERAL INTERMETALLIC ALLOYS AND SUPERALLOYS. DATA IS PRESENTED FOR NiAl (7), Fe-40% Al-1%Hf (2), Ni_3Al ALLOY IC-264 (13), Ti-14%Al-21%Nb (15), TiAl (16), DIRECTIONALLY SOLIDIFIED EUTECTIC NiAl+34%Cr (12), WASPALOY (17), AND SINGLE CRYSTAL NASAIR 100 (18).

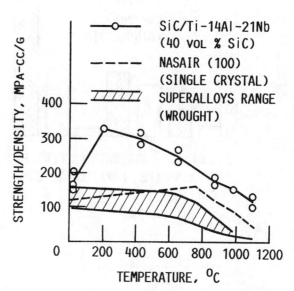

FIGURE 2. - TENSILE BEHAVIOR OF A SiC/Ti_3Al+Nb COMPOSITE COMPARED TO SUPERALLOYS.

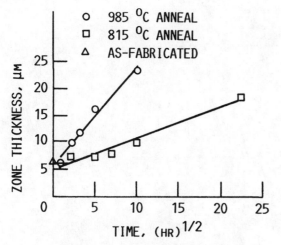

FIGURE 3. - EXTENT OF INTERDIFFUSION BETWEEN FIBER AND MATRIX UPON HIGH TEMPERATURE EXPOSURE.

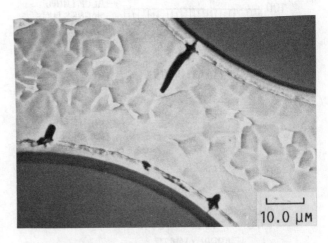

FIGURE 4. - REACTION ZONES, MATRIX DEPLETION, AND CRACKS DEVELOPED IN SiC/Ti_3Al+Nb COMPOSITE AFTER CYCLIC EXPOSURE AT 985 °C.

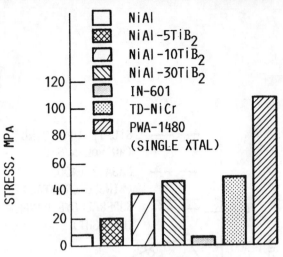

FIGURE 5. - STRENGTH PROPERTIES OF NiAl-TiB$_2$ PARTICULATE COMPOSITE.

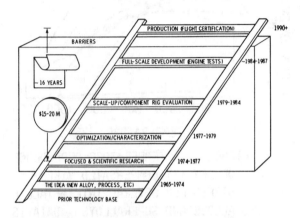

FIGURE 6. - TECHNOLOGY TRANSFER LADDER TO OVERCOME BARRIERS TO INTRODUCTION OF A NEW MATERIAL INTO AIRCRAFT ENGINES, USING THE OXIDE DISPERSION STRENGTHENED ALLOY, MA 6000 AS AN EXAMPLE.

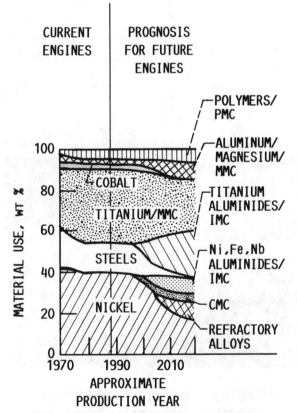

FIGURE 7. - TRENDS IN AIRCRAFT GAS TURBINE MATERIALS USE AND ESTIMATE OF NEW MATERIAL ENGINE INCORPORATION AND GROWTH.

REACTION KINETICS BETWEEN FIBER AND MATRIX

COMPONENTS IN METAL MATRIX COMPOSITES

M.W. Kopp[*], J.K. Tien[*] and D.W. Petrasek[**]

[*]Center for Strategic Materials
Henry Krumb School of Mines
Columbia University
New York, New York 10027
[**]NASA-Lewis Research Center
21000 Brook Park Road
Cleveland, Ohio 44135

Abstract

Interdiffusion and interdiffusion controlled intermediate phase formation in metal matrix composites can be of interest for the prolonged application of these systems at high temperatures. Methods are discussed that address the kinetics of interdiffusion in systems that exhibit solid solution intermixing or the formation of a third intermediate phase at the fiber/matrix interface, or both. The tungsten fiber reinforced niobium and tungsten fiber reinforced superalloy systems are employed as model systems for experimentation and discussion. In an effort to impede interdiffusion, the concept of ion implanted diffusion barriers have been examined. Preliminary results on the feasibility and effectiveness of ion implanted diffusion barriers are presented.

Introduction

The current search for advanced materials to surpass the high temperature capabilities of the present single crystal superalloys has concentrated attention on metal matrix composites (MMCs) (1,2,3). The inherent advantages of composite materials, however, do not come without some potential liabilities. For example, when one considers composites for ambient temperature applications, factors such as fiber/matrix adhesion must be considered. When one extends the question to short term high temperature applications, thermal-mechanical compatibility must also be considered. For long term high temperature applications, thermally activated processes such as interdiffusion and interfacial reactions between the reinforcing fibers and the matrix can become challenging concerns. These latter issues have been the focus of recent work at Columbia University and the NASA-Lewis Research Center (3,4).

Interdiffusion promoted effects on metal-matrix composite properties has been documented. Phenomena of this type include dissolution of the reinforcing fibers, formation of intermediate phases at the fiber/matrix interface, and, in the case of refractory metal fibers, the diffusion triggered recrystallization of the heavily worked fibers. These phenomena and their effect on composite properties must therefore be understood before the advantages of metal matrix composites can be fully utilized.

Results and Discussion

The Tungsten Fiber Reinforced Niobium System. In order to gain the most fundamental understanding of interdiffusional kinetics in fiber reinforced MMCs, we chose the single phase, or entirely mutually soluble, W/Nb system for initial study. Planar and cylindrical interface diffusion couples of W/Nb were fabricated. The cylindrical interface composites employed 218CS commercial tungsten wire and were made using the arc-spray technique developed at the NASA-Lewis Research Center (5). The planar interface couples were made by diffusion bonding of bulk materials. These diffusion couples were annealed in vacuum at 1800K and 1500K for 50 to 500 hours. Composition profiles for the annealed couples were obtained using a calibrated EDS microprobe at 1 micron intervals.

Composition dependent interdiffusion coefficients were obtained from the profiles of the planar interface composites using Boltzmann-Matano analysis. This technique allows the simple extraction of diffusion coefficients by setting the flux balance point to x=0, the Matano interface. Diffusion coefficients can then be determined numerically from the composition profile such that:

$$D(C,T) = \frac{-1}{2t} \left[\frac{dC'}{dx}\right]^{-1} \int_{C_0}^{C} x \, dC' \qquad (1)$$

Experimentally determined diffusion coefficients are shown in Figure 1. For this system in this temperature range the interdiffusion coefficient may be assumed to be independent of composition. The magnitude of the coefficients is considerably higher than what was expected based on Arrhenius extrapolations from the higher temperature data of Hehemann and Leber (6).

In order to assess the accuracy of the experimentally determined interdiffusion coefficients numerical solutions to Fick's second law (Eqn. 2) were employed. The code originally developed by Tenney and Unnam (7) was adapted for this purpose. These numerical solutions calculate composition

$$\frac{\delta C}{\delta t} = \frac{1}{r^n}\frac{\delta}{\delta r} r^n D(C,T)\frac{\delta C}{\delta r} \qquad (2)$$

profiles for planar, cylindrical, or spherical geometry composites (n=0, n=1, n=2) with finite boundary conditions.

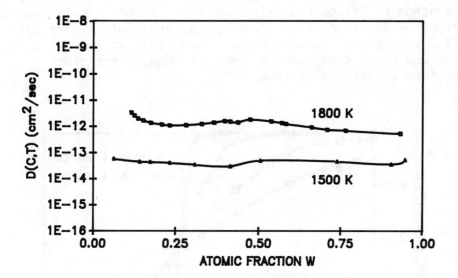

Figure 1 - Experimentally determined interdiffusion coefficients for the W/Nb system at 1500K and 1800K.

Figure 2 illustrates the good agreement between an experimentally determined composition profile for a cylindrical geometry composite and one that was calculated with interdiffusion coefficients determined from a planar interface diffusion couple. The good agreement between experimental and calculated composition profiles led to efforts to employ these same numerical solutions in order to predict the long term interdiffusional behavior of these composites.

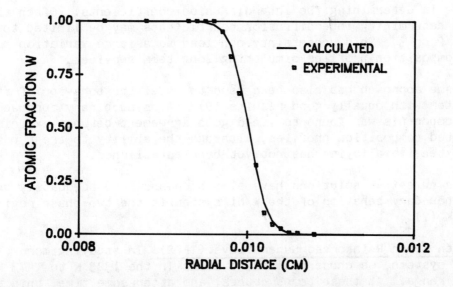

Figure 2 - Experimental and calculated composition profiles for a cylindrical geometry W/Nb composite annealed at 1500K for 500 hours.

The ability of the numerical solutions which were employed to handle finite boundary conditions also allows for the consideration of long term diffusion where the diffusion fields of individual fibers begin to overlap. Figure 3 shows a calculated time progression of the concentration profile of a cylindrical geometry W/Nb composite at 1500K. This hypothetical composite contains 200 micron diameter tungsten fibers making up approximately 40 volume percent of the composite. Recrystallization has been shown in this and other systems to follow the diffusion field of the matrix material in the refractory metal fibers (8). Accordingly, one would expect significant recrystallization of the tungsten fiber in one year at 1500K. Similar calculations and predictions can be done, of course, for other time-temperature conditions of interest.

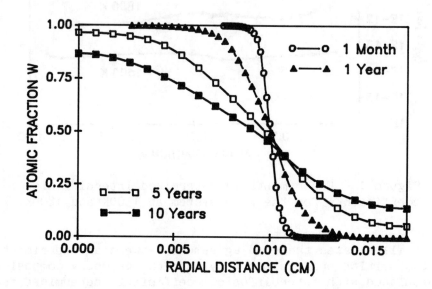

Figure 3 - Long term progression of composition profiles of a cylindrical geometry W Nb composite at 1500K.

The ability to make predictions in this manner does require a great deal of care in determining the interdiffusion coefficients. Often the accuracy in determining interdiffusion coefficients may be limited to within a factor of 2 or 3. This uncertainty can lead to a large variation in predicted composition profiles simulating long term services.

This same approach has also been used to model the behavior of the W/Nb1Zr system with equally good results (9). A pseudobinary approach of W and Nb1Zr components was found to yield good agreement between experimental and calculated composition profiles, although the ability to treat phases with more extensive alloying has not yet been established.

Similar numerical solutions have also been shown to adequately describe the moving boundary behavior of the W/Ni system in the two-phase regime above 1343 K (7).

<u>Tungsten Fiber Reinforced Superalloys (TFRS)</u>. In studying more complicated systems, we chose the TFRS systems in the 1323 K to 1473 K temperature range. At these temperatures, and after some time, this system exhibits the formation of a brittle reaction zone at the fiber/matrix interface. Matrix alloys that were initially studied included FeCrAlY, Waspaloy, Incoloy 907, Incoloy 903, and 316 stainless steel (10,11). These composites were fabricated using a 100 micron diameter ST300 thoriated

tungsten wire (1.5% thoria) by a powder metallurgical technique. These composites were annealed in vacuum for 50 to 500 hours. Figure 4 illustrates the growth of the "reaction zone" phase at the fiber/matrix interface. The formation of the reaction zones was found to be long range diffusion controlled (10) and the kinetics to be highly matrix chemistry dependent (11).

In order to quantitatively assess the kinetics of reaction zone formation, the moving boundary equations were solved simultaneously with Fick's second law, yielding the expression for the diffusion coefficient of the reaction zone phase (Eqn. 3). Assuming the interfacial concentrations

$$D_{rz}(T) = 1/2 \ [(C_{f|rz}-C_{rz|f})/(C_{rz|f}-C_{rz|m})] \ K_f(T)^{1/2} K_{rz}(T)^{1/2} \tag{3}$$

(where $C_{f|rz}$ would be the W concentration on the fiber side of the fiber/reaction zone interface, for example) can be considered roughly constant for the different matrix alloys studied, this result indicates that the diffusion coefficient is proportional to the product of the roots of parabolic rate constants for the total growth of the reaction zone and the component of the reaction zone which displaces the fiber (12).

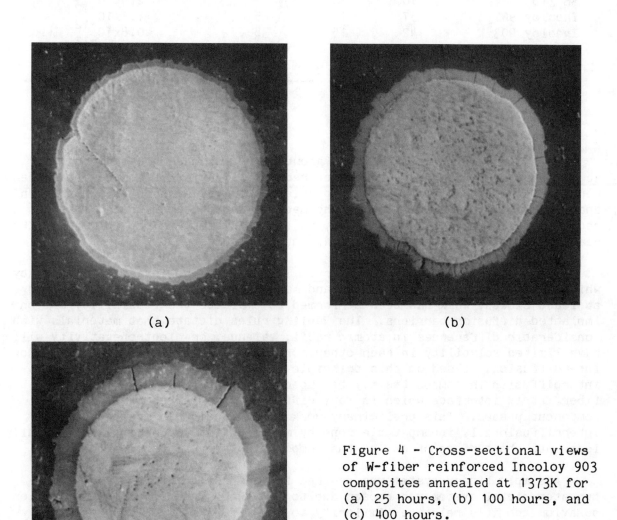

Figure 4 - Cross-sectional views of W-fiber reinforced Incoloy 903 composites annealed at 1373K for (a) 25 hours, (b) 100 hours, and (c) 400 hours.

By ranking the different alloys studied in order of these kinetic constants, it was determined that the kinetics of reaction zone growth decreased with decreases in Fe and Co content, and the corresponding increase in Ni content. Based on this kinetic data from conventional superalloy matrices, a new matrix alloy (Alloy 89) was designed and fabricated which totally eliminated Fe and Co. The kinetic constants were determined for this composite and the results for all of the matrix materials studied are shown in Table 1. As is evident, by eliminating these stable reaction zone formers, the kinetics of reaction zone formation have been greatly reduced.

Table 1. Product of the square roots of the parabolic rate constants for various TFRS composites annealed at 1366K and their corresponding Fe, Co, and Ni contents in atomic percent.

Matrix	Fe	Co	Ni	$[K_{rz}^{1/2} K_f^{1/2}]$ (cm^2/sec)
FeCrAlY	71	0	0	3.5×10^{-12}
SS 316	70	0	12	2.9×10^{-12}
Incoloy 907*	57	13	25	1.7×10^{-12}
Incoloy 903	42	15	38	0.8×10^{-12}
Waspaloy	0	13	56	0.3×10^{-12}
Alloy 89	0	0	66	0.05×10^{-12}

* Annealed at 1373K.

This decrease in reaction zone growth kinetics may, however, be accompanied by increasing the rate of recrystallization of the W-fibers when considering long term composite properties. Work by Larsson (13) has shown that increasing the Ni content in the matrix leads to accelerated fiber recrystallization. This may be due, at least in part, to the ability of the reaction zone to limit the flux of Ni into the tungsten fiber.

Ion Inplanted Diffusion Barriers. In an effort to investigate means by which interdiffusion between fiber and matrix components can be halted, or at least slowed, work has been performed to evaluate the feasibility of ion implanted diffusion barriers. The Pauling rules dictate that materials with considerable differences in atomic radii, valence, or electronegativity will have limited solubility in each other, or will have little driving force for interdiffusion. Based on this principle, it might be expected that interdiffusion in composites may be impeded by creating a layer at the fiber/matrix interface which is very different than either of the two component phases. This preliminary work is an attempt to create such a interdiffusionally incompatible zone by means of ion implanting large alkali ions into the surface of one of the components prior to fabrication.

A simple binary system was desired for a model for this work. The tungsten-nickel system was chosen due to it's well understood interdiffusion behavior and it's relative similarity to the TFRS system which is of technological interest. Diffusion couples were fabricated employing ion implanted diffusion barriers of sodium, calcium, barium, and potassium. These alkali and alkali earth ions were implanted into bulk tungsten substrates and subsequently coated by an ion-beam sputtered nickel film of approximately 100 microns in thickness. Each diffusion couple received an equal dose of the respective barrier ions of 6.4×10^{11} ions/cm^2 implanted with an accelerating potential of 190 keV. The substrate temperature during the subsequent nickel deposition was estimated to be approximately 700K over the 70 hour deposition time. These diffusion couples, along with couples fabricated without an implanted layer, were annealed in vacuum at 1500K for 50 hours. Composition profiles were obtained for each diffusion couple were obtained by an EDS microprobe at 1 micron intervals across the interface.

The composition profiles for the sodium and calcium barriers diffusion couples are plotted along with the profiles of the non-implanted and as-received couples in Figures 5 and 6, respectively. The evident "saturation" of tungsten completely through the nickel "matrix" can be attributed to the very significant grain boundary diffusion induced by the characteristic microstructure of the sputtered nickel film. In spite of this problem which makes quantitative analysis of the efficacy of these barriers impossible, it is possible to make some preliminary observations for the effect of the barriers relative to the tungsten-nickel standard. As Figure 5 shows, it appears that the implanted calcium barrier did decrease the kinetics of interdiffusion in the tungsten-nickel system, although this

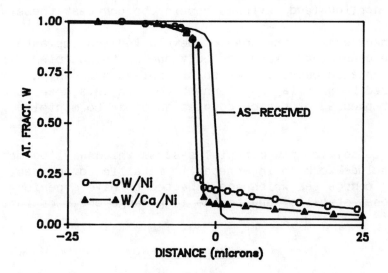

Figure 5 - Composition profile for tungsten-nickel diffusion couples with and without ion implanted calcium layer.

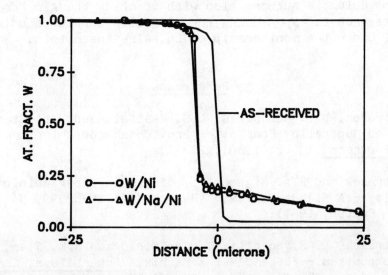

Figure 6 - Composition profile for tungsten-nickel diffusion couples with and without ion implanted sodium layer.

may be due at least in part to inhibiting the short circuit paths along the grain boundaries. The effect of the sodium barrier, illustrated in Figure 6, appears to be to promote interdiffusion in this system. Barium and potassium barriers were also found to promote interdiffusion, but to an even greater extent.

A possible explanation of this behavior may be that the distortion produced by the larger barium and potassium ions may be triggering a dynamic recovery at this elevated temperature. This process might be eliminated if lower ion doses are used. Similarly, variation of ion dose and accelerating potential may prove to yield better results for the calcium barrier by tailoring the calcium atom density and the thickness of the affected zone.

Conclusion

Understanding interdiffusion between metal matrix composite components and its affect is essential in terms of material selection for long term, high temperature applications. At very high temperatures the W/Nb (W/Nb1Zr) and the TFRS composites do show certain interdiffusional intermixing and intermediate phase formation, respectively. Diffusion mechanics have been successfully applied (and reviewed herein) to forecast these happenings.

Preliminary work in the area of developing ion implanted diffusion barriers is encouraging. Although only one of the barriers initially tested showed a moderate success, variation of implant process parameters may lead to an optimized barrier layer. Current work in this area at Columbia is utilizing hot-pressed diffusion couples in order to minimize grain boundary effects.

Efforts to develop new material systems which will be more chemically stable may include work on intermetallic/metallic composites, intermetallic/intermetallic composites which are expected to have superior interdiffusional stability due to the long range order of intermetallic phases.

Acknowledgment

The authors from Columbia University wish to thank the NASA-Lewis Research Center for sponsoring the research on the tungsten fiber reinforced niobium and tungsten fiber reinforced superalloy systems through grants NASA-NAG-3-410 and NASA-NAG-3-720, which were monitored by Donald Petrasek. The Columbia authors also wish to thank the Air Force Office of Scientific Research for sponsoring the diffusion barrier studies under grant AFOSR-86-0312 under the monitorship of Dr. Alan Rosenstein.

References

1. D.W. Petrasek, D.L. McDaniels, L.J. Westfall and J.R. Stephens, "Fiber Reinforced Superalloy Composites Provide an Added Performance Edge," Metals Progress, 130(2)(1986), 27-31.

2. D.W. Petrasek and R.A. Signorelli, "Tungsten Fiber Reinforced Superalloys--A Status Review," (Report NASA TM-82590, NASA-Lewis Research Center, 1981).

3. D.W. Petrasek, R. Signorelli, T. Caulfield and J.K. Tien, "Supercomposites - FRS and Metal Matrix," Superalloys, Composites and Ceramics, eds. J.K. Tien and T. Caulfield, (New York, Academic Press), in press.

4. M.W. Kopp and J.K. Tien, "Interdiffusional Instability in High Temperature Metal Matrix Composites," (Paper presented at the 9th Riso International Symposium, Roskilde, Denmark, 5-9 September, 1988).

5. L.J. Westfall, "Tungsten Fiber Reinforced Superalloy Fabrication by an Arc-Spray Process", (Report NASA TM-86917, NASA-Lewis Research Center, 1985).

6. R.F. Hehemann and S. Leber, "Chemical Diffusion in the Columbium-Tungsten System," Trans. AIME, 236 (1966) 1040-1044.

7. D.R. Tenney and J. Unnam, "Numerical Analyses for Treating Diffusion in Single-, Two-, and Three-phase Binary Alloy Systems," (Report NASA TM-78636, NASA-Langley Research Center, 1978).

8. D.W. Petrasek and J.W. Weeton, "Alloying Effects on Tungsten Fiber Reinforced Copper Alloy or High Temperature Alloy Matrix Composites," (Report NASA TN-D-1568, NASA-Lewis Research Center, 1966).

9. M.W. Kopp, "Long Term Interdiffusional Behavior in Tungsten Fiber Reinforced Niobium-base Matrix Composites," (Masters thesis, Columbia University, 1987).

10. T. Caulfield, R.S. Bellows and J.K. Tien, "Interdiffusional Effects between Tungsten Fibers and an Iron-Nickel-Base Alloy," Met. Trans. A, 16A (1985) 1961-1968.

11. J.K. Tien, T. Caulfield and Y.P. Wu, "High Temperature Reaction Zone Growth in Tungsten Fiber Reinforced Composites: Part II--Matrix Chemistry Effects," Met. Trans. A, in press.

12. T. Caulfield and J.K. Tien, "High Temperature Reaction Zone Growth in Tungsten Fiber Reinforced Composites: Part I--Application of the Moving Boundary Equations," Met. Trans. A, in press.

13. L.O.K. Larsson, Metal Matrix Composites for Turbine Blades in Aero-engines, (Goteborg, Sweeden: Chalmers Institute of Technology, 1981), 1-41, F1-F30.

INDUSTRIAL SCALE PROCESSING AND ELEVATED TEMPERATURE

PROPERTIES OF $Ni_3Al-Cr-Zr-B$ ALLOYS[*]

V. K. Sikka

Metals and Ceramics Division
Building 4508; MS 083
Oak Ridge National Laboratory
P. O. Box 2008
Oak Ridge, Tennessee 37831

E. A. Loria

Metallurgical Consultant
1828 Taper Drive
Pittsburgh, Pennsylvania 15241

Abstract

This study is the first to evaluate the elevated temperature properties of two of the strongest $Ni_3Al-Cr-Zr$ alloys when made in sufficient quantities and processed to products simulating industrial-scale operations. The technologies considered are powder processing, melting techniques, near-net-shape processes, and cold and hot working of cast products. The tensile properties from ambient to 1100°C and creep rupture properties in the range of 704 to 1093°C are presented. These properties are most sensitive to the grain size differences produced by the various processing techniques. The results compare favorably with those obtained on commercial superalloys and reveal good potential for elevated temperature structural use.

---

[*]Research sponsored by the Division of Materials Sciences, U.S. Department of Energy, under contract DE-AC05-840R21400 with Martin Marietta Energy Systems, Inc.

Introduction

Alloys based on the ductile-ordered intermetallic compound Ni_3Al are being considered for a range of structural applications (1-3). These include gas and steam turbines, automotive pistons, turbochargers and valves, heating elements for appliances, aircraft fasteners, gas and oil well tubular products, and components for corrosive environments. Acceptance will depend on the development of the processing technology and the mechanical properties of various products for these applications. With this in mind, a study was conducted on the effect of processing scaled up quantities of two Ni_3Al-Cr-Zr-B alloys by six different methods and determining their elevated temperature tensile, creep, and rupture properties. The two alloys (IC-218 and IC-221) provide improved high-temperature strength via solid solution-hardening effects of 0.8 or 1.7 wt % Zr, while the 7.8 wt % Cr alleviates dynamic embrittlement by changing the oxidation process. To accommodate these elements, the aluminum content is reduced from 11.3 to 8.5 wt % and boron remains at 0.02 wt %. The IC-218 and IC-221 alloys are primarily ordered structure (γ'-phase) with small amounts of the disordered structure (γ-phase) at room temperature. The fraction of disordered structure in these alloys increases at high temperatures (greater than 1000°C). On the other hand, IC-50 (11.3Al-0.6Zr-0.02B) has only the ordered structure (γ'-phase) up to its melting point, but it lacks the high-temperature ductility of the chromium-containing alloys.

Materials and Processes

This study is the first to evaluate materials made in sufficient quantities and processed to products simulating industrial-scale operations. Because of expected restricted hot workability and its negative effect upon a number of possible applications, various processing techniques have been explored for the determination of elevated temperature properties. These include powder metallurgy (PM) techniques utilizing 125 to 250 kg of powder from two atomizing sources which were either hot isostatic pressed (HIP) or consolidated under atmospheric pressure (CAP) and then extruded to bar stock. Also, small isothermally forged disks and rapid omnidirectional compacted (ROC) disks were produced from the two PM processes. Melting and casting 50- to 250-kg air-induction melt (AIM) or vacuum-induction melt (VIM) heats as well as duplexing via electroslag remelting (ESR) provided cast billet and hot extruded products. Cast tube hollows of 125-mm OD and 25-mm wall were produced from argon-induction melting of 250-kg heats as well as 15-kg castings and twin roller cast sheet of 200- to 250-mm width and 1.5- to 2.0-mm thickness were produced from 50-kg AIM heats. Flowcharts describing the status of each of these processes investigated by the Oak Ridge National Laboratory (ORNL) in conjunction with industry have been presented (4).

For testing purposes, sheet specimens were used principally to reduce machining cost. This was the case for the VIM-extruded, ESR-extruded, AIM-extruded, cast billet-cold rolled, twin-roller sheet, and PM-extruded specimens. Although extruded products were sheet or round bar, sections from these were cold rolled and annealed to a thickness of 0.76 mm. Round specimens of 6.4-mm diam were prepared from cast billet as well as 3.2-mm-diam subsize rounds which were cut from an isothermal and an ROC disk prepared from the PM products. Tensile tests were conducted on four Instron machines, using sheet specimens of 0.76-mm thickness by 25-mm gage length. In all cases, a constant crosshead rate was used rather than strain rate. Most commonly used was 8.3×10^{-4}/s (because of ASTM requirements), but 3.3×10^{-3}/s was used on some tests. Several tests were conducted at a range of strain rates to explore superplasticity and effects on strength and ductility. All creep tests were constant load tests. Those on sheet specimens were conducted using dead-load creep machines. Creep tests on 6.4-mm-diam round

specimens were conducted using a lever-arm machine with lever-arm ratios of
12:1 or 20:1. For all sheet specimens, strain was measured by extension of
the pull rod by a mechanical dial gage. For round specimens, an attached
strain-averaging extensometer provided strain measurements. For all cases,
strain-time data were computer plotted and analyzed for minimum creep rate
and other quantities such as start of second-state creep, start of third-
stage creep, etc. Ruptured specimens were measured for fracture strain and
reduction of area.

Results

The grain size of six products of IC-218 vary between 9 and 21 µm for
three different PM processes, cast and hot extruded and cast and cold-
rolled sheet versus 727 µm in cast tube hollow, per Figure 1. The tensile

Figure 1 - Optical microstructure of IC-218 alloy specimens processed by six
different processes. (a) Cast tube hollow, grain size = 727 µm; (b) cast and
hot extruded, grain size = 11 µm; (c) cast and cold-rolled sheet, grain size
= 16 µm; (d) powder extruded, grain size = 16 µm; (e) powder isothermal disk,
grain size = 9 µm; and (f) powder ROC disk, grain size = 21 µm. Magnification: 500×.

properties, plotted from ambient to 1100°C, show a band of values for 0.2%
yield strength, ultimate strength, and total elongation, per Figure 2.
Yield strength is one of the more fundamental properties of a structural
alloy. It is a property to be maximized and, in the case of elevated temperature applications, to be invariant with temperature. The yield strength
of the five fine-grain products remains stable within a band of 550 to
750 MPa up to 600°C, then declines gradually within 400 to 600 MPa at 800°C
then declines sharply to below 100 MPa at 1100°C. The coarse-grain tube
hollow has the lowest yield strength of all products at room temperature of
450 MPa, which rises gradually to the highest value of all products of
680 MPa at 800°C and then declines to still the highest value of 250 MPa at
1100°C. The ultimate strengths of the five fine-grain products decline
gradually within a band of 1250 to 1500 MPa at room temperature to 520 to

650 MPa at 800°C, while the coarse-grain cast product which has a much lower strength of 750 MPa at room temperature maintains it to 800°C. All products then decline in the same manner to 1100°C with the latter still exhibiting the highest strength. The total elongation of all six products is within 20 to 38% at room temperature and the range then widens between 7 and 35% at 800°C in five products, followed by a sharp rise to values exhibiting superplasticity at 1000°C. The 20% elongation of the cast tube hollow only declines to 15% at 800°C and shows no superplasticity at 1075°C. Unexpectedly, the 22% elongation in cast and cold-rolled sheet product, having a fine-grain size of 16 μm, declines continuously to a nil ductility value at 800°C and then rises sharply to a superplastic value of 260% at 1000°C.

The IC-221 alloy was processed by similar techniques used for IC-218 alloy. The same trends in grain size and tensile were observed in IC-221. When compared under the same conditions with IC-218 (i.e., fine-grain PM-extruded product), the yield and tensile strength of IC-221 are higher than IC-218, see Figure 3. The IC-221 alloy maintains its yield strength of 635 MPa up to 800°C, while IC-218 declines slightly from 525 to 435 MPa at 800°C, then both decline rapidly to 100 MPa at 1000°C. Ultimate strength declines gradually from 1500 MPa for IC-221 versus 1400 MPa for IC-218 to 690 MPa versus 550 MPa at 800°C and then to the same 120 MPa at 1000°C. Elongation values are the same at 35% at room temperature, with a gradual decrease to a minimum of 11% at 800°C in IC-218 compared to an unexpected deep decline from 28% at 600°C to 5% at 800°C in IC-221. Then, there is the usual steep ascent at 1000 to 1100°C. It is noteworthy that the room temperature ductility of the various IC-218 and IC-221 products is sufficient for conventional mill material handling and shipping operations.

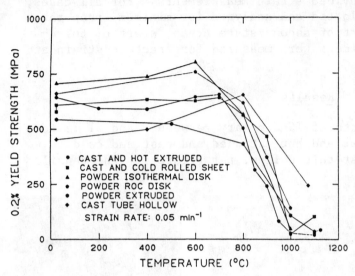
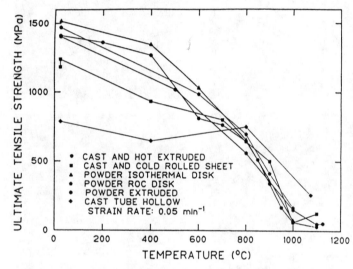
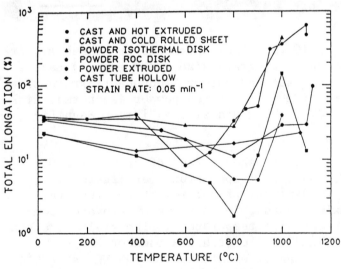

Figure 2 - Tensile properties comparison for IC-218 alloys processed by six different processes. All tests were conducted at a strain rate of 0.05/min (8.3×10^{-4}/s).

The effects of grain size on tensile properties of Ni_3Al produced in small (laboratory) quantities have been extensively studied (5-6). Reducing the grain size has been found to significantly increase the room-temperature strength with relatively little effect on the ductility. Our results on industrial-scale processing of larger quantities reveal superior yield and ultimate strength from ambient to 700°C for fine-grain products versus coarse-grain cast products, but the latter remain stable and gradually increase to the highest strength values at 800°C. This change is brought about by the increase in lattice resistance to slip and the decrease in the effectiveness with which grain boundaries impede slip as temperature rises. Above 950°C, a fine-grain product produced by either extrusion of cast billet or extrusion of powder shows superplastic behavior, whereas coarse-grain cast product shows higher strength and low ductility (2-3).

Creep tests over a temperature range of 649 to 871°C (1200 to 1600°F) and times from 10 to 12,464 h provided data for analysis using the Larson-Miller parameter, see Figure 4. Plotting creep-rupture strength versus parameter value produces a narrow band of values, represented by a single line of negative slope, for the five fine-grain products and a separate single line for the coarse-grain castings of IC-221. At each stress level, the parameter is significantly higher in the case of the latter, which translates to a higher rupture strength at the same parameter value. On the other hand, the corresponding values for total elongation and reduction of area for the fine-grain products are significantly higher than those for castings over the entire parameter range, and the data, in each case, are represented by a single

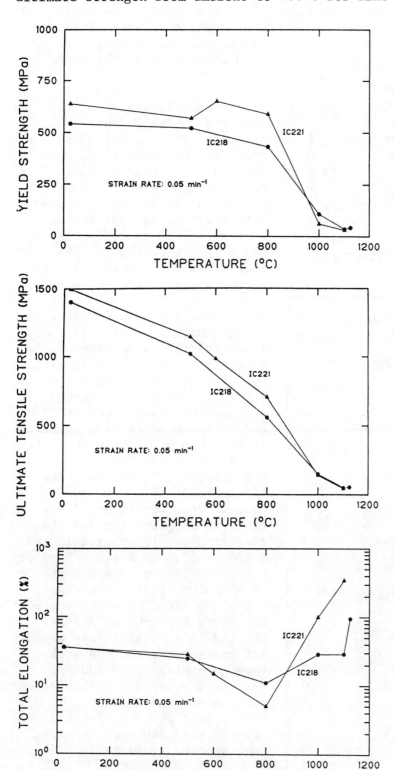

Figure 3 - Comparison of elevated temperature tensile properties of IC-221 versus IC-218 extruded products prepared under the same conditions.

positive slope line. Also, it is apparent that fine-grain material from cast billets provides higher values than fine-grain material from powder billets so that separate lines are drawn differentiating these two fine-grain products. In general, the same trends are observed in comparing IC-218 fine-grain products with very coarse-grain (6600 μm) cast billet, per Figure 5. The plotting of the data for fine-grain products of IC-218 reveals that separate lines separating 14- to 16-μm grain size from 7- to 10-μm grain size are drawn which show that the former has a higher rupture strength but lower elongation and reduction of area than the latter at the same parameter.

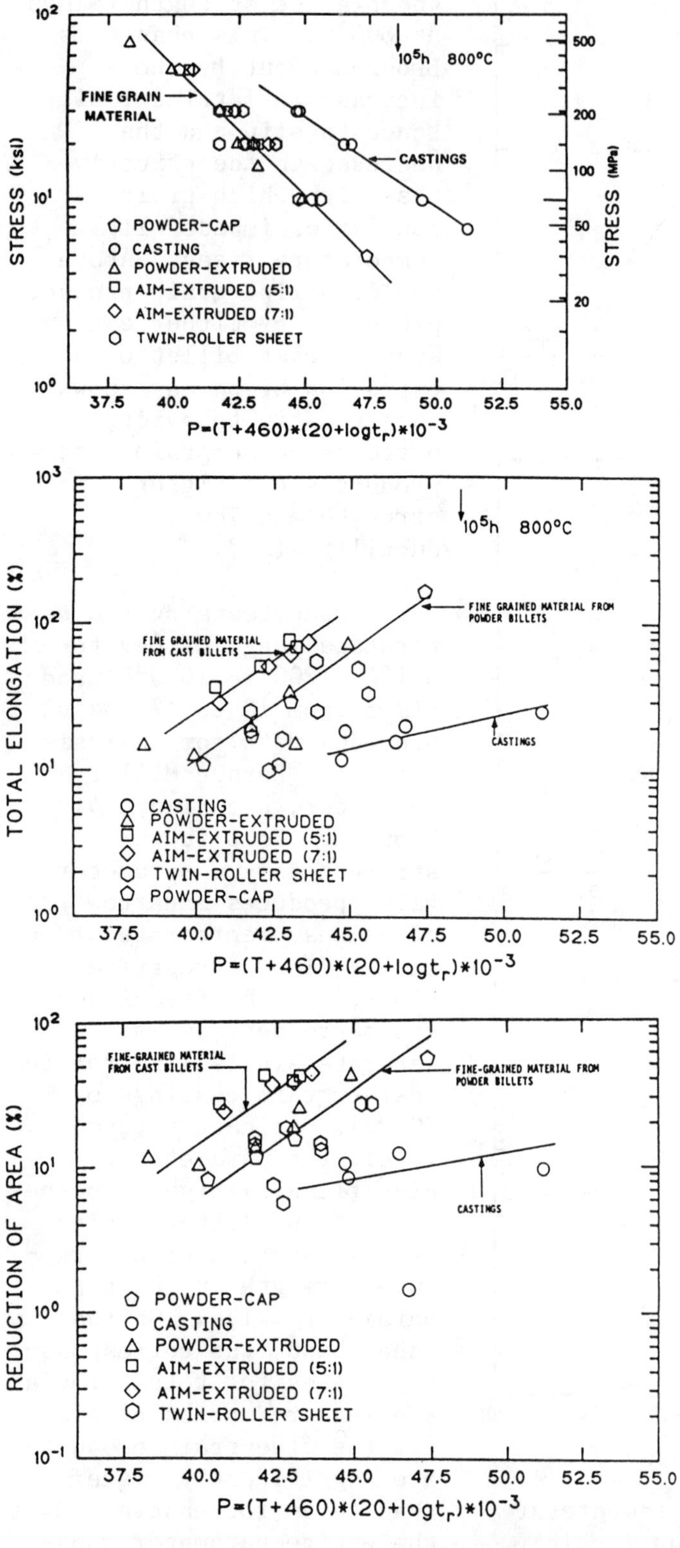

From this analysis, the 100- and 1000-h rupture strengths at 649, 732, 816, and 982°C (1200, 1350, 1500, and 1800°F) are listed in Table I for the various products of IC-218 and IC-221 and compared with electroslag cast IC-50. Once again, it is obvious that a coarse-grain cast structure has much superior rupture strength than fine-grain wrought material. The diffusional creep mechanism operative in the test temperature range of this investigation is believed to be responsible for such a significant grain-size effect. The 100- and 1000-h creep rupture strength of the various IC-218 and IC-221 products are compared in Figure 6 with those published for A-286, N-155, V-57, Waspaloy, and IN-100 over the 649 to 982°C (1200 to 1800°F) range. Some IC-218 and IC-221 products are superior and the rest are closely comparable to A-286, V-57, and N-155. Some of our products are even comparable to Waspaloy within this range, but above 816°C (1500°F) where the above Fe-Ni-Cr alloys are not used and Ni_3Al alloys would not be recommended, the nickel-base superalloys (IN-100 and Waspaloy) provide superior creep rupture strength. Again, in all cases, cast products provide higher creep rupture strengths than wrought products.

Figure 4 - Larson-Miller parameter of creep rupture properties of IC-221.

The Arrhenius plot of the steady-state creep rate versus temperature, Figure 7, for several products of both alloys yields an activation energy for creep of 81.7 kcal. This value

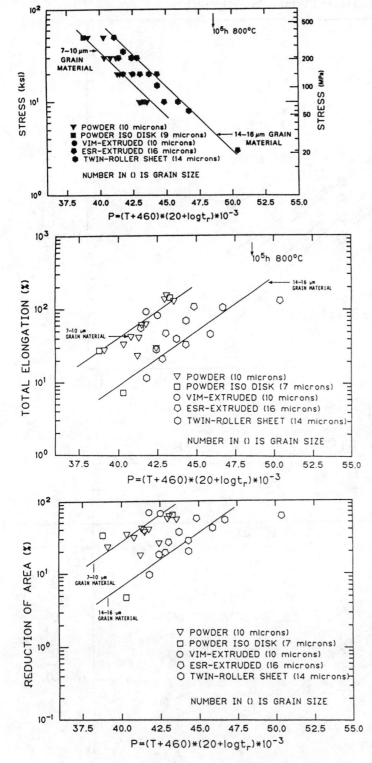

Figure 5 - Larson-Miller parameter of creep rupture properties of IC-218.

is higher than the published values of self-diffusion of nickel (68 kcal) and γ-iron (70 kcal) but still suggests that the rate-controlling processing in high-temperature creep is diffusion controlled. Analysis of the creep data via the Nabarro-Herring creep mechanism, Figure 8, provides reasonable agreement between predicted and observed values for different grain size materials, and the grain size dependence of creep at high temperature is explained by the Nabarro-Herring creep mechanism which involves lattice diffusion under the action of applied stress.

Discussion

This paper has presented property data from scaled up quantities of two high-strength Ni_3Al-CrZr-B alloys prepared by various techniques. The tensile and creep properties are sensitive to grain-size differences resulting from the various processing techniques, but the alloys still provide an attractive alternative to certain superalloys. The current status of the processes being explored at ORNL in conjunction with industry is that powder processing is closest to commercialization and ideal for fabricating complex shapes. Nickel aluminide can be melted within specifications by the simplest method such as AIM. Secondary melting such as electroslag remelting produces ingots of most desirable grain structure for forging and extruding and of excellent surface quality. Near-net-shape processes from molten metal offer the best fabrication possibility via the melting route for a variety of shapes. These processes offer the best chance of success for utilization of nickel aluminides in various applications suggested for them. Hot-working processes such as hot forging and hot rolling for the cast products are still not fully developed. Hence, cold processing followed by heat treatment currently has the best chance of success. Among the hot-processing methods, hot extrusion has been investigated most thoroughly and has immediate potential for commercialization.

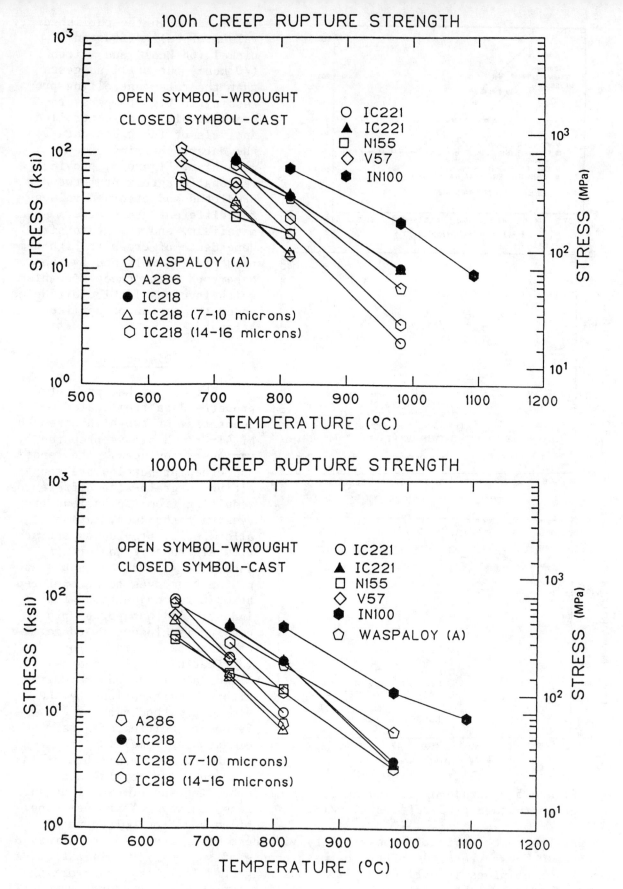

Figure 6 - Comparison of 100- and 1000-h creep rupture strength of IC-221 and IC-218 with A-286, N-155, V-57, IN-100, and Waspaloy.

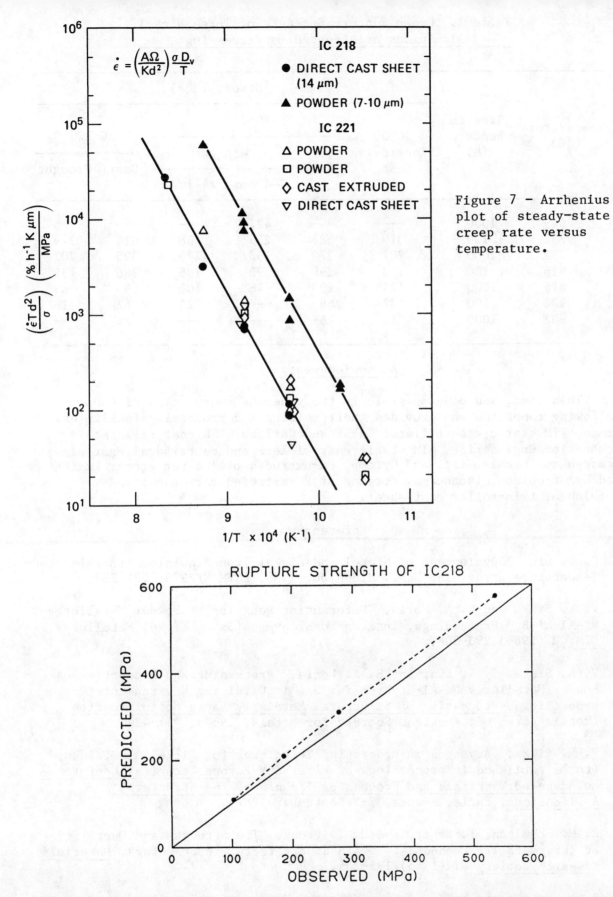

Figure 7 - Arrhenius plot of steady-state creep rate versus temperature.

Figure 8 - Plot showing predicted and observed rupture strength of IC-218 alloy. Values for 15-μm grain size were predicted using Nabarro-Herring equation and data on 10-μm grain size material. Comparison is for data in the temperature range of 732 to 816°C and for 100- and 1000-h rupture strength.

Table I. Creep Rupture Strength of Three Nickel Aluminides as Affected by Processing

Temperature (°C)	Time to Rupture (h)	Stress (MPa)					
		IC-50 Electroslag Cast	IC-218			IC-221	
			Cast	Wrought		Cast	Wrought
				7–10 μm	14–16 μm		
649	1000	---	---	427	---	---	655
732	100	310	586	255	538	614	379
732	1000	207	379	138	276	393	207
816	100	172	290	93	186	296	138
816	1000	124	193	48	103	193	69
982	100	37	69	---	23	67	16
982	1000	14	26	---	---	24	---

Acknowledgments

This study was made possible by the generous cooperation of the following companies who provided their melting and processing facilities: Armco, VIM static cast billets; Cabot and CarTech, ESR cast billets; Combustion Engineering, AIM static cast billets and centrifugal cast pipe; Homogeneous Metals, Inc. and Cytemp, PM-extruded billet/bar stock; Ladish and Wyman-Gordon, isothermal forging of PM-extruded bar stock; and Bethlehem, twin-roller cast sheet.

References

1. C. T. Liu, "Development of Nickel and Nickel-Iron Aluminides for Elevated-Temperature Structural Use," Micon 86, 1988, ASTM STP979, 222-237.

2. V. K. Sikka and E. A. Loria, "Deformation Behavior of Powder Metallurgy Ni_3Al-Hf-B," Proceedings, International Symposium on Nickel Metallurgy, CIM, 2 (1986) 293-308.

3. V. K. Sikka, C. T. Liu, and E. A. Loria, "Processing and Properties of Powder Metallurgy Ni_3Al-Cr-Zr-B for Use in Oxidizing Environments," proceedings, Processing of Structural Metals by Rapid Solidification (Metals Park, OH: American Society for Metals, 1987), 417-427.

4. V. K. Sikka, "Advances in Processing Techniques for Nickel Aluminides," (to be published in proceedings of First ASM Europe Technical Conference on Advanced Materials and Processing Techniques for Structural Applications, Paris, France, 7-9 September 1987).

5. E. M. Schulson, I. Baker, and H. T. Frost, "The Strength and Ductility of Intermetallic Compounds: Grain Size Effects," proceedings, Materials Research Society 81 (1985) 195-205.

6. R. N. Wright and V. K. Sikka, "Elevated Temperature Tensile Properties of Powder Metallurgy Ni_3Al Alloyed with Cr and Zr," (submitted for publication in J. Mater. Sci.).

Directionally Solidified /Single Crystal Alloys

Best Paper Award

The following paper "Creep Deformation Anisotropy in Single Crystal Superalloys," by P. Caron, Y. Ohta, Y.G. Nakagawa and T. Khan was selected by the Honors and Awards Subcommittee of the Seven Springs International Symposium Committee as the Best Paper of the Sixth Symposium. The criteria for judging all papers was based on technical excellence, originality, and pertinence to the superalloy and gas turbine industries.

CREEP DEFORMATION ANISOTROPY

IN SINGLE CRYSTAL SUPERALLOYS

P. Caron*, Y. Ohta**, Y.G. Nakagawa** and T. Khan*

* Office National d'Etudes et de Recherches Aérospatiales (ONERA)
B.P. 72, 92322 Châtillon Cedex, France
** IHI Research Institute
Toyosu, Koto-ku, Tokyo 135, Japan

Abstract

The anisotropic behaviour in creep of a number of single crystal superalloys (CMSX-2, alloy 454, MXON and CMSX-4) was investigated in the temperature range 760-1050°C as a function of the γ' particle size. The particle size is shown to have a spectacular effect both on the creep behaviour and stress rupture life at intermediate temperatures (760-850°C). In this temperature range, a γ' particle size of about 0.5 µm which results in optimum creep strength for the [001] orientation drastically reduces the stress rupture life for the [$\bar{1}$11] orientation. However, a very fine γ' particle size ($\approx$ 0.2 µm) leads to a dramatic improvement for the [$\bar{1}$11] orientation, which becomes the strongest. In the CMSX-2 alloy, for instance, the creep anisotropy between [001] and [$\bar{1}$11] can be totally eliminated for a γ' particle size of about 0.3 µm. The relationship between the creep behaviour of various single crystals and their deformation microstructures was analyzed in detail. At higher temperatures (950-1050°C) the effect of orientation and the particle size on creep strength is considerably reduced.

Introduction

Single crystal alloys for blade applications have now become an industrial reality. Because of their highly anisotropic properties, these materials pose unique challenges for use as structural materials. The orientation of single crystal blades is [001], which provides the best combination of properties that are required in the hot section of an aircraft turbine. Although the blade is essentially subjected to centrifugal loading along the [001] axis, some multiaxial stresses are generated locally due to the complex shape of the blade root and, in some cases, because of the presence of intricate cooling schemes in the airfoil section. Moreover, temperature gradients between different parts of the blade cause high thermal stresses along various directions. An important requirement for a proper and efficient design of single crystal blades is the development of appropriate constitutive models. Extensive work is being carried out both in Europe and USA to develop such models. The micro-mechanical approach for modelling requires in particular that the complex anisotropy of plastic deformation be fully identified and characterized. One of the major objectives of this paper is to perform a detailed study of the creep deformation anisotropy of various single crystal superalloys in the temperature range 760-1050°C as a function of precipitation heat treatments. A detailed investigation was carried out on the CMSX-2 alloy, but some results were also obtained on other alloys (CMSX-4, alloy 454, MXON) in order to propose a coherent explanation regarding the anisotropic creep behaviour of the single crystal superalloys.

Experimental procedures

The compositions of the different studied alloys are presented in the Table I. The CMSX-2 and CMSX-4 alloys were developed by Cannon-Muskegon Corporation (1,2) whereas alloy 454 was developed by United Technologies (3). The CMSX-4 alloy was chosen, since it contains rhenium, an element not commonly used in superalloys. The MXON experimental alloy was developed by ONERA (4). Dendritic single crystal rods or plates were grown by the withdrawal process using a seed or a selector. The crystallographic orientations of the single crystals were checked by the Laue X-ray back reflection technique. In the case of CMSX-2 and alloy 454, the selected crystal orientations fully covered the stereographic triangle. For the other alloys, the single crystals were grown only in the [001] and [$\bar{1}$11] orientations.

Table I. Chemical compositions (wt.%) of the single crystal alloys

Alloy	Ni	Co	Cr	W	Mo	Re	Al	Ti	Ta
CMSX-2	Base	5	8	8	0.6	-	5.5	1	6
CMSX-4*	Base	9.2	6.6	5.9	0.6	3.3	5.6	0.9	6.2
MXON	Base	5	8	8	2	-	6.1	-	6
Alloy 454	Base	5	10	4	-	-	5	1.5	12

* Chemical analysis at IHI

All single crystals were first heat treated at temperatures around 1300°C, followed by air quenching, in order to solutionize the γ/γ' eutectic pools and to partially homogenize the dendritic segregation. The CMSX-4 single crystals were subjected to a two-step solutioning heat treatment recommended by Cannon-Muskegon to reduce the volume fraction of residual γ/γ' eutectic pools. These single crystals were then subjected to different precipitation heat treatments resulting in various γ' particle mean sizes, ranging from 0.2 to 0.54 µm.

Constant load stress-rupture tests were performed in air in the temperature range 760-1050°C. Some creep tests were interrupted before rupture in order to prepare samples for transmission electron microscopy (TEM) examinations. Thin foils were prepared by twin-jet electropolishing using a solution of 45% butylcellosolve, 45% acetic acid and 10% perchloric acid at 0°C and 25 V. These foils were examined both at IHI and ONERA.

Experimental results

Creep properties at intermediate temperatures (760-850°C)

Previous work has shown that the creep behaviour of single crystal superalloys is strongly dependent on the orientation at intermediate temperatures (5-8). Another study by Caron and Khan (9) on the CMSX-2 alloy had demonstrated that the creep behaviour of [001] oriented single crystal could be modified by varying the size (and perhaps the morphology) of the strengthening γ' precipitates. It was therefore decided to undertake a detailed investigation on the effect of heat treatments on the creep behaviour of selected Ni-based single crystal alloys in various orientations.

Three different precipitation heat treatments were applied to the CMSX-2 single crystals in order to produce γ' sizes of 0.23, 0.3 and 0.45 μm respectively. Stress rupture lives as a function of orientation and heat treatments are shown in the stereographic triangles of Figure 1. The creep behaviour at 760°C for the three principal orientations [001], [$\bar{1}$11] and [011] is illustrated by the typical creep curves of Figure 2.

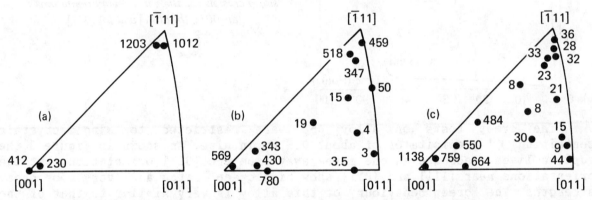

Figure 1 — Orientation dependence of the stress rupture life (in hours) at 760°C and 750 MPa of CMSX-2 single crystals, as a function of γ' precipitate size : (a) 0.23 μm, (b) 0.3 μm and (c) 0.45 μm.

In the [001] orientation, the longest creep lives are obtained with the largest precipitates (0.45 μm). We have previously shown that if the precipitate size is much larger than 0.5 μm the stress rupture lives are again substantially lowered at intermediate temperatures (10). A reduction of the precipitate size increases both the amplitude of the primary creep and the secondary creep rate, thereby resulting in a poor stress rupture life. On the other hand, the [$\bar{1}$11] oriented crystals containing γ' precipitates of 0.45 μm exhibit extremely poor creep strengths and large elongations to rupture due to an almost total absence of strain hardening.

Contrary to the case of [001] specimens, decreasing the γ' size led to a spectacular increase of the creep strength of the [$\bar{1}$11] crystals: a particle size of 0.23 μm, the smallest achieved in this investigation, led to a 30-fold increase in the stress rupture life at 760°C compared with that obtained with a particle size of 0.45 μm. The [$\bar{1}$11] single crystals containing the smallest particles exhibited very low creep rates and the rupture occurred after a small elongation. The single crystals oriented far away from the [001] and [$\bar{1}$11] exhibit short rupture lives, irrespective of the precipitation heat treatments, but the effect is much more drastic near the [$\bar{1}$11] orientation than near the [001] orientation. The creep behaviour is however dependent on the γ' size in the case of [011] oriented crystals: specimens containing 0.3 µm particles do not exhibit any strain hardening and the rupture occurs rapidly after large elongations, whereas those having the largest particles showed some strain hardening and lower elongations.

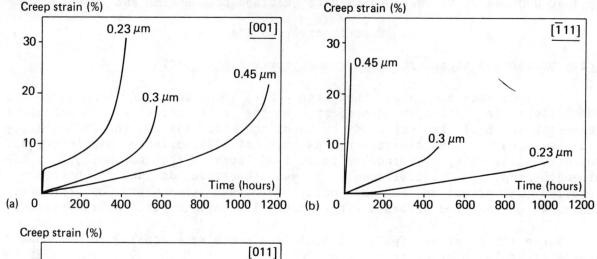

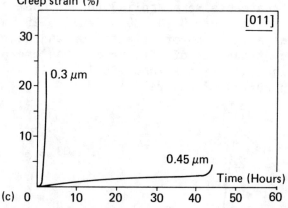

Figure 2 — Effect of γ' size on the creep behaviour at 760°C and 750 MPa of CMSX-2 single crystals for the three main orientations : (a) [001], (b) [1̄11] and (c) [011].

The creep tests on alloy 454 were restricted to single crystals containing γ' precipitates of about 0.5 µm in size. As shown in Figure 3 the rupture lives decrease as one moves away from the [001] orientation, but the orientations near [1̄11] and [011] show high creep rates and very poor creep strengths. The creep behaviour of this alloy is very similar to that of the CMSX-2 alloy containing 0.45 µm γ' particles. It is also worth mentioning that in both these alloys the stress rupture properties do not show any drastic drop up to misorientations of about 15° around [001].

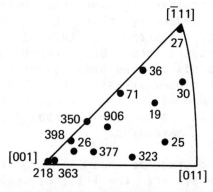

Figure 3 — Effect of orientation on the stress rupture life (in hours) at 760°C and 758 MPa of alloy 454 single crystal (γ' size = 0.5 µm).

In order to confirm the spectacular effect of the precipitation heat treatment on the creep behaviour of [1̄11] oriented crystals, creep tests were run at 760°C and 750 MPa on two other single crystal alloys, MXON and CMSX-4 containing γ' precipitates of two different sizes. Again, in the MXON alloy the highest creep strength along the [1̄11] axis was obtained with the finest γ' precipitates. Indeed, for the [1̄11] orientation at 760°C and 750 MPa the rupture life of specimens containing 0.2 µm precipitates is over 1800 hours, whereas the life with 0.36 µm precipitates drops to about 50 hours. For the

[001] oriented crystals the decrease in γ' precipitate size from 0.36 to 0.2 μm results in a substantial increase in the primary creep amplitude and a two-fold decrease in creep life as already observed in the CMSX-2 alloy.

The effect of precipitate size on creep is much more subtle in the case of Re-bearing alloy CMSX-4. Typical creep curves for the [001] and [īll] CMSX-4 specimens are shown in Figure 4. Increasing the γ' size from 0.22 to 0.34 μm again leads to a two-fold increase in the creep life along [001], by reducing the primary creep strain amplitude and lowering the secondary creep rate. The effect of γ' size on the creep life of [īll] crystals was however much less pronounced than for CMSX-2 and MXON alloys: [īll] single crystals of CMSX-4 containing precipitates with mean sizes of 0.22 and 0.34 μm showed comparable stress-rupture lives, although their creep behaviours were significantly different. So, by reducing the γ' size from 0.34 to 0.22 μm, the time for 1% creep increased by a factor of ten, and the minimum creep rate was divided by the same factor. Concurrently, the elongation to rupture became ten times smaller, the net result being that the rupture lives were not significantly affected by the γ' particle size. However, when the particle size in this alloy is about 0.5 μm, the creep strength of [īll] crystals is significantly reduced compared to the [īll] crystals containing smaller precipitates, but it remained much higher than those of the [īll] CMSX-2 and MXON crystals containing precipitates of comparable sizes.

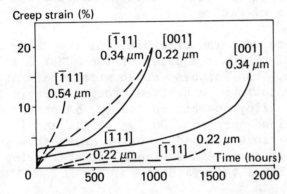

Figure 4 — Effect of γ' size on the creep behaviour at 760°C and 750 MPa of [001] and [īll] CMSX-4 single crystals.

The creep lives at 850°C and 500 MPa of the CMSX-2 and MXON single crystal alloys are still anisotropic, but the effects of orientation and heat treatments are less pronounced than at 760°C. For the CMSX-2 alloy containing 0.3 μm precipitates, the strongest orientation is [īll], while the [011] oriented crystals exhibited poorest creep lives. Increasing the γ' size to 0.45 μm improved the strength of [001] specimens by a factor of two, whereas the creep life of [īll] crystals was reduced by a factor of five. The creep tests performed on the MXON alloy confirmed the beneficial effect of decreasing the γ' size on the creep lives of [īll] single crystals, as shown in Figure 5. By reducing the precipitate size from 0.36 to 0.2 μm, the creep life of [īll] specimens was improved by a factor of ten, while preserving a high ductility.

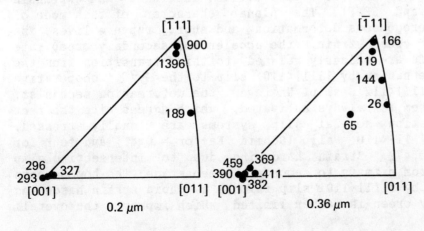

Figure 5 — Effect of orientation and γ' size on the stress rupture life at 850°C and 500 MPa of MXON single crystals.

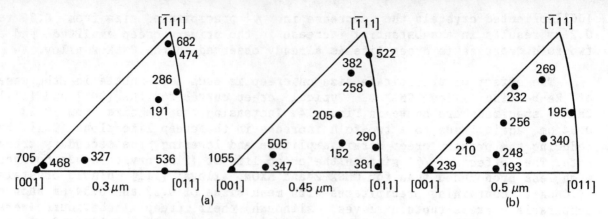

Figure 6 — Effect of orientation and γ' size on the high temperature stress rupture life of single crystals : (a) CMSX-2, 1050°C, 120 MPa ; (b) alloy 454, 980°C, 200 MPa.

Creep properties at high temperature (980-1050°C)

Creep tests were performed at 980°C and 200 MPa on alloy 454 single crystals containing 0.5 μm precipitates and at 1050°C and 120 MPa on CMSX-2 single crystals containing 0.3 and 0.45 μm γ' particles. The creep lives are reported in the stereographic triangles of Figure 6.

In this high temperature creep range, the creep life of the alloy 454 did not depend on the crystal orientation. In the case of the CMSX-2 alloy containing the largest precipitates, the highest creep strength is still obtained along [001], while other orientations exhibited creep lives in the range 150-500 hours. For the CMSX-2 alloy containing 0.3 μm γ' particles, the highest creep strength was obtained around [001], [$\bar{1}$11] and [011]. Orientations away from the main directions showed a weaker creep strength. The general creep behaviour at 1050°C of the CMSX-2 single crystal alloy is however much less anisotropic than at lower temperatures.

Discussion

The typical deformation microstructures generated during primary creep at 760°C and 750 MPa of [001] and [$\bar{1}$11] oriented CMSX-2 single crystals containing γ' precipitates with sizes in the range 0.22-0.45 μm are shown in Figures 7 and 8. In the [001] single crystals containing the largest precipitates, the deformation operated by homogeneous {111}<110> multiple slip in the matrix (Figure 7a), leading rapidly to the formation of dense dislocation networks at the γ/γ' interfaces. Reducing the γ' size to 0.22 μm promoted extensive cooperative shearing of the γ/γ' structure by {111}<112> slip (Fig. 7b). The precipitates are sheared by 1/3<112> dislocations with the creation of superlattice intrinsic and extrinsic stacking faults (SISF and SESF). The planar character of this mode of deformation promotes heterogeneous deformation and short rupture lives. The increase in the primary creep strain, the accelerated secondary creep rate and the reduced creep life are clearly related to this transition from the γ' by-passing Orowan mechanism by {111}<110> slip to the γ/γ' cooperative shearing mechanism by {111}<112> slip. Whatever the deformation mechanism, more than one slip system is always activated, which agrees with the fact that in the [001] orientation several slip systems are equally stressed, respectively eight for {111}<110> slip (Schmid Factor = 0.41) and four for {111}<112> slip (S.F. = 0.47). Strain hardening due to intersecting slip causes the transition from primary to secondary creep stage. The homogeneous character of deformation by {111}<110> slip favours a rapid strain hardening and the extent of primary creep is rather limited, which improves the overall creep behaviour.

 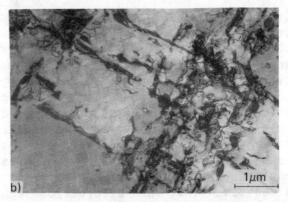

*Figure 7 — Dislocation structures during primary creep at 760°C and 750 MPa of [001] CMSX-2 single crystals :
(a) γ' size = 0.45 μm, ε = 0.2 % ; (b) γ' size = 0.23 μm, ε = 2 %.*

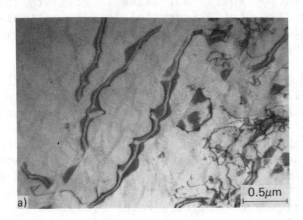 

*Figure 8 — Dislocation structures during primary creep at 760°C and 750 MPa of [$\bar{1}$11] CMSX-2 single crystals :
(a) γ' size = 0.23 μm, ε = 0.6 % ; (b) γ' size = 0.45 μm, ε = 1.5 %.*

Analysis of the creep deformation microstructures in the [011] CMSX-2 single crystals deformed at 760°C shows that the largest γ' particles favoured homogeneous (111)[$\bar{1}$01] *single slip* in the matrix. Moreover, a lot of precipitates are individually sheared by (111)[$\bar{2}$11] slip with the creation of superlattice stacking faults. When the γ' size is decreased to 0.3 μm, the deformation occured by (111)[$\bar{2}$11] extensive planar single slip through the γ/γ' structure. In this orientation, the Schmid Factors for these slip systems are the same as for the [001] orientation. Although the [011] orientation favours, theoretically, deformation on two equally stressed slip systems, the rotation of the single crystals induced by the creep deformation tends to move the tensile axis away from [011] and favours single slip. In both cases, the creep life of the single crystals is short due to the absence of strain hardening. The specimens containing the smallest precipitates, however, always showed the highest creep rate due to the operation of extensive heterogeneous (111)[$\bar{2}$11] slip.

The same heterogeneous deformation mechanism is involved during the primary creep of [$\bar{1}$11] CMSX-2 single crystals containing the smallest precipitates (0.23 μm) (Fig. 8a). The deformation is however limited to the (111) primary slip plane. These specimens showed a better creep strength than the corresponding [001] crystals, which deform by the same creep deformation mechanism. The Schmid Factor for these {111}<112> slip systems is clearly smaller for [$\bar{1}$11] than for [001] (respectively 0.31 and 0.47). The smaller resolved shear stress thus justifies the lower creep rate exhibited by the [$\bar{1}$11] oriented single crystals, even if the deformation operates by single slip, as observed during the initial stage of creep. This argument was previously advanced to explain the optimum creep strength of [$\bar{1}$11] Mar-M 247 single crystals at 774°C (6). Increasing the precipitate mean size to 0.3 μm promoted Orowan bowing of matrix dislocations between the γ' particles, instead of shearing by {111}<112> slip, thereby resulting in an increase in the creep rate. During the primary creep, the deformation is inhomogeneous, some areas being free of dislocations, which suggests that there is a strong resistance to deformation by Orowan by-passing. An increase in the γ' size to 0.45 μm decreases the Orowan stress, which leads to a homogeneous deformation structure in the matrix (Fig. 8b). Despite the fact that [$\bar{1}$11] is a multiple slip orientation (six equally stressed {111}<110> slip systems with F.S. = 0.275), careful analysis of the dislocations in near-[$\bar{1}$11] oriented crept specimens showed that the primary creep deformation operated mainly by coplanar slip in the primary slip plane (111) along the [$\bar{1}$10] and [$\bar{1}$01] directions (8). This deformation mode causes only a weak strain hardening, which explains the high creep rates observed in this case. The absence of secondary creep stage seems to indicate that the first activated slip system remains predominant throughout the life of the [$\bar{1}$11] single crystals. The relatively low value of the Schmid Factor would not allow to activate the conjugate slip systems contrary to the case of [001] crystals, where multiple slip is operative at the beginning of the primary creep stage. No cube slip was observed for this orientation at 760°C.

T.E.M. analysis performed on MXON, CMSX-4 and alloy 454 single crystals crept at 760°C showed the same overall features as in CMSX-2. For instance, decreasing the γ' size to 0.22 μm in the CMSX-4 alloy promotes heterogeneous deformation by {111}<112> planar slip instead of homogeneous {111}<110> slip in the matrix. This creep mechanism transition gave rise to the same effects as in CMSX-2, i.e. increase of the primary creep strain along [001] and decrease of the secondary creep rate along [$\bar{1}$11]. In the case of [$\bar{1}$11] CMSX-4 single crystals with the finest precipitates, the fracture surface was planar and parallel to (111). The low elongation to rupture which resulted in a short rupture life was therefore due to the planar nature of deformation. On the other hand, the alloy 454 single crystals containing precipitates with a mean size of 0.5 μm deformed mainly by {111}<110> slip in the matrix, which is in accordance with the good creep behaviour of the [001] specimens and the very poor creep strength along [$\bar{1}$11], as explained in the case of the CMSX-2 crystals containing the largest precipitates.

The dislocation structures observed after creep at 850°C and 500 MPa in the CMSX-2, CMSX-4 and MXON [001] single crystals are homogeneous independently of the size of the precipitates. The deformation operated mainly in the γ matrix by {111}<110> slip. Moreover, some γ' particles were shown to be individually sheared by {111}<112> slip or by pairs of ½<110> dislocations. Extensive shearing of the γ/γ' structure by {111}<112> slip was never observed at 850°C and 500 MPa, which explains the limited amplitude of the primary creep strain even in the case of [001] crystals containing the smallest precipitates. The better creep strength of the [$\bar{1}$11] CMSX-2 and MXON single crystals containing the finest precipitates, compared to the corresponding [001] oriented specimens, may be explained by the reduced value of the Schmid Factor for {111}<110> slip, i.e. 0.275 vs. 0.41. Again, the [011] orientation is weak due to operation of single slip. The improvement of the creep strength of the [001] CMSX-2 single crystals as the precipitate size increases from 0.3 to 0.45 has been shown previously to be related to the reduction of the kinetics of dislocation climb during the secondary creep stage.[11] In the case of [$\bar{1}$11], the reduction of the creep strength when the γ' size increases can be interpreted using arguments similar to those advanced to explain the creep behaviour at 760°C. However, the major difference at 850°C, compared with 760°C, is that the deformation occurs in the matrix by {111}<110> slip, even in the case of the finest precipitates.

The main feature observed during the high temperature creep of nickel-based superalloys is the rapid coarsening of the strengthening γ' particles.

Several studies have shown that, in [001] single crystal superalloys where the γ/γ' misfit is negative at the testing temperature, γ' rafts develop normal to the tensile axis. The low creep rates and long rupture lives observed in these cases are generally thought to be due to this particular morphology of the γ' phase, which inhibits dislocation climb around them. Such γ' rafts normal to the [001] stress axis have previously been observed in CMSX-2 (9), CMSX-4 (2) and MXON (4) single crystals during creep at high temperature. Additional observations made on a [$\bar{1}$11] oriented CMSX-2 single crystal crept at 1050°C show that the precipitates apparently coarsen parallel to the three {001} planes, which results in irregular shaped coarse γ' particles. In view of the fact that, in general, the creep strengths are not stronly dependent on orientation one may question the so-called beneficial role played by the rafted γ' morphology for [001] oriented specimens at high temperature. Certainly, careful further work is required to elucidate this problem.

Conclusions

This investigation shows some spectacular effects of the crystal orientation and heat treatments on the creep behaviour and strength of a number of single crystal superalloys. The salient features can be summed up as follows:

1. At intermediate temperatures (760-850°C), the creep behaviour of nickel-base single crystal superalloys is extremely sensitive to the orientation and γ' precipitate size. For a γ' size in the range 0.35-0.5 µm, the highest creep strength is obtained near [001], whereas orientations near the [111]-[011] boundary of the standard stereographic triangle exhibit very short creep lives. When the γ' size decreases to 0.2 µm, the longest creep lives are exhibited, in decreasing order, by the crystals oriented near [$\bar{1}$11], [001] and [110]. The anisotropy in creep between the [001] and [$\bar{1}$11] orientations can therefore be reduced by appropriate precipitation heat treatments. The creep strengths, however, remain extremely poor near the [011] orientation.

2. At high temperatures (980-1050°C), the creep behaviour of the single crystal superalloys is much less sensitive to orientation and γ' size than at intermediate temperatures. The [001] oriented single crystals develop a rafted γ' structure normal to the tensile stress axis, while the γ' precipitates coarsen rather irregularly in the [$\bar{1}$11] specimens.

References

1. K. Harris, G.L. Erickson, and R.E. Schwer, "CMSX Single Crystal, CM DS & Integral Wheels, Properties and Performance", High Temperature Alloys for Gas Turbines and Other Applications 1986, ed. W. Betz & al. (Dordrecht, Holland: D. Reidel Publishing Company, 1986), 709-728.

2. K. Harris, G.L. Erickson, and R.E. Schwer, "Development of an Ultra High Strength Single Crystal Superalloy CMSX-4 for Small Gas Turbines" (Paper presented at the 1983 TMS-AIME Fall Meeting, Philadelphia, Pennsylvania, 3 October 1983).

3. M. Gell, D.N. Duhl, and A.F. Giamei, "The Development of Single Crystal Superalloy Turbine Blades", Proc. of the Fourth International Symposium on Superalloys, ed. J.K. Tien & al. (Metals Park, OH: ASM, 1980), 205-214.

4. T. Khan, P. Caron, and C. Duret, "The Development and Characterization of a High Performance Experimental Single Crystal Superalloy", Proc. of the Fifth International Symposium on Superalloys, ed. M. Gell, C.S. Kortovich, R.H. Bricknell, W.B. Kent, and J.F. Radavich (Warrendale, PA: The Metallurgical Society of AIME, 1984), 145-155.

5. B.H. Kear, and B.J. Piearcey, "Tensile and Creep Properties of Single Crystals of the Nickel-Base Superalloy Mar-M200", Trans. TMS-AIME, 239 (1967) 1209-1215.

6. R.A. MacKay, and R.D. Maier, "The Influence of Orientation on the Stress Rupture Properties of Nickel-Base Superalloy Single Crystals", Met. Trans. A, 13A (1982) 1747-1754.

7. M.R. Winstone, "The Effect of Orientation on the Properties of Single Crystals of the Nickel Superalloy Mar-002" (NGTE Memorandum M81017, 1981).

8. P. Caron, T. Khan, and Y.G. Nakagawa, "Effect of Orientation on the Intermediate Temperature Creep Behaviour of Ni-Base Single Crystal Superalloys, Scripta Met., 20 (1986) 499.

9. P. Caron, and T. Khan, "Improvement of Creep Strength in a Nickel-Base Single Crystal Superalloy by Heat Treatment", Mat. Sci. and Eng., 61 (1983) 173-184.

10. T. Khan, P. Caron, D. Fournier, and K. Harris, "Single Crystal Superalloys for Turbine Blades", Matériaux et Techniques, 10-11 (1985) 567-578.

11. P. Caron, P.J. Henderson, T. Khan, and M. McLean, "On the Effects of Heat Treatments on the Creep Behaviour of a Single Crystal Superalloy", Scripta Met., 20 (1986) 875-880.

HIGH PERFORMANCE SINGLE CRYSTAL SUPERALLOYS

DEVELOPED BY THE d-ELECTRONS CONCEPT

N.Yukawa, M.Morinaga, Y.Murata, H.Ezaki, and S.Inoue[*]

Toyohashi University of Technology, Toyohashi, Aichi 440, Japan

*Numazu College of Technology, Numazu, Shizuoka, 410, Japan

Abstract

New single crystal superalloys of Ni-10at%Cr-12at%Al-Ti-Ta-W-Mo-Re have been developed by the d-electrons concept, which was devised on the basis of molecular orbital calculations of Ni based alloys. The process for the design of single crystal superalloys introduced in our previous study was modified to a more simple one and applied to the present study. The alloys were designed so as to have an optimum combination of the properties such as high temperature strength and ductility, hot corrosion resistance, density and cost of the alloys, while holding good fabricability and phase stability. The d-orbital energy level (Md) of alloying elements was utilized in order to predict and control the occurence of the undesired eutectic γ' and the topologically close packed (TCP) phases in alloys. The alloying effect of Cr, Ti, Al, Re and Co on the properties of single crystal alloys was investigated systematically. Based on these results several single crystal superalloys were designed. Some of them have well balanced properties, compared to the various alloys so far developed. In addition, several important factors were discussed for strengthening single crystal superalloys.

Introduction

A number of single crystal superalloys have been developed in several countries during the past decade (1). However, these development seems to be relied on several empirical methods. Further advances in the method of alloy design are necessary for the future improvement in the temperature capability of single crystal turbine blades and vanes (1,2).

Recently we have developed a new d-electrons concept which is based on the theoretical calculations of the electronic structure of alloys (3-5), and applied to the design of superalloys (6-8). In a previous study, a new process was proposed for the design of single crystal superalloys (8). In this process, fabricating limitations of single crystals are considered in addition to the phase stability limitations of alloys. Following this design process, several single crystal superalloys with a long creep-rupture life and high corrosion resistance were developed successfully.

In the present study, a more simplified method was devised and applied to the design of high performance single crystal superalloys. Some developed alloys exhibited an optimum balance in the properties which are required for the high temperature applications.

Experimental Procedure

Ni-Cr-Al-Ti-Ta-W-Mo-Re-Co alloys with a variety of compositions were investigated together with some reference alloys listed in Table 1.

Table 1 Chemical compositions of reference alloys.

Alloy	Ni	Cr	Al	Ti	Ta	W	Mo	Re	Co	Country
PWA 1480	bal.	10.0	5.0	1.5	12.0	4.0	–	–	5.0	U.S.A.
NASAIR 100	bal.	9.0	5.8	1.2	3.3	10.5	1.0	–	–	U.S.A.
CMSX-2	bal.	8.0	5.6	1.0	6.0	8.0	0.6	–	4.6	U.S.A.
CMSX-4	bal.	6.6	5.6	1.0	6.5	6.4	0.6	3.0	9.6	U.S.A.
MXON	bal.	8.0	6.1	–	6.0	8.0	2.0	–	5.0	France
TMS-1	bal.	5.5	5.2	–	5.1	16.6	–	–	7.5	Japan
TMS-12·1	bal.	6.6	4.6	–	8.1	13.5	–	–	–	Japan
SC-53A	bal.	6.9	5.5	–	5.9	5.9	5.1	–	–	Japan

Both poly- and single-crystal specimens of these alloys were prepared. Poly-crystal specimens were prepared by a tri-arc furnace. Single crystal specimens were grown by the withdrawn process at a growth rate of 10cm/h and a temperature gradient of about 120°C/cm. The single crystal bars of 11mm in diameter was obtained, and its axial direction was near the [001] direction; the deviation from this direction was less than 10° for all the crystals used for the present investigation.

A series of experiments was carried out with these specimens after heat-treatments, 1300°C/4hrs, B.A.C. + 1050°C/16hrs + 850°C/48hrs, A.C.; here, B.A.C. means the blast air cooling.
1) Solidus and solvus temperatures and melting range of alloys, ΔT, were measured using a differential thermal analysis (DTA). The magnitude of heat-treatment window (H.T.W.) that is defined by the difference, ($T_{solidus} - T_{\gamma' solvus}$), were then determined.
2) Volume fraction of the γ' phase and the partitioning ratios of alloying elements between the γ and γ' phases were measured using an EPMA. This analysis was carried out with the residues extracted from as-grown specimens.

The extraction was performed using a potentiostatic apparatus under the condition of a constant potential in an 1% ammonium sulfate - 1% tartaric acid aqueous solution.

3) Hot corrosion resistance was examined using single crystal specimens (5×10×1mm), surface-coated with an Na_2SO_4 - 25%NaCl salt by the amount of 20mg/cm^2. The thermogravimetric measurement was performed at 900°C in the flow of air (50cc/min). The hot corrosion index (H.C.I.) was then defined as the weight gain of the specimen (mg/cm^2) after 20 hrs exposures, and used as a measure of the hot corrosion resistance. The corrosion resistance of alloys increases with decreasing this index.

4) Creep-rupture strength was tested at 1040°C under a constant load of 14kgf/cm^2.

5) Tensile strength was also examined at 982°C.

6) Precipitation behaviors of the γ', eutectic γ', α-W and topologically close packed (TCP; μ or σ) phases were investigated. In order to examine phase stability with respect to the α-W and TCP phases in the γ phase, both poly- and single-crystal specimens were aged for 500hrs at 982°C after the solution-treatment at 1300°C.

7) Densities of alloys were computed according to the Hull's regression equation (9).

8) Cost of alloys was calculated using the following equation:

$$\text{Cost of alloy} = (0.7(\text{wt\%Ni}) + 1.3(\text{wt\%Cr}) + 0.35(\text{wt\%Al}) \\ + 2.0(\text{wt\%Ti}) + 35.0(\text{wt\%Ta}) + 6.0(\text{wt\%W}) \\ + 0.5(\text{wt\%Mo}) + 100.0(\text{wt\%Re}) + 3.0(\text{wt\%Co}))/100, \text{yen/g}, \quad (1)$$

where the coefficient in this equation indicates the price of each metal (yen/g) in June 1987 in Japan. The price might be changed by many factors, but for comparison it was fixed in this study.

Alloy Design

In the alloy design devised in our previous study (8), two types of limitations were considered. The one is a limitation for the fabrication, and the another is a limitation for the phase stability which can be treated by the d-electrons concept. Following this design process, the optimum compositional field was obtained for the single crystal superalloys of Ni-10at%Cr-12at%Al-1.5at%Ti-Ta-W-Mo. The result is represented in the coordinates of Ta and W+(Mo) content, as shown Fig.1.

In the figure, the fabricating limitation was shown by the two lines, $\Delta T \leq 50°C$ and $H.T.W. \geq 20°C$. The former limitation is necessary to grow single crystal superalloys, and the latter limitation is required to permit a solution-treatment process.

The phase stability limitation was shown by three lines. Two of them were expressed by the average d-orbital level of alloying transition metal, $\overline{Md}$, and that is defined as,

$$\overline{Md} = \Sigma X_i (Md)_i. \quad (2)$$

Here, X_i and $(Md)_i$ are the atomic fraction and Md value of i element, respectively. Table 2 shows Md values together with the values of the another parameter, bond order between M and Ni (called Bo), which will be discussed later.

The microstructural observation was performed for the alloys which were solidified in a controlled way through the single crystal growth or the constant cooling (5°C/min) in the DTA experiment. It was found that the eutectic γ′ phase disappeared by the solution-treatment at 1300°C when its volume fraction was less than 2%. (In the case of the arc-melted specimens, the boundary increased to 4%.) This condition could be expressed using an inequality relationship, $\overline{Md}t \leq 0.985$, where $\overline{Md}t$ is the $\overline{Md}$ value calculated from the total alloy composition.

Table 2 List of Md and Bo values.

	Element	Md	Bo		Element	Md	Bo
	Ti	2.271	1.098		Zr	2.944	1.479
	V	1.543	1.141	4d	Nb	2.117	1.594
	Cr	1.142	1.278		Mo	1.550	1.611
	Mn	0.957	1.001		Hf	3.020	1.518
3d	Fe	0.858	0.857	5d	Ta	2.224	1.670
	Co	0.777	0.697		W	1.655	1.730
	Ni	0.717	0.514		Re	1.267	1.692
	Cu	0.615	0.272	Others	Al	1.900	0.533
					Si	1.900	0.589

The another $\overline{Md}\gamma$ was calculated from the composition of the residual γ matrix following the same process described in the previous study (8). Microstructural observation of alloys aged for 500hrs at 982°C revealed that the TCP phase (in this case, μ phase) could be suppressed when $\overline{Md}\gamma \leq 0.93$, and hence this is the another limitation for phase stability.

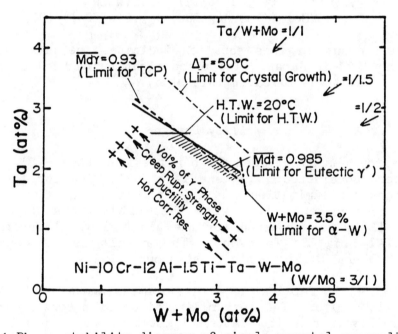

Fig.1 Phase stability diagram of single crystal superalloys.

In the previous study, the condition, $\overline{Md}\gamma(Ta+W) \leq 0.105$, was used as the limitation for the precipitation of the α-W phase. In the subsequent study, however, it was found that W+Mo ≤ 3.5at% was the better limitation for Ni-10at%Cr-12at%Al-Ti-Ta-W-Mo alloys, then it was used in this study.

As can be seen in Fig.1, both lines, $\overline{Md}t=0.985$ and $\overline{Md}\gamma=0.93$, were overlapped, and may be considered practically as one line. Therefore, only the $\overline{Md}t$ was employed in the present alloy design in order to represent the phase stability. This criterion seems to be applicable for the alloys containing 10∼13at% Cr. When the Cr content decreases to 9∼7at%, the $\overline{Md}t$ increases to 0.986, as is seen in CMSX-4.

In the figure, trend of the changes in alloy properties with Ta and W+Mo contents is also illustrated. The vol% of the γ´ phase increased with Ta content, and its maximum value was 67vol%. Creep-rupture strength increased with W+Mo content. By contrary, ductility decreased with W+Mo content. In addition, hot corrosion resistance increased with Ta content. From the above trend, it was concluded that these properties could be controlled by the Ta/W+Mo ratio. (This ratio was calculated in at%.) In the present study, alloys were designed in the range of Ta/W+Mo atomic ratio from 1/1 to 1/2. The optimum compositional field of the alloys is indicated by the shaded area in Fig.1.

The process of alloy design employed in this study is shown in Fig.2. W/Mo atomic ratio was kept at 3/1 in the calculation. Density and cost of alloys were also calculated together with the γ´ vol%, the compositions of the both γ and γ´ phases, and the $\overline{Md}_\gamma$ value.

For the development of single crystsl superalloys, the following five major properties were considered: 1) creep-rupture life, 2) creep-rupture elongation, 3) H.C.I., 4) density, 5) materials cost. An optimum combination of these properties was searched for in the present design.

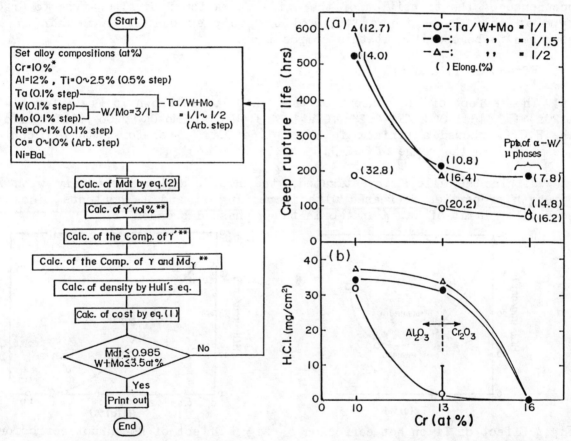

Fig.2 Flow chart for the design of single crystal superalloys.
* In this study, Cr content was changed up to 16%.
**For the calculation procedure, see Ref.(8).

Fig.3 Effect of Cr on (a) the creep-rupture strength and (b) the hot corrosion resistance of Ni-10~16Cr-12Al-1.2Ti-1.2~2.7Ta-1.2~2.9W-0.4~1.0Mo (at%) alloys.

Alloying Effects on the Properties of Single Crystal Superalloys

The alloying effects of elements on the properties of single crystal superalloys were investigated for the designed alloys and some reference alloys. Here, only the relevant results are outlined. The detailed explanation on the hot corrosion resistance will be given elsewhere (10).

Effect of Cr and Ta/W+Mo atomic ratio

Fig.3 shows the effect of Cr content on the creep-rupture properties and the hot corrosion resistance of Ni-10∼16Cr-12Al-1.2Ti-1.2∼2.7Ta-1.2∼2.9W-0.4∼1.0Mo (at%) alloys. These experimental alloys had different Ta/W+Mo atomic ratios of 1/1, 1/1.5, and 1/2.

As shown in Fig.3 (a), creep-rupture life decreased with Cr content and with Ta/W+Mo atomic ratio except for the alloys with higher Cr content. Rupture elongation indicated by the number in parenthesis decreased with Cr content except for some alloys with the Ta/W+Mo atomic ratio of 1/2, but increased with Ta/W+Mo atomic ratio. As is shown in the figure, the precipitation of the α-W and μ phases was observed in the 16at%Cr alloys. Relatively small elongation of these alloys may be attributed to the precipitation of these phases. Fig.3 (b) shows the change of H.C.I. with Cr content. The H.C.I. decreased with Cr content, and became zero at 16at%Cr. The H.C.I. also decreased with Ta/W+Mo atomic ratio. Its decrease was remarkable for the alloys with the Ta/W+Mo atomic ratio of 1/1. Data was scattered at 13at%Cr, as shown by an error bar. This may be due to the occurence of the transition of the oxide formation from Al_2O_3-type to Cr_2O_3-type at this Cr content (11), and the complex oxidation mechanism was supposed to be operated at this composition.

Effect of Ti and Al

The effect of Ti content on the H.C.I. is shown in Fig.4 for Ni-10∼12Cr-11.5∼13Al-0∼2.5Ti-1.2∼4.0Ta-1.3∼3.5W-0∼0.6Mo-0∼5.2Co (at%) alloys. The H.C.I. showed a minimum at about 1.5at%Ti. Therefore, the Ti content was varied in the range of 1.2∼1.5 at% in this design.

Within the alloys investigated, the H.C.I. changed scarcely with Al. Strength properties decreased with increasing Ti and Al contents since the solubility limit of Ta, W and Mo in the γ phase decreased.

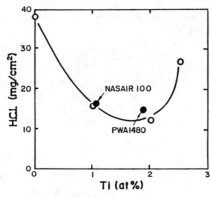
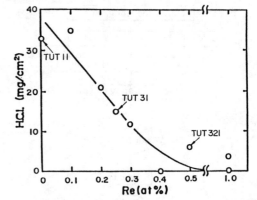

Fig.4 Effect of Ti on hot corrosion resistance of Ni-10∼12Cr-11.5∼13Al-0∼2.5Ti-1.2∼4.0Ta-1.3∼3.5W-0∼0.6Mo-0∼5.2Co (at%) alloys.

Fig.5 Effect of Re on hot corrosion resistance of Ni-9∼12Cr-11∼12Al-1.2∼2.5Ti-1.1∼2.7Ta-1.9∼3.8W-0.64∼0.67Mo-0.1∼1.0Re (at%) alloys.

Effect of Re

Fig.5 shows the effect of the Re addition on the H.C.I. of Ni-9∼12Cr-11∼12Al-1.2∼2.5Ti-1.1∼2.7Ta-1.9∼3.8W-0.64∼0.67Mo-0∼1.0Re (at%) alloys. The addition of Re to the alloys up to 1at% reduced the H.C.I. remarkably. However, the Re addition more than 0.25at% into the alloys containing 10at%Cr caused the phase instability. The eutectic γ' phase never disappeared after solution-treatments, and also the excess precipitation of the α-W phase took place. As the result, rupture life and elongation decreased greatly. Thus, the maximum amount of Re was considerd to be 0.25at%.

Effect of Co

The effect of Co on the single crystal superalloys was investigated by substituting Co for Ni in TUT 92 and CMSX-4. The substitution was made so that the alloys have the same $\overline{M}dt$ value or the same total W+Mo content as TUT 92. The results are summarized in Table 3. The chemical compositions, W+Mo content, Ta/W+Mo atomic ratio, $\overline{M}dt$, cost, density, creep-rupture properties and the H.C.I. were listed in the table.

Table 3 Effect of Co on the properties of TUT 92 and CMSX-4.

Alloy	Composition, wt%								
	Ni	Cr	Al	Ti	Ta	W	Mo	Re	Co
TUT 92	bal.	8.7	5.4	1.2	6.2	7.1	1.2	0.8	–
TUT 921	bal.	8.7	5.4	1.2	6.2	7.1	1.2	0.8	8.8
TUT 922	bal.	8.7	5.4	1.2	5.7	6.5	1.1	0.8	8.9
CMSX-4	bal.	6.6	5.6	1.0	6.5	6.4	0.6	3.0	9.6
CMSX-4 (0Co)	bal.	6.5	5.6	1.0	7.1	7.0	0.7	3.0	–

Alloy	W+Mo at%	Ta/W+Mo AT.ratio	$\overline{M}dt$	Dens. g/cm^3	Cost yen/g	Creep-Rup. life(h)	El.(%)	H.C.I. mg/cm^2
TUT92	3.1	1/1.5	0.985	8.62	4.1	950	10.0	13.3
TUT921	3.1	1/1.5	0.990	8.60	4.3	680	8.7	6.0
TUT922	2.8	1/1.5	0.985	8.54	4.1	570	17.0	3.9
CMSX-4	2.5	1/1.1	0.986	8.78	6.5	610	19.5	0.0
CMSX-4 (0Co)	2.8	1/1.1	0.986	8.86	6.5	950	17.3	15.2

For TUT 921, containing 8.8wt%(9at%)Co but the same W+Mo content (3.1at%) as TUT 92, a considerable decrease was observed in the rupture life, the rupture elongation and H.C.I. index. This was interpreted as due to the eutectic γ' phase which was retained even after heat-treatments probably owing to the higher $\overline{M}dt$ value (0.990) than the critical value of 0.985 (see Fig.1).

Also, as is seen in TUT 922, when the Co addition was made under the same $\overline{M}dt$ value, rupture life and the H.C.I. further decreased, whereas the rupture elongation increased. By contrary, the elimination of 9.6wt%(9.9at%)Co from CMSX-4, resulted in the increase of rupture life and H.C.I., but caused the slight decrease of rupture elongation.

From these results it is concluded that the addition of Co decreases the solubility limit of Ta, W and Mo in the alloy, which led to the decrease of the strength properties. On the other hand, hot corrosion resistance of the alloy increased to some extent with the addition of Co.

The Properties of Developed Single Crystal Superalloys

The compositions and the properties of some developed alloys are shown in Table 4. All of these alloys consisted of homogeneous microstructure after the heat-treatment. Except for TUT 321, no precipitation of the μ and α-W phases was observed after the aging for 500hrs at 982°C or after the creep-rupture test.

Creep-rupture life increases with increasing W+Mo content and with decreasing Ta/W+Mo atomic ratio. The addition of Re drastically increases creep-rupture life and hot corrosin resistance, but this also causes the increase of density and materials cost.

Table 4 Chemical compositions and the properties of developed alloys.

Alloy	Composition, wt%								W+Mo at%	Ta/W+Mo AT.ratio	Dens. g/cm^3	Cost yen/g	Creep-Rup.		H.C.I. mg/cm^2
	Ni	Cr	Al	Ti	Ta	W	Mo	Re					life hrs	El. %	
TUT 11	bal.	8.6	5.4	1.0	8.1	6.2	1.1	–	2.7	1/1	8.60	3.9	282	15.9	33.5
TUT 31	bal.	8.6	5.4	1.0	7.9	6.0	1.0	0.8	2.6	1/1	8.65	4.6	1700	7.0	15.4
TUT 321*	bal.	8.6	5.4	1.0	7.7	5.9	1.0	1.5	2.6	1/1	8.71	5.3	674	14.4	6.4
TUT 22	bal.	8.7	5.4	1.2	7.5	5.7	1.0	0.2	2.5	1/1	8.56	3.8	219	16.6	22.7
TUT 52	bal.	8.7	5.4	1.2	6.9	6.5	1.1	0.2	2.7	1/1.25	8.57	3.7	574	9.9	27.5
TUT 82	bal.	8.7	5.4	1.2	6.3	7.2	1.3	0.2	3.1	1/1.5	8.58	3.5	652	7.4	36.4
TUT 92	bal.	8.7	5.4	1.2	6.2	7.1	1.2	0.8	3.1	1/1.5	8.62	4.1	950	10.0	13.3

* Precipitation of the μ and α-W phases was found after the aging and creep-rupture test.

In Fig.6, five properties of the developed alloys, TUT 31 and 92, are illustrated using a pentagonal axis, and compared with the data of typical reference alloys. It is apparent that TUT 92 exhibited the more balanced properties than the reference alloys. As for the reference alloys, it is found that creep-rupture life increases with W+Mo content. Poor hot corrosion resistance of MXON (12) and TMS (13,14) alloys may be attributed partially to lack of Ti. As for TMS alloys, low Cr and high W+Mo contents, and low values of Ta/W+Mo atomic ratio probably cause the very poor hot corrosion resistance though they exhibited excellent strength properties. The addition of 3wt%Re in CMSX-4 results in excellent hot corrosion resistance in spite of the reduced Cr content of 6.6wt%. However, it is penalized by the increase of cost and density.

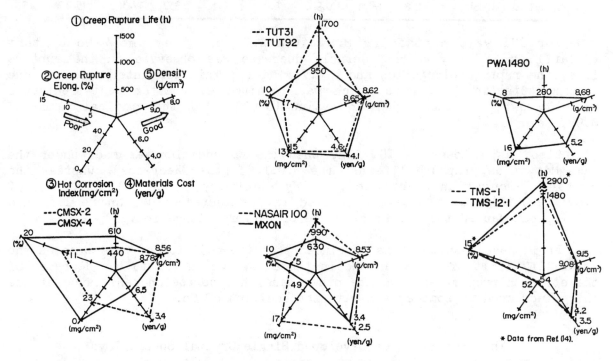

Fig.6 Comparison of the five properties of typical developed alloys with those of the reference alloys.

Discussion

It was demonstrated that a new method utilized the $\overline{Mdt}$ is a nice tool for the design of single crystal superalloys. In Fig.7 measured solvus temperatures are plotted as a function of $\overline{Mdt}$ for Ni-10/11Cr-12/14Al-0/2Ti-1∼4Ta-1.25∼3.50W-0.81∼0.88Mo-0∼7.5Co (at%) alloys and some of reference alloys. The vol% of the eutectic γ′ phase in alloys solidified at a constant cooling rate of 5°C/min is also shown in the figure. The solvus temperature increases linearly with $\overline{Mdt}$ up to 0.985 and appears to stay constant for higher $\overline{Mdt}$. Vol% of the eutectic γ′ increases with $\overline{Mdt}$. As stated before, to devoid of this phase after solution-treatments the critical amount should be less than 2% and this condition is achieved around the $\overline{Mdt}$ value of 0.985. So, this criterion for $\overline{Mdt}$ was employed in the present design.

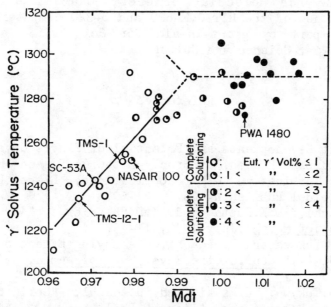

Fig.7 Relation between volume % of the eutectic γ′ phase and γ′ solvus-temperature of Ni-10/11Cr-12/14Al-0/2Ti-1∼4Ta-1.25∼3.50W-0.81∼0.88Mo-0∼7.5Co (at%) alloys and the reference alloys.

Many guidlines based on the concepts of physical metallurgy have been proposed for the design of high strength superalloys. For example, the lattice misfit between the γ and γ′ phases was often taken into account in the alloy design (13). As described in Fig.1, however, the trend of the change in alloy properties with Ta and W+Mo contents implied that the solution strengthening of the γ phase is the most crucial factor for strengthening single crystal superalloys. The effect of the γ/γ′ lattice misfit on the strengthening, of course, can not be neglected, but this appears to be subsidiary compared to the strengthening effect of the γ phase through the solutioning of refractory metals such as W, Mo and Re. This effect may be related to the magnitude of bond order (Bo). As shown in Table 2, these elements have high covalent bond-strength with Ni, and they are preferably partitioned to the γ phase and effective in strengthening it. The elimination of Co resulted in the decrease of $\overline{Mdt}$, which allows the further solutioning of W, Mo and Re into the matrix and strengthens the alloy. This is in good agreement with the experimental results for MXON alloy (12). Utilizing this concept, it is possible to tailor single crystal alloy compositions for specific applications.

Conclusions

Based on an electronic theory, a new design process for single crystal supperalloys was devised and applied to the development of high performance single crystal alloys. It was concluded that the d-electrons concept makes the alloy design more efficient and accurate compared to the currently used methods standing on several empirical rules and many trial-and-error experiments.

Acknowledgements

We would like to express sincere thanks to Prof. H. Adachi of Hyogo University of Teacher Education and to Prof. M.Kawakami of Toyohashi University of Technology for their considerable advice, and to Messrs. K.Matsugi, S.Yasui, A.Suzuki, M.Nishimura, S.Samori, A.Baba, T.Tange and C.Ikeda for their experimental assistance. We also acknowledge the Computer Center, Institute for Molecular Science, Okazaki National Research Institute, for the use of the HITAC M-680 and S-820 computers. This research was supported in part by grant-in-aid for Scientific Research from the Ministay of Education, Science and Culture.

References

1. T.Kahn, "Recent Developments and Potential of Single Crystal Superalloys for Advanced Turbine Blades, " Proc. of the Conf. on High Temperature Alloys for Gas Turbines and Other Applications, Liege, Belgium, Oct. 6-9, 1986, (D. Reidel Publishing Co.), 21-50.
2. M.Gell, D.N.Duhl, D.K.Gupta and K.D.Sheffler, "Advanced Superalloy Airfoils," Jour.of Metals, July, 1987, 11-15.
3. M.Morinaga, N. Yukawa and H. Adachi, "Alloying Effect on the Electronic Structure of Ni_3Al," Jour.of the Phys. Soc. of Japan, 53(2) (1984), 653-663.
4. M.Morinaga, N.Yukawa, H.Ezaki and H.Adachi, "Solid Solubilities in Transition-Metal-Based f.c.c. Alloys," Phil. Mag., A, 51(2) (1985), 223-246.
5. M.Morinaga, N.Yukawa, H.Ezaki and H.Adachi, "Solid Solubilities in Nickel-Based f.c.c. Alloys," Phil. Mag., A, 51(2) (1985), 247-252.
6. M.Morinaga, N.Yukawa, H.Adachi and H.Ezaki, " New PHACOMP and its Application to Alloy Design , " Superalloys 1984, ed. M.Gell et al. (Warrendale, PA: The Metallurgical Society, 1984), 523-532.
7. N.Yukawa and M. Morinaga, " New Approach to the Design of Superalloys," Proc.of the 2nd Japan-U.S. Seminar on Superalloys, Dec. 7-11, 1984, Susono-shi, Shizuoka-Pref., Japan Inst. of Metals, 37-48.
8. N.Yukawa, M.Morinaga, H.Ezaki and Y.Murata , " Alloy Design of Superalloys by the d-Electrons Concept, " Proc. of the Conf. on High Temperature Alloys for Gas Turbines and Other Applications, 1986, 935-944.
9. F.C.Hull,"Estimating Alloy Density," Metal Progress, (1969), Nov., 139-140.
10. M.Kawakami et al., "Study on Hot Corrosion of Single Crystal Superalloys inNa_2SO_4-NaCl by Thermo-Gravimetric Measurement, "Proc. of the 1988 MRS International Meeting on Advanced Materials, (Material Research Society), 1988, May 30-June 3, Tokyo, to be published.
11. F.S.Petit, G. H. Meier, "Oxidation and Hot Corrosion of Superalloys," Superalloys 1984., 651-687.
12. T.Khan, P.Caron, C.Duret, " The Development and Characterization of a High Performance Experimental Single Crystal Superalloy," Superalloys 1984, 145-155.
13. M.Yamazaki, " Development of Nickel-Base Superalloys for a National Project in Japan," Proc. of the Conf. on High Temperature Alloys for Gas Turbine and Other Applications, 1986, 945-954.
14. T.Yamagata, " Alloy Design for Nickel-base Superalloys and Titanium Alloys," International Symp. on Basic Technologies for Future Industries -Materials Development and Technology Innovation-, 1988, March, 22-25, Kobe, to be published.

SECOND-GENERATION NICKEL-BASE

SINGLE CRYSTAL SUPERALLOY

A. D. Cetel and D. N. Duhl

Materials Engineering
Pratt & Whitney
400 Main Street
East Hartford, CT 06108

Abstract

Significant increases in high temperature strength capability have been achieved in a second-generation nickel-base single crystal alloy designed for advanced military and commercial turbine airfoil applications. This new alloy, designated PWA 1484, offers a 50°F improvement in metal temperature capability (creep-rupture strength, thermal fatigue resistance, and oxidation resistance) over PWA 1480, the first nickel-base single crystal turbine alloy to enter production. PWA 1484 represents the first use of rhenium in a production single crystal superalloy. The alloy properties have been fully characterized and demonstrate an outstanding combination of high-temperature creep and fatigue strength as well as excellent oxidation resistance. Production processing of single crystal PWA 1484 castings has shown that the alloy has good castability and a wide solution heat treatment range. The superior properties of PWA 1484 have been confirmed through extensive engine testing.

Introduction

As operating temperatures of gas turbine engines have steadily increased, the demands placed on turbine airfoils have escalated dramatically. To meet these accelerating demands on the capability of high-temperature materials, significant breakthroughs in alloy design and processing have been made (1,2,3).

The directional solidification casting process, pioneered by Pratt & Whitney in the 1960's, has led to the development of a series of columnar grain and single crystal alloys designed for turbine airfoil applications, with temperature capabilities more than 100°F higher than their conventionally cast predecessors. PWA 1480 (4), the first production single crystal turbine airfoil alloy, entered service in 1982 and has successfully accumulated more than five million flight hours in commercial and military engines.

To obtain further improvements in alloy capabilities needed to meet the turbine airfoil requirements for advanced commercial and military engines for the 1990's, alloy development efforts at Pratt & Whitney have been focused on defining a second-generation single crystal alloy (5). The demanding goals established for this program were a 50°F improvement in airfoil metal temperature capability compared to PWA 1480, in an alloy that is microstructurally stable after long-time exposure, and has good castability.

Alloy Goals and Design Approach

The basic design goals established for the second-generation single crystal alloy were as follows:

1. 50°F improvement in airfoil capability relative to PWA 1480
   a. Creep-rupture strength
   b. Coated and uncoated oxidation resistance
   c. Thermal fatigue resistance

2. Other critical properties
   a. Absence of secondary phase precipitation after long-time static and dynamic thermal exposure
   b. Intermediate to elevated temperature high cycle fatigue strength comparable to PWA 1480

3. Ease of processing
   a. Castability equivalent to PWA 1480
   b. Solution heat treatment range greater than PWA 1480

To meet these challenging goals, an extensive alloy development effort was undertaken using the knowledge gained in over twenty years of cast superalloy development (1). It was found that to meet the high-temperature strength goals, a significant increase in refractory metal content compared to the 16 weight percent (w/o) tantalum plus tungsten contained in PWA 1480 was required. These studies also showed that it was necessary to maintain a high volume fraction (60 to 65%) of γ' with a solvus temperature in excess of 2350°F to achieve the strength goals. In these studies, rhenium was found to be an especially potent and necessary solid solution strengthening agent to achieve the required ultra high strength levels. To produce a high volume fraction of γ' precipitate, which was relatively resistant to coarsening at high temperatures and had a high solvus temperature, it was determined that the alloy under development should have at least 5.5 w/o aluminum and a high level of tantalum.

Previous alloy development efforts had determined that raising the refractory content above a critical level, however, could lead to the precipitation of refractory-rich secondary phases such as alpha tungsten, as well as mu or sigma phases. These phases which can form following long-time elevated temperature exposure, robbing the alloy of its prime solid solution strengthening agents, have been observed to compromise alloy properties and thus must be avoided.

To achieve the creep strength goal of a 50°F improvement over PWA 1480, which is a threefold increase in life, while maintaining microstructural stability, a series of alloys was screened at 36 ksi at 1800 and 1850°F to define the optimum balance of the refractory elements: molybdenum, tantalum, tungsten, and rhenium. The alloy with the optimum combination of properties resulting from these screening trials has been designated PWA 1484. The composition of PWA 1484 is compared to that of PWA 1480 in Table I. As shown in the table, PWA 1484 contains 3 w/o rhenium and is the first production single crystal superalloy to employ this element as an alloying addition. To ensure maintenance of microstructural stability at the high (20 w/o) refractory content of PWA 1484, the chromium level of the new alloy was lowered to 5 w/o, compared to 10 w/o in PWA 1480.

Table I. PWA 1484 Composition and Heat Treatment

	\multicolumn{10}{c}{Chemistry (weight percent)}
	Ni	Cr	Ti	Mo	W	Re	Ta	Al	Co	Hf
PWA 1480	Bal	10	1.5	-	4	-	12	5	5	--
PWA 1484	Bal	5	-	2	6	3	8.7	5.6	10	0.1

PWA 1484 Heat Treatment: 2400°F (4 hours) + 1975°F (4 hours) + 1300°F (24 hours)

Alloy Properties

Microstructure and Heat Treatment

The composition of the new alloy was designed to have a large range between its incipient melting temperature (2440°F) and its γ' solvus temperature (2370°F). The cobalt content of PWA 1484 was increased by 5 w/o relative to PWA 1480 to help expand the heat treatment range, partially offsetting the effect of the reduced chromium. As shown in Figure 1, a nominal 2400°F solution heat treatment for 4 hours results in almost complete solutioning of the coarse as-cast γ' in the alloy and their reprecipitation in a uniform array of fine ($\leq 0.3\mu$) cuboidal particles.

The microstructure of PWA 1484 in the as-solutioned and fully heat treated conditions is presented in Figure 2. PWA 1484 retains a very fine γ' particle size after full heat treatment which is believed to result from rhenium additions that retard γ' coarsening (6).

The heat treatment of PWA 1484, as shown in Table I, consists of a three-step process employing a 2400°F for 4 hour solution heat treatment. This solution heat treatment dissolves the coarse as-cast γ' present in the alloy which is reprecipitated as a fine regular array of cuboidal particles upon controlled cooling from 2400°F. Following the solution heat treatment cycle, a 1975°F for 4 hour cycle is used to bond the coating to the alloy, as well as to produce an optimum γ' size and distribution. To enhance intermediate temperature yield strength, the PWA 1484 heat treatment incorporates a 1300°F for 24 hour age which precipitates ultra fine γ' ($\leq 0.1\mu$) between the larger particles as seen in Figure 3.

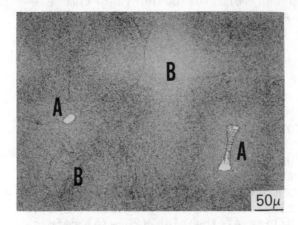

GAMMA/GAMMA PRIME EUTECTICS (A) AND UNRESOLVED GAMMA PRIME PRECIPITATES (B) IN GAMMA MATRIX

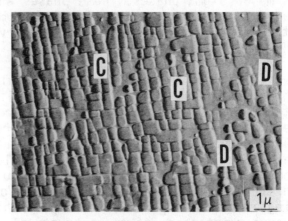

FINE GAMMA PRIME PRECIPITATES (C) IN GAMMA MATRIX (D)

Figure 1 - Typical microstructure of solution heat treated PWA 1484.

AS-SOLUTIONED 2400°F (4HRS)

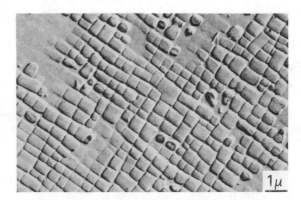

FULLY HEAT TREATED 2400°F (4HRS) + 1975°F (4 HRS) + 1300°F (24 HRS)

Figure 2 - PWA 1484 retains fine γ' microstructure after full heat treatment.

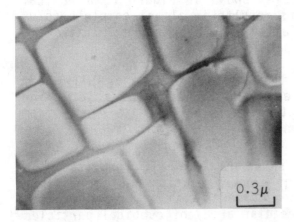

NO AGE

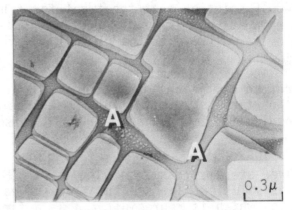

704°C (1300°F) FOR 24 HOURS

Figure 3 - 1300°F aging heat treatment precipitates an additional small volume fraction of ultra fine γ' precipitates (A) in PWA 1484.

Long-Term Microstructural Stability

The PWA 1484 composition was designed to be free of secondary phase instabilities after long-time elevated temperature exposures. Metallographic evaluation of specimens furnace exposed between 1600 and 2000°F for 1000 hours as well as specimens creep tested for up to 2000 hours between 1600 and 2000°F (Figure 4) has confirmed the excellent microstructural stability of the alloy.

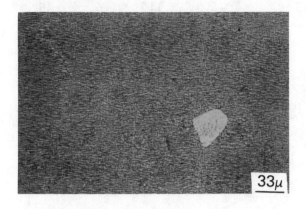
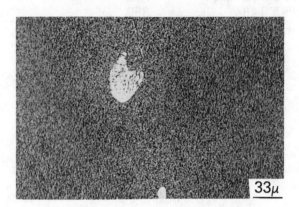

FURNACE EXPOSURE AT 1800°F
FOR 1000 HOURS

CREEP TESTED AT 1800°F/25 ksi
FOR 1997 HOURS

Figure 4 - PWA 1484 microstructure is free of secondary phase instabilities after long-time static and dynamic exposure at elevated temperatures.

Creep-Rupture Strength

Creep-rupture testing of PWA 1484 at 36 ksi indicates that the alloy offers almost a fourfold increase in life over PWA 1480 at this stress, which corresponds to a 70°F metal temperature advantage, easily surpassing the program goal of a 50°F improvement (Table II). At a stress of 36 ksi, PWA 1484 exhibits longer creep and rupture lives when tested at 1850°F than does PWA 1480 in testing at 1800°F, clearly demonstrating that PWA 1484 offers more than a 50°F improvement in creep-rupture capability over PWA 1480. The density of PWA 1484 is 0.323 lb/in^3; and on a density corrected basis, PWA 1484 has a 65°F metal temperature advantage over PWA 1480. Testing at other temperature/stress conditions demonstrates that PWA 1484 offers an even greater improvement over PWA 1480 with advantages in life ranging from 4X at 1700°F/50 ksi to greater than 8X at 2000°F/18 ksi. Comparing the creep strength of the two alloys on the basis of stress for time to 1% creep in 300 hours versus temperature (Figure 5), it can be seen that PWA 1484 retains greater than a 50°F temperature advantage relative to PWA 1480 over a wide temperature range (1500 to >2000°F). Since the initial screening trials, several hundred additional creep-rupture tests have been performed on specimens machined from cast bars, as well as from solid blades, verifying the initial data. These tests have been conducted on specimens from more than a dozen large commercial heats ($\geq$2000 lb).

A study of the effect of cooling rate from the solution heat treatment temperature on strength has shown that the creep-rupture strength of PWA 1484 can be maximized by producing a γ' particle size in the range of 0.25 to 0.35μ (7). This optimum size range can be readily achieved using the full heat treatment cycle for the alloy with a rapid cooling rate from the solution heat treatment temperature, and then applying a 1975°F for 4 hour coating diffusion cycle followed by a 1300°F for 24 hour age.

Table II. Creep-Rupture Data at 36 ksi

	1800°F		1850°F	
	Rupture Life (hours)	Time to 1% Creep (hours)	Rupture Life (hours)	Time to 1% Creep (hours)
PWA 1480	90	36	30	12
Program Goal	270	110	90	36
PWA 1484	350	135	117	45

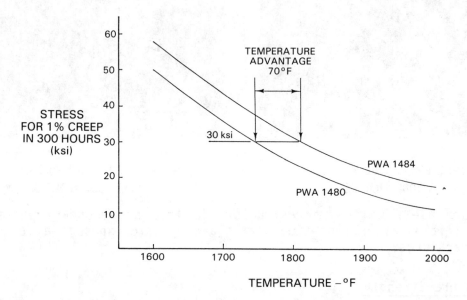

Figure 5 - PWA 1484 exhibits a significant creep strength advantage relative to PWA 1480 over a wide range of temperatures.

Tensile Properties

The tensile properties of PWA 1484 have been extensively characterized in a series of tests between room temperature and 2000°F (Figure 6). The yield strength of PWA 1484 at temperatures below 1200°F is approximately 140 ksi which is similar to that of other single crystal alloys (8). At intermediate to elevated temperatures ($\geq$1400°F), the yield strength of PWA 1484 is superior to most other single crystal alloys. Despite its high yield strength, PWA 1484 exhibits excellent tensile ductility from room temperature to 2000°F (Figure 6) with typical ductilities greater than 12% at all temperatures.

Oxidation and Hot Corrosion Resistance

Burner rig oxidation tests conducted on PWA 1484 at 2100°F, in both the uncoated and overlay coated conditions, indicate that the alloy possesses oxidation resistance even better than that of PWA 1480 which has excellent oxidation resistance. Based on metal recession data (Figure 7), uncoated PWA 1484 specimens oxidize at about half of the rate of PWA 1480 specimens under severe conditions (temperatures $\geq$2000°F). On the basis of time to 5 mils of metal loss at 2100°F, PWA 1484 demonstrates more than a 60% life improvement over PWA 1480 (Figure 7). Testing conducted between 2050 and 2150°F indicates that PWA 1484 offers a 65°F improvement in metal temperature capability over PWA 1480, surpassing the program goal of a 50°F improvement in oxidation resistance.

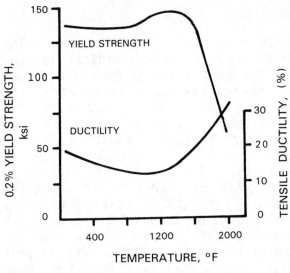

Figure 6 - Tensile properties of PWA 1484.

Figure 7 - PWA 1484 uncoated oxidation resistance at 2100°F is superior to that of PWA 1480.

The excellent oxidation resistance of PWA 1484 is attributable to its higher aluminum content compared to PWA 1480, its relatively high tantalum content, and the fact that it does not contain titanium.

In addition to the uncoated tests, a series of burner rig oxidation tests with a NiCoCrAlHfSiY overlay coating were conducted between 2000 and 2150°F. These tests, some of which ran for more than 6000 hours at 2000°F, demonstrated the superior oxidation performance of the overlay coating on PWA 1484, with lives 70% greater than on PWA 1480 (Figure 8). This translates to a 70°F advantage in metal temperature capability over PWA 1480, surpassing the program goal of a 50°F improvement.

The alloy has also been evaluated in a series of ducted burner rig hot corrosion tests using a synthetic sea salt mixture to induce hot corrosion. These tests, which were conducted both isothermally and cyclically, have shown that PWA 1484 has hot corrosion resistance similar to PWA 1480 and other high-strength turbine airfoil alloys.

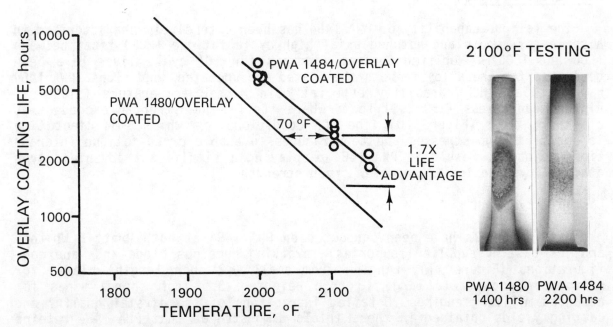

Figure 8 - Overlay coating exhibits significant increase in oxidation life on PWA 1484 compared to PWA 1480.

Thermal Fatigue Resistance

The thermal fatigue capability of the overlay NiCoCrAlHfSiY coated alloy was evaluated in a series of strain controlled thermal-mechanical fatigue (TMF) tests using hollow tube specimens. The strain and temperature components of the cycle employed were out-of-phase (maximum tensile strain at minimum temperature - Cycle I). Specimens were cycled between 800 and 1900°F and the tests were run until cracking was extensive, resulting in a 50% load drop. As shown in Figure 9, PWA 1484 specimens exhibited almost a 2X life advantage over PWA 1480 at a total strain range of 0.5%. This translates to a 65°F advantage in TMF temperature capability, which corresponds closely to the creep-rupture advantage exhibited by PWA 1484 over PWA 1480. Studies have indicated that elevated temperature TMF strength is related to alloy creep strength. It is believed that an alloy with increased creep strength is better able to resist the creep relaxation that occurs due to repeated cycling, which in turn increases the mean tensile stress imposed on the test specimen (9). The higher the mean tensile stress that develops, the lower the resulting TMF life.

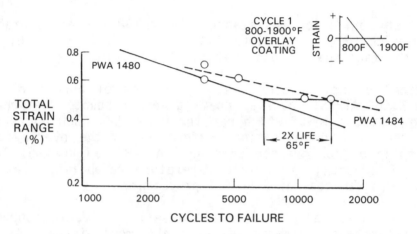

Figure 9 - PWA 1484 thermal fatigue life at 1900°F (T_{max}) is superior to PWA 1480.

High Cycle Fatigue

The fatigue capability of PWA 1484 has been extensively characterized in a series of smooth and notched axial high cycle fatigue (HCF) tests between 1100 and 1800°F. Modified Goodman diagrams for 10^7 cycle lives have been generated for the alloy in both the coated and uncoated conditions. PWA 1484 has excellent HCF capability especially in higher temperature (≥1600°F), higher mean stress (≥50 ksi) tests where its fatigue strength exceeds that of other alloys (Figure 10). The HCF conditions, in which sufficient time is spent at high stresses and temperatures to enable creep-fatigue interactions to occur, result in PWA 1484 displaying a significant advantage over other alloys due to its superior creep strength.

Castability

Casting trials have been conducted on PWA 1484 at both Pratt & Whitney and its casting supplier foundries, involving numerous blade and vane configurations (Figure 11), ranging from small (≤2 inch length) blades for helicopter engines to large (6 inch height x 3.5 inch chord) vanes for advanced military engines. Detailed inspections of the crystal quality and casting yields obtained in these trials demonstrated that PWA 1484 retains the excellent castability displayed by PWA 1480. Extensive experience gained by Pratt & Whitney's casting suppliers over the past four years in scaling up their casting processes, in preparation for production applications for PWA 1484, has confirmed the alloy's excellent castability. Post casting processing of PWA 1484 is similar to that for PWA 1480.

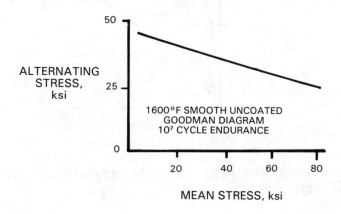

Figure 10 - High cycle fatigue strength of PWA 1484.

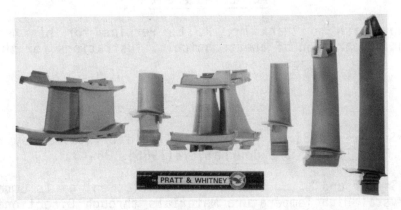

Figure 11 - PWA 1484 has been cast in a wide variety of airfoil configurations.

Engine Test Evaluation

Numerous engine tests of PWA 1484 turbine blades and vanes, in both military and commercial engines, has been successfully conducted over the last three years. This testing has confirmed the results of the laboratory characterization effort and demonstrated the excellent performance of PWA 1484. Growth measurements on PWA 1484 turbine blades have confirmed the low growth predicted by laboratory creep testing (Figure 12). Visual and metallographic evaluation of airfoils, engine tested for thousands of cycles under severe conditions, has shown them to be in excellent condition. Based on the successful engine test results, PWA 1484 will be employed as a turbine airfoil alloy in many advanced commercial and military engines. It is scheduled to enter production by the end of 1988.

Summary and Conclusions

A second-generation single crystal alloy, designated PWA 1484, has been identified which possesses more than a 50°F improvement in turbine airfoil capability compared to the strongest production single crystal alloys. The alloy displays good castability, has a large solution heat treatment range, and retains its microstructural stability after long-time thermal exposures. Extensive engine test evaluation of PWA 1484 has confirmed the excellent combination of properties displayed by the alloy in laboratory testing. The alloy is expected to enter production by the end of 1988.

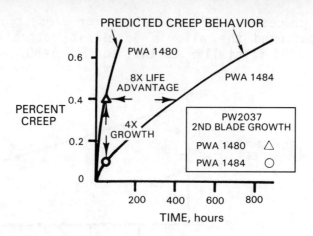

Figure 12 - PWA 1484 creep strength advantage confirmed in 2100 cycle PW2037 engine test.

Acknowledgement

The authors wish to thank Mr. R. L. Perkins for his metallographic assistance and preparation of the technical illustrations for this paper.

References

1. F. L. VerSnyder and B. J. Piearcey, "Single Crystal Alloy Extends Service Life Four Times," SAE Journal, 74(1966), 36-43.

2. F. L. VerSnyder and M. E. Shank, "The Development of Columnar Grain and Single Crystal High Temperature Materials through Directional Solidification," Materials Science and Engineering, 6(1970), 213-240.

3. M. Gell and D. N. Duhl, Advanced High Temperature Alloys, (Metals Park, OH: American Society for Metals, 1985), 41-49.

4. M. Gell, D. N. Duhl and A. F. Giamei, Superalloys 1980, Proceedings of the Fourth International Symposium on Superalloys, (Metals Park, OH: American Society for Metals, 1985), 205-214.

5. M. Gell, D. N. Duhl, D. K. Gupta and K. D. Sheffler, "Advanced Superalloy Airfoils," Journal of Metals, (1987), 11-15.

6. L. S. Lin, A. F. Giamei and R. E. Doiron, Proceedings of the 38th Annual Meeting of ESMA, (Baton Rouge, LA: Claitors, 1980), 330.

7. A. D. Cetel and D. N. Duhl, Proceedings of the 2nd International SAMPE Metals and Metals Processing Conference, (Covina, CA: SAMPE, 1988).

8. R. V. Miner, R. C. Voigt, J. Gayda and T. P. Gabb, "Orientation and Temperature Dependence of Some Mechanical Properties of the Single-Crystal Nickel-Base Superalloy Rene N4: Part 1, Tensile Behavior," Metallurgical Transactions, 17A(1986), 491-496.

9. P. P. Norris, N. E. Ulion and D. N. Duhl, "Thermal Fatigue Crack Propagation in Single Crystal Nickel-Base Superalloys," (Report NADC-82169-60, Final Report NADC Contract N62269-83-C-0329, 1987).

ENHANCED RUPTURE PROPERTIES

IN ADVANCED SINGLE CRYSTAL ALLOYS

S.M. Foster, T.A. Nielsen, and P. Nagy

Advanced Materials and Processes Department
Williams International
Walled Lake, MI 48088

Summary

The rupture lives of two single crystal alloys, CMSX-2 and CMSX-4G, were improved by extended thermal processing. The chemistries of the alloys were similar except for an addition of rhenium for high temperature stability in CMSX-4G. Two high temperature aging cycles were given to each alloy which resulted in γ' platelet formation and varying degrees of γ' coarsening. The effect of γ' coarsening on the rupture and tensile properties was investigated. The results for each alloy were compared to the ONERA aging cycle which produced a fine cuboidal γ' morphology in both alloys. It was found that overaging the γ' in CMSX-2 occurred more readily than in CMSX-4G.

Additionally, for CMSX-2, crystallographic orientations versus tensile and rupture properties from each aging cycle were evaluated. Regardless of the aging cycle, the [111] crystallographic orientation had the best rupture properties and the [001] orientation had the best tensile properties. Mechanical property testing in both portions of the study showed that extended thermal cycles resulted in increased rupture lives due to γ' platelet formation and coarsening with only minor decreases in tensile strength.

Introduction

Advanced limited life gas turbine engines require high strength and rupture resistant turbine blades. Previous work in the area of single crystal superalloys has shown a large number of variables contributing to tensile and rupture behavior. (1,3,8,9) The most predominant variables were heat treatment, crystallographic orientation, and chemistry. The effect of high temperature/extended aging cycles and chemistry were evaluated to determine interactive effects on tensile and rupture properties in Cannon-Muskegon alloys CMSX-2 and CMSX-4G. High stress levels for the rupture tests were selected to produce the relatively short rupture lives, that are typical for limited life engines.

The two aging cycles investigated were based on previous work conducted at Williams International in developing an activated diffusion bonded multiple alloy turbine rotor with single crystal airfoils. These cycles were compared against the typical ONERA heat treatment to provide a common reference point. In addition to the study of aging cycles and chemistry modifications on single crystal properties, the effect of aging cycles on the mechanical properties of various crystallographic orientations was also evaluated for CMSX-2.

Experimental Procedure

Alloy Selection

CMSX-2 was chosen for evaluation since it was a well characterized single crystal alloy, that had been successfully used in other studies. CMSX-4G, a derivative of CMSX-2 was included due to reports of its increased creep rupture properties. The main difference in the chemistries of CMSX-2 and CMSX-4G was the addition of approximately 3 percent rhenium (Re). A small amount of hafnium (Hf) and additional cobalt (Co) was also added to the CMSX-4G alloy. The Re was added to reduce the coarsening rate by decreasing the diffusion kinetics at the γ/γ' interface, which results in increased high temperature microstructural stability. (10) Table I lists the chemistries of the two alloys.

Table I. Alloy Chemistries.

Alloy	Re	Hf	Al	Ti	Cr	W	Co	Mo	Nb+Ta	Ni
CMSX-2	0	0	5.66	1.02	7.9	7.9	4.7	0.6	6.05	Bal.
CMSX-4G	2.93	0.10	5.45	0.98	6.22	6.53	9.54	0.64	6.52	Bal.

Material Procurement

Earlier studies have shown that rupture properties vary with crystallographic orientation, (1,9) therefore the CMSX-2 evaluation included four major crystallographic orientations ([001], [011], [111], and [112]). The CMSX-4G portion of the study included only the [111] orientation, since the [111] had the best rupture properties behavior in the CMSX-2 study.

The CMSX-2 and CMSX-4G single crystal test material was produced by Howmet using the seeded Bridgeman technique. Different casting solidification rates were required to produce the different orientations needed for each alloy. Since the test material exhibited only minor differences in either the primary or secondary dendrite spacings, casting effects were considered negligible, which allowed a reasonable basis for comparing all the test material.

To avoid recrystallization, the CMSX-2 and CMSX-4G single crystal slabs and bars were solutioned immediately after casting. The solutioning cycles were different, due to the chemistry variations and different γ' solvus temperatures. A minimum of 99.7% of the γ/γ' eutectic was put into solution for both alloys.

The test material was Laue X-ray inspected. All of the tested specimens were within 10 degrees of the desired crystallographic orientation.

Material Test Plan

Two high temperature aging cycles were developed as part of an activated diffusion bonding study to produce multiple alloy turbine rotors. The effects of the rotor bond cycles on the rupture lives and tensile properties of the CMSX-2 and CMSX-4G were investigated to determine maximum rotor capabilities. These two cycles were compared to the standard ONERA aging cycle. The complete thermal processing sequences were as follows:

    a. Solution Treatment
    b. HIP: 1135°C/207 MPa/3 hrs.
    c. Age

The aging cycles compared in this study were:

o ONERA Heat Treatment (3) - 1050°C/16 hrs/furnace cool plus 850°C/48 hrs/furnace cool.

o Cycle #1 - 1165°C/6 hrs/furnace cool plus 760°C/16 hrs/furnace cool

o Cycle #2 - 1165°C/53 hrs/furnace cool plus 760°C/16 hrs/furnace cool.

Most of the material was HIP'ed after solutioning because the activated diffusion bonding cycles included HIP'ing as part of the bonding operation. However, the material given the ONERA cycle was HIP'ed prior to solutioning per its standard processing cycle.

Results/Discussion

Aging Treatment and Orientation Effects on CMSX-2 Mechanical Properties

After the aging treatment, testing was conducted on the specimens in the [001], [011], [111], and [112] crystallographic orientations. The [001] orientation exhibited the highest tensile strength in all aged conditions and test temperatures. Previous work by other investigators confirm these findings. (9) The results of the CMSX-2 tensile tests are shown in Figure 1. For each crystallographic orientation, there was a reduction in tensile strength from cycle #1 to cycle #2, due to overaging caused by the longer aging cycle.

Rupture testing on CMSX-2 showed that the [111] orientation had the best properties. The rupture results are presented in Table II. The data shows that cycle #1 produced the best rupture properties, cycle #2 had slightly lower rupture properties, and the ONERA cycle had the shortest rupture lives.

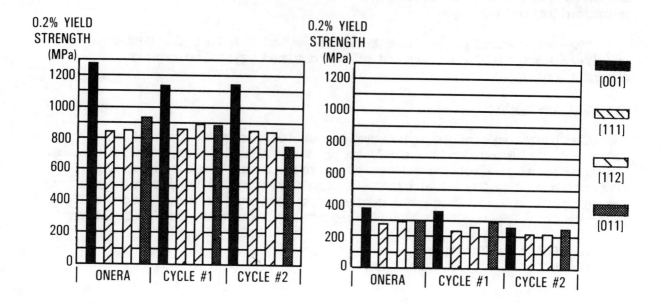

(a.) 0.2% Yield Strength (MPa) vs. Crystallographic Orientation and Aging Cycle for CMSX-2 at 760°C.

(b.) 0.2% Yield Strength (MPa) vs. Crystallographic Orientation and Aging Cycle for CMSX-2 at 1095°C.

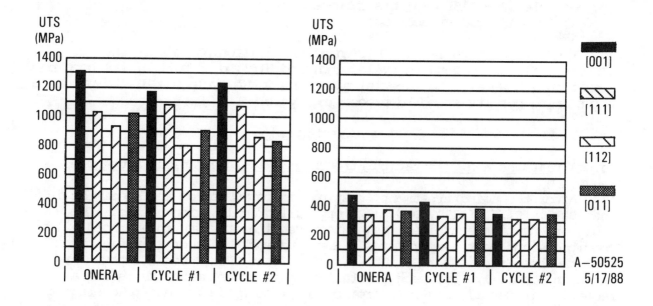

(c.) Ultimate Tensile Strength (MPa) vs. Crystallographic Orientation and Aging Cycle for CMSX-2 at 760°C.

(b.) Ultimate Tensile Strength (MPa) vs. Crystallographic Orientation and Aging Cycle for CMSX-2 at 1095°C.

Figure 1. Tensile and Yield Strength vs. Crystallographic Orientation and Heat Treatment for CMSX-2.

Table II. Rupture lives in hours, for all of the tested orientations of CMSX-2 at 927 and 1095 degrees C.

Orientation	Heat Treatment	Temp.(°C)	Stress (MPa)	Rupture Time (Hrs)
[001]	ONERA	927	345	64.3
[001]	Cycle #1	927	345	134.5
[001]	Cycle #2	927	345	72.1
[011]	ONERA	927	345	66.2
[011]	Cycle #1	927	345	121.7
[011]	Cycle #2	927	345	65.0
[111]	ONERA	927	345	98.4
[111]	Cycle #1	927	345	153.1
[111]	Cycle #2	927	345	114.3
[112]	ONERA	927	345	68.5
[112]	Cycle #1	927	345	110.2
[112]	Cycle #2	927	345	91.7
[001]	ONERA	1095	117	56.0
[001]	Cycle #1	1095	117	293.1
[001]	Cycle #2	1095	117	81.1
[011]	ONERA	1095	117	54.8
[011]	Cycle #1	1095	117	189.3
[011]	Cycle #2	1095	117	88.9
[111]	ONERA	1095	117	63.7
[111]	Cycle #1	1095	117	328.2
[111]	Cycle #2	1095	117	117.6
[112]	ONERA	1095	117	81.7
[112]	Cycle #1	1095	117	270.7
[112]	Cycle #2	1095	117	61.2

Figures 2a, b, and c show the γ' morphology of CMSX-2 in all three conditions. The ONERA cycle produced a fine cuboidal γ', while cycles #1 and #2 produced platelet γ' morphologies. The higher temperature and shorter time of Cycle #1 produced a fine platelet type γ' morphology that had the same width as the ONERA cycle's cuboidal γ'. The extended aging time in Cycle #2 coarsened the platelets in both length and width. Similar results on platelet coarsening were shown by Nathal. (8) The actual γ' dimensions are shown in Table III.

The platelet type γ' morphologies were more effective at impeding dislocation motion, because the dislocations must shear the γ' rather than circumventing as is possible with cuboidal structures. This resulted in increased rupture lives.[1] The finer γ' platelet morphology characteristic of Cycle #1 had better rupture properties than Cycle #2's coarse γ' platelet, since it had a higher γ/γ' surface area which provided more barriers to dislocation motion. It was concluded that cycle #2 coarsened and overaged the γ', which resulted in the shorter rupture lives.

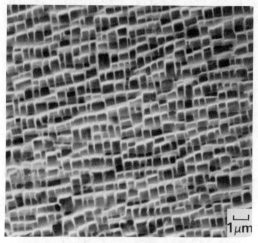

(a.) CMSX-2 with the ONERA Aging cycle. 3800X.

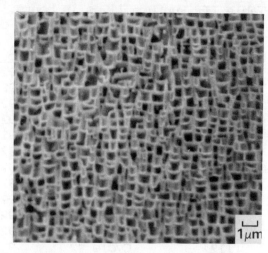

(d.) CMSX-4G with the ONERA Aging cycle. 3800X.

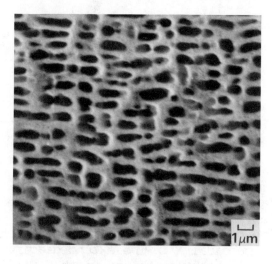

(b.) CMSX-2 with Aging Cycle #1. 3800X.

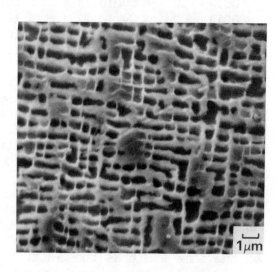

(e.) CMSX-4G with Aging Cycle #1. 3800X.

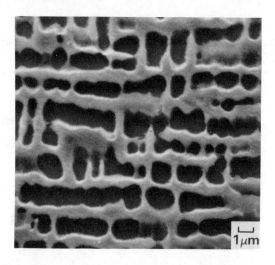

(c.) CMSX-2 with Aging Cycle #2. 3800X.

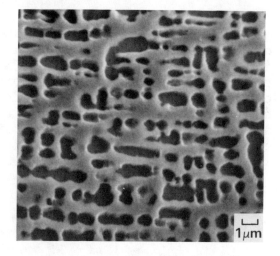

(f.) CMSX-4G with Aging Cycle #2. 3800X.

Figure 2. Photomicrographs of γ' Morphology as a Function of Aging Cycle and Alloy. γ' elongated in the [001] direction.

Aging Cycle Effects on CMSX-4G Compared to CMSX-2

The aging cycle response of CMSX-4G in the [111] orientation was compared to CMSX-2 by observing the growth and morphology change in the γ' and by measuring the resultant mechanical properties. The two cycles produced dramatic differences on the γ' morphology, which resulted in changes in tensile and rupture properties. The γ' morphologies of the two alloys that form after each of the aging cycles are shown in Figure 2. Table III lists the approximate γ' size and morphology for CMSX-2 and CMSX-4G.

The ONERA aging cycle produced a fine cuboidal γ' morphology in CMSX-4G as observed in CMSX-2. These microstructures served as a baseline for comparison of structures produced from the other cycles for both alloys.

The CMSX-4G photomicrographs show that the Re addition tended to retard both the platelet formation and subsequent coarsening. As was shown in CMSX-2, Cycle #1's higher temperature/short time aging cycle promoted γ' platelet formation in CMSX-4G. However, the platelet formation appeared to be retarded due to the Re addition. Cycle #2's increased aging time promoted longitudinal growth of the γ' platelet rather than thickening of the platelet as seen in CMSX-2. It was observed that for all the aging cycles, the CMSX-4G containing Re had a finer γ' platelet structure. The Re addition has apparently inhibited the coarsening of the γ' platelets, that occurred in CMSX-2.

Table III. Approximate γ' size and morphology for each alloy and aging cycle.

	ONERA	Cycle #1	Cycle #2
CMSX-2			
width	0.5um	0.5um	1.0um
length	0.5um	1.5um	4.0um
l/w	1.0	3.0	4.0
CMSX-4G			
width	0.5um	0.5um	0.75um
length	0.5um	1.0um	2.5um
l/w	1.0	2.0	3.3

The longest rupture lives for CMSX-4G were achieved with aging cycle #2, indicating that CMSX-4G has more stability and can be subjected to prolonged thermal treatments without overaging which was not the case for CMSX-2. Table IV shows the rupture lives, in hours, for both alloys in all aged conditions. CMSX-4G has longer rupture times than CMSX-2 for either high temperature aging cycle. The CMSX-4G rupture life is nearly doubled at 927°C for cycle #2, compared to the ONERA aging cycle.

The tensile strength of CMSX-4G and CMSX-2 resulting from all of the aging cycles is shown in Figure 3. CMSX-4G had higher tensile strength than CMSX-2 in any of the aging cycles. The superior tensile strength of CMSX-4G was inherent to the alloy chemistry, since the tensile properties were better than CMSX-2 regardless of the aging cycle.

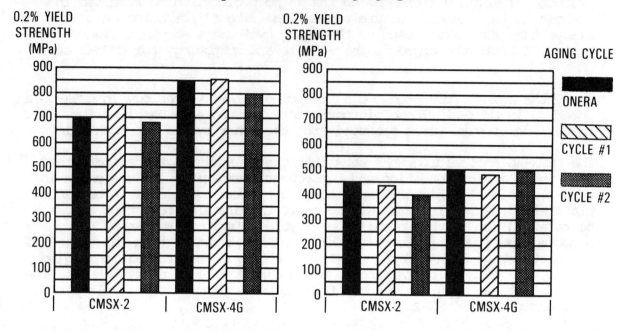

(a.) 0.2% Yield Strength (MPa) vs. Alloy and Aging Cycle at for the [111] Orientation at 870°C.

(b.) 0.2% Yield Strength (MPa) vs. Alloy and Aging Cycle for the [111] Orientation at 983°C.

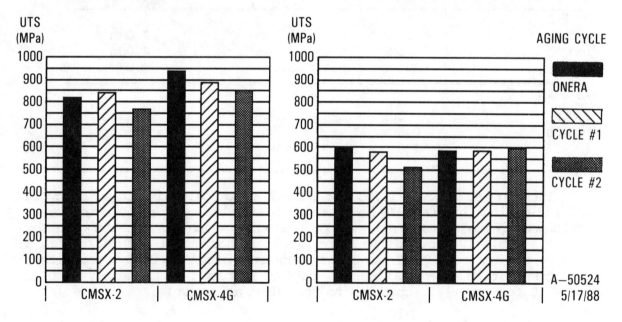

(c.) Ultimate Tensile Strength (MPa) vs. Alloy and Aging Cycle for the [111] Orientation at 870°C.

(d.) Ultimate Tensile Strength (MPa) vs. Alloy and Aging Cycle for the [111] Orientation at 983°C.

Figure 3. Tensile Strengths of CMSX-4G and CMSX-2 vs. Aging Cycle.

Table IV. Rupture Lives as a Function of Aging Cycle and Alloy.

Heat Treatment	Temperature °C	Stress (MPa)	Rupture Time (Hrs.) CMSX-4G	CMSX-2
ONERA	927	483	15.0	11.6
CYCLE #1	927	483	18.5	17.0
CYCLE #2	927	483	29.0	15.6
ONERA	1095	117	165.7	63.7
Cycle #1	1095	117	208.5	328.2
Cycle #2	1095	117	427.3	117.6
ONERA	1150	83	*	51.7
Cycle #1	1150	83	*	135.3
Cycle #2	1150	83	*	72.5

Conclusions

1. The morphology of the γ' in single crystal CMSX-2 and CMSX-4G alloys was modified by thermal processing. An aging temperature of approximately 1050°C produced fine cuboidal γ'. Increasing the aging temperature to 1165°C caused the γ' to have a platelet morphology.

2. The aging cycles had a significant effect on the mechanical properties of single crystal CMSX-2 and CMSX-4G alloys due to the resultant γ' morphology and coarsening rate. Time at temperature and composition dictated the rate of γ' platelet formation and subsequent coarsening.

3. For CMSX-2, the [111] orientation had better rupture properties than the [001], [011], or [112], regardless of aging cycle. The [001] orientation had better tensile properties than the [011], [111], or [112], regardless of aging cycles in CMSX-2.

4. For CMSX-2, the fine γ' platelet morphology produced in cycle #1 optimized rupture properties. Overaging associated with cycle #2 coarsened the platelet γ', decreasing rupture properties. However, both of these aging cycles resulted in increased rupture properties compared to the ONERA cycle.

5. For CMSX-4G, the rupture properties were improved with cycle #2. CMSX-4G exhibited platelet lengthening with the longer thermal cycle resulting in increased rupture properties.

6. The CMSX-4G chemistry modifications resulted in better rupture and tensile strengths than found in CMSX-2. The Re additions in CMSX-4G resulted in a much more stable γ' structure, which allowed longer aging cycles without coarsening and overaging the γ' platelets.

Acknowledgments

The authors wish to express their appreciation to Mr. Ken Harris of Cannon-Muskegon Corporation for his technical assistance and many discussions throughout the coarse of this evaluation, Mr. Kent Perkins for the metallography, and Mr. Dan Jones for the S.E.M. photomicrographs.

References

1. R.A. Mackay and L.J. Elbert, " The Development of γ-γ' Lamellar Structures in a Nickel Base Superalloy during Elevated Temperature Mechanical Testing," Metall. Trans. Vol 16A, 1985, 1969-1982.

2. S.L. Draper, D.R. Hull and R.L. Dreshfield, "Observations of Directional Gamma Prime Coarsening During Engine Operation," (NASA Technical Memorandum 100105), Lewis Research Center, 1987.

3. P. Caron and T. Khan, "Improvement of Creep Strength in a Nickel-base Single Crystal Superalloy by Heat Treatment," Material Science and Engineering. 61 (1983) 173-184.

4. R.P. Dalal, C.R. Thomas, and L.E. Dardi, "The Effect of Crystallographic Orientation on the Physical and Mechanical Properties of an Investment Cast Single Crystal Nickel-Base Superalloy," Superalloys 1984. M. Gell et. al. eds., AIME, New York, 1984, 185-197.

5. A. Fredholm and J.L. Strudel, "On the Creep Resistance of Some Nickel Base Single Crystal," Superalloys 1984. M. Gell et. al. eds., AIME, New York, 1984, 211-220.

6. D.D. Pearson, F.D. Lemkey, and B.H. Kear, "Stress Coarsening of γ' and its Influence on Creep Properties of a Single Crystal Superalloy," Superalloys 1980. J.K. Tien et. al. eds., (Metals Park, OH: American Society for Metals, 1980), 385-394.

7. R.A. Mackay, R.L. Dreshfield, and R.D. Maier, "Anisotropy of Nickel-Base Superalloy Single Crystals," Superalloys 1980. J.K. Tien et. al. eds., (Metal Park, OH: American Society for Metals, 1980), 385-394.

8. M.V. Nathal, "Effect of Initial Gamma Prime Size on the Elevated Temperature Creep Properties of Single Crystal Nickel Base Superalloys," Metall. Trans., Vol. 18A, 1987, 1961-1970.

9. D.M. Shah and D.N. Duhl, "The Effect of Orientation, Temperature, and Gamma Prime Size on the Yield Strength of a Single Crystal Nickel Base Superalloy," Superalloys 1984. M. Gell et.al. eds., AIME, New York, 1984, 105-114.

10. K. Harris, private communication with authors, Cannon-Muskegon Corporation. 1987 and 1988.

11. R.A. MacKay, private communication with authors, NASA-Lewis Research Center. 1988.

FORMATION OF TOPOLOGICALLY CLOSED PACKED PHASES IN

NICKEL BASE SINGLE CRYSTAL SUPERALLOYS

R Darolia, DF Lahrman, RD Field

Engineering Materials Technology Laboratories

GE Aircraft Engines
Cincinnati, Ohio 45215
and
R Sisson
Worcester Polytechnic Institute
Worcester, MA 01609

ABSTRACT

The formation of topologically closed packed (TCP) phase was studied for a series of nickel base superalloys containing rhenium. The alloys were found to form three types of TCP phases: rhombohedral mu, tetragonal sigma and a relatively unknown orthorhomic phase, P. The crystal structures of these phases are closely related. They often coexist in the alloy, even within the same precipitate. The segmented appearance of the precipitates is due to a "basket weave" morphology, consisting of ribbons of precipitates, overlapping at 90 degree angles. Computer programs such as PHACOMP which are used to predict alloy stability do not necessarily provide good correlation with TCP formation in Re containing alloys. The paper describes in detail the nature and formation of TCP phases in one specific alloy composition.

INTRODUCTION

The development of advanced turbine nickel base superalloys has progressively produced more complex alloy compositions in the quest for higher strengths and operating temperatures. With this increase in complexity comes an increase in the difficulty of predicting and controlling the formation of topologically close packed (TCP) phases. These phases have long been associated with property reduction as sites for crack initiation, or simply by robbing the matrix of solid solution strengthening alloying elements. In many instances, in alloys prone to TCP phase formation, short time properties (tensile properties) are not affected, but the long time creep rupture properties are severely degraded. Understanding the formation of these phases is particularly important, since the alloy developer is generally working on the edge of phase instability, in order to achieve the best balance of properties.

Today's superalloys contain high levels of refractory elements, such as Mo, W, Re, and Ta, in order to increase high temperature creep and rupture properties. These elements act as solid solution strengtheners in both the gamma and gamma prime phases. Re is a particularly potent strengthener, residing mainly in the gamma phase and is thought to retard gamma prime coarsening. High levels of these alloying elements can make the alloys prone to TCP phase formation. The control of TCP phases by judicious selection of strengthening elements is required. This has prompted the current investigation, in which TCP phase formation has been studied for a series of alloys containing Re. The nature of the phases, their relationship with the matrix and with each other, as well as the role of Re, has been investigated.

EXPERIMENTAL PROCEDURE

Several single crystal superalloy compositions were evaluated by GE Aircraft Engines in a study of the formation of TCP phases as a function of composition and exposure conditions. This paper describes results of such an investigation on a superalloy designated "Alloy 800", the composition of which is listed below in wt%:

Ni	Cr	Co	Mo	W	Ta	Ti	Al	Re
63.6	8.0	7.5	1.5	4.0	5.0	1.5	5.8	3.0

The fully solutioned and aged specimens were exposed at 1600, 1800, 2000, 2100 and 2200°F for times up to 1000 hours to determine the occurrence of TCP phases. The TCP phases were identified using a variety of analytical electron microscopy (AEM) techniques. Crystallographic analysis was conducted using both selected area diffraction (SAD) and convergent beam electron diffraction (CBED) techniques. Energy dispersive spectroscopy (EDS) of characteristic X-rays was used for compositional analysis. In addition, scanning electron microscopy (SEM) was used to evaluate the morphology and distribution of the extracted precipitates.

RESULTS

Table I lists the temperature and time combinations which led to precipitation of the TCP phases. A time temperature transformation (TTT) curve, constructed from the data presented in Table 3, is given in Figure 1.

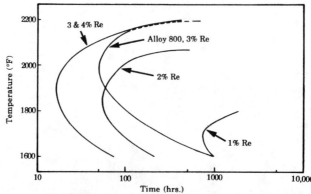

Figure 1. Comparison of TTT curve for formation of TCP phases in Alloy 800 with those of alloys 240 through 243[1].

Table I: Time for Start of TCP Phases at Various Temperatures of Exposure for Alloy 800.

Time Hours	1600°F	1800°F	2000°F	2100°F	2200°F
20	N	N	N	N	N
35	N	N	T	N	N
50	N	N	T	N	N
75	N	N	TCP	T	N
100	N	T	TCP	T	N
500	N	TCP	TCP	-	-
1000	T	TCP	TCP	-	-

N = None, T = Trace Amounts of TCP, TCP = minor to moderate amounts of TCP.

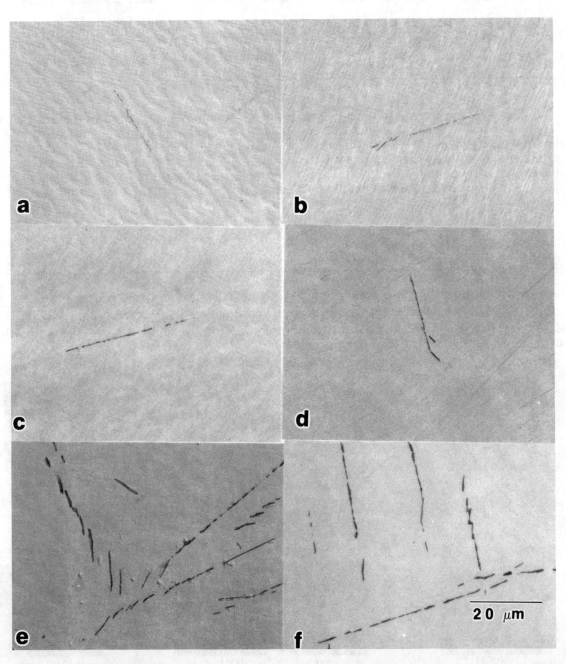

Figure 2. Optical micrographs showing the morphological development of the TCP phases at 2000°F a) 35 hrs., b) 50 hrs., c) 75 hrs., d) 100 hrs., e) 500 hrs., and f) 1000 hrs.

Also included in Figure 1 are TTT curves for other Re containing alloys (1). Figure 2 shows the morphology of the TCP phases for several exposure times at 2000°F. It is apparent that they increase in quantity and size as the time is increased. The precipitates in Figure 2 have a segmented morphology which was observed in several alloys at all stages of development of the TCP phases.

TEM observations revealed the presence of three different TCP phases in the Re containing superalloys: rhombohedral mu, tetragonal sigma, and a relatively unknown orthorhombic phase, P. The P phase was first observed by Rideout et al (2) in the Cr-Ni-Mo system. Its crystal structure was determined by Shoemaker et al (3) to be orthorhombic and is very closely related to sigma. Crystal structure information for the three phases is given in Table II.

Table II: Crystal Structure Information on Mu, Sigma, and P

TCP Phase	Space Group	Pearson Symbol	Lattice Parameter (Å)
Sigma	$P4_2/mnm$	tP30	a=9.3, c=4.86
P	Pnma	oP56	a=17.2, b=4.86, c=9.2
Mu	$R\bar{3}m$	hR13	a=9.0, a=30.8

The P and sigma phases were predominant, and appeared as platelets with {111}gamma habit planes. Electron diffraction data for sigma and P are given in Figure 3. In many cases, sigma and P were found to coexist in the same platelet. As observed in the optical and SEM micrographs, the platelets were often segmented in nature. Figure 4a shows a TEM micrograph of a sigma/P precipitate along with selected area diffraction patterns from each phase. The orientation relationships between the two phases and the gamma matrix were found to be:

$[100]_{sigma}$ ∥ $[100]_P$ $[1\bar{1}0]_{sigma}$ ∥ $[102]_P$ ∥ $[\bar{1}12]_{gamma}$

$[110]_{sigma}$ ∥ $[10\bar{2}]_P$ ∥ $[110]_{gamma}$ $(001)_{sigma}$ ∥ $(010)_P$ ∥ $(1\bar{1}1)_{gamma}$

Thus, the habit plane for the platelets in terms of the TCP phase crystal structure is $(001)_{sigma}$ and $(010)_P$. The relationship between the sigma and P phases is depicted graphically in Figure 4b. The P phase was found to exist in two different variants within the same platelet, with the (010) planes parallel for the two variants, and $[100]_{P1}$ parallel to $[001]_{P2}$.

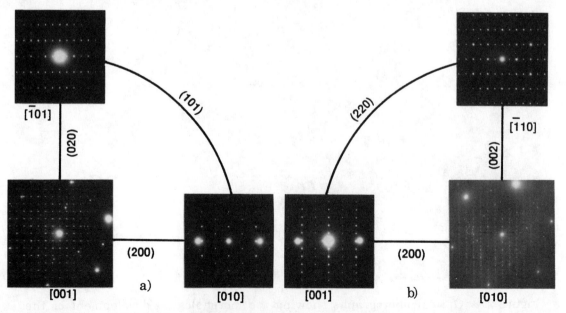

Figure 3.   Selected area diffraction patterns from a) sigma precipitate, b) P precipitate in Alloy 800.

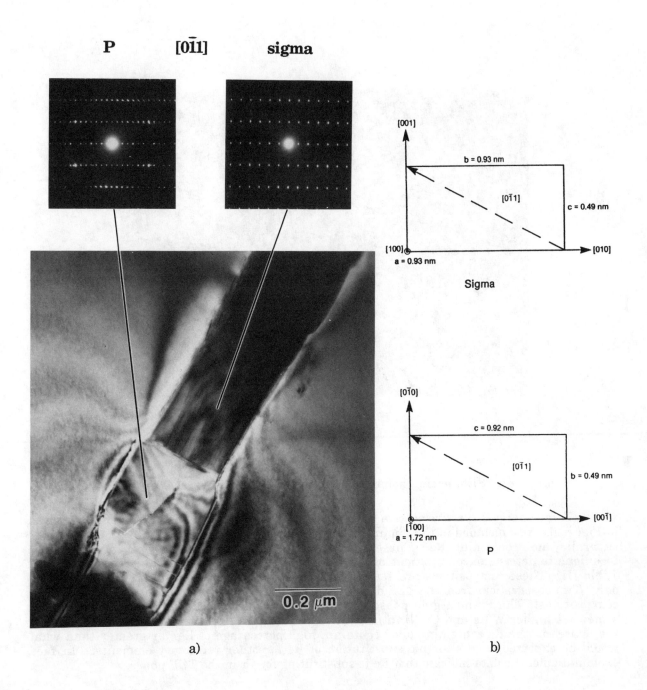

Figure 4. a) TCP precipitate consisting of P and sigma sections.
b) relationship between the sigma and P crystal structures.

The morphology of the segmented plates is shown in Figure 5. In Figure 5a, an SEM micrograph is given of a specimen which was electropolished to show the TCP particles in three dimensions. This particle was extracted and placed on a carbon grid for TEM analysis. As shown in Figure 5b, the TCP phase has a "basket weave" morphology, consisting of ribbons of primarily P with some sigma, at 90 degree angles. The growth direction of the ribbons was found by diffraction to be <100> and [001] for sigma and P, respectively.

Figure 5a. SEM micrograph of TCP precipitate after 75 hrs. at 2000°F.

X-ray energy dispersive spectra (EDS) taken from the TCP phases in the AEM are given in Figure 6. Also included in this Figure is a spectrum taken from a sigma plate in a Ni base superalloy not containing Re. These spectra were analyzed using a peak intensity ratio technique to determine compositions of the phases. Results from these analyses are given in Table III. These compositions are semiquantitative and are included for rough comparison only. Two observations from the EDS data are of interest. Firstly, the phases have very similar compositions. The P and sigma phases are essentially indistinguishable, while the mu has somewhat higher W/Re and Co/Ni ratios, but still has quite a similar composition. Secondly, the phases in the Re containing alloy contain a high percentage of Re, far greater than what would be expected from a simple substitution of Re for other refractory elements. In fact, examination of the data indicate that Re is substituting for Cr in the TCP phase.

Table III: Results From Analysis of EDS Spectra From TCP Phases in At %

Element	P or Sigma	Mu	Sigma without Re
Ni	32	25	22
Re	19	18	-
W	8	11	5
Mo	6	7	6
Cr	25	25	54
Co	10	15	12

Figure 5b. TEM micrograph of extracted TCP precipitate after 75 hrs. at 2000°F. Note the crystallographic relationships between the various segments of the precipitate.

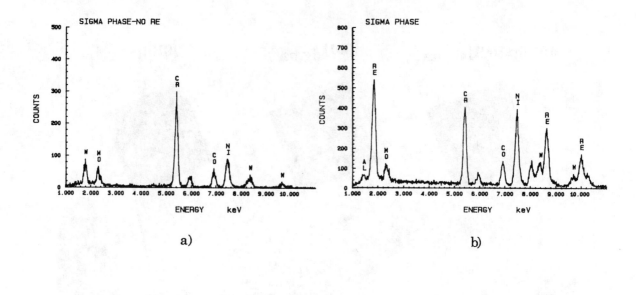

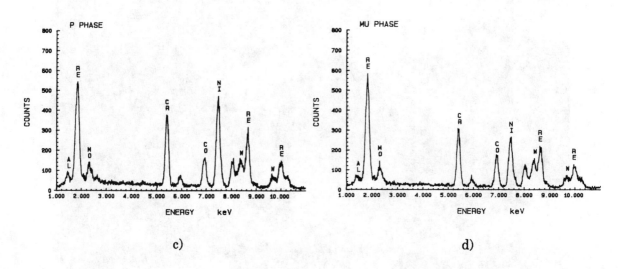

Figure 6. EDS spectra from TCP phases (a) Sigma phase in an alloy not containing Re, (b) sigma, (c) P, and (d) mu phase in an alloy containing Re.

DISCUSSION

Computer programs such as PHACOMP (N_V) are generally used to predict the overall tendency to form TCP phases. Recently, a new PHACOMP (Md) was introduced (4) for better predictive capability. It is interesting to note that the PHACOMP N_V and Md values of the alloys listed in Table IV are essentially the same, even though they behave differently with respect to TCP phase formation. This is not surprising, because the calculation of the alloy stability numbers such as N_V or Md ignores the synergistic effects of the alloying elements on their partitioning behavior. As an example, it is generally assumed that Re partitions mostly to the gamma phase. In reality, Re can partition to gamma prime. As shown in Figure 7, the addition of 2 atomic (6 weight) percent W causes up to 20% of the Re present in the alloy to partition to gamma prime (5). It is quite clear that, though the calculation of N_V and Md for phase stability is a useful guide as a first approximation, total reliance or these calculations is not recommended.

Table IV: Compositions of Alloys in Figure 1 (wt%) and Their Nv and Md Valves

Alloy	Cr	Co	Mo	W	Ta	Ti	Al	Re	Ni	Nv	Md
800	8.0	7.5	1.5	4.0	5.0	1.5	5.8	3.0	63.6	2.19	0.9244
240[1]	7.5	5.0	2.0	3.0	12.0	1.0	5.0	1.0	63.5	2.15	0.9233
241[1]	7.5	5.0	2.0	2.0	12.0	1.0	5.0	2.0	63.5	2.15	0.9211
242[1]	7.5	5.0	2.0	1.0	12.0	1.0	5.0	3.0	63.5	2.14	0.9209
243[1]	7.5	5.0	2.0	0.0	12.0	1.0	5.0	4.0	63.5	2.14	0.9197

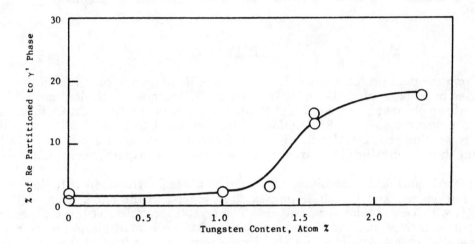

Figure 7. Plot showing synergistic effects of W and Re on gamma/gamma prime partitioning of Re.

The similarity of the compositions of the P and sigma phases, and their orientation relationship within the platelets, can be understood if one examines the crystal structures of the phases shown in Table II. Sigma is tetragonal, with a=0.93 nm and c=0.49 nm. P is orthorhombic with a=1.72 nm, b=0.49 nm, and c=0.92 nm. As can be seen from these parameters, the 'a' parameter of the P unit cell is approximately twice that of sigma and the two phases are very closely related crystallographically (3). Thus, the energy difference between the two phases is quite small, and they are found to coexist in the microstructure with almost identical compositions. Due to the difficulty in distinguishing between the two phases in the AEM, the precise conditions under which P or sigma preferentially precipitate were not determined.

It is interesting to consider the origin of the segmented plate morphology observed in the optical micrographs. One possible explanation for the morphology is the existence of a "precursor" phase, which is continuous. The original continuous phase may become segmented or serve as a nucleation site for other precipitates, thereby maintaining a common growth plane. However, examination of specimens aged for short times revealed the presence of segmented plates even at the earliest stages of precipitation. The other possibility is that the segments are actually connected in three dimensions, and that the segmented appearance represents a growth morphology within the habit plane. This may be due to slow growth of the TCP phase in the gamma prime phase in which the concentration of TCP forming elements is low, yielding a "swiss cheese" morphology of a plate with holes, or some other growth morphology inherent in the TCP structure itself.

As observed in Figure 5, the segmented plates seen in three dimensions display a "basket weave" morphology, consisting of ribbons of precipitate, overlapping at 90 degree angles. Although these precipitates maintain the {111}gamma habit plane determined by their orientation relationship with the matrix, the "basket weave" morphology is determined by the crystallography of the TCP phases themselves. Thus, the 90 degree angles between the ribbons is a manifestation of the [001] growth direction for P or <100> growth direction for sigma. It should be noted that these directions are essentially equivalent, parallel to the 0.92nm lattice parameter of the two unit cells. The fact that only one such direction exists in the P structure, the predominate phase, probably gives rise to the ribbon morphology. The precipitates appear to have grown as "solid state dendrites" within the {111}gamma habit plane. This is thought to be caused by the preferred growth direction of the platelets combined with the diffusion fields of the TCP forming elements within the matrix. Thus, a perturbation along the side of a ribbon of TCP phase would extend into a region with a higher concentration of TCP forming elements (ie. outside the depleted zone surrounding the precipitate) and become stable, in a solid state analogy to dendritic growth at a liquid/solid interface. The slow diffusion rates of the TCP forming elements, particularly Re which is a major constituent of the precipitates in these alloys, promotes this growth mechanism. In the case of P phase precipitates, the growth of one of these "secondary arms" involves the formation of a low energy interface to create a variant with the [001] growth direction at 90 degrees to that of the original precipitate. The precipitates are heavily faulted, and the formation of such an interface would not be difficult. The 4-fold symmetry of the sigma (001) axis precludes the necessity of such a variant.

SUMMARY

From the data presented here, it is clear that Re plays a special role in the formation of TCP phases in Ni base superalloys. It is a potent TCP former which does not act similarly to other refractory elements, such as W and Mo, but appears to substitute for Cr in the TCP phase. Computer programs such as PHACOMP which are used to predict alloy stability do not necessarily provide good correlation with TCP formation in these Re containing alloys and it seems that the way in which Re is treated in such programs requires reevaluation.

It is evident that Re containing alloys can be unstable with respect to three types of TCP phases: mu, sigma, and P. These phases often coexist in the alloy, _even within the same precipitate_, with very similar compositions. The crystal structures of these phases are closely related, particularly sigma and P, and it is likely that such multiple phase TCP precipitates exist in non-Re containing alloys as well. Thus, "sigma" formation , in these alloys at least, is more complicated than is traditionally believed.

Lastly, the origin of segmented plates of TCP phase involves a growth mechanism inherent in the phases themselves. Upon examination in three dimensions, it is found that they grow on the {111}gamma habit planes as "solid state dendrites" with preferred growth directions defined by the crystallography of the TCP phases. This also may be promoted by diffusion gradients around the growing precipitates. The relative detrimental effects of "segmented" versus continuous plates is not known.

ACKNOWLEDGMENTS

The authors wish to sincerely acknowledge the contributions of the following people: Dr. Stanley T. Wlodek for SEM analysis of the TCP phases, Mr. James C. Nickley and Mr. John Snyder for their work at various stages of the investigation.

REFERENCES

1. S. Chin and D.N. Duhl, "Evaluation of Advanced Single Crystal Superalloy Compositions", Final Report NADC 78120-60, (1980).

2. S. Rideout, W.D. Manly, E.L. Karmen, B.S. Lament and P.A. Beck, J. Metals/Trans AIME, Oct. (1951), 872-876.

3. D.P. Shoemaker, C.B. Shoemaker and F.C. Wilson, Acta Cryst., 10 (1957), 1-14.

4. M. Moringa, N.Yukawa, M. Adachi and H. Ezaki, "New PHACOMP and its Applications for Alloy Design", Superalloys 1984 Eds. M. Gell, et. al., (American Society of Metals, Metals Park, Ohio 1984), 523-532.

5. M.L. Frey and P. Aldred, "Effect of Rhenium on the Properties of Superalloys", General Electric Report No. R75AEG189, (1975).

THE INFLUENCE OF HIGH THERMAL GRADIENT CASTING, HOT ISOSTATIC

PRESSING AND ALTERNATE HEAT TREATMENT ON THE STRUCTURE AND

PROPERTIES OF A SINGLE CRYSTAL NICKEL BASE SUPERALLOY

L. G. Fritzemeier

Rocketdyne Division, Rockwell International
6633 Canoga Avenue
Canoga Park, CA 91303

SUMMARY

The results of a program to improve the cyclic properties of the single crystal superalloy PWA 1480 are reported. The program objective was to reduce or eliminate casting porosity as fatigue initiation sites by the application of improved commercial casting process parameters and hot isostatic pressing. An alternative to the standard PWA 1480 coating and aging heat treatment cycle was also evaluated for potential mechanical property improvement in high pressure hydrogen environment. Higher thermal gradient casting was found to provide a reduction in dendrite arm spacing, reduction in overall casting porosity density and, most importantly, a reduction in pore size. Tensile properties of PWA 1480 were not significantly affected by casting gradient while stress rupture lives were increased. Improvements in low cycle fatigue and high cycle fatigue lives were achieved as a result of the reduced pore size and improved homogeneity. The alternative heat treatment provides a slight increase in tensile ductility in hydrogen environment, but no apparent benefit to cyclic properties. The most dramatic improvements to material properties were provided by hot isostatic pressing. Increased high cycle fatigue life is directly attributed to elimination of casting pores as crack initiation sites. Increased stress rupture lives may be attributed to a delay in the onset of tertiary creep due to the absence of pre-existing voids and an increased volume fraction of fine gamma prime due to improved solution heat treatment. Hot isostatic pressing has now been instituted as a production requirement for the next generation of single crystal blades for the Space Shuttle Main Engine.

Introduction

Single crystal (SC) superalloys are currently under development as a potential replacement for directionally solidified MAR-M246 + Hf as turbine blades in the Space Shuttle Main Engine (SSME) high pressure turbopumps. The operating conditions in the SSME, however, are significantly different from the conditions in airbreathing gas turbine engines for which SC alloys were developed.(1) Consequently, the primary deformation modes of SSME turbine blades are low cycle fatigue (LCF) and high mean stress high cycle fatigue (HCF) compounded by hydrogen environment embrittlement rather than creep/stress rupture assisted by oxidation and sulphidation, for which the alloys were developed. The stress rupture, low cycle fatigue, high cycle fatigue and hydrogen environment embrittlement resistance of the candidate SC alloys are superior to DS MAR-M246, however, significant improvements in fatigue properties have been projected due to the application of advanced processing methods.(1,2) In the absence of intentional carbides in most modern SC superalloys, interdendritic casting porosity has been found to be the primary initiation site for fatigue failures. Reduction in the density and, more importantly, size of the inherent porosity can be expected to substantially improve fatigue lives from the view point of increased crack initiation times.

Increased casting thermal gradient has been shown to provide reduced dendrite arm spacing (DAS),(1) improved chemical homogeneity(3) and reduction in interdendritic pore size.(4) The influence of increased thermal gradient on SC nickel base superalloys was first demonstrated in the early 1970's,(5) but the technology to apply significant improvements on a commercial scale has only recently been developed.(6) Hot isostatic pressing (HIP) has also been recognized as a valuable tool for improving the fatigue capability of cast superalloys.(7) Early attempts at applying the technology to SC superalloys met with only limited success due to problems unique to the SC structure. Typical HIP temperatures and pressures produced plastic flow around closing pores, leading to local recrystallization. Fatigue cracks then initiate at the isolated secondary grains, negating the benefits due to elimination of porosity.(8) Additionally, low carbon single crystal superalloys contain many alloying elements with high affinities for carbon. Extremely careful control of HIP autoclave environments must be exercised to avoid significant surface carburization.(9) Despite these early concerns, development of a SC HIP process has been spurred on for the SSME program due to the potential benefits in improved fatigue lives. Alternative heat treatments for several superalloys,(10) have been shown to provide improvements in hydrogen environment embrittlement resistance.(11) Since creep is not a primary mechanism in SSME turbine blade deformation, alternatives to the PWA 1480 heat treatment have been evaluated for improved hydrogen environment properties.

The program described herein was undertaken to evaluate the benefits in material properties due to high thermal gradient casting, hot isostatic pressing and alternate heat treatments for PWA 1480.

Experimental Procedure

The PWA 1480 test materials for this program were cast in commercial facilities, representing the range of available thermal gradients. Commercial facilities were specified to facilitate transfer of results to the SSME program without the need to scale up from laboratory experiments. Materials representing three casting thermal gradients were obtained in the form of 1.25 to 1.6 centimeter diameter cylindrical rods. Primary crystallographic orientation of the cast bars was maintained within 10 degrees of the <001> direction. All of the test material received standard PWA 1480 solution heat treatment. Low and intermediate thermal gradient cast materials were tested in the standard heat treat condition; 1080C/4 hours plus 871C/32 hours. The intermediate and high thermal gradient materials were tested with an alternative heat treatment of 1010C/2 hours plus 871C/48 hours to evaluate possible improvements in hydrogen environment embrittlement. The intermediate gradient material was also tested in the hot isostatic pressed (HIP) plus standard heat treated condition. A viable HIP schedule has been devised which avoids the microstructural pitfalls associated with the densification of SC superalloys. The process was developed by Rocketdyne for application to PWA 1480 and has been found to be generic to other single crystal superalloys with the adjustment of HIP temperature profiles. Details of the process are restricted by U. S. Patent Secrecy Order.

Mechanical test samples were low stress crush ground from the fully processed castings. Standard straight gage section samples of 0.625 cm diameter were employed for all tests. Tensile and low cycle fatigue tests were conducted according to ASTM standards with extensometry attached to the specimen gage section. Tensile tests were conducted at 20C in air and 34.5 MPa hydrogen gas to verify the viability of the various processes. Low cycle fatigue tests were conducted at 538C, 0.33 Hz and 2.0% total strain range. Stress rupture tests were conducted at 871C and 550 MPa initial stress. High cycle fatigue tests were conducted in load control at room temperature and 843C, at a stress ratio of R = 0.47 at 30 Hz. The cyclic test parameters were chosen to represent conditions of interest for SSME turbopump turbine blades.

Optical metallography was employed to document the density and size of casting porosity and DAS. Statistics of the pore distribution were calculated from quantitative metallography measurements. Gamma prime morphology and distribution were evaluated by scanning electron microscopy. Optical and SEM fractography were conducted on all failed samples, with special emphasis placed on characterization of fatigue initiation sites.

Results

Representative photomicrographs of the casting porosity and etched microstructures are shown for the respective materials in Figure 1-3, with average area fraction of pores, average pore sizes and maximum pore sizes also given. The observed density and size of the interdendritic porosity are reduced as the thermal gradient is increased. In addition, pore morphology changes qualitatively to a lower aspect ratio, more circular cross section

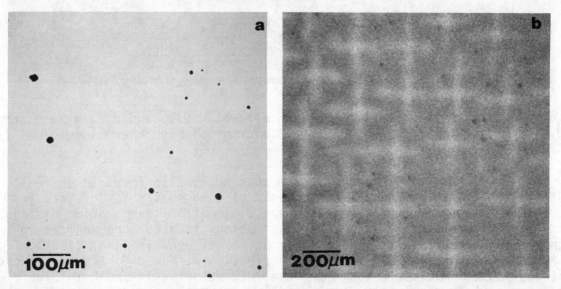

Figure 1. a)Typical porosity; area percent = 1.01 and d_{avg} = 32 micron; and b)typical microstructure; DAS = 446 micron; of low thermal gradient cast PWA 1480.

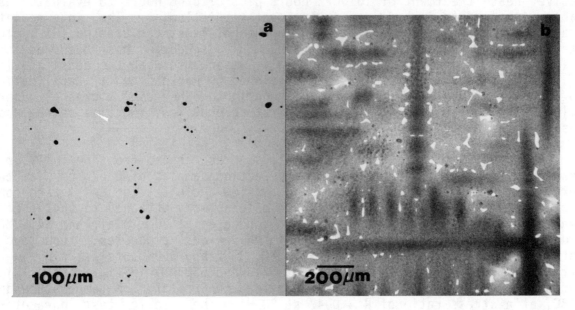

Figure 2. a)Typical porosity; area percent = 0.36 and d_{avg} = 20 micron; and b)typical microstructure; DAS = 324 micron; of intermediate thermal gradient cast PWA 1480.

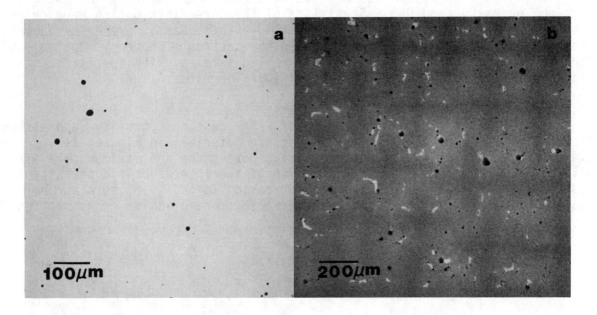

Figure 3. a)Typical porosity; area percent = 0.30 and d_{avg} = 14 micron; and b)typical microstructure; DAS = 222 micron; of high thermal gradient cast PWA 1480.

as pore size decreases. Dendrite arm spacing is also found to decrease as casting gradient is increased. An increase of homogeneity is inferred from a reduction in the amount of casting eutectic. The final gamma prime distribution between dendrite and interdendritic regions is also more uniform in the high gradient material. The influence of the alternative heat treatment is to force a more uniform gamma prime distribution with a slightly larger precipitate size. Microstructure of the HIP material is shown in Figure 4. Porosity has been completely eliminated and internal recrystallization has been avoided by the HIP process. The final microstructure is better homogenized due to increased time at the solution heat treatment temperature during the HIP process.

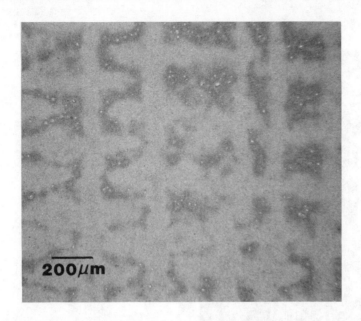

Figure 4. Microstructure of hot isostatically pressed PWA 1480 shows elimination of casting porosity and improved alloy homogeneity.

Tensile properties for the various materials are presented in Table I. Room temperature yield and ultimate strengths are unaffected by process parameters, except for increased ultimate strengths observed for the alternate heat treated material. The alternate heat treatment also provides a slight benefit in ductility in both air and high pressure hydrogen environments. Stress rupture results are given in Table II. Short time stress rupture life increased as thermal gradient increased and increased for HIP relative to non-HIP material. A SEM photomicrograph of a failed, low gradient cast, stress rupture sample is shown in Figure 5. Stress rupture failures initiate at interdendritic casting porosity and associated eutectic islands which link up by ductile overload in the intervening regions. Low cycle fatigue test results, presented in Figure 6, indicate slightly increased life as thermal gradient increased, for the strain range tested. LCF failures in the intermediate temperature regime initiate primarily at surface connected porosity and are Stage I, with the {111} fracture plane often encompassing the entire specimen. High cycle fatigue test results are presented in Figure 7. HCF life increases with increasing casting thermal gradient at

TABLE I. Typical Room Temperature Tensile Properties

Material Condition	Atm.	0.2% Yield (MPa)	Ultimate (MPa)	R. A. (%)	Elongation (%)
Low Grad/Std Heat Treat	Air	1024	1077	12.5	11.7
Int. Grad/Std Heat Treat	Air	1020	1179	5.6	4.0
Int. Grad/Alt Heat Treat	Air	992	1282	7.0	8.0
Int. Grad/HIP/Std Heat Treat	Air	1010	1122	9.2	12.5
Int. Grad/Std Heat Treat	H_2*	979	986	2.0	2.0
Int. Grad/Alt Heat Treat	H_2	1020	1089	4.0	3.0
Int. Grad/HIP/Std Heat Treat	H_2	970	1007	3.2	3.1
High Grad/Alt Heat Treat	Air	1082	1211	10.3	10.3

*H_2 = 34.5 MPa Hydrogen Atmosphere

TABLE II. 871C, 550MPa Stress Rupture Lives

Material Condition	Time To Failure (Hours)	Elongation (%)
Low Gradient/Std Heat Treat	Not Tested	
Int. Gradient/Std Heat Treat	12.5	11.4
Int. Gradient/HIP/Std Heat Treat	61	10.6
High Gradient/Alt. Heat Treat	63.4	14.6

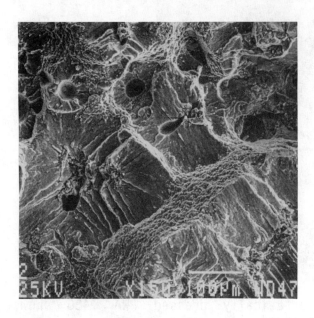

Figure 6. Low gradient cast stress rupture fracture surface.

Figure 7. Low cycle fatigue life ranges at 538C and 2.0% strain range.

Figure 7. High cycle fatigue curves at R = 0.47, 30 Hz and a)843C and b)24C.

both room temperature and 870C. Alternative heat treatment was found to have no discernable influence on fatigue life. Those test results are, therefore, included in the representative fatigue curves. Representative fractographs are presented in Figure 8. HCF failures initiate predominantly at internal interdendritic casting porosity at 870C. Propagation is Stage II at 870C until overload is reached or until the crack meets the specimen surface, when crack growth becomes Stage I. Room temperature propagation is predominantly Stage I from near surface pore initiation. HCF life increases as initiating pore size decreases. HCF curves for HIP material are compared to the non-HIP material in Figure 9. The greatest benefit in fatigue life is obtained at room temperature. Failures of the HIP material initiate at the specimen surface, eutectic islands or at isolated small carbides, as shown in Figure 8d.

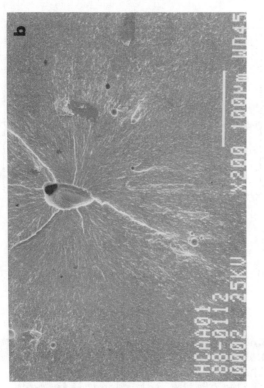

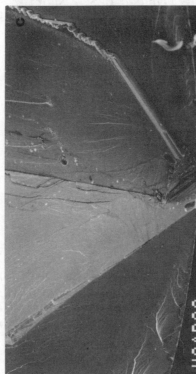

Figure 8. Representative high cycle fatigue fractographs. a) and b) high gradient cast, 843C, R = 0.47, maximum stress = 793 MPa, N_f=924,869; c) low gradient cast, 24C, R = 0.47, maximum stress = 793 MPa, N_f=111,789 and d) HIP intermediate gradient cast, 24C, R = 0.47, maximum stress = 896 MPa and N_f=6.95 X 10^6

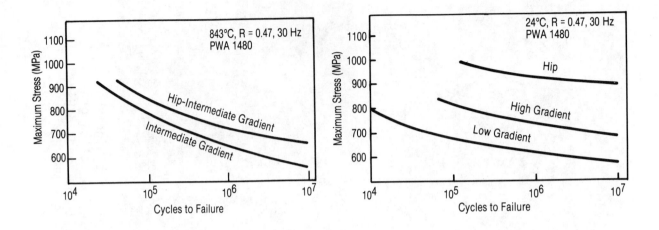

Figure 9. Comparison of fatigue life of HIP versus non-HIP PWA 1480 at a) 843C and b) 24C.

Discussion

Higher thermal gradient single crystal casting has been found to provide benefits in terms of microstructure and improved cyclic loading properties. Increased casting thermal gradient reduces the DAS, provides improved chemical homogeneity and reduces the size and density of the interdendritic porosity. The improved homogeneity and reduced pore size are both direct consequences of the reduced DAS. Shorter diffusion distances within the solidification front reduce the amount of solute rejected from each dendrite core and reduce the local magnitude of interdendritic eutectic formed. These shorter diffusion distances also improve the homogenization provided during a conventional heat treatment cycle. Casting porosity is formed by the inability of the liquid metal to flow into the dendritic interstices and subsequent shrinkage as the remaining liquid freezes. Finer DAS reduces the volume of liquid in the interdendritic region, resulting in a finer pore size. Pore morphology is also changed to a more spherical shape as surface tension reduces the surface area to volume ratio.

Tensile properties at room temperature are virtually unaffected by the microstructural improvement due to high gradient casting. The primary benefit is manifested as an increase in stress rupture capability at high stresses. There are several factors which contribute to this improvement. The reduction of the amount of casting eutectic increases the free concentration of the alloying elements responsible for the formation of fine gamma prime. A slightly increased volume fraction of the optimum size precipitates is then produced. In addition, the gamma prime size in both dendritic and interdendritic regions is more uniform and closer to the optimum size for creep resistance. The onset of tertiary creep is also delayed due to the reduction of porosity size and density in the initial microstructure. Crack nucleation must then take place on a broader scale, prior to linkup and net section loss leading to failure.

Low cycle fatigue lives were found to increase with increased casting thermal gradient. The relatively high strain range employed produces a significant amount of macroscopic strain. Under these conditions, crack initiation can occur in slip bands, reducing the importance of internal defects as initiation sites. The improved LCF capability in this test regime is primarily due to the improvement in strength and/or ductility afforded by the improved alloy homogeneity. High cycle fatigue lives are very clearly controlled by crack initiation at internal defects; especially at low temperature. Increased HCF lives with increased casting thermal gradient are a direct consequence of reduced interdendritic pore size and lower aspect ratio morphology. This benefit has recently been analytically predicted through linear elastic fracture mechanics calculations.(12) It is apparent, though, that return on investment for increasing thermal gradient will diminish due to practical limitations on production casting gradients.

Hot isostatic pressing, when properly applied, provides a significant improvement in high cycle fatigue life, especially at lower temperatures and long lives. At elevated temperatures stress rupture becomes a contributing factor in the final failure and reduction in initiating defect size is not as significant. Again, the increased cyclic life can be partially attributed to the removal of porosity as fatigue crack initiation sites since no appreciable affect of HIP was found on tensile properties. The high temperature, high mean stress fatigue capability in the creep/fatigue regime may also be improved slightly by the increased stress rupture life obtained due to improved solution heat treatment and pore removal. Alternative heat treatment had no significant impact on the cyclic properties of PWA 1480 since defect tolerance, rather than bulk mechanical properties, is the overriding factor in HCF life. This conclusion is consistent with recent results on the fatigue capability of rafted microstructures.(2)

Conclusions

High thermal gradient casting is an effective method for improving the homogeneity and reducing the size, amount and aspect ratio of interdendritic porosity in single crystal nickel base superalloys. Primary benefits due to improved thermal gradients are manifested as increased stress rupture and increased fatigue lives. A diminishing rate of return on improved properties attributed to high thermal gradient casting places practical limits on its use for production hardware.

Alternative heat treatment does not significantly affect the cyclic properties of PWA 1480 since fatigue life is predominantly controlled by crack initiation at microstructural defects. Hot isostatic pressing is an effective method of improving the cyclic properties of single crystal superalloys especially in the long life regime. Increased time at the solution heat treat temperature during HIP improves the alloy homogeneity and provides a significant increase in stress rupture life. Elimination of casting porosity shifts fatigue crack initiation sites to less severe defects such as carbides and, therefore, significantly improves fatigue life. Improved casting homogeneity and reduction of the initial pore size due to high gradient casting improves the homogenization due to solution heat treatment and increases process parameter windows for the HIP process, reducing the possibility of recrystallization due to pore closure.

Acknowledgements

Portions of this program were funded by Rocketdyne discretionary funds, the Space Shuttle Main Engine program and NASA Technology contract NAS3-24646. Discussions with J. D. Frandsen were instrumental in developing the HIP process. The author is indebted to A. K. Thompson for his efforts in microstructural and fractographic characterization.

REFERENCES

1. W. T. Chandler, "Materials for Advanced Rocket Engine Turbopump Turbine Blades," (NASA CR-174729, Lewis Research Center, 1983.

2. T. Khan et. al., "Single Crystal Superalloys for Turbine Blades: Characterization and Optimization of CMSX-2 Alloy" (Paper presented at the 11th Symposium on Steels and Special Alloys for Aerospace, Paris Air Show, Le Bourget, 6 June 1985).

3. M. McLean, <u>Directionally Solidified Materials for High-Temperature Service</u> (London, The Metals Society, 1983).

4. D. D. Pearson, D. L. Anton and A. F. Giamei, "High Thermal Gradient Superalloy Crystal Growth" (Presented at NASA Conference on Structural Integrity and Durability of Reusable Space Propulsion Systems, Cleveland Ohio, June 4-5, 1985).

5. J. K. Tien and R. P. Gamble, "The Suppression of Dendritic Growth in Nickel-Base Superalloys during Unidirectional Solidification," <u>Materials Science and Engineering</u>, 8(1971), 152-160.

6. M. J. Goulette, P. D. Spilling and R. P. Arthey, <u>Proceedings of the Conference on In-situ Composites III</u> (Lexington MA., Ginn and Co. 1979).

7. D. J. Kenton, "HIP - Nemesis of Shrinkage in Investment Castings," <u>Diesel and Gas Turbine Progress</u>, March 1976.

8. T. E. Strangman, M. Fujii and X. Nguyen-Dink, "Development of Coated Single-Crystal Superalloy Systems for Gas Turbine Applications," <u>Superalloys 1984</u>, ed. M. Gell et. al. (Warrendale, PA: The Metallurgical Society, 1984), 795-804.

9. R. L. Dreshfield, "Application of Single Crystal Superalloys for Earth-to-Orbit Propulsion Systems", (Presented at the 23rd Joint Propulsion Conference, San Diego, CA, June 29-July 2, 1987).

10. T. Khan and P. Caron "Effect of Heat Treatments on the Creep Behaviour of a Ni-Base Single Crystal Superalloy", (Paper presented at the 4th RISO International Symposium on Metallurgy and Materials Science, Denmark, 1986).

11. Rocketdyne Division unpublished research.

12. S. Majumdar and R. Kwasny, "Effects of a High Mean Stress on the High-Cycle Fatigue Life of PWA 1480 and Correlation of Data by Linear Elastic Fracture Mechanics" (NASA CR-175057, Argonne National Laboratory, November 1985.)

THE EFFECT OF TEMPERATURE ON THE DEFORMATION STRUCTURE OF

SINGLE CRYSTAL NICKEL BASE SUPERALLOYS

M. Dollar[‡] and I.M. Bernstein[‡]

Carnegie Mellon University, Pittsburgh, PA 15213

‡ now at the Illinois Institute of Technology, Chicago, IL 60616

ABSTRACT

The temperature dependence of the yield and flow stress was analyzed in the superalloys PWA 1480 and CMSX-2. From measured dislocation densities at different strains and temperatures, flow stresses were accurately predicted from an extended Copley and Kear model (5). Differences in the strain response of the two alloys were described.

I. INTRODUCTION

Nickel-base superalloys offer desirable creep strength, thermal fatigue stress, oxidation resistance, and hot corrosion resistance [1,2], making them important materials for gas turbine engine applications. Recently, single-crystal nickel-base superalloys have been developed, in order to remove grain boundary strengthening elements, providing an increase in the incipient melting temperature [3,4]. This permits solution heat treatments to be carried out at higher temperatures which both improves materials homogeneity and dissolves coarse primary particles.

The ability to use these superalloys at progressively higher temperatures has encouraged a number of recent studies designed to fully characterize their high temperature properties [3,4]. What is lacking are a sufficient number of attempts to understand the fundamentals of the deformation processes in superalloys, particularly the effects of temperature on superalloy deformation and properties. Earlier efforts were at the best incomplete [5,6]. This perceived gap has led to the present research. Its aim has been to clarify the role of temperature on the development of deformation structure and tensile properties in PWA 1480 and CMSX-2, produced by Pratt and Whitney Aircraft and Cannon-Muskegon Corporation, respectively, both being considered for use in the space shuttle main engine [7].

II. EXPERIMENTAL PROCEDURES

The chemical compositions of the superalloys investigated in the present study are given in Table I. They were provided in the form of bars, both with longitudinal orientations within 10° of [001]. Tensile samples were then machined in accordance with ASTM specification, selecting only bars with orientations $5° \pm 1°$ from [001], to minimize any influence of variations in orientation. The materials were heat treated according to schedules reported in Table II.

Table I. Alloy Compositions (wt. pct.)

	Al	Co	Cr	Ta	Ti	W	Mo	Ni
PWA 1480	4.8	5.3	10.4	11.9	1.3	4.1	---	bal.
CMSX-2	5.6	4.6	8.0	6.0	1.1	8.0	0.6	bal.

Tensile testing was conducted on an Instron machine at temperatures from $20°C$ to $800°C$ at an initial strain rate of $1.2 \times 10^{-3} \, s^{-1}$. Interrupted tensile tests to selected strain levels were also carried out to study the development of deformation structure.

TABLE II. The heat treatment procedures for PWA 1480 and CMSX-2 superalloys

PWA 1480	CMSX-2
1285°C / 4h	1315°C / 3h
1080°C / 4h	1050°C / 16h
870°C / 32h	850°C / 48h

Discs for transmission electron microscopy (TEM) were prepared from undeformed heat treated materials, as well as from tensile samples strained to different strain levels at different temperatures. The discs were cut perpendicular to the tensile axis, as well as parallel to {111} planes. Thin foils were prepared by twin-jet electropolishing and examined at 120kV, in a Phillips 420 microscope.

III. EXPERIMENTAL RESULTS

A. Microstructure

The general characteristics of the macrostructure and microstructure produced by the heat treatments reported in Table II in both alloys have previously been presented (CMSX-2 [8], PWA 1480 [6]).

In both, a high volume fraction of 65% cuboidal, ordered particles in a matrix were observed, with an edge length of 500 ± 100 nm (400 ± 150 nm in PWA 1480). TEM failed to reveal the presence of γ/γ' interfacial dislocations, in accordance with the reported small γ/γ' misfit at room temperature of 0.14% for CMSX-2 [9] and 0.28% for PWA 1480 [6].

B. Mechanical Properties

The true values of the yield stress, flow stress and ultimate tensile stress at 1% strain, versus temperature for PWA 1480 are shown in Fig.1. The yield stress at room temperature of about 1000 MPa decreases quite significantly with increasing temperature until 400°C, above which the yield stress then increases, peaking at 730°C. Beyond 730°C, the strength again drops. The other stresses behave similarly, (Fig.1), compatible with the strong work hardening of the alloy over all strains and temperatures examined.

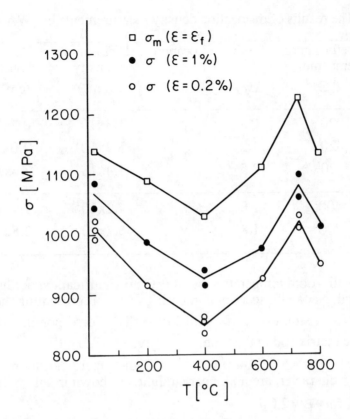

Figure 1. The effect of temperature on the yield stress, the flow stress for the strain of 0.01 and the ultimate tensile stress in PWA 1480.

The yield stress at room temperature for CMSX-2 is slightly smaller than PWA 1480 and in addition there is no drop in strength beyond this temperature. In fact, the yield stress is the same at 20 and 400°C, with a peak yield stress peak at 730°C (as in PWA 1480). In contrast with PWA 1480, much less work hardening is observed in CMSX-2.

C. *Deformation structures*

The development of dislocation structure at different stress levels was investigated in PWA 1480 at 20, 400 and 730°C in [001] oriented foils, unless indicated otherwise. This is the most appropriate orientation to observe simultaneously the deformation behavior of both phases in [001] single crystals which contain particles with their faces parallel to the [001] growth direction.

At room temperature and a strain of 0.24%, significant dislocation activity was already present in the matrix (Fig.2), in contrast with the γ' phase, where dislocations were observed only occasionally. The matrix dislocations were determined to have a Burgers vector of <110> and usually were present as loosely coupled pairs, suggesting that they are superdislocations trapped in the matrix at the earliest stages of plastic deformation.

With increasing strain the dislocation density increases in both the matrix and the precipitates, but is still substantially higher in the matrix, as illustrated in Fig. 3, for a strain of 1.2%. Dislocation densities were measured in the matrix for strains of 0.24, 1.2, and 3.5% (the strain to failure), and are given in Table III.

The dislocation structure in a foil cut parallel to {111} planes after 1.2% strain is shown in Fig.4. The use of this orientation permits superdislocations to be shown on their slip planes and are seen to be straight <110> superdislocations of screw character (i.e. parallel to <110> directions). By the use of weak-beam dark field imaging the mean separation of superpartials was found to be 4.8 ± 0.8 nm.

TABLE III. The results of dislocation density measurements in PWA 1480 superalloy

Deformation temperature (°C)	Strain (%)	Location	Dislocation density (cm^2)	Standard deviation (cm^2)
20	0.24	γ	1.3x10^{10}	0.7x10^{10}
20	1.2	γ	5.2x10^{10}	2.0x10^{10}
20	3.5	γ	1.4x10^{11}	3.6x10^{10}
400	1.8	γ	4.2x10^{10}	1.7x10^{10}
400	1.8	γ'	5.7x10^{10}	2.8x10^{10}

In contrast with room temperature, at 400°C no significant superdislocation trapping in the matrix was found, as can be seen by comparing Fig. 5, representing the typical dislocation structure at 400°C for a strain of 1.8% with Fig. 3. The corresponding dislocation densities measured in both the matrix and in the γ' phase are given in Table III.

In contrast with both 20 and 400°C, at 730°C, the superdislocations while still exhibiting predominantly screw character, are no longer straight, as shown in Fig. 6, representing a typical dislocation structure for $\varepsilon = 2.0\%$.

The development of the dislocation structure in CSMX-2 was investigated at room temperature only. At the earliest stages of plastic deformation, numerous dislocations were found in the matrix, with occasional superdislocation pairs observed in the precipitates. Detailed analysis indicates that, as in PWA 1480, the matrix dislocations are most likely superdislocations trapped in the disordered phase.

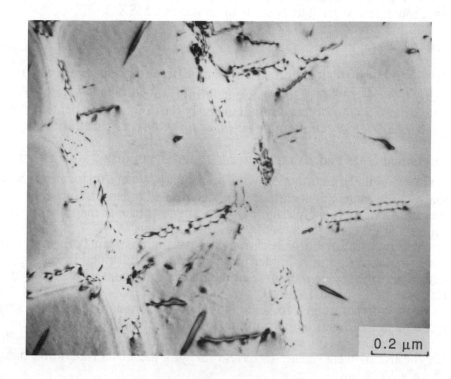

Figure 2. TEM micrograph, PWA 1480, T_{def}= 20°C, ε_P=0.24%, the foil normal is [001].

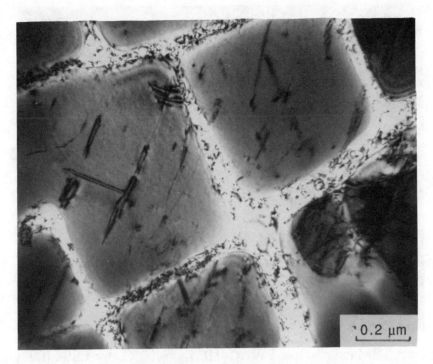

Figure 3. TEM micrograph, PWA 1480, T_{def}= 20°C, ε_P=1.2%, [001] foil.

Figure 4. TEM micrograph, PWA 1480, T_{def}= 20°C, ε_P=1.2%, [111] foil.

At higher strains the comparative dislocation structure in both alloys is remarkably different. In CMSX-2 the disocation distribution is more uniformly distributed through both the γ and γ'. There appear to be no barriers to prevent the superdislocations from moving over appreciable distances, suggesting the mean free path of superdislocations is significantly greater in CMSX-2 than in PWA 1480.

The mean separation of superpartials was also measured for CMSX-2 and found to be 4.6 ± 0.3 nm in CMSX-2, not significantly different from PWA 1480.

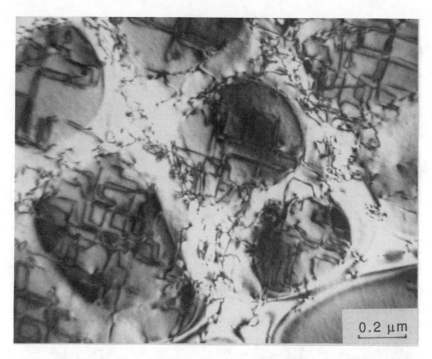

Figure 5. TEM micrograph, PWA 1480, $T_{def}=400°C$, $\varepsilon_P=1.8\%$, [001] foil.

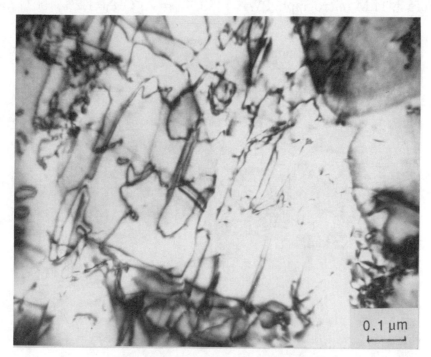

Figure 6. TEM micrograph, PWA 1480, $T_{def}=730°C$, $\varepsilon_P=2.0\%$, [001] foil.

IV. DISCUSSION

A reasonable approach for a discrete two phase single crystal superalloy is to try to calculate its yield stress from the yield stresses of its constituent single phases. For example with PWA 1480 appropriate data is available [5,10]. The yield stress, σ_P, versus temperature of a single-phase alloy of chemical composition similar to that of γ' in PWA 1480, is shown in Fig.7 (curve c), as is the yield stress, σ_M, as a function of temperature (curve a) of a commercial, solution hardened Hasteloy X alloy, of chemical composition similar to that of the γ phase in PWA 1480 (for both, the critical resolved shear stresses (CRSS) for {111}<110> primary octahedral slip are multiplied by the reciprocal of the Schmid factor M = 0.41, the value appropriate for the <001> orientation [5,11]). The experimental values of the yield stress for PWA 1480 are shown in Fig. 7 as well. It is apparent that the use of any conventional model for yielding in two-phase alloys [12] can not allow a quantitative prediction of the yield behavior of the superalloy, since the resultant stress, as calculated by conventional models, would never exceed that of the harder phase. Any successful explanation has to account for the strong strengthening of the superalloy in comparison to its ordered phase.

Models have been developed by Huther and Reppich [13] and by Copley and Kear [5], specifically for alloys strengthened by large, coherent, ordered particles. The latter authors considered both the leading and trailing dislocations (forming a superdislocation) in the balance of forces, and the additional strengthening of a superalloy arises from the energy that must be supplied as a superdislocation passes from the disordered matrix into the ordered particle. The CRSS, σ_C, is given by:

$$\tau_C = \left(\frac{E_A}{2b}\right) - \left(\frac{T}{br}\right) + [0.41(\tau_M + \tau_P)] \quad (1)$$

where: E_A = antiphase boundary energy, b = dislocation Burgers vector, T = dislocation line tension, r = particle radius, τ_M = CRSS of γ and τ_P = CRSS of γ'.

We can apply Eq. (1) to calculate first the yield stress of PWA 1480 at room temperature, and then to model the temperature dependence of the yield stress following Copley and Kear's procedure for the MAR M200 superalloy [5].

At room temperature: G = 57 GPa (G is estimated as G_{Ni} parallel to {111}<110> [14]), b = 2.5 x 10^{-8} cm, and r = 0.15 x 10^{-4} (as measured from TEM results). T was calculated from T = $Gb^2/2$ [15], E_A was calculated from the relationship developed in reference [16]:

$$E_A = \frac{c G b^2}{2 \pi d} \quad (2)$$

where: c = constant equal to unity for screw dislocations [16], and d = the mean distance between superpartials, measured to be 4.8 nm. A value of E_A = 118 mJ/m^2 was estimated. Using τ_M = 180 MPa and τ_P = 115 MPa from [5], the CRSS of PWA 1480 predicted by Eq.(1) then equals 317 MPa. Thus, the predicted yield stress:

$$\sigma_C = M^{-1} \tau_C \quad (3)$$

equals 773 MPa for Schmid factor M=0.41 compared to the experimental value of 1013 MPa.

To model the temperature dependence of the yield stress for [001] oriented PWA 1480 crystals, the τ_M and τ_P versus T relationships were used exclusively by assuming that the other parameters appearing in Eq. (1) were temperature-independent. The resultant yield stress-temperature relationship obtained (M = 0.41) is shown in Fig. 7 as $\tau_C = \tau_1 + \tau_2$ (curve d).

The above approach successfully predicts to within almost 10% why at temperatures above 400°C superalloys in general, and PWA 1480 in particular, are strengthened in excess of the intermetallic compounds forming their precipitates. However, the approach does not explain why the experimental yield stress at room temperature is much higher than predicted, and also why there is a significant drop of the yield stress with increasing temperature till 400°C.

To better model this region of the yield stress-temperature regime, we have made use of our TEM studies, which showed the existence of a substantial dislocation density in the matrix slightly above yield at room temperature. In as much as Eq. (3) was derived under the implicit

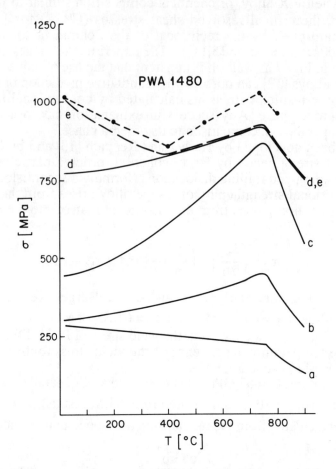

Figure 7. The plot of yield stress versus temperature curves (a) τ_M - yield stress of γ (b) τ_P - yield stress of γ' (c) $\tau_1 = .41(\tau_M + \tau_P)$ (d) τ_C (Eq. 3) (e) σ_Y of PWA 1480. Experimental data represented by dots.

assumption that a moving superdislocation, passing from the matrix into the γ' phase does not interact with other dislocations, it can not reflect the dislocation accumulation in the matrix which inevitably strengthens the superalloy. From conventional work hardening theories [17,18], the additional strengthening can be given as:

$$\delta\sigma = M^{-1} \alpha G b \rho^{1/2} \qquad (4)$$

where: α = constant, typically about 0.5 and ρ = total dislocation density.

The yield stress, σ_Y, can be now postulated to consist of two components: σ_C (Eq. (3)) and $\delta\sigma$ (Eq. (4)), thus:

$$\sigma_Y = M^{-1}\left\{\left(\frac{E_A}{2b}\right) - \left(\frac{T}{br}\right) + [0.41(\tau_M + \tau_P)] + \alpha G b \rho^{1/2}\right\} \qquad (5)$$

Taking $\rho = 1.3 \times 10^{-10}$ cm^{-2} at yield (Table III) and fitting the value of $\alpha = 0.6$, we now estimate the yield stress at room temperature to be 1010 MPa, in excellent agreement with the experimental yield stress. Clearly this exactness results from the choice of $\alpha = 0.6$, a very reasonable value in agreement with many work hardening theories. Since above 400°C no pronounced dislocation accumulation was observed in the matrix, the incremental strengthening given by Eq. (4) is negligible and we assume that it decreases from 98 MPa at 20°C to zero at 400°C. Using this approach, curve e in Fig. 7 is obtained, and is in excellent agreement with experiment.

To test the generality of the approach used for PWA 1480 the yield stress versus temperature relationship for CMSX-2 was also calculated. The experimental data is re-shown in Fig. 8 and the yield stress is calculated from Eqs.(1-3), using measured $r = 0.18 \times 10^{-4}$ cm and $d = 4.6$ nm, calculating $E_A = 123$ mJ/m^2, and taking the same $\tau_M(T)$ and $\tau_P(T)$ relationships, G and b as for PWA 1480. The predicted σ_C versus T relationship for [001] oriented crystals (M=0.41) is shown in Fig. 8 and agrees with experiment to within 11% over the entire temperature range, in contrast to PWA 1480.

It is clear, and supported by TEM results, that in CMSX-2, unlike PWA 1480, strength increases due to work hardening do not occur. While at small strains some superdislocations are trapped in the matrix of CMSX-2, the tendency is much less pronounced than for PWA 1480. For instance, at room temperature the dislocation density in the matrix of CMSX-2 at 0.8% strain is significantly lower than that in the matrix of PWA 1480 at 0.24% strain.

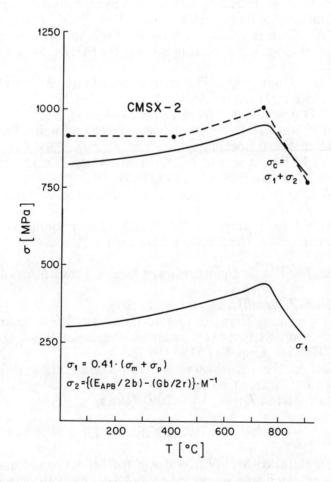

Figure 8. Plot of σ_Y versus temperature. The σ_C curve is predicted for CMSX-2. Experimental data represented by dots.

V. CONCLUSIONS

1. PWA 1480 and CMSX-2 superalloys are strengthened in excess of their constituent phases. This is a consequence of the fact that energy must be supplied as a superdislocation passes from the disordered matrix into the ordered particle. If this results in the trapping of superdislocations in the matrix, an additional increment of strengthening ensues.
2. The yield stress versus temperature relationships in both superalloys can be reasonably reproduced from existing or extended phenomenological models, taking account of both theoretical considerations and structural observations.
3. The development of dislocation structure is different in the two superalloys (as is their ductility and work hardening). The reasons for such behavior is yet to be established.

REFERENCES

1. R.F.Decker and T.S.Sims, in: *The Superalloys*, T.S.Sims and W.C.Hagel (eds.), John Wiley and Sons, New York 1972, p.33
2. T.S.Sims, in: *Superalloys '84*, Proc.5th International Symposium on Superalloys, M.Gell et al. (eds.), The Metallurgical Society Publication, New York 1984, p.399
3. M.Gell, D.N.Duhl, A.F.Giamei, in: *Superalloys '80*, Proc.4th International Symposium on Superalloys, J.K.Tien et al. (eds.), American Society for Metals, Metals Park Ohio 1980, p.205
4. M.Gell, D.N.Duhl, D.K.Gupta and K.D.Sheffer, *J. of Metals*, **39**, 11 (1987)
5. S.M.Copley and B.H.Kear, *Trans. of AIME*, **239**, 984 (1967)
6. W.W.Milligan and S.D.Antolovich, *Metall.Trans.*, **18A**, 85 (1987)
7. W.T.Chandler, Final Report - Materials for Advanced Rocket Engine Turbopump Turbine Blades, Rockwell International Rocketdyne Division, Canoga Park, CA, 1984
8. C.L.Baker, J.Chene, I.M.Bernstein and J.C.Williams, *Metall.Trans.*, **19A**, 73, (1988)
9. P.Caron and T.Khan, *Mat. Sci.Eng.*, **61**, 173 (1983)
10. B.J.Piercay and R.W.Smashey, *PWA Report 65-018*, Pratt and Whitney Aircraft, North Haven Conn. 1965
11. D.M.Shah and D.N.Duhl, in: *Superalloys '84*, Proc.5th International Symposium on Superallys, M.Gell et al. (eds.), The Metallurgical Society Publication, New York 1984, p.105
12. R.W.K.Honeycombe, *The Plastic Deformation of Metals*, Edward Arnoldt Publ., London 1984, p.252
13. W.Huether, B.Reppich, *Z.Metallkunde*, **69**, 628 (1978)
14. J.F.Nye, *Physical Properties of Crystals*, Oxford University Press, London 1957, p.148
15. D.Hull and D.J.Bacon, *Introduction to Dislocations*, Pergamon Press, Oxford 1984
16. J.T.M. De Hosson, *Mat.Sci. Eng.*, **81**, 515 (1986)
17. D.Kuhlmann-Wilsdorf, in: *Work Hardening in Tension and Fatigue*, A.W.Thompson (ed.), The Metallurgical Society Publication, New York 1976, p.1
18. D.Kuhlmann-Wilsdorf, *Metall.Trans.*, **16A**, 2091 (1985)

VI. ACKNOWLEDGEMENTS

The authors are grateful to Mr. W.S.Walston for his technical assistance and many fruitful discussions. This research was supported by NASA-Lewis Research Center under the technical direction of Dr. R. Dreshfield.

INTERMEDIATE TEMPERATURE CREEP DEFORMATION IN CMSX-3 SINGLE CRYSTALS

Tresa M. Pollock and A.S. Argon

Massachusetts Institute of Technology, Cambridge, MA

Abstract

Creep deformation in <001> oriented single crystals of CMSX-3 has been studied at temperatures around 850°C at a stress level of 552 MPa. Detailed observations of the evolution of the dislocation structure during creep have been made. TEM stereo pairs show creep deformation in the matrix phase, leaving the γ' precipitates undeformed. In previously undeformed crystals there is a brief incubation period during which dislocations spread from widely spaced sources, bowing out between the precipitates in the narrow matrix gaps. This Orowan bowing process is the major source of creep resistance. In the early stages of primary creep, over large regions dislocations all have the same Burgers vector in spite of the fact that there are eight equivalent slip systems. Later in the primary transient dislocations with other Burgers vectors interpenetrate from neighboring regions, gradually building up a three dimensional nodal network, which is characteristic of steady state creep. At 850°C static recovery processes are sluggish, indicating that dynamic recovery may be important at this temperature and at lower temperatures. Finite element analysis shows that the constraint of the non-deforming precipitates leads to the buildup of a negative pressure in the matrix channels normal to the applied stress. The negative pressure provides load support and reduces the effective stresses driving the steady state creep process in the matrix. The γ' precipitates are increasingly stressed as the creep process progresses, and under the conditons studied here, they are occasionally sheared in the later stages of steady state creep.

Introduction

Nickel base alloys have long been used in high temperature applications because of their exceptional creep resistance at high fractions of their melting temperature. Based on the bulk properties of the constituent γ and γ' phases, continuum estimates of the creep strength greatly underestimate the actual properties of these two phase alloys. It is clear that while the presence of the high volume fraction of ordered precipitates must be altering the development of the creep substructure relative to what is observed in the bulk solid solution materials, that the actual scale of the γ' phase domains is also of key importance. For this reason a detailed study of the development of the dislocation structure during creep has been undertaken. The temperature range of 800°C to 900°C has been selected since in this range the γ' precipitates are stable with respect to rafting in the duration of a normal laboratory test. An understanding of the creep process at these intermediate temperatures should provide a meaningful starting point for dealing with the creep problem at higher temperatures where the precipitates are changing shape.

Experimental Procedures

CMSX-3 single crystals of $<001>$ orientation and nominal composition listed in (1) were used in this investigation. Samples were given a solution treatment of 1293°C/2 hours + 1298°C/3 hours. A two step aging treatment of 1080°C/4 hours + 870°C/16 hours produces 65-70 volume percent cuboidal gamma prime precipitates, 0.45μm in average diameter.

Prior to testing, samples were electropolished and thermocouples were welded to the sample surface for accurate temperature measurement. Creep experiments were run in a resistance heated vacuum furnace, with four LVDTs mounted on the loading rods for measurement of creep strains. The majority of the creep tests were at a temperature of 850°C (T/T_m=0.7) and a stress of 552 MPa. To assess the effects of static recovery on the mechanical response of the material, samples were deformed to steady state creep, unloaded to 10% of the original load, allowed to anneal for varying amounts of time, and reloaded to record the transients.

For TEM observations, the samples were cooled under load to preserve the dislocation structure produced by creep. Foils were cut from planes normal to the direction of stressing and were prepared by conventional jet polishing techniques. Stereo techniques were utilized to obtain a three dimensional view of the dislocation arrangements. With a double tilt holder stereo pairs were obtained by maintaining a constant diffraction condition while tilting an average of 10-20°. The average foil thickness in the vicinity of the observations was 1 to 1.5 times the precipitate diameter. In-situ annealing experiments were performed using a single tilt high temperature heating stage.

Experimental Results

Mechanical Response of CMSX-3 in the Creep Regime

A typical creep curve for 850°C/552 MPa is shown in Figure 1. A short primary transient is followed by steady state creep at a rate of 2.5×10^{-8}/s. There is also a long period of accelerating tertiary creep, but this aspect of the creep problem will not be dealt with here.

Closer inspection of the response of the sample immediately after loading reveals an incubation period prior to the primary transient, as shown in Figure 2. At 850°C/552 MPa the incubation period is approximately 600 seconds in length. At 800°C the incubation is 1.6×10^4 seconds, while at 900°C it is immeasurably short.

In Figure 3 are the results of a recovery test at 850°C/552 MPa. After steady state creep was reached, the load was reduced by 90% and the sample was annealed

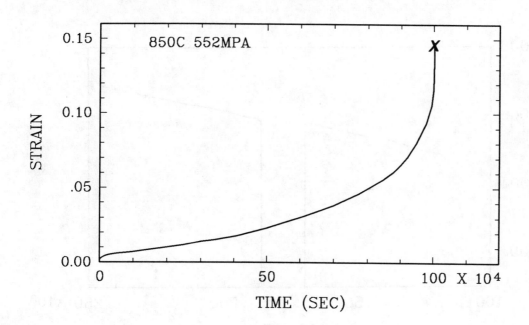

Figure 1 - Typical creep curve for < 001 > oriented CMSX-3 at 850C/552MPa.

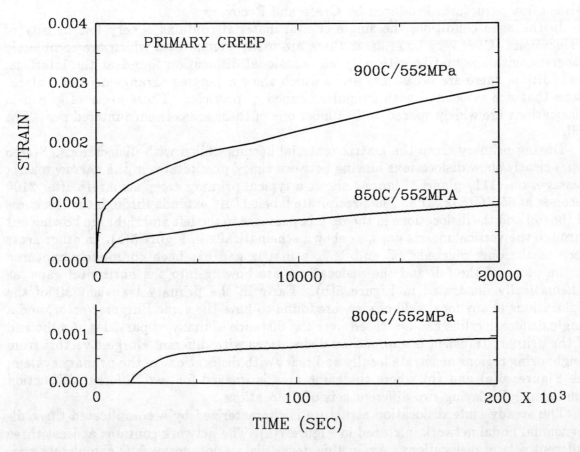

Figure 2 - Details of the primary transient. Note the incubation periods at 800 and 850°C.

for 1.62×10^4 seconds (4.5 hours). Upon reloading a very slight transient was observed, during which a strain of only 2.1×10^{-4} was accumulated.

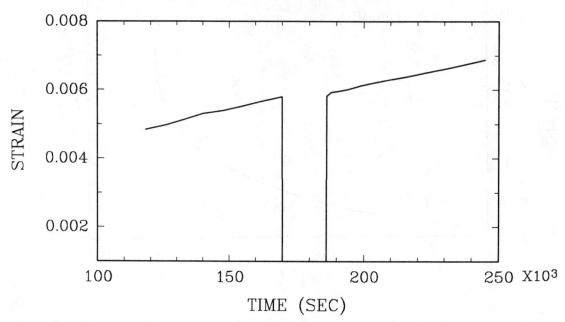

Figure 3 - The effect of static recovery on the reloading transient of a sample which was deformed to steady state creep, unloaded for 1.6×10^4 seconds, and then reloaded.

Dislocation Structures Produced by Creep and Recovery

In the aged condition, the single crystal material contains a very low density of dislocations. Over very large areas there are many precipitates which are completely coherent with the matrix with only an occasional dislocation found at the interface. In addition, there are occasional areas which show a tangled arrangement of dislocations that are associated with irregular shaped γ' particles. These areas of grown-in dislocations are widely spaced and at most one of these areas is encountered per TEM foil.

During primary creep the matrix material becomes filled with dislocations. Stereo pairs clearly show dislocations bowing between the γ' precipitates in the narrow matrix passages on {111} planes. Figure 4 shows a typical primary creep structure after 2100 seconds at 850°C/552 MPa. The precipitate labeled "A" extends through the thickness of the foil and the dislocations in the matrix passages to the left and right are bowing out through the vertical matrix gaps, as shown schematically in Figure 5(a). In other areas such as the ones marked "B" and "C", a matrix gap has been completely captured in the plane of the foil and the dislocations are bowing into the horizontal gaps, as schematically illustrated in Figure 5(b). Early in the primary transient all of the dislocations in any local field of view are found to have the same Burgers vector, and a single dislocation line can be traced over the distance of many γ' particles. As the end of the primary transient is approached dislocations with different Burgers vectors from neighboring regions penetrate locally and react with dislocations of the primary system; see Figures 6(a) and (b) where the same area is imaged for two different diffraction conditions, displaying two different sets of dislocations.

The steady state dislocation structure is characterized by a complicated three dimensional nodal network, pictured in Figure 7(a). The network contains at least three different sets of dislocations. Again this dense dislocation network is completely contained in the matrix, and there are very few dislocations in the matrix passages which are not associated with the network. The gamma prime precipitates are occasionally observed to be sheared by pairs of dislocations in the later stages of steady state creep.

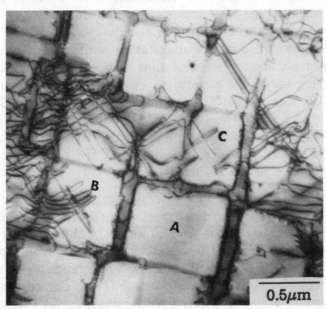

Figure 4 - Dislocation structure during primary creep. Dislocations in the matrix on either side of "A" are bowing through vertical gaps, while near "B" and "C" a horizontal matrix gap with dislocations is contained in the plane of the foil.

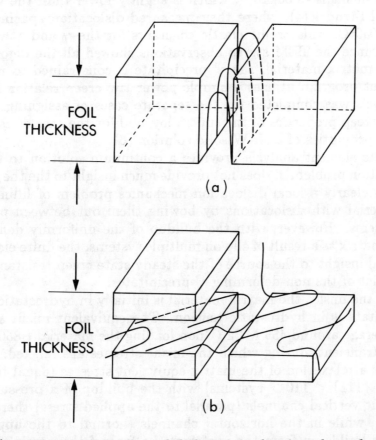

Figure 5 - (a) Schematic 3-D view of dislocations bowing through vertical gaps and (b) a horizontal gap contained in the foil with dislocations bowing through the gap captured in the plane of the foil.

After creeping to steady state a sample was bulk annealed under no stress for 750 hours. From Figure 7(b), it is obvious that the overall density of dislocations has decreased and the network has coarsened. Although the network appears to be more planar, it is not necessarily at the γ/γ' interface. In-situ static annealing at 850°C and 900°C with a heating stage also shows a gradual decrease in the local dislocation density associated with the three dimensional network at 850°C for annealing times of 2 to 3 hours, but the kinetics of this dislocation annihilation process at these short times at 850°C do not necessarily suggest that static recovery is fully responsible for this short time clearing up of the structure. However, at 900°C the rate of decrease of dislocation density does suggest that some diffusion controlled static recovery is operative. It is necessary to be careful in interpreting the results obtained with thin foils, since dislocations are able to escape from the free surface. A more thorough analysis of the in-situ static recovery processes and the problem of surface effects in thin films will be presented in the future.

Finite Element Model of the Two Phase Microstructure

A finite element model of the γ/γ' microstructure was constructed to observe the changing distribution of stress, strain and pressure in the structure due to the creep process. The model used generalized plane strain elements, which permit the development of a uniform strain in the third dimension. The symmetry of the problem allows one quarter of the precipitate plus matrix to be modeled. The top and right boundaries are constrained to remain plane to satisfy overall symmetry in the microstructure. The finite element mesh is shown in Figure 8(a). Misfit was incorporated into the model by assigning to the γ and γ' their respective coefficients of thermal expansion. The coefficients were estimated using the data of Grose and Ansell (2) and Giamei et al. (3). At 850° the misfit is -3.05×10^{-3}, which is slightly lower than the value reported by Fredholm and Strudel (4), where they measured dislocation spacings after aging of CMSX-2 at 1050°C. Anisotropic elastic constants for the γ and γ' were also used in the analysis. Since the TEM stereo observations showed all the deformation to be occurring in the matrix material, the γ' precipitate is constrained to remain elastic. The finite element program utilizes a simple power law creep relation in its calculations. The analysis was completed for the separate cases of assigning to the matrix the macroscopic creep properties of the superalloy, and for the case of assigning to the matrix the creep properties of a Ni-W solid solution (6).

Since the finite element analysis provides a continuum solution to the two phase material deformation problem, it does not provide much insight to the the primary creep process, which is clearly a local dislocation mechanics problem of filling the initially undeformed material with dislocations by bowing them out between precipitates on selected slip systems. However, with the buildup of the uniformly dense dislocation networks in the matrix as a result of slip on multiple systems, the finite element analysis can provide useful insight to the source of the steady state creep resistance that comes from the constraint of the non-deforming γ' precipitates.

As a result of the misfit, the matrix material is initially in hydrostatic compression, while the precipitate is in hydrostatic tension. The equivalent misfit stresses in the matrix are on average around 455 MPa. Upon loading the material (850°C/552 MPa) there is a rapid transient, during which the misfit stresses are relaxed. As the creep proceeds, there is a relaxation of the matrix equivalent stresses (equal to the resolved shear stresses on $\{111\} < 110 >$ systems) with the buildup of a pressure gradient in the matrix. In the vertical channels (parallel to the applied stress) there is a positive pressure buildup, while in the horizontal channels (normal to the applied stress) a negative pressure builds up from the center of the face of the precipitate, providing load support. Decreasing the matrix creep resistance leads to shorter relaxation times and an increased macroscopic creep rate which develops at long times $t/\tau > 5$. (The macroscopic creep rate is calculated from the displacements of the upper boundary of the mesh with respect to the lower, and τ is a relaxation time equal to the time to

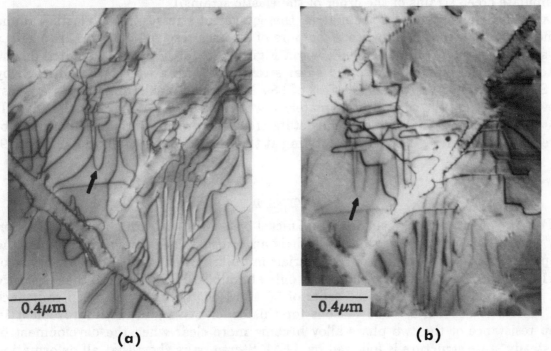

Figure 6 - Dislocations in (b) with a Burgers vector different from the primary dislocations in (a) begin to interpenetrate and form nodes. The two sets of dislocations are imaged for (a) $g=\bar{2}00$ and (b) $g=020$. The arrows mark a common reference point in each micrograph.

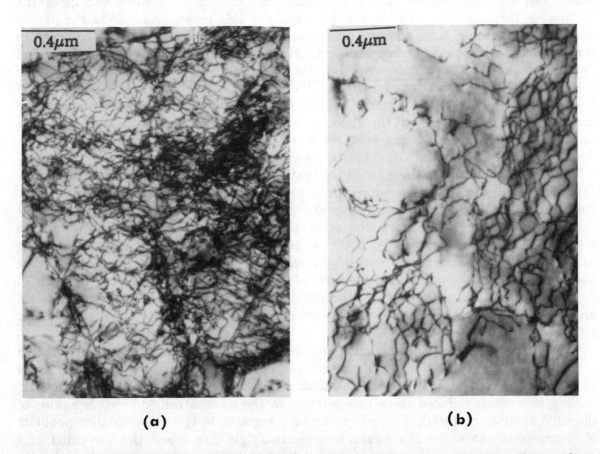

Figure 7 - (a) Characteristic steady state dislocation structure with a dense three dimensional nodal network in the matrix. (b) Coarsening of the three dimensional network following 750 hours of bulk static annealing. The scale in these two micrographs is slightly different.

accumulate creep strains of the order of the elastic strains.)

As the creep continues in the matrix, the equivalent stresses in the elastic gamma prime precipitate continue to rise. Contours of stresses resolved onto $\{111\}<110>$ slip systems in the precipitate are shown in Figure 8(b) for $t/\tau=485$. As noted on the contour plot, the maximum resolved shear stress is 281 MPa which when divided by the Schmid factor gives a tensile stress of 689 MPa. The maximum stresses are just below the corner of the cube, approximately 1/4 of the distance along the cube edge. It is interesting to note that late in steady state creep the precipitates are sometimes observed to be sheared by dislocations entering at this very location, as shown in Figure 9.

Discussion

For a given creep rate, a single crystal nickel base alloy with a high volume fraction of γ' has a tensile creep resistance which is an order of magnitude higher than the creep resistance of either the γ or γ' materials in bulk. For example, for a creep rate of 2.5×10^{-8}/s the creep resistance of Ni_3Al is only about 60-70 MPa (5), and the resistance of a Ni-W solid solution is on the order of 30-40 MPa (6), while the creep resistance of CMSX-3 is 550 MPa. The reasons for this order of magnitude improvement in creep resistance of the two phase alloy become more clear when the development of the steady state structure is followed by TEM. Stereo pairs show that all deformation is occurring in the matrix. Apparently at the stress levels studied here or at lower stresses, the difficulty of pushing a pair of normal γ phase dislocations into the ordered gamma prime phase is so high that the dislocations are confined to the matrix. Since the matrix in the aged condition contains so few dislocations, the development of the plastic response requires generation of dislocations from limited sources and spreading by bowing them out between the precipitates. For this reason, an incubation period may be required and large areas of dislocations of a single Burgers vector are seen in the early stages of primary creep. For a matrix passage with a thickness of 60 nm and deformation on a $\{111\}<110>$ system, the Orowan stress is 167 MPa, which gives a tensile deformation resistance of 408 MPa. This is clearly the major contribution to the deformation resistance of the two phase alloy. The quantitative aspects of the kinematics of the spreading of the dislocations and the resistance of the gamma prime to shearing by precipitates have not yet been treated, but will be necessary for the development of a model for the constitutive creep response of the material.

As the dislocations continue to fill the material during the primary transient, dislocations spreading from widely spaced sources in neighboring regions begin to interpenetrate and react, thereby preventing large crystallographic rotations. Eventually a dense three dimensional nodal network, characteristic of steady state creep, is built up in the matrix. At temperatures around 850°C annealing shows the static recovery process to be sluggish, producing only very slight reloading transients. At this temperature and at lower temperatures, a combination of static and dynamic recovery appears to be required for the maintenance of the steady state creep process. At temperatures around 900°C and higher the steady state creep process is probably controlled by static recovery.

As the steady state creep process proceeds, the constraint of the non-deforming precipitates leads to the buildup of a negative pressure in the matrix passages normal to the applied stress. This provides additional load support for the creeping matrix, leading to a much reduced creep rate relative to the unconstrained creep behavior of the solid solution material. This effect can be compared to the more familiar problem of compressing a thin disk of material between two rigid dies; where the flow constraint due to wall friction produces an increased load carrying ability of the disk due to pressure buildup. This problem has been discussed in an earlier paper (7).

In the later stages of steady state creep, as the stresses in the matrix relax and the stresses in the non-deforming γ' continue to rise, the precipitates may be sheared by

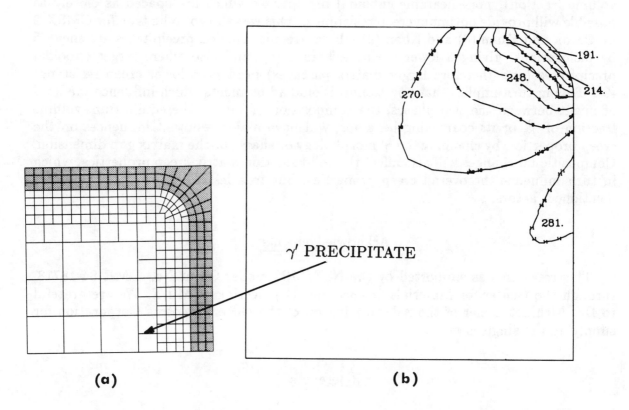

Figure 8 - The γ/γ' finite element mesh (a), with the shaded region corresponding to the matrix and the unshaded to the precipitate. The resolved shear stresses on $\{111\} <110>$ systems in the precipitate are shown in (b).

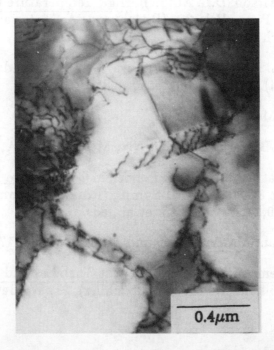

Figure 9 - Dislocations shearing the γ' precipitate in the later stages of steady state creep. Note that the point of entry corresponds to the point of maximum resolved shear stresses in Figure 8.

dislocations. For the conditions studied here this was only an occasional event, but for higher stresses it is expected that this would be observed more frequently.

These results suggest that for alloys with a high volume fraction of γ' (at a constant volume fraction), non-shearable cuboidal precipitates which are spaced as closely as possible will provide optimum creep resistance. This was shown to be true for CMSX-2 by Caron and Khan (8) and Khan (9) where irregular shaped precipitates are sheared by dislocations, giving a lower creep resistance (8), and also where larger cuboidal precipitates (and therefore larger matrix passages) produce a lower creep resistance (9). This in turn implies that any compositional adjustments which influence the level of misfit between the two phases, the composition of the γ', the equilibrium volume fraction of γ', or its coarsening behavior, will have major resultant influences on the creep properties, by changing the γ' morphology or shape, or the matrix gap dimension. Compositional changes will also affect the solid solution matrix creep properties, which in turn influence the overall creep properties, but to a lesser extent than the above mentioned factors.

Acknowledgements

This research was supported by the NSF/MRL under Grant No. DMR 84-18718, through the Center for Materials Science and Engineering at M.I.T. We are grateful to Dr. Mehmet Doner of the Allison Division of the General Motors Corporation for supplying the single crystals.

References

1. K.L. Harris, G.L. Erickson, and R.E. Schwer in Superalloys 1984 (Warrendale, PA : The Metallurgical Society, 1984), 227, M. Gell, et al., eds.

2. D.A. Grose and G.S. Ansell, Met. Trans. 12A, 1981, 1631.

3. A.F. Giamei, D.D. Pearson, D.L. Anton in High Temperature Ordered Intermetallic Alloys, MRS Symp. Proc. Vol. 39, (Pittsburg, PA: Materials Research Society, 1986), 293, C.C. Koch et al., eds.

4. A. Fredholm and J.L. Strudel in Superalloys 1984, (Warrendale, PA :The Metallurgical Society, 1984), 211, M. Gell, et al., eds.

5. D.M. Shah, Scripta Met. 17, 1983, 997.

6. W.R. Johnson, C.R. Barrett and W.D. Nix, Met. Trans. 3A, 1972, 963.

7. A.S. Argon, A.K. Bhattacharya, T.M. Pollock in Constitutive Relations and Their Physical Basis, Proc 8th RISO Intl. Symp., (Roskilde, Denmark: Riso National Laboratory, 1987), 39, .S.I. Anderson, et al., eds.

8. P. Caron and T. Khan, Materials Sci. and Engr. 61, 1983, 173.

9. T. Khan in High Temperature Alloys for Gas Turbines and Other Applications 1986, (Dordrecht, Holland: D. Riedel Publishing), 21, W. Betz, et al., eds.

THE EFFECT OF HYDROGEN ON THE DEFORMATION AND

FRACTURE OF PWA 1480.

W.S. Walston, N.R. Moody‡, M. Dollar*, I.M. Bernstein* and J.C. Williams

Carnegie Mellon University, Pittsburgh, PA 15213
‡ Sandia National Laboratories, Livermore, CA 94550
* Illinois Institute of Technology, Chicago, IL 60616

ABSTRACT

The effect of internal hydrogen on mechanical properties and fracture behavior is being studied in the single crystal nickel-base superalloy PWA 1480. In particular, plane strain fracture toughness and tensile tests have been performed in hydrogen-free samples and hydrogen gas phase charged samples in two different sample orientations. The mechanical properties as well as the fracture process were found to be very dependent on orientation. Hydrogen did not affect the yield strength but reduced the uniform tensile elongation more than 75%. However, hydrogen reduced the tensile reduction-in-area and the fracture toughness by only 10%. Cleavage of the eutectic γ' as well as the role of hydrogen trapping will be discussed. These results will be analyzed in terms of the effects that orientation, microstructure and hydrogen have on the fracture process.

I. INTRODUCTION

The presence of a hydrogen-rich environment in the space shuttle main engine (SSME) provides a basis for the study of the effect of hydrogen on PWA 1480. PWA 1480 is a single crystal nickel-base superalloy and a candidate alloy for use as turbine blades in the high pressure fuel turbopump of the SSME. The effect of high pressure hydrogen and hydrogen-enriched steam on mechanical properties such as creep, low cycle fatigue and tensile properties has previously been studied for PWA 1480.[1-3] However, a detailed analysis of how hydrogen affects the deformation and fracture behavior has not been performed. This is the aim of the current study with a particular focus being to correlate observed hydrogen effects with the role of microstructural heterogeneities as trapping centers for hydrogen. PWA 1480 is being utilized as a model material to gain a better understanding of high γ' volume fraction superalloys in general. As a beginning to understanding the effect of hydrogen on the deformation and fracture behavior of PWA 1480, tensile tests and fracture toughness tests have been performed with samples containing a uniform concentration of internal hydrogen in two different orientations. The effect of hydrogen on the ductility and strength of this alloy was studied through the tensile tests, while the fracture toughness tests provided a means to study the crack growth behavior. This paper is concerned with the results of tests conducted at room temperature. The effect of higher temperatures on the deformation behavior has been previously reported[4] and the effect of internal hydrogen at these higher temperatures will be studied in the near future.

II. EXPERIMENTAL PROCEDURES

The composition of PWA 1480 is given in Table I. Single crystal bars were received from TRW Metals Division in the solutionized condition. The material was in the form of rectangular castings, 1.6 cm x 6.4 cm x 15.2 cm, with the solidification axis in the long direction. A standard heat treatment was performed which consists of a solutionizing treatment at 1288°C for 4 hours followed by water quenching, an aging treatment at 1080°C for 4 hours followed by air cooling and a final aging treatment at 875°C for 32 hours followed by air cooling.

Table I. Alloy Composition (wt. pct.)

Al	Co	Cr	Ta	Ti	W	Ni
4.8	5.3	10.4	11.9	1.3	4.1	bal.

Tensile tests and fracture toughness tests were performed with the loading direction parallel to the <001> orientation and a transverse orientation which was determined to be <130>. All hydrogen containing samples were gas phase charged at Sandia National Laboratories for 15 days at 350°C and a pressure of 103.5 MPa . Hydrogen content was analyzed by Luvak, Inc. by vacuum hot extraction at 900°C.

III. RESULTS

A. Microstructure

The microstructure of PWA 1480 is similar to other high γ' volume fraction alloys[5-7] having a dendritic macrostructure with interdendritic porosity and eutectic γ' as shown in Figure 1. The dendritic structure has the following dimensions: core diameter ~150 μm; arm length

~400-500 μm; and spacing ~200 μm. The pores were fairly spherical with an average size of 5.9 μm and an average spacing of 49.7 μm. However, pores an order of magnitude larger have also been observed. The volume fraction of porosity is .0012 which corresponds with the results of Milligan and Antolovich.[8] Eutectic γ' is also found in the interdendritic region and has an irregular shape. The volume fraction of this phase is .023 with an average size of 13.5 μm. It is important to note that the amounts of porosity and eutectic γ' remain fairly constant within each single crystal bar but can vary from bar to bar. Figure 2 illustrates the γ/γ' microstructure for 65% volume fraction of γ' with an average size of about 0.4 μm. As previously reported,[8] there are isolated areas in the interdendritic regions where the γ' is about 1 μm. The average γ/γ' mismatch has been measured by Bowman[9] using extracted γ' and was found to be 0.28 percent.

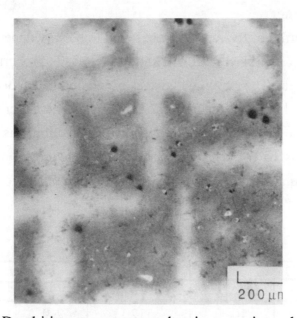

Figure 1. Dendritic macrostructure showing porosity and eutectic γ'.

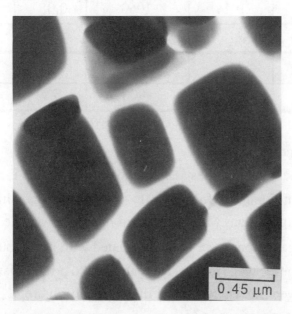

Figure 2. γ/γ' microstructure of PWA 1480.

B. *Hydrogen Trapping*

The measured hydrogen content was 30,000 appm (~3 at. pct.). This value is significantly higher than an estimated lattice concentration of 5,000 appm which was based upon data from a low γ' volume fraction alloy. In addition, after desorption for 5 hours at 300°C there is still a significant amount of hydrogen (17,000 appm) in the material. The high hydrogen concentration after charging and desorbing suggest the presence of strong trapping sites.

C. *Mechanical Behavior*

1. *Tensile Properties*

The tensile properties of PWA 1480 were found to be very dependent upon orientation as well as to the presence of hydrogen. Table II shows the tensile properties for both orientations with and without hydrogen. The <130> transverse orientation had about the same yield strength as the <001> orientation, but there was no work hardening. The major effect of orientation was seen in the ductility. The total elongation in the <001> direction was only 2.9%, while in the transverse direction it was 24.3%. The presence of internal hydrogen did not affect the strength or reduction-in-area, but dramatically reduced the elongation to failure for both orientations.

Table II. Tensile Properties

[001]	σ_{ys} (MPa)	σ_{UTS} (MPa)	e_t (%)	R.A. (%)
uncharged	1001	1120	2.9	3.1
charged	1014	1060	0.38	2.7
transverse				
uncharged	929	929	24.3	25.3
charged	906	906	6.9	21.2

2. *Tensile Fractography*

The differences in these properties as a function of orientation is reflected in the associated fractography, as shown in Figure 3. In the <001> tensile samples with and without hydrogen, 10-20 μm diameter cleavage regions were surrounded by smaller ductile voids around a micron in size. These large cleavage regions were often associated with pores, but more importantly there were many cases where the cleavage regions were not next to pores. Plateau etching revealed that these cleavage regions were the eutectic γ' as illustrated in Figure 4. In some cases the pores seemed to act as initiation sites, however in many cases there was no identifiable microstructural feature with the initiation site. In all cases, the river markings led back to the edge of a cleavage facet and never to the interior.

The entire transverse fracture surface consists of the small ductile voids which were seen in the <001> samples. Thus, the only difference between the two orientations is the presence of the cleavage facets in the <001> orientation. Plateau etching also shows that there is a relationship between these voids and the γ/γ' microstructure. The voids on the fracture surface usually consist of two or three γ' particles with the edge of the void representing a tear ridge along various γ/γ' interfaces.

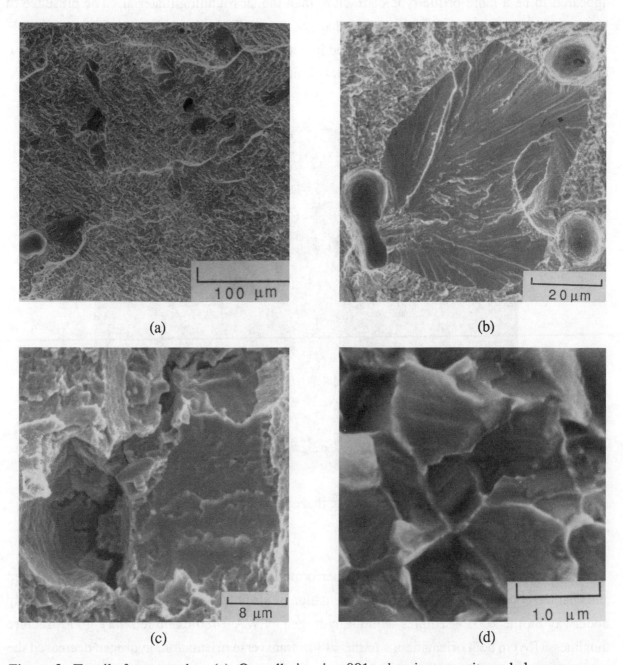

Figure 3. Tensile fractography. (a) Overall view in <001> showing porosity and cleavage. (b) <001> cleavage facet. (c) <130> with hydrogen showing brittle area. (d) Small ductile voids present in all specimens.

The presence of internal hydrogen had little effect on the tensile fractography. The only difference observed between the uncharged and charged samples showed up in the cleavage facets. In the [001] orientation, cracking within the cleavage regions was seen in the hydrogen charged samples only. Furthermore, there were small, flat regions in the charged transverse samples as seen in Figure 3(c) which were about the same size as the cleavage facets seen in the <001> samples. These regions were not seen in the transverse samples without hydrogen. They weren't the classic cleavage facets with river markings as were seen in the <001> samples, but appeared to be a more brittle type of fracture than the surrounding material. The presence of hydrogen did not seem to affect the size or shape of the small ductile voids.

Microcracks have also been observed in both orientations which change directions at 90° about every 0.5 μm suggesting that the crack is following the γ/γ' interface.

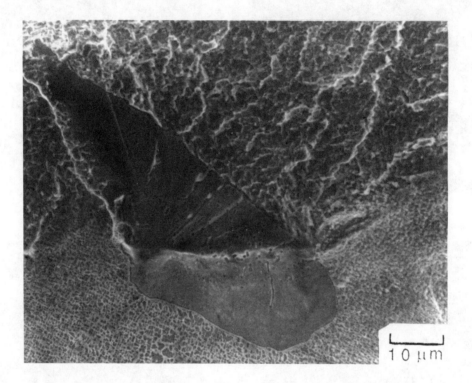

Figure 4. Plateau etch showing that eutectic γ' are the cleavage facets

3. Fracture Toughness Properties

Fracture toughness tests have been performed in the <130> orientation in the uncharged and charged conditions and in the <001> orientation in the charged condition. Testing is currently underway on the <001> samples without hydrogen. PWA 1480 exhibited very good fracture toughness (K_Q) in both orientations. In the <130> transverse orientation, hydrogen decreased the values of K_Q from 133 to 101 MPa·m$^{1/2}$. The value of K_Q in the charged <001> sample was 94 MPa·m$^{1/2}$. As a result of these high values and the morphology of the single crystal bars, valid toughness data could not be obtained.

4. Fracture Toughness Fractography

Fractographic analysis has only been performed on the transverse samples to date. The uncharged samples fractured predominantly along crystallographic planes while hydrogen causes a tortuous, less crystallographic crack path. In contrast to the tensile specimens, there were no cleavage regions observed in any samples. It was also noted that in all cases pores acted as initiation sites. Most of the pores had cracks either emanating from them or passing through them linking them to other pores. There were no other microstructural features which seemed to act as crack initiation sites.

IV. DISCUSSION

A. *Hydrogen Trapping*

The large amount of hydrogen in the material after the initial gas phase charging indicates the presence of at least weak traps. Even after desorption for several hours at temperatures of 150°C and 300°C, the concentration of hydrogen was an order of magnitude greater than has been found in nickel and austenitic stainless steels.[10,11]. It appears that the presence of internal hydrogen embrittled the eutectic γ' and weakened the γ/γ' interface and both can be considered trap sites, but the most obvious site for strong trapping is the pores.

More can be learned from the above observations by comparing them with a previous study on a similar alloy CMSX-2.[5] CMSX-2 is a single crystal nickel-base superalloy with approximately the same volume fraction porosity as PWA 1480. CMSX-2 contained some eutectic γ', but not to the same extent as PWA 1480. Similar concentrations of hydrogen were found in CMSX-2 after cathodically charging thin disks. Desorption experiments showed that the pores was a strong trapping site which the authors estimated to have a binding energy of 0.7 to 0.8 eV. Further desorption experiments are necessary to determine an accurate binding energy for the porosity in PWA 1480.

In CMSX-2, after desorbing for a short time at 300°C there was very little hydrogen left whereas there was still a large amount of hydrogen in PWA 1480 after desorbing for several hours at 300°C. Based upon microstructural comparison given it doesn't seem that the trapping in PWA 1480 should be that much stronger than in CMSX-2. Both alloys contain similar trapping sites which should have similar trap binding energies. The γ/γ' mismatch is larger in PWA 1480 so the trapping strength of this interface may be slightly greater. The trapping characteristics of the eutectic γ' are unknown and cannot be compared. There may be more trap sites in PWA 1480 due to the much higher volume fraction of eutectic γ' and the widely varying porosity distribution. The average volume fraction of porosity in these single crystals is known to vary widely from bar to bar. A possible explanation for the differences could be that the bars selected for desorption actually had quite different volume fractions of porosity. A further explanation may be that PWA

1480 simply has more available sites for trapping due to the larger γ/γ' misfit and the higher eutectic γ' volume fraction.

B. *Tensile Specimens*

The major difference in the tensile properties of the two orientations was manifested in the tensile elongations, however at this point there is no explanation for this large difference. The determination of active slip systems is currently underway as are TEM studies which may explain the observed differences.

Internal hydrogen had no effect on the strength of PWA 1480 however hydrogen slightly decreased the reduction-in-area and markedly decreased the total elongation in both orientations. The hydrogen charged tensile samples exhibited less elongation of the gage length, but as much reduction in cross-section diameter as the uncharged samples. These observations suggest that hydrogen localized the plastic flow.

It was observed that only some cleavage facets initiated pores while other facets didn't have any obvious microstructural initiation site. These observations suggest that the stress field surrounding the pores do not cause cleavage of the eutectic γ'. If it was the stress field surrounding the pore, then every cleavage facet would be initiated at a pore. It is not known exactly what is causing the cleavage initiation. One observation that may help is that in every instance the river markings led back to the edge of the facet suggesting a nucleation site at the interface of the eutectic γ'.

It should also be noted that the cleavage facets were only observed in the tensile specimens and never in the fracture toughness specimens. Taking into account that the tensile specimen undergoes a combination of stress and strain prior to failure and that the fracture toughness specimen principally experiences only a stress state in the crack tip region prior to failure, one may conclude something about the requirements for cleavage of the eutectic γ'. Orowan's[12] original proposal that a critical value of tensile stress is required to produce cleavage fracture doesn't explain the observations in this study. It seems that some critical strain or combination of stress and strain is necessary to produce cleavage as found by Lewandowski and Thompson[13,14] in pearlitic 1080 steel.

Plateau etching revealed that there is a relationship between the ductile voids on the fracture surface and the γ/γ' microstructure. The ductile voids are on the order of 1 micron, while the size of the γ' is about 0.4 μm. Close observation of the plateau etching micrographs suggests that the fracture occurs along the γ/γ' interface. The actual void consists of two or three γ' particles with the edge of the void representing a tear ridge along various γ/γ' interfaces. This observation can be correlated with the occurrence of microcracks and with the development of the dislocation structure during deformation of PWA 1480[4]. The microcracks appear to be occurring along the γ/γ' interface. This hypothesis is supported by the deformation observations shown in

Figure 5. In particular, the dislocation density in the matrix was found to be significantly higher than that in the precipitates for all strains. This promotes strain localization in the matrix and may result in premature plastic failure due to strain exhaustion, as manifested by the occurrence of microcracks following γ/γ′ interfaces.

(a)   (b)

Figure 5. (a) Microcracks in tensile specimen. (b) Accumulation of dislocations in γ matrix at small strain, $\varepsilon_p = 0.24\%$ (from Dollar and Bernstein[4]).

C. *Fracture Toughness*

Preliminary test results indicate that PWA 1480 has very good fracture toughness. There does not appear to be any effect of orientation on the fracture toughness values, at least in the two orientations tested. Although the tests on the <001> samples without hydrogen have not been completed, it may be assumed that this value will be close to the uncharged <130> value. This is based upon similar values in the two orientations for the charged samples and also the results of short rod fracture toughness tests.[15] If this is the case, then hydrogen decreases the fracture toughness value by about 25%. The presence of internal hydrogen also affected the crack growth behavior in the transverse samples. It will be interesting to see if hydrogen has the same effect on the <001> orientation. In contrast to the tensile samples, the pores seem to act as fracture initiation sites in all fracture surfaces examined to date. Completion of the fractography and J integral tests to determine valid fracture toughness values are part of the future research.

V. CONCLUSIONS

1. The pores present in PWA 1480 act as very strong trapping centers for hydrogen while the eutectic γ' and the γ/γ' interface may serve as weaker traps.
2. The presence of internal hydrogen localized the plastic flow in the tensile samples as evidenced by a large change in elongation and a correspondingly small change in reduction-in-area.
3. The cleavage facets seen on the tensile fracture surfaces are the eutectic γ' in the microstructure. The stress field surrounding each pore is not responsible for initiating these cleavage facets, however it is not known what is initiating the cleavage.
4. It appears that a critical value of tensile stress is not sufficient to initiate cleavage, but that a critical strain or combination of stress and strain may be necessary.
5. Tensile fracture is occurring along the γ/γ' interface or more likely through the γ phase. This is supported by plateau etching, the observation of microcracks and the study on the development of the deformation structure.
6. PWA 1480 exhibited very good fracture toughness values with and without hydrogen.
7. The fracture initiation site in all of the fracture toughness samples was the porosity.

VII. REFERENCES

1. D.P. Deluca and B.A. Cowles, *AFWAL-TR-84-4167*, Feb. 1985.
2. Cowles et. al., *Adv. Earth-to-Orbit Propulsion Tech. 1986*, Huntsville, AL, NASA CP 2437, pp. 727-748.
3. N.L. Weeks and J. Mucci, *NASA FR 19269*, June 1986.
4. M. Dollar and I.M. Bernstein, *Acta Met.*, in press.
5. C.L. Baker, J. Chene, I.M. Bernstein and J.C. Williams, *Met. Trans.*, **19A**, 73, (1988)
6. B.J. Piearcey, B.H. Kear and R.W. Smashey, *Trans. ASM*, **60**, 634, (1967)
7. M.V. Nathal, R.D. Maier and L.J. Ebert, *Met. Trans*, **13A**, 1767, (1982)
8. W.W. Milligan and S.D. Antolovich, *Met. Trans.*, **18A**, 85, (1987)
9. R.R. Bowman, *M.S. Thesis*, Georgia Inst. of Tech., Atlanta, GA, 1986.
10. W.M. Robertson, *Met. Trans.*, **8A**, 1709, (1977)
11. Y. Shehu, P. Menut, J. Chene and M. Aucouturier, *Proc. 3rd Int. Congress on Hydrogen and Materials,* Paris, 1982, p. 889.
12. E. Orowan, *Trans. Inst. Engrs. Shipbuilders Scot.*, **89**, 165, (1945)
13. J.J. Lewandowski and A.W. Thompson, *Met. Trans.*, **17A**, 1769, (1986)
14. J.J. Lewandowski and A. W. Thompson, *Acta Met.*, **35**, 1453, (1987)

ACKNOWLEDGEMENTS

The authors are grateful to A.W. Thompson for many fruitful discussions. This research was supported by NASA-Lewis Research Center under the technical direction of Dr. R. Dreschfield.

AN ATOM-PROBE STUDY OF SOME FINE-SCALE MICROSTRUCTURAL FEATURES IN Ni-BASED SINGLE CRYSTAL SUPERALLOYS

D. BLAVETTE*, P. CARON** and T. KHAN**

* Faculté des Sciences de Rouen,
Laboratoire de Microscopie Ionique
UA 808 CNRS, UFR des Sciences & Techniques
B.P. 118, 76134 MONT SAINT AIGNAN CEDEX, FRANCE

** Office National d'Etudes et de Recherches
Aérospatiales (ONERA)
B.P. 72, 92322 CHATILLON CEDEX, FRANCE

ABSTRACT

Some fine-scale microstructure of nickel-based single crystal superalloys have been investigated by means of atom probe techniques. The beneficial effect as well as the role of Re additions in the creep resistance of CMSX-2 and PWA 1480 alloys have been studied. Atom probe analyses show that very fine clusters, 10 Å in size, are present in the matrix. Re additions also modify the partitioning of other elements. Other experiments conducted on a Re-free CMSX-2 alloy exhibit the occurence of a concentration gradient in the precipitates near the phase boundary.

INTRODUCTION

Nickel base superalloys for turbine blades have constantly been improved through chemistry modifications and innovative processing techniques. Modern single crystal superalloys like CMSX-2 and PWA 1480 derive their excellent creep performance from the presence of a large volume fraction of γ' precipitates (70 %). High temperature properties strongly depend upon the lattice misfit which is in turn controlled by the phase composition. It is important to insist here that the misfit is closely related to the local concentrations of various elements in the vicinity of the γ/γ' interface. It is therefore of utmost importance to characterize the γ and γ' phases chemically as well as the interface.

A number of refractory elements such as W, Ta, Mo and more recently Re are added in nickel base superalloys in order to improve their mechanical strength. The element Re for instance has been shown to be a potent strengthener in creep of single crystal materials. However, the physical reasons behind the improvement in strength of Re containing superalloys have not been elucidated. The role played by these elements can be more subtle than just solid solution strengthening. The performances of superalloys are in effect, intimately dependent on the solute distribution in both γ and γ' phase. For instance, the stacking fault energy as well as the antiphase boundary energy are probably two major parameters which determine the temperature capability of nickel base alloys. Both parameters are directly dependent one hand, upon the occurence of a short range order in the γ solid solution and on the other hand, on the long range order in γ' particles.

The atom-probe techniques are well suited for the investigation of these fine scale microstructural parameters. This instrument is probably the most sensitive microanalytical tool currently available. Quantative informations can be obtained for a large variety of elements, including for the light species (B, H). The spatial resolution of the apparatus may be varied from 5 to 25 Å at the specimen surface. In addition, the depth resolution attains one atomic layer. This means that very fine γ' particles may be chemically analysed, small clusters (10 Å) may be detected and interphase boundaries may be investigated as well. In addition, by making an atom-probe analysis of the (001) superlattice planes of the ordered γ' phase, one is able to estimate the occupancy frequencies of sites for each element, these parameters being deduced from the average composition of both types of planes (mixed and pure planes). In this paper, we will try to focus the attention on the information provided by this technique and discuss the possible consequences of fine scale features on the high temperature properties of superalloys.

EXPERIMENTAL

Principle of atom-probe

The FIM - atom-probe combines an ultra high resolution

field ion microscope with a mass spectrometer of single atom sensitivity. The Field ion microscope provides an atomic resolution image of a specimen. The atom-probe is capable of performing chemical analysis with equal sensitivity for all elements and a spatial resolution of better than 1nm.

The principle of these techniques is based on field ionisation of a rare gas near the specimen surface (FIM) and controlled field evaporation of atoms. Both physical processes require a high electrical field. This is obtained by applying a positive voltage to the specimen prepared in the form of a sharply pointed needle. Atom probe investigations of a material may be interpreted as a layer by layer analysis along a cylinder whose the axis is parallel to the tip axis. The diameter of this cylinder is equal to the size of the area which is selected by the detector. More details may be found in a recently published paper [1].

Although atom-probe analyses do not necessitate the knowledge of some calibration parameters, care must be taken with the experimental conditions. The details of optimum operating conditions which are required to obtain reliable compositional data are discussed elsewhere [2].

RESULTS

Role of Re additions in improving creep resistance

Two base alloy chemistries, designated CMSX-2 and PWA 1480, were selected in this investigation. Rhenium was partly or entirely substituted for W in these reference alloys. The chemical compositions (at %) of various alloys are shown in Table I. The basic intention here is to study the role of Re in creep and not to propose optimized alloy chemistries for which further work is required in order to avoid phase instability. Single crystals were grown parallel to the (001) direction by the seeded technique under a temperature gradient of 250°C/cm at a withdrawal rate of 15 cm/hour. All single crystal specimens for creep tests were within 5° from (001) and given a solutioning and homogenizing treatment between 1280 and 1325°C for 15 hours followed by air quenching and precipitation heat treatments at 1050°C/16h/AC + 850°C/48h.

Alloy	Al	Ti	Cr	Ni	Co	Ta	W	Re
CMSX-2	12.21	1.25	9.22	Bal.	5.08	1.99	2.61	-
CMSX-2-4W-4Re	12.21	1.25	9.22	Bal.	5.08	1.99	1.30	1.29
CMSX-2-0W-5Re	11.96	1.23	9.03	Bal.	4.98	1.95	-	1.58
PWA 1480	11.26	1.90	11.68	Bal.	5.15	4.03	1.32	-
PWA 1480-Re	11.26	1.90	11.68	Bal.	5.15	4.03	-	1.31

Table I. Nominal compositions of single crystal superalloys (at %)

Creep tests were run in the temperature range 850-1000°C and the results are reported in Table II. Clearly, the creep strength of the Re-modified alloys, despite the slightly lower volume fraction of the γ' phase, is significantly higher than that of the base alloys at all temperatures. However, the improvement is most significant around 950°C both in the time for 1 % creep and in stress rupture life.

Temperature (°C)	Stress (MPa)	CMSX-2 t_A (h)	CMSX-2 t_B (h)	CMSX-2 4W-4Re t_A (h)	CMSX-2 4W-4Re t_B (h)	PWA 1480 t_A (h)	PWA 1480 t_B (h)	PWA 1480-Re t_A (h)	PWA 1480-Re t_B (h)
850	500	102	382	123	498	88	356	64	497
950	240	141	390	265	646	125	354	220	580
1 000	200	60	176	106	315	52	118	71	223

Table II. Effect of Re on creep properties
of the (001) single crystal alloys
t_A : Time to 1 % creep
t_B : Rupture life

Composition at %	Al	Ti	Cr	Ni	Co	Ta	W	Re
Matrix	3.1	0.6	25.5	59.6	8.6	0.1	2.5	0
Precipitates	16.7	3.0	2.4	70.5	3.2	3.0	2.4	0

Table III. Phase composition in CMSX-2

Composition at.%	Al	Ti	Cr	Ni	Co	Ta	Re
Matrix (33 000 ions)	4.76 ± 0.12	0.2 ± 0.02	20.12 ± 0.22	63.89 ± 0.26	7.5 ± 0.15	0.4 ± 0.03	3.59 ± 0.1
Precipitates (18 000 ions)	16.59 ± 0.27	1.79 ± 0.25	3.5 ± 0.14	71.7 ± 0.34	3.2 ± 0.13	2.49 ± 0.11	0.52 ± 0.052

Table IV. Phase composition in
the CMSX-2-0W-5Re material

The mass spectrum of γ phase of the CMSX-2-0W-5Re alloy is given in figure 1. This histogramm exhibits the main features of the matrix : while γ' precipitates are Al, Ti, Ta rich, the γ solid solution is a Cr, Co rich phase. Re element appears in the γ spectrum in two charge states : Re^{2+} and Re^{3+}. Phase composition of both phases are given in Table IV. It is clear that the Re partitions preferentially to the matrix. As shown in the Table IV, the partitioning ratio of this element is close to 7 ($C_\gamma/C_{\gamma'}$). Phase composition in the standard CMSX-2 alloy is given in Table III for comparison.

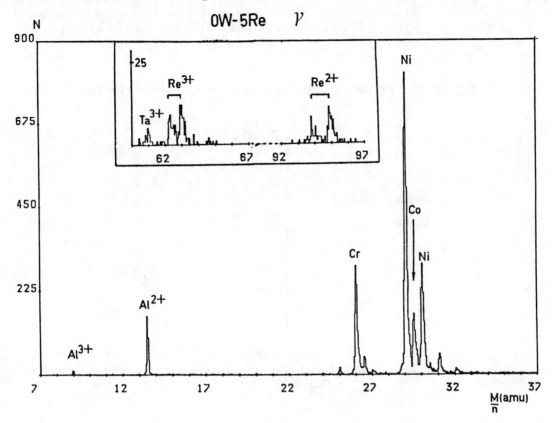

Figure 1 : Mass spectrum of γ phase in the CMSX-2-0W-5Re alloy.

Figure 2 is a synthetic representation of atom-probe data shown in Tables III, IV, V, for the three CMSX-2 versions. This graph simply uses the well known lever rule :

$$C_n = f\, C_{\gamma'} + (1 - f)\, C_\gamma$$

$C_{\gamma'}$, C_γ are respectively the composition of γ' and γ, C_n is the nominal composition of the material and f the volume fraction.

Atom-probe results are reported here for the standard CMSX-2 as well as for two Re-modified versions. The fact that all experimental data for a given material are aligned along a single line proves that atom-probe analyses are consistent with the bulk composition of the material. The figure also clearly shows in addition that the Re addition reduces the partitioning ratios of Al and Cr elements ($C_{\gamma'} - C_\gamma$ decreases). However, no detectable change is observed for the volume fraction. The slope of the "best fit" line is close to 65 % for each material.

Composition at %	Al	Ti	Cr	Ni	Co	Ta	W	Re
Matrix	4.63	0.23	21.94	60.4	7.60	0.52	1.82	2.87
Precipitates	17.22	1.87	3.23	70.0	3.27	2.99	0.90	0.41

Table V. Phase composition in CMSX-2-4W-4Re

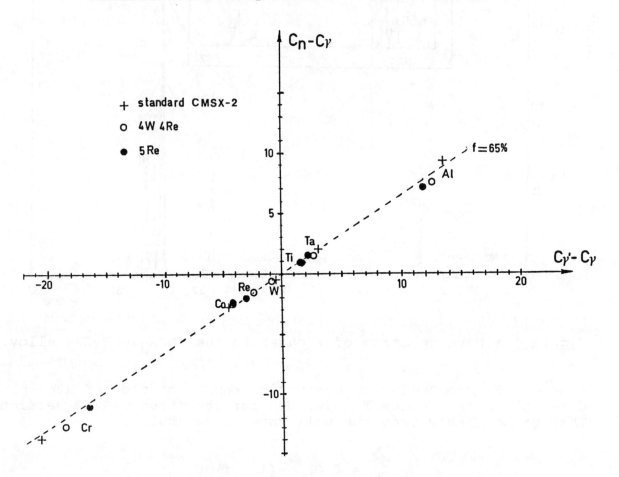

Figure 2 : Volume fraction diagramm. C_n, C_γ, $C_{\gamma'}$ are respectively the nominal composition of the three materials and the composition of both γ and γ' phases.

Among the fine-scale information provided by atom-probe techniques, the preferential sites of solute elements in the γ' ordered phase are of great interest. By performing an investigation of (001) superstructure planes of γ', one is able to estimate the long range order in this phase [3]. The preferential sites of various elements may be deduced from the mass spectra of both the Al rich mixed planes and the Ni rich pure planes. The results as deduced from these spectra (figure 3) may be qualitatively interpreted as follows : Ti, Ta, W as well as Re are preferentially located in mixed planes. These elements substitute for Al sites (cube corners). In contrast, the presence of Cr and Co in both types of planes suggests that these atoms preferentially substitute for Ni sites. These results are in qualitative agreement with the orientation of solubility lobes in ternary phase diagramms (Ni Al X) [4].

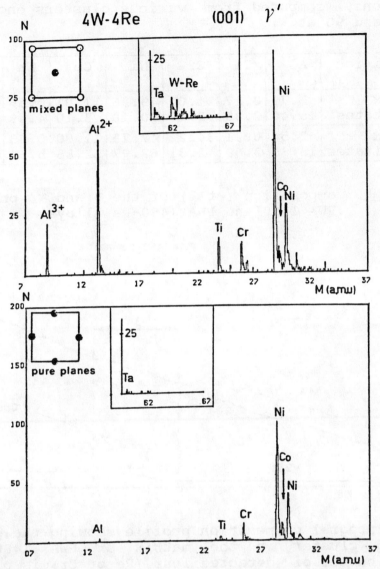

Figure 3 : Mass spectra of both types of (001) planes (mixed and pure planes) in the ordered γ' phase (CMSX-2-4W-4Re).

In order to establish a firm support for the observed improvement in creep resistance brought about by Re additions, it is important to investigate the spatial distribution of this element in the FCC solid solution.

The experiments we conducted on Re containing alloys revealed interesting informations. The matrix of both CMSX-2 and PWA 1480 superalloys was found to contain Re enriched clusters. The fine scale fluctuations of Re observed in the γ matrix of the PWA 1480 alloy are illustrated by figure 4. The phase composition of the PWA 1480 versions, as shown in Table VI, confirms that Re atoms segregate preferentially to the matrix. The rhenium profile clearly exhibits a small enriched region extending over 13 Å. The lateral resolution being of the same order as the apparent size of the cluster, the observed composition (as given by the curve slope) has to be considered as an integrated measurement of the small cluster and the surrounding matrix. The actual Re concentrations, computed from various clusters encountered, are between 20 and 90 at %.

Composition at %	Al	Ti	Cr	Ni	Co	Ta	W	Re
γ Matrix	2.79	0.17	34.03	49.26	11.01	0.56	2.15	0
γ precipitates	15.40	2.85	2.17	70.10	3.10	4.59	1.68	0
γ' Matrix	2.54	0.31	34.22	47.79	11.00	0.33	0	3.71
γ' precipitates	15.35	3.09	2.31	69.97	2.88	5.64	0	0.76

<u>Table VI</u>. Composition (at %) of the γ and γ' phases in PWA 1480 and PWA 1480-Re Alloys

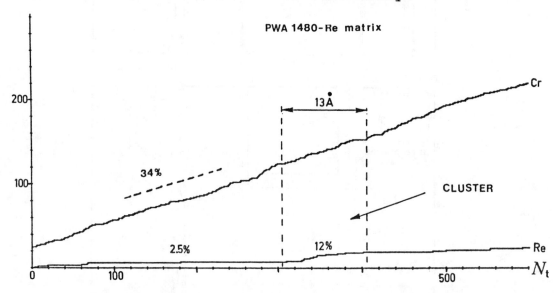

<u>Figure 4</u> : Integral composition profile showing the presence of a Re enriched cluster in the matrix of PWA 1480 alloy. The cumulative number of detected ions (Re or Cr) is plotted versus the total number of analysed atoms (N_t). The slope of each curve gives the local concentration. N_t is proportional to the probed depth.

γ/γ' interface

Atom-probe techniques are also particularly suitable for the study of γ/γ' interfaces [5]. Figure 5 presents a concentration profile obtained in the vicinity of the phase boundary of the standard CMSX-2 material. This alloy was subjected to the standard two-stages precipitation treatment which consists in heat treating at 1050°C for 16 hours plus air quenching and then at 850°C for 48 hours. However, the γ' size distribution was found to remain monomodal [6].

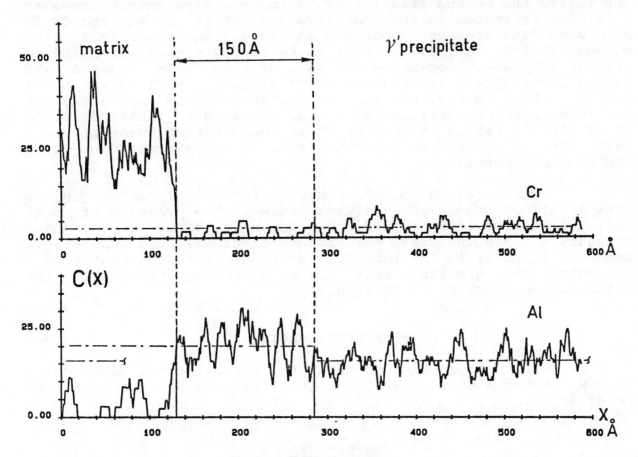

Figure 5 : Concentration profile showing the transition between the matrix and a precipitate in the standard CMSX-2 alloy. This figure clearly exhibits an Al enriched Cr depleted shell in the vicinity of the interface.

The profile clearly exhibits the occurence of a local concentration gradient in γ', near the interface. The width of the Al enriched region (or Cr depleted zone) is in good agreement with the observed increase of the γ' volume fraction due to the second precipitation heat treatment at 850°C for 48 hours. Indeed, the atom probe measurements show that this second heat treatment increaseas the volume fraction of the γ' phase by 5% compared with the single stage heat treatment performed at 850°C only. This suggests that the occurence of this enriched region around γ' particles is due to a non-equilibium phenomenon. Although the physical mechanisms involved are not completely elucidated, it is thought that this zone is due to the growth of initially formed particles at 1050°C (first precipitation stage) during the second heat treatment at 850°C.

CONCLUSION

The few examples described here show the importance of atom-probe techniques in the investigation of very fine scale features of superalloys. The investigations conducted on Re containing single crystal materials provide a physical basis for the observed improvement of the stress rupture properties of these materials. The experiments demonstrate that Re additions do not modify the volume fraction of γ' to any large extent. However a slight decrease in the partitioning ratios of Al and Cr is observed. The studies conducted on long range order in γ' also showed that W, Ta, Ti as well as Re occupy the Al sites in the ordered γ' phase. Atom-probe analyses of both CMSX-2 and PWA 1480 alloys clearly revealed the occurence of small Re clusters in the matrix. These Re enriched regions probably act as more efficient obstacles against dislocation motion compared to isolated solute atoms in a γ solid solution. It is thought that these clusters must therefore play an important role in creep properties of Re containing superalloys.

With regard to the study of the γ/γ' interface in the CMSX-2 alloy, the microanalyses reveal the presence of an Al enriched region, 150 Å in width, in the vicinity of the phase boundary. Since the creep performance of nickel base superalloys depends, in part, on the lattice misfit, we postulate that such a concentration gradient must have a strong effect on the mechanical properties of these alloys.

ACKNOWLEDGEMENTS

The authors wish to thank the Direction des Recherches, Etudes et Techniques (DRET) for its financial support (contract nr 84.1230).

REFERENCES

[1] D. BLAVETTE and A. MENAND "La sonde atomique et la microscopie ionique à émission de champ en Sciences des Matériaux," Annales de chimie 11 (1986) 321 - 384.

[2] D. BLAVETTE, A. BOSTEL and J.M. SARRAU "Atom probe microanalysis of a nickel base superalloy," Met. Trans 16 A (1985) 1703 - 1712.

[3] D. BLAVETTE et A. BOSTEL "Phase composition and long range order in γ' phase of a nickel base single crystal superalloy CMSX-2 : an atom probe study," Acta Met., 32 pp (1984) 811 - 816.

[4] S. OCHIAI, Y. OYA, T. SUZUKI "Alloying behaviour of Ni_3Al, Ni_3Ga, Ni_3Si and Ni_3Ge," Acta Met. 32 (1984) 289.

[5] D. BLAVETTE and A. BOSTEL "FIM atom-probe investigation of the interphase boundary of a nickel-base superalloy," Surf. Sci. Lett, 177 (1986) L994.

[6] P. CARON and T. KHAN "Improvement of creep strength in a nickel-base single-crystal superalloy by heat treatment," Mat. Sci. ang Eng. 61 (1983) 173

SOLID SOLUTION STRENGTHENING OF Ni$_3$Al SINGLE CRYSTALS

BY TERNARY ADDITIONS

F. E. Heredia and D. P. Pope

Department of Materials Science and Engineering
University of Pennsylvania
Philadelphia, Pa. 19104.

Abstract

A systematic approach intended to determine the mechanism by which ternary additions increase the strength of Ni$_3$Al is presented. Uniaxial tensile and compressive flow stress measurements at constant strain rate were performed on single crystals of Ni$_3$Al containing additions of Hf, Ta, Zr, and B, and also on binary Ni-rich Ni$_3$Al. Data were collected over temperatures ranging from liquid nitrogen to 1100 °C, for different orientations within the standard unit triangle, different amounts of ternary additions as well as a function of the sense of the applied uniaxial stress. In agreement with previous studies, the yield strength of binary Ni$_3$Al presented a positive temperature dependence up to about 800 °C. The critical resolved shear stress (CRSS) for the {111}<110> slip was found to be orientation dependent, mostly over the positive temperature range (200 °C up to 800 °C). Additions of ternary elements are found to enhance the CRSS of octahedral slip with much less effect on the CRSS for cube slip. The temperature of the maximum CRSS as well as the orientation of the stress axis for which the tension-compression asymmetry is zero are both found to be composition dependent. The value of the octahedral CRSS measured at low temperatures exhibits a strong dependence on both composition and orientation of the stress axis. The tension-compression asymmetry, taken as a measurement of the rate of cross slip, is affected by the ternary additions, specially Ta and Hf, and mostly at intermediate temperatures. At low temperatures the asymmetry is almost zero in most cases. For orientations in which the tension-compression asymmetry was found to be zero, the increase of the octahedral CRSS with ternary additions is believed to be a consequence of a lattice parameter/modulus mismatch mechanism. For orientations in which the asymmetry is non zero the effect of ternary additions on dislocation core configuration also needs to be considered.

Superalloys 1988
Edited by S. Reichman, D.N. Duhl,
G. Maurer, S. Antolovich and C. Lund
The Metallurgical Society, 1988

Introduction

A positive temperature dependence of the flow stress is well documented, both for polycrystalline and single crystalline Ni_3Al (1). This unusual behavior has been attributed to thermally and stress-activated cross-slip from the (111) to the (010) plane of the leading $1/2(111)[\bar{1}01]$ screw superpartial comprising a $(111)[\bar{1}01]$ superdislocation. This cross-slip can be driven by anisotropy of the antiphase boundary (APB) energy, resolved shear stress on (010), or elastic anisotropy (1,2). These effects have been combined in a theory which explains the observed dependence of the CRSS for $(111)[\bar{1}01]$ slip on both orientation and sense of the applied uniaxial stress (3-6).

Many ternary elements are soluble in Ni_3Al and some result in remarkable strengthening. Curwick was the first to study their effect in single crystals (7) and recently there has been a great deal of research performed on polycrystalline material (8). The current work was initiated on single crystalline, ternary Ni_3Al-based alloys to separate the effects of Fleischer-type solid solution strengthening from effects arising from cross-slip considerations.

Experimental Procedure

Tension/compression specimens where machined by grinding from single crystal bars of binary nickel rich Ni_3Al and Ni_3Al containing ternary additions of Hf, Ta, Zr and B. We tested two compositions of each, Hf, Ta, and Zr, and three compositions of B. The detailed chemical composition of the single crystal bars is shown in Table I. The orientation of the specimens in the standard [001]-[011]-[$\bar{1}$11] unit triangle is shown in Fig.1. Schmid Factors for octahedral slip are listed in Table II. Experimental details are described in ref (6).

Table I. Alloy Composition of the Ni_3Al + X Single Crystal Bars

	Composition (at%)		
Material (addition)	at% Ni	at% Al	at% X
Binary	76.60	23.40	none
1 Ta	74.30	24.70	1 Ta
2.5 Ta	74.50	23.00	2.5 Ta
1 Hf	75.80	23.00	1.0 HF+0.2 B
3.3 Hf	74.80	21.80	3.28 Hf
0.3 Zr	76.01	23.63	0.26 Zr
1 Zr	76.19	22.80	1.04 Zr
0.2 B	77.15	22.63	0.195 B
0.7 B	76.92	22.56	0.67 B
1 B	76.52	22.08	0.98 B

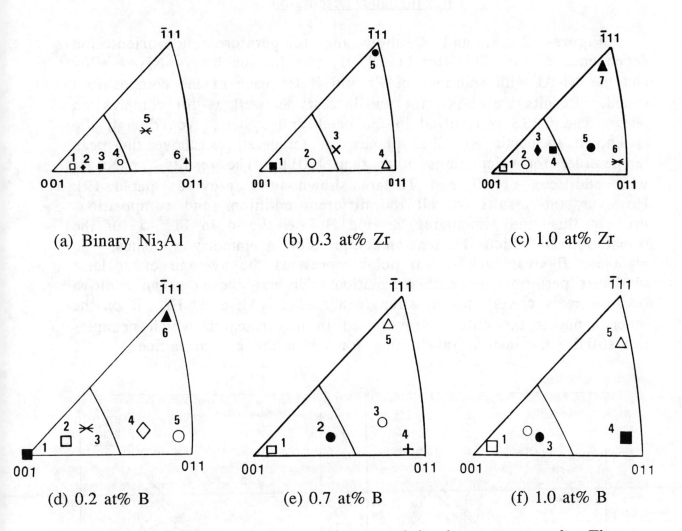

Figure 1 - Orientation of the samples tested in the present study. The numbers refer to the orientation in table II.

Table II. Schmid Factors on the Octahedral System for the Orientations Tested

	OCTAHEDRAL SYSTEM (111)[$\bar{1}$01]						
	Orientation						
Material	1	2	3	4	5	6	7
Binary	0.449	0.463	0.491	0.5	0.479	0.429	
1.0% Ta	0.44	0.478	0.489	0.492			
2.5 % Ta	0.435	0.47					
1.0% Hf	0.431	0.474					
3.3% Hf	0.453	0.475	0.489	0.478	0.435	0.412	
0.3% Zr	0.455	0.493	0.499	0.455	0.306		
1.0% Zr	0.428	0.476	0.486	0.49	0.486	0.44	0.33
0.2% B	0.419	0.474	0.489	0.489	0.443	0.307	
0.7% B	0.435	0.498	0.455	0.442	0.496		
1% B	0.442	0.486	0.493	0.451	0.4		

Results and Discussion

Figures 2, 3, and 4 show the temperature and orientation dependence of the CRSS for (111)[$\bar{1}$01] slip for the binary Ni$_3$Al alloy and the Ni$_3$Al with additions of Zr and B for each of the compositions tested. Results are shown for tensile tests as well as for compression tests. The CRSS is resolved in the octahedral system, even though slip is known to occur on (001) planes for temperatures above the peak temperature for orientations other than [001]. The results for Ni$_3$Al with additions of Hf and Ta are shown in a previous paper (9). However, the results for all the different additions and compositions used in this study, including Zr and B, are shown in Fig. 5 for the orientation in which the tension/compression asymmetry vanishes. In all these figures, each datum point represents the average of at least two tests performed on each orientation. Zr was chosen as an additive because it is a very potent strengthener of Ni$_3$Al, as is Hf. B on the other hand, is the only additive used in this research which occupies interstitial sites instead substituting for Al in the crystal lattice.

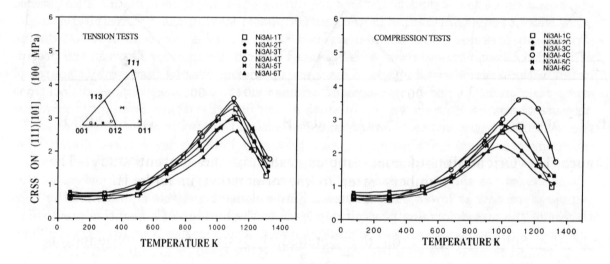

Figure 2 - CRSS for (111)[$\bar{1}$01] slip of binary Ni$_3$Al as a function of temperature and orientation of the stress axis.

As pointed out elsewhere (3) the value of the CRSS at temperatures below the peak temperature depends on the orientation of the tension/compression axis, and additions of Zr and B make the dependency more dramatic, as did the additions of Hf and Ta. Note that the CRSS at low temperature for the 0.7 at% B alloys is approximately 75 MPa in tension, about 100 MPa in compression, but the CRSS of the binary Ni$_3$Al is only about 25 MPa in both tension and compression.

The magnitude of the CRSS is increased in the range of positive temperature dependence by additions of 1 at% Zr and 0.7 at% B, but it is increased by additions of 1 at% B by approximately 150 MPa over the entire range of temperatures studied. In contrast, the value of the CRSS was not seriously affected by additions of 0.3 at% Zr and 0.2 at% B, and only at 77 K do these two additions change the magnitude of the CRSS: 0.3 at% Zr has a "softening effect" lowering the value of the CRSS by an average of about 18 MPa, and 0.2 at% B increases the CRSS at 77 K by approximately 15 MPa on average, depending on the orientation tested.

Comparing these results with those previously obtained for additions of 1 and 3.3 at% Hf and 1 and 2.5 at% Ta, it is noticed that the effect of 0.7 and 1 at% B on the octahedral CRSS are the strongest

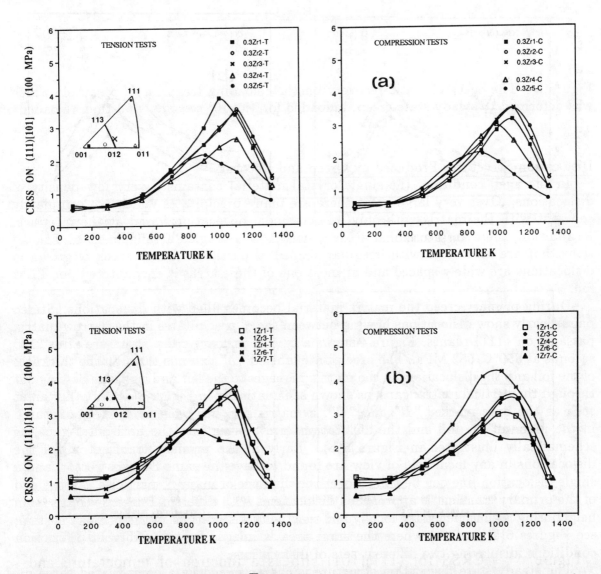

Figure 3 - CRSS for $(111)[\bar{1}01]$ slip as a function of temperature and orientation of the stress axis for $Ni_3(Al, Zr)$ with additions of (a) 0.3 at% Zr, and (b) 1.0 at% Zr.

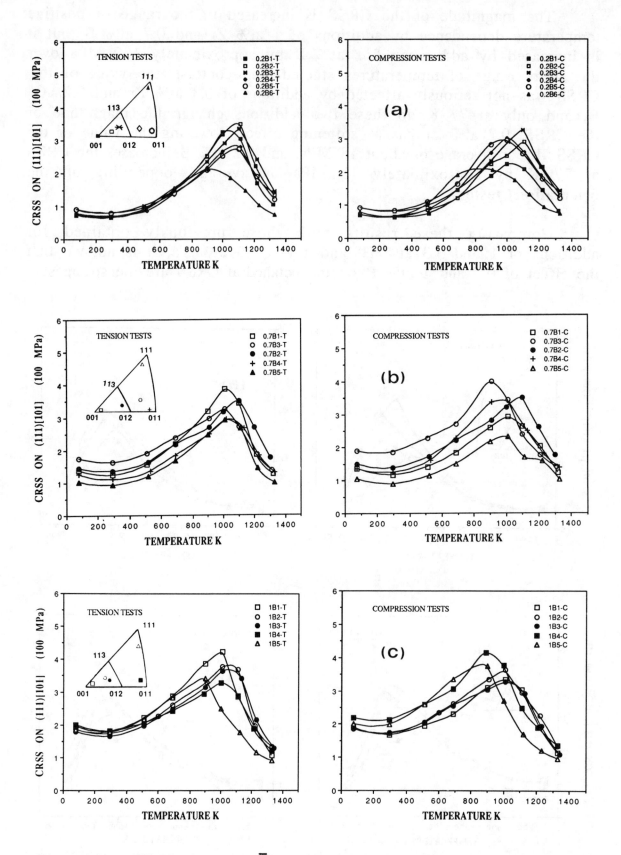

Figure 4 - CRSS for (111)[$\bar{1}$01] slip as a function of temperature and orientation of the stress axis for Ni_3Al + B: (a) 0.2 at% B; (b) 0.7 at% B; (c) 1.0 at% B

at low temperatures but addition of 3.3 at% Hf has the most effect at the peak temperature, especially for orientations near [001]. The effect of the ternary additions on the tension/compression asymmetry follows the same trends discussed for the increase of the CRSS. The additions of 0.3 at% Zr and 0.2 at% B slightly increase the asymmetry over that of binary Ni_3Al. The remaining additions all increase the asymmetry but the increase is similar for each case. The sense of the asymmetry follows the same trends as predicted by the Paidar *et al.* model (3) and verified later by Ezz (4) and Umakoshi (5) on single phase single crystals and by Heredia and Pope in a two phase single crystalline alloy (6).

It is important to note at this point that Zr as well as Hf, Ta and Nb (4) occupy the Al sites in the ternary alloy Ni_3Al + X, but B occupies interstitial sites and therefore the effect of lattice friction plays an important role on the strengthening mechanism of the alloys. Among the compositions tested, 0.7 at% B produces the most accentuated effect. 1 at% B raises the magnitude of the CRSS qualitatively similar to the effect of the substitutional type elements.

In Figure 5 the CRSS for octahedral slip as a function of temperature and composition for the orientations showing zero tension/compression asymmetry is presented. Data for 5 at% Ta and 4.3 at% Nb are from Umakoshi *et al.* (5) and S. Ezz *at al* (4) respectively. Data for the other Ta and Hf alloys have been reported by the present authors earlier (9). It is clearly seen that ternary additions do strengthen Ni_3Al at temperatures as low as 77 K with only one exception, 0.3 at% Zr. The 2.5 and 5 at% Ta alloys show the highest CRSS at the peak temperature, and the highest strengthening rate over the range of positive temperature dependence. The peak temperature, the temperature where the maximum CRSS occurs, decreases as the amount of ternary addition increases, with the exception of Ta. Boron, even though it increases the value of the CRSS at low temperatures, as mentioned previously, does not increase its magnitude at the peak temperature. The high temperature value of the RSS (above the peak temperature) is much less dependent on the composition, because slip actually occurs on the cube system in this temperature range (10).

The orientation for which the tension/compression asymmetry vanishes also shows a compositional dependency: as the amount of ternary element is increased, the orientation for T=C moves toward [001] from the [$\bar{1}$13]-[012] great circle. The T=C orientation of the binary Ni_3Al is the closest to this circle.

Finding the orientation for zero asymmetry is important because for these orientations the rate of cross-slip does not depend on the

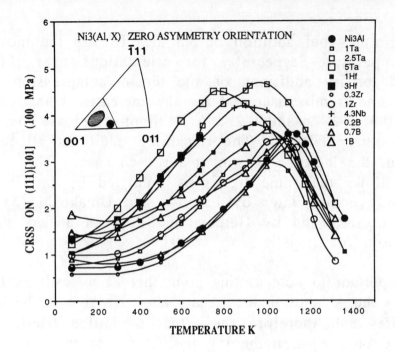

Figure 5 - The CRSS for octahedral slip as a function of temperature and composition for orientations which show zero tension/compression asymmetry.

dislocation core width. For such a condition the expression for the activation enthalpy of cross-slip is greatly simplified (3). For this particular case, assuming that the difference in APB energy on the (111) and (001) planes is small compared with the resolved shear stress (RSS) on the (001) plane, equation [23] in Paidar *et al.* (3) has been shown previously (9) to reduce to:

$$T \ln \Delta T_i = A - B (T_{i, pb})^{1/2}, \qquad (1)$$

where ΔT_i is the difference between the RSS at a given temperature with respect to the RSS at 77 K for the same ith ternary addition; $T_{i, pb}$ is the RSS on the primary (111) plane in the direction of the [$\bar{1}$01] Burgers vector, also for the same ith ternary addition, and A and B are constants.

In Fig. 6, $T \ln \Delta T_i$ is plotted against $(T_{i,pb})^{1/2}$. A fairly narrow band is obtained in which twelve different compositions and four different temperatures are represented. A slight deviation from the band is found for the boron-containing material which as mentioned earlier has different alloying characteristics. The fact that all the different compositions follow this pattern indicates that the difference in the APB energies between (111) and (001) planes do not depend on temperature or composition. Thus, for the particular case of zero asymmetry orientations, ordinary solid solution strengthening effects seems to be most important in determining the CRSS for $(111)[\bar{1}01]$ slip.

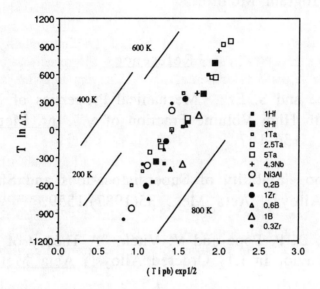

Figure 6 - $T \ln \Delta \tau_i$ as a function of composition and RSS. The straight lines indicate the ranges over which data are presented for each temperature

Conclusions

1. Additions of ternary elements to Ni_3Al increase the CRSS for octahedral slip but have little effect on the RSS for cube slip.

2. In the present study, Hf and Ta are the elements that show the strongest effects on the CRSS for the $(111)[\bar{1}01]$ slip.

3. Additions of more than 0.7 at% Boron markedly increase the value of the octahedral CRSS at low temperatures.

4. The orientation for which the tension/compression asymmetry vanishes is composition dependent.

5. The core dissociation effect is an important mechanism for the tension/compression asymmetry but where it disappears, solid solution strengthening effects are the dominant ones.

Acknowledgments

The authors wish to express their appreciation to the Office of Naval Research for financial support of this study (Grant Nº 5-21233), Dr. Donald Polk, Program Monitor.

References

1. D. P. Pope and S. Ezz, "Mechanical Properties of Ni_3Al and Ni-base Alloys with High Volume Fraction of γ'", Int. Metals Rev., 29(1984), 136-167.

2. M. H. Yoo, "Stability of Superdislocations and Shear Faults in $L1_2$ Ordered Alloys", Acta Met., 35(1987),1559.

3. V. Paidar, D. P. Pope and V. Vitek, "A Theory of The Anomalous Yield Behavior in $L1_2$ Ordered Alloys", Acta Met., 32(1984), 435.

4. S. Ezz, D. P. Pope and V. Paidar, "The Tension/Compression Flow Stress Asymmetry in Ni_3(Al, Nb) Single Crystals", Acta Met., 30(1982), 921.

5. Y. Umakoshi, D. P. Pope and V. Vitek, "The Asymmetry of the Flow Stress in Ni_3(Al, Ta) Single Crystals", Acta Met., 32(1984), 449.

6. F. E. Heredia and D. P. Pope, "The Tension/Compression Flow Stress Asymmetry in a High γ' Volume Fraction Ni-Base Alloy", Acta Met., 34(1986),279.

7. L. R. R. Curwick, "Strengthening Mechanisms in Ni-Base Superalloys" (Ph. D. thesis, University of Minnesota, 1972), 93.

8. Y. Mishima, S. Ochiai, M. Yodogawa and T. Suzuki, "Mechanical Properties of Ni_3Al with Ternary Addition of Transition Metal Elements", J. Japan Inst. Metals, 27(1986), 41-50.

9. F. E. Heredia and D. P. Pope, "Strengthening of Ni_3Al by Ternary Additions" (Paper presented at the MRS Fall Meeting, Boston, Mass., 5 December 1986, pp. 213-220 in the Procd. MRS Symposium, Vol 81, 1987).

10. S. Ochiai, S. Miura, Y. Mishima and T. Suzuki, "High Temperature Yielding of Ni_3(Al, Ti) Single Crystals", J. Japan Inst. Metals, 51(1987), 608-615.

HIGH SPEED SINGLE CRYSTAL CASTING TECHNIQUE

Shogo Morimoto, Akira Yoshinari
Hitachi Research Lab., Hitachi Ltd.
832 Horiguchi Katsuta, Ibaraki, 312 Japan

Eisuke Niyama
Tohoku Univ.
Aoba Aramaki Sendai, Miyagi, 980 Japan

Abstract

Production of a single crystal with a <100> direction with good yield and at a higher speed is an important factor for improving the productivity in a mold withdrawal unidirectional solidification method. This factor was investigated both in solidification calculations and casting experiments. It was found that casting conditions, such as the pouring temperature and the mold heating temperature, are important in the production of a single crystal with a <100> direction; and by optimizing these conditions, <100> single crystals were produced with good yield. In order to increase the speed of production of a single crystal, attempts were made at providing a high temperature gradient at the solidification front by two stage heating and regulation of convection by a magnetic field. Two stage heating was effective for increasing the temperature gradient solely without raising the temperature of the molten metal as a whole, and facilitated the production of a single crystal airfoil having a complicated shape. On the other hand, the application of a magnetic field had a deleterious effect and produced typical freckles.

Introduction

Single crystal castings of Ni-base superalloys exhibit excellent characteristics at high temperature[1,2]. They have begun to be used as buckets or nozzles of jet engines for aircraft, and are expected to be used as buckets and nozzles for heavy duty gas turbines in the near future.

The characteristics such as the high temperature strength and thermal fatigue strength of a single crystal casting of a Ni-base superalloy are excellent when the casting is crystallized in the direction of <100>[3]. It is therefore essential for bringing the characteristics of a single crystal into full play to make the direction of the main stress of the airfoil the <100> direction of the crystal.

The direction of crystallization of a single crystal casting is controlled by a seeding method or a selector method. Although it is difficult to directly control the direction of a single crystal by a selector method, it is suitable for industrial production in the respect that it is a comparatively simple manufacturing method. Accordingly, factors for controlling the direction of crystallization by the selector method were investigated by calculation and experiments.

On the other hand, single crystal castings produced by a mold withdrawal unidirectional solidification method are much more expensive than those produced by an ordinary casting method because the solidification time is so long as to lower the productivity. In order to apply a single crystal airfoil to heavy duty gas turbines, improvement of the productivity is therefore essential and development of a solidification controlling technique is required.

One of the factors for improving the productivity is a higher withdrawal rate. For this purpose, attempts were made at providing a high temperature gradient at the solidification front and regulation of convection. As a method of providing a high temperature gradient, a two stage heating process was investigated in which the temperature of the bottom of the susceptor was raised[4]. This was because it was believed that the two stage heating process would increase the temperature gradient solely without raising the temperature of the molten metal as a whole. As a method of regulating convection, application of a static magnetic field was investigated. This method is called an MCZ method and is adopted for production of a high-quality Si single crystal[5]. We supposed that if the solidification speed was increased, the solidification front would become unstable and the convection produced was regulated by the magnetic viscosity[6] and, hence, the production of new crystal and the deterioration of crystallinity would be regulated.

The results of these investigations are reported in this paper.

Method for Research

Calculation Method

Solidification calculations were carried out both on a model starter and on a model airfoil having a simplified two-dimensional shape. Fig. 1 shows the calculation model. The calculation model of the airfoil had a two-dimensional plate shape, while the starter had a two-dimensional columnar shape. Calculation was carried out by an explicit finite difference scheme[7]. On the assumption that the heat transfer between the mold surface and the atmosphere was of a Newtonian type, an apparent heat transfer coefficient was used. The coefficients used in the calculation are shown in Table 1. Thermo-physical values used in the calculation were assumed to be constant irrespective of the temperature. When the mold withdrawal rate was calculated, the apparent heat transfer coefficient and atmospheric temperature were varied as a function of time and distance from the chillplate.

Table 1 Physical properties used in the calculation

		standard	variation
Density (Metal)	(g/cm)	8.9	
Specific heat (Metal)	(cal/g.deg)	0.11	
Thermal conductivity (Metal)	(cal/cm.°C.s)	0.07	
Liquidus	(°C)	1390	
Solidus	(°C)	1350	
Latent heat of solidification	(cal/g)	75	
Density (Mold)	(g/cm)	2.4	
Specific heat (Mold)	(cal/g.deg)	0.25	
Thermal conductivity (Mold)	(cal/cm.°C.s)	0.003	
Time increment	(s)	0.03	
Heat Transfer Coefficient (cal/cm.s.deg)	Casting-Chill	0.01	
	Mold-Chill	0.005	
	Mold-Susceptor Chill-7 cm	0.001	
	7 cm-24 cm	0.0015	
	Mold-Atmosphere	0.00075	
Susceptor Temperature (°C)	Upper Heater	1500	1500-1650
	Lower Heater	1600	1400-1650
Heater Height (cm)	Upper Heater	17	17-24
	Lower Heater	7	0-7
Chillplate Temperature	(°C)	20	
Withdrawal Velocity of Mold	(cm/h)	20	

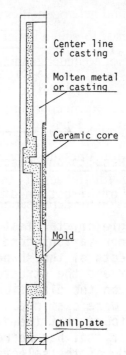

Fig. 1 Calculation model.

Casting Experiment

A single crystal of a Ni-base superalloy was cast by the apparatus shown in Fig. 2 in a vacuum of about 10^{-4} Torr. The susceptor in Fig. 2 was divided into two parts so as to control the temperature at the top and at the bottom separately from each other. A magnetic field for convection was applied from the outside of the furnace wall by a DC electromagnet. The magnetic field intensity was 2000 G at maximum at the center of the furnace. Table 2 shows the chemical composition of the alloys used in the experiments. The pouring temperature, the mold heating temperature, the mold withdrawal rate, etc., were varied in the casting experiments. The temperature of the mold was measured by a platinum thermocouple.

In the convection regulation experiments, a test piece of a bar having a 15 mm diameter was cast. After casting, the mold was withdrawn downwardly for effecting unidirectional solidification while a magnetic field was being applied.

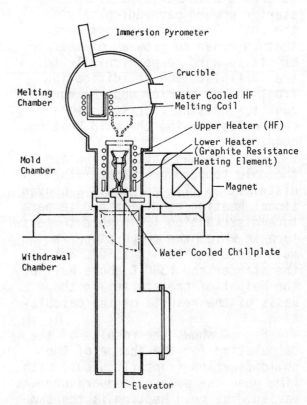

Fig. 2 Schematic structure of single crystal casting furnace.

Table 2 Chemical composition of S.C. Alloys (w%)

Alloy	Cr	Mo	W	Co	Ta	Al	Ti	Hf	Ni
NASAIR-100	9.0	1.0	10.5		3.2	5.7	1.2		bal
CMSX-3	8.0	0.6	8.0	4.6	6.0	5.6	1.0	0.1	bal
TMS-12	6.6		12.8		7.7	5.2			
TMS-26	5.6	1.85	10.9	8.2	7.7	5.1			bal

Results and Discussion

Control of Direction of Growth

Single crystal test pieces were cast under various conditions, and the effects of the shape of the selector and the structure of the starter on the direction of a single crystal were examined. As a result, it was found that the direction of a single crystal has no relation to the shape of the selector and that the structure of the starter is important. More specifically, in order to obtain the <100> direction of a single crystal, it is important to grow a columnar crystal in the starter upward perpendicularly to the chillplate. It was considered that in order to grow a columnar crystal upward perpendicularly to the chillplate the solidification front in the starter must be horizontal, and single crystals were cast while varying the mold heating conditions.

Fig. 3 shows the distribution of the mold temperature. Temperature distribution (1) represents a conventional heating condition and temperature distribution (2) a mold temperature in which the temperature is maintained at 1300 to 1350°C below the starter and 1390°C above half of the height of the starter on the basis of the results of the calculation.

Fig. 4 shows the results of the calculation for the change of the solidification front (liquidus) with time when the pouring temperature was varied. The mold heating is the same as temperature distribution (1) shown in Fig. 3. From the results of the calculation, it is not expected that the columnar crystal will grow sufficiently from the chillplate to the portion above the starter (entrance of the selector) because the solidification front protrudes downwardly

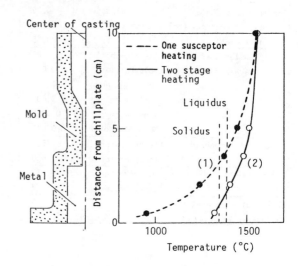

Fig. 3 Temperature profile of mold.

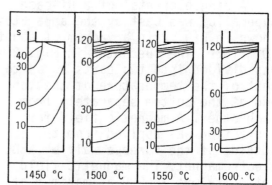

Fig. 4 Effect of pouring temperature on the location of solidification front vs. time elapsed.

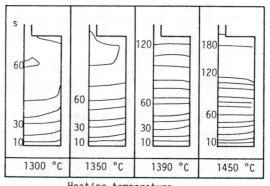

Fig. 5 Effect of mold temperature on the location of solidification front vs. time elapsed.

to a great extent at a pouring temperature of 1450°C. When the pouring temperature is 1500°C, the solidification front in the vicinity of the center line becomes substantially horizontal. If the pouring temperature is raised further, the area of horizontal solidification is enlarged, but the solidification front protrudes downwardly at the outer peripheral portion which is in contact with the mold.

Fig. 5 shows the results of the calculation for the change in the solidification front with time when the mold temperature was varied. The mold heating temperature was constant in the furnace. The solidification front was horizontal at about 1/3 of the height of the starter when the mold heating temperature was 1300°C, and at about 2/3 of the height of the starter when the mold heating temperature was 1350°C (solidus). When the mold heating temperature is higher than 1390°C (liquidus), the solidification front is horizontal over the entire area of the starter and a columnar crystal is expected to grow upward perpendicular to the chill.

From the above results, it was found that even if the pouring temperature is low, the mold cools the molten metal and it is impossible to obtain a completely horizontal solidification front. Since the raise of the mold heating temperature was considered to be most effective for obtaining a horizontal solidification front, temperature distributions (1) and (2) were investigated by both calculation and by casting experiment.

Fig. 6 shows the results of the calculation of the change in the solidification front with time and the macrostructure of the starter when the mold heating temperature was varied in accordance with Fig. 3. When the heating temperature is in agreement with temperature profile (1), the solidification front protrudes downwardly due to the chill from the mold. From observations of the macrostructure, it is clear that a large columnar crystal exists at the outer peripheral portion, which in fact corresponds to the results of the calculation. Thus, it was found that the heating condition in accordance with temperature profile (1) cannot produce a horizontal solidification front.

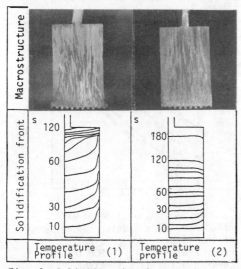

Fig. 6 Solidification front and macrostructure of starter.

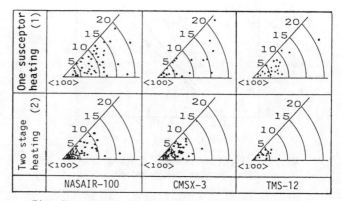

Fig. 7 Distribution of orientation from <100>.

As a result of the calculation using temperature profile (2), it was found that the solidification front is horizontal at any position unlike the adoption of temperature profile (1). The columnar crystal of the starter grows upward perpendicularly to the chillplate, which in fact agrees well with the results of the calculation.

Fig. 7 shows the directions of the single crystals cast under mold heating conditions (1) and (2). If it is assumed that single crystals having a deviation of not greater than 10° from the <100> direction are good products, great improvement in the yield was observed in all alloys used when mold heating condition (2) was adopted.

Provision for High Temperature Gradient

1) Two-Stage Heating (Calculation)

In order to increase the temperature gradient at the solidification front, it is most effective to raise the mold heating temperature. Fig. 8 shows the temperature gradient when the mold heating temperature is raised while making the temperature within the susceptor constant. The temperature gradient becomes higher in proportion to the rise in the mold heating temperature. The rise of the mold heating temperature brings about problems such as: (1) dimensional changes due to deformation of the mold and the core, (2) changes in the composition of the molten metal due to evaporation of the alloying element such as Cr, and (3) reaction of the mold and core with the molten metal. It is, therefore, very useful to increase only the temperature gradient at the solidification front without raising the temperature of the cast as a whole. As one measure, a two-stage heating process in which the temperature of the bottom of the susceptor is raised was investigated.

Fig. 9 shows the change in the temperature gradient at the center of the casting when the temperature of the upper heater is constantly 1500°C and the temperature of the lower heater having a height of 70 mm is varied.

When the temperature of the lower heater is not lower than 1500°C, the temperature gradient increases in proportion to the temperature of the lower heater, but the increase in the temperature gradient is slightly smaller than that in the case of maintaining the whole susceptor at a uniform higher temperature.

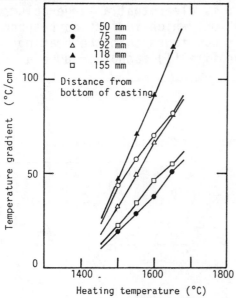

Fig. 8 Effect of heating temperature on the temperature gradient.

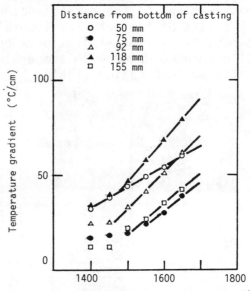

Fig. 9 Effect of lower heating temperature on the temperature gradient.
Upper heater ; 1500°C
Lower heating unit ; Height 70 mm

Fig. 10 shows the temperature distribution of the casting when the mold is withdrawn 5 cm and 10 cm under the conditions shown in Fig. 9. In proportion to the increase in the temperature of the lower heater, the temperature change in the vicinity of the solidification front becomes large rapidly. However, when the temperature of the lower heater is excessively high, namely 1650°C, the temperature distribution of the molten metal is reversed.

Fig. 11 shows the rate of increase in gradient in the casting when the temperature of the upper heater is constantly 1500°C and the temperature of the lower heater is constantly 1600°C while the height of the lower heater is varied. When the height ratio is 0, the temperature of the whole susceptor is constantly 1500°C. The larger the height ratio of the lower heater, the larger the temperature gradient obtained.

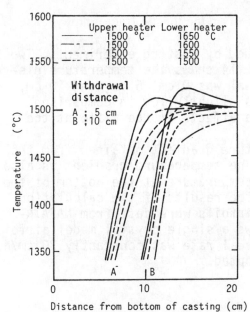

Fig. 10 Temperature profile of casting at several heater temperatures.

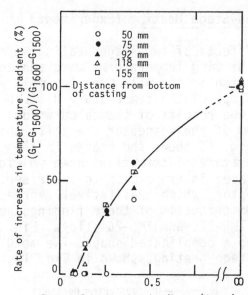

Fig. 11 Rate of increase in temperature gradient vs. heater ratio.
G_L ; Temperature gradient (Lower=L)
G_{1500} ; (1500°C const.)
G_{1600} ; (1600°C const.)

Fig. 12 shows the change in the temperature gradient with the withdrawal distance of the model airfoil mold under two heating conditions. The temperature gradient is suddenly lowered at the airfoil tip and at the enlarged portion directly above the platform of the casting. However, the change in the temperature gradient is almost the same under both heating conditions.

Fig. 13 shows the temperature distribution in the casting when the mold is withdrawn 5 cm, 10 cm, and 15 cm under the same heating conditions shown in Fig. 12. In spite of having almost the same change in the temperature gradient, a large difference is observed in the temperature distribution in the casting. When the temperature at the bottom of the susceptor is raised by the lower heater, the temperature of the molten metal is 1500°C to 1520°C, while the temperature of the molten metal is as high as 1580°C when the whole susceptor is uniformly heated to 1600°C. Thus, it was found that the two-stage heating process in which the temperature at the bottom of the susceptor is raised is very effective in increasing the temperature gradient solely without raising the temperature of the casting as a whole.

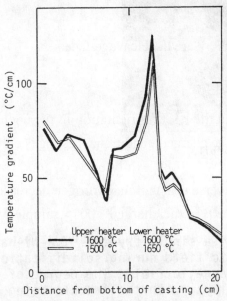

Fig. 12 Temperature gradient of model airfoil at two heating conditions.

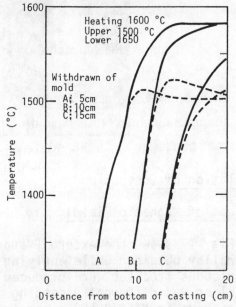

Fig. 13 Temperature profile of casting.

2) Two-Stage Heating (Experiment)

Effects of two-stage heating were examined by casting experiments. Two mold heating temperatures were adopted. Fig. 14 shows the temperature distribution in the mold when the mold withdrawal was 0 cm, 5 cm, and 10 cm. The slope of the temperature distribution is slightly low in comparison with the results of the calculation, but the effects of the raise at the bottom of the susceptor are still observed.

Fig. 15 shows the change in the temperature gradient obtained from the temperature distribution shown in Fig. 14. The temperature gradient is also increased in proportion to the rise in the temperature at the bottom of the susceptor, which qualitatively agrees with the results of the calculation.

On the basis of these findings, model airfoils were cast from NASAIR-100, TMS-12, and TMS-26 alloys. Fig. 16 shows a single crystal model airfoil having a complicated shape. The mold withdrawal rate was constantly 20 cm/h. Two-stage heating, shown in Fig. 14, was adopted.

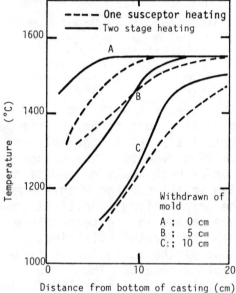

Fig. 14 Temperature profile of mold.

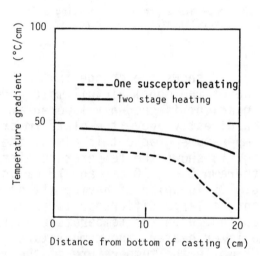

Fig. 15 Change of temperature gradient accompanying withdrawal of casting.

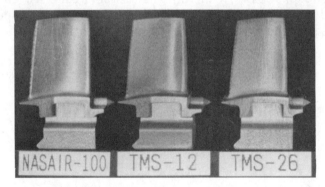

Fig. 16 Single crystal airfoil.

Regulation of Convection

(Effect of Magnetic Field)

Fig. 17 shows the external appearance of a single crystal of a Ni-base superalloy obtained while applying a magnetic field during solidification. Conspicuous freckles were produced by the magnetic field. The degree of freckling was different depending on the alloys, with the freckle being more conspicuous in NASAIR-100. Even in the test pieces with freckles, no change was observed in the composition either in the direction of growth or in the transverse direction, and no change was observed in the structure of either the vertical section or the cross section due to the magnetic field.

Fig. 18 shows the position at which freckling was produced. Freckling was constantly produced on the left-hand side of the casting relative to the magnetic line of force irrespective of the position in the susceptor and the configuration of the castings.

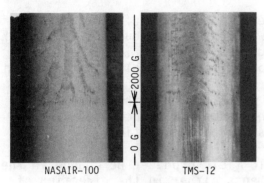

Fig. 17 Freckle

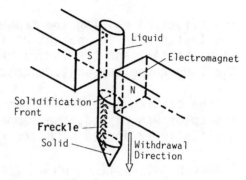

Fig. 18 Generation of freckle.

It is believed that freckling is produced due to the segregation of low-melting point elements[8]. However, since freckling in the experiment was produced due to the magnetic field irrespective of the configuration of the casting, it was not caused by the segregation of an element. We consider the cause of freckles to be as follows:

A thermoelectromotive force is generated on the basis of a difference in the molten metal and the mold, the molten metal and the casting, or the mold and the casting, and produces a circuit for a current. When a current flows in this circuit, the magnetic field interferes, so that the molten metal is caused to flow in a constant direction. At the solidification front with which the flow of the current comes into contact, freckles are produced. Since the main constituents of the mold are zircon sand and silica, the mold becomes electrically-conductive in the vicinity of the solidification temperature of the alloy, thereby causing a current to flow on the basis of the thermoelectromotive force.

Clarification of the mechanism for producing freckles including the establishment of the above-described theory remains to be solved in subsequent studies.

Conclusions

Solidification calculations and casting experiments were carried out for the purpose of improving the accuracy of the direction of growth and increasing the withdrawal speed in a mold withdrawal unidirectional solidification method, and the following conclusions were reached.

1. In order to produce a single crystal of the <100> direction, it is necessary that the starter has a columnar structure.

2. In order to grow a columnar crystal in the starter, it is desirable to raise the pouring temperature and heat the mold above the liquidus of the alloy. If the mold heating temperature is low, the columnar crystal grows obliquely due to the chill from the mold, thereby increasing the difference in the angle between the direction of the vertical axis and the <100> direction of the single crystal.

3. The two-stage heating process in which the temperature at the bottom of the susceptor is raised is very effective for increasing the temperature gradient without raising the casting temperature.

4. The temperature gradient at the solidification front increases in proportion to the temperature of the lower heater at the bottom of the susceptor. However, when the temperature of the lower heater exceeds a predetermined temperature, the temperature distribution in the molten metal is reversed so that the temperature of the lower portion becomes higher.

5. The effect of raising the temperature at the bottom of the susceptor is manifested when the height of the lower heater is about 10% that of the upper heater, and the larger the height ratio of the lower heater is, the larger the temperature gradient becomes at the soldification front.

6. In the casting experiments, when the temperature at the bottom of the susceptor was raised, the temperature gradient was increased, which qualitatively agreed with the results of the calculation.

7. Although the application of a magnetic field produced no change in the major components of the alloying elements either in the direction of growth or in the horizontal plane of the castings, freckles which are characteristic of a single crystal casting were produced.

8. The freckles due to the application of the magnetic field were constantly produced on the left-hand side of the castings relative to the magnetic lines of force irrespective of the configuration of the castings. However, when the mold withdrawal rate was high, the freckles were rarely produced.

9. It is suggested that the freckles were produced by the forced circulation due to the interference of the magnetic field with the thermoelectromotive force which had been produced between the molten metal and the mold, the molten metal and the casting, or the mold and the casting.

Acknowledgement

This work was performed under management of Research and Development Institute of Metals and Composites for Future Industries as part of the R&D Project of Basic Technology for Industries sponsored by Agency of Industrial Science and Technology, MITI.

Reference

1) B.H.Kear, B.J.Piearcey; Trans. Metal.Soc.of A.I.M.E.239('67),1207
2) F.L.Versnyder, M.E.Shank; Mater.Sci.Eng. 6('70),213
3) J.S.Higginbotham; Mater.Sci.Tech. 2('86),442
4) R.J.Naumann; J. of Crystal Growth 58('82),554
5) H.Kimura,M.F.Harvey,D.J.O'Connor; J. of Crystal Growth 62('83)523
6) G.M.Oreper,J.Szekely; J. of Crystal Growth 67('84) 405
7) S.Morimoto,A.Yoshinari,E.Niyama; Superalloys '84 p177
8) A.F.Giamei,B.H.Kear; Metall.Trans. 1('70),2185

EFFECT OF CHEMISTRY MODIFICATIONS AND HEAT TREATMENTS ON THE MECHANICAL PROPERTIES OF DS MAR-M200 SUPERALLOY

Zheng Yunrong*, Wang Yuping*, Xie Jizhou*,
Pierre Caron**, and Tasadduq Khan**

*Institute of Aeronautical Materials, Beijing (BIAM)
**Office National d'Etudes et de Recherches Aerospatiales (ONERA)

Abstract

Hf and Zr can promote the formation of eutectic ($\gamma+\gamma'$), MC_2, and Ni_5M phases. In the alloy with equal atomic percent Zr and Hf, the solubility of Zr in eutectic γ' is lower than that of Hf, and Zr content in Ni_5M is much higher than Hf. This distribution of Zr is beneficial to the formation of Ni_5M and lowers the strengthening efficiency of Zr. A pretreatment of 1130°C/3hr efficiently eliminates Ni_5M and, as a consequence, increases the incipient melting temperature of the alloy. The precipitation treatment of 1100°C/4hr leads to cuboidal γ' precipitation of about 0.5 μm size and causes the Hf-containing alloy to have a much higher creep strength than the Hf-free alloy in the temperature range of 760 to 1050°C. The Hf-containing alloy showed greater LCF (low cycle fatigue) life in comparison to the Hf-free alloy. A similar tendency was found when Zr was either partially or totally substituted for Hf. A higher rate of solidification facilitates enhanced LCF life due to a finer structure and more perfect orientation. Surface slip analysis showed that intersection of two sets of slip in adjacent grains occurred in the Hf-free and Hf-containing alloys, but cracking at the columnar grain boundary easily took place in the Hf-free alloy. The number of surface cracks of LCF specimens and their length per unit area are much higher in the Mar-M200 alloy. MC cracking preferentially occurs at long rod-shaped carbides perpendicular to the stress axis, and then propagates through the interdendritic region. The Hf-containing alloy cracks along the crystallographic planes by separation of slip bands.

Introduction

Mar-M200 is an early superalloy which has good high temperature strength; however, it was found that premature rupture of this alloy occurred at intermediate temperature. Later, a potential problem (i.e., "ductility valley") that occurs with many high-strength cast superalloys at intermediate temperature was recognized. Hafnium (Hf) additions and directional solidification techniques were considered as better ways to solve this problem. Directionally solidified (DS) superalloys were first introduced by Pratt & Whitney for military engines in the late 1960's and for commercial engines around 1974. Since that time, effort was devoted more specifically to the development of single crystal (SC) superalloys because it was thought that no significant improvement in properties was possible with the DS alloys. In spite of the advances made in the development of new SC alloys, attention has been rediverted toward the DS alloys for economic and technical reasons.

As we know, Hf improves transverse mechanical properties of DS superalloys, but makes stress-rupture life above 980°C drop (1). In order to overcome this disadvantage, a modified heat treatment (2) suitable for SC superalloys was introduced into DS Mar-M200+Hf. On the other hand, Zr and Hf have similar alloying behavior (3); and recent work (4) has proved that Zr can be substituted for Hf. Accordingly, BIAM and ONERA conducted a joint program to explore the effect of Hf and Zr additions, as well as a modified heat treatment, on the tensile, creep, and fatigue strength of the DS Mar-M200 superalloy. This paper describes some of the results, with particular emphasis on the low cycle fatigue behavior of such alloys.

Experimental Procedures

The Mar-M200 alloy and its modified versions (Hf and/or Zr) were cast into master ingots at BIAM. Their chemical compositions are listed in Table I. A number of DS rods were then prepared at both BIAM and ONERA. The former used a ISP2/III DS furnace and the latter did a high gradient DS set. The high rate solidification process used at BIAM results in a cooling rate (gradient x withdrawal rate) of 25°C/min, whereas the cooling rate at ONERA was 10°C/min. Various heat treatments were used for the different alloys (Table II). Pretreatment of 1130°C/3hr for Zr-containing alloys was performed to eliminate existing Ni_5Zr phase in order to increase the incipient melting temperature of the alloy. Microstructural analysis of as-cast and heat treated specimens was conducted using an optical microscope, SEM, and EPMA.

Table I. Chemical Composition of Alloys

Alloy	Cr	Co	Al	Ti	Nb	W	Hf	C	B	Zr	Ni
Mar-M200	8.56	9.80	5.05	2.03	1.03	11.20	--	0.15	0.015	0.0003	base
Mar-M200+Hf	8.88	9.95	5.13	2.16	1.02	12.20	1.78	0.13	0.015	0.035	base
Mar-M200+Zr	8.56	9.80	5.05	2.03	1.03	11.20	--	0.15	0.015	0.70	base
Mar-M200+Zr +Hf	8.56	9.80	5.05	2.03	1.03	11.20	0.80	0.15	0.015	0.40	base

Tensile and creep tests were performed at a temperature range of 20 to 1050°C, and specimens were machined into standard cylindrical bars. A large number of fully reversed strain-controlled LCF tests (R=-1, frequency 0.33 Hz, total strain range 1.0 to 1.4%) were performed on a MTS servo-hydraulic fatigue testing machine for various DS alloys at 760°C by using cylindrical and plate specimens. The plate specimens were prepolished before testing in order to reveal slip trace and crack initiation sites. Some LCF tests were interrupted at different cycles before failure for transmission electron microscopic and metallographic analysis to understand the effect of Hf on deformation mechanisms and crack initiation. For ruptured specimens, fracture surfaces were observed by SEM.

Experimental Results

Microstructure of Alloys

<u>As-Cast Microstructure</u>. Specimens given rapid solidification have a finer dendritic structure. Results of quantitative metallography have shown that the dendritic densities of BIAM and ONERA bars are $1391/cm^2$ and $706/cm^2$, respectively. The directional diversion from (001) is less than 6° and 16° for BIAM and ONERA bars. In the as-cast Mar-M200 alloy, there are MC carbide, eutectic ($\gamma+\gamma'$), and minor M_3B_2 and M_2SC phases. In addition to these phases, there are also MC_2 and Ni_5M in the Hf- and/or Zr-containing alloys. Both elements promote the formation of eutectic ($\gamma+\gamma'$) and the intermetallic compound Ni_5M. Moreover, the effect of Zr is more intense. The volume percents of eutectic ($\gamma+\gamma'$) and Ni_5Hf in the Hf-containing alloy are 10.53 and 0.23%, whereas eutectic ($\gamma+\gamma'$) and Ni_5Zr in the Zr-additive alloy are 10.38 and 0.71%. Ni_5Zr appears around eutectic ($\gamma+\gamma'$) and in cellular form (Fig. 1), as does Ni_5Hf. Hf and Zr can change script MC carbide into blocky carbides, but they cannot influence the amount of MC carbide distinctly.

The composition of some phases was measured by EPMA, and the results are listed in Table III. The composition of MC only represents the metallic 'M' half of the carbide. Table III shows that in an alloy with equal atomic percent of Zr and Hf, the solubility of Zr in eutectic γ' is lower than that of Hf, and the Zr content in Ni_5M is much higher than Hf. The above characterization promotes the formation of Ni_5Zr in Zr-containing alloys. This implies that we cannot substitute Zr for Hf in the same atomic fraction, and a rather low content of Zr is beneficial.

<u>Heat Treated Microstructure</u>. A pretreatment of 1130°C/3hr is performed to promote the transformation of Ni_5Zr into MC_2 and to restrict incipient melting during solution treatment. After 1130°C/3hr, Zr-rich MC_2 formed around the eutectic ($\gamma+\gamma'$) (Fig. 2). The 1230°C/2hr treatment is basically a full solid solution treatment for Mar-M200, but treatment in the temperature range of 1205 to 1225°C gives an incomplete solid solution for the four alloys. During solid solution treatment, secondary MC_2 is scattered in the vicinity of the eutectic ($\gamma+\gamma'$) (Fig. 3). Precipitation treatment at 1100°C/4hr leads to cuboidal γ' of about 0.5 μm.

Tensile and Creep Properties

Ultimate tensile strength (UTS), 0.2% yield strength (YS), and creep life were measured (Tables IV and V). Table IV shows that both alloys have almost the same UTS, 0.2% YS, and elongation in the temperature range of R.T. to 760°C; but at higher temperature (950°C), these properties in the Hf-containing alloy are rather higher than those in the Hf-free alloy. The effect of heat treatment is not obvious in both alloys.

It can be seen in Table V that the Hf-containing alloy has much longer rupture life than that of the Hf-free alloy in the temperature range of 760 to 1050°C. Previous investigation indicated that Hf additions to the DS Mar-M200 alloy improved the transverse stress-rupture life and elongation at intermediate temperature; however, Hf is not beneficial at high temperature, especially above 1000°C in an alloy given an incomplete solid solution treatment. The present work shows that by using heat treatments II and III, satisfactory results can be obtained.

Table II. Heat Treatment (H.T.) of Alloys

H.T.	Temperature/Time	Application
I	1205°C/2hr, A.C.+870°C/32hr, A.C.	Mar-M200 Mar-M200+Hf
II	1230°C/2hr, A.C.+1100°C/4hr, A.C. +870°C/20hr, A.C.	Mar-M200
III	1225°C/2hr, A.C.+1100°C/4hr, A.C. +870°C/20hr, A.C.	Mar-M200+Hf
IV	1130°C/3hr, A.C.+1210°C/3hr, A.C. +1100°C/4hr, A.C.+870°C/20hr, A.C.	Mar-M200+Zr
V	1130°C/3hr, A.C.+1220°C/3hr, A.C. +1100°C/4hr, A.C.+870°C/20hr, A.C.	Mar-M200+Zr+Hf

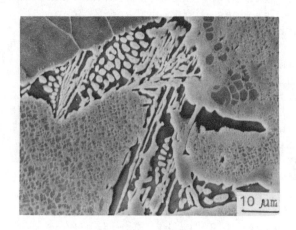
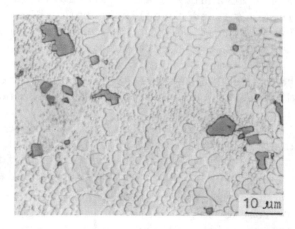

Figure 1 - Morphology of Ni_5Zr in Mar-M200+Zr.

Figure 2 - Zr-rich MC_2 in Mar-M200+Zr.

Low Cycle Fatigue (LCF) Properties

Figure 4 shows the results of axial tension-compression LCF tests for both alloys. As shown, the Hf-containing alloy has a longer fatigue life. The life of some specimens of the Hf-containing alloy is longer by about one order of magnitude than the Hf-free alloy, and BIAM bars have longer lives than ONERA bars due to the different directionally solidified conditions.

A similar tendency was found when Zr was either partially or totally substituted for Hf. The role of Zr in improving LCF is not as distinct as Hf, but the Zr-containing alloy is better than the Hf-free alloy in LCF.

Observation of LCF Cracking

Systematic observation was conducted for a set of plate specimens at 760°C, 1.4% total strain range, and 0.33 Hz. The number N and length L of surface cracks per unit area as well as the average length were measured for about half life and ruptured specimens. The results are listed in Table VI. It was found that the cracking tendency of the Hf-containing alloy is

Table III. Phase Composition in Alloy, Atomic %

Alloy	Phase	Ti	Al	Cr	Ni	W	Co	Nb	Hf	Zr
Mar-M200	MC₁ *	43.29	0.10	1.85	6.84	13.92	0.97	32.16	0.88	--
	Eutectic γ'	4.68	13.02	5.49	66.84	1.42	7.84	0.68	0.30	--
Mar-M200 + 0.60Hf	MC₁ *	40.52	0.07	1.30	5.96	14.19	0.95	26.46	10.55	--
	MC₂ *	21.74	0.40	2.24	10.26	6.29	2.18	19.47	37.42	--
	Ni₅Hf	4.53	1.65	5.14	61.88	1.41	9.05	1.93	14.40	--
	Eutectic γ'	3.70	11.95	7.15	64.01	2.07	8.70	0.70	1.72	--
Mar-M200 + 0.45Zr	MC₁ *	47.77	0.00	0.96	3.41	13.91	0.42	26.88	0.30	6.34
	MC₂ *	11.30	0.18	1.45	7.95	5.84	1.33	21.93	1.17	48.86
	Ni₅Zr	0.71	1.59	2.25	66.19	0.26	7.05	0.53	0.24	21.17
	Eutectic γ'	4.61	11.91	4.96	67.73	1.72	7.49	0.62	0.00	0.95
Mar-M200 + 0.26Zr+0.27Hf	MC₁ *	43.82	0.06	0.91	3.92	13.28	0.54	27.39	6.50	3.59
	MC₂ *	19.23	0.05	1.26	7.42	6.54	1.39	21.80	21.52	20.79
	Ni₅(Zr, Hf)	1.79	2.45	2.79	67.24	0.44	7.42	0.43	5.75	11.71
	Eutectic γ'	3.54	13.63	6.55	63.98	2.07	8.19	0.50	0.88	0.65

* Composition of MC only represents metallic radical 'M'

Table IV. Tensile Properties of DS Mar-M200 Alloy

Temperature °C	D.S. Condition	H.T.	Hf-free alloy				Hf-containing alloy			
			0.2%Y.S. (MPa)	U.T.S (MPa)	δ (%)	ψ (%)	H.T.	0.2%Y.S (MPa)	U.T.S (MPa)	δ (%)	ψ (%)

Temperature °C	D.S. Condition	H.T.	0.2%Y.S. (MPa)	U.T.S (MPa)	δ (%)	ψ (%)	H.T.	0.2%Y.S (MPa)	U.T.S (MPa)	δ (%)	ψ (%)
R.T.	BIAM	II	1007	1115	5.85		III	1024	1217	5.63	
		I	1001	1197	10.8		I	1091	1284	8.4	
		I	989	1238	11.1	13.5	I	1106	1256	8.7	9.6
650	BIAM	II	927	1119	11.1		III	977	1173	8.7	
760	BIAM	II	1030	1109	6.5		III	1078	1121	11.2	
		I	967	1205	8.8	14.0	I	1091	1317	7.0	10.0
		I	990	1227	9.2	14.5	I	1106	1332	7.2	10.6
	ONERA	I	886	1107	6.2	7.0	I	920	1118	4.7	8.7
950	BIAM	II	437	559	16.1		III	555	641	22.5	

Table V. Creep Properties of DS Mar-M200 Alloy

Test Conditions		Hf-free alloy				Hf-containing alloy		
Temperature °C	Stress (MPa)	H.T.	Time for 1% creep (h)	Rupture life (h)	Elongation (%)	H.T	Time for 1% creep (h)	Rupture life (h)	Elongation (%)

Temperature °C	Stress (MPa)	H.T.	Time for 1% creep (h)	Rupture life (h)	Elongation (%)	H.T	Time for 1% creep (h)	Rupture life (h)	Elongation (%)
760	750	II	13	351	8.6	III	5	1300	---
850	450	II	45	197	19.6	III	162	523	24.7
950	200	II	125	251	6.8	III	228	625	26.5
1050	100	II	104	173	13.6	III	156	254	21.2

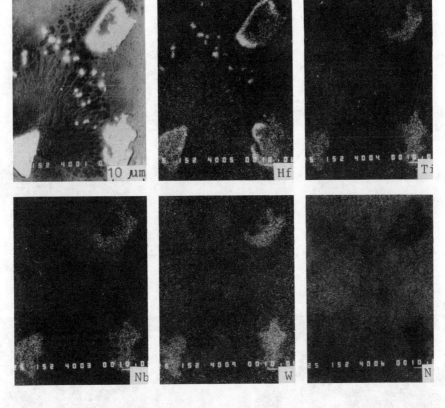

Figure 3 - Element distribution in MC_2 for Mar-M200+Hf after heat treatment I.

Table VI. Strain-Controlled LCF Results for Plate Specimens

Alloy	Specimen No.	Cycle N_f	Result	Density of Cracks N/mm^2	Length of Cracks $L(mm/mm^2)$	Average Length $\overline{L}(mm)$
Mar-M 200	4T	21	unruptured	2.30	0.12	0.0534
	6B	57	ruptured	3.89	0.22	0.0561
Mar-M 200+Hf	3T	203	unruptured	0	0	0
	3B	315	ruptured	0.07	0.0064	0.0875

very low. After testing, the surface of the Mar-M200 specimens was rougher due to deformation than that of Mar-M200+Hf. Surface slip analysis showed that intersection of two sets of slip bands in adjacent grains occurred at the columnar grain boundary. For Mar-M200, microcracks preferentially formed (Fig. 5). Although similar slip band intersections took place in Mar-M200+Hf, the grain boundaries did not crack. The cracking pattern of both alloys is much different within the grains. Mar-M200 has predominantly interdendritic cracks (Fig. 6a), but Mar-M200+Hf has interdendritic and transdendritic cracks. The latter is very flat with straight segments having some steps, as shown by the arrow in Figure 6b. It means that this kind of crack is related to crystallographic cracking caused by slip. At high strain amplitude, extensive MC cracking was observed in the Hf-free alloy during the first few cycles. The MC cracking preferentially occurs at long rod-shaped carbides perpendicular to the stress axis (Fig. 7a). The Hf-containing alloy has higher resistance to MC cracking due to the presence of blocky carbides. Even if some carbides crack, the crack size is very small (Fig. 7b).

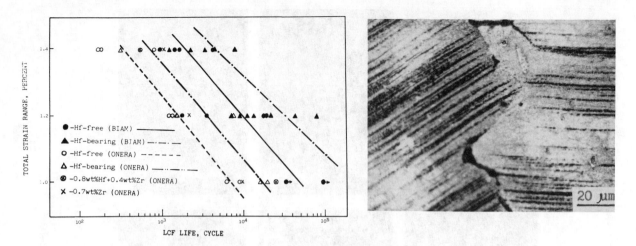

Figure 4 - LCF properties of alloys.

Figure 5 - Intersection of two sets of slip in specimen 6B.

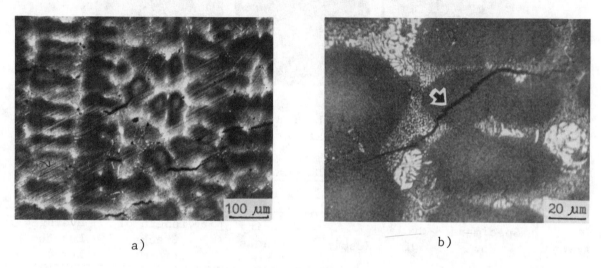

Figure 6 - Intragranular cracking pattern of two alloys: (a) specimen 6B, and (b) specimen 3B.

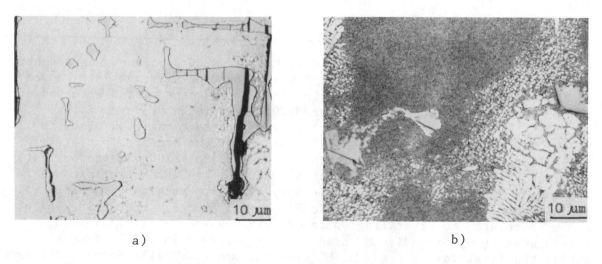

Figure 7 - MC carbide cracking of Hf-containing and Hf-free alloy: (a) specimen 6B, and (b) specimen 3T.

The microscopic characterization of fatigue rupture is very different for the two alloys. At the fracture surface of Mar-M200, the following pattern often exists: a cluster of MC carbides crack which then propagates through the interdendritic region, as clearly indicated by the fatigue striation and stops at dendritic arms (Fig. 8a). In the late period of fatigue, dendrites are separated by a number of local fatigue cracking zones. In the Hf-containing alloy, cracking takes place along the crystallographic plane of heavy slipping and the fracture surface appears to be a flat zone corresponding to the segment shown by the arrow in Figure 6b. On the flat fracture, there are some slip bands (Fig. 8b).

A TEM examination of thin foils taken from the fatigue specimens was carried out. The results showed that there were a lot of stacking faults caused by long distance slip in the Hf-free alloy (Fig. 9a), but dislocations were difficult to spread into a stacking fault and prone to twining in the Mar-M200+Hf alloy (Fig. 9b). This demonstrated that the Hf-containing alloy had more uniform deformation than the Hf-free alloy. The effect of Zr on the LCF deformation mechanism is very similar to Hf.

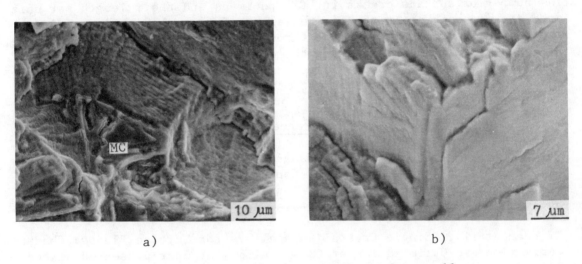

Figure 8 - Fatigue fracture surface of two alloys:
(a) specimen 6B, and (b) specimen 3B.

Figure 9 - Dislocation structure in both alloys: (a) Mar-M200, $\epsilon t=1.2\%$, 1700 cycles; and (b) Mar-M200+Hf, $\epsilon t=1.2\%$, 44484 cycles.

Conclusions

1. Both Hf and Zr can promote the formation of eutectic γ', MC_2, and Ni_5M phases. Zr promotes the formation of Ni_5M to a greater extent than Hf. The solubility of Zr in Ni_5M is higher than that of Hf, but its solubility in γ' is lower than Hf. As a result, the strengthening efficiency of Zr is not as strong as Hf.

2. Ni_5Zr in a Zr-containing alloy can be eliminated by pretreatment at 1130°C/3hr; therefore, incipient melting can be avoided below 1235°C.

3. The Hf-containing alloy showed greater LCF life compared to the Hf-free alloy. A similar tendency was found when Zr was either partially or totally substituted for Hf. High rate solidification enhances LCF life further.

4. The creep life for DS Mar-M200+Hf at 760°C and 1050°C was increased with a 1225°C/2hr + 1100°C/4hr + 870°C/20hr heat treatment.

5. The number of surface cracks in LCF specimens and their length per unit area are much higher in the Hf-free alloy. The cracks mainly originate from MC carbides, then propagate through the interdendritic region. The Hf-containing alloy cracks along the crystallographic planes. This proved that the Hf-containing alloy has a more uniform deformation mode and higher interdendritic coherent strength.

References

1. Wang Luobao, Chen Rongzhang and Wang Yuping, "Influences of Directionally Solidified Techniques and Hf Content on a Nickel-Base Superalloy," The Production and Application of Less Common Metals, (The Metals Society and Chinese Society of Metals, 1982), Book II, Paper 43, 1-17.

2. T. Khan et al., "Single Crystal Superalloys for Turbine Blades: Characterization and Optimization of CMSX-2 Alloy," (Paper presented at the 11th Symposium on Steels and Special Alloys for Aerospace, Paris Air Show Le Bourget, 6 June 1985).

3. D. N. Duhl, "Enhancement of Transverse Properties of Directionally Solidified Superalloys," USPO 3, 700, 433, Oct. 24, 1972.

4. Zheng Yunrong, Cai Yulin and Ruan Zhongci, "Investigation of Zr as a Substitute for Hf in Cast Ni-Base Superalloys," Acta Metallurgica Sinica, 23(5)(1987), A430-433.

A HAFNIUM-FREE DIRECTIONALLY SOLIDIFIED

NICKEL-BASE SUPERALLOY

Dongliang Lin (T. L. Lin) and Songhui Huang
Department of Materials Science and Engineering
Shanghai JIao Tong University, Shanghai, China

Chuanqi Sun
Institute of Aeronautic Materials
Beijing, China

Abstract

This paper provides a review of current efforts on design of a hafnium-free directionally solidified nickel-base superalloy with good castability, post-casting transverse ductility and improved creep strength in our laboratories. Emphasis is being placed on the effect of alloy modifications on castability and the improvement of creep strength by increasing solid solution temperature.

Introduction

Although the benefits of directionally solidified (DS) alloys have been amply realized, there is still much room for improvement. Hafnium was added to DS MAR-M200 to prevent grain boundary cracking of hollow-blade castings during solidification and to provide good post-casting transverse ductility (1,2,3). However, the addition of hafnium had led to the formation of hafnium containing inclusions which affected casting yields and lowered the incipient melting temperature, which in turn limited possible increases in solution temperature for further strengthening DS nickel-base superalloys by improving their creep strength. In order to overcome the shortcomings of hafnium containing DS superalloys, hafnium should be minimized or eliminated and, perhaps, replaced with other grain boundary strengthening elements, and a hafnium-free superalloy with good castability and post-casting transverse ductility developed.

In recent years, a DS nickel-base superalloy, DZ-3 has been developed in our laboratories based on a cast nickel-base superalloy, K3. This alloy was developed in our laboratories in the early 1960's and has been widely used since then in China for first and second turbine blades and vanes in various kinds of aircraft engines (4). As a DS alloy, the composition of DZ-3 was modified from that of alloy K3 as shown in Table 1. It has been shown that the advantages of directionally solidified superalloy, DZ-3 over conventional cast superalloy, K3 are better creep properties, greater thermal fatigue life, greater rupture life, and greater rupture ductility, which are comparable to those of DS alloy PWA 1422 (5). However, in many instances, cracking of hollow airfoils is associated with thin walls present in the castings of DZ-3. By proper positioning of the core in the mold and close matching of the core and wax tooling, this difficulty can be partly overcome. The addition of hafnium to DZ-3 alloy can also reduce the tendency for grain boundary cracking. In order to completely prevent grain-boundary cracking of hollow blade casting during directional solidification and to provide good post-casting transverse ductility, an alloy modification of DZ-3 has been investigated and results compared with the effects of hafnium additions. A new hafnium-free DS alloy called DZ-4 was developed and its composition is listed in Table 1. A large number of DS hollow-airfoils have been successfully cast with DZ-4 alloy (Fig. 1).

This paper provides a review of the alloy design with emphasis placed on the effect of alloy modifications on castability and the improvement of creep strength by increasing the solution temperature.

Table 1. Composition of K3, DZ-3, and DZ-4

	C	Cr	Co	W	Mo	Al	Ti	Ti+Al	B	Ce	Zr	Ni
K3	0.11	10.0	4.5	4.8	3.8	5.3	2.3		0.01	0.01	0.1	Bal
	0.18	12.0	6.0	5.5	4.5	5.9	2.9		0.03	0.03		
DZ-3	0.07	9.5	4.5	4.8	3.8	5.2	2.3		0.015	0.01	0.1	Bal
	0.15	11.0	6.0	5.8	4.6	5.9	2.9		0.03			
DZ-4	0.10	9.0	5.5	5.1	3.5	5.6	1.6	>7.6	0.012			Bal
	0.16	10.0	6.0	5.8	4.2	6.4	2.2		0.025			

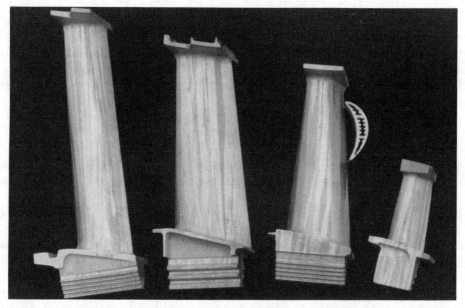

Fig. 1 Hollow airfoils directionally solidified in DZ-4.

Castability Improvement by Alloy Modification

It has been shown that Zr and Ti are the elements which tend to enlarge the melting and solidification range of alloys, $\Delta T = T_L - T_S$, where T_L, T_S are the liquidus and solidus temperatures of an alloy.

The composition of the experimental alloys studied to assess castability are shown in Table 2. Alloy 1 is DZ-3, the baseline alloy. In alloys 2 to 4, Zr is eliminated and 0.3 to 0.8 weight percent (w/o) Ti is replaced by 0.6 to 0.9 w/o Al. In alloys 6 to 10, various amounts of hafnium from 0.3 to 2.0 w/o are added to DZ-3. The cracking tendency on both inner and outer surfaces of the hollow-blade castings was selected as a measure of alloy castability. A typical crack on the surface of a DZ-3 hollow-blade is shown in Fig. 2. Five sections (I, II, III, IV, and V) of a hollow-blade with a minimum of 0.8 mm and a maximum of 3.3 mm in thickness were chosen as the location to examine for the presence of cracks using penetrant inspection and optical microscopy. The degree of castability was divided into four classes: A - no cracks; B - no more than 2 cracks with a length of less than 5 mm in any of the sections; C - no cracks on the outer surface of all five sections, but cracks on the inner surfaces; and D - cracks on inner and outer surfaces of all five sections.

The results of the castability tests listed in Table 3 indicate that the Zr-free alloys 2, 4 and 5 where some of the Ti has been replaced by Al and alloy 10 with 2% Hf display superior castability compared to the other

Table 2. Chemical Composition of Experimental Alloys, Weight Percent

Alloy No.	C	Cr	Co	W	Mo	Al	Ti	B	Ce	Zr	Hf	Ni
1	0.14	10	5.0	5.4	4.3	5.4	2.6	0.015	0.01	0.1		Bal
2	0.14	10	6.0	5.9	3.8	6.0	1.8	0.015				Bal
3	0.14	10	5.0	5.4	4.3	5.4	2.5	0.015				Bal
4	0.17	10	6.0	5.9	3.8	6.3	2.3	0.015				Bal
5	0.10	9	5.0	5.4	3.8	5.3	1.6	0.015				Bal
6	0.14	10	5.0	5.4	4.3	5.4	2.6	0.015	0.01	0.1	0.3	Bal
7	0.14	10	5.0	5.4	4.3	5.4	2.6	0.015	0.01	0.1	0.6	Bal
8	0.14	10	5.0	5.4	4.3	5.4	2.6	0.015	0.01	0.1	1.0	Bal
9	0.14	10	5.0	5.4	4.3	5.4	2.6	0.015	0.01	0.1	1.5	Bal
10	0.14	10	5.0	5.4	4.3	5.4	2.6	0.015	0.01	0.1	2.0	Bal

Table 3. Castability of Experimental Alloys

Alloy No.	Solidification Process	Volume Percent	Castability Rank	Yield (A Rank)
1	DS	57.9	D	0
2	DS	57.4	A	100
3	DS	57.9	C	50
4	DS	63.4	A	100
5	DS	52.8	A	100
6	DS	57.8	B	25
7	DS	58.1	B	40
8	DS	58.6	B	60
9	DS	59.5	B	75
10	DS	60.37	A	100
2	CC	57.9	A	100

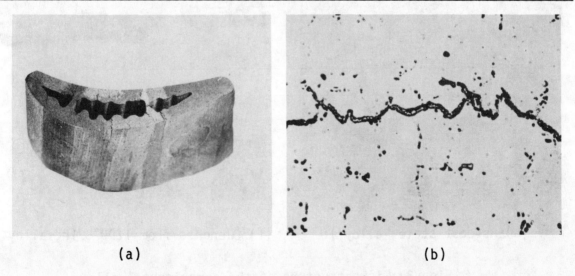

(a)   (b)

Fig. 2 A typical crack in a hollow blade of alloy DZ-3 (a) and its micrograph at higher magnification (b).

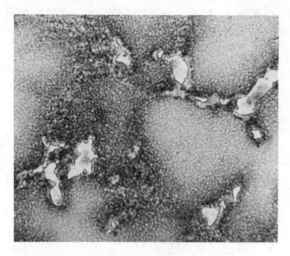
(a) As-cast state, alloy 1

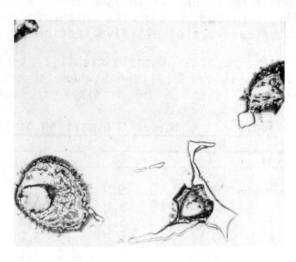
(b) Quenched from 1250°C/1h, alloy 1

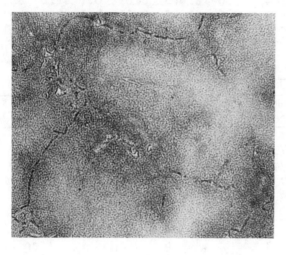
(c) As-cast state, alloy 2

(d) Quenched from 1290°C/1h, alloy 2

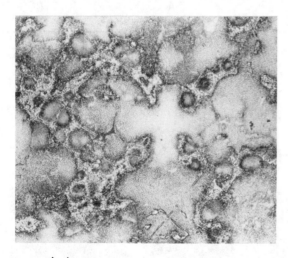
(e) As-cast state, alloy 10

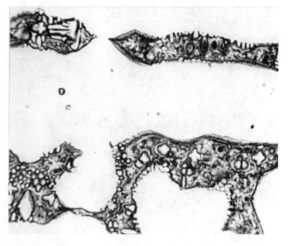
(f) Quenched from 1290°C/1h, alloy 10

Fig. 3 Microstructure of the experimental alloys.

alloys. The differential thermal analysis and quenching method were used to determine the melting and solidification range of the alloys and some results are shown in Table 4. Both liquidus and solidus temperatures of alloy 2 (DZ-4) are higher than those of alloy 1 (DZ-3) especially during cooling where the solidus temperature of alloy 2 is 30°C higher than that of alloy 1. The solidus temperature for alloy 10 (DZ-3 with 2% Hf) is 26°C lower than that of alloy 1 and 56°C lower than that of alloy 2. Since there is no $\gamma+\gamma'$ eutectic in alloy 2, as shown in Fig. 3c, its incipient melting temperature is greater than 1290°C, while alloy 1 starts to incipiently melt at 1170°C (Fig. 3b and 4) because it contains 2.0 to 4.7 volume percent of $\gamma+\gamma'$ eutectic (Fig. 3a). This suggests that grain-boundary cracking of hollow-blade castings can be prevented by the elimination of $\gamma+\gamma'$ eutectic (6). However, in hafnium containing alloys, the $\gamma+\gamma'$ eutectic increases as the Hf content increases. When an alloy contains greater than 1.5 w/o Hf, the amount of the $\gamma+\gamma'$ eutectic is 24 volume percent as in the case of alloy 10 (Fig. 3e) which contains 2 w/o Hf. Therefore, grain boundary cracking can be avoided by producing a large amount of $\gamma+\gamma'$ eutectic in DS superalloys. In summary, an improved understanding of the solidification process, the local chemistry and microstructure that control grain boundary strength would be useful to explain these phenomena. A hafnium-free DS superalloy, DZ-4, was developed to meet the requirements for hollow-blade castings based on the compositions of alloys 2 and 4.

Table 4. Differential Thermal Analysis

Alloy No.	Heating			Cooling		
	Liquidus Temperature	Solidus Temperature	Melting Range	Liquidus Temperature	Solidus Temperature	Melting Range
1	1356°C	1296°C	60°C	1347°C	1266°C	81°C
2	1365	1309	56	1356	1296	60
10	1339	1260	79	1335	1240	95

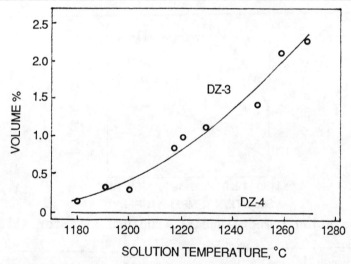

Fig. 4 Amount of incipient melting phase as a function of solution temperature for DZ-3 and DZ-4.

Creep Strength and Rupture Life Improvement by Increasing Solid Solution Temperature

It has been shown that solution treatments at temperatures sufficiently high to homogenize the alloy and dissolve the coarse γ' and eutectic $\gamma+\gamma'$ constituent for reprecipitation in the form of a uniform fine γ' dispersion will further strengthen DS nickel-base alloys by improving their creep strength. J. J. Jackson and his co-workers (7) found a threefold increase in creep rupture life at 982°C under 220 MPa in DS MAR-M200 + Hf when the amount of fine γ' increased from about 30 volume percent to approximately 45 volume percent. The role of high temperature solution treatments in increasing the creep strength and rupture life of DS nickel-base superalloys has been investigated systematically by Dongliang Lin (T. L. Lin) and his co-workers (8,9) and it was found that the secondary creep rate at 760°C is related to the size a, center departure L or particle spacing λ and volume fraction V_f of fine γ' particles as follows:

$$\varepsilon \alpha \lambda^2/a \text{ or } \varepsilon \alpha a/V_f^{2/3}\left(1-V_f^{2/3}\right)^2 \text{ for alloy DZ-3}$$

and $\quad \varepsilon \alpha L^2/a$ or $\varepsilon \alpha a/V_f^{2/3} \quad$ for alloy DZ-17G and DS René 80

The smaller the size and the higher the volume fraction of fine γ', the lower the secondary creep rate. The relation between rupture life t_f and secondary creep rate ε was found to fit the following expression:

$$t_f \varepsilon^\alpha = C\varepsilon_f$$

where α and C are alloy constants. The extension of creep rupture life was found to be due to a decrease of the secondary creep rate and an extension of the secondary creep stage. Raising the solution temperature will increase the volume fraction and decrease the size of the fine γ' which leads to a lower secondary creep rate and extends rupture life. The limitation for increasing the solution temperature is the low incipient melting temperature for alloys containing $\gamma+\gamma'$ eutectic. As mentioned above, DZ-4 has its incipient melting point greater than 1290°C (Fig. 3b and 4), so it is possible to further increase its creep strength and extend its rupture life by increasing its solution temperature. Fig. 5 and 6 show that either rupture life or tensile strength at 760°C and 980°C for DZ-4 can be significantly improved by increasing the solution temperature to 1270°C.

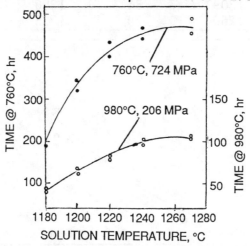

Fig. 5 Solution temperature vs. rupture life for alloy DZ-4.

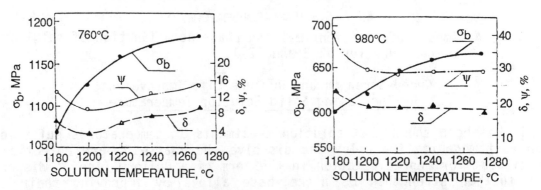

Fig. 6 Solution temperature vs. tensile properties at 760°C and 980°C for DZ-4.

The size, a, and volume fraction, V_f, of fine cuboidal γ' were determined by area measurements on representative electron photomicrographs and then the separation of γ' particles was calculated by the method described in Ref. 5.

Specimens of DZ-4 were solution treated at 1180, 1200, 1220, 1240, and 1270°C for 2 h/AC. All specimens for creep rupture testing were aged for 16 hours at 870°C after the solution treatment. The morphology and size of γ' in solution treated DZ-4 are shown in Fig. 7. The as-cast coarse γ' was dissolved gradually with increasing solution temperature and fine cuboidal γ' was reprecipitated during subsequent cooling, while undissolved γ' coalesced. At 1220°C, the as-cast coarse γ' was completely dissolved in the dendritic regions and only a few retained as-cast γ' particles existed in the interdendritic regions. Above 1230°C, a fine uniform cuboidal γ' could be found after solution treatment. The volume fraction V_f and size a (the side of a cuboid) of fine γ' after solution treatment at various temperatures, followed by aging at 870°C for 16 hours, are shown in Fig. 8. The largest volume fraction V_f and the smallest fine γ' size can be obtained by solutioning above 1270°C, which is the solution temperature to obtain the optimum rupture life and tensile strength at 760°C and 980°C for DZ-4. Therefore, solution temperature can be raised from 1210-1230°C for DZ-3 to 1260-1280°C for DZ-4.

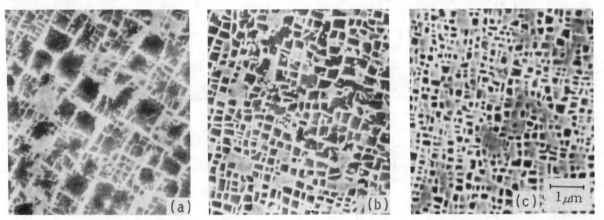

Fig. 7 γ' phase morphology in interdendritic regions for DZ-4 after 1180°C (a), 1220°C (b), and 1290°C (c) solutioning and 870°C/16h aging.

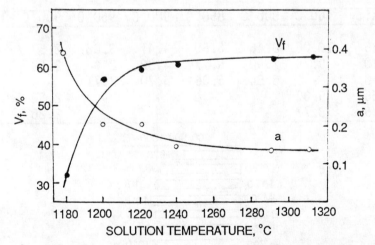

Fig. 8 Solution temperature vs. volume fraction, V_f, and size, a, of fine γ'.

Mechanical Properties of DZ-4

Additional benefits in mechanical properties and phase stability of DZ-4 can be achieved with low values of Nv (Nv = 2.16). The mechanical properties shown below are for specimens given a standard heat treatment, e.g., 1220°C/4h/AC + 870°C/32h/AC. DZ-4 presently under development has a higher incipient melting temperature (>1290°C). One can therefore increase the solution temperature to provide a more homogenous and uniform distribution of fine γ' which will lead to further increases in alloy strength.

Tensile Strength

As shown in Table 5, the tensile strength and ductility of DZ-4 are comparable with those of advanced DS superalloys.

Table 5. Tensile Properties

Alloy	20°C			760°C			980°C		
	UTS MPa	YS MPa	EL %	UTS MPa	YS MPa	EL %	UTS MPa	YS MPa	EL %
DZ-4	1059	947	6.0	1187	996	6.0	647	466	20
PWA 1422	1108	941	6.1	1187	961	9.4	608	470	21
DZ-3	971	-	5.0	1167	-	15	666	-	21
K3	912	-	5.1	-	-	-	-	-	-

Stress Rupture Strength

With superior rupture strength (Table 6) and reasonable density (Table 7), DZ-4 displays superiority in specific stress rupture strength and temperature capability over other alloys as shown in Table 7 and Fig. 9.

Table 6. 100-Hour Stress Rupture Strength (MPa)

Alloy	760°C	800°C	850°C	900°C	950°C	980°C	1000°C	1040°C	1090°C
DZ-4	840	677	520	353	245	206	181	142	78.5
DZ-3	755	628	490	343	235	206	177	137	78.5
K3	-	520	392	294	216	171	147	-	-
PWA 1422	755	-	-	-	-	200	-	127	-
Mar-M200	741	-	-	-	-	-	-	118	-

Table 7. Specific Stress Rupture Strength

Alloy	Density	760°C	800°C	850°C	900°C	950°C	980°C	1000°C	1040°C
DZ-4	8.15	10.06	8.46	6.50	4.41	3.06	2.57	2.26	1.77
DZ-3	8.10	9.75	7.90	6.17	4.32	2.96	2.59	2.20	1.72
K3	8.10	-	6.54	5.06	3.70	2.71	-	1.85	-
PWA 1422	8.56	9.00	-	-	-	-	2.38	-	1.52
DS-IN100	7.75	8.77	-	-	-	-	2.26	-	-

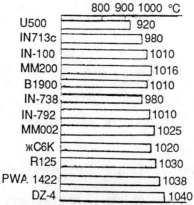

Fig. 9 Temperature capability of alloys at 137 MPa for 100 hours.

Creep Strength

Creep curves at 760°C and 980°C are shown in Fig. 10 and display low secondary creep rates, long rupture life, and high rupture ductility. It should be emphasized that the primary creep strain of DZ-4 at intermediate temperature and high stress is very low, similar to that of DZ-3 (10) and unlike DS MAR-M200 which exhibits relatively large primary creep strains at high stresses and temperatures around 760°C (11).

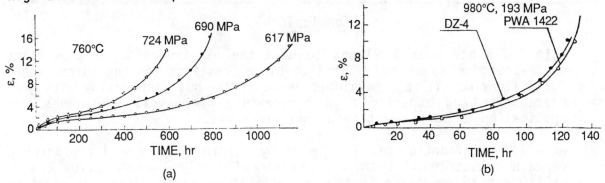

Fig. 10 Creep curves at 760°C for DZ-4 (a) and at 980°C for DZ-4 and PWA 1422 (b).

Thermal Mechanical Fatigue (TMF)

Thermal fatigue cracks in many advanced cooled turbine blades initiate in the coating at the blade leading edge and propagate into the superalloy. Fig. 11 shows the results of a laboratory stress-controlled TMF test that simulates blade cracking. The results for low cycle fatigue (LCF) in Fig. 11 show that the LCF life is greater than that for TMF. If the operating stress for a turbine blade is 196 MPa, its thermal fatigue life will be more than 10,000 cycles.

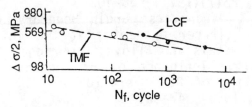

Fig. 11 Curves of LCF and TMF at 980°C.

Transverse Stress Rupture Properties

The results for transverse rupture strength (Table 8) for DZ-4 show that the 100-hour transverse rupture strength is 85% of the longitudinal strength at intermediate temperatures and 90% at higher temperatures. Compared to alloy DZ-3, more Al has been added to DZ-4 replacing some Ti and Zr has been eliminated. DZ-4 has a much longer rupture life and much higher rupture ductility which are the same levels as those of alloys with hafnium, e.g., DS M002 and PWA 1422 (Fig. 12).

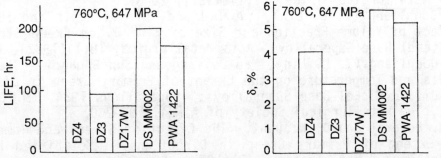

Fig. 12 Transverse rupture life and ductility of various alloys.

Table 8. Transverse Stress Rupture Strength of DZ-4

Orientation	Rupture Strength for 100 hours, MPa			
	760°C	850°C	980°C	1040°C
Transverse	674	441	181	128
Longitudinal	804	520	206	142
K = σ_T/σ_L	0.84	0.85	0.87	0.90

Conclusions

1. Alloy DZ-4 has been developed to meet the requirements for preventing grain-boundary cracking of hollow-blade castings during directional solidification. It is a hafnium-free DS nickel-base superalloy with good castability and superior high temperature mechanical properties for application as hollow turbine blades and vanes.

2. An alloy modification of DZ-3, made by eliminating Zr and replacing a certain amount of Ti with Al, can significantly improve alloy castability due to reduction in the solidification range ΔT and elimination of the formation of $\gamma+\gamma'$ eutectic. This leads to the prevention of grain-boundary cracking in hollow-blade castings.

3. DZ-4 has a higher incipient melting temperature (>1229°C) which permits an increase in solution temperature and provides an increase in rupture strength and tensile strength with increasing solution temperature up to 1270°C.

References

1. D. N. Duhl and C. P. Sullivan, "Some Effects of Hafnium Additions on the Mechanical Properties of a Columnar-Grained Nickel-Base Superalloy," J. Metals, 23(7)(1971), 38-40.
2. B. H. Kear et al., "Transverse Grain Boundary Strengthening in a DS Nickel-Base Alloy," (Paper presented at the International Conference on the Strength of Metals and Alloys, Cambridge, Aug. 1973), 134-138.
3. Wang Yuping et al., "Influence of Hf Content on Microstructure and Mechanical Properties of DS Superalloy DZ 22," (Proceedings on Cast Superalloy, China, 1986), 47-52.
4. Yin Keqing and Li Qijuan, "Test of K3 Superalloy in Cast State," (Proceedings on Cast Superalloy, Institute of Aeronautic Materials, Beijing, China, May 1982), 1-41.
5. Sun Chuanqi et al., "The Properties of DZ 3 Superalloy," Superalloys Handbook, (China, 1980), 940-947.
6. Sun Chuanqi et al., "A Study of the Castability of DZ3 DS Superalloy," Aeronautic Materials, 3(1984), 1-6.
7. J. H. Jackson et al., "The Effect of Volume Percent of Fine γ' on Creep in DS Mar-M200 + Hf," Met. Trans., 8A(10)(1977), 1615-1620.
8. Lin Dongliang (T. L. Lin), "Role of Solution Treatment in Improving the Creep Strength of a Directionally Solidified Nickel-Base Superalloy," Acta Metallurgica Sinica, 17(1)(1981), 26-37.
9. Lin Dongliang (T. L. Lin), Yao Deliang, Lin Xiangjin and Sun Chuanqi, "Effect of Volume Fraction and Size of Fine γ' on Creep Strength of a DS Nickel-Base Superalloy," Acta Metallurgica, 18(1)(1982), 104-114.
10. Lin Dongliang (T. L. Lin), Yao Deliang and Sun Chuanqi, "The Effect of Stress and Temperature on the Extent of Primary Creep in Directionally Solidified Nickel-Base Superalloys," Superalloys 1984, (M. Gell et al. eds., The Metallurgical Society of AIME), 187-200.
11. G. R. Leverat and D. N. Duhl, "The Effect of Stress and Temperature on the Extent of Primary Creep in Directionally Solidified Nickel-Base Superalloys," Met. Trans., 2(3)(1971), 907-908.

THE PROCESSING AND TESTING OF A HOLLOW

DS EUTECTIC HIGH PRESSURE TURBINE BLADE

R.G. Menzies, C.A. Bruch, M.F. Gigliotti*, J.A. Smith, and R.C. Haubert

General Electric Co.
Evendale, Ohio
*Schenectady, New York

ABSTRACT

A program to assess the feasibility of Directionally Solidified Eutectic (DSE) High Pressure Turbine (HPT) blades is described. The overall objectives of the program were to define the materials, design and fabrication technology required for the application of eutectic composites in HPT blade components and, through engine testing, to demonstrate their applicability in advance aircraft gas turbine engines.

NiTaC-14B a DSE alloy, has a combination of mechanical properties judged to be superior for turbine blade applications over all other known eutectics. It has a potential temperature advantage of 225°F over the conventionally cast superalloy Rene' 80 and 100°F over the directionally solidified (DS) alloy Rene' 150. A NiTaC-14B blade for testing in a demonstrator engine was designed by significantly modifying the existing DS Rene' 150 blade. The cooling air was markedly reduced and the pitchline airfoil temperature predicted to increase by 100°F. This allowed an effective evaluation of NiTaC HPT blade capabilities under realistic engine conditions.

Hollow HPT blades for test evaluations were cast as close to final external dimensions as practical. Complex internal air cooling passages were cast to size through the use of a new alumina core material/processes developed by General Electric under Air Force sponsorship. A shell mold developed by General Electric was used to form the external blade surfaces. However, because of surface defects in castings made with this mold, the blade was cast oversize to provide sufficient stock for removal of such defects.

HPT blades were finish machined and then subjected to a series of bench tests. The results were combined with blade life analysis results to assess the adequacy of the blade design for engine testing. Subsequently, eight uncoated blades were successfully engine tested.

The program demonstrated that the mechanical properties of the selected NiTaC alloy are very good and that although processing is difficult the manufacture of engine quality hollow DS eutectic blades is feasible.

Introduction

In the 1970's, several Directionally Solidified Eutectic (DSE) systems were identified as promising blade alloys [1]. One such eutectic system, NiTaC, developed by General Electric, consists of aligned, high-strength tantalum carbide (TaC) fibers in a DS gamma-gamma prime, nickel-base matrix. The NiTaC system offers greatly improved rupture strength and other mechanical property improvements over current DS and single crystal alloys. In 1977, the high potential of the NiTaC eutectic system was demonstrated through the successful engine test of solid, uncooled NiTaC-13 low pressure turbine blades. NiTaC alloy development continued concurrently, resulting in significant strength improvements over NiTaC-13 while retaining an excellent balance of ductility, toughness and environmental resistance.

The highest payoff for DS NiTaC eutectics occurs in hollow high pressure turbine (HPT) blade applications. It was expected that the complex internal geometry of HPT blades could be achieved by the cored casting process; however, readily removable cores that could endure the severe environment for casting NiTaC alloys were not available. Consequently, General Electric conducted a core development program under U. S. Air Force sponsorship that identified an injection-molding process for producing the leachable Koralox® alumina core which could survive the casting cycle without deformation or deleterious reactions.

In light of the potential offered by DS NiTaC eutectics, the development of improved NiTaC alloys and the identification of an alumina ceramic that had high potential for fulfilling core requirements, a program was established by the U. S. Air Force to fabricate and evaluate hollow NiTaC HPT blades. The overall objectives were to develop the materials, design, and fabrication technology required for the application of DS NiTaC eutectics in HPT blade components and, through engine testing, to demonstrate their applicability in advanced aircraft gas turbine engines. This ambitious program required the combining of a new alloy system, a new core system and a new internal blade design. To accomplish the objectives, work in the following areas was carried out: 1) microstructural characterization; 2) turbine blade design and analysis; 3) material property data acquisition; 4) core and mold technology development; 5) casting development; 6) machining development; and 7) blade hardware evaluation and engine test. These areas will now be reviewed.

Microstructural Characterization

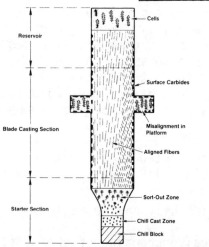

Fig. 1 Schematic Microstructure of NiTaC Blade Casting

NiTaC-14B is a hypereutectic alloy. After an initial chill cast zone, the alloy adjusts to the eutectic composition by the precipitation of coarse carbides in a sort out zone as shown in Figure 1. Aligned fiber growth then proceeds until near the top of the casting where the growth becomes cellular. As noted by Gigliotti and Henry [2] significant chemical segregation occurs throughout the length of the casting, which reduces the rupture strength toward the top of the casting. For this reason, NiTaC blades are normally cast tip down, as shown in Figure 1, to ensure that the best rupture properties are achieved in the airfoil. In addition, a reservoir of metal is placed on top of the casting to limit the amount of chemical segregation, to prevent the onset of cellular growth in the blade dovetail and to ensure that the dovetail has adequate mechanical properties.

Another microstructural aspect of the NiTaC system is that blocky surface carbides precipitate at both the internal and external blade surfaces. These carbides may be up to 30 mils deep and were considered undesirable. Accordingly, the eutectic blade was cast 30 mils oversize so that the carbides on external surfaces could be removed. Analysis and mechanical testing showed that surface carbides on the inside of the blade would not compromise its performance in the planned engine test.

Blade Design and Analysis

Initial studies were performed to determine the potential payoffs of NiTaC-14B and to establish realistic design goals. As a result of these studies, the following goals were set: 1) The pitch line bulk temperature must be more than 100°F higher than the existing DS Rene' 150 blade design; 2) The air cooling flow must be significantly less than the Rene' 150 design; 3) The predicted rupture and low cycle fatigue (LCF) life must be equal to or greater than the Rene' 150 design; 4) The external blade configuration must be interchangeable with the Rene' 150 blade design and; 5) The internal design should be of reduced complexity to allow a more rugged core and greater blade castability.

A preliminary configuration for the eutectic blade was designed and analyzed. By reducing the cooling flow by approximately 15% compared to the Rene'150 blade, the bulk temperature of the NiTaC-14B blade was predicted to rise 100°F. A computer model to calculate the blade natural frequencies for flexural and torsional vibration modes was then constructed using geometric properties of the airfoil cross sections, estimated bulk temperature distributions, and NiTaC-14B elastic modulus data. Analyses were conducted both for laboratory bench test conditions (room temperature and no centrifugal loading) and for engine design point conditions (elevated metal temperatures and centrifugal stiffening effects). Because of some uncertainties in the modeling, the engine test frequencies for the eutectic blade were predicted by comparing the calculated frequencies for the eutectic and Rene' 150 blades with the actual measured bench and engine test frequencies of the Rene' 150 blades. The predicted frequencies were very similar to the Rene' 150 blade; hence, the eutectic blade configuration was considered acceptable for the blade flex and torsional modes.

Enlarged (10X) aluminum models of the pitch and tip section were made to check certain chordwise bending frequencies not predictable by the computer program used. One resonant frequency was found to fall in the engine operating range, and additional stiffening ribs were added to the design to overcome this problem.

The mechanical stresses due to centrifugal loading, gas bending and airfoil untwist were calculated. The higher density of NiTaC-14B over the existing DS alloy resulted in slightly higher predicted stresses in the eutectic blade. Using these predicted stresses, the target bulk metal temperatures required to meet the life goals were then calculated at various airfoil radial spans. Available LCF data indicated that the maximum edge temperatures should be 2100°F. The preliminary blade cooling design was then tailored to produce the desired temperatures. Having achieved a preliminary blade design, a finite element heat transfer computer program was then run to analyze the blade temperatures in detail. The actual combustor gas temperature profiles obtained during engine tests were used in this analysis. The eutectic blade was predicted to run at essentially the same airfoil root temperature as the Rene' 150 blade, but approximately 100°F hotter in the upper airfoil spans.

To obtain detailed rupture and LCF life predictions, a finite stress analysis program was run, inputting the calculated stresses together with the detailed temperature distributions from the heat transfer analysis. Additional data on elastic constants, thermal coefficients, yield point, stress rupture, creep and thin wall property derates were also entered into the program. The minimum rupture life of the final eutectic blade design was predicted to be approximately 3 times that of the Rene' 150 blade.

The limiting LCF location was calculated to be at the 80% span leading edge nose hole with a life equivalent to the Rene' 150 blade. Furthermore, the Rene' 150 blade was calculated to be limiting in HCF, LCF, and rupture all at the 15% airfoil span; whereas the eutectic blade was HCF limited at the 15% span, LCF limited at the 80% span, and rupture limited at the 50% span. Thus, the eutectic blade was a more balanced design and should have greater capability since it avoids possible interaction of HCF, LCF, and rupture at the limiting life locations.

In summary, the eutectic blade was designed to require 38% fewer film cooling holes, use 15% less cooling air, operate at 110°F higher pitchline bulk temperature and have equivalent LCF capability and 3 times the rupture life of the DS Rene' 150 blade. Were the comparison made with a conventionally cast rather than the DS Rene' 150 design, the advantages of the eutectic design would be even greater, with an estimated 34% savings in overall cooling flow.

Material Property Data Acquisition

To support the design effort, extensive mechanical and physical properties of heat treated NiTaC-14B were measured in laboratory tests. These properties included stress and creep rupture, tensile, HCF, LCF (strain controlled and load controlled), sustained peak LCF, thermal shock and impact. Typical properties for NiTaC-14B are shown in Table I. As can be seen in the longitudinal (casting) direction, mechanical properties exhibited an approximately 225°F superiority over conventionally cast Rene' 80 which is widely used in GE production engines. In the transverse direction, mechanical properties of the two alloys were essentially equivalent. The most surprising result from the testing was the excellent oxidation resistance of NiTaC-14B, which is comparable with some coatings. Additional testing was performed to evaluate the effects on properties of some microstructural defects in castings, microstructural alterations during pre-test thermal exposures, and a selected external overlay coating for environmental protection.

Surface carbides did degrade properties but not drastically. It was calculated that for a 30 mil stress rupture sheet specimen at 18 ksi, bare NiTaC-14B still showed a 230°F advantage over Codep coated Rene' 80, even when taking the greater density of NiTaC-14B into account. The HCF capability was also estimated to be equivalent to current DS or mono alloys. An early test of one blade confirmed that there was more than adequate HCF capability. It was therefore concluded that surface carbides on internal surfaces could be tolerated.

Pre-test thermal exposure did degrade stress rupture and impact properties, but these effects were not unique to eutectics and were not cause for concern. However, the significant degradation in HCF properties due to crack initiation in the overlay coating was judged to present a risk of blade failure during engine testing that was unnecessary to take because NiTaC-14B had adequate oxidation resistance for the planned engine testing. Therefore, the use of the coating was abandoned.

TABLE I. Summary of Round Bar Properties of Solutioned and Aged NiTaC-14B

	Longitudinal	Transverse	CC Rene' 80
Temperature for 100 hours rupture life at 20 ksi	2081°F	1898°F	1840°F
Temperature for 100 hours HCF life (10^7 cycles) at A=00, alternating stress = 60 ksi	1850°F	1410°F	1370°F
Temperature for 100 hours LCF life (10^4) cycles at A=00, alternating pseudo stress = 70 ksi, strain control	1800°F	Est. at least equivalent to R' 80	72°F
Room temperature ultimate tensile strength	205 ksi	155 ksi	148 ksi
Depth of average penetration after 100 hours at 2075°F in Mach 1 dynamic oxidation	--	2.9 mils	24 mils
Depth of average maximum penetration after 100 hours at 1700°F 5 ppm salt	--	10 mils	4 mils

Core and Mold Development

Core Development

Koralox, an all alumina core, was developed at GE by Klug and Pasco (3, 4). The core has a high surface density to resist penetration by the molten alloy and impart a good surface finish to the blade interior. The core also has a porous interior, which allows it to be crushed as the alloy solidifies and shrinks, thereby decreasing the likelihood of casting cracks. In addition, the internal porosity of the core allows it to be readily leached in caustic soda at rates dependent upon the percentage porosity, but at least equivalent to the removal rates of current silica cores.

Much early core development work was done at GE Corporate Research and Development Center and the process was then transferred to Sherwood Refractories, Inc. (SRI) a subsidiary of TRW for scale up. Initially, SRI produced cores of the Rene' 150 blade design and, when tooling was completed, adapted the Koralox process to the production of cores for the NiTaC blade.

The eutectic blade core, very similar to that shown in Figure 2, was designed for airfoil-tip-down castings. The T-bar was used to suspend the core in the mold, and the print outs were used to grip the tip of the core in the mold, as shown in Figure 3. Cores were fired on alumina setters in a covered molybdenum retort which had a piping system for bathing the core surface with dry hydrogen. The retort was placed in a hydrogen furnace and heated at a specified and relatively high rate to 3235°F. After a 2-hour hold time, the furnace was allowed to cool.

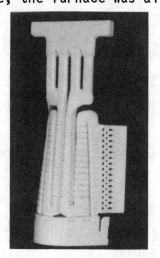

Fig. 2 Typical HPT Blade Core

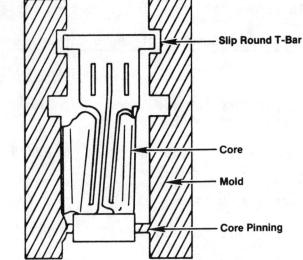

Fig. 3 Schematic of Core Pinning

Three serious problems were encountered: excessive shrinkage, distortion, and cracking. The injection molding formulation for the Koralox core contains alumina and carbon which react at high temperatures (3). If the carbon is prematurely oxidized at low temperatures due to a relatively high dew point or too slow a heating rate, then excessive core shrinkage takes place. As a result of ongoing improvements to ensure a sufficiently dry hydrogen environment, core shrinkage gradually decreased to about 2.0% and became more consistent. However, small process variations affected core shrinkage significantly and further improvements in control are necessary.

It was found that the furnace hearth sagged causing in turn the retorts, setter blocks, and cores to distort. This was significantly reduced by ensuring that the Mo retort always rested on a flat surface in the hydrogen furnace. During firing, it was also found that some of the thin sections of the core slumped under gravity; to alleviate this problem, cross members called tie bars were added to join and mutually support thin sections. These tie bars were added by modifying the core injection die and were removed by grinding after core firing.

Core cracking occurred during firing due to excessive shrinkage and distortion. However, reducing the shrinkage and distortion to acceptable limits also reduced the cracking. As a result of the above modifications, core quality steadily improved.

Blade Wax Injection

Because of the poor core quality produced during the early part of the program, a large proportion of the early castings were scrapped due to core kiss-out. This was attributed to cracking of the cores during injection of blade wax patterns. It thus became imperative to produce wax patterns containing uncracked cores. For this purpose, three methods for producing blade wax patterns with eutectic design cores were evaluated: 1) injection molding at low pressure; 2) pouring molten wax into the wax injection die; and 3) pouring molten wax into a silicone rubber die. The cored blade wax patterns cast in the rubber mold were the most successful because painstaking care had been taken to obtain acceptable wax-pattern wall thicknesses. It was, however, recognized that injection molding was the preferred technique and refinement continued until a sufficient number of satisfactory patterns were eventually produced. Wax patterns of the starter and reservoir sections were then added to the injected blade wax pattern, as shown in Figure 1.

Shell Mold Fabrication

The shell molds developed at GE for casting DS eutectics were made of silica-bonded alumina. The mold construction was of two inner (or face coats) and four outer coats using coarser particles (5). The placement of the core within the mold required attention to the difference in thermal expansion behavior of the mold and core. Figure 4 shows a schematic of dilatometer traces of a Koralox core and an alumina base mold. The core material behaves essentially as fully dense alumina. However, the mold undergoes a net contraction as it is heated between about 1800°F and 2800°F. Core placement and pinning must allow for this difference in thermal behavior in order to properly locate the core in the casting. Thus, the core pinning was designed to allow slippage between the mold and core as in Figure 3. The core tip was clamped by the mold, and pockets were created in which the T-bar could slide. After preparation of the green mold, it was dewaxed by microwave heating. The cored mold was then fired to 1832°F for 2 hours. Heating and cooling rates during this firing step were controlled to a rate 180°F/hr.

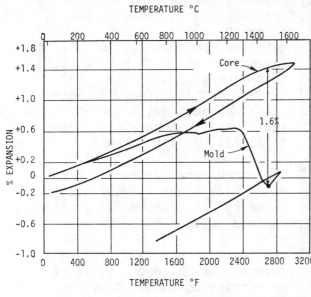

Fig. 4 Core and Mold Expansion

The cored molds were then visually inspected and x-rayed to identify any cores which were improperly pinned or cracked. Mold cracks were repaired with alumina-silica slurry.

Casting Development

The directional solidification furnace was a Bridgman design capable of producing three single blade castings simultaneously. The furnace tank contained 3 high thermal gradient, resistance heated furnaces with individual mold withdrawal units (Figure 5). NiTaC-14B charges were placed in separate shell molded melt cups with a small hole in the bottom sealed with a Ni metal plug. Resistance heated furnaces were placed around each of the melt cups and the assemblies were placed atop the three directional solidification furnaces.

The casting procedure consisted of positioning the molds on pull rods and raising them into the individual furnaces, loading the upper (melting) furnaces with NiTaC-14B charges and closing the chamber. The furnace chamber was evacuated and the individual furnaces heated to 1830°F in vacuo. At this point, the furnace chamber was backfilled with argon-10% carbon monoxide, and the furnace temperatures raised. After holding the mold at temperature to assure conversion to a stable microstructure, the upper furnaces were energized to melt the charge. The molten NiTaC alloy dissolved the Ni plug at the bottom of the melt cup and then quickly filled the cored mold below.

After the charge had melted and filled the mold, the DS furnaces were set for a 3137°F wall temperature and mold withdrawal was initiated. The mold withdrawal rate was held constant through solidification of the airfoil section and then reduced gradually as solidification progressed through the platform, shank, and dovetail of the blade. This rate reduction was done to maintain a constant solidification velocity and carbide fiber alignment. The progression of the solidification front from a section of low cross sectional area to one of high cross sectional area was observed to first reduce and then increase the local solidification rate; the withdrawal rate was decreased to counter this.

Fig. 5 DS Casting Facility

After casting, the mold was removed, the casting cut, and the core leached in caustic soda solution in an autoclave. After leaching, detailed non-destructive evaluation was performed on each casting to determine acceptability. The evaluations included: 1) internal visual inspection; 2) external visual and fluorescent penetrant inspection; 3) airfoil wall thickness measurements via ultrasonics; 4. microstructural examination for MC-carbide fiber alignment; 5) grain boundary orientation; and, 6) x-ray radiography. A total of 64 castings were examined, of which 17 were considered acceptable for engine testing and 32 were adequate for non-engine component testing.

Machining Development

The first finishing step, removal of excess stock by electro-discharge machining (EDM), was the most critical because internal datum points had to be used to EDM the external airfoil and platform top. During this step, external datum points were established for finish machining of the remaining external surfaces. It was quickly learned that the cores had not been uniformly positioned in the castings, and that it was advantageous to divide the castings into 3 lots to reduce both the number of fixturing setups required and wall thickness deviation from design intent. Establishing fixturing setups for each lot required the use of sophisticated equipment, including a validator and CAD/CAM equipment in an iterative procedure. After each setup was established, using the blade with the average deviation from design intent for the lot, all blades in that lot were machined. This procedure was continued until all airfoils were fully machined.

In the next step, two-piece tip caps were machined from NiTaC-14B, and the blades and tip caps were given a gamma prime solutioning heat treatment. The tip-caps were then brazed in position, inspected, and rebrazed when necessary.

The airfoils and platform tops were then hand ground to finish dimensions, measured ultrasonically for wall thicknesses, etched, and inspected for external defects. During inspection, it was found that many surface carbides were not fully removed which was due, in part, to unequal metal removal from both sides of the airfoil. An unexpected type of defect detected was cracking in the airfoil root on the concave pressure side. Most of the cracks, however, were shallow and were removed by light hand grinding. There was evidence to indicate that these cracks had formed during finishing and not during casting.

Finishing thereafter proceeded more smoothly and included conventional low-stress grinding of the dovetail, EDM of the shank and platform undersides, EDM of cooling holes and special blade features, brazing of balls in the shank to complete the internal cooling system, final inspection, testing of the cooling system in each blade, and shot peening of the dovetails.

Blade Hardware Evaluation and Engine Test

During bench testing, one DSE blade was used to determine nodal patterns during resonance at frequencies in and beyond the engine operating range. The same blade was used to determine strain distributions at all resonant frequencies expected in the engine operating range. Resonant frequencies for the 8 DSE blades selected for engine testing were subsequently determined and showed little scatter. Analysis of the data showed that the vibrational characteristics of the DSE blade were very similar to the Rene' 150 blade. HCF tests of 5 bare DSE blades at the anticipated engine operating temperature were completed prior to engine testing and the results exceeded previous results for the Rene' 150 blade. Hence, these results demonstrated adequate HCF strength of the eutectic blades for engine testing.

Results from nodal pattern and strain distribution measurements were used, in a complex analysis, to determine a single optimum position for the strain gage on all engine test eutectic blades. The analysis also established the stress limits for various resonances that could be expected during engine operation. Bench tests to determine the ultimate-tensile and low cycle fatigue strengths of the blade dovetail also gave satisfactory results.

In early engine testing, data from the strain gages showed that vibratory stresses in the eutectic blades were well below the established limit for the first flex mode. Subsequently, all strain-gage signals were lost and no further stress measurements were possible. Testing was continued with only optical pyrometer measurements of the airfoil surface temperatures on all blades. Engine testing was intermittent and after each test run, all blades were inspected using a borescope.

It was obvious during these inspections that the tips of the eutectic blades were being oxidized, particularly at the trailing edge. Despite this, the engine test was completed as planned.

Testing included 180 accelerated mission test cycles. The total running time was about 184 hours, including 2,290 full thermal cycles with 46 hours at gas stream temperatures within $100°F$ of maximum operating conditions, and over 2 hours above the design-rated level. The eutectic blades, with a reduced cooling-flow design, operated at temperatures at least $80°F$ above the Rene' 150 blade.

Post-test visual examination of the eutectic blades revealed that all squealer tips were severely oxidized, especially at the trailing edge. In addition, one blade had a small (0.030") elongated hole through the airfoil wall and another had a series of cracks in the airfoil wall. Microscopic evaluations indicated that the tips ran at a temperature considerably hotter, about $2200°F$, than the analytically predicted $2100°F$. The excessive temperature was at least partially due to obstruction of tip-cap cooling holes by oxidized surface-carbide particles that were blown from internal cooling cavities. Flow tests of 6 blades after engine testing showed an average loss in tip-cooling flow of about 11 percent, with one blade losing about 37 percent.

Distress of the other two blades mentioned were also surface carbide related. In one blade, oxidation of a cluster of surface carbides through the wall created an elongated hole. In the other blade, the wall was extremely thin, and this combined with surface carbides led to a series of short, jagged cracks. Despite the distress of the eutectic blades, all successfully completed the severe accelerated mission test. Plans have been made to further test two of the blades in another build of the demonstrator engine.

Conclusions

The Rene' 150 blade design was successfuly modified to take advantage of the very good mechanical properties of NiTaC-14B. The measured properties show that the alloy has an approximately 225°F superiority over conventionally cast Rene' 80. A core and mold system suitable for casting NiTaC-14B was developed but further improvements in control of core firing and core location during casting are required. The casting procedure was refined to provide in-situ melt/pour and to ensure the requisite aligned fibers in the blade airfoil and dovetail. Finish machining operations were also successfuly developed. The successful engine test demonstrated the improved performance provided by NiTaC-14B in the HPT blade application.

Acknowledgements

This work was sponsored by the US Air Force Materials Laboratory under contract F33615-77-C-5200. D. R. Beeler and T. Fecke were Technical Program Managers for the Air Force. There were many other GE contributors to this program and their work is gratefully acknowledged as is that of B. Ferg and A. Mihelsic of Sherwood Refractories Inc.

References

1. J. L. Walter, M.F.Gigliotti, B. F. Oliver and H. Bibring, Conference on In-Situ Composites-III, Ginn Custom Publishing, Lexington, Ma. 1979.

2. M. F. Gigliotti and M. F. Henry, Conference on In-Situ Composites-II, p. 253, Xerox Individualized Publishing, Lexington, Ma., 1976.

3. F. J. Klug and W. D. Pasco, United States Patent 4,108,672.

4. W. D.Pasco and F. J. Klug, United States Patent 4,186,885.

5. C. D. Greskovich, MFX Gigliotti and P. Svec, Trans. Brit. Ceram. Soc., 1978, $\underline{77}$ 3, pps 98-103

® Koralox is a Registered Trademark of the General Electric Company

Advances in Processing

SUPERALLOY RECYCLING 1976 - 1986

John F. Papp

Bureau of Mines
2401 E Street, North West
Washington, D.C. 20241

Abstract

About 55 million pounds of clean and contaminated superalloy scrap were processed in 1986; about 92 pct (50 million pounds) went to domestic buyers, and 8 pct (4.3 million pounds) was exported. About 93 pct (4.0 million pounds) of the exported material was refinery-destined grindings and 7 pct (0.3 million pounds) was vacuum-melting-grade superalloy scrap. Of the 55 million pounds of superalloy scrap processed in 1986, about 70 pct (38.5 million pounds) was recycled into the same superalloy, 20 pct (11 million pounds) was downgraded, and 10 pct (5.5 million pounds) was sold to refineries. The average element content of superalloy scrap processed in 1986 was about 44 pct Ni, 16 pct Cr, 5 pct Co, 2 pct Cb, less than 1 pct each of Mn and Ta, and nil for Re. The remaining 30 pct was primarily Al, Fe, Mo, Ti, W, and other minor constituents.

The major changes in the superalloy recycling industry since 1976 were the introduction of premelted superalloy scrap as a material supply source and increased use of closed loop recycling agreements among forger-scrap processor-alloy producer-engine manufacturer groups. Since 1976 Inconel 718 has become the predominantly produced superalloy.

Introduction

The Bureau of Mines has long been interested in recycling as part of its minerals program. Chromium and superalloys have been the subject of previous Bureau studies because chromium is a critical and strategic metal and superalloys represent a strategic use for chromium (1-5). The objective of this study was to characterize the superalloy recycling industry. An industry structure was determined; major superalloy producing, consuming, and processing companies were identified; and superalloy material flow was estimated. Information was collected through personal interviews and site visits to companies that volunteered to participate. Data were collected both by Bureau of Mines commodity specialists and by industry analysts contracted by the Bureau. Data were organized and presented in material flow circuit diagrams.

The superalloy recycling industry was found to be composed of scrap generators, scrap dealers, superalloy processors, and scrap consumers. Scrap is generated when superalloys are produced, cast or wrought into semifinished products, and cut or ground into finished products, and when finished products become obsolete. Superalloy scrap is collected and processed by scrap collectors, scrap dealers, wholesale scrap dealers, and superalloy scrap processors. Superalloy scrap is sorted, cleaned, sized, and certified for chemical composition by a superalloy scrap processor before it re-enters the use cycle as a superalloy. The numerous material flow relationships between scrap industry and scrap generator, and among scrap collectors, scrap dealers, wholesale scrap dealers, superalloy scrap processors, and scrap brokers obscure the quantity of superalloy scrap available for recycling, the quantity downgraded, and the quantity exported. A superalloy industry diagram based on material flow between processing steps was constructed. A previous Bureau of Mines study (IC 8821, reference 3) of the superalloy industry was used as a reference to which 1986 data were compared. Thus the terminology, industry structure, and material flow patterns used in this study are similar to those of IC 8821.

Results

Companies

No major change in superalloy producing, consuming, and recycling companies was found to have taken place since 1976. Only a few company name changes have taken place. About 75 companies were identified as significantly involved in the U.S. superalloy industry. They were classified by their major role in the superalloy industry as a superalloy producer (AP), casting producer (CP), forger (F), end user (EU), gas turbine manufacturer (GTM), metals producer (MP), product manufacturer (PM), scrap dealer (SD), or scrap processor (SP). Table 1 lists these industries.

Table 1.--Companies and Organizations identified as part of this study.

Company	Activity code
Abex Corp., New York, NY.	AP, CP
Abex Research, Mahwah, NJ.	CP
Air Force Materials Laboratory, Dayton, OH.	EU
AiResearch Manufacturing Co. (see Garrett Corp).	GTM, PM
Allegheny Ludlum Industries, Special Metals Div., Lockport, NY.	AP
Atlas Metals & Iron Corp., Denver, CO.	SD, SP
Brush Wellman Inc., Cleveland, OH.	CP
Cameron Iron Works Inc., Houston, TX.	F
Canon-Muskegon Corp., Muskegon, MI.	AP, CP

Company	Codes
Carpenter Technology Corp., Reading, PA	AP
Certified Alloy Products, Inc., Long Beach, CA	AP, CP
Chromalloy Corp., St. Louis, MO	AP, PM
Cytemp Specialty Steel, Division of Cyclops Corp., Pittsburgh, PA	AP
Degussa Electronics Inc., Vallejo, CA	AP
Detroit Diesel, Allison Division, Detroit, MI	GTM, PM
Duraloy Co., Division of Blaw-Knox Corp., Scottdale, PA	AP
Eaton Corp., Cleveland, OH	PM
Electralloy Corp. (see Michael Kral Industries)	
Electrometals (see Degussa)	
Elkem Metals Co., Pittsburgh, PA	MP, AP
Ford Motor Co., Aeronutronic Div., Detroit, MI	GTM, PM
Garrett Corp, Airesearch, Torrance, CA	AP
General Electric Co., Cincinnati, OH	GTM, PM
General Motors Corp. (see Detroit Diesel)	GTM, PM
Haynes International Inc., Kokomo, IN	AP
Howmet Turbine Components Corp., Plymouth, MI, Dover, NJ, Norfolk, VA	AP, CP
Inco Ltd., Toronto, Canada	MP
Inco Alloys International Inc., Huntington, WV	AP
Ireland Alloys Inc., Houston, TX	SD, SP
Kaydon Ring & Seal Inc., Baltimore, MD	CP
Koppers Co, Sprout Waldron, Muncie, IN	CP
Ladish Corp., Cudahy, WI	F
LMC Metals Corp., San Jose, CA	SD, SP
Martin-Marietta Corp., Bethesda, MD	AP, PM
Michael Kral Industries, Electralloy Corp., Oil City, PA	AP, SP
Kokomo Tube Co., Peru, IN	AP, CP
Monico Alloys Inc., Los Angeles, CA	SD, SP
National Aeronautics and Space Administration, Cleveland, OH	EU
Norco Alloys Corp., Farmington Hills, MI	SD, SP
Northeast Alloys & Metals Inc., Utica, NY	SD, SP
Outokumpu Oy, Kokkola, Finland	MP
PCC Airfoils Inc., Minerva, OH	AP
Powmet Inc., Rockford, IL	SD, SP
Pratt & Whitney (see United Aircraft)	
Precision Castparts Co., Inc., Portland, OR	AP, CP
Precision Rolled Products Inc., Florham, NJ	AP, MP
Quaker Alloy Inc., Myerstown, PA	CP
Rainbow Metals Inc., Charlotte, NC	SD, SP
Reading Alloys Inc., Reading, PA	MP
Rolls Royce Inc., Miami, FL	GTM, PM
S. Wilkoff & Son Co., Cleveland, OH	SD, SP
Samuel Keywell, Inc., Detroit, MI	SD, SP
Samuel Zuckerman and Co., Front Royal, VA	SD, SP
Schnitzer Steel, Portland, OR	SD, SP
Shieldalloy Metallurgical Corp., Newfield, NJ	AP, MP
Solar Turbines Inc., San Diego, CA	GTM
Special Metals Corp., New Hartford, NY, Princeton, KY	AP, CP
Spectrum Alloys Inc., Los Angeles, CA	SD, SP
Stoody Deloro Stellite, Inc., Industry, CA	CP
Suissman and Blumenthal, Inc., Hartford, CT	SD, SP
Techni Cast Corp., South Gate, CA	CP
Teledyne Allvac, Monroe, NC	AP
Textron Lycoming, Stratford, CT	GTM
TRW, Inc., Cleveland, OH	AP, CP
Unico Alloys Inc., Columbus, OH	SP
United Aircraft Corp., Pratt & Whitney, East Hartford, CT	GTM, PM
United Airlines, Chicago, IL	EU
United Alloys Inc., Los Angeles, CA	SD, SP
Universal Metals Inc., Worchester, MA	SP
Utica Alloys, Utica, NY	SP
Vac Air Alloys, Frewsburg, NY	SD, SP
Venango Metallurgical Products, Oil City, PA	CP
Wells Mfg. Co., Skokie, IL	AP

Westinghouse Electric Corp., Pittsburgh, PA.................. GTM, PM
Wisconsin Centrifugal Inc., Waukesha, WI.................... AP, CP
Wyman-Gordon Co., Worchester, MA............................ F

Scrap Disposition

Superalloy materials resulting from the production and manufacturing process were classified as product, scrap, or waste. Superalloy scrap was further subdivided into solids, turnings, or grindings. Superalloy scrap domestically processed was subdivided into scrap that was used domestically and that exported, and into scrap graded for superalloy use, graded for other alloy use, and graded for refinery use. Figures 1 and 2 show the distribution of domestically processed superalloy scrap in 1986 based on these categories.

Figure 1 shows that a small amount of processed superalloy scrap was exported and that most of it was refinery-grade material. Figure 2 shows that most of the superalloy scrap processed was returned to superalloy use, some was used in other alloys, and the remainder required refining before reuse.

Production

The relative fraction of superalloy production by grade changed significantly for both wrought and cast superalloys. For both types of superalloy processing, Inconel 718 has become the dominant grade. Wrought superalloy production in 1986 declined about 30 pct compared to that of 1976, while cast superalloy production increased about 10 pct. Table 2 shows the changes in production by alloy class and fabrication method from 1976 to 1986.

Table 2.--Alloy production in 1976 and 1986 by alloy class and by fabrication method.

Alloy designation	Quantity (million pounds)		Percent of total	
	1976	1986	1976	1986
WROUGHT NICKEL-BASE ALLOYS				
Waspaloy....................	10	1.8	11.1	3
Inconel 718.................	10	27.0	11.1	45
Inconel 600 series..........	20	15.0	22.2	25
Inconel X750 and X751.......	6	3.0	6.7	5
Other.......................	44	13.2	48.8	22
Total..................	90.0	60.0	NAp	NAp
INVESTMENT CAST NICKEL-BASE ALLOYS				
Inconel 713 and 713C........	5.0	6.8	21.6	26.1
B-1900 and B-1900+Hf........	2.0	0.7	8.6	2.7
Rene 77.....................	2.0	2.2	8.6	8.5
IN 738......................	1.5	3.8	6.5	14.6
Inconel 718.................	1.0	6.5	4.3	25.0
Other.......................	11.7	6.0	50.4	23.1

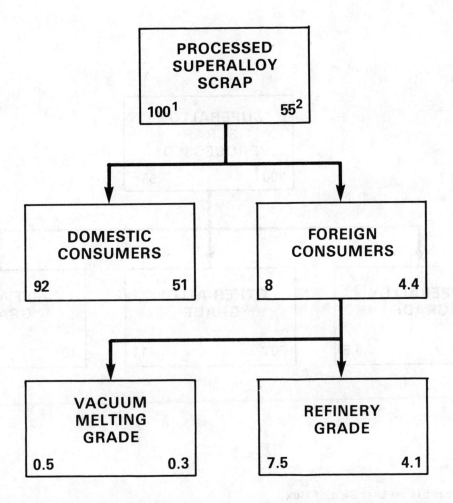

[1] Percent Is on Left Side of Box.
[2] Million Pounds Is on Right Side of Box.

Figure 1.--Domestically processed superalloy scrap material flow circuit for foreign and domestic use.

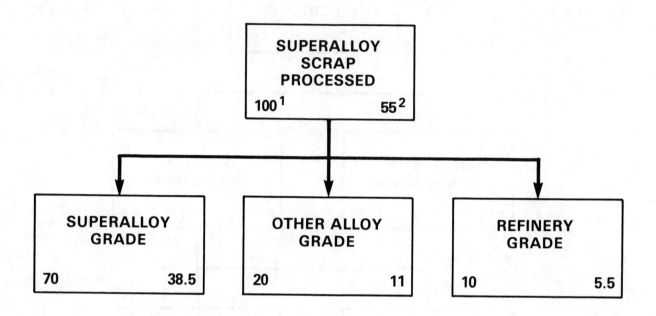

[1]Percent Is on Left Side of Box.
[2]Million Pounds Is on Right Side of Box.

Figure 2.--Domestically processed superalloy scrap material flow circuit for end uses by grade.

```
    Total..............     23.2       26.0           NAp         NAp
                       ----------------------------------------------------
                       ----------------------------------------------------
    Grand total..........   113.3      86.0           NAp         NAp
```

NAp Not Applicable. 1976 data from IC 8821.

The decline of wrought superalloy production was thought possibly to have resulted from several factors including (1) a decline in chemical industry (rather than turbine engine industry) demand for nickel-base alloys, (2) increased use of powder-metallurgy-produced parts, (3) increased use of casting, and (4) greater end-user efficiency. Greater end-user efficiency results in lower use and stocking, and therefore lower demand for parts and engines. From 1976 to 1986, the major civilian commercial aircraft changed from the three-engine 727 to the two-engine 737. This change may have resulted in reduced demand for original and replacement parts. Cooperative maintenance agreements wherein one engine repair company services many carriers may also have contributed to greater end-user efficiency by reducing the need for each carrier to stock parts.

Demand for superalloys, as for other metals, is cyclic. It is likely that 1976 and 1986 simply fell on different parts of the demand cycle.

The increased use of one superalloy grade suggests that recycling of prompt scrap should be easier because there would be more markets for the processed scrap. A disadvantage results from disuse of previously popular grades. Obsolete scrap of unpopular grades, although technologically recyclable, would not be in demand. Such scrap would, therefore, more likely be downgraded. Where possible, it is industry practice to use excess scrap of one grade to produce another grade.

The shift to Inconel 718 was thought to have resulted from (1) the shift away from cobalt caused by the high cobalt prices of 1979, and (2) the fact that General Electric used Inconel 718 widely in their engines, and they manufacture a large and increasing share of engines. Inconel 718 is a nickel-iron superalloy that can be produced using low-carbon, low-nitrogen ferrochromium in place of chromium metal for the required chromium units. This allows a cost advantage because ferrochromium is less expensive than chromium metal.

Material Flow Circuits

Assumptions required to produce material flow circuits for wrought and cast superalloy recycling included (1) the available obsolete scrap in 1986 was equal to 1976 finished product, (2) purchased scrap was first supplied from prompt scrap, then from obsolete scrap, and (3) half of the balance of solid scrap was exported and the remainder was downgraded.

Casting. Figure 3 shows that the casting superalloy industry material flow in 1986 was essentially the same as in 1976. The 1986 flow circuit shows that product yield per unit of raw material consumed in 1986 increased compared to that of 1976. It was thought that casting nearer to finished dimensions was the reason for improved product yield. The decreased use of scrap in 1986 was thought to have resulted from the imposition of more stringent chemical specifications by engine manufacturers that excluded previously used scrap. More stringent chemical specifications were imposed to reduce contamination by fixture alloys that include bismuth, lead, and tin.

Wrought. Figure 4 shows that the wrought superalloy industry material flow circuit in 1986 was essentially the same as in 1976 except for the addition of premelted scrap as a feed material. It was found that scrap premelted by argon-oxygen decarburization was widely used for vacuum induction melted alloys, allowing the alloy producers to use materials that

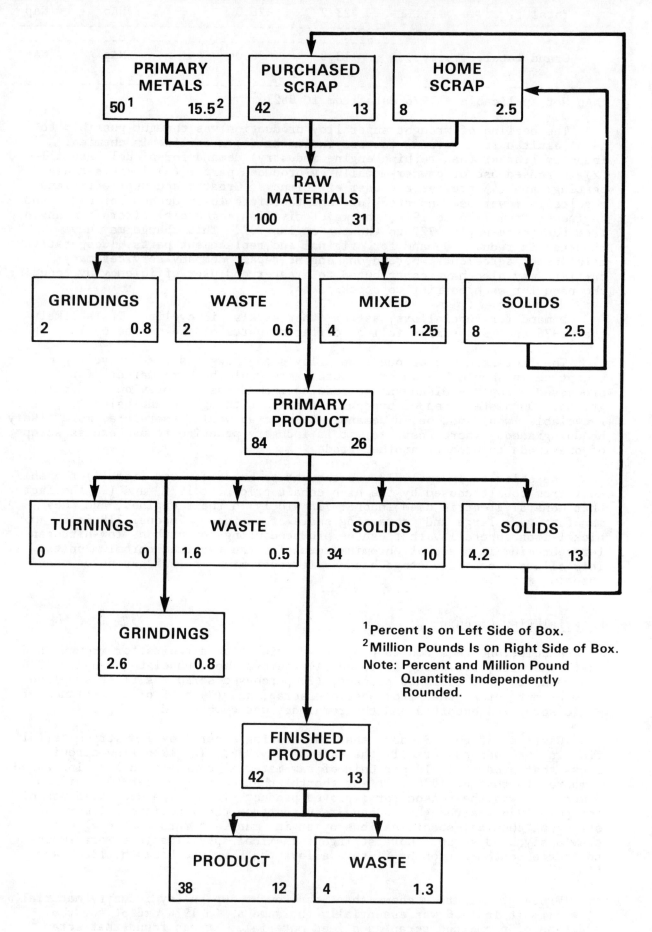

Figure 3.--Cast superalloy material flow circuit in 1986.

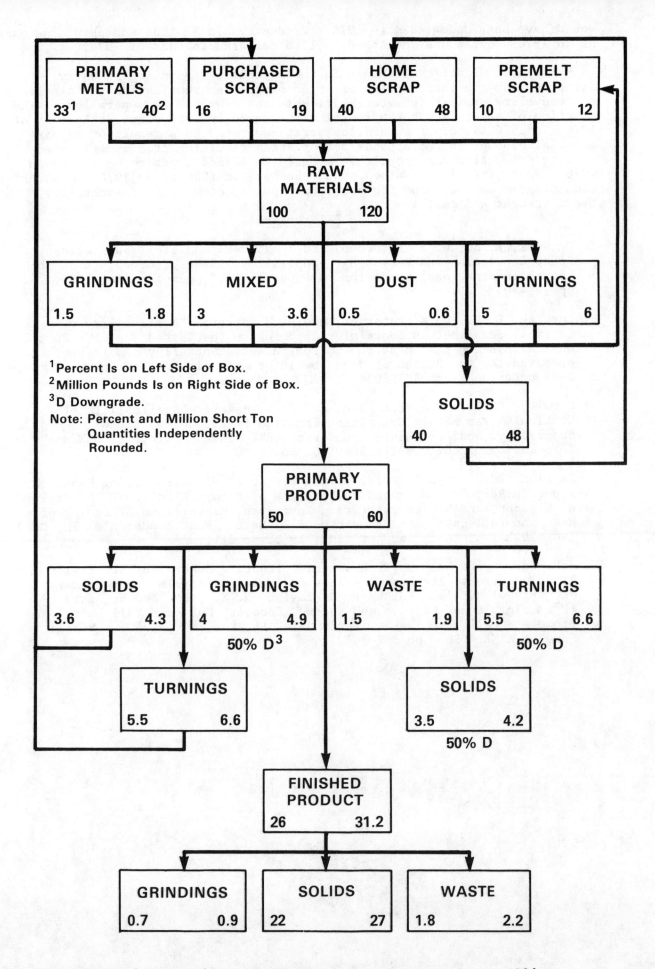

Figure 4.--Wrought superalloy material flow circuit in 1986.

would have been downgraded in 1976. Product yield in 1986 was about the same as in 1976. Scrap use increased in 1986 compared to that of 1976.

The most significant change in the superalloy recycling industry structure was the introduction of argon-oxygen decarburization refining into the recycling process for wrought nickel- and cobalt-base superalloys. The fraction of superalloy scrap recycled increased in 1986 compared to that of 1976. In 1986 49 pct of superalloy scrap generated as a result of primary material production and about 9 pct of scrap resulting from semifinished product production was recycled, compared to 47 and 3 pct respectively in 1976. Waste fraction remained about the same in 1986 as in 1976. A material flow diagram was produced for each year and for each type of superalloy, i.e., wrought and cast.

References

1. Curwick, L. R., W. A. Petersen, and J. J. deBarbadillo. Superalloy Scrap Generation and Recycling. Superalloys 1980. Proceedings of the Fourth International Symposium on Superalloys. Champion, PA, 1980, pp. 21-30.

2. Curwick, L. R., W. A. Petersen, and H. V. Makar. "Availability of Critical Scrap Metals Containing Chromium in the United States. Part 1: Superalloys and Cast Heat and Corrosion Resistant Alloys." (Report prepared for the Bureau of Mines by International Nickel Company, Inc. in 1979 under contract J0188056.)

3. Curwick, L. R., W. A. Petersen, and H. V. Makar. "Availability of Critical Scrap Metals Containing Chromium in the United States. Superalloys and Cast Heat- and Corrosion-Resistant Alloys." BuMines Information Circular 8821, 1980, 51 pp.

4. Kapalan, R. S. "An Overview of the Bureau of Mines Recycling Research. Paper in Recycle and Secondary Recovery of Metals." Ed. by P. R. Taylor, H. Y. Sohn, and N. Jarrett (Proc. Int. Symp. Recycle and Second. Recovery Met. and Fall Extr. and Process Metall. Meet., Fort Lauderdale, FL, Dec. 1-4, 1985). Metall. Soc. of AIME, 1985, pp.3-11.

5. Atkinson, G. B., and D. P. Desmond. "Treating Superalloy Scrap with Zinc To Increase Its Leaching Rate." Paper in Recycle and Secondary Recovery of Metals." Ed. by P. R. Taylor, H. Y. Sohn, and N. Jarrett (Proc. Int. Symp. Recycle and Second. Recovery Met. and Fall Extr. and Process Metall. Meet., Fort Lauderdale, FL, Dec. 1-4, 1985). Metall. Soc. of AIME, 1985, pp.337-348.

THE INVESTIGATION OF MINOR ELEMENT ADDITIONS ON OXIDE
FILTERING AND CLEANLINESS OF A NICKEL BASED SUPERALLOY

S. O. Mancuso, F. E. Sczerzenie, and G. E. Maurer

Special Metals Corporation
New Hartford, New York 13413
USA

Summary

Effects of minor element additions on filtering were studied by examining filtered and unfiltered melts of UDIMET® Alloy 700B (U-700B) with small additions of hafnium, rhenium and yttrium. Comparisons were made by sectioning filters, EB-button raft measurement, and sessile drop melting. Results provided evidence that these additions influence the filtration of molten metal. The examinations of filter cross sections also provided a perspective which supports oxide flotation as the basic function of the filter system.

Introduction

High performance superalloys designed for fatigue resistance in turbine engine applications demand ultra-clean material processing. Liquid metal filtration, using ceramic foam filters, is an advancing technology which responds to this demand. An understanding of the phenomena involved in the filtering process is necessary to develop improved procedures. During the observation of current filtering practices, the presence of minor elements in superalloys appeared to influence the characteristics of filtration and filter efficiency.

Several authors including Sutton[1,2,3,4], Apelian[5] and Ali[6] have reported varying results on the effects of alloy composition and filtering techniques on inclusion removal and filter efficiency. This study was intended to examine filtration phenomena in present filtering processes and to determine how alloy composition affects filtration.

Experimental Design

U-700B is a nickel based superalloy which contains aluminum and titanium for precipitation hardening similar to alloys now being manufactured and processed using filtration technology. The alloy contains no rhenium, hafnium or yttrium and it was therefore used to make a systematic investigation of

®UDIMET is a registered trademark of Special Metals Corporation

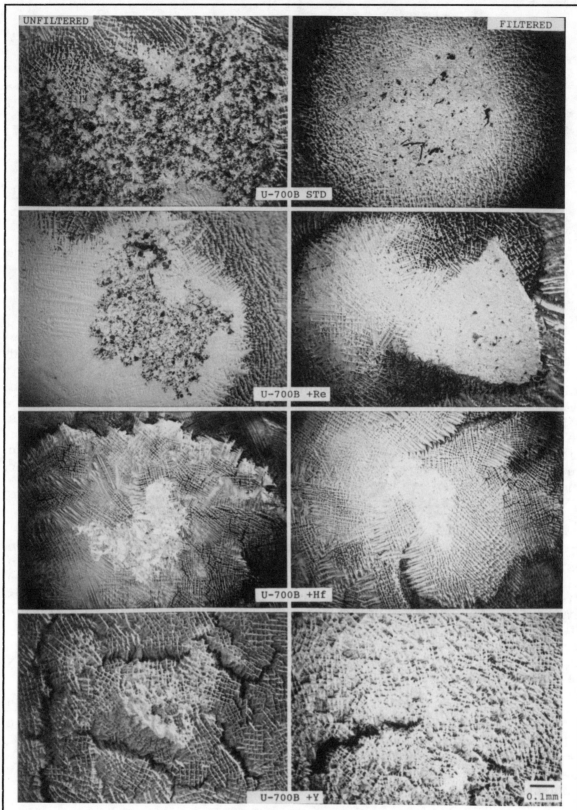

Figure 1. Scanning electron micrographs of EB button rafts from tops of filtered and unfiltered ingots. Materials with low atomic number elements, e.g., aluminum, appear dark, and high atomic numbers, e.g., hafnium, appear light.

the effects of these elements on the removal of oxide inclusions by filtration. Rhenium, hafnium, and yttrium were chosen for study because they are minor elements used in recent years in advanced investment cast and cast-wrought alloys.

Table I Specific Oxide Areas and particle sizes for each of the unfiltered and filtered alloy modifications.

ALLOY	RAFT OXIDE	UNFILTERED RATING cm²/kg	PARTICLE (μm) TYPICAL	LARGEST	FILTERED RATING cm²/kg	PARTICLE (μm) TYPICAL	LARGEST
STD	TOP Al	0.362	1-200	400	0.060	1-100	200
	BOT Al	0.050	1-200	400	0.091	1-100	200
	AVE Al	0.206	1-200	400	0.076	1-100	200
+Re	TOP Al	0.14	1-200	400	0.038	1-10	300
	BOT Al	0.05	1-200	400	0.057	1-10	100
	AVE Al	0.10	1-200	400	0.048	1-10	300
+Hf	TOP Hf	0.036	1-200	700	0.015	1-10	10
	Al	0.006	1-100	700	0.004	1-10	100
	BOT Hf	0.122	1-200	800	0.071	1-10	10
	Al	0.035	1-100	300	0.011	1-10	400
	AVE Hf	0.079	1-200	800	0.043	1-10	10
	Al	0.021	1-100	700	0.008	1-10	400
+Y	TOP Y	0.05	1-200	500	0.014	1-10	10
	BOT Y	0.125	1-200	500	0.028	1-10	10
	AVE Y	0.088	1-200	500	0.021	1-10	10

Rhenium was included in this study because difficulties were encountered in the past in EB cleanliness evaluations of Re bearing alloys. Experience has shown that oxides are difficult to sweep to the center of the button for subsequent analysis. About 1.0% rhenium is present in commercial alloys. Hafnium and yttrium were included because they are strong oxide formers and are expected to reduce Al_2O_3 and other less stable oxides. Hafnium concentrations of 1.0% are typical of several hafnium bearing alloys now manufactured. Yttrium has been used in a wide range of concentrations from as much as 1% in coating materials to as little as 0.05% in wrought superalloys. A concentration of 0.5% Y was used in this series.

Eight experimental heats with different minor element additions were vacuum induction melted (VIM) and poured with and without filters. The same lots of virgin raw materials were used for all melts in order to obtain consistent melt quality. The modifying elements were added to the base charge in a magnesia ram crucible. The filtered melts were poured through alumina pour cup assemblies containing high density partially stabilized zirconia (HDPSZ) foam filters. The filters were disks 76mm φ x 25mm (3" φ x 1") thick with an open pore size of 8 pores per cm (20 pore per inch). The filter assemblies were pre-heated to 1090°C (2000°F) for one hour before pouring. The pre-heated assemblies were transferred to the VIM furnace at the time of pour to minimize heat loss. The poured materials were collected in 75mmφ x 412mm (3"φ x 16.5") cast iron molds incorporating hot top insulation to improve ingot soundness. Ingots and pour assemblies were allowed to cool in vacuum before they were removed from the VIM furnace for further operations. The ingots were removed and sectioned for remelting in an EB button furnace. Procedures for EB button melting are described elsewhere[7]. The rhenium modified unfiltered heat presented a problem due to insufficient filling of the hot top and, subsequently, the top of the ingot was sampled by sectioning off center to avoid the primary pipe. Cleanliness was assessed at the top and bottom of each ingot. Oxide rafts from filtered and unfiltered experimental alloys are shown in Figure 1. An oxide area measurement of each raft was made by a planimeter trace of the elemental x-ray maps from a scanning electron microscope. Oxides of Al, Hf, Y, Mg and Zr were detected. The latter two were only found in trace amounts and were not included in the measurements. The areas were initially normalized by the area fraction of particles in the raft, measured by a line intercept method.

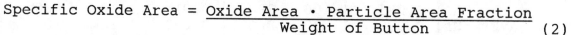

$$\text{Particle Area Fraction} = \frac{\text{Particle Intercepts}}{\text{Total Intercepts}} \qquad (1)$$

These areas were again normalized by the weight of the EB button to give a specific oxide area, cm^2/kg.

$$\text{Specific Oxide Area} = \frac{\text{Oxide Area} \cdot \text{Particle Area Fraction}}{\text{Weight of Button}} \qquad (2)$$

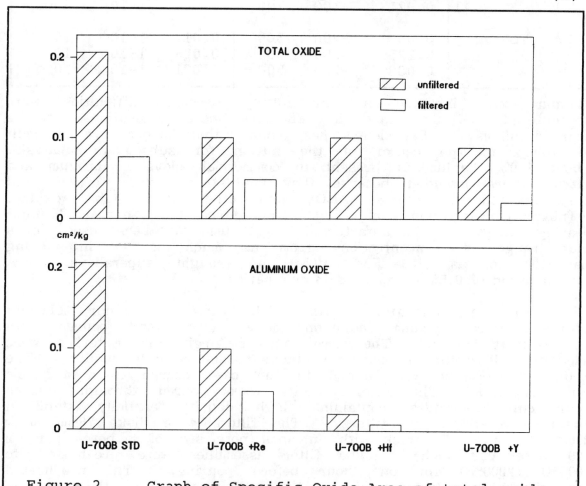

Figure 2. Graph of Specific Oxide Area of total oxide raft and aluminum oxide raft for unfiltered and filtered melts of each experimental alloy.

The filter/pour cup assemblies were disassembled and the filters were sectioned, mounted and metallographically polished. The polished surfaces were then coated with carbon to provide a conductive layer for examination in the SEM. The filter cross

Figure 3. Differences in particle sizes in rafts of unfiltered and filtered standard grade and Hf-Modified U-700B.

sections were examined for oxide particles attached to filter spine surfaces, imbedded in the molten alloy within the filter and at the top surfaces of the filter. The size of the oxide particles were measured and their chemistries were determined by Energy Dispersive Spectroscopic (EDS) microanalysis.

Samples were taken from each alloy for chemistry including O and N analysis, Differential Thermal Analysis (DTA), and sessile drop melting.

Experimental Results and Discussion

A summary of the raft oxide data obtained in the SEM is provided in Table I. Figure 2 compares the cleanliness before and after filtration. All four alloys were cleaner after filtration. For the standard alloy and the rhenium modified U-700B only Al_2O_3 particles were found in the rafts. The hafnium modified alloy produced 80% HfO_2 particles and 20% Al_2O_3 particles in the rafts of both the unfiltered and filtered materials. The yttrium modified alloy produced Y_2O_3 particles and only a trace of Al_2O_3 particles in the rafts of the unfiltered and filtered melts. The amount of Al_2O_3 particles was significantly reduced by the minor additions of hafnium and yttrium, Figure 2. Cleanliness improvements were achieved before filtration by the additions of rhenium, hafnium or yttrium. The specific oxide areas were averaged for top and bottom ingot samples. Without filtering, the average specific oxide area for rafts of the modified alloys were approximately half of those for the rafts of the standard alloy. The rhenium modified alloy

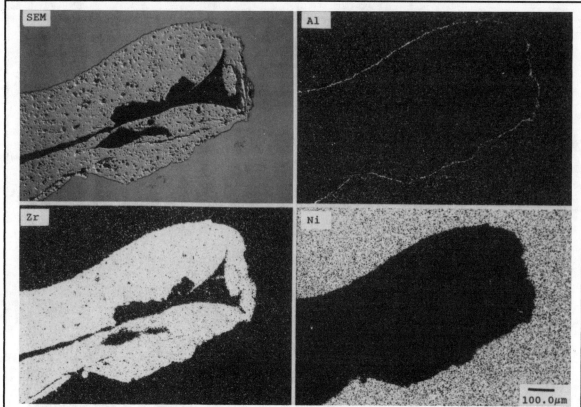

Figure 4. SEM photo and elemental x-ray maps of metal and filter solidified after pour. Example was taken from standard grade U-700B melt.

result was unexpected and may be due to problems in sampling the inadequately hot topped unfiltered ingot. Figure 3 demonstrates the reduction in typical particle size with filtration. This is important because studies have shown that particle size of inclusions is related to alloy fatigue properties[8]. The inclusion size in low cycle fatigue is inversely proportional to the number of cycles to failure.

There appears to be four mechanisms by which ceramic foam filters remove oxides from molten metal: 1) <u>Mechanical sieving</u> which is simply the blockage of large particles by small holes, 2) <u>Physio-chemical attraction</u> which involves the tendency of oxide particles to cluster and attach themselves to each other and the filter spines, 3) <u>Tundish flotation enhancement</u> which results from the decrease in molten metal turbulence in the pouring vessel and 4) <u>Hydrodynamic separation</u> by which particles are trapped in eddies within the torturous filter channels. Previous literature strongly suggests that the primary mechanism for filtration is the physio-chemical removal. Filtering efficiency relationships such as presented by Apelian[3]

$$\eta = \frac{C_i - C_o}{C_i} = 1 - \exp\left[\frac{-K_o L}{U_m}\right] \qquad (3)$$

where:

η = the effiency
C_i = conc. before filtration
C_o = conc. after filtration

U_m = melt approach velocity
K_o = kinetic parameter
L = length of filter

imply that the ability for physio-chemical attraction to occur is a function of the time the oxide particles are in close vicinity of the filter material. The evidence gathered in the present study suggest that this mechanism may not be the primary cause of oxide removal.

If physio-chemical attraction was the primary cause for oxide removal, one would expect to find large residual particles in the filter cross sections. Cross sectioned filters from pour cup assemblies, after casting, contained metal solidified in the cavities and on top of the filters. In Figure 4, the SEM image and elemental maps illustrate the typical oxide film observed at the interface between the metal and filter spines. In filters from standard and rhenium modified U-700B, Al_2O_3-MgO spinel material was found in the interface film. At the pour temperature used in this experiment (1650°C), MgO used to stabilize the zirconia, may be reduced to form the more stable Al_2O_3 from elemental aluminum in the melt due to the differences in oxide stability as indicated by the Relative Free Energy of Formation of the oxides at pour temperature, Table II. In the filters used in the hafnium modified and the yttrium modified alloys, hafnia and yttria respectively were found at the interfaces. Hafnium and yttrium form more stable oxides and can reduce the ZrO_2 or the MgO from the filter material. No evidence of particle adhesion was observed at the metal/filter interfaces. The width of the films observed at the filter interface were in the order of 1-2µm which were smaller than the particles eliminated from the rafts of the filtered material. The oxide film from the filter spines of the yttrium modified alloys was yttria, yet the particles eliminated from the rafts of the filtered samples of this alloy were alumina.

Table II Relative Free Energy of Formation for oxides of interest at pour temperature, 1650°C.

OXIDE	$\Delta F°$ (Kcal)
2 MgO	-162.6
2/3 Al_2O_3	-170.3
ZrO_2	-175.4
HfO_2	-180.6
2/3 Y_2O_3	-202.4

Figure 5 shows that large and small particles of oxides including Al_2O_3 were found in the layer of metal frozen on top of the filters. Surface tension of the liquid metal prevents this material from passing through the filter at the end of the pour and it therefore solidifies at the top of the filter disk. The size and chemistry of these particles are similar to those of the type of particles missing from the raft of the filtered melts and is a result of flotation during pouring. The flotation of these particles is improved by the change in flow within the pour cup when the filter is present. The observation of this phenomena suggests that this flotation enhancement plays a major role in the separation of particles from the pour stream. It may result from a combination of factors including a more constant metal head in the pour cup and reduced turbulence at the bottom of the cup.

While oxide flotation was previously considered a supplementary effect of filtration, it appears that filters are very valuable in controlling turbulence that keeps fine oxides in suspension. If this observation is correct, the filtering efficiencies predicted by equation (3) are still valid, since the same factors which promote physio-chemical attraction (low melt velocities and thick filter thickness) also increase the ability for oxide particles to float out in the tundish.

A 125mm³ sample was taken from each alloy for melting in a sessile drop furnace using a substrate made from the same material as the filters (Figure 6). The alloy samples along with the ceramic substrates were

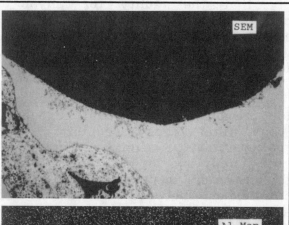

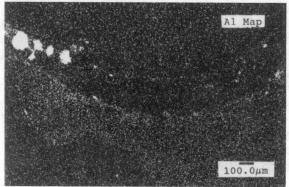

Figure 5. SEM photo and elemental maps showing oxide particles at top of filter.

heated in vacuum to a temperature of 167°C (300°F) above the liquidus. The metal droplet and the substrate interfaces were

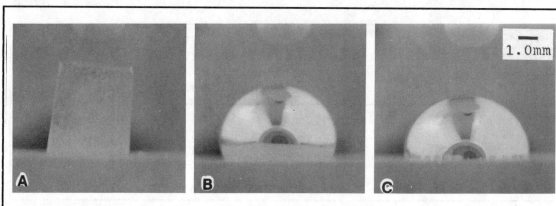

Figure 6. Three photos showing stages of the sessile. Zirconia substrate is at the bottom, metal specimen is placed on the substrate, and the object at the top of photo is the thermocouple. (A) Before melting (B) At melting (C) Metal reaction with substrate

examined in the SEM. Figure 7 shows a typical film found on the interface of the metal sessile drop and its substrate and these films are similar to those found on the filter interfaces. For the samples of the standard and rhenium modified alloy a film of Al_2O_3+MgO was observed. The interface of hafnium modified alloy drop contained a film of HfO_2 and the interface of the yttrium modified alloy drop contained a film of Y_2O_3. The widths of these interfaces were 5-10μm. The oxide products found at these interfaces are from reactions between the MgO stabilized ZrO_2 substrates and the stable oxide formers in the

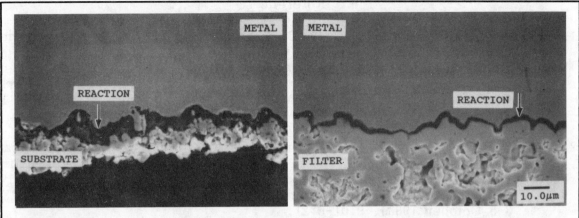

Figure 7. Interfacial reaction films found in both the sessile drop specimens and the filter after pour. (U-700B)

molten alloy. This supports the hypothesis that the films found on the filter interfaces are formed by reactions between the filter material and the molten alloy during pouring.

Conclusions

- Thin oxide films are formed on the filter spines during pouring.

- The minor addition of the elements hafnium or yttrium enhance the cleanliness of the melt by reducing Al_2O_3 inclusions. While the unfiltered Re modified alloy appeared cleaner, this was judged to be due to a sampling problem.

- Filtration of U-700B with minor additions of hafnium or yttrium significantly reduces the amount of Al_2O_3 found in the melts.

- No particles were found adhering to filter spines in any of the poured filters. Large and small particles of Al_2O_3 were observed at the top layer of the metal in the filter cross sections. Small particles were found in the metal solidified between filter spines but no particles were found in the filter-metal interfaces.

- Particle adhesion or physio-chemical attraction does not appear to play a major role in the reduction of oxide particles. The major proportion of the oxide reduction is due to enhanced flotation of oxides above the filter.

Recommendations

- The further study of the effects of other minor elements and the definition of minimum concentration for effective filtration may introduce the development of cleansing elements for ultra-clean superalloys.

- The study of tundish design is recommended to take advantage of the flotation improvement benefits achieved by the use of filters in pour systems.

References

1. W. H. Sutton, J. C. Palmer and J. R. Morris, "Development and Evaluation of Improved High Temperature Ceramic Foam for Filtering Investment Casting Alloys", (Paper presented at the Sixth World Conference on Casting Meeting, Washington, D. C., 12 October 1984), 15:01-15:18.

2. W.H. Sutton, J. R. Morris, "Development of Ceramic Foam Materials for the Filtration of High Temperature Investment Casting Alloys", (Paper presented at the 31st Annual Meeting, Investment Casting Industry, Dallas, TX, 3-5 October 1983), 9:01-9:21.

3. W. H. Sutton and S. O. Mancuso, "Evaluation of Filter Performance in Purifying MM-200+Hf Melts by the EB-Button Test", Electron Beam Melting and Refining - State of the Art 1985, (Bakish ed., Bakish Materials Corporation, 4-6 November 1985), 123-146.

4. W. H. Sutton and S. O. Mancuso, "Evaluation of Filter Performance in Purifying Superalloy Melts by the EB-Button Test", Electron Beam Melting and Refining - State of the Art 1984, (Bakish ed., Bakish Materials Corporation, 8-10 November 1984), 347-364.

5. D. Apelian and W. H. Sutton, "Utilization of Ceramic Filters to Produce Cleaner Superalloy Melts", (Paper presented at the Fifth International Symposium on Superalloys, Sevens Springs, PA, 7-11 October 1984), 421-432.

6. S. Ali, R. Mutharasan and D. Apelian, "Physical Refining of Steel Melts by Filtration", Metallurgical Transactions, 16B, (12)(1985), 725-742.

7. W. H. Sutton and I. D. Clark, "Development of an EB-Melting Test to Evaluate the Cleanliness of Superalloys", Electron Beam Melting and Refining - State of the Art 1983, (Bakish ed., Bakish Materials Corporation, 6-10 November 1983), 36-60.

8. D. R. Chang, D. D. Krueger and R.A. Sprague, "Superalloy Powder Processing, Properties and Turbine Disk Applications", (Paper presented at the Fifth International Symposium on Superalloys, Sevens Springs, PA, 7-11 October 1984), 245-273.

EVALUATION OF ELECTRON-BEAM COLD HEARTH REFINING (EBCHR) OF VIRGIN AND REVERT IN738LC

P.N. Quested*, M. McLean* and M.R. Winstone[†]

*Division of Materials Applications, National Physical Laboratory,
Teddington, Middx, TW11 0LW, UK

† Propulsion Department, Royal Aerospace Establishment,
Pyestock, Farnborough, Hants. UK

Summary

A collaborative programme involving 12 UK industrial and research organisations[#] has assessed the effects of EBCHR on the chemistry, microstructure, casting performance and mechanical behaviour of both virgin and revert casts of IN738LC. Small differences in chemistry and microstructure of the alloy variants do not significantly influence either the casting performance or the mechanical behaviour. Consequently, the work provides no commercial incentive for the use of EBCHR on cast superalloys of the purity levels considered in this investigation. Acceptable castings were obtained from all four bar stocks, the most expensive of which was about ten times the cost of the cheapest.

[#] AE Turbine Components Ltd
Cameron Iron Works Ltd
Glossop Superalloys Ltd
INCO Alloys Ltd
INCO Engineered Products Ltd
National Physical Laboratory
Rolls-Royce plc – Aero and Industrial/Marine Divisions
Ross and Catherall Ltd
Royal Aerospace Establishment
Ruston Gas Turbines Ltd
TI Reynolds Ltd
Vickers Precision Components Ltd

Introduction

It is well established that the mechanical behaviour of superalloys can be profoundly influenced by the presence of traces of various residual solutes and contaminating gases[1] and that the occurrence of porosity in cast superalloys can be associated with increases in nitrogen content[2]. The quality of VIM bar stock for investment casting is an important factor in optimising the service performance of cast components. It has been claimed that EBCHR of VIM bar-stock can lead to significant improvements in the yield of difficult-to-cast parts, particularly for revert superalloys[3,4]. However, there is little systematic evidence of the effects of EBCHR on chemistry, microstructure, castability and mechanical performance[5]. The present investigation considers the use of EBCHR secondary melting to improve the purity and cleanliness of a superalloy of two quite different levels of starting purity.

a) Examination of the trends in the fall of stress rupture life and rupture ductility with increasing trace element concentration shows that the relative effects are greatest at the lowest concentrations[1]. This suggests that significant improvements in mechanical performance may result from reducing the concentrations of detrimental trace elements to even lower levels than can be achieved in virgin alloy melts. Consequently one objective was to examine whether a superalloy, prepared by EBCHR of a virgin cast prepared to the highest VIM specification, had any advantages over existing cast superalloys.

b) The main detrimental effects of reverting nickel-base superalloys appear to be associated with accumulations of nitrogen, oxygen, and silicon originating from melt/mould and melt/environment interactions. These can lead to a variety of problems that limit the use of revert alloys for critical components - eg. microporosity, hot tearing, reduced fatigue resistance, decreased ductility, increased scatter in rupture life, etc. It is believed that these phenomena are consequences of inclusions, such as TiN and Al_2O_3, that remain stable within the melt and that can be removed by sophisticated EBCHR processing. A second objective of the programme was to evaluate EBCHR when applied to a 100% revert cast prepared using current commercial practice.

The work was carried out on the alloy IN738LC which is extensively used in industrial and marine gas turbines. The programme was organised by the UK Superalloy Panel and the characterisation of the various alloy variants, which was shared by the participating organisations, included duplication and cross-referencing to minimise systematic errors in various mesurements. The plan of the study is summarised in the flow diagram shown in Figure 1.

Alloy Production

The two initial bar-stocks for the programme were produced in the UK by vacuum induction melting (VIM) using the best current industrial practice on different types and grades of starting materials.

a) A high purity virgin cast of IN738LC, weighing approximately 600 kg, was prepared using the highest purity raw materials that could be obtained consistent with economic requirements. Except for zirconium, only primary raw materials were used. The base charge in the furnace, consisting of Ni, Cr, Co, Ta, Mo, W and Nb, was melted in a high alumina crucible while the more reactive constituents (Al, Ti, B, Zr, Mg) were added late in the process. The charge was cast into 75 mm diameter steel moulds using tea-pot pouring and passing the metal through two tundishes in order to minimise the transport of inclusions from the melt. The total time for the complete process of 14 hours included an extended melt-hold period to ensure the evaporation of metalloids.

b) A revert cast weighing 500 kg was prepared from cleaned scrap of known pedigree produced during investment casting in a single foundry. A commercial manufacturing route was followed using a crucible of spinel dry-rammed refractory powder (70% Al_2O_3, 30% MgO nominal). The alloy was teemed into 75 mm diameter steel moulds through a refractory tundish incorporating an alumina reticulated foam filter and a head box.

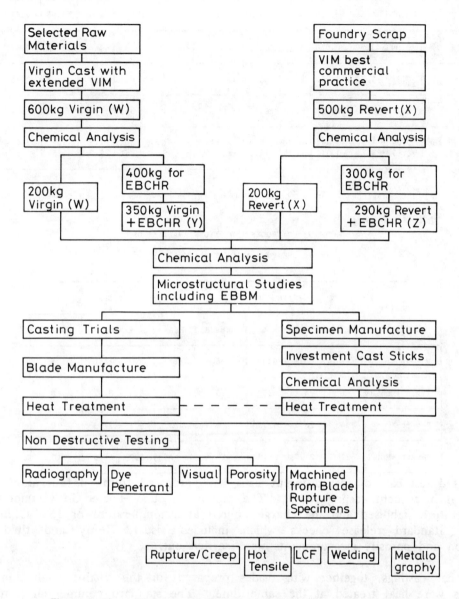

Figure 1 Flow Diagram of Technical Programme

About 400 kg of the virgin alloy (W) and 350 kg of the revert alloy (X) were reprocessed by EBCHR by a specialist U.S. subcontractor. The EBCHR process, which has been described previously[6], used a single gun with a maximum available power of 250 kW at 30 kV and the electron beam was programmed to melt the ingot, sweep the melt contained in a water-cooled copper hearth in order to isolate any floating inclusions and control the solidification of the ingot. No attempt was made to adjust the alloy compositions during EBCHR to compensate for losses of the more volatile constituents such as chromium because of the dangers of introducing associated gaseous contaminants. In order to conserve material during EBCHR, the virgin alloy (W) was remelted first to produce a refined alloy (Y). The skull formed on the water-cooled hearth by the virgin alloy was retained and used in processing the revert VIM alloy (X) into the refined version (Z). It was estimated that the first 50 kg of alloy Z would be mixed with some of the alloy W skull; this material was not used in subsequent characterisation.

TABLE 1: CHEMICAL ANALYSIS OF MAJOR CONSTITUENTS
IN ALLOY VARIANTS OF IN738LC

Alloy	C	Al	B	Co	Cr	Fe	Mo	Nb	Ni	Ta	Ti	W	Zr
Virgin W	0.09	3.44	0.010	8.6_1	15.9_5	0.04	1.77	0.92	bal	1.76	3.55	2.64	0.038
Virgin + EBCHR Y	0.09	3.50	0.010	8.5_7	15.8_0	0.06	1.77	0.93	bal	1.68	3.54	2.61	0.036
Revert X	0.11	3.40	0.012	8.8_0	16.1_7	0.11	1.75	0.90	bal	1.77	3.46	2.76	0.037
Revert + EBCHR Z	0.11	3.42	0.011	8.7_7	16.1_2	0.11	1.74	0.89	bal	1.70	3.48	2.74	0.036
Specification	0.09 -0.13	3.2 -3.7	0.007 -0.012	8.0 -9.0	15.7 -16.5	0.10 max	1.5 -2.0	0.6 -1.1	bal	1.5 -2.0	3.2 -3.7	2.4 -2.8	0.03 -0.08

The following minor constituents were analysed and found to be below the specified levels.

Si, Mn, P, S, Ag, As, Bi, Ca, Cd, Cu, Ga, In, Hf, Mg, Pb, Sb, Se, Sn, Te, Tl, V, Zn.

TABLE 2: NITROGEN AND OXYGEN ANALYSES OF
THE FOUR ALLOY VARIANTS

HEAT	NITROGEN			OXYGEN		
	MEAN	RANGE	NUMBER	MEAN	RANGE	NUMBER
Virgin W	6	5.6 - 6.9	5	12	8.3 - 14.0	5
Virgin + EBCHR Y	6	4.9 - 7.7	10	6	3.3 - 8.5	10
Revert X	30	28.2 - 30.5	10	12	9.6 - 17.3	10
Revert + EBCHR Z	27	24.7 - 28.8	10	7	3.9 - 11.5	10

Tapered test bars of two different sizes were prepared from all four alloy variants by an identical investment casting route. The metal was super-heated in vacuum to 1560°C, the temperature stabilised and the charge poured at a temperature of 1530°C into ceramic moulds. Standard release checks which included visual, X-ray and fluid penetrant examination and porosity measurements were carried out.

The bar castings, together with blades prepared for the casting evaluation, from all four alloys were heat treated at the same time. The standard commercial heat treatment for IN738LC was used; viz. a solution treatment of 2h at 1120°C in vacuum followed by an argon fan quench and a precipitation treatment of 24 hours at 845°C.

Characterisation

a) <u>Chemical analysis</u>

The chemical analyses for all four alloys are compared in Tables 1 and 2 with the original alloy specification. The values quoted in Table 1 represent the findings of 7 analysts from the collaborating organisations for the major constituents in the four bar-stocks. However, the results for oxygen and nitrogen obtained by the inert gas fusion method showed a wide scatter. This inconsistency has led to an independent investigation into the gas fusion technique aimed at producing reference standards and a UK protocol for gas analysis of superalloys. In subsequent discussion the figures quoted are for gas analysis from the single laboratory that performed the greatest number of analyses showing little scatter and which were generally consistent with the results from other laboratories (Table 2).

There are small differences in the major element compositions of the VIM virgin and revert alloys the latter having higher Cr, Co, Fe and W and lower Ti. However, both fall within the specified range for IN738LC adopted for this programme. The nitrogen level in the revert alloy X at 30 ppm was significantly higher than the 6 ppm measured in the virgin material (W). The oxygen contents of alloys W and X were both 12 ppm.

After EBCHR there was a marginal reduction in nitrogen content for the revert heat (30 ppm for X to 27 ppm for Z) but a significant change in oxygen content for both virgin and revert alloys. Changes in major element concentrations after EBCHR were small the largest effect being a reduction in Cr. However, the changes detected were close to the limits of accuracy of X-ray fluorescence spectrometry and EBCHR had no significant effect on the major alloy chemistry. No effects on the concentrations of volatile trace elements were noted the values being close to the reporting limits.

No significant changes in chemistry between the bar-stock and cast forms of the alloys were detected.

b) Microstructure

Polished sections of both the bar-stock and the cast bars were examined by optical microscopy; both etching with Marbles reagent and use of interference film microscopy enhanced contrast of various features. Microporosity was determined by different participants in the programme using established in-house procedures utilising quantitative metallography.

Examination of the bar-stock showed the carbides in the revert alloy X to be small and blocky rather than having the acicular, or mixed blocky/acicular, form characteristic of the other alloy variants (Figure 2). The detailed structure of the carbides in the revert alloy was complex generally exhibiting oxide and nitride centres which were not readily apparent in the other materials. Differences in the carbide shape and structure were less pronounced in the castings than in the bar stock. However, the castings produced from revert alloy X generally had a more equiaxed grain structure than those prepared from alloys W, Y and Z (Figure 3).

The highest levels of porosity in the bar castings occurred in the revert material (X) although this was at a very low level. There was negligible porosity in castings produced from the other three alloys.

c) Electron beam button melting

Samples of each bar-stock weighing approximately 700 gm were electron beam button melted (EBBM) to provide a measure of the total inclusion content. The alloy cleanliness was assessed by measuring the extent of the inclusion-rich raft on the button surface. Detailed examination of inclusions by scanning electron microscopy using back-scattered electron images, cathodoluminescence and X-ray mapping were also carried out. A full account of this phase of the characterisation has been published previously[6] and only the most important aspects are reported below.

Typical cap areas for the four alloys are shown in Figure 4. There was a remarkable consistency in the results obtained using a standard cycle in a single machine for runs carried out over several months. There is a clear reduction in cap size between the revert alloy X and EBCHR version Z. Alloy X exhibits both a rough central raft of oxide particles (~ 100 μm diameter) and a larger surrounding region of (~ 5 μm diameter) TiN inclusions in a nickel matrix. The oxide zone is much smaller in alloy Z which still retains TiN cubes. In the virgin alloy W the cap contained oxides but few nitrides while the EBCHR version Y only exhibited carbides although the cap was larger than that obtained for the original VIM material. Chakravorty et al[7] discuss the differences in EBBM analysis associated with different machine geometries.

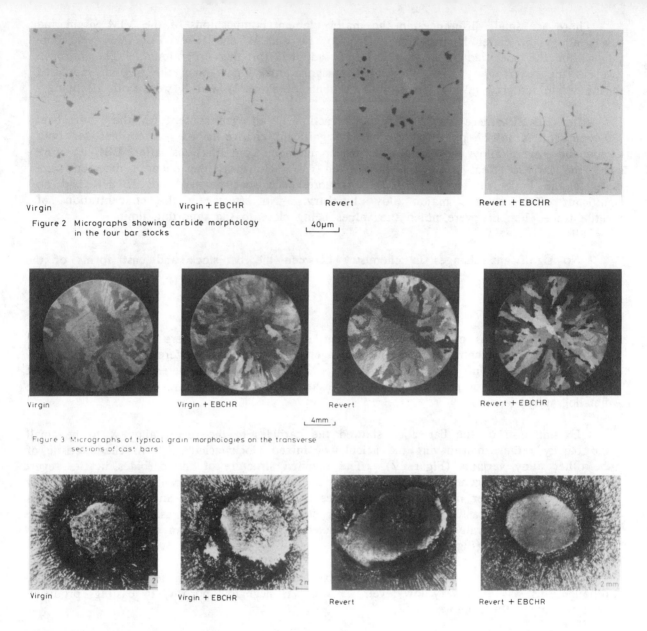

Figure 2 Micrographs showing carbide morphology in the four bar stocks

Figure 3 Micrographs of typical grain morphologies on the transverse sections of cast bars

Figure 4 Micrographs of inclusion rich caps on buttons prepared by EBBM of the four bar stocks

Mechanical Tests

The mechanical test programme was shared by 7 of the collaborating laboratories and included a series of callibration tests to ensure compatability of results. In all cases the inter-laboratory variation was within the expected experimental scatter.

Tensile Tests

Tensile specimens from each alloy were tested over the temperature range 20°C to 900°C. Tests were concentrated around 700-800°C to identify any ductility trough or peak in strength. Figure 5 summarises the results. Most of the data points are the average of 2 tests. A sharp peak in the proof stress and tensile strength occurred at 750°C and above this temperature the strengths of all the alloys were similar. However at lower temperatures there were some small, but significant, differences. The virgin alloy W had a consistently higher proof stress, while alloy Z had a low proof stress. The revert alloy X had a lower tensile strength than the other alloys. It was noted that EBCHR increased the low temperature ductility of both alloys. This was most noticeable for the revert + EBCHR alloy Z and resulted in a significant increase in tensile strength.

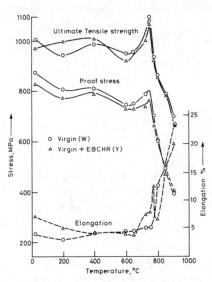

Figure 5a Tensile data for cast bars prepared from Virgin and Virgin + EBCHR alloys

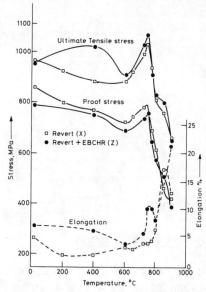

Figure 5b Tensile data for cast bars prepared from Revert and Revert + EBCHR alloys

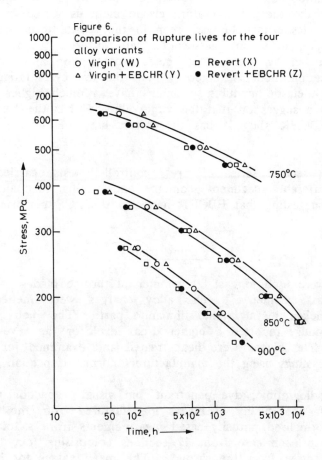

Rupture Tests

The results of stress rupture tests at 750°C, 850°C and 900°C with stresses chosen to give rupture lives in the range 50–10000 hours are plotted in Figure 6. All the data are grouped within a tight scatter band and it is not possible to identify a clear advantage for any one material. However, there is a difference between the virgin and revert alloys. Both virgin alloys (W and Y) tend to lie at the top of the scatter band, with the revert alloys (X and Z) towards the bottom. No advantage of EBCHR can be identified: the ductility increase noted in the tensile test data was not observed under creep conditions.

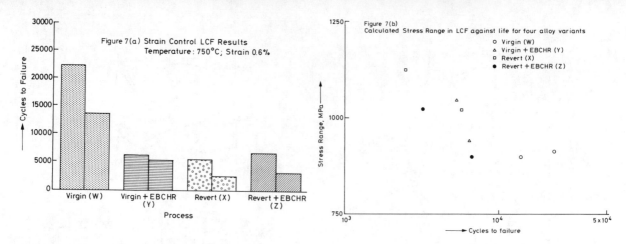

Low Cycle Fatigue Tests

Duplicate strain control fatigue tests were performed on each alloy at 750°C using a strain range of ± 0.3% and a frequency of 1 Hz. The fatigue life of the virgin alloy W was more than twice those of the other materials, the revert alloy X falling slightly below the EBCHR materials (Y and Z). This result must be treated with caution because it was also observed that there was a considerable variation in elastic modulus between test pieces. For example, the room temperature elastic modulus varied from 189 - 238 GPa. A similar variation in elastic modulus was seen in the tensile tests and is believed to be the result of crystallographic texture aassociated with different grain morphologies (Figure 3) and to be compounded by a small test piece cross-section and large grain size. To remove the dependence on modulus, Figure 7 plots the low cycle fatigue data in terms of stress range. The high elastic modulus specimens have a much higher stress range, and a shorter life. There is a suggestion that the virgin alloy W has the best fatigue properties, while the revert + EBCHR alloy Z has the lowest life. However, the differences are quite small.

The results of a second series of stress controlled tests, carried out in a second laboratory using a different specimen geometry and testing machine show a different ranking of materials, suggesting that EBCHR increases the LCF performance of both virgin and revert materials.

Casting trials

The four alloys were used to cast 80 commercial turbine blades which are known to have a difficult-to-cast geometry. The alloy charges were melted in a disposable silica-lined crucible backed by a mullite/sillimanite part. The melt pouring temperature was 1490°C. The foundry reported a higher dross level for the revert alloy X than for the other materials. The blades were heat treated and examined for any differences in the soundness of the castings using the manufacturers normal inspection procedures.

The results of radiographic, dye penetrant and visual inspection failed to show any significant difference in the quality of blades manufactured from the four alloy variants. Figure 8 shows the normalised porosity-based measurements from a total of 44 blades for the four alloys taken in each case from 12 sections comprising four from the root, five from the aerofoil and three from the shroud. The mean values for blades manufactured from typical revert and virgin heats are included for reference. All the blades cast from the four alloys under consideration would be accepted on the basis of the manufacturers criteria for microporosity levels. By contrast, a proportion of the blades from the comparison production revert cast would be rejected. Surprisingly, the lowest microporosity levels were obtained from blade prepared from the revert alloy which is the opposite of the trend observed for the rod castings.

Little difference in the microstructures of the various blades was noted. Thus, similar carbide morphology, extent of γ/γ' eutectic and γ' distribution were detected at equivalent sections of blades prepared from the different alloys. The largest microstructural differences were associated with different positions in the blade (eg. root, aerofoil, shroud) rather than with alloy and, consequently, can be considered to indicate sensitivity to cooling rate rather than composition.

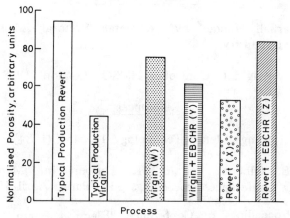

Figure 8 Mean Cumulative Porosity for samples cut from blades

Creep rupture specimens machined from the lower aerofoil sections of the blades were tested at 545 MPa/760°C and 230 MPa/930°C. The company requirement is for minimum lives of 60 and 20 h and mean lives of 87 and 28 h for these two test conditions respectively. For the four alloys under consideration, all tests exceeded the specified <u>average</u> life by at least 80%.

The welding characteristics of the cast versions of all four alloys was assessed using both the cast rod and blade materials. A range of welding procedures was considered, viz hard facing with Stellite 12 and lap-, butt- and patch-welds to Nimonic 75 using a range of filler alloys (INCO625, HASTELLOY X, NIMONIC 75). There was no evidence that EBCHR led to significant differences in welding characteristics. However, more crack-free welds were observed when using the revert casts (X,Z) rather than the virgin materials (W,Y).

Discussion and Conclusions

The present study shows that EBCHR can result in small but significant changes in the composition and microstructure of both virgin and revert IN738LC. In particular, there is a reduction in the oxygen concentration, present in the form of oxides in the bar-stock, though surprisingly little change in nitrogen level; some changes in carbide morphology were also apparent. The bar-castings produced from these materials exhibit some changes in both the grain shape and the level of porosity. However, any changes in the mechanical properties of the bar-castings or the quality of blades produced from the four materials are marginal. The mechanical properties of all the alloys were near the top of the scatter band for commercial heats of IN738LC. Some minor differences between the various alloys were observed, but no clear benefit could be attributed to either the use of a carefully selected, high purity virgin heat or to the application of the EBCHR process. EBCHR did give a slight increase in the low temperature ductility and a modification of elastic modulus associated with a change in grain texture, but these factors are unlikely to be of engineering significance.

It is estimated that the cheapest bar-stock (alloy X) which was prepared by remelting foundry scrap was about ten per cent of the cost of the most expensive (alloy Y) which was prepared from selected primary raw materials and electron beam cold hearth refined. Clearly, the present results can not justify this additional expenditure. The revert alloy considered in this study showed acceptable casting performance and mechanical behaviour. It is possible that a material with particularly poor casting characteristics would show greater benefits from EBCHR although the present study can not confirm this speculation.

However, bearing in mind that the revert alloy considered here was prepared to current commercial practice using 100% scrap, it may be more cost-effective to select or blend appropriate scrap alloys rather than incur the expense of a secondary remelting process.

Acknowledgement

This programme was partially funded by the Metals Sectoral Advisory Committee of the Department of Trade and Industry.

Copyright C Controller HMSO, London, 1988

References

1. M. McLean and A. Strang, Metals Technology **11**, 454 (1984).

2. P.N. Quested, T.B. Gibbons and G.L.R. Durber, in "Materials Substitution and Recycling", AGARD Conference Proceedings 356, Paris, April 1984.

3. J.R. Mosher, in Proceedings of Conference "Electron Beam Melting and Refining - State of Art", edited by R. Bakish, Reno, NV 1986. p.277.

4. M. Krehl and J.H.C. Lowe, in Proceedings of Conference" Electron Beam Melting and Refining - State of Art", edited by R. Bakish, Reno, NV (1986). p.286.

5. T.B. Gibbons, Proceedings of International Conference on Electron Beam Welding and Melting, Paris, p.545 (1970).

6. C d'A. Hunt, J.H.C. Lowe and S.K. Harrington in Proceedings of Conference "Electron Beam Melting and Refining - State of Art 1985" edited by R. Bakish, Reno, NV (1985) p.58.

7. S. Chakravorty, P.N. Quested and M. McLean, Proceedings of 1st European Conference of ASM International on "Advanced Materials and Processing", edited by T. Khan, Paris (in press).

ELECTRON BEAM COLD HEARTH REFINEMENT PROCESSING OF

INCONEL* ALLOY 718 AND NIMONIC* ALLOY PK50

S. Patel+, I.C. Elliott+, H. Ranke++, H. Stumpp++

+ Inco Alloys Limited, Holmer Road, Hereford, U.K.
++ Leybold AG, Wilhelm-Rohn-Strasse 25, 6450 Hanau 1, West Germany

Abstract

The use of Electron Beam Cold Hearth Refining (EBCHR) to produce very clean nickel based superalloys has been the subject of a number of investigations in the past, aimed at providing material with enhanced properties for components such as aero-engine gas turbine discs. In these, the useful working life is largely limited by the presence of certain defects which would ultimately initiate fracture. Such undesirable features can include nitrides, carbonitrides and oxides, the quantity and morphology of which is generally referred to as the cleanness of an alloy. This paper describes the EB refining of two established disc alloys melted in a 250 kw Leybold EB furnace equipped with two KSR 250 guns, INCONEL alloy 718 and NIMONIC alloy PK50 (similar composition to Waspaloy**) and describes the EB melting process performance and resulting metallurgical structures, chemistry, forgeability, cleanness levels and mechanical behaviour .

* Trademark of the Inco family of Companies.
** Trademark of the United Technology Corporation.

Introduction

There is an increasing demand nowadays from manufacturers of aero-engine gas turbines to be supplied with nickel based superalloys which will perform satisfactorily at higher stresses and temperatures. It is therefore necessary to investigate methods of achieving extremely high cleanness levels in these alloys by reducing the size and quantity of non metallic low density inclusions (LDI's) which are primarily responsible for initiating fatigue cracks in critical components such as turbine discs (1).

Electron Beam Cold Hearth Refining (EBCHR) is proposed as a production means for superalloy refining especially relating to LDI reduction (2). It is a sensitive process during which optimum melting conditions need to be achieved in order to obtain the highest quality material. Any deviation from these optimum conditions can lead to problems with the most critical parameters, and may result in some of the undesirable features ending up in the material.

In order therefore to evaluate and optimise the process, ingots of VIM + ESR INCONEL alloy 718 and VIM + ESR NIMONIC alloy PK50 were melted in an EBCHR furnace and the resulting material processed and assessed against material produced using conventional methods.

Electron Beam Furnace Design

Several investigations have been performed in the past to try and achieve refined superalloys with superior material qualities especially with regard to an improvement of behaviour under higher stresses. It is well known that stress related behaviour closely correlates with the size and quantity of low density inclusions and high levels of the latter can lead to a significant reduction in gas turbine component lifetimes.

Electron Beam Cold Hearth Refining is known to be a production means for superalloy refining especially in relation to LDI reduction (3).

In order to evaluate this process, pre-refined ingots of VIM + ESR INCONEL alloy 718 and VIM + ESR NIMONIC alloy PK50, supplied by Inco Alloys Limited were melted in a horizontally fed EB furnace installed in the laboratory area at Leybold in Hanau.

The arrangement of the EB guns, trough and crucible in an EBCHR process is shown schematically in Figure 1. The refining effect is based on LDI flotation. Oxide particles, which float to the surface of the trough pool, are restricted by means of a water cooled mechanical skimmer from entering the crucible.

A medium sized sate of the art, EB furnace manufactured at Leybold in which EBCHR processes can be performed on a pilot scale is shown in Figure 2. Because of its modular design, different set up modifications are possible which allow the performance of other EB processes such as horizontal and vertical dripmelting for the refining of refractory metals. Bar shaped feedstock can be fed via a horizontal or vertical feeding system whereas granules can be fed by means of a flanged vibration feeding system.

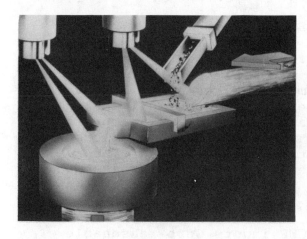

Figure 1 - Schematic showing the arrangement of EB guns, trough and crucible in EBCHR process.

Figure 2 - A medium sized state of the art EB furnace manufactured at Leybold.

The furnace is equipped with two EB guns, type KSR 250/35 with a maximum power of 250 kw at 35 kv. The installed EB power amounts to 300 kw. The furnace evacuation is done by means of a baffled oil diffusion pump having a nominal pumping capacity of 30,000 l/s.

To guarantee continuous melting operation without intermediate venting, the horizontal bar feeding chamber can be separately vented and recharged during melting. The capacity of the bar feeding chamber is three bars with a maximum length of 800 mm and a maximum diameter of 120 mm each.

In order to control the homogeneity with respect to the chemistry and structure of the final ingot, a reproducible beam pattern control is absolutely necessary. This is achieved using computer system BD 564. It contains a flexible u-computer system which allows simultaneous control of up to five EB guns. Fifty different programmes, each consisting of up to 64 different preprogrammed patterns such as circles, various areas and inclined lines can be stored. Beside pattern frequency, the position and dwell time can also be programmed. The application of the computerised deflection control leads to well defined process parameters which allow melting performances of a high quality.

For improved quality control, the furnace is additionally equipped with a residual gas analyser, a pyrometer for temperature indication of the melt surface and a sample taking mechanism by means of which samples can be extracted from the melt pool for chemistry evaluation purposes.

In these tests the dimensions of the installed trough were as follows: 450 mm length x 150 mm width x 50 mm depth. The crucible diameter was 200 mm, resulting in EB melted ingots weighing approximately 250 kg. The influence of the melting speed on the ingot structure was studied by varying the melt rate in the range between 40 kg/hr and 130 kg/hr. The melt rates of the INCONEL alloy 718 and NIMONIC alloy PK50 ingots which were forged and subsequently evaluated and whose results are outlined below were 110 kg/hr and 125 kg/hr respectively.

Forgeability of EBCHR alloys

The EB melted material was processed through a conventional route which included forging of the material from 200 mm dia. to 100 mm dia., slab forging the 100 mm dia. material to achieve a reduction in width of 3 to 1 and then carrying out all the mechanical tests on the resultant material. The main aspects of interest were the "metallurgical" structure achieved to assess whether there was any segregation, the forgeability of the alloys, the change in chemistry, the cleanness levels as shown by Electron Beam Button Samples and the mechanical properties obtained.

The surfaces of the as-cast EBCHR ingots were reasonable and comparable to those of ingots melted by conventional methods (Figure 3). Top and bottom slices from the as-cast ingots were

Figure 3 - 200 mm dia., as-cast EBCHR ingot of NIMONIC alloy PK50

taken and macroetched to ascertain the structure. It can be seen from Figure 4 that good structures were obtained with no evidence of serious macro segregation or undesirable features.

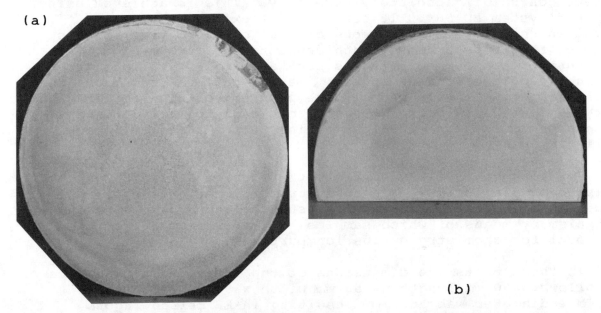

Figure 4 - 200 mm dia. macro-slices of (a) INCONEL alloy 718 and (b) NIMONIC alloy Pk50 showing no serious macrosegregation.

The chemistries of the two alloys after EB melting were analysed and are shown in tables 1 and 2. As expected, there is

Start Stock		EB Melted
0.036%	C	0.033%
18.57%	Cr	17.63%
17.60%	Fe	18.00%
53.50%	Ni	53.84%
0.55%	Al	0.59%
3.05%	Mo	3.13%
5.06%	Nb	5.40%
1.10%	Ti	1.03%
5ppm	S	5ppm
70ppm	N	40ppm
7ppm	O	5ppm
37ppm	Mg	-

Table 1 - Chemical analysis of INCONEL alloy 718 before and after EB melting.

Start Stock		EB Melted
0.02%	C	0.010%
19.30%	Cr	18.72%
Bal	Ni	Bal
1.29%	Al	1.30%
13.50%	Co	13.50%
4.20%	Mo	4.38%
3.16%	Ti	3.24%
4ppm	S	4ppm
20ppm	N	20ppm
7ppm	O	4ppm
40ppm	Mg	-

Table 2 Chemical analysis of NIMONIC alloy PK50 before and after EB melting.

a loss of chromium in both alloys. It is greater in INCONEL alloy 718 (0.94%) than in NIMONIC alloy PK50 (0.58%). This could have been due to the greater rate at which the INCONEL alloy 718 was melted. Magnesium was reduced in both cases from around 40 ppm to virtually zero. The achievable nitrogen level in nickel based alloys depends amongst other things on the titanium content. NIMONIC alloy PK50 is a high titanium (~ 3.16%) and consequently low nitrogen (20 ppm) containing alloy. There was no detectable change in its nitrogen level after remelting. In the case of INCONEL alloy 718 which has a lower titanium content (~ 1.10%) and therefore higher nitrogen (70 ppm), there was indeed a significant reduction in nitrogen content from 70 ppm to 40 ppm. This is an important result as the lower nitrogen level means fewer titanium nitrides and carbonitrides contributing to a "cleaner" alloy. The oxygen level of the INCONEL alloy 718 showed a reduction from 7 ppm to 5 ppm whereas the NIMONIC alloy PK50 showed a reduction from 7 ppm to 4 ppm. The trace elements such as Sb, Zu, Pb and Sn were also favourably reduced.

The forgeability of the alloys depends very much on the chemistry and it was therefore both interesting and encouraging to note that both forged well. The magnesium and sulphur contents are especially important. It is considered crucial to obtain a Mg:S ratio of greater than 1 to achieve the optimum elevated temperature ductility in these alloys and especially so in the case of INCONEL alloy 718. In both cases however virtually all the magnesium was lost upon EB remelting and therefore forgeability problems might have been expected. This has, in the past, been considered to be one of the drawbacks of the EBCHR process. Nevertheless, none of the anticipated problems were encountered and both alloys forged without any difficulties. Macroslices from the forged bars (Figure 5) once again show very good structures with no adverse macro segregation.

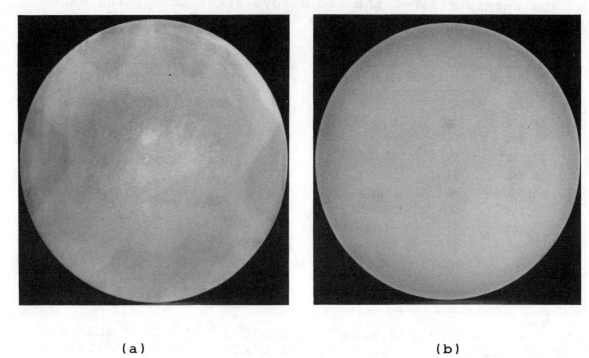

(a)   (b)

Figure 5 - Macroslices from 100 mm dia. forged bars of (a) INCONEL alloy 718 and (b) NIMONIC alloy PK50.

Cleanness and Mechanical Properties

Electron Beam Button melting is a means of assessing the cleanness of an alloy with regards to LDI content (4). A 2 lb sample is melted under vacuum by an electron beam into a water cooled copper crucible. The LDI's float to the surface of the button and form a cap, the evaluation of which gives an indication of the relative cleanness of the alloy.

Electron Beam Buttons were melted from the EB melted material as well as material used as starting stock. In the case of INCONEL alloy 718 (Figure 6) the aluminium X-ray maps of the button from the VIM + ESR starting stock material and the button from the VIM + ESR + EBCHR material show that there is an agglomeration of Al-rich particles on both buttons but that the one on the latter is very much smaller. This result corresponds well with reduction in oxygen level shown by the chemical analysis (Table 1). There is no single solid cap of nitride particles on either the button from the starting stock material or that from the EB melted material. Instead there are some very small (mainly <5 um) Ti-rich particles scattered on dendrite arms in the central region of the buttons and once again their number appears to have reduced after EB melting more or less in direct proportion to the reduction in nitrogen content as indicated by chemical analysis (70 to 40 ppm).

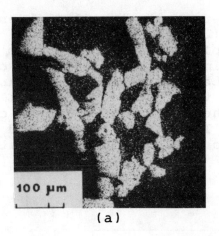 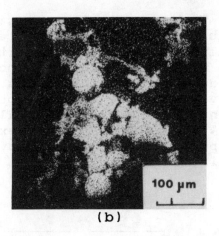

Figure 6 — Al X-ray maps of EB button rafts of (a) start stock and (b) EBCHR INCONEL alloy 718 showing a reduction in oxides after EB remelting.

NIMONIC alloy PK50 also shows an indication of reduction in oxides after EB melting. The button from the VIM + ESR + EBCHR material contained a very small agglomeration of purely Al-rich oxide particles (Figure 7) and absolutely no trace of any other particles anywhere on the button surface. The button from the starting stock material, however, showed several purely Al-rich as well as Al + Mg spinels scattered around the central region of the button. An estimation based on a detailed analysis of both buttons showed that there was at least a four fold decrease in the volume fraction of oxides after EB melting. This is a greater reduction than was anticipated on the basis of the chemical analysis (Table 2).

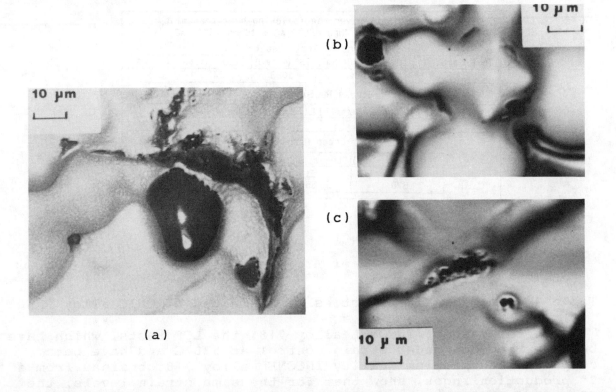

Figure 7 — SEM backscattered micrographs of EB button rafts showing (a) the only oxide particles found on EBCHR NIMONIC alloy PK50 and (b) and (c) two typical regions from start stock NIMONIC alloy PK50 button showing several oxide particles.

The mechanical properties of interest are tensile, stress rupture, creep and low cycle fatigue (LCF). Good tensile results were obtained despite the complete loss of magnesium upon EB melting and the expected detrimental effect of this on the ductility of the alloys. It can be seen from Tables 3 and 4 however that the results obtained are very good and especially so in the case of INCONEL alloy 718 where the R of A figure of 62.7% at 650°C was deemed excellent. The stress rupture and creep results for both alloys are also well within specification requirements.

All samples were given the following heat treatment; 1hr 980°C OQ + 8hrs 720°C FC 50/60°C/hr to 620°C. Hold 8hrs AC.				
Tensile Test results				
Test Temp. (°C)	0.2% PS (N/mm2)	TS (N/mm2)	Elong (%)	R of A (%)
20	1213	1410	22.6	45.8
650	1017	1166	27.7	62.7
Stress Rupture Results				
Test Temp. (°C)	Stress (N/mm2)	Life to Rupture (hrs)	Elong. (%)	
650	760	74	28	

Table 3 - Mechanical Properties of EB melted INCONEL alloy 718

All samples were given the following heat treatment; 4hrs 1020°C OQ + 4hrs 850°C AC + 16hrs 760°C AC				
Tensile Test results				
Test Temp. (°C)	0.2% PS (N/mm2)	TS (N/mm2)	Elong (%)	R of A (%)
535	781	1083	24	26.1
Stress Rupture Results				
Test Temp. (°C)	Stress (N/mm2)	Life to Rupture (hrs)	Elong. (%)	
730	550	35	31	
Creep Results				
Test Temp. (°C)	Stress (N/mm2)	TPS at 100hrs		
670	510	0.04		
670	555	0.05		

Table 4 - Mechanical Properties of EB melted NIMONIC alloy PK50

In the case of INCONEL alloy 718, the LCF tests, which have been carried out under strain control at 538°C and have been compared with premium quality INCONEL alloy 718 obtained from a 9" production ingot, show that for the same strain levels, the EB melted INCONEL alloy 718 performed consistently better (Table 5). Both materials had gone through a similar processing route which involved slab forging on a laboratory scale.

Due to a lack of suitable material being available at the time, samples from a cut-up of a forged disc were used as a reference for the EB melted NIMONIC alloy PK50.

The results show that whereas at the highest strain levels the EB material performed better, at lower strain levels the material from the forged disc gave longer lives (Table 6). These results for the NIMONIC alloy PK50 should perhaps be viewed with the fact in mind that the two materials had not gone through a similar processing route and that the material from the forged disc would have received more controlled work compared to the EB melted material which had been slab forged on a laboratory scale. More work is being carried out in this area to provide comparable data.

Strain %	Cycles to Failure	
	Reference	EB Melted
0.55	21070	42160*
0.75	12350	32250
0.85	5196	13680
0.95	5923	8225
1.02	2004	2173
1.10	1526	3025
* Sample unbroken		

Strain %	Cycles to Failure	
	Reference	EB Melted
0.60	123100*	122900*
0.75	9167	5897
0.85	6175	3697
0.95	7420	4275
1.02	2347	2642
1.10	1195	2106
* Sample unbroken		

Table 5 - LCF results for reference and EB melted INCONEL alloy 718.

Figure 6 - LCF results for reference and EB melted NIMONIC alloy PK50.

Despite the experimental nature of the EB melts, it has been demonstrated that with these two alloys forgeable ingot may be obtained with good structures. Moreover, electron beam control and patterning has been developed such that the controlled solidification of the ingot avoids the deleterious segregation observed in many previous exercises. Chemical and mechanical properties of the EBCHR forged material indicate that the process control in the hearth region may not have been optimum. However, further to the encouraging structure and forging properties observed, reductions in nitrogen and oxygen appear to be reflected in preliminary LCF results. Further work is being carried out to establish the optimum hearth refining conditions to maximise this effect and to apply the processing information thus developed to more defect sensitive alloys, as these are the alloys where EBCHR can be most effective in producing high quality material for rotating components in gas turbine engines.

Conclusions

1. The EBCHR process is a sensitive one, which, under optimum process parameters can be used to produce excellent quality, "clean", Ni-base superalloys.

2. The structures obtained for both INCONEL alloy 718 and NIMONIC alloy Pk50 were very good with no unacceptable macrosegregation.

3. The forgeability of both alloys was remarkably good considering the complete loss of magnesium upon EB melting.

4. There was a reduction in oxygen levels for both alloys. A similar reduction in nitrogen levels was observed for INCONEL alloy 718.

5. These chemical analysis results were reflected in the comparison between EB raft test buttons from starting stock material and the EB melted material.

6. The changes in chemistry had no detrimental effect on the mechanical properties. LCF properties of EB melted INCONEL alloy 718 were better than those of premium quality production material of the alloy.

8. The work reported in this paper was done on a purely experimental basis. The results however are clearly encouraging and form a base for future work.

References

1. J.K. Tien and E.A. Schwarzkopf
"Assessing the Needs for EB Refining of Superalloys".
Electron Beam Melting and Refining, State of the Art, 1983.

2. E.E. Brown, et al.
"The Influence of VIM Crucible Composition, Vacuum Arc Remelting and Electroslag Remelting on the Non metallic Inclusion Content of MERL 76". Fourth International Symposium on Superalloys, 1980.

3. C. d'A Hunt, et al.
"Electron Beam, Cold-Hearth Refining Furnace for the Production of Nickel and Cobalt Base Superalloys."
Electron Beam Melting and Refining, State of the Art, 1983.

4. C.E. Shamblen, S.L. Culp and R.W. Lober.
"Superalloy Cleanliness Evaluation Using the EB Button Melt Test." Electron Beam Melting and Refining, State of the Art, 1983.

THE MAGNESIUM PROBLEM IN SUPERALLOYS

A. Mitchell, M. Hilborn, E. Samuelsson and A. Kanagawa
Dept. of Metals and Materials Engineering
The University of British Columbia
Vancouver, B.C., Canada, V6T 1W5

INTRODUCTION

Magnesium in Superalloys

Magnesium is added to nickel-based superalloys to improve high temperature ductility. This property is important both from the manufacturing point of view, particularly in terms of open-die forgeability, and also from the in-service standpoint where rupture ductility is of primary importance. Further, it has been suggested that magnesium improves the low cycle fatigue strength[1,2].

Vacuum Induction Melting

A typical VIM cycle starts with the charging of either virgin raw materials or scrap. Following the pumping down, the furnace contents are melted. When the metal is entirely molten and the bath stable, desulphurizing or deoxidizing additions are made. Finally, reactive elements are added, and when the bath composition is correct the melt is cast[3].

Additions of aluminum and alkaline earth (AE) or rare earth (RE) metals decrease the activity and content of oxygen during the melting of a virgin heat. The increase of oxygen in the revert heat is probably due to pick-up from the refractories. In both cases, however the AE or RE additions help with the removal of oxygen-containing inclusions by the suggested mechanism of globurizing the alumina clusters[4].

Calcium, magnesium, and rare earth metals are also strong desulphurizing agents. Figure 1 shows that the sulphur content can readily be reduced from ~ 150 ppm to ~ 10 ppm. The sulphide products deposit on the crucible walls. Hence, it is critical that the correct amount of desulphurizing agent can be predicted for each charge[4].

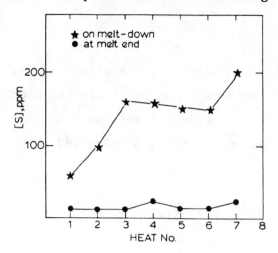

Figure 1: Reduction in melt sulphur content [S] over vacuum induction melting period[4].

Alexander[4] suggests that the removal of magnesium from VIM melts is controlled by liquid mass transfer while Fu et al[5] suggest that the controlling steps are liquid mass transfer and the evaporation reaction. However, the latter were forced, for lack of better data to assume that the activity of magnesium in nickel alloys is equal to the content. Clearly, additional data is required for more accurate analysis of the evaporation process during Vacuum Induction Melting.

Vacuum Arc Remelting

Fu et al[5] investigated the evaporation of magnesiuim from a nickel-base superalloy during VAR. They found that magnesium evaporates mainly from the electrode tip. It was suggested that the evaporation rate is limited by the transport of magnesium through the melt to the surface. However, the analysis of the problem is limited by lack of fundamental data.

Electro Slag Remelting

Ichihashi et al[6] investigated the ability of the ESR process to retain titanium, aluminum, and magnesium in the metal using different slag compositions. They show that increasing the contents of CaO and MgO will increase the content of magnesium in the ingot, as seen in Figure 2. They also suggest that titanium or aluminum can be lost to the slag through following reactions:

$$3MgO + 2\underline{Al} \rightleftarrows Al_2O_3 + 3\underline{Mg}$$
$$2MgO + \underline{Ti} \rightleftarrows TiO_2 + 2\underline{Mg}$$

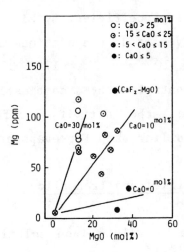

Figure 2: Effect of MgO content in the ESR slag on Mg content in produced ESR ingots[6].

Simultaneously, these reactions would increase the magnesium content in the alloy. It may be noted that proper control of the reactive element content is dependent on a delicate balance between slag and metal chemistry. Ichihasi[6] also found that magnesium can not be retained in the metal unless the slag is protected from air oxidation by an inert atmosphere, such as argon. Generally, the conclusions of Chen et al.[7] agree with those of Ichihashi[6].

Electron Beam Melting

The Electron Beam melting process has recently emerged as another alternative for superalloy refining[8]. The process, usually applied in the form of a Cold-Hearth Refining (EBCHR) furnace, offers more flexibility as to melting speed and ingot shape than the VAR process, since the heat source is independent of the charge material. However, the large area to volume ratio in the EBCHR process makes evaporation of alloying elements unavoidable. Hence, knowledge of the evaporation parameters of alloying elements is important if desired alloy compositions are to be produced.

Mitchell[9] and Herbertson[10] have discussed the evaporation of magnesium from superalloys during EBCHR refining. Mitchell[9] suggests that the evaporation rate is controlled by the evaporation step and gives an evaporation rate constant of 2.5×10^{-7} m s^{-1} at 1700°C. On the other hand, Herbertson[10] claims that the evaporation rate is controlled by bulk mass transfer and gives a Langmuir evaporation rate constant of 1.5×10^{-1} m s^{-1} at 1700°C. It appears that Herbertson[10] has arrived at this value assuming an activity coefficient for magnesium in the liquid superalloy of unity. Using the suggested activity coefficient by Mitchell[9] of 10^{-3} for magnesium in liquid nickel, one arrives at an evaporation rate constant of 1.2×10^{-4} m s^{-1} at 1700°C and 3.6×10^{-4} m s^{-1} at 1700°C[11]. This would mean that magnesium evaporation is controlled by a combination of liquid mass transfer and evaporation. Hence, the activity coefficient is important in determining the rate-controlling step for evaporation during EBCHR refining.

Magnesium and Calcium in Nickel and Iron Alloys

The thermodynamic data of dilute iron solutions are better known than those for nickel solutions[12,13]. Therefore, when data for nickel alloys are lacking, many investigators apply those for iron as a best approximation. While this may be a valid approach for many systems, in the case of magnesium and calcium it is probably far from correct.

While the Mg-Fe phase diagram is incomplete, the binary systems Mg-Fe and Mg-Ni are essentially opposites[14,15]. Nickel-magnesium exhibits complete solubility in the liquid phase and two intermetallic compounds in the solid phase, whereas iron-calcium has an extensive miscibility gap in the liquid phase and no solid intermetallic phases. The situation is similar for the calcium-nickel and calcium-iron systems. The Ca-Fe phase diagram is even more tentative than the Mg-Fe diagram[16], but it is well known that the solubility of calcium in liquid iron is only 0.032 weight percent[17]. As in the nickel-magnesium system, nickel-calcium exhibits several solid intermetallic compounds and complete solubility in the liquid[18].

In terms of activity data one can then expect $\gamma_{Ca}/\gamma_{Mg} > 1$ in iron as in these systems the like atom attractions are stronger than the unlike atom attractions. In contrast, in nickel solutions γ_{Ca}/γ_{Mg} ought to be smaller than 1 as the intermetallic compounds indicate unlike atom attraction. This is also confirmed by Meysson and Rist[19] for nickel and iron-rich solutions containing calcium.

de Barbadillo[20] also discusses the influence of other alloying elements on the solubility of magnesium in nickel alloys. For instance, iron and chromium decreases the magnesium solubility. While insufficient data are available de Barbadillo manages to propose a tentative ternary liquid phase diagram for the Fe-Ni-Mg system. Figure 3 reproduces estimated activity data based on the limited data for the binary Ni-Mg system[21] and data for the system Ni-Fe-Mg[22].

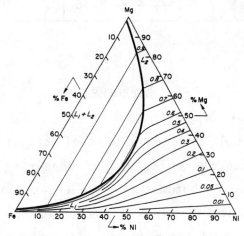

Figure 3: Estimated activity of magnesium in the Fe-Ni-Mg system[20].

Magnesium and Calcium

No experimental study of the deoxidation equilibria of Mg and Ca in nickel-base alloys has been found in the literature. Some work has been done on these elements in various iron-based solutions[23-26]. While these data may be useful in establishing deoxidation equilibria in nickel-base alloys the different solubility of Mg/Ca in the two solvents cannot be forgotten and must be taken into account in using the data of Tables I & II.

TABLE I: Deoxidation Equilibria in Iron Alloys

Deoxidation Reaction	log K	K	T (°C)	Ref.
$2Al + 3O \rightleftarrows Al_2O_3$ (s)		1.8×10^{13}	1600	28
		7.9×10^{13}	1600	27
		1.7×10^{13}	1600	23
$Ca + O \rightleftarrows CaO$		6.2×10^{5}	1600	23
		1.1×10^{6}	1600	24,25
$Mg + O \rightleftarrows MgO$		5.5×10^{5}	1600	24,25

TABLE II: Deoxidation Equilibria in Nicel Alloys

Deoxidation Reaction	log K	K	T (°C)	Ref.
$2Al + 3O \rightleftarrows Al_2O_3$ (s)	$\frac{60795}{T} - 18.81$	4.48×10^{13}	1600	13
$Ca + O \rightleftarrows CaO$ (s)	$\frac{27959}{T} - 6.59$	2.16×10^8	1600	13
$Mg + O \rightleftarrows MgO$ (s)	$\frac{26009}{T} - 7.38$	3.2×10^6	1600	13

EXPERIMENTAL

The technique used to determine the activity coefficient of Mg in nickel-base alloys was the novel one of measuring vapour pressure with an atomic-absorption spectrophotometer. The experimental details are given in reference 29. The system was calibrated using both pure magnesium and Mg + Sn alloys. The equilibrium constant for the reaction:

$$Mg(g) \rightleftarrows [Mg]_{Ni}$$

was determined as:

$$\log K = 2.7 \text{ at } 1450°C$$

The Raoultian activity coefficient of Mg at 1450°C was found to be 1.2 (± 0.2) $\times 10^{-2}$, with $e_O^{Mg} = -28$. The addition of Al, Fe and Cr did not substantially change these values in the concentration ranges up to 50% Ni, as shown in Figure 4.

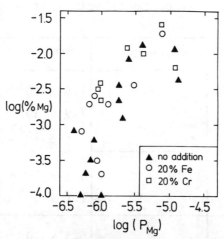

Figure 4: Comparison between data for alloys containing no major additions and alloys containing 20% Fe or 20% Cr.

DISCUSSION

Using the data given above we may compute the losses of Mg giving various process steps and compare these with the practically observed values. Since $[Mg]_{Ni}$ must also include that in equilibrium with MgS and MgO, we are able at the same time to estimate the levels of Mg in the three forms $[Mg]_{Ni}$, MgS and MgO at the various process stages.

The solubility product of MgO is extremely small, and in a system contained in pure MgO, we can assume that all oxygen is present as solid MgO. Therefore the maximum Mg present in this form is aprpoximately 10 ppm for a typical superalloy. In a VIM melt with an addition of 150 ppm Mg, and 10 ppm S, almost all the magnesium is $[Mg]_{Ni}$.

A comparison of Langmuir evaporation rates and bulk diffusion coefficients in VIM indicates that the latter controls the evaporation process. A surface depletion rate of 1.5×10^{-5} g cm^{2}s^{-1} is estimated for the melting conditions of a 5 tonne VIM melt of Inconel 718, equivalent to a bulk loss of 5 ppm Mg/min. This value accounts well for the "Mg-fade" observed during melting and casting.

Using the thermodynamic data generated above, we can also calculate the expected Mg contents of Inconel 718 melted by ESR, through a slag containing MgO, as shown in Figure 5. These values also compare well with observed practice and indicate that after ESR all the ingot Mg is as MgO or MgS.

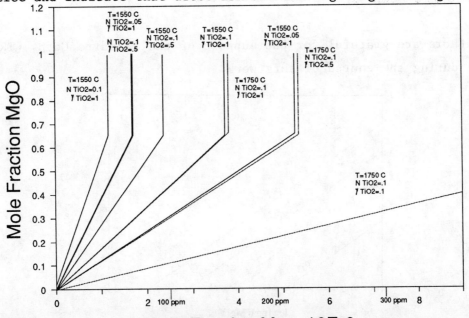

Figure 5: Calculated Equilibria for:
$2[Mg] + (TiO_2) \rightleftarrows 2MgO + [Ti]$ for ESR slags based on $40CaF_2 : 30CaO : 30Al_2O_3$ and alloy Inconel 718

The loss-rate of Mg from the same alloy melted by VAR may also be calculated. In an ingot of 500 mm φ, melted under typical industrial conditions, the loss rate is 100 ppm Mg/sec which indicates that all $[Mg]_{Ni}$ is removed by VAR, leaving only the equilibration of Mg from MgS and MgO as the ingot Mg content.

The solubility product of MgS is of considerable interest, as indicated by several workers, because of its impact on mechanical properties. Secondary precipitation of MgS has not yet been observed in Inconel 718, but the effect of Mg on mechanical properties must be through the elimination of grain-boundry S as MgS particles. The MgS carried in the liquid alloy system dissociates (Figure 6), but at room temperature it is likely tht all of the magnesium present in a VAR alloy is in the form of MgS or MgO. The same is true of EB-melted material. In view of the dissociation behaviour of MgS, and the high capacity for Mg removal in both EB and VAR, it is essential that the alloys processed this way should have a very low sulphur content. It appears that the dissociation of MgS will not permit the retention of more than 10 ppm sulphur as MgS at any content of Mg in Inconel 718 which is compatible with VAR or EB melting. It is to be noted that an intermediate de-sulphurizing step, such as ESR, will make this requirement possible. We also note that at a sufficiently-low sulphur content (<5ppm), adequate intermediate-temperature ductility may be obtained with magnesium contents at the residual levels of VAR or EB.

ACKNOWLEDGEMENTS

The authors are grateful for the support of the Teledyne Corp. (TRAP Programme) during the course of this work.

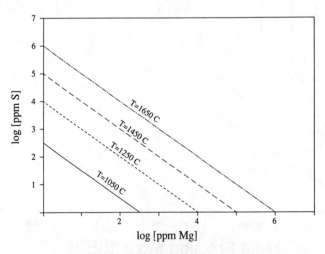

Figure 6: Reaction Equilibria for:
$[Mg] + [S] = MgS(s)$ in Inconel 718

REFERENCES

1. C.T. Sims, W.C. Hagel, eds.: "The Superalloys", John Wiley & Sons Inc., 1972.

2. J.M. Moyer: Superalloys 1984, Prof. Fifth Int. Symposium on Superalloys, eds.: M. Gell et al., 1984, pp. 443-454.

3. R.S. Cremiso: "Melting" in "The Superalloys", eds.: Sims et al., John Wiley & Sons Inc., 1972, pp. 373-401.

4. J. Alexander: Material Science and Technology, Feb. 1985, V.1, pp. 167-170.

5. J. Fu, H. Wang, D. Wang, E.P. Chen: Proc. 7th ICVM, 1982, Tokyo, Japan.

6. H. Ichihashi, R. Baba, T. Ikeda: Proc. Vacuum Metallurgy Conf. 1984, pp. 75-82, Published by the Iron and Steel Soc., Eds.: G.K. Bhat, L.W. Lherbiev.

7. C.X. Chen, R.F. Gao, W.X. Zhao: 8th Intl. Conf. on Vacuum Metl., Linz, Austria, Sept. 30 - Oct. 4, 1985, V.2, pp. 1046-1052.

8. C. D'A. Hunt, J.H.C. Lowe, S.K. Harrington: Electron Beam Melting and Refining, Conf. Proc. Nov. 4-6, '85, Part I, pp. 58-70, Ed. R. Bakish.

9. A. Mitchell, K. Tagaki: Proc. Vacuum Metallurgy Conf. 1984, pp. 55-59, Published by the Iron and Steel Soc., Eds.: G.K. Bhat, L.W. Lherbier.

10. J. Herbertson: Conf. Proc.: Electron Beam Melting and Refining-State of the Art, 1986, pp. 19-29; Ed.: R. Bakish.

11. G.H. Geiger, D.R. Poirier: "Transport Phenomena in Metallurgy", Addison Wesley, 1973, pp. 560-567.

12. G.K. Sigworth, J.F. Elliott: Metal. Sci., V. 8, 1974, pp. 298-310.

13. G.K. Sigworth, J.F. Elliott, G. Vaughn, G.H. Geiger: Metallurgical Society of CIM, Annual volume featuring Molybdenum, 1977, pp. 104-110.

14. A.A. Nayeb-Hashemi, J.B. Clark, L.J. Swartzendruber: Binary Alloy Phase Diagrams, ed. T.B. Massalski, Am. Soc. Metl., 1986, p. 1076.

15. A.A. Nayeb-Hashemi, J.B. Clark: ibid, p. 1529.

16. W.G. Moffatt: "Handbook of Binary Phase Diagrams", Genium Publ. Corp., 1984.

17. D.L. Sponseller, R.A. Flinn: Trans. Met. Soc. AIME, V. 230, 1964, pp. 876-888.

18. W.G. Moffatt: ibid ref. 57, p. 628.

19. N. Meysson, A. Rist: Revue de Metallurgie, V. 62, 1965, pp. 1127-31.

20. J.J. de Barbadillo: "Magnesium and Calcium Treatment of Nickel-Base Alloys", Presented at American Vacuum Society Conf. 1975 (Available from the author at Inco Alloys Int., Huntington, W. VA).

21. N.G. Schmahl, P. Sieben: "The Physical Chemistry of Metallic Solutions and Intermetallic Compounds", Chemical Publishing Co. Inc., New York, 1960, pp. 268-290.

22. P.K. Trojan, R.A. Flinn: Trans. Am. Soc. Met., V. 54, 1961, pp.. 549-566.

23. S. Gustafsson, P-O. Mellberg: Scand. J. of Met., V. 9, 1980, pp. 111-116.

24. M. Joyant, C. Gatellier: "Influence d'une addition de calcium ou de magnesium sur la solubilite de l'oxygene et du soufre dans l'acier liquid", IRSID PCM-RE 1108, May 1984.

25. M. Nadif, C. Gatellier: ibid, PCM-RE 1108 Bis, June 1985.

26. N.A. Voronova: "Desulphurization of Hot Metal by Magnesium", Int. Magnesium Ass. and AIME, 1983, pp. 67-102.

27. A. McLean, H.B. Bell: J. Iron Steel Inst., Feb. 1965, pp. 123-130.

28. R.J. Fruehan: Met. Trans., V. 1, Dec. 1970, pp. 3403-3410.

29. Samuelsson, E., Magnesium in Liquid Nickel Solutions, Ph.D. Thesis, University of British Columbia, 1988.

LIQUID METAL TREATMENTS TO REDUCE MICROPOROSITY

IN VACUUM CAST NICKEL BASED SUPERALLOYS

R.E. Painter and J.M. Young

Department of Metallurgy and Materials,
University of Birmingham, Birmingham, B15 2TT, UK

Synopsis

Results from the present work suggest that additions of nucleants during remelting for vacuum casting can reduce microporosity levels. The mechanism controlling porosity reduction is consistent with reducing dissolved gas levels. The treatments involving additions of titanium nitride and alumina are somewhat radical as they lead to the formation of additional phases not originally designed for the alloy. The effect of these phases on mechanical properties was not ascertained. Argon purging was shown to reduce microporosity levels without the need to form deleterious phases and is suggested as a suitable mechanism for reducing microporosity levels.

Introduction

Cast components often contain porosity which is responsible for a degradation in properties compared to wrought components (assuming no other difference in microstructure). The porosity may be classified as gross shrinkage porosity or microporosity. Whilst the former is invariably localised and can be controlled by correct design of mould feeding and gating systems, control of the more evenly dispersed microporosity is difficult.

Vacuum melting, whilst capable of producing metal with a low impurity content will not eliminate microporosity and in some circumstances could increase levels (e.g., by increased gas evolution). The high strength nickel based superalloys are vacuum cast to produce many components using the investment casting technique. Freedom from microporosity in cast components is not always achieved and limits on microporosity levels are normally included in specifications covering manufacture. Hot isostatic pressing is also routinely used on some components to ensure product integrity.

The reasons for microporosity formation in vacuum cast superalloys are not clear, although melt cleanliness is known to effect the levels. Increased amounts of nitrogen have been shown to give an increased incidence of microporosity in MAR M002 (1) and IN939 (2) whilst additions of silicon reduced microporosity levels in MAR M002 (1). Recycling of foundry scrap (reverting) has also been shown to increase microporosity which has been associated with an increase in nitrogen levels (3).

Control of microporosity in revert melts is normally achieved by dilution of the scrap with virgin alloy. Modification of the composition during remelting is not considered feasible owing to the absence of a carbon boil that is responsible for refining in primary vacuum induction melting to produce bar stock. The object of the present work was to investigate the effects of liquid metal treatments aimed at promoting gas removal during remelting and thereby reduce microporosity levels.

Experimental Details

All melts were produced from IN100 bar stock of the composition shown in Table I. Billets measuring 35 mm in diameter and 90 mm in length were

Table I. Composition of IN100 Used for Melting

Element	Ni	Co	Cr	Ti	Al	Mo	V	C
Weight %	bal	14.8	9.59	4.9	5.62	2.97	0.99	0.18

Element	Mg	Si	Fe	S	B	Zr	O	N
ppm	17	<500	600	40	120	400	15	10

melted in a laboratory induction furnace under vacuum (10^{-4} mbar) using high purity recrystallised alumina crucibles. A new crucible was used for each melt to prevent contamination from previous melts.

Once molten the melt was held for 5 minutes at 1500°C after which the required addition was made. The melt was then held for a further 5 minutes before pouring into heated ceramic moulds at 1000°C. Moulds were made by the investment technique using a zirconium silicate based slurry with zircon stucco for the face and secondary coats, and graded molochite for subsequent back-up coats. For the cast specimen a simple tapered cylindrical shape (carrot test sample outlined in British Standard HC100) was chosen in order to obtain good feeding. The dimensions of the carrots were 22 mm diameter over a length of 28 mm then tapering to 14 mm diameter to give an overall length of 75 mm. Two vertically orientated carrots that were joined at the top by a square feeder incorporating a pouring cup constituted a mould. Heating of the mould was carried out in situ using a resistance heated ceramic furnace manufactured in a similar manner to the moulds.

Initial melting concentrated on the effects of nitrogen additions in the form of high nitrogen chromium (designated as Cr(N)) and titanium nitride (TiN). The castings produced were as follows:

(1) base material with no addition;
(2) base + 300 ppm N as TiN;
(3) base + 1000 ppm N as TiN;
(4) base + 100 ppm N as Cr(N);
(5) base + 300 ppm N as Cr(N).

During solidification the furnace pressure was increased to 800 mbar by admitting high purity argon. The castings were allowed to cool to around 700°C before removing them from the furnace.

Subsequent melting was aimed at reducing any microporosity brought about from the previous melts using the following treatments:

(1) base material with no addition;
(2) base + 300 ppm N as Cr(N);
(3) base + 300 ppm N as Cr(N) + 1000 ppm N as TiN;
(4) base + 300 ppm N as Cr(N) + argon bubbling;
(5) base + 300 ppm N as Cr(N) + 300 ppm O as Al_2O_3.

Immediately after pouring the furnace pressure was increased to 1013 mbar by admitting high purity argon. Castings were allowed to cool to 700°C before removing them from the furnace.

Microstructural evaluation was carried out on transverse sections taken from identical positions in all castings. Quantitative metallography was carried out on unetched polished sections (0.25 µm finish) using the Kontron SEMIPS system. This allowed characterisation of porosity and carbides in terms of area fractions, shapes and sizes, along with statistical testing for significant differences in the distribution of phases. A significance level of 0.95 was used with two tests that do not assume a normal distribution of data (Kolmogoroff-Smirnoff and Mann-Whitney U tests [4]). Interference film metallography was used to identify compositional changes within the primary carbide phase as described elsewhere [5].

Results and Discussion

Results for the initial melts are shown in Fig. 1 from which it is apparent that microporosity increased with increasing additions of high nitrogen chromium, but decreased with increasing amounts of TiN. Statistical analysis of the results found significant differences between all porosity levels. The microporosity has been expressed as a normalised value to the base cast which is >1 for an increase and <1 for a decrease in microporosity. The results are interesting as they suggest microporosity is sensitive to the form of nitrogen in the melt rather than the total amount present. High nitrogen levels will not necessarily result in an increase in microporosity.

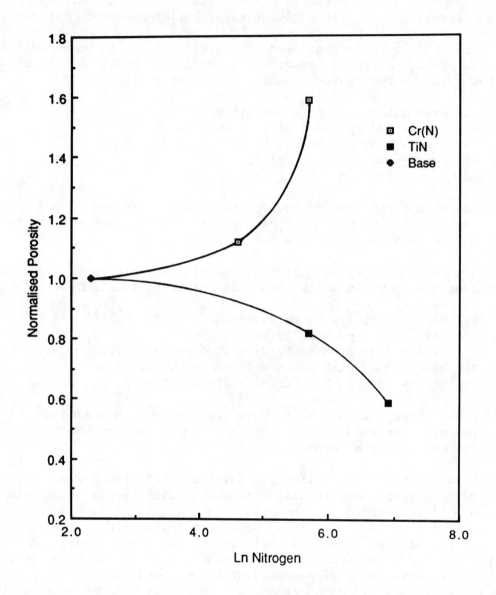

Fig 1 Effect of nitrogen additions on microporosity levels for IN100. Note that porosity levels are normalised to the base cast. Levels greater than 1 indicate an increase in porosity compared to the base cast.

Both additions resulted in the appearance of primary MC carbides containing a second phase. Interference film microscopy suggested the second phase to be based on Ti(C,N) or TiN. Laser ionised mass analysis confirmed the presence of TiN and Ti(C,N) in the high nitrogen chromium castings (although Ti(C,N) was by far the dominant phase), whilst the predominant phase was TiN in casts having TiN additions. This suggests that TiN was stable in the melt whereas the high nitrogen chromium increased the dissolved nitrogen content of the melt to promote TiN and Ti(C,N) formation during solidification.

Average grain sizes were also reduced for both additions as shown in Table II, although as with microporosity, the amount of refinement was dependent on the form of addition rather than the total amount of nitrogen in the melt. Additions of nitrogen in the form of TiN gave a reduced grain size compared to an identical addition in the form of high nitrogen chromium.

Table II. Effect of Nitrogen Additions on Grain Size in IN100 Determined by the Line Intercept Method

Melt addition	Amount of N addition, ppm	Grains mm^{-1}
Base melt	None	7.0
Cr(N)	100	6.4
Cr(N)	300	5.8
TiN	300	4.7
TiN	1000	3.2

As the amount of grain refinement appears to be influenced by the presence of TiN, the indications are that TiN survived in the melt to act as a nucleant. Additions of high nitrogen chromium also resulted in some TiN formation, although the dominant second phase was Ti(C,N). Hence, the reduced amount of TiN resulted in some refinement of the grain size compared to the base cast, but was limited by the necessity to form a nucleant during solidification.

Previous work has associated the presence of TiN to an increase in microporosity (6) which appears to contradict the present results for additions of TiN. However, TiN was also identified in casts having high nitrogen chromium additions, albeit in a limited amount. This suggests that the presence of TiN alone is insufficient to predict whether the melt is prone to microporosity. If the TiN was present in the melt before pouring it may even be beneficial in reducing microporosity. Alternatively, if it forms during solidification, then apart from promoting some grain refinement, it will not assist in reducing microporosity.

Results for subsequent melts having treatments aimed at reducing microporosity are shown in Fig. 2. Once again, as the nitrogen level of the melt was increased by additions of high nitrogen chromium, microporosity in the cast sample increased. Having previously observed that additions of TiN reduced microporosity levels in the base cast, an addition of TiN to a high nitrogen melt was made. The effect was an overall decrease in microporosity thereby supporting the results for the initial melts. Microporosity was also decreased by bubbling argon through the melt whilst additions of alumina, although, only giving a marginal decrease in microporosity, did reduce levels significantly when compared to the cast containing a nitrogen addition but no alumina.

The effect of pressure during solidification was also investigated. Increasing the pressure from 800 mbar to 1013 mbar resulted in a decrease in microporosity for both the base cast and melts having nitrogen additions in

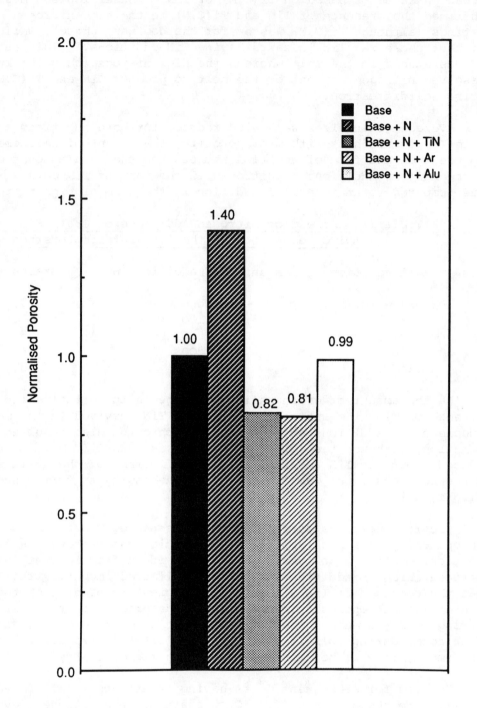

Fig 2 Effect of melt treatments on microporosity levels for IN100. Note that porosity levels are normalised to the base cast. Levels greater than 1 indicate an increase in porosity compared to the base cast.

the form of high nitrogen chromium as shown in Table III.

Table III. Effect of Increasing Pressure during Solidification on Microporosity levels in IN100

Melt addition	Amount of N addition, ppm	Pressure, mbar	Total area fraction porosity, %
Base melt	None	800	0.295
Base melt	None	1013	0.176
Cr(N)	300	800	0.468
Cr(N)	300	1013	0.247

The mechanism responsible for reducing microporosity is not clear, as indeed the mechanism governing microporosity formation in the base cast or virgin alloy is similarly not clear. Several hypothesis have been proposed to explain the increase in microporosity in revert casts, that presumably may also be applied to microporosity in virgin alloy. By definition they all suggest microporosity is caused by insufficient liquid metal to compensate for the volume reduction accompanying solidification. For example, Bachelet and Lesoult (7) proposed that liquid metal is prevented from feeding the solidification front by carbides blocking interdendritic channels. Changes in primary MC carbide shape from Chinese script to blocky (that is characteristic of revert casts) were thought to assist in promoting microporosity. Rupp et al (8) suggested a difference in the volume fraction solid as a function of temperature was responsible for differences in revert and virgin IN100. For a given temperature below the liquidus, revert alloys had less volume fraction liquid and hence a reduced permeability of the solidifying structure. Assuming fluidity of the metal was governed by temperature alone, at a given temperature the decreased permeability in the revert alloy would inhibit feeding when compared to the virgin alloy. Measurements of solidification enthalpy for virgin and revert have been made by Lecomte-Beckers and Lamberigts (9). Results indicated that revert casts of IN100 had a lower solidification enthalpy which, for a given fraction solid, would amount to a reduced fluidity of the remaining liquid. An increase in microporosity levels was attributed to a reduction in viscosity due to a reduced enthalpy that limited feeding.

All of the proposed mechanisms are concerned with apparent differences in behaviour of revert and virgin alloys that are related to changes in chemical composition. Changes in carbide morphology, volume fraction liquid/solid or solidification enthalpy are all assumed to result as nitrogen and oxygen contents increased, although no direct measurements of viscosity or permeability have been made. The major difference in composition of revert alloys is an increase in nitrogen and oxygen levels. An alternative mechanism based on the rejection of dissolved gas during solidification has not been considered, even though microporosity increases with increasing gas levels.

The results from the present work suggest that rejection of gas during solidification could be responsible for microporosity in superalloys. Increasing the dissolved gas content of the melt by additions of high nitrogen chromium resulted in an increase in microporosity. The increase could be reduced by increasing the pressure during solidification and this is a well known ploy that is often practised industrially. Indeed, the time taken before the vacuum is released in the mould chamber is often due to mechanical restrictions rather than metallurgical considerations. A limiting factor which could govern the ability to form gas bubbles is the amount of dissolved gas and the availability of a suitable substrate for nucleation. Adding high nitrogen chromium would have increased the amount of dissolved nitrogen. During solidification some of the nitrogen would combine with titanium and carbon to form TiN and Ti(C,N) which could act as nucleation sites for gas bubble formation. As the metal would be semi-solid, escape of the gas would be hindered and bubbles would become entrapped. Adding a suitable nucleant to the melt before casting would allow gas bubble formation and give sufficient time for the bubbles to leave the melt. The latter situation is thought to prevail for additions of TiN that effectively reduced the amount of microporosity in the base cast as well as for casts having additions of high nitrogen chromium. Even though a suitable nucleant for gas bubble nucleation during solidification was present in the form of TiN, gas rejection would be limited as the amount of dissolved gas was minimised, even though the overall nitrogen content of the casting would be significantly greater than in the base cast.

It is also known that high chromium alloys (e.g., IN939) are less prone to microporosity than low chromium alloys (e.g., IN100). As chromium content increases the solubility of nitrogen would be expected to increase and hence the tendency to form microporosity by rejection of dissolved gas would decrease.

Other additions or melt treatments promoting gas bubble formation would similarly be expected to reduce microporosity and this was also observed. Argon bubbling and additions of alumina both reduced microporosity in melts having nitrogen additions as high nitrogen chromium. Argon bubbling under vacuum is a well known method of reducing gas levels during steelmaking to below the levels obtained after the carbon boil during primary refining (10). The argon bubbles act as nucleation sites for further gas bubble nucleation or diffusion of dissolved gas to pre-existing argon bubbles that allows transport from the melt. Additions of alumina were thought to increase the dissolved oxygen content of the melt and promote a carbon boil as alumina is unstable in liquid superalloys under vacuum. The formation of gas bubbles resulting from the boil would facilitate removal of nitrogen similar to nitrogen reduction in primary refining.

Conclusions

Microporosity formation in nickel based superalloys has been shown to be influenced by nitrogen additions to the melt. Additions of high nitrogen chromium increased microporosity whilst additions of TiN reduced microporosity. The results indicate that microporosity is not a function of total nitrogen content, but the form in which the nitrogen is present. Adding TiN to melts having high nitrogen chromium additions also reduced microporosity levels indicating that melt treatments to reduce microporosity were possible during secondary melting.

The effect of these phases on mechanical properties was not ascertained. As microporosity levels were reduced then mechanical properties would be expected to increase. However, whether the presence of, for example, titanium nitride would deleteriously effect properties is uncertain. Likewise, melt treatments to promote gas bubble formation also resulted in reduced microporosity levels, while increasing the presence during solidification similarly reduced microporosity. Argon purging was shown to reduce microporosity levels without the need to form deleterious phases and is suggested as a suitable mechanism for reducing microporosity levels. Reductions in microporosity are consistent with microporosity formation being caused by rejection of dissolved gas during solidification.

References

1. G.L.R. Durbar, S. Osgerby and P.N. Quested, Metals Technology, 1984, 11. 129.

2. R.E. Painter and J.M. Young, "The Role of Nitrogen on Microstructure of Recycled Nickel-Based Superalloys", 242, Proc. of the 25th Annual Conf. of Metallurgists-1986, Nickel Metallurgy, Vol. II, Toronto, Ed. by E. Ozbek and S.W. Marcuson, Canadian Institute of Mining and Metallurgy Montreal, 1986.

3. D.A. Ford, P.R. Hooper and P.A. Jennings, "Foundry Performance of Reverted Alloys for Turbine Blades", 51, Proc. of the Conf. on High Temperature Alloys for Gas Turbines and Other Applications 1986, Liege, 1986, Ed. W. Betz et al, D. Reidal Publ. Co., Dordrect, 1986.

4. L. Sachs, "Applied Statistics - A Handbook of Techniques", 291, Springer-Verlag, New York, 1982.

5. P.N. Quested and K.W. Raine, Metals and Mat., 1981, October, 37.

6. M. Lamberigts, J. Lecomte-Beckers and J.M. Drapier, "Reverting Foundry Nickel-Base Superalloys", Proc. of Conf. on High Temperature Alloys for Gas Turbines and Other Applications 1986, Liege, 1986, 777, Ed. W. Betz et al, D. Reidal Publ. Co., Dordrect, 1986.

7. E. Bachelet and G. Lesoult, "Quality of Castings of Superalloys", 665, Proc. of the Conf. on High Temperature Alloys for Gas Turbines, Liege, 1978, Ed. D. Coutsouradis et al, Applied Science Publishers Ltd., London, 1978.

8. S. Rupp, J. Massol and Y. Bienvenu, "Effect du Recyclage sue les Structures de Fonderie et la Santa Interne de L'Alliage IN100; Etude de Laboratoire", 20-1, Proc. of the AGARD Conf. Nos. 356 on Materials Substitution and Recycling, Vimerio, Portugal, 1983, AGARD, 1984.

9. J. Lecomte-Beckers and M. Lamberigts, "Microporosity Formation in Nickel Based Superalloys in Relation with their Solidification Sequence, 745, Proc. of the Conf. on High Temperature Alloys for Gas Turbines and Other Applications 1986, Liege, 1986, Ed. W. Betz et al, D. Reidal Publ., Co., Dordrect, 1986.

10. "Fundamentals of Metallurgical Processes", 297, Ed. L. Coudurier et al, Pergamon Press, Oxford, 1985.

COMPOSITIONAL CONTROL AND OXIDE INCLUSION LEVEL COMPARISON OF PYROMET®718

AND A-286 INGOTS ELECTROSLAG REMELTED UNDER AIR VS. ARGON ATMOSPHERE

D. D. Wegman

Process Research and Development Group
Carpenter Technology Corporation
P.O. Box 14662, Reading PA, 19612-4662 USA

Summary

The effect of electroslag remelting (ESR) Pyromet 718 and A-286 alloys under an argon atmosphere and its resulting effect on ingot compositional control and oxide inclusion level were investigated using a laboratory consumable electrode furnace. Chemical composition data indicate that an argon cover during ESR rendered the slag a closed system where oxidation/reduction reactions were interdependent. This result provides potential for improved compositional control and lower ingot oxygen levels compared to air-ESR. Reasonable correlation between ingot oxygen content and volume percent inclusion level was attained and data show that low oxygen argon-remelted ingots exhibited a lower inclusion content than air-remelted ingots of higher oxygen content. These preliminary results demonstrate the improved process control of slag reactions provided by argon-ESR has the potential to improve the compositional control and "cleanliness" of superalloy ingots and consequently their mechanical properties.

Introduction

Melting processes that can provide precise compositional control, superior "cleanliness," and control of solidification structure are of great interest as pressure mounts to improve the mechanical property performance of superalloys in critical applications. Electroslag remelting (ESR) is a common refining method, but can encounter difficulty in controlling ingot composition of superalloys containing high levels of reactive elements (Ti,Al) due to molten metal/slag and slag/air chemical reactions. In addition, if proper slag deoxidation practices are not followed, ingot oxygen levels can increase. The use of an argon cover over the slag bath during ESR to improve process control over slag reactions was investigated and the results are presented.

Experimental Procedure

Six inch (150 cm) rd electrodes of Pyromet 718 and A-286 were electroslag remelted into eight inch (200 cm) rd ingots in Carpenter Technology's laboratory consumable electrode furnace. For comparison, melting trials were conducted under both air and argon atmospheres. During argon remelting, the exit gas v/o oxygen content from the melting chamber was maintained at less than 2%. Pyromet 718 ingots were remelted at $\sim$215 lb/h using an "A" slag of 36% CaF_2, 29% CaO, 3% MgO, and 32% Al_2O_3. Pyromet A-286 ingots were remelted at $\sim$150 lb/h using a "B" slag of 32% CaF_2, 30% CaO, 3% MgO, 34% Al_2O_3, and 1% SiO_2. Melting parameters for 718 ingots (1, 2, 3) and A-286 ingots (4, 5, and 6) were:

Ingot 1 - air atmosphere + "A" slag + 6% TiO_2 addition + 20g Al/10 min deoxidation treatment
Ingot 2 - argon atmosphere + "A" slag
Ingot 3 - argon atmosphere + "A" slag + 3% TiO_2 addition
Ingot 4 - air atmosphere + "B" slag + 3% TiO_2 addition
Ingot 5 - argon atmosphere + "B" slag
Ingot 6 - argon atmosphere + "B" slag + 6% TiO_2 addition.

Electrode composition was determined at top and bottom locations prior to remelting while remelted ingot compositions were determined from A, B, C and X locations as diagrammed in Figure 1. Oxygen analyses were determined via a combustion infrared detection method with a precision of $\pm$ 5 ppm at levels below 40 ppm and $\pm$ 10 ppm at levels above 40 ppm. Sulfur levels were measured with a $\pm$ 1-2 ppm level of precision.

Metallographic specimens adjacent to compositional analysis specimens from the center of the A, B, C, and X disc locations of each remelted Pyromet 718 and A-286 ingots were removed to measure oxide inclusion content. Specimens were hot isostatically pressed (HIP'd) at 2125°F/3h/15ksi to eliminate shrinkage porosity from solidification. Each sample was automatically polished and rated for oxide inclusion content on three parallel planes using a Leitz TAS quantitative image analysis microscope. An average of 200 fields were rated to generate each data point. Image analysis used a screen magnification of 2620x and determined volume percent oxide content, average particle diameter, and average particle area. Confidence limits of $\pm$95% also were determined by the microscope for each data point. The mean ($\bar{x}$) and standard deviation (s) for the inclusion content of each ingot were calculated and levels of significant difference between mean volume percent inclusion levels of the respective ingots determined. Linear correlation coefficients (r) between volume percent inclusion content and oxygen content were also calculated. SEM analysis was performed on randomly selected primary carbides containing

TABLE I

SELECTED PYROMET 718 AND A-286 ELECTRODE AND INGOT ELEMENTAL COMPOSITIONS

Pyromet 718

Ingot 1 (Air + 6% TiO_2 + Al Deoxidant Addition)

Element	Electrode X	Electrode A	Ingot X	Ingot C	Ingot B	Ingot A
S (PPM)	10	20	7	7	7	7
Ti	0.94	1.00	0.94	0.98	1.00	1.07
Al	0.58	0.62	0.45	0.50	0.50	0.48
O (PPM)	<5	<5	33	29	38	46

Ingot 2 (Argon)

Element	Electrode X	Electrode A	Ingot X	Ingot C	Ingot B	Ingot A
S (PPM)	10	40	<5	<5	5	6
Ti	1.02	1.00	0.81	0.87	0.95	1.07
Al	0.56	0.58	0.65	0.63	0.60	0.51
O (PPM)	<5	<5	13	20	24	8

Ingot 3 (Argon + 6% TiO_2)

Element	Electrode X	Electrode A	Ingot X	Ingot C	Ingot B	Ingot A
S (PPM)	10	10	7	8	9	8
Ti	1.01	1.02	0.94	0.98	0.97	1.08
Al	0.53	0.54	0.54	0.56	0.57	0.49
O (PPM)	22	<5	12	15	20	<5

All Pyromet 718 ingots remelted utilizing a 36% CaF_2, 29% CaO, 3% MgO, 32% Al_2O_3 slag.

Pyromet A-286

Ingot 4 (Air + 3% TiO_2)

Element	Electrode X	Electrode A	Ingot X	Ingot C	Ingot B	Ingot A
Si	0.22	0.22	0.27	0.23	0.22	0.23
S (PPM)	<10	<10	<5	<5	<5	<5
Ti	2.02	2.15	1.81	1.62	1.78	2.02
Al	0.23	0.23	0.29	0.30	0.22	0.25
O (PPM)	11	<5	43	35	34	22

Ingot 5 (Argon)

Element	Electrode X	Electrode A	Ingot X	Ingot C	Ingot B	Ingot A
Si	0.28	0.22	0.29	0.26	0.23	0.23
S (PPM)	<10	<10	<5	<5	<5	<5
Ti	2.01	2.06	1.67	1.81	1.95	2.19
Al	0.28	0.31	0.50	0.46	0.42	0.27
O (PPM)	<5	<5	7	9	10	5

Ingot 6 (Argon + 6% TiO_2)

Element	Electrode X	Electrode A	Ingot X	Ingot C	Ingot B	Ingot A
Si	0.18	0.19	0.31	0.24	0.23	0.23
S (PPM)	NA	NA	<5	<5	8	<5
Ti	2.02	2.13	1.87	2.03	2.08	2.24
Al	0.35	0.34	0.41	0.35	0.36	0.28
O (PPM)	NA	NA	<5	5	10	<5

All Pyromet A-286 ingots remelted utilizing a 32% CaF_2, 30% CaO, 3% MgO, 34% Al_2O_3, 1% SiO_2 slag.

All values are weight percent
NA - Not analyzed

TABLE II

PYROMET 718 AND A-286 OXIDE INCLUSION DIAMETER AND AREA DATA

Ingot	Diameter (μm) Mean ($\bar{X}$)	Diameter (μm) Std. Dev. (s)	Area (μm^2) Mean ($\bar{X}$)	Area (μm^2) Std. Dev. (s)
1	0.978	0.138	0.949	0.398
2	0.945	0.154	0.866	0.267
3	1.001	0.161	1.107	0.384
4	0.736	0.128	0.508	0.189
5	0.569	0.056	0.364	0.205
6	0.690	0.113	0.482	0.171

Figure 1 - Ingot Disc Locations Sampled for Argon and Air ESR Study

inclusions from the 3-C and 6-A ingot locations. Visual metallographic inspection of the specimens was performed and electron microprobe analysis performed on three isolated inclusions observed at the 3-X ingot location.

Experimental Results

The nominal analysis (w/o) of the VIM Pyromet 718 and arc-AOD A-286 electrode material used in this study was as follows:

	C	Mn	Si	P	S	Cr	Ni	Mo	V	Cb	Ti	Al	B	Fe
Pyromet 718	0.050	0.10	0.13	0.010	0.001	18.50	52.50	3.10	----	5.30	1.00	0.60	0.004	Bal.
Pyromet A-286	0.030	0.35	0.22	0.015	0.001	14.50	24.60	1.10	0.24	----	2.05	0.30	0.005	Bal.

Table I lists the elemental levels which were altered by remelting for Pyromet 718 ingots 1, 2, and 3. Data of ingot 1 (air cover) show Al decreased 0.10-0.15% from electrode levels and oxygen increased from <5 ppm to as high as 46 ppm. A slight increase in Ti was also observed at the ingot top. Sulfur was lowered a slight amount by remelting. Ingot 2 (argon cover) revealed an overall loss of Ti, but increase in Al. The largest loss of Ti and highest increase in Al was at the X location, but this difference diminished as remelting progressed until there was a slight increase of Ti and loss of Al observed at the A location. A smaller increase in oxygen content was observed compared to ingot 1. Sulfur level was also lowered. Ingot 3 (argon cover + TiO_2 slag addition) showed much improved compositional control compared to ingot 2. Only a slight loss of Ti and increase in Al were observed. Oxygen levels were comparable to ingot 2. Sulfur level also remained low.

Table I also lists the elemental levels that were altered by remelting of Pyromet A-286 ingots 4, 5, and 6. Ingot 4 (air) shows a significant loss of Ti at all locations during remelting and a slight increase of Al. Oxygen level increased significantly at all locations during melting while sulfur was maintained at low levels and Si was essentially unchanged. Ingot 5 (argon) also showed a significant overall loss of Ti during melting, but with a corresponding increase in Al content. The condition was most pronounced at the X location and diminished as remelting progressed until a slight increase in Ti was observed at the A location. Oxygen content increased only a slight amount compared to ingot 4. Sulfur remained at very low levels and Si level was basically unchanged. Ingot 6 (argon + 6% TiO_2) showed a significantly smaller Ti loss and Al increase compared to ingot 5 although a slight increase of Si was observed. Oxygen and sulfur levels were maintained at very low levels.

Figure 2 lists the volume percent inclusion data for ingots 1-6. Given the very low measured volume fractions, scatter in the data is minimal for a specified location. Table II lists the average particle diameter and area of the inclusions and shows that averages for both Pyromet 718 and A-286 were in the vicinity of 1 micron or less. Correlation coefficients (r) between volume percent inclusion content and oxygen content for the respective Pyromet 718 and A-286 data were calculated and determined to be 0.769 and 0.673 respectively.

Image analysis showed the great majority of inclusions to be located within Ti-rich carbides and/or carbonitrides. Figures 3 and 4 show respective SEM photomicrographs of a carbide cluster and isolated carbide containing inclusion nuclei from the 3-C ingot location. For the inclusion shown in Figure 4, energy dispersive spectrographic (EDS) analysis with the SEM indicated the carbide/inclusion center to be very high in Ti with significant amounts of Al and Mg while an analysis of the outer shell

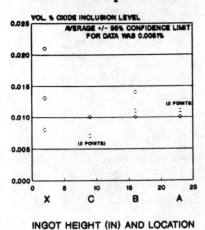
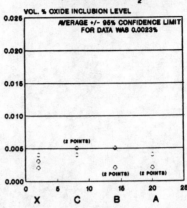
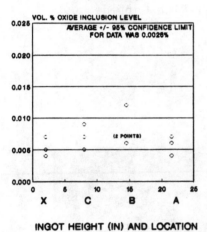
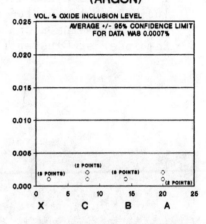
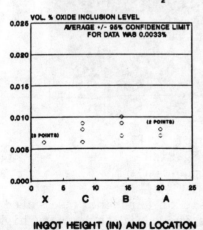
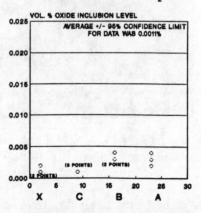

Figure 2 – Volume Percent Inclusion Data for Laboratory Pyromet 718 and A-286 8" Rd. Ingots Electroslag Remelted under Argon and Air Atmosphere

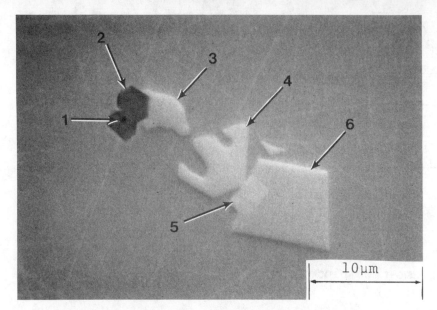

Figure 3 - SEM Photomicrograph of Typical Primary Carbide Cluster Observed In Pyromet 718 (Ingot 3-C Location). Note Al-Mg-Rich Inclusion (1) Located Within Ti-Rich Carbide (2). Columbium-Rich Carbides Constitute Remainder of Cluster (3,4,5,6).

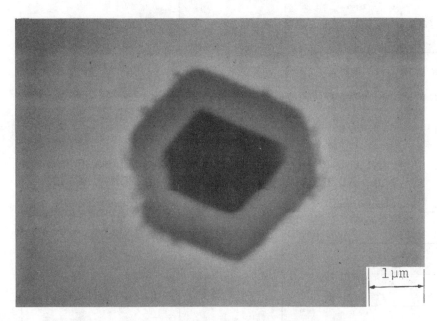

Figure 4 - SEM Photomicrograph of Ti-Rich Primary Carbide Containing Al-Mg-Rich Inclusion Nucleus (Ingot 3-C Location).

revealed only very high Ti levels. Analysis of a randomly selected primary carbide in A-286 with an inclusion nucleus showed it to be very high in Ti with small levels of Al and Mg. In both 718 and A-286, inclusions were found only at the center of Ti-rich primary carbides. However, in both 718 and A-286, Ti-rich carbides were frequently observed within larger Cb-rich carbides. Microstructural analysis also showed inclusions were also more frequent in 718 than in A-286 and more frequent in air remelted ingots than argon remelted ingots. Metallographic observation also revealed three small inclusions enriched in Al, Ca, and Mg at the X location of ingot 3.

Discussion

Compositional Control

Examination of Pyromet 718 ingots 1, 2, and 3 revealed important insights into slag reactions and their effect on ingot composition during ESR. Ingot 1 demonstrated that even with TiO_2 and Al deoxidation slag additions, a significant loss of Al occurred. Further refinement of slag compositions through TiO_2 and deoxidant additions can provide improved ingot compositional control as is evidenced by routine air remelting of 718 production ingots. However, the increased oxygen content of ingot 1 compared to the electrode content indicated that even with a slag deoxidation addition of Al, the ESR operation increased the ingot oxygen content. The loss of Ti and increase in Al in ingot 2 demonstrates that the argon cover over the slag during ESR isolated it from the air and rendered the slag a closed system where the slag oxidation/reduction reactions became interdependent. Similar behavior was observed in Ti stabilized stainless steels by Schwerdtfeger et al.(1). The Ti and Al compositional variations can be explained by the reaction:

$$3Ti + 2(Al_2O_3) \rightleftharpoons 4Al + 3(TiO_2) \qquad (1)$$

as discussed by Pateisky et al(2). This reaction occurred as the slag attempted to reach an equilibrium state during remelting. The interrelationship between Ti and Al and low ingot oxygen level also suggest that the oxidation/reduction reactions of Ti and Al with FeO were minimized and that low slag oxygen potentials were obtained during ESR under argon(1,3). In argon-ESR, the isolation of the slag from an air atmosphere and the interdependence of slag reactions favors a low slag oxygen potential which consequently produces a remelted ingot of lower oxygen content compared to air-ESR (compare ingots 1 and 2)(4). Data of Ingot 3 demonstrated that when a slag of near equilibrium composition was used in conjunction with an argon cover during ESR, improved control of ingot Ti and Al levels and lower ingot oxygen levels can be obtained compared to ingots remelted via air-ESR. Remelting under an argon cover also maintained very low sulfur levels and did not adversely affect the desulfurizing capability of the slag.

The A-286 data of ingots 4, 5, and 6 essentially confirmed the trends observed in Pyromet 718 heats, but also revealed that control of ingot composition can be extremely difficult during air-ESR of alloys with a high Ti/Al ratio as shown by ingot 4. Ingot 4 also shows the substantial increase of ingot oxygen content after air-ESR just as did Pyromet 718 ingot 1. Ingot 5 (argon) data reflect the reduction of Al_2O_3 in the slag to form TiO_2 as the slag attempted to reach an equilibrium state. Ingot oxygen content only increased a slight amount after remelting under an argon atmosphere. Ingot 6 (argon + 6% TiO_2 slag addition) data showed improved compositional control of Ti and Al compared to ingot 5 while maintaining a low ingot oxygen content. Utilization of slightly higher TiO_2 content in the slag during ESR under argon should improve compositional control of Ti and Al while maintaining low ingot oxygen levels. Comparison of ingots 3 and 6 demonstrate that the proper slag TiO_2 content to maintain compositional control will vary according to alloy composition, particularly Ti and Al content(2). This demonstrated that the argon-ESR process is capable of remelting Ti-Al containing superalloys with improved compositional control while maintaining low ingot oxygen and sulfur levels.

Ingot Inclusion Level

An important implication of the lower ingot oxygen content obtained through argon-ESR is its effect on ingot inclusion content. A comparison of the calculated inclusion volume percent averages showed that the difference between the two 718 ingots remelted in argon and air was significant above a 95% confidence level. This trend was also observed in the A-286 ingots. The statistically significant lower inclusion levels of the argon remelted heats indicated that the lower oxygen content produced a direct reduction in the inclusion levels. The three small Al-Ca-Mg-rich inclusions at the X location of ingot 3 can be attributed to entrapped slag from the start up of the melt at that location and do not indicate a refining problem with the slag.

Figure 5 shows bar charts where the average of the three measured inclusion levels are compared with the measured oxygen content for each location. Previous investigations(5,6) have found reasonably good correlations between oxygen level and oxide inclusion content for oxygen levels above 40 ppm. Given both the low measured oxygen levels (precision of $\pm$ 5-10 ppm) and the very low measured inclusion levels, the degree of correlation of this study is considered quite good.

It is interesting to note that for a given oxygen level, the inclusion level of Pyromet 718 ingots is higher than that of A-286. Factors contributing to this difference could include both varying oxygen solubilities of the 718 and A-286 matrices and varying densities of the oxide inclusions depending upon the oxide formed. Microstructural analysis also showed that lower oxygen levels in argon remelted ingots did reduce the number of inclusions compared to air remelted ingots. Therefore, the data indicate that a given oxygen level can produce varying inclusion levels depending on the alloy system and oxide species formed during melting and solidification.

Work by Fox et al. (7) on alloy 718 also showed that Ti-rich carbides in 718 nucleated on small aluminum oxide inclusions with smaller amounts of Ca and Mg also detected. Detection of Al and Mg at the center of Ti-rich carbides in both 718 and A-286 supports this work and indicates that Al and Mg oxide inclusions were nucleation sites for Ti-rich carbide precipitation during solidification. Thus, if the inclusion level is lowered, it is quite possible that the average size of Ti-rich carbides could be reduced since their nucleation could be delayed during solidification. Mitchell and Tripp (8) have addressed the potential deleterious effect which TiN/oxide clusters can exert on superalloy mechanical properties. This topic is of considerable importance since a uniform primary carbide distribution imparts optimum mechanical properties to an alloy component. The preliminary data of this study suggest argon-ESR has the potential to achieve oxygen levels of < 10 ppm in superalloy ingots and consequently very low oxide levels. This could contribute to improved carbide distributions by minimizing the interaction between TiN's and oxides and improve alloy performance. The potential for improved alloy "cleanliness," and more uniform carbide distributions makes the argon-ESR process worthy of continued investigation. Also, further refinements in slag composition and deoxidation/desulfurization additions could produce very "clean" ingots with mechanical properties superior to either conventional VAR or air-ESR ingots.

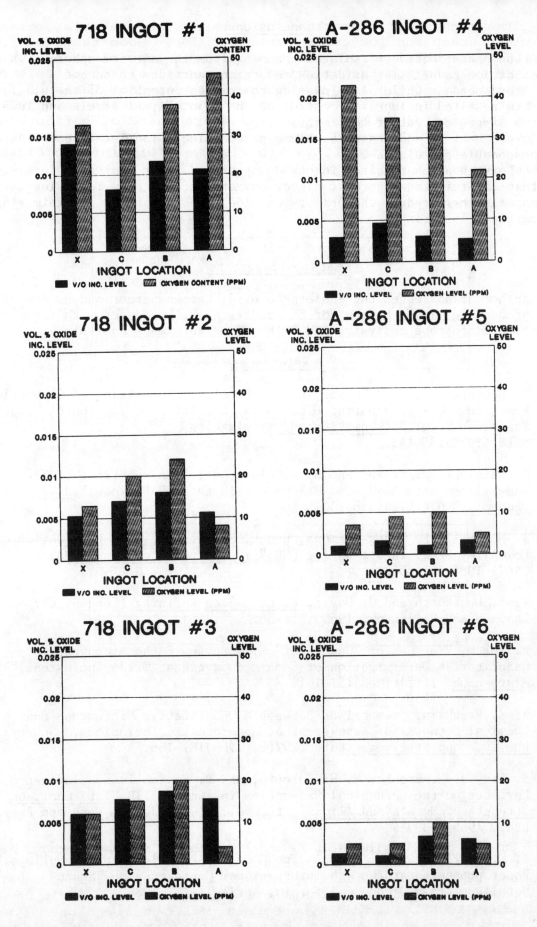

Figure 5 - Comparison of Oxygen and Average Volume Percent Inclusion Levels in Laboratory Pyromet 718 and A-286 8" Rd. Ingots Electroslag Remelted under Argon and Air Atmosphere

Conclusions

1. Argon-ESR effectively isolated the slag from the air atmosphere. This condition caused slag oxidation/reduction reactions to become interdependant while also lowering the oxygen potential of the slag. This resulted in improved control of the ingot Ti, Al levels and lowered the oxygen content of the ingot.
2. Oxygen content and measured volume percent inclusion level demonstrated reasonable correlation.
3. Low oxygen argon-remelted ingots demonstrated lower inclusion levels than air-remelted ingots of higher oxygen content. Low inclusion levels may also help reduce the primary carbide size and assist in minimizing carbide clustering.

Acknowledgements

The author acknowledges the assistance of J. Leibensperger and D. Benzel in ESR processing, J. W. Bowman for SEM analysis, B. L. Messersmith for electron microprobe analysis, and D. Kauffman for image analysis.

References

1. K. Schwerdtfeger, W. Wagner, and G. Pateisky, "Modelling of Chemical Reactions Occurring During Electroslag Remelting: Oxidation of Titanium Stainless Steel," Ironmaking and Steelmaking, 5 (3)(1978)135-143.

2. G. Pateisky, H. Biele, and H. J. Fleischer, "The Reactions of Titanium and Silicon with Al_2O_3-CaO-CaF_2 Slags in the ESR Process," J Vac Sci Technol, 9(6)(1972)1318-1321.

3. A. Mitchell, F. Reyes-Carmona, and E. Samuelsson, "The Deoxidation of Low-Alloy Steel Ingots During ESR," Trans Iron Steel Inst Jpn, 24(7)(1984)547-556.

4. W. E. Duckworth and G. Hoyle, Electro-Slag Refining (London, UK: Chapman and Hall Ltd., 1969), 51-56.

5. R. Roche, "The Use of "Quantimet" Microscope in the Micrographic Quantitative Determination of Combined Oxygen at Oxide Inclusions," The Microscope, 16(2)(1968)151-161.

6. A. G. Franklin, "Comparison Between a Quantitative Microscope and Chemical Methods for Assessment of Non-Metallic Inclusions," Journal of the Iron and Steel Institute, 207(2)(1969)181-186.

7. S. Fox, J. W. Brooks, M. H. Loretto, and R. E. Smallman, "Influence of Carbides on the Mechanical Properties in Inconel 718,"Int. Conf. on Strength of Metals and Alloys, 7th, vol.1 (Oxford, UK: Pergamon Press, Ltd., 1985), 399-404.

8. A. Mitchell and D. W. Tripp, "The Case for Low Nitrogen Superalloys," Paper presented at the 9th International Conference on Vacuum Metallurgy, San Diego, California, April 11-15, 1988.

DETERMINATION OF THE SOLIDIFICATION BEHAVIOUR

OF SOME SELECTED SUPERALLOYS

U. Heubner and M. Köhler
VDM Nickel-Technologie AG
Werdohl, W-Germany

B. Prinz
Metallgesellschaft AG
Frankfurt, W-Germany

Summary

Equilibrium solidification ranges and melt equilibria of the alloys 625, 718, 825 hMo, X, X-750 and 80 A have been determined. After isothermal holding and quenching of alloy samples the partition coefficients of the alloy components Ni, Cr, Mo, Fe, Cb and Ti as a function of temperature have been evaluated by means of microprobe measurements. There is no identical segregation behaviour of the individual alloy components in the above mentioned alloys. The partition coefficients nevertheless can be described by a linear function of temperature. The segregation behaviour of Cr and Cb in alloy 718 during quasi-isothermal solidification is in accordance with a mathematical model which was developed previously for binary alloys.

The influence of increasing solidification rates on dendrite arm spacing as well as on microsegregation has been studied by directional solidification experiments. The primary dendrite arm spacing is related to the cooling rate by a power function. However, a pronounced reduction of segregation can only be achieved at the cost of significant extension of the dendrite arm spacing.

Introduction

The development and application of new casting technologies in the manufacturing of semi-finished products of nickel base alloys demands a more precise knowledge of the solidification behaviour of these complex multi-component systems. The usually apppplied multi-stage hot forming of ingots allows working towards an optimum microstructure by means of forming and annealing operations. With the aim of cost reduction, continuous casting has already been introduced to the manufacturing of nickel base alloys which includes the advantage of steady state solidification. However, with new near net shape casting technologies the potential to influence the microstructure by extensive forming and annealing operations will be reduced considerably.

Because of this development a better knowledge of the proceedings during solidification gains more importance. At an early stage already, mathematical models had been developed to describe the solidification behaviour of binary alloys. These models differentiated between micro- and macrosegregation. Basic research in this field was done by Scheuer (1), Scheil (2), Brody and Flemings (3), Clyne and Kurz (4), as well as by Oeters and coworkers (5). For the description of segregation in multi-component systems, as they are to be found in superalloys, no mathematical models have been published so far. Therefore, the process parameters governing the solidification have to be verified experimentally.

Table 1 shows the alloys which have been studied with regard to their solidification behaviour. The solid solution hardening nickel base alloys 825 hMo, X and 625 have been examined. Alloys 718, X-750 as well as 80 A have been included as examples of age-hardenable materials. In addition to the principal alloy components Ni, Cr, Mo and Fe the segregation behaviour of Cb and Ti was examined.

Table 1: Chemical composition of the experimental alloys, wt%

Alloy	Ni	Cr	Mo	Fe	Cb	Ti	Al	other
825 hMo	bal.	20.0	6.3	28.4	–	0.80	0.1	2.3 Cu
X	bal.	21.0	9.0	19.0	–	–	–	0.56 W, 1.0 Co
625	bal.	21.8	8.9	2.0	3.4	0.19	0.18	
718	bal.	18.2	3.2	18.4	5.4	0.93	0.64	
X-750	bal.	15.7	–	7.9	1.0	2.6	0.8	
80 A	bal.	19.9	–	–	–	2.3	1.5	

Examination of the equilibrium solidification behaviour

Equilibrium solidification is usually described by means of a phase diagram which refers to extremely slow cooling. In case of binary systems thermal analysis is preferred because knowledge of both the individual basic components and the liquidus and solidus temperatures allows the design of a phase diagram and provides a measure for the segregation to be expected, characterized by the width of the liquidus-solidus temperature interval. In multi-components system this procedure cannot be applied. Therefore, the method of isothermal holding in the heterogeneous area between the liquidus and solidus temperatures was introduced to determine the equilibrium solidification behaviour. Subsequent to the adjustment of the equilibrium at a previously defined temperature, the sample is quenched in water.

During quenching, the residual melt, which is in the state of equilibrium with the precipitated solid crystal, solidifies with the shape of extremely fine dendrites. By means of microprobe analysis both phases were subsequently analyzed with respect to their composition. Several experiments within the equilibrium solidification interval allow an exact determination of the partition coefficients of the individual alloy components as a function of temperature and composition.

Figure 1 shows a typical microstructure as it is to be found in this experimental procedure. Coarse primary crystals are surrounded by a dendritic solidified residual melt. The analysis of the concentration profiles across the coarse primary crystals verified that the equlilibrium was reached completely in the solid crystal.

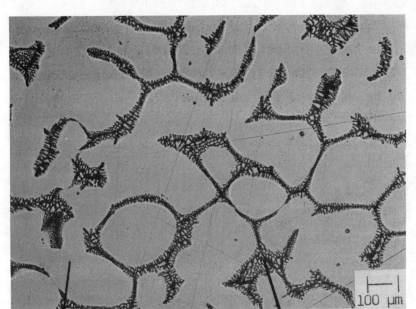

Figure 1:

Microstructure of alloy 718 after isothermal holding at 1310°C and water quenching

S : primary crystal   L : residual melt

The microprobe analysis showed, that during solidification, the primary crystals become enriched with nickel and iron in all alloys under consideration. The same applies for chromium. There are, however, exceptions. As demonstrated in <u>Fig. 2</u> it is only in alloy X that chromium segregates preferentially into the residual melt and the chromium concentration in the residual melt does not vary during solidification of alloy 80 A. In all alloy compositions examined so far, molybdenum, titanium and particularly columbium are enriched to a high degree in the residual melt. Columbium segregation is shown in <u>Fig. 3</u>.

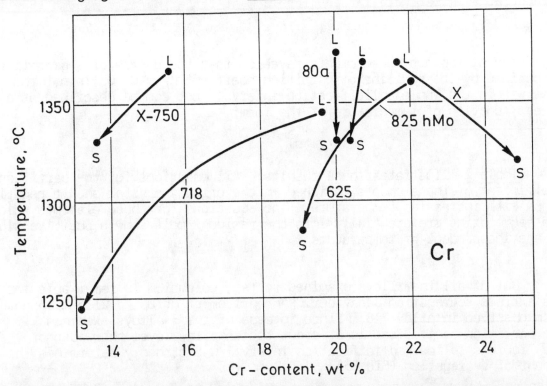

Figure 2: Chromium equilibrium concentration in the residual melt of various nickel base alloys as a function of temperature

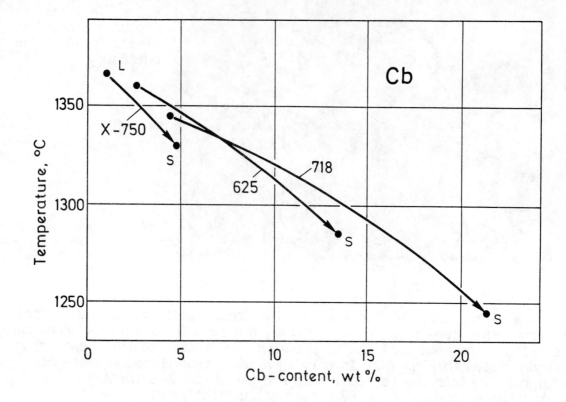

Figure 3: Columbium equilibrium concentration in the residual melt of various nickel base alloys as a function of temperature

The microprobe measurements revealed that the degree of segregation as defined by the equilibrium partition coeffient $K_o = C_s/C_l$ is not the same within the whole solidification interval, but can be described as a linear function of temperature of the form:

$$K_o = a + bT$$

Table 2 illustrates these results. All equations for the partition coefficient of the elements enriched in the primary crystal show a negative slope with increasing temperature. The equations for those elements which are segregating preferentially into the residual melt show a positive slope or are independent of temperature.

Out of all the alloys examined so far, columbium is remarkable for its extreme segregation behaviour. An enrichment of up to 21.5 % Cb has been observed in alloy 718. Since in case of the 3 alloys examined the partition coefficient is at about the same low level the width of the liquidus-solidus solidification interval L-S strongly determines the amount of segregation (Fig. 3).

Table 2: Equilibrium-partition-coefficients ($K_o = C_s / C_L$) as a function of temperature in the solidification range (T in °C)

	T_L °C	T_s °C	Ni K_o	Fe K_o	Cr K_o	Mo K_o	Cb K_o	Ti K_o
alloy 825	1370	1330	$2.08-7.50 \cdot 10^{-3}T$	$10.29-6.75 \cdot 10^{-3}T$	n.d.	$-5.47+4.50 \cdot 10^{-3}T$	n.d.	$-4.79+3.75 \cdot 10^{-3}T$
alloy X	1370	1320	$9.04-5.80 \cdot 10^{-3}T$	$2.87-1.20 \cdot 10^{-3}T$	$0.36-0.40 \cdot 10^{-3}T$	$0.64+0.8 \cdot 10^{-3}T$	n.d.	n.d.
alloy 625	1360	1285	$4.13-2.27 \cdot 10^{-3}T$	$7.45-4.67 \cdot 10^{-3}T$	$3.54-1.87 \cdot 10^{-3}T$	$0.18+0.53 \cdot 10^{-3}T$	$-1.29+1.20 \cdot 10^{-3}T$	$-0.59+0.80 \cdot 10^{-3}T$
alloy 718	1345	1245	$1.72-0.50 \cdot 10^{-3}T$	$5.50-3.2 \cdot 10^{-3}T$	$5.64-3.40 \cdot 10^{-3}T$	0.8*	0.25*	$2.00-1.20 \cdot 10^{-3}T$
alloy X-750	1367	1330	$2.85-1.35 \cdot 10^{-3}T$	$10.39-6.76 \cdot 10^{-3}T$	$4.38-2.43 \cdot 10^{-3}T$	n.d.	$-6.68+5.14 \cdot 10^{-3}T$	$-1.40+1.35 \cdot 10^{-3}T$
alloy 80 A	1375	1330	$3.14-1.56 \cdot 10^{-3}T$	n.d.	$2.22-0.89 \cdot 10^{-3}T$	n.d.	n.d.	$-2.56+2.22 \cdot 10^{-3}T$

* $\hat{=}$ independent of temperature

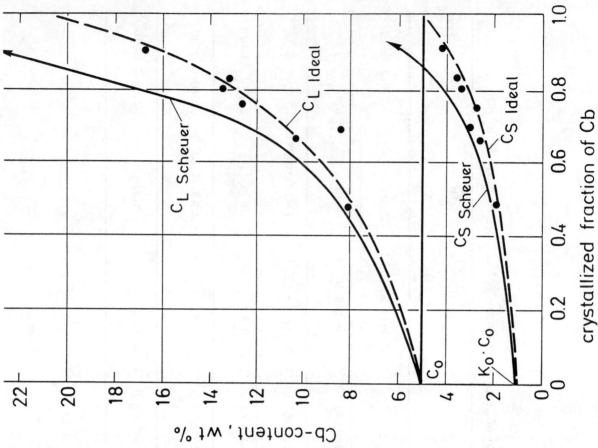

Figure 4 and Figure 5:
Variation of chromium and columbium concentrations in the crystal and in the residual melt of alloy 718 with progressing solidification (calculation according to E. Scheuer and the ideal solidification model).

Mathematical description of the solidification processes

Under real solidification conditions the residual melt will be segregated stronger than under ideal solidification conditions, since a complete equalization of the chemical composition by diffusion according to the equilibrium in the solid crystal and the residual melt cannot be obtained. As mentioned above, mathematical models for the solidification processes of binary alloys have already been developed and shall now be applied to the complex multi-component systems. This is, however, done on the basic assumption that the individual alloy components do not influence each other with respect to segregation behaviour.

Assuming a complete mixing of composition in the liquid and no equalization by diffusion at all in the crystal, the course of the change of chemical composition in the residual melt can be determined according to E. Scheuer (1) as a function of the crystallized fraction fs:

$$C_L = C_0 (1 - fs)^{(K_0 - 1)}$$
$$C_S = K_0 C_0 (1 - fs)^{(K_0 - 1)}$$

In case of ideal solidification conditions with a complete balance of composition by diffusion in both the solid and the residual melt the element concentration will always change according to the following equations:

$$C_L = \frac{C_0}{1 - fs + K_0 fs}$$
$$C_S = \frac{C_0 \cdot K_0}{1 - fs + K_0 fs}$$

Since on this assumption the element concentrations do not depend on the solidification rate, the results obtained by means of isothermal holding experiments are expected to follow this rule.

Figures 4 and 5 show, as an example, the element concentrations of Cr and Cb in alloy 718 as a function of the crystallized fractions for both assumptions. As to be expected, the experimental data correspond well with the model for ideal solidification. Under real solidification conditions the element concentration in the residual melt will be found somewhere between both lines, since some equalization by diffusion can take place in the crystal as soon as a concentration gradient is built up at the solidification front.

The influence of the cooling rate on the microstructure and segregation

In order to analyze the microstructure obtained at different cooling rates, samples of alloys 625, 718 and X were cooled down in a temperature controlled tube furnace. The temperature gradient at the solidification front was about 20 K/min. The cooling rates ranged from 24°C/min to 1°C/min. A solidified microstructure is given in Figure 6 for alloy 625. The transverse microsection clearly shows the primary dendrite arm spacing which had been determined as a function of the cooling rate. As shown in Fig. 7 and to be expected, the primary dendrite arm spacing is becoming smaller with increasing cooling rates.

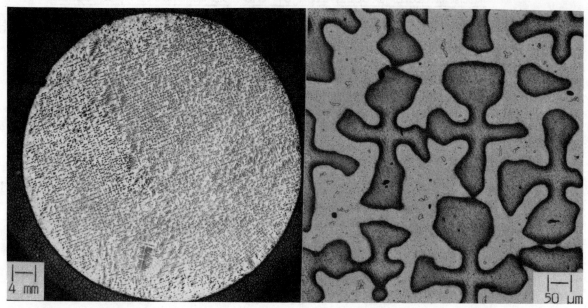

Figure 6: Transverse microsection of an alloy 625 sample, which had been directionally solidified. Cooling velocity was 12°C/min.

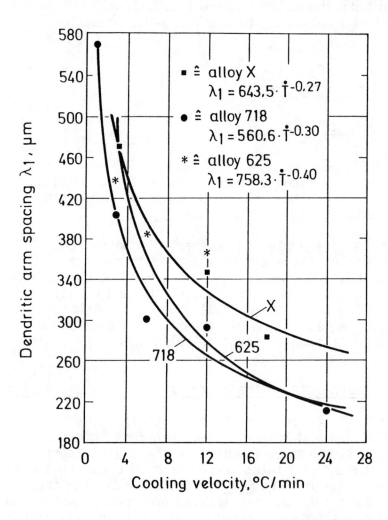

Figure 7: Comparison of the dendrite arm spacing of different nickel base alloys as a function of the cooling velocity

- ■ ≙ alloy X $\lambda_1 = 643.5 \cdot \dot{T}^{-0.27}$
- ● ≙ alloy 718 $\lambda_1 = 560.6 \cdot \dot{T}^{-0.30}$
- * ≙ alloy 625 $\lambda_1 = 758.3 \cdot \dot{T}^{-0.40}$

The influence of the cooling rate $\dot{T}$ on the primary dendrite arm spacing λ_1 can be described by the emperical equation indicated in Fig. 7.

In order to assess the amount of microsegregation as a function of the cooling rate, concentration profiles across the dendrite arms have been determined by microprobe analysis. Some of the results having been obtained on alloy 625 are shown in Table 3.

Table 3: Segregation behaviour of alloy 625 at controlled rates of cooling

$\frac{dT}{dt}$ $\frac{°C}{min}$	λ_1 µm		Ni	Cr	Fe	Mo	Cb
12	367	C_S	65.0	23.0	2.37	8.0	1.6
		C_L	59.0	20.0	1.62	11.0	6.5
		$K_E = \frac{C_S}{C_L}$	1.10	1.15	1.50	0.73	0.25
3	436	C_S	66.0	22.9	n.d.	7.7	2.0
		C_L	61.0	20.4	n.d.	10.3	6.3
		$K_E = \frac{C_S}{C_L}$	1.08	1.12	n.d.	0.75	0.32

The influence of the cooling rate can be seen from the fact that the slowly cooled sample has effective partition coefficients which are closer to the ideal value 1. However, reduced segregation can only be achieved at the cost of significant extension of the diffusion distance, here characterized by the dendrite arm spacing (Fig.7).

Further microprobe analyses allowed to assess how different diffusion annealing treatments result in a reduction of segregation. Table 4 summarizes the results of a 20 and 50 hrs homogenization annealing at 1200°C (2192°F) in correlation to the dendrite arm spacing.

Thus, under identical annealing conditions, an extension of the diffusion distance of 19 % results in a remarkably less reduced segregation profile. If the concentration profile has to be equalized to a similar extent, the extension of the diffusion distance of 19 % demands an increase in the diffusion annealing time of about 150 %, provided the temperature is the same.

Table 4: Influence of diffusion annealing treatments on the reduction of segregation in controlled cooled alloy 625 samples

λ_1 µm	diffusion annealing treatment		Ni	Cr	Fe	Mo	Cb
367	20 h, 1200°C	C_S	63.1	22.2	2.17	8.6	2.8
		C_L	61.6	20.8	1.83	9.6	3.6
		$K_E = \frac{C_S}{C_L}$	1.02	1.07	1.19	0.9	0.78
		reduction of segr.	80 %	53 %	62 %	63 %	68 %
436	20 h, 1200°C	C_S	64.6	22.9	-	8.5	3.1
		C_L	61.3	20.6	-	10.4	4.5
		$K_E = \frac{C_S}{C_L}$	1.05	1.11	-	0.82	0.69
		reduction of segr.	38 %	8 %	-	28 %	59 %
	50 h, 1200°C	C_S	63.5	22.5	-	7.8	3.0
		C_L	62.5	21.8	-	9.2	3.5
		$K_E = \frac{C_S}{C_L}$	1.02	1.03	-	0.85	0.86
		reduction of segr.	75 %	75 %	-	40 %	81 %

Conclusions

1. During quasi-isothermal solidification of the nickel base alloys 625, 718, 825 hMo, X, X-750 and 80 A the alloy elements Mo, Ti and Cb will be enriched in the residual melt. Ni and Fe are segregating into the primarily solidifying crystals. Depending on the type of alloy Cr will be enriched either in the residual melt or on the primary crystals.

2. During quasi-isothermal solidification the degree of segregation as defined by the equilibrium partition coefficient $K_o = C_s/C_l$ is, in most cases, not constant within the solidification range but varies with temperature according to the equation $K_o = a + bT$.

3. Within the scope of the alloys examined, liquidus temperatures T_L varied between 1345 and 1375°C (2451 to 2507°F) only, whereas solidus temperatures T_s showed a much broader range from 1245°C to 1330°C (2273 to 2426°F).

4. In all of the alloys examined so far, columbium is distinguished by its extreme segregation behaviour. The final degree of columbium segregation is largely depending on the solidification interval ΔT, i.e. increasing from alloy X-750 ($\Delta T = 37$°C) over alloy 625 ($\Delta T = 75$°C) to alloy 718 ($\Delta T = 100$°C).

5. The segregation of Cr and Cb during quasi-isothermal solidification of alloy 718 is in accordance with the ideal model for binary alloys. Also E. Scheuer's model for extreme non-equilibrium solidification (no equalization of composition at all in the primarily solidifying crystals) may be applicable to superalloys and the segregation occuring under real conditions will probably lie between both models.

6. The primary dendrite arm spacing is related to the cooling rate by a power function $\lambda_1 = cT^{-d}$, the parameters c and d depending on the type of alloy. With decreasing cooling rate the partition coefficients are approaching the ideal value 1. However, a pronounced reduction of segregation can only be achieved at the cost of significant extension of the dendrite arm spacing, i.e. the diffusion distance in subsequent homogenization treatments.

7. The reduction of segregation in diffusion annealing is mainly influenced by the width of dendrite arm spacing. Therefore, it is most important to create a narrowly spaced and uniform dendrite pattern during solidification whereas the amount of microsegregation is less important.

Acknowledgement

This work was promoted by the German Federal Ministry for Research and Technology (03 ZG 201 6). The responsibility for this report is with the authors.

References

1. E. Scheuer, "Zum Kornseigerungsproblem", Z. Metallkunde 23 (1931), pp 237 - 241.

2. E. Scheil, "Bemerkungen zur Schichtkristallbildung", Z. Metallkunde 34 (1942), pp 70 - 72.

3. H. D. Brody and M. C. Flemings, "Solute Redistribution in Dendritic Solidification", Trans. TMS-AIME 236 (1966), pp 615 - 624.

4. T. W. Clyne and W. Kurz, "Solute Redistribution During Solidification with Rapid Solid State Diffusion", Met. Trans. 12 A (1981), pp 965-971.

5. F. Oeters and M. Seidler, "Experimentelle Untersuchungen zur horizontalen Makroseigerung bei Erstarrung mit heterogener Schicht", Arch. Eisenhüttenwesen 48 (1977), pp 527 - 531.

EFFECTS OF CHEMISTRY ON VADER PROCESSING OF

NICKEL BASED SUPERALLOYS

P. W. Keefe, F. E. Sczerzenie and G. E. Maurer

Special Metals Corporation
New Hartford, New York

ABSTRACT

VADER's unique solidification mechanisms produce continuous cast ingots characterized by a relatively fine equiaxed grain structure. The understanding of these mechanisms and how they are affected by various alloy and process modifications is essential to the optimization of VADER casting. In addition, this information will enable the formulation of new alloys which can fully utilize the capabilities of VADER.

This study was designed to elucidate how alloy variations alter the response of VADER casting. Alloy chemistries were designed to alter the volume of interdendritic fluid and the amount of carbide phase present during the final stages of solidification. Five heats of modified UDIMET® Alloy 720 were VIM/VAR/VADER processed. They included three heats containing 0.01, 0.03 and 0.06% carbon and two heats containing 1% of a carbide stabilizer, niobium , with 0.01 and 0.02% carbon. Analysis included characterization of ingot solidification and determination of the structural and chemical uniformity of the ingot at various radial locations.

It was observed that the 1% addition of a carbide stabilizer effectively reduced chemical and structural gradients near the ingot surface. Alloys with intermediate carbon levels of 0.02 and 0.03% produced relatively uniform ingots while alloys with high and low levels, .06 and .01%, exhibited substantial structural gradients. Intermediate carbon in combination with the addition of a carbide stabilizer produced an ingot which exhibited less of a chemical gradient than a typical VAR ingot of the same size.

Observations were made that increased the understanding of VADER solidification mechanisms and clearly defined useful parameters to develop alloys specifically for VADER processing.

®UDIMET is a registered trademark of Special Metals Corporation

INTRODUCTION

Vacuum arc double electrode remelting (VADER) is a unique casting technology that is capable of producing uniform fine grained ingots of various superalloys, including P/M type alloys that could not be made using conventional cast-wrought technology. What makes VADER unique is that casting is carried out at temperatures slightly below the liquidus. Semimolten metal, dispersed with solid nuclei, is dripped into the mold where solidification proceeds with the formation and growth of fine equidirectional dendrites. VADER's characteristic fine grained cellular structure results not from rapid solidification but from the simultaneous growth of many evenly dispersed solid nuclei in a low temperature gradient. A relatively slow rate of solidification avoids the formation of microporosity associated with other fine grained casting techniques (1,2) by allowing the intercellular flow of liquid metal.

Previous literature (3-6) comparing the microstructure of VADER ingots to those of ESR and VAR indicates that VADER ingots, are in general, more homogeneous. However, sharp chemical and structural variations have been observed near the surface of VADER ingots. These variations are the result of an interaction between the solidifying ingot and the rapidly cooled solid ingot surface, whereby microporosity in this outer skin is filled by the drawing of interdendritic fluid from the semi-liquid region inside the solid-mushy zone interface.

It follows that the observed surface inconsistencies can be controlled by increasing the resistance to intercellular liquid flow. Previous work indicates that minor chemistry variations can effectively alter the permeability of the mushy zone (7). It is anticipated that increasing carbon and/or increasing the stability of the carbides formed will reduce chemical and structural gradients by decreasing this permeability. The following report explores the effectiveness of these modifications in controlling chemical and structural gradients in VADER ingots.

EXPERIMENTAL PROCEDURE

Five 200 Kg heats with chemistries given in Table I, were VIM/VAR/VADER withdrawal cast as 20 cm diameter ingots. All alloys were modifications/versions of the high strength nickel based superalloy UDIMET Alloy 720. Alloys 1 and 3 are low and high C versions, respectively. One percent niobium was added to Alloys 4 and 5 for carbide stabilization. Alloy 4 was a low C version and 5 had an intermediate carbon level; both alloys had a reduced level of chromium in order to prevent the formation of sigma phase.

Transverse sections were cut from each ingot at a location representative of the steady state portion of the continuous cast ingot (approximately 50 cm from the bottom of the ingot). Center-to-edge micros were prepared for the determination of structural uniformity and phase identification. Chemical analysis and differential thermal analysis (DTA) were subsequently performed at several radial locations for each heat.

TABLE I. ALLOY CHEMISTRIES (Wt%)

#	Ni	Co	Cr	Mo	W	Nb	Ti	Al	B	Zr	C
1	BAL	14.9	17.9	3.1	1.2	---	5.0	2.5	.031	.030	.007
2	BAL	14.9	17.9	3.1	1.2	---	5.0	2.5	.032	.030	.033
3	BAL	14.9	17.9	3.1	1.2	---	5.0	2.5	.031	.031	.061
4	BAL	14.5	14.6	3.0	1.2	1.0	4.9	2.5	.031	.029	.010
5	BAL	14.5	14.6	3.0	1.2	1.0	4.9	2.5	.030	.030	.023

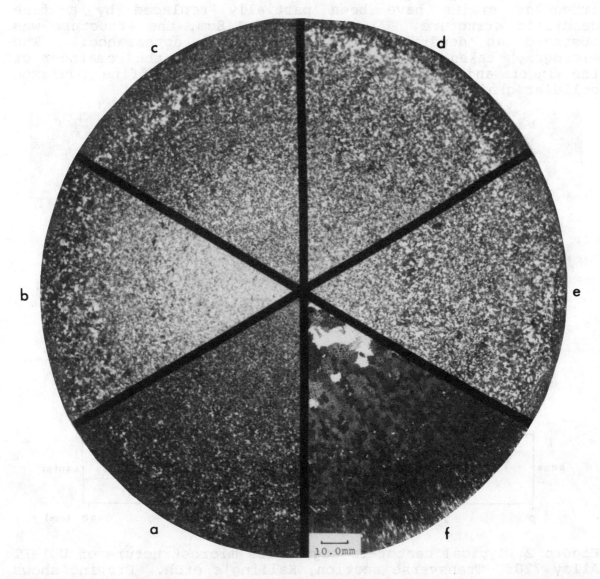

Figure 1. VADER ingot macrostructures, etched in HCl and hydrogen peroxide. a) Alloy 3 (high C) b) Alloy 2 (standard U-720) c) Alloy 1 (low C) d) Alloy 4 (low C and carbide stabilized) e) Alloy 5 (carbide stabilized) f) U-720 VAR ingot.

RESULTS

A comparison of the as cast VADER macrostructures (Figure 1) revealed a definite effect of carbon on the structural uniformity of the ingot. The low carbon heat (Alloy 1) showed the most severe grain size variation while Alloys 2 and 3 exhibited only slight variations. The effect of carbide stabilization was also obvious. The carbide stabilized low carbon heat, Alloy 4, showed much less grain size variation than the low carbon heat, Alloy 1. Likewise, the carbide stabilized heat, Alloy 5, was more uniform than the standard heat, Alloy 2. Alloy 5 produced the most uniform grain structure of all the heats observed.

Typical center-to-edge microstructures of VADER ingots are exhibited in Figure 2. At a depth of about 0.6cm beneath the ingot surface, the structure consists entirely of fine irregularly shaped grains. At 1.2cm into the ingot, these irregular grains have been partially replaced by a fine dendritic structure. At a depth of 1.8cm, the structure was observed as dendritic with a cellular appearance. The micrograph taken at 5 cm is representative of the remainder of the ingot and shows the characteristic VADER fine equiaxed cellular grain structure.

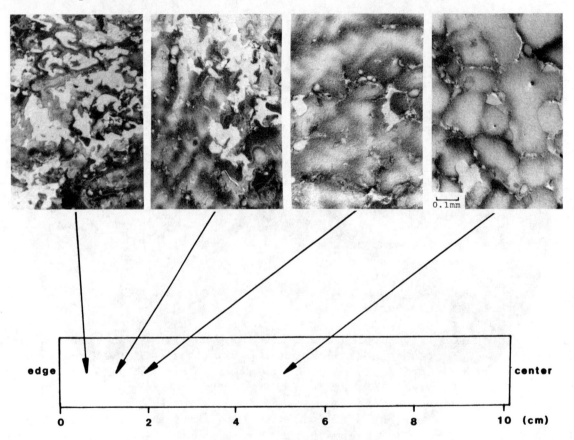

Figure 2. Typical center-to-edge VADER microstructure of UDIMET Alloy 720. Transverse section, Kalling's etch. Drawing shows approximate location of photomicrographs.

The variation of titanium with respect to location (Figure 3) further emphasizes the effect of carbide stabilization on ingot uniformity. These heats, Alloys 4 and 5, exhibit noticeably flatter chemical profiles than their non-carbide stabilized counterparts. As expected, increasing carbon content from a low level, 0.007%, to an intermediate level, 0.033%, dramatically reduced the chemical gradient. However, increasing carbon to a high level, 0.061%, resulted in an obvious chemical gradient which was not observed in the macrostructure. Again the carbide stabilized heat, Alloy 5, exhibited the most uniform center-to-edge chemistry, comparable to a typical VAR ingot of the same size.

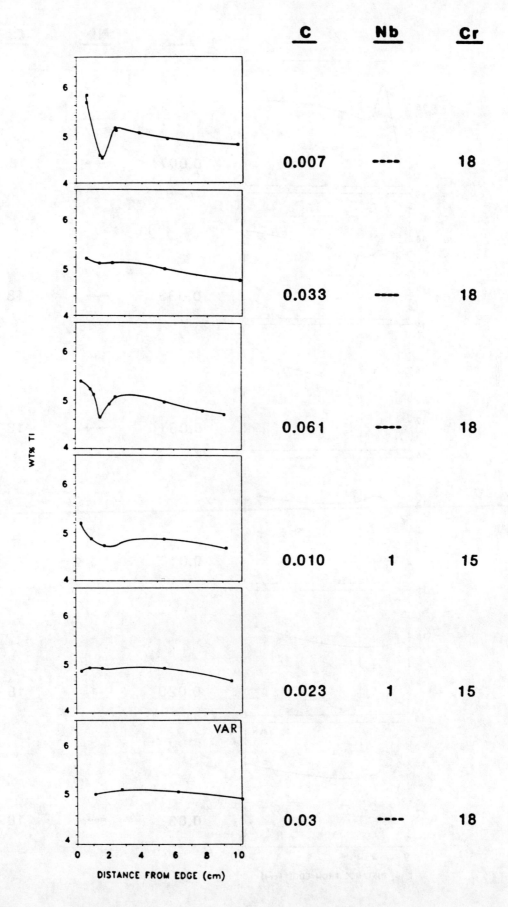

Figure 3. Titanium concentrations of VADER ingots as a function of center-to-edge location. The titanium profile for a VAR ingot is also given for comparison.

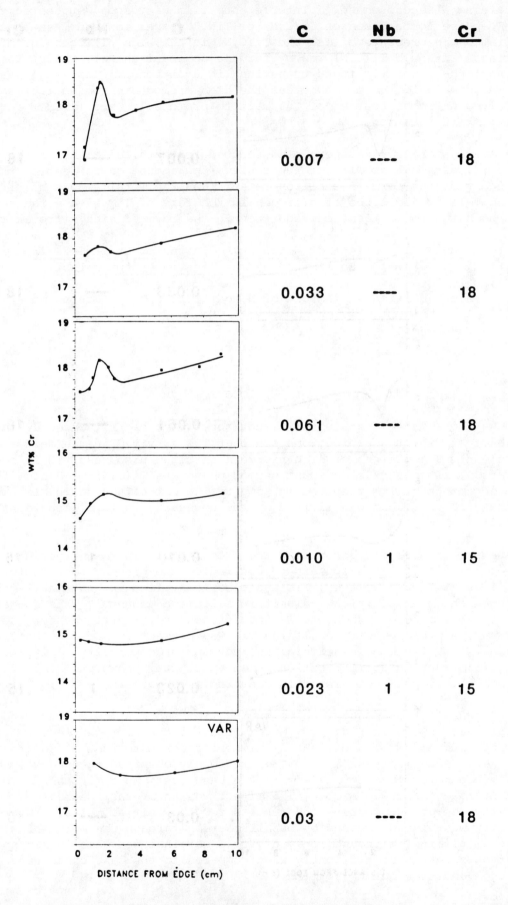

Figure 4. Chromium concentrations of VADER ingots as a function of center-to-edge location. The chromium profile for a VAR ingot is also given for comparison.

The variation of chromium content, Figure 4, exhibited the same trends as titanium but in the opposite direction (i.e, high levels of chromium were observed where low levels of titanium existed). This is due to the relative increase in chromium content as titanium-rich intercellular liquid metal moves toward the surface of the ingot, leaving the chromium-rich primary cellular dendrites in the interior.

The micrographs in Figure 5 represent typical cast microstructures of Alloys 1, 3, 4 and 5 (photos A, B, C and D respectively). SEM-EDX analysis of each ingot revealed that all alloys contained phases similar to VAR cast UDIMET Alloy 720, including eutectic gamma-gamma prime, primary borides, titanium-rich MC-type carbides and small amounts of eta, and that these structures were not noticeably affected by changes in carbon content. The precipitation of titanium-rich globular gamma prime and acicular eta phases (light phases in Figure 5) was greatly increased by the addition of niobium. In addition to increased amounts of globular gamma prime and eta, carbide stabilized alloys also exhibited MC-type carbides containing equal amounts of titanium and niobium. In general, all ingots contained porosity comparable to VAR ingots.

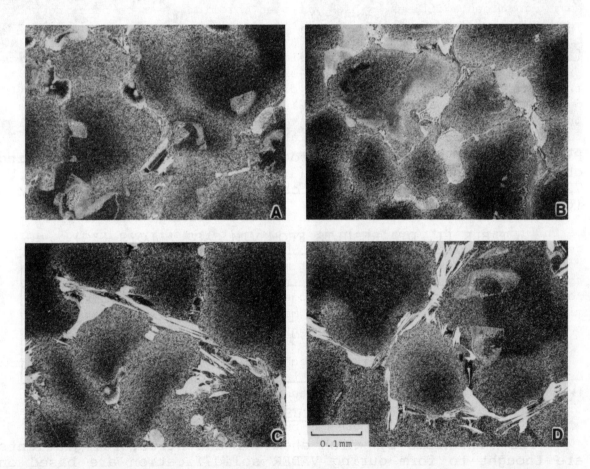

Figure 5. Midradius microstructures. A) Alloy 1 (low C) B) Alloy 3 (high C) C) Alloy 4 (low C and carbide stabilized) D) Alloy 5 (carbide stabilized).

The DTA results for each alloy, in the VIM cast condition, are presented in Table II. The data from Alloys 1, 2 and 3 shows that carbon depresses the solidus temperature, T_s, but does not alter the liquidus, T_l, or incipient melting points,

I.M.. As a result the melting range, T_l-T_s, increases with carbon. Alloys 4 and 5, however, do not demonstrate this carbon effect. Both alloys exhibit a melting range of about 90°C.

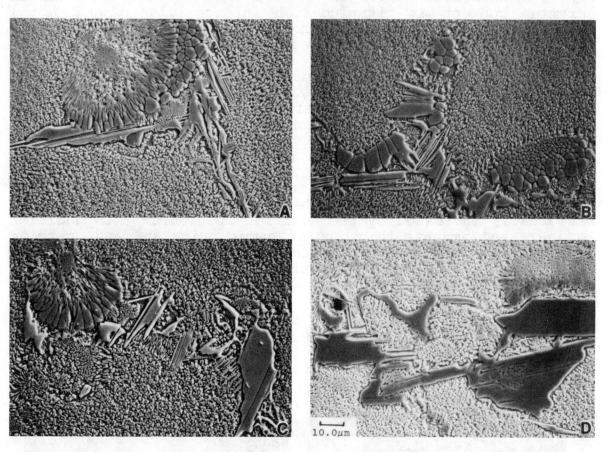

Figure 6. Midradius SEM micrographs of VADER ingots showing observed phases. A) Alloy 1 (low C) B) Alloy 2 (standard U-720) C) Alloy 4 (low C and carbide stabilized) D) Alloy 5 (carbide stabilized).

TABLE II. DTA RESULTS FROM VIM CAST ALLOYS (°C)

ALLOY #	T_l	T_s	T_l-T_s	I.M.
1	1331	1242	89	1172
2	1334	1233	101	1158
3	1332	1222	110	1168
4	1338	1246	92	1173
5	1333	1246	87	1172

DISCUSSION

The mechanisms by which chemical and structural gradients are thought to form during VADER solidification are based on the flow of intercellular liquid to the surface. The driving forces for this material transport include shrinkage of the solid surface, centrifugal forces and thermal gradients. VADER casting basically consists of dripping semimolten metal into a rotating water-cooled mold. Material in contact with the mold wall is rapidly cooled, forming a thin solid layer and establishing a steep thermal gradient. As this skin shrinks, micropores are formed which draw liquid from the contained solid+liquid mass. Further shrinkage results in the separation of the skin from the mold wall causing a reduction in the thermal gradient and a decrease in the rate of solidification. The advancement of the solid solid+liquid interface is greatly reduced, effectively increasing the extent of liquid flow by extending the time for material transport.

Current technology is not capable of eliminating these thermal and solidification gradients. However, chemical gradients may be reduced by other means, such as reducing melting ranges or by impeding liquid flow. Narrower melting ranges reduce gradients by limiting the amount of liquid available for filling the interdendritic gaps at the rapidly cooled surface. Higher concentrations of carbides or other precipitates stable during solidification reduce chemical gradients by impeding the flow of liquid in the intercellular region.

Additions of carbon increase both the concentration of carbides and the melting range. The combined effects of these opposing properties are evident in Alloys 1, 2 and 3. Alloy 1 contained an insufficient concentration of carbides; as a result, intercellular liquid flow was not effectively inhibited during solidification. Even with the benefit of a relatively narrow melting range, a steep chemical gradient was observed. At high levels of carbon (Alloy 3) the increased carbide concentration was incapable of overcoming the effect of a relatively broad melting range. The carbon level of Alloy 2 produced sufficient carbides to control liquid flow without deleteriously increasing the melting range; as a result, a relatively smooth chemical profile was produced.

Carbide stabilization was expected to increase the range of temperatures at which carbides could impede liquid flow. This particular study employed niobium as a carbide stabilizer; however, it is anticipated that other elements such as tantalum or vanadium could also be effectively used. The improvement in chemical uniformity through the addition of niobium is observed in the comparisons of Alloy 1 with 4 and Alloy 2 with 5. Both low carbon Alloys, 1 and 4, exhibited steep chemical gradients due to insufficient amounts of carbides; however, the carbide stabilized alloy produced a noticeably smoother chemical profile. Comparing the chemical uniformity of Alloys 2 and 5 further demonstrated the benefit of carbide stabilization. The addition of niobium also eliminated the detrimental effect of carbon on the melting range. It appears that the formation of stable carbides effectively removes carbon from solution, thereby eliminating the melting range effect.

Although the addition of niobium enhanced the VADER processibility, it also resulted in an increase in the formation of globular gamma prime and eta. The effect of these phases on mechanical properties has not been determined, but it is expected that, if they cannot be eliminated by thermomechanical processing, they will decrease the strength of the alloy by virtue of tying up gamma prime formers. A simple chemical equivalence calculation shows that only about 0.25 wt% niobium is required to form NbC with all of the available carbon. The niobium-bearing carbides observed contained approximately equal amounts of titanium and niobium.

Therefore, it is possible that about 0.13% niobium would be sufficient to stabilize the carbides. This would effectively reduce the tendency of the alloy to form globular gamma prime and eta as well as stabile the alloy with respect to sigma formation. PHACOMP calculations suggest that a niobium content of about 0.15% would be stable at chromium levels on the order of 17.5%.

CONCLUSIONS

The chemical and structural uniformity of VADER withdrawal cast ingots can be altered by alloy modifications. Modifications, which affect the relative amounts and stability of precipitates or affect the amount or characteristics of the liquid during solidification, can greatly influence the relative structural and chemical uniformity of the ingot.

Insight into the relationships of alloy chemistry and VADER uniformity clearly demonstrates the potential for alloy design to optimize the process.

REFERENCES

1. J. R. Brinegar, L. F. Norris, and L. Rozenberg, "Microcast-X Fine Grain Castings - A Progress Report," Superalloys 1984, ed. M. Gell et al (Warrendale, PA, TMS-AIME, 1984), 23-32.

2. B. D. Ewing, K. A. Green, "Polycrystalline Grain Controlled Castings for Rotating Compressor and Turbine Components," Superalloys 1984, ed. M. Gell et al (Warrendale, PA, TMS-AIME 1984), 33-42.

3. W. J. Boesch, G. E. Maurer, and C. B. Adasczik, "VADER - A New Melting and Casting Technology," High Temperature Alloys for Gas Turbines - 1982, ed. R. Breenetaud et al, (The Netherlands, D. Reidel Publishing Company, 1982), 823-838.

4. K. O. Yu et al, "A Comparison of VIM, VAR and VADER Superalloy Ingot Structures," Proceedings of the Vacuum Metallurgy Conference - 1986, ed. L. W. Lherbier et al, (Warrendale, PA, Iron and Steel Society Inc), 167-174.

5. Y. E. Vasil'ev et al, "Investigation of Vacuum Arc, Double Electrode Remelting (VADER)," Steel in The USSR, 17 (Jan. 1987), 30-33.

6. J-L Xu et al, "Summary Abstract: Comparison of Ingot Structure of Vacuum Arc Remelting - Vacuum Arc Double-Electrode Remelting Processes: Study of the Possibility of Using Cast Titanium Directly as an Isothermo-forging Billet," J Vac. Sci. Technol., A5(4)(1987), 2682-2683.

7. L. Ouichou, F. Lavlud, and G. Lesoult, "Influence of the Chemical Composition of Nickel-Base Superalloys on Their Solidification Behavior and Foundry Performance," Superalloys 1980, ed. J. K. Tien et al (Warrendale, PA, TMS-AIME, 1980), 234-244.

EFFECT OF HIP PARAMETERS ON FINE GRAIN CAST ALLOY 718

Patty Siereveld and John F. Radavich
Purdue University, W. Lafayette, IN 47907

Tom Kelly
General Electric Co., Cincinnati, OH 45202

Gail Cole
Howmet Turbine Corp., La Porte, IN 46350

Robert Widmer
Industrial Materials Tech., Inc., N. Andover, MA 01845

Abstract

A study was undertaken to determine the effects of HIP parameters which would close porosity and economically reduce segregation. Cast 718 samples with boron levels of 30 ppm were HIPed at 2050°F/15 ksi/3 hr, 2000°F/30 ksi/6 hrs, and 1950°F/45 ksi/9 hrs. The HIP temperature range of 2050°F to 1950°F was selected as the temperature range within which grain growth could be controlled. HIP pressures of 15-45 ksi were selected as those available in commercial HIP furnaces.

The results of the study showed that porosity was closed by all HIP cycles used. The degree of segregation varied inversely with HIP temperatures used, but residual chemical segregation was still present even after the highest temperature HIP cycle. The amount of Laves decreased as the HIP temperature increased.

The room temperature tensile properties are sharply reduced when the HIP temperature is below 2000°F due to insufficient Cb diffusing from interdendritic to dendritic areas.

Superalloys 1988
Edited by S. Reichman, D.N. Duhl,
G. Maurer, S. Antolovich and C. Lund
The Metallurgical Society, 1988

INTRODUCTION

Components of fine grain cast 718 are being considered for various applications where LCF properties are critical. Regular castings of alloy 718 normally are large grained and in addition to porosity contain varying amounts of segregation which require longtime and high temperature homogenization practice [1,2,3]. The normal HIP cycle uses temperatures of 2125°F to 2175°F at 15 ksi for 3-5 hours. Because of the inherently large grains, grain growth has not been a serious problem but closure of casting porosity was the prime consideration.

Recent developments by the Howmet Turbine Components Corp. in casting technology have resulted in the production of fine grain castings of alloy 718 by a process called Microcast X. Due to the rapid solidification of the Microcast process, cast 718 with grains of ASTM 3-5 are obtained. However, the very rapid solidification of this process results in decreased segregation but extensive porosity. Both porosity and segregation need to be minimized by HIP cycles without sacrificing LCF life due to grain growth.

A study was undertaken to determine the effects of HIP parameters which would close porosity and economically reduce segregation. Cast 718 samples with boron levels of 30 ppm were HIPed at 2050°F/15 ksi/3 hrs, 2000°F/30 ksi/6 hrs, and 1950°F/45 ksi/9 hrs. The HIP temperature range of 2050°F to 1950°F was selected as the temperature range within which grain growth could be controlled. HIP pressures of 15-45 ksi were selected as those available in commercial HIP furnaces.

STRUCTURAL CHARACTERIZATION TECHNIQUES

The structural responses of the as-cast and HIPed samples were evaluated by optical and scanning electron microscopy of electropolished and etched surfaces. Some chemical analyses were carried out with the dispersive X-ray analyzer attached to the SEM. Identification of phases was carried out on electrolytically extracted residues of selected samples and comparing the diffraction patterns to known standard patterns available at Micro-Met.

Additional heat treatments were carried out to determine the response of the HIPed structures to possible post HIP commercial cycles or heat treatments. Other non-conventional heat treatments were given to various samples to "TAG" the residual segregation by precipitation of δ or γ' phases.

RESULTS

SEM

The as-cast structural pattern as a composite-like structure with interdendritic segregation consisting of MC and Laves phases formed during solidification surrounded by precipitation of δ, γ', and γ phases during the cool down from temperature. The porosity is mostly found in the interdendritic regions in the as-cast condition and was closed by all HIP cycles and is easily revealed by the preparation techniques.

Sample 1 - As-Cast

Figure 1 show the wide variations of segregation present. At low magnification, the high Cb areas are very bright and consist of large islands of eutectic-like Laves phase with small white MC particles interdispersed in the Laves regions. The dendrites are the dark areas and contain very low amounts of Cb. Larger and more discrete MC particles are found scattered in the dendritic areas.

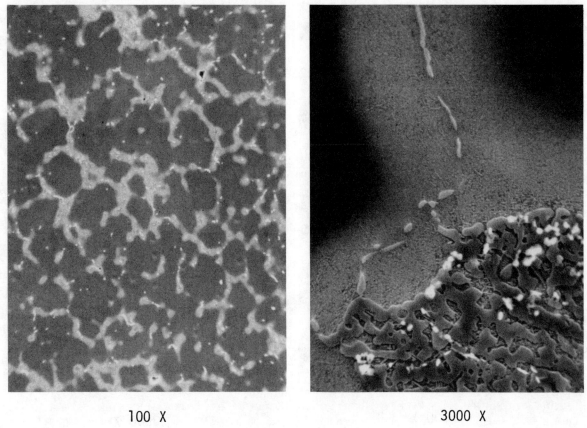

100 X   3000 X

Fig.1. As Cast

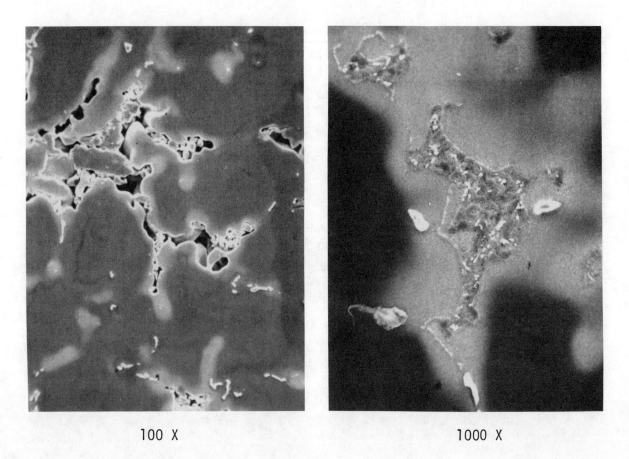

100 X   1000 X

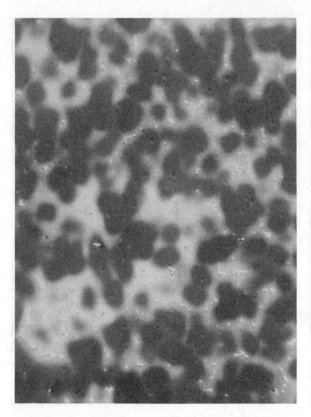

100 X

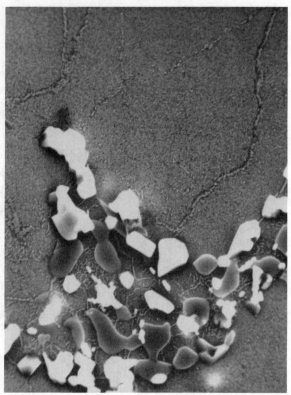

3000 X

Fig. 2. 2050°F HIP

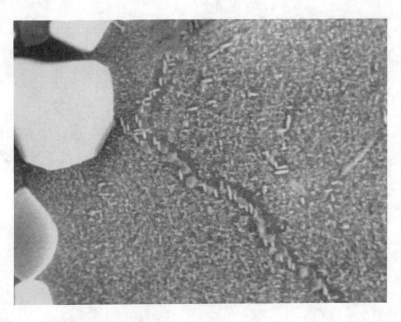

10,000 X

The Laves island is surrounded by plates of γ' which become smaller as the distance increases from the Laves islands into the low Cb dendrites. The dendrite areas show no evidence of precipitation as can be detected within the resolution of the SEM. The porosity appears to be associated with the interdendritic regions and heavy Laves precipitation.

The grain boundaries of the as-cast material show precipitation which appears to be a Laves phase. This identification is based on the similar preparation behavior of the grain boundary phase as that of the Laves islands.

Sample 13 - 2050°F/15 ksi/3 hr HIP

HIPing at 2050°F produces a less distinct black and white picture of segregation as compared to that of the as-cast condition, Figure 2. The white areas indicating the presence of a Cb rich precipitation extend further into the dendrite regions. This phase is probably γ'. No plates of large δ phase are present.

The Laves islands appear to have been broken up and an extensive large MC precipitation is formed in the Laves regions. Subgrain networks are seen extending into the dendrite areas.

Sample 23 - 2000°F/30 ksi/6 hrs HIP

Figure 3 shows more contrast in the black and white segregation pattern than that of Sample 13. Fewer and smaller MC particles are present in the Laves areas, but more than in the original as-cast condition. The Laves islands show less tendency to have broken down. Subgrain boundaries are present and plates of γ' appear near the Laves particles.

Sample 33 - 1950°F/45 ksi/9 hrs HIP

Figure 4 shows more Laves islands remaining after HIPing at this temperature, and the Laves islands are not showing any tendency to break down as seen in Figure 2. The amount of the MC formation in the Laves regions is much less than found at 2050°F or 2000°F but more than present in the as-cast condition. δ phase plates are growing from and near the Laves particles. The subgrain boundaries show small γ' precipitation as does areas around the δ phases.

Characterization of Effects of Cooling Rate

Two samples of as-cast 718 were given a heat treatment of 3 hours at 1925°F. One sample was furnaced cooled and the other was water quenched. Figure 5 shows the resultant phases when the sample is furnace cooled. Delta plates are growing from the Laves phase and γ' precipitation is found in the Laves region. Porosity is very evident in the Laves areas. On the other hand when the sample is water quenched, only the MC and Laves phases are found as seen in Fig. 6. It appears that the appearance of δ and γ' phases is a function of the cooling rate through the temperature range of 1600°F–1900°F.

Effect of Post HIP Heat Treatment -- 1925°F/1 hr + 1400°F/5 hrs

Samples with the different HIP cycles were given 1925°F/1 hr + 1400°F/5 hrs thermal treatments. Figure 7 shows the changes of diffusivity of the segregation pattern due to the density of the precipitation of the γ' phase at 1400°F. As the

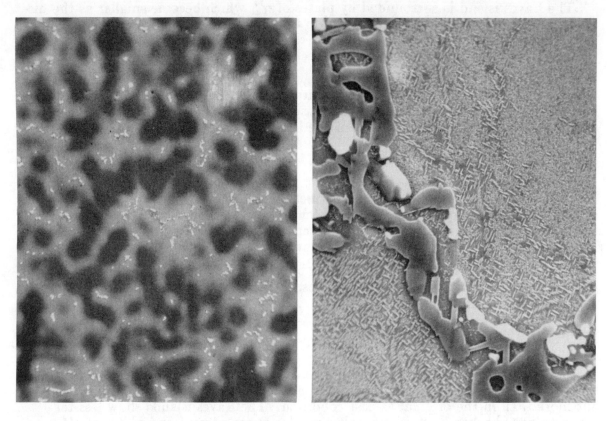

100 X   3000 X

Fig. 3. 2000°F HIP

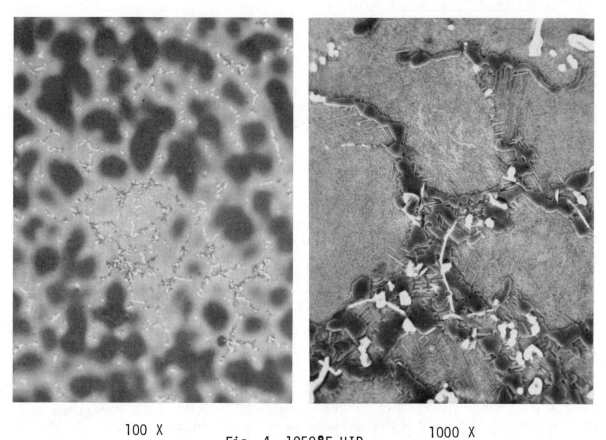

100 X   1000 X

Fig. 4. 1950°F HIP

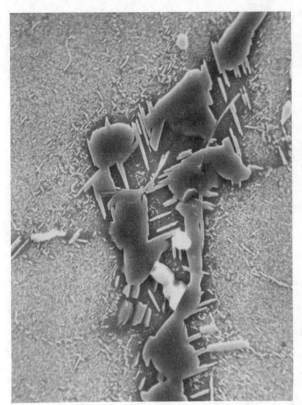 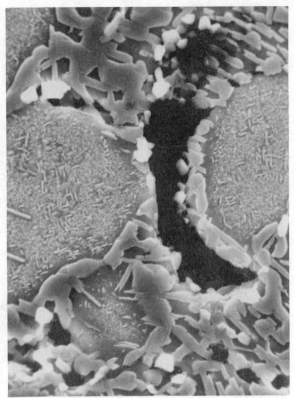

3000 X   3000 X

Fig. 5. As Cast + 1950°F/3 hr/FC

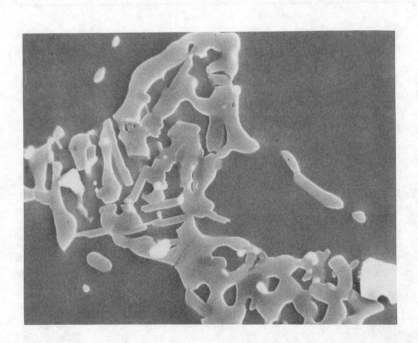

3000 X

Fig. 6. As Cast + 1950°F/3/WQ

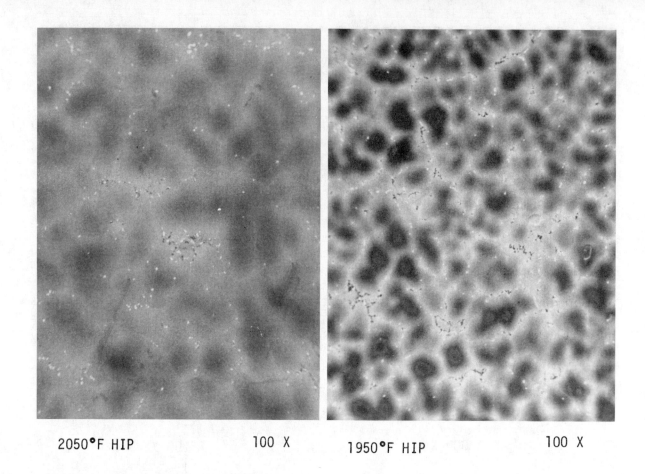

2050°F HIP   100 X   1950°F HIP   100 X

Fig. 7. 1950°F/1h + 1400°F/5h

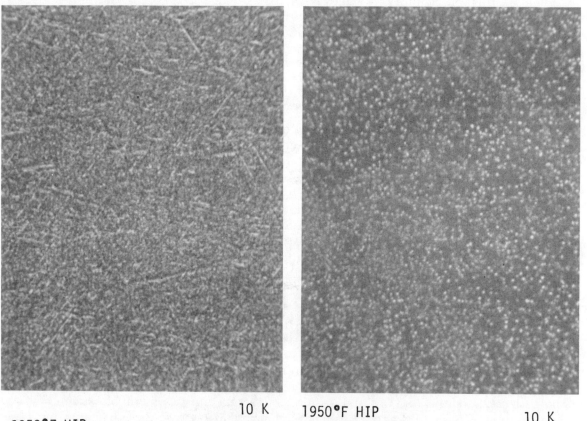

2050°F HIP   10 K   1950°F HIP   10 K

Fig. 8. 1925°F/1h + 1400°F/5h

temperature of the HIP cycle is decreased, the contrast of the white and black pattern increases signifying less precipitation occurring in the dendritic areas. High magnification studies of the white areas show that the density of the precipitation in the white or high Cb areas decreases with decreasing HIP temperature. Figure 8 clearly shows that the dark areas precipitate fewer particles when the HIP temperatures decrease. The precipitation in the dark or dendrite regions tend to show more spheroidal shapes when the density of the precipitation is less. This precipitate may be a γ' phase due to the low Cb contents in the dendritic areas.

X-Ray Diffraction Results

The results of the X-ray diffraction studies of the extracted residues from selected samples show the as-cast phases are predominately MC and Laves phases. When the material is HIPed at increasingly higher temperatures, and longer times, the amount of the MC phase increases relative to the Laves phase.

Effect of HIP Cycles on Mechanical Properties

The effect of the HIP cycles on the room temperature tensile properties is shown in Table II. Although porosity may be closed and grain growth does not occur at temperatures as low as 1900°F, a limitation of using a low temperature HIP cycle is that Cb diffusion from the interdendritic to the dendrites occurs very slowly. Dendrites which contain less than 4% Cb do not precipitate the strengthening γ/γ' phases which results in lower tensile properties as seen in Table II. Thus, to ensure adequate Cb diffusion, HIPing should not be carried out below 2000°F.

TABLE II. Microcast X Mechanical Test Summary

Serial Number	HIP Parameters	Boron Level	Room Temp. Tensile Data*			
			UTS	YS	EL	RA
13	2050°F/15 ksi/3h	.003	169.4	147.4	12.1	17.9
23	2000°F/30 ksi/6h	.003	168.6	145.6	12.5	19.7
33	1950°F/45 ksi/9h	.003	156.1	137.2	6.7	12.0
1	1900°F/25 ksi/3h	.003	146.5	123.6	5.2	5.5

*HT = 1950°F/1h + 1325°F/8h + 1150°F/8h

CONCLUSIONS

1. Porosity was found to be concentrated in the interdendritic regions and is associated with the Laves phase. The amount of porosity decreased with increasing boron levels. All HIP cycles closed the porosity in the cast material.

2. The degree of segregation remaining after HIPing varied inversely as the HIP temperature, i.e., less segregation with higher temperatures. The segregation pattern in the fine grain material was similar to that found in conventionally cast materials except it was on a much smaller scale due to the finer DAS. It consisted predominately of Laves and MC phases formed during the high temperature portion of the solidification surrounded by δ, γ' and γ phases formed

temperature portion of the solidification surrounded by δ, γ' and γ phases formed during the cool down, while the dendritic regions were precipitation free.

3. X-ray studies of the as-cast material showed mainly strong MC and Laves phases. The amount of the Laves phase decreased with higher HIP temperatures while at the same time the amount of MC apparently increased. It appears that carbon was made available for more MC formation from either dissolved matrix carbon or carbon freed by primary MC breakdown. Since Laves phases were found to contain approximately 25% Cb, dissolution of the Laves phase would enrich the interdendritic areas in Cb. EDAX analyses showed all interdendritic areas to contain much higher Cb and lower Fe + Cr compared to the dendritic areas.

4. All of the HIPed samples when subjected to the "TAG" treatment for residual segregation determination confirmed that segregation is present in amounts inversely proportional to the temperature of HIPing.

5. Cb does not diffuse readily below 2000°F which results in a lack of sufficient Cb for γ' and γ precipitation during aging. The room temperature tensile properties are severely degraded.

REFERENCES

1. John F. Radavich and Robert G. Carlson, "Microstructural Characterization of Cast 718," T. M. No. TM85-512, 10/2/85.

2. John F. Radavich and Shifan Tian, "Effects of Composition on Microstructure of Cast Alloy 718," 1986 Research.

3. Susan Jones, "Effects of Composition on the Structure and Mechanical Properties of Cast Alloy 718," M.S. Thesis, Purdue University, May 1987.

HISTORY OF CAST INCO 718

Oren W. Ballou   and   Melvin W. Coffey

Casting Design Manager   and   Process Engineering Manager
Production Development Engineering
Large Structurals Business Operation
Precision Castparts Corporation

Abstract

PCC entered the Large Structurals Casting business for aircraft engines in 1965. Large castings were 24 inches in diameter with a ship weight of 50 pounds and made of IN-718. By the late 1970's, large castings were 50 inches or more in diameter with a ship weight of 350 pounds. During that time period dimensional requirements and part quality became much more stringent. In order to comply with these new requirements the processing technology had to change. X-ray film requirements moved from 4084 square inches of film to 14,000 square inches. On similar castings Hot Isostatic Pressing (HIP) was introduced to improve part internal quality. Dimensional quality was improved with better process control, the incorporation of target machining and coordinate measuring machines.

After more than 20 years as the work-horse alloy of large structural castings IN-718 is being gradually replaced. Higher engine operating temperatures are necessitating the introduction of alloys such as Rene 220C, C263, IN-939, and MM509. PCC has several programs in place to evaluate the castability, weldability and mechanical properties of these alloys.

HISTORY OF CAST INCO 718

PCC entered the Large Structural Casting business for aircraft engines in 1965. Large castings were 24" diameter with a ship weight of 50 pounds. The primary alloy was INCO 718. The general requirements were:

>   Tolerances: ± .005 "/"
>   Wall Thickness: .100 minimum with ± .025 tolerance
>   X-Ray Quality: Shrink-plate 4; gas-plate 5
>   Penetrant Quality: .030 Non-INterpretable
>   .060 Acceptable
>   Internal Quality: No limits.

The aircraft engine design engineers used this criteria in their designs for the next 15 years. The cost effectiveness and the design freedom of structural castings out-weighed any associated quality or weight penalties. INCO 718 allowed the Designers to increase the aircraft engine operating temperatures from the 400 alloy series of 900 °F to the 1200 °F range.

FIGURE I

TYPICAL LARGE CASTING

1960's era actual example 28" dia. ± .080 wall thickness .050/ .120

8 strut stand-ups .030R

In the late 1970's the size of large casting more than doubled and the quality requirements were increased. They typical casting may weigh as much as 350 pounds and be 50 inches or more in diameter. The general requirements are:

        Tolerances: ± .003 "/" To Datums
        Wall Thickness: .050 Minimum with ± .015
        X-Ray Quality: Shrink-plate 1 or less; gas-plate 3
        Penetrant Quality: .015 Non-Interpretable
                                .030 Rejectable
        Internal Quality: Equal to the surface requirements

These new Quality requirements were incorporated at the same time as the inspection frequency and coverage was increased. Initially 3 castings were fabricated into a frame that today is cast as one piece. The initial three castings required a total of 4084 square inches of x-ray film. The current one piece casting requires over 14,000 square inches of film.

In order to comply with these new requirements the processing technology had to change. Hot Isostatic Pressing (HIP) was added to the process. This operation did improve quality but not to the degree required by the engine manufacturers. It was also determined that H.I.P. reduces the weldability of INCO 718, particularly in heavier sections. The welding problems associated with HIPped 718 were heat affected zone (HAZ) cracks and out-gassing occurred because Argon was forced into such minute voids as non detected porosity and concentrations of Laves during the HIP cycle. There were several corrective actions implemented but the most cost effective was to improve the as cast skin, reduce the Laves concentrations and lower the HIP temperature.

The processes implemented to improve the HIP ability of cast INCO 718 had a minimum impact on eliminating the H.A.Z. cracks. Several process improvements, such as extended pre-HIP anneals, chemistry control and welder techniques reduced the HAZ cracks but the most consistent, cost effective method is to restrict post HIP welding. PCC has a vast amount of data which illustrates that HIP is very effective in eliminating all weld associated cracking.

Several process innovations and improvements have been incorporated to enhance the HIPability of INCO 718. The challenge was to improve the process and resulting quality without significantly increasing the costs. These process innovations and more rigid controls did improve the NDT quality to the degree the customer required.

TYPICAL LARGE CASTING
1980'S ERA

The new dimensional quality requirements were addressed with better controls and the incorporation of target machining. The dimensional quality of large structural castings has always been good but the methods and tools to confirm this on a part by part basis was not available, at least in a cost effective method. Small cast hardware types of fixtures were not effective tools to measure large structures and they were very expensive. Coordinate Measuring Machines in conjunction with targeting machines, both computer controlled, are very effective tools for dimensional control with a permanent record of the selected dimensions. The human element has been removed from the part by part dimensioning of large structural production castings.

After more than 20 years as the work horse alloy of large structural castings, INCO 718 is being gradually replaced. The new large generation aircraft engines are being designed to operate at temperatures higher than 1200°F. Most of the high temperature resistant alloys are non weldable which makes their ease of use difficult. These alloys include Rene 220, C263, IN-939, and MM509. PCC has several programs in place to test these alloys in structural castings. Tests include castability, weldability, mechanical properties and gating.

The results of current studies show these alloys to have fair to good castability when compared against each other and IN718 using standardized castability specimen. Fill and shrink behavior are good, based on the same specimen. However, standardized hot tear specimen data shows a wide range of results. C263 is the least prone to hot tearing followed by IN718, Rene 220C, MM509 with IN939 being very hot tear prone.

Mechanical property data of cast and heat treated specimen compared to wrought heat treated specimen varies. Alloys C263 and IN939 were similar in that, room temperature ultimate strengths were higher for the cast specimen compared to wrought specimen. Yield strengths were similar. Data from 1450°F tensile test showed similar trends. Mechanical property data for cast MM509 was similar to wrought data from literature. Wrought alloy Rene 220C mechanical property data is not available for comparison to cast data.

In terms of weldability, C263 and Rene 220C are as workable as IN718. However, IN939 and MM509 are difficult to weld. The reason appears to be either carbide or secondary phase precipitation in the matrix and grain boundaries due to either solidification time or section thickness. It is suspected that the metallurgical factors contributing to the hot tear property of these two alloys also contribute to their weldability behavior. In order to address the latter problem, structural castings made from these two alloys must be either made defect free, defect tolerant, or subject to repair techniques other than welding.

SUMMARY

To summarize PCC's position on the use of high temperature resistant alloy applications in structural investment castings:

Producing complex shapes in these alloys appears possible as long as good foundry practice, gating techniques, and chemistry control are used. Casting mechanical properties can be expected to be good. Casting defect repair can be a problem and more work in this area needs to be done.

The long range outlook for large structural castings indicate that the major changes will be in thinner walls, (as low as .020) better profile tolerances (approaching ± .001 "/" in the "new" structural alloys.

SKIN EFFECT OF Hf-RICH MELTS AND SOME ASPECTS IN ITS

USAGE FOR Hf-CONTAINING CAST NICKEL-BASE SUPERALLOYS

Zheng Yunrong and Li Chenggong

Institute of Aeronautical Materials, Beijing, China

Abstract

Hf-rich melts have a skin effect when Hf-containing cast nickel-base superalloys melt or solidify at temperature range 1250-1290°C. In this interval the interdendritic liquid maintains perfect interconnection and can overflow to surface. It has been proved that Hf content in skin liquid is four times more than its average content in the alloy. The liquid has superior fluidity and wettability, therefore it has a capability of self brazing and a recovery effect on cracks. Using Hf-rich melts as a brazing material has successfully brazed cast superalloy. By means of diffusion treatment, residual Ni_5Hf in lapping zone transforms into secondary $MC_{(2)}$. As a result, in lapping zone a reasonable structure like Mar-M200+Hf was obtained. High chemical activity of the skin liquid is a disadvantage. At the surface where the skin liquid contacts shell mold, a thin layer of HfO_2 can form.

Introduction

Today, Hf-containing cast nickel-base superalloys have been used for blade material of advanced gas turbines. Many papers concerning the effect of Hf on superalloys were published. Early works were focused on microstructure and mechanical properties (1) (2). Recently. metallurgists pay more attention to the effect of Hf in solidification field and determine the distribution and segregation of Hf during solidification of superalloys (3) (4).

Our works have shown that Hf widens the solidification range of superalloy, but it can narrow down the range between temperature lost the interdendritic capillary feeding and solidus, as well as can decrease the liquid volume percent nesessary for linking the interdendritic pools in the late solidification, therefore can cause beneficial effect on castability and weldability of alloy(5). In this paper, we will explore some characteristics of Hf-rich melts during solidification and melting.

Materials and Procedures

Chemical composition of testing materials is listed in Table I.

Table I. Chemical Composition of Alloys, wt.-%

Alloy	Co	Cr	W	Mo	Nb	Al	Ti	Hf	Zr	C	B	Ni
K5	9.85	9.53	4.82	3.80	--	5.65	2.40	--	0.083	0.14	0.023	base
K5H	9.58	9.36	4.74	3.74	--	5.73	2.32	1.41	0.115	0.13	0.023	base
K3H	4.99	10.41	4.87	3.96	--	6.14	2.72	5.62	--	0.13	0.016	base
K19H	11.95	5.65	10.15	1.95	2.42	5.28	1.26	1.27	0.045	0.10	0.08	base
DZ22	9.93	8.61	12.55	--	1.06	5.21	1.98	1.96	0.078	0.14	0.016	base

Specimens of K5H superalloy for isothermal solidification were heated to 1370°C, soaked for 15 minutes, then cooled to different solidified temperature and quenched into water after 5 minutes isothermal duration. Some specimens were directionally solidified into dendritic structure and then quenched.

The plates of alloy K5H with 5 mm thickness were polished by metallography and then melted by tungsten inert gas(TIG) welding in order to observe the Hf-rich liquid film in molten and heat-affected zone.

Alloy K19H was put into a vacuum furnace and heated to 1260-1270°C. Collecting incipient melting drops and solidifying it, we analyzed microstructure of the button ingot.

The morphology and structure of ingot surface touching shell mold was determined by SEM, EDAX, EPMA and X-ray diffraction.

Using Hf-rich skin melts brazed Hf-free K3H and then lapped specimens experienced diffusion treatment to remove residual Ni5Hf phase.

Experimental Results

Change of Hf Concentration During Solidification

Onset formation of γ dendrite takes place at 1350°C for alloy K5H. As temperature decreases, dendrite grows very rapidly and dendritic skeleton basically forms at 1310°C. 7 vol.% liquid still exists in the interior of alloy and the melts at the interdendrite remain interconnected at 1250°C. There is still 0.5 vol.% liquid below 1180°C. This part of liquid does not completely solidify until Ni_5Hf forms at 1130°C.

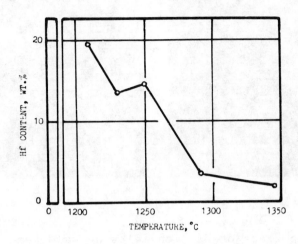

Figure 1 -- Relationship between Hf content of melts and temperature of isothermal solidification for alloy K5H.

Change of Hf concentration in the liquid is very obvious during solidification. The results were given in Fig.1. It can be seen that below 1290°C Hf content in interdentritic liquid increases rapidly, but from 1250 to 1230°C Hf concentration in molten pools decreases slightly. This is relative to precipitation of Hf-rich $MC_{(2)}$ carbide within the temperature interval.

Skin Effect of Hf-Rich Melts

A thin Hf-rich layer can be observed near the weld bead and heat-affected zone when plates of alloy K5H are melted by TIG. Hf content in this layer is four times more than its average content in the alloy. At solidified surface a river pattern caused by overflowing and spreading of Hf-rich melts along interdendritic capillary can be observed, Fig.2a. Metallographic analysis revealed a typical eutectic structure of Ni-Hf in river zone, Fig.2b.

Skin liquid film was analyzed by EPMA and its composition was compared with that of dendrite and interdendrite, as shown in Table II. As we know from the Table, the composition of liquid film is closer to that of interdendrite. This indicates that the liquid film is to overflow to surface during the late period of solidification. Fig.3 shows the profile of Hf.

Table II. Composition of Skin Melts in Alloy K5H, wt.%

Location	Cr	Al	Ni	Ti	Mo	W	Hf	Co
Skin melt	9.04	4.64	58.45	3.93	3.85	4.89	6.55	8.84
Interdendrite	9.83	5.18	62.22	3.38	4.29	3.39	3.06	8.65
Dendrite	9.37	5.09	64.50	1.90	3.59	5.84	0.00	9.71

A dark copper-color layer was found at the surface of internal shrinkage cavity of Hf-containing ingots and its content of Hf was identified as 3 to 7wt.% by EPMA. This is another important evidence for skin effect of Hf.

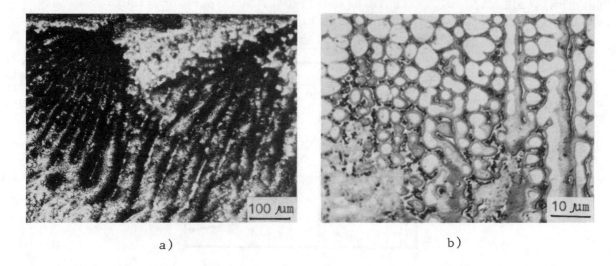

Figure 2 -- Flowing trace of Hf-rich melts at weld bead surface for K5H, a) river pattern and b) Ni-Hf eutectic at river zone.

In heat affected zone the melting point of interdendrite is 160°C lower than that of dendrite. Hf-rich liquid film appears earlier than liquid spherule which means heavy melting. The film and spherule can be observed under SEM, as shown in Fig.4.

It can be noticed that Hf-rich melts spread on surface and then form liquid drops when alloy K19H is put into a vacuum furnace and heated to 1260 -1270°C. After collecting the drop and solidifying it, a typical eutectic structure consisted of eutectic γ', Ni_5Hf, M_3B_2, MC and γ was found in the button ingot, Fig.5. Quantitative metallographic analysis has show that the volume percent of Ni_5Hf is 1.2% in alloy, but 8.1% in microingot. EDAX result has proved that average content of Hf in drop is three times as high as that in alloy.

Influence of Skin Effect on Weld Cracking

Alloy K5 and K5H were welded by TIG without filler. Crack length L and numbers N per unit area in both molten and heat affected zone were measured by quantitative metallography. The results are given in Table III. L and N in Hf-free superalloy are a magnitude order higher than ones in Hf-containing alloy. Using Hf-rich melts as a filler of weld can also decrease weld cracking.

Hf decreases tendency to weld cracking, because Hf-containing alloy has perfect connection of interdendrite and skin liquid film has a self recovery effct on formed cracks. There are a lot of distinct cracks in HAZ for alloy K5 (Fig.6a), but some cracks are welded by Hf-rich liquid, only residual mark of crack remains for K5H (Fig.6b).

Application of Hf-Rich Melts to Brazing

Ni_5Hf phase melts at 1160°C for Hf-containing superalloy and the Hf-rich melts formed have superior fluidity and wettability. As metioned above, skin liquid mainly contains Ni and Hf. We prepared a series of alloys in which content of Hf is more than 10 percent. Fig.7 shows the microstructure of a typical alloy and it is similar to Fig.2b. Hf-free K3H alloy was brazed by

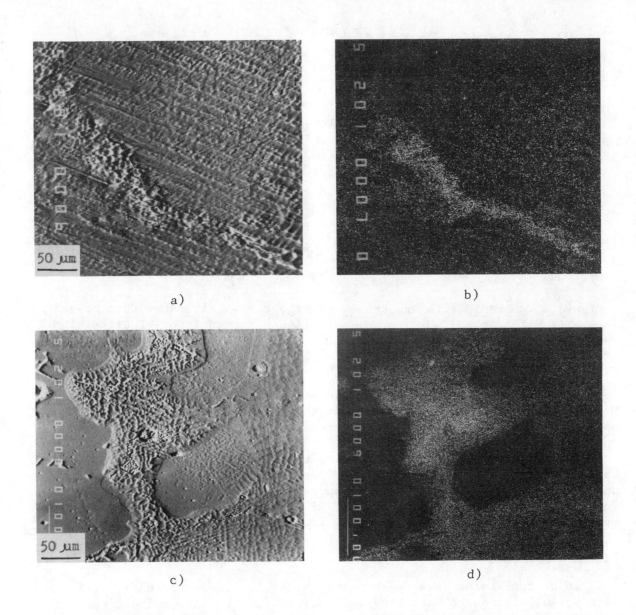

Figure 3 -- Distribution of Hf, a), b) surface of weld bead and c), d) HAZ for alloy K5H.

Table III. Comparison of Sensitivity to Weld Cracking

Alloy	Weld bead zone		HAZ	
	L(mm/mm^2)	N(number/mm^2)	L(mm/mm^2)	N(number/mm^2)
K5	0.20	0.17	0.20	0.54
K5H	0.018	0.02	0.025	0.057

this material under vacuum. In weld condition Ni$_5$Hf still exists in lapping zone. Microstructure in this region is similar to incipient melting structure of Hf-containing alloy, Fig.8.

Hf is a strong former of γ' and MC carbide. It can easily resolve into γ' and form Hf-rich secondary MC$_{(2)}$ by the reaction of Ni$_5$Hf + γ (C) $\longrightarrow$ MC$_{(2)}$ + γ. After diffusion treatment, Ni$_5$Hf can be eliminated and reasonable

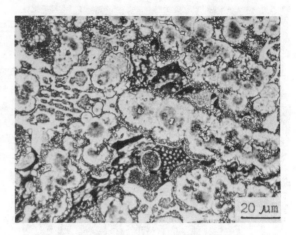

Figure 4 -- Liquid film and spherule of alloy DZ22.

Figure 5 -- Microstructure of skin liquid button ingot of alloy K19H.

lapping structure can be obtained, Fig.9. It is noticed that there is no Ni$_5$Hf, but there are a lot of eutectic $\gamma + \gamma'$ and block MC$_{(2)}$ in lapping zone. This kind of microstructure is familiar to us in Mar-M200+Hf alloy.

As compared to Ni-Cr-Si-B, Hf-rich brazing material has no brittleness, because Ni$_5$Hf phase is not brittle. Vicker hardness of the phase is 480kg/mm^2 and much lower than ones of boride and silicide. On the other hand, Si is considered as a impurity element, but Hf is a beneficial one in superalloys. Hf diffused into γ' will strengthen it further.

Formation of HfO$_2$

In Hf-containing investment castings, it is not in the early, but in the late solidification to form HfO$_2$ most easily. In the late solidification, skin liquid can react with shell mold and form a layer of HfO$_2$.

HfO$_2$ often grows as small dendrites at surface of castings. A continuous layer of HfO$_2$ was revealed after slight polishing, Fig.10a. The microphotograph is not clear due to bad electrical conductivity of oxide. Evaparating gold on oxide makes dendrites of HfO$_2$ more destinct, Fig.10b. A large area polished in Fig.10a was analyzed by EDAX and the result was illustrated in Fig.11. X-ray diffraction has proved that HfO$_2$ exists at the surface of castings for alloy K3H, K5H and DZ22. Fig.12 shows X-ray diffraction diagram of HfO$_2$ for K3H. It is obvious that oxide layer consists of pure HfO$_2$.

SEM and EPMA experiments have demostrated that HfO$_2$ appears as grained shape when Hf content in alloy is less than 1.0 wt.%, but as Hf content in alloy increases to 2.0 wt.%, a continuous layer of HfO$_2$ forms. The compositions are same in both grained and continuous HfO$_2$. Hf content of HfO$_2$ is as high as 80wt.%; concentration of oxygen is about 12 wt.%; Ni, Co and Ti contents are 3.38, 0.66 and 3.53 wt.% respectively and the amounts of other elements are very low. The characteristic of its composition corresponds to the result of Fig.11.

It is found that the scale of HfO$_2$ is discontinuous for alloy K5H with 1.7 wt.% Hf. Surface area in which HfO$_2$ exists exactly matches the spreading

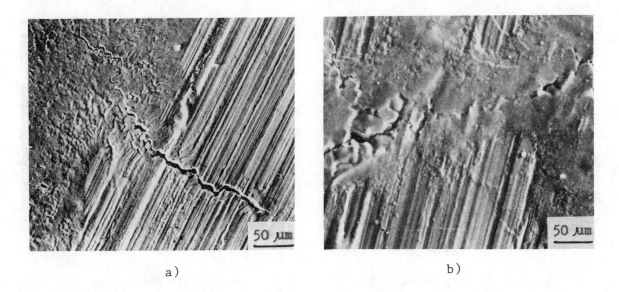

Figure 6 -- Cracking and self recovery of crack at HAZ
a) K5, b) K5H.

zone of Hf-rich melts in Fig.3. This proved relationship between formation of HfO_2 and Hf-rich melts.

Discussion of Results

In specimens of isothermal solidification and quenching at 1250°C, dendrites of primary γ are enveloped by Hf-rich eutectic nets. It is predicted that skeleton of dendrites is brazed by Hf-rich melts. The idea using the melts as brazing material originates from the fact mentioned above. Later results have demostrated that Hf-rich melts can really braze cast nickel base superalloy.

Ni_5Hf is a unhelpful phase in superalloy, although it is not brittle. The target is to eliminate the phase from lapping zone. Fortunately, Ni_5Hf is very unstable. Our work has revealed (6) that Ni_5Hf can be removed by means of 1150°C/8h treatment in two ways: the reaction of $Ni_5Hf + \gamma (C) \longrightarrow MC_{(2)} + \gamma$ and the solid solution of Ni_5Hf into γ'. In this paper the above results was proved further. Up to now, we both apply the low melting point characteristic of Ni_5Hf to brazing and obtain optimum microstructure of lapping. The benefit of this method is to avoid the troubles that Ni-Cr-Si-B brazing material brings large amount of Si, B and forms brittle phases in lapping zone.

The skin effect of Hf-rich melts is a practical phenomenon of physical metallurgy. During solidification of weld bead, the effect plays benefical role in self-brazing and makes existing solidification cracks healed. The fact that weld crack in Hf-containing alloy obviously reduces is related to the characteristic of Hf-rich melts. On the other hand, skin effect would bring disadvantageous influence on inverstment castings. Hf-rich skin liquid film having high activity contacts with shell mold and forms HfO_2. High density HfO_2 may be a trouble for using revert material, therefore the layer of HfO_2 must be removed by grinding or blast-sanding. As revert material is remelted, residual HfO_2 can be filter out through ceramic foam filters.

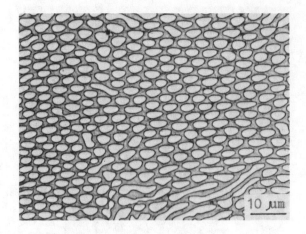

Figure 7 -- Microstructure of Hf-rich brazing material.

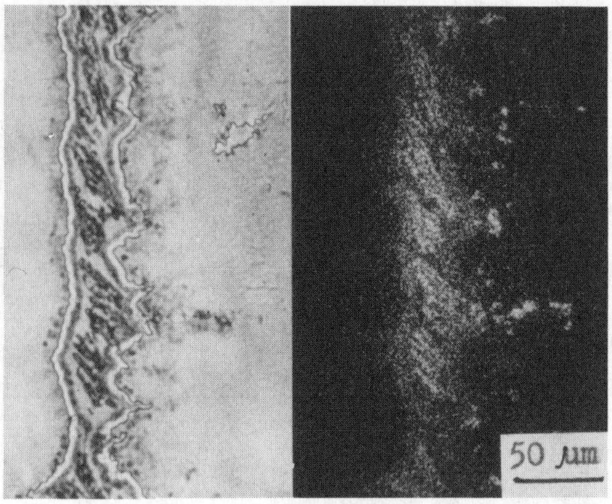

Figure 8 -- Microstructure of lapping zone and Hf profile.

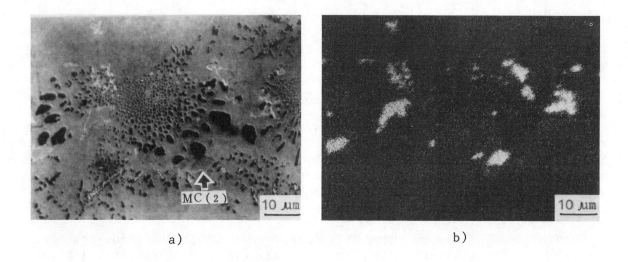

a)   b)

Figure 9 -- MC$_2$ carbide and eutectic $\gamma + \gamma'$ formed by diffusion treatment, a) morphology and b) Hf distribution.

Conclusions

1. Hf-rich melts have a skin effect when Hf-containing cast nickel-base superalloys melt or solidify at temperature range 1250-1290°C. The melts flow out from interdendritic capillary filled by molten liquid and have superior fluidity. Hf content in skin liquid is four times more than its average content in the alloy.

2. The interdendritic liquid maintains improved interconnection before eutectic γ' forms and liquid volume percent necessary for keeping the interconnection is about 7 vol.% at least. Moreover, Hf-rich skin melts have the capability of self brazing and the recovery effect on cracks. These are rea-

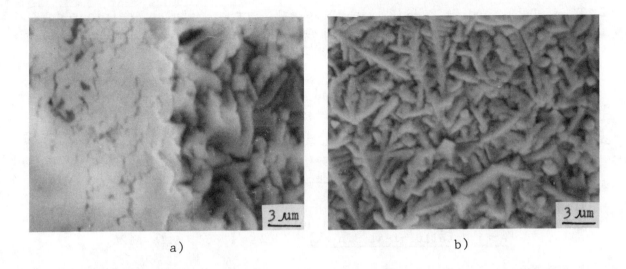

Figure 10 -- Morphology of HfO_2 for alloy K3H, a) local polishing and b) evaporating gold at surface.

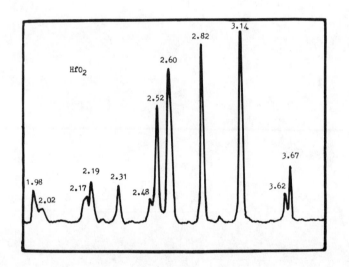

Figure 12 -- X-ray diffraction diagram of HfO_2 for K3H.

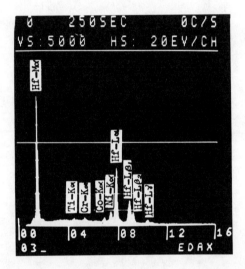

Figure 11 -- EDAX result of HfO_2 for K3H.

sons that Hf-containing superalloys show low sensitivity to cast or weld cracking.

3. Hf-rich melts as a brazing material have shown advantages of low brittleness and reasonable lapping microstructure.

4. Skin liquid can react with shell mold and form a layer of HfO_2.

References

1. J. M. Dahl, W. F. Danesi, and R. G. Dunn, " The Partitioning of Refractory Metal Elements in Hafnium-Modified Cast Nickel-Base Superalloys ", Met-

allurgical Transactions, 4 (4) (1973), 1087-1096.

2. D. N. Duhl, and C. P. Sullivan, " Some Effect of Hafnium Additions on the Mechanical Properties of a Columnar-Grained Nickel-Base Superalloy", Journal of Metals 23 (7) (1971), 38-40.

3. R. Sellamuthu, and A. F. Giamei, " Measurement of Segregation and Distribution Coefficients in MAR-M200 and Hafnium-Modified MAR-M200 Superalloys ", Metallurgical Transactions A , 17A (3) (1986), 419-428.

4. R. Sellamuthu, H. D. Brody, and A. F. Giamei, " Effect of Fluid Flow and Hafnium Content on Macrosegregation in the Directional Solidification of Nickel Base Superalloys ", Metallurgical Transactions B, 17B (6) (1986),347-356.

5. Zheng Yunrong, "Behaviour of Hf in Solidification of Cast Ni-Base Superalloys ", Acta Metallurgical Sinica, 22 (2) (1986), A119-A124.

6. Zheng Yunrong, Wang Luobau and Li Chenggong, " Influence of Carbon Content and Solidification Condition on Incipient Melting of DS Superalloy Mar-M200 + 2Hf ", (Paper presented at the First ASM Europe Technical Conference, Paris, 7-9 September 1987).

SPRAY-FORMED HIGH-STRENGTH SUPERALLOYS

K.-M. Chang and H.C. Fiedler

GE Corporate Research and Development
Materials Research Laboratory
P.O. Box 8
Schenectady, NY 12301

Abstract

Five commercial nickel-base superalloys, Rene'95, AF115, AF2-1DA, Astroloy, and MERL 76, were spray-formed into disk preforms. Metallurgical evaluation and high-temperature deformation studies were performed on the as-sprayed materials. Spray forming offers these superalloys the advantages of rapid solidification: segregation-free, uniform structure with a fine grain size. The preforms demonstrated good forgeability even without superplasticity in the as-sprayed condition. All alloys were press-forged and supersolvus-annealed. High-temperature strengths and stress rupture properties were compared, and AF115 shows the best temperature capability.

Introduction

Spray forming is a new process that combines rapid solidification and high deposition rate. This process involves atomizing a stream of molten metal into droplets with high-speed gas, with the droplets being collected on a substrate before solidifying. As compared to the conventionally cast ingot, the spray-formed deposit, or preform, has little segregation and small grain structure. The process is considered as one of the least cost methods to produce high-quality, rapidly solidified (RS) materials (1).

For the past twenty years, the advantages of rapid solidification for high-strength superalloys has been realized (2). Powder metallurgy (P/M) processing is currently the standard route to high-performance aerospace components, such as compressor and turbine disks in advanced aircraft engines. The high inherent cost of powder processing limits the applications of RS alloys. Spray forming offers less cost because of the absence of collecting and handling powders, and also results in a low oxygen content in the preforms.

In the investigation to be reported, five commercial P/M superalloys, including Rene'95, AF115, AF2-1DA, Astroloy, and MERL 76, were prepared by spray-forming. The tensile and stress rupture properties of the preforms were examined after hot die press forging. In addition to their metal processing characteristics, alloys are compared for their performance capability.

Experimental Procedures

Table I lists the designation and the nominal compositions of 5 superalloys employed for this study. Also included in Table I are the density and the volume fraction of strengthening precipitates estimated for each alloy. Because of the high precipitate content (48 - 61%), these alloys can only be processed through rapid solidification (3).

Table I. Chemical Compositions of Five Commercial P/M Superalloys

ALLOY:	RENE'95	AF115	AF2-1DA	MERL76	ASTROLOY
Ni	BALANCE	BALANCE	BALANCE	BALANCE	BALANCE
Co	8.00	15.00	10.00	18.50	17.00
Cr	13.00	10.70	12.00	12.50	15.00
Mo	3.50	2.80	2.75	3.20	5.00
W	3.50	5.90	6.50	---	---
Al	3.50	3.80	4.60	5.00	4.00
Ti	2.50	3.90	2.80	4.40	3.50
Nb	3.50	1.70	---	1.40	---
Ta	---	---	1.50	---	---
Hf	---	0.75	---	0.40	---
Zr	0.05	0.05	---	0.06	0.045
C	0.060	0.050	0.040	0.020	0.020
B	0.010	0.020	0.015	0.020	0.030
PPT. v%	49%	55%	52%	61%	48%
DENSITY (lb./in^3)	0.298	0.301	0.301	0.283	0.289

Figure 1　Disk preform of Rene'95 made by spray forming.

One 20 kg heat of each alloy was prepared by vacuum induction melting (VIM) and cast into a 100-mm-diameter, chilled copper mold under an argon atmosphere. The ingots were conditioned and remelted by induction heating in a magnesia crucible. A zirconia nozzle was cemented into the bottom of the crucible, and a disk of the alloy being melted was inserted into a recess in the nozzle. Spraying began when the disk melted, and nitrogen was used as the atomizing gas. The atomized stream was deposited on a rotating disk of cordierite. A typical deposit is shown in Figure 1.

Isothermal compression tests were carried out at forging temperatures to study the hot deformation behavior of spray-formed materials. A closed-loop servohydraulic machine equipped with a resistance furnace was controlled by a constant strain rate signal generator. Cylindrical coupons cut from the preforms were tested in air with a 60% reduction in height.

A billet about 75 mm by 75 mm by 90 mm was machined from the preforms for press forging. Forging was carried out on a 300-ton press with the die blocks preheated to 900°C (1652°F). The press speed was selected at 7 in/min. The forging procedures included an initial upset with 42% reduction and a final press with 43% reduction. In all cases, the forging temperature was kept below the precipitate solvus of the alloys.

Metallographic samples were prepared by using conventional mechanical grinding and polishing procedures. An etchant consisting of 10 ml HCl, 10 ml HNO_3, and 30 ml H_2O_2 revealed the grain structure. Subsize round tensile specimens of 0.10-in.-gauge diameter were machined and low-stress ground for both tensile and rupture testing. Tensile tests were performed in a vacuum chamber with an initial strain rate of 0.05 per minute. Stress rupture tests were run under an argon atmosphere by using a lever-arm machine with a constant load.

Results

Microstructure

All compositions were spray-formed successfully except AF2-1DA. When cut into halves, a large crack was observed at the center of the AF2-1DA preform. The crack developed during the cooling after solidification. A second attempt with a careful control of cooling still resulted in a crack. The other four alloys that were processed with the same spraying condition did not show any cooling crack.

The as-deposited microstructure (Figure 2) consists of equiaxed grains with no indication of chemical segregation. The grain size is about 25 to 40μm, depending upon the alloy composition. A certain degree of annealing after rapid solidification is expected to cause some grain growth before the preform is cooled. The cooling rate is also reflected by the formation of precipitate particles inside the grains and along the grain boundaries (Figure 2.b).

Figure 3 shows the relative density values of a slab cut through the center of the Rene'95 preform. Low density regions are found along the bottom horizontal row and the center line of rotation. Porosity as seen in Figure 2.a is responsible for the low density in both regions, but the formation mechanism may differ. One porosity is caused by the deposition on the cold surface of collector; the other is a consequence both of centrifugal forces and of nonuniform deposition parallel to the face of the collector disk. All pores are isolated and do not connect with the free surface. The forging brought the preforms essentially to full density; the porosity can also be removed by hot isostatic pressing (HIP). All spray-formed preforms contained less than 35 ppm oxygen. There is a substantial pickup of nitrogen that is, the atomizing gas, to the amount of 100 to 150 ppm, depending on the alloy being atomized.

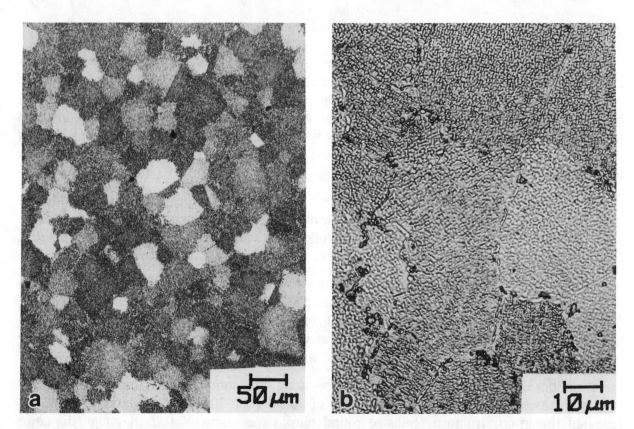

Figure 2   Metallography of as-sprayed Rene'95: a.) a low magnification showing the uniform equiaxed grain structure with some porosity; b.) a high magnification showing the cooling precipitates inside grains and along grain boundaries.

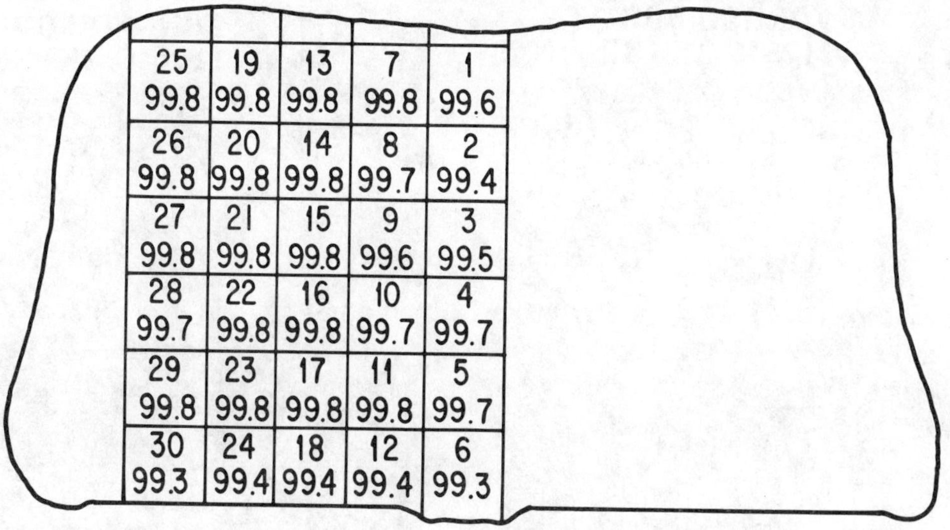

25	19	13	7	1
99.8	99.8	99.8	99.8	99.6
26	20	14	8	2
99.8	99.8	99.8	99.7	99.4
27	21	15	9	3
99.8	99.8	99.8	99.6	99.5
28	22	16	10	4
99.7	99.8	99.8	99.7	99.7
29	23	17	11	5
99.8	99.8	99.8	99.8	99.7
30	24	18	12	6
99.3	99.4	99.4	99.4	99.3

Figure 3   Density measurements on the cross section of a Rene'95 disk preform. 100% density equals 8.30 g/cc.

Superalloy structures are determined by the forging conditions and the subsequent heat treatments. A supersolvus solution treatment was applied to all preforms so that the property comparison will be predominantly the chemistry effect. Fully recrystallized grain structures were developed for every alloy.

Forgeability

One of the significant advantages of RS materials is the hot workability provided by their fine and homogeneous structure. Figure 4 shows various Rene'95 coupons tested in compression with a 60% reduction. The conventionally cast coupon suffers a serious cracking problem, while both P/M and spray-formed coupons can sustain the deformation satisfactorily.

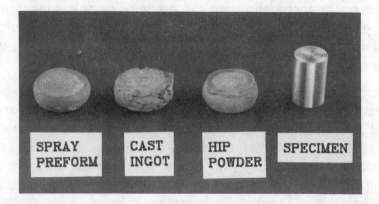

Figure 4   Improved forgeability of spray formed materials as illustrated by isothermal compression testing.

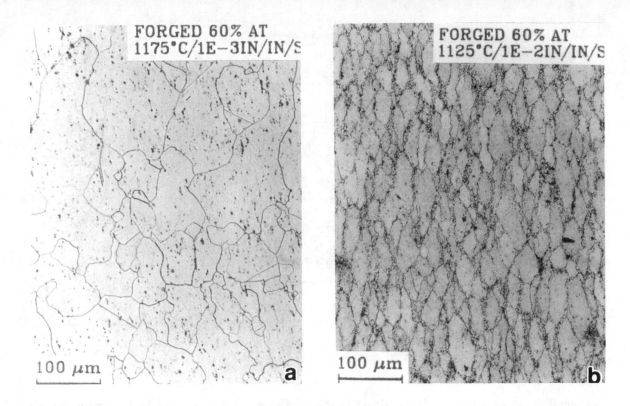

Figure 5   Metallography of as-forged Rene'95 preforms forged at: a.) supersolvus; b.) sub-solvus.

The hot deformation behavior of sprayed preforms was studied at different temperatures and different strain rates. The flow curves indicate that the as-deposited preforms never can reach a steady state, i.e., a constant stress for a given strain rate at a fixed temperature. Metallographic examination agrees with the above observation. Figure 5 shows the as-deformed microstructures of Rene'95 after 60% reduction in height. When the preform is deformed above the precipitate solvus of the alloy, the grains grow continuously during the deformation, and the slow strain rate generates a large grain size. In contrast, when the temperature is below the precipitate solvus, dynamic recrystallization takes place along the grain boundaries. However, the recrystallization can never be completed within a 60% reduction. As a result, a duplex grain structure was developed during forging below the precipitate solvus. Such a grain structure has been called the "necklace" structure, which shows some unique properties.

The stress for hot working relates to the grain size directly, illustrated by Figure 6, in which the stress at 30% strain is plotted as a function of strain rate for different Rene'95 materials tested at 1100°C. The lower working stress required for RS materials means better forgeability. The strain rate sensitivity, defined as the slope of the curves in Figure 6, can suggest the occurrence of superplasticity (4). Unlike the P/M material, which shows a strong strain rate dependent flow stress at slow strain rates, the spray-formed materials behaves similar to the cast material. No superplastic regime is found in the as-sprayed preform; the grain size is not fine enough as suggested by the metallography (Figure 5.b). However, superplasticity can be easily achieved for the spray-formed superalloys with some thermal mechanical processing (TMP).

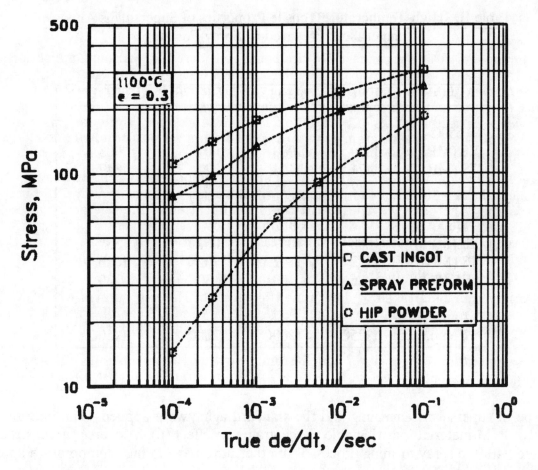

Figure 6 Comparison of strain rate dependence of flow stress for Rene'95 made from different processes.

Property Comparison

To eliminate the influence of forging operation on the alloy properties, alloy forgings were solutioned above their precipitate solvus. The solution temperatures are listed in Table II. A fully recrystallized, equiaxed grain structure was developed for every alloy. A standard aging treatment, 760°C for 16 hours, was then applied subsequently.

Table II lists the tensile properties of four spray-formed superalloys measured at 650 and 760°C, respectively. AF115 stands out as the strongest alloy, followed by Rene'95, MERL 76, and then Astroloy. The difference becomes pronounced for the 760°C tensile strength. The alloy strength order can be rationalized by the volume fraction and the chemical composition of strengthening precipitates. A high precipitate content, as well as a high level of refractory alloying addition, are the important factors for high-temperature strength (5).

Alloy temperature capability was evaluated by the stress rupture test at 663°C/758MPa and at 760°C/620MPa. The stress rupture test data for each preform forging was normalized by the Larson-Miller parameter, (absolute temperature × [log(rupture life) + 20]), and the results are plotted in Figure 7. The same relationship as found for alloy strength is observed for temperature capability. However, the chemistry of alloy matrix is believed to play a role as important as that of precipitates, especially at high temperatures.

Figure 7 also includes a data point of P/M Rene'95 tested at 760°C/620MPa. This P/M alloy was HIP'ped and forged under conditions similar to the spray-formed preform. The

Table II. High-Temperature Tensile Properties of Supersolvus Annealed Superalloys Prepared by Spray-Forming

ALLOY:	RENE'95	AF115	MERL76	ASTROLOY
650C				
YS (MPa)	1040	1080	1007	903
TS (MPa)	1471	1464	1327	1347
EL (%)	12	15	21	27
760C				
YS (MPa)	1008	1049	1007	887
TS (MPa)	1152	1242	1102	1040
EL (%)	26	30	30	30
Anneal Temp.	1175C	1200C	1200C	1150C

same type of supersolvus annealing and the standard aging were applied. The inferior rupture life of P/M material is attributable to the grain size effect (6). A coarser grain structure associated with the spray forming process is another advantage in high-temperature applications.

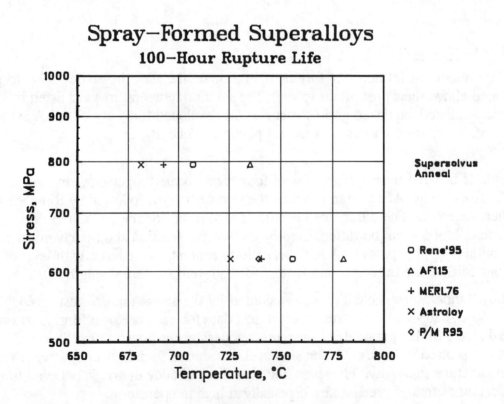

Figure 7   Comparison of stress rupture capability measured in spray-formed, high-strength superalloys.

Conclusions

A. Spray forming has been successfully applied to high-strength superalloys and offers an economical route for RS materials. The preforms show good forgeability and a homogeneous structure.

B. The spray-formed superalloys exhibit a unique structure with beneficial characteristics. The smaller grain size of the spray-formed alloy relative to a cast structure improves forgeability, and the coarser grain size relative to the P/M structure improves the high-temperature capability.

C. From the results of five commercial alloys studied, the alloy containing high refractory metal additions like AF115, show the most potential for high-temperature applications.

Acknowledgement

The authors gratefully acknowledge the direct assistance of experimental work of E.H. Hearn and T.F. Sawyer. Many thanks are also due to C. Canestraro, and P.L. Dupree for mechanical testing, and to R.G. Trimberger and R.O. Auer for metal processing. Helpful discussions with Dr. M.F. Henry and Dr. M.G. Benz are greatly appreciated.

References

1. H.C. Fiedler, et al., "The Spray Forming of Superalloys," Journal of Metals, 1987, 28-33.

2. M.M. Allen, R.L. Athey and J.B. Moore, "Application of Powder Metallurgy to Superalloy Forgings," Metals Engineering Quarterly, 1969, 20.

3. F. Sczerzennie and G.E. Maurer, "Developments in Disc Materials," Materials Science and Technology, 3(1987), 733-742.

4. T.E. Howson, W.H. Couts, Jr. and J.E. Coyne, "High Temperature Deformation Behavior of P/M Rene'95," Superalloys 1984, M. Gell et al., eds., TMS-AIME, 1984, 275-284.

5. R.E. Duttweiler, "The Role of Refractory Elements in Powder Metal Superalloy Development," Refractory Alloying Elements in Superalloys, ed. J.K. Tien and S. Reichman, ASM, 1984, 77-86.

6. D.J. Deye and W.H. Couts, "Super Waspaloy Microstructure and Properties," MiCon 78, H. Abrams, et al., eds., ASTM STP 672, 1979, 601-615.

EVALUATION OF THE POTENTIAL OF LOW PRESSURE PLASMA SPRAYING

AND SIMULTANEOUS SPRAY PEENING FOR PROCESSING OF SUPERALLOYS

J V Wright

Combustion Technology and
Engineering Centre
Lucas Aerospace Limited
Burnley UK

J E Restall

Royal Aerospace Establishment
Farnborough
UK

Abstract

The potential of low pressure plasma spraying with or without simultaneous peening has been assessed as a means of producing superalloy materials. The equipment and techniques utilised are described and some of the limiting features identified. The quality of the deposits obtained has been investigated using a variety of metallurgical techniques and the structural differences attributable to simultaneous peening are described in detail. Mechanical testing, primarily tensile at ambient and elevated temperature, has been carried out to characterise the properties of the deposits obtained. Effort has been devoted primarily to work on conventional sheet materials such as Nimonic alloy 263 and to work on oxide dispersion strengthened alloys. Superior levels of tensile strength were obtained at room temperature on the sprayed and spray peened superalloys relative to ingot processed stock. Spray joining of $FeCrAlY_2O_3$ - type alloys was achieved with negligible loss in tensile strength providing that post spraying heat treatments were carried out.

Introduction

Plasma spraying is routinely used as a method of depositing metallic or ceramic materials on component surfaces to restore worn or damaged areas and to provide a variety of coatings for engineering purposes. Work has also been performed on the direct forming of components by the plasma spray route. A further extension of spray forming of materials proposed and patented by Singer [1] is that of simultaneous spray deposition and peening of metals (SSP). This process consists of the spray deposition of metal by various means on the surface of a substrate whilst concurrently bombarding the deposit with a stream of high velocity, hard, rounded particles (e.g. shot-peening balls). The peening process plastically deforms the deposit as it is being built up thereby enhancing the physical and mechanical properties. Further details are given in the literature [1].

A programme of work has been undertaken at Lucas Aerospace Limited utilising an experimental facility designed to carry out plasma spraying or spray-peening processes in a low pressure, controlled atmosphere chamber. The choice of the reduced pressure processing offers significant benefits such as:-

(i) The increased length of the plasma flame, typically 250 mm to 350 mm, provides a more homogeneous distribution of particles in the flame and a higher mean particle velocity and temperature.

(ii) The low pressure environment reduces the level of oxide contamination in the deposit.

The potential of the SSP route was assessed by comparing the structure and properties of low pressure plasma sprayed (LPPS) material produced with or without simultaneous peening.

Equipment and Processing

The facility required considerable design and procurement effort as SSP techniques are entirely experimental. It was designed to fulfil the requirements of low pressure plasma spraying, with or without concurrent peening, and to have sufficient flexibility to enable equipment modifications to be carried out easily. The equipment is shown schematically in Fig. 1 and comprises a water cooled vacuum chamber in which a plasma gun, workpiece holder and mechanical peening unit are mounted. Associated power supplies, motors for linear and rotational manipulation of the workpiece and for the shot slinger, and the powder feed system were located outside the chamber. The shot peening facility is capable of delivering high carbon steel shot at velocities up to 30 metres per second. The shot is directed onto the workpiece by a tapered chute, enabling 100% coverage to be achieved over a 100 mm square target on each traverse.

For commissioning purposes a number of powders readily available within Lucas were utilised. A series of parametric surveys were carried out to identify the more critical factors controlling deposit efficiency and integrity. As the work progressed, areas of potential interest became more clearly identified and a wider range of materials was produced for metallographic characterisation and for mechanical tests where appropriate. Powder materials have included Stellite [2] 31 and Nimonic [3] AP1. More recently work has been done on the powder processed sheet superalloys Nimonic [3] 86 and 263, and also on powder having a similar composition to the iron-based oxide dispersion strengthened Incoloy [3] MA 956 alloy.

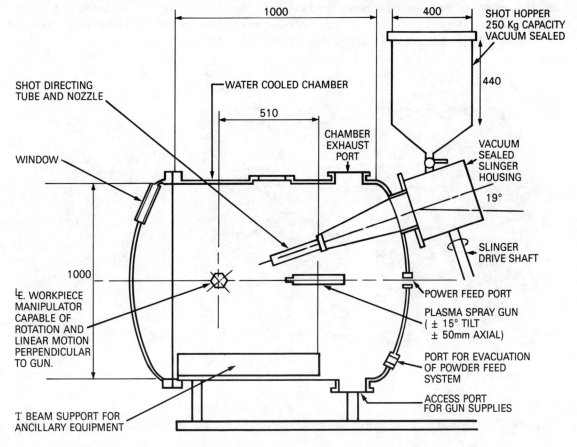

Figure 1 — Low pressure simultaneous spray peening chamber
(Dimensions in mm)

The sequence of operations for low pressure plasma spraying is well established ([4]) and will not be described further. Simultaneous peening was usually introduced after a brief period (< 30s) of LPPS deposition to permit the build up of a well bonded initial deposit on a shaped mandrel or other type of former. On completion of the spraying or spray peening operation the deposit was stripped from the mandrel by machining or chemical dissolution.

Experimental Assessment

Initial Trials

Plasma spraying trials were carried out initially using readily available spray powders to identify and optimise the critical parameters involved in LPPS. The critical factors in achieving acceptable deposit efficiencies were found to be carrier gas flow rate and particle size. The gas flow in conjunction with the gun nozzle geometry regulates the injection of powder into the mainstream of the plasma gas jet. The influence of powder particle size on deposit quality is illustrated in Fig. 2 for MAR M002 alloy powder deposited under LPPS conditions. Spraying was carried out at a chamber pressure of 45 mbar.

Prior to SSP trials an assessment of the peening intensity was made using Almen 'A' test strips. The total arc height of the strips was measured after peening to provide comparative data for various operating conditions of the slinger. The shot, made from 1.0% carbon steel was supplied in the clean and dry condition and was heat treated in hydrogen prior to use. A high energy sifter was used to separate the overspray powders from the shot so that shot could be re-used in subsequent SSP trials.

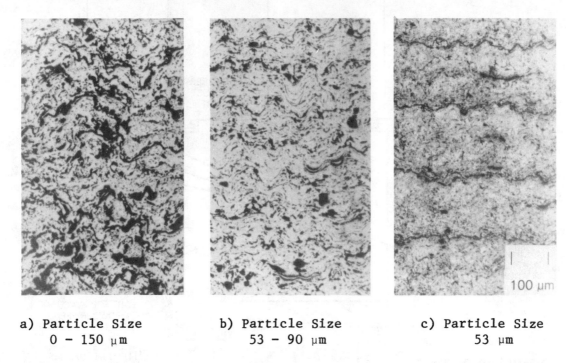

a) Particle Size 0 - 150 μm

b) Particle Size 53 - 90 μm

c) Particle Size 53 μm

Figure 2 - Effect of particle size on structure in LPPS MAR M002 deposits

Materials produced by SSP had a comparable microstructure to that seen in LPPS samples but contained lower levels of porosity. Using the image analyser (quantimet-type) technique, the inclusion count (i.e. total porosity plus inclusions) was measured at 1.8% for SSP material compared with 4.8% for the LPPS material. In both LPPS and SSP samples the major element concentrations were in accordance with the analysis of the powder. The etched microstructure revealed particles which had passed through the plasma flame without melting and became embedded in the plasma spray deposit. In SSP deposits the larger particles had been severely deformed by impact with the peening shot. Typical microstructures obtained for LPPS and SSP Stellite 31 alloy deposits are shown in Fig. 3 (a) and (b) respectively. The powder size range used for the spraying of these deposits was -150 + 75 μm. In Fig. 3 (b) the larger unmelted artefacts have effectively been eliminated and the structure is more homogenous.

Residual stress levels in spray peened deposits can be controlled by varying the peening intensity and temperature. The effect was demonstrated by slitting longitudinally tubes which had been spray deposited on a cylindrical mandrel. The change in diameter was a measure of the residual stress in the tube wall and showed that the tensile stresses in an LPPS deposit 1mm thick were typically three times those in the corresponding SSP deposit.

LPPS/SSP of Nickel-Base Superalloys

Interest in plasma spraying materials such as Nimonic alloys 263 or 86 lies in the potential for spraying thin section (sheet) components as one piece constructions, thus obviating the need for complex and expensive forming and welding operations. The work formed part of a collaborative programme with Rolls-Royce Limited. In the first instance it was necessary to establish the practicability of producing high integrity material by the plasma spray route and then to explore any benefits arising from spray peening.

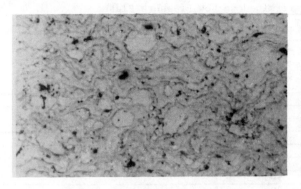 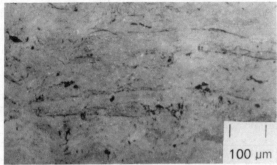

a) Deposit produced by LPPS   b) SSP deposit using same size of powder

Figure 3 - Microstructure of LPPS and SSP Stellite 31 alloy

Extensive parametric surveys were carried out on Nimonic 263 powder to determine optimal plasma spray conditions. The conditions identified as producing a sound deposit were used as baseline conditions for the spraying of all subsequent superalloy powders. The powder was supplied in two size ranges, 12 - 35 μm and 35 - 70 μm, the finer powder being used for LPPS deposition whereas a mixture of the two particle sizes was used for the manufacture of SSP deposits.

Metallographic examination of the Nimonic 263 deposits showed that both the SSP and the LPPS material had a relatively low level of porosity Figure 4. The layered structure is typical of plasma spray deposited material and, as expected the layer thickness was greater with the coarser powder. The influence of simultaneous peening on the layer profile can clearly be seen, producing a series of linked smooth facets rather than the more familiar 'splat' structure of LPPS deposits. Deposits up to 9 mm thick were produced to enable a number of miniature bar tensile test pieces to be manufactured. Data for material processed by LPPS or SSP routes are compared in Table I with corresponding data on ingot processed material. Following heat treatment at 1100° or 1150°C, the LPPS or SSP material showed improved levels of ductility and excellent proof and ultimate strengths. The strength advantage of the LPPS/SSP material was maintained at 600°C, but at 800°C there was a significant reduction in strength relative to wrought bar. In order to optimise the properties of alloy 263 deposits further work is required to evaluate the interplay between LPPS/SSP and heat treatment.

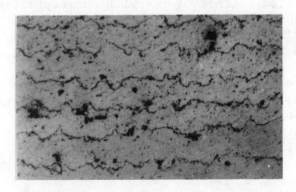 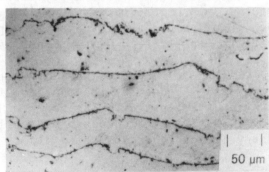

LPPS Nimonic Alloy 263 12 - 35 μm   SSP Nimonic Alloy 263 12 - 70 μm

Figure 4 - Influence of simultaneous peening on layer profile

For the limited amount of work carried out on Nimonic alloy 86 powder trends similar to those for Nimonic 263 were observed.

Table I. Tensile Test Data Determined on Low Pressure Plasma Sprayed and Simultaneously Spray Peened Nimonic Alloys 263 and 86

Material	Processing Route	Heat Treatment	0.2% P.S. MPa	UTS MPa	Elong. %	Reduction in Area %	Hv	Data Ref.
Nimonic Alloy C263 at Room Temperature	Wrought Bar	Solu. Treat & Age[1]	600	985	27	24	265	
	SSP	1100°C 1h AC	690	907	9	7	280	
		1150°C 1h AC	678	975	16	11	285	
		Solu. Treat & Age[1]	883	1048	9	7	353	
	LPPS	Solu. Treat & Age[1]	877	1180	16	16	348	
Nimonic Alloy C263 at 600°C	Wrought Bar	Solu. Treat & Age[1]	490	820	43	50	-	3
	SSP	Solu. Treat & Age[1]	905	937	7	6		
	LPPS	Solu. Treat & Age[1]	906	975	5	5		
Nimonic Alloy C263 at 800°C	Wrought Bar	Solu. Treat & Age[1]	460	587	15	26		3
	SSP	Solu. Treat & Age[1]	-	449	5	5		
	LPPS	Solu. Treat & Age[1]	-	357	11	15		
Nimonic Alloy 86 at Room Temperature	Wrought Sheet	1150°C 15m AC	438	873	45	-	200	3
	SSP	1100°C 1h AC	665	902	15	13	269	
		1150°C 1h AC	645	925	20	16	272	
	LPPS	1150°C 1h AC	-	-	-	-	271	
Nimonic Alloy 86 at 800°C	Wrought Bar	1150°C 2/4h AC	210	400	60	-		3
	SSP	1150°C 1h AC		298	7	5		

Note: [1] - 15-90 min. (depending on section) at 1150°C, W.Q. + 8h 800°C, AC

LPPS/SSP of Oxide Dispersion Strengthened Material

Incoloy alloy MA 956 develops its high temperature strength by combining an elongated lamellar grain structure with a uniform dispersion of yttria. However, the strength and structural stability of MA 956 are degraded by fusion welding processes. Both LPPS and SSP processing of powder $FeCrAlY_2O_3$ materials offer the possibility of manufacturing components directly and may circumvent the problems associated with joining.

For proprietary reasons the powder made available by the supplier had slightly different chemistry to alloy MA 956 and is designated $FeCrAlY_2O_3$. The $FeCrAlY_2O_3$ particles processed by mechanical attrition were more irregular than the gas atomised Nimonic 263/86 powders. To compensate for the reduced flow characteristics of the $FeCrAlY_2O_3$ it was necessary to reduce the powder feed rate or increase the carrier gas flow rate to the plasma gun.

Tensile tests were again carried out at room and elevated temperature to compare the properties of spray deposited $FeCrAlY_2O_3$ bar with corresponding data on Incoloy MA 956 and $FeCrAlY_2O_3$ material processed by powder consolidation and manufactured into plate. Tests were also carried out on deposits heat treated at temperatures in the range 1100°C - 1300°C, Table II.

High levels of proof and ultimate strengths were obtained at room temperature. However, the most consistent results were obtained when spraying the finer cuts of powder. Spray deposits processed from <35 μm $FeCrAlY_2O_3$ powder also performed well after heat treatment in miniature Charpy impact tests absorbing almost 13 Nm without fracture.

Table II. Tensile Test Data Determined on Low Pressure Plasma Sprayed and on Simultaneously Peened (FeCrAlY$_2$O$_3$) Material

Material and Condition	Heat Treatment	Test Temp. °C	0.2% P.S. MPa	UTS MPa	Elong. %
Incoloy MA 956 - Wrought Plate	As Supplied	20	550	680	16
FeCrAlY$_2$O$_3$ - Wrought Plate	As Supplied	20	525	620	15
FeCrAlY$_2$O$_3$ LPPS	As Deposited	20	–	765	–
	1100°C 1h AC	20	797	875	3.0
	1150°C 1h AC	20	785	900	4.0
	1300°C 1h AC	20	–	840	20
FeCrAlY$_2$O$_3$ SSP	As Deposited [1]	20		761	3
	1100°C 1h AC	20	810	860	2
	1150°C 1h AC	20	825	900	4.5
	As Deposited [1]	600	–	323	15
	As Deposited [1]	800	–	105	19

[1] Data determined at RAE, Pyestock on spray deposit material provided by Lucas

Spray Joining of Incoloy MA 956 Sheet

The joining of two sheets of MA 956 sheet (0.9 mm) with a spray deposit of FeCrAlY$_2$O$_3$ material was investigated. Problems to be overcome included joint design, bonding of MA 956 with plasma spray deposit, application of preheat and optimisation of spray deposition without developing oxide debris or porosity in substrate or deposit material. It was important that all traces of surface oxide be removed from the joint area immediately prior to mounting of the MA 956 alloy in the plasma spray chamber. Also, the duration of preheat to a surface temperature of approximately 800°C should be kept to a minimum. A typical joint profile, and microstructure after heat treatment are shown in Fig. 5.

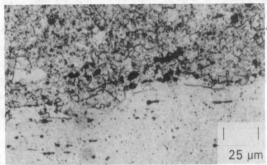

Figure 5 Joint profile and interfacial structure of MA 956 spray joined with LPPS FeCrAlY$_2$O$_3$

The tensile properties of flat strip joints are compared in Table III for MA 956 alloy measured by the supplier and by Lucas Aerospace.

Table III Tensile Test Data Determined on Incoloy MA 956 Sheet Unwelded and Low Pressure Plasma Spray Joined with FeCrAlY$_2$O$_3$

Material and Condition	Heat Treatment	Test Temp. °C	0.2% P.S. MPa	UTS MPa	Elong. %	Failure Locn.	Number of Tests
Unwelded MA 956	As recd. (Inco)	20	550	650	10	-	
	As recd. (Lucas[1])	20	-	629 ± 25	10 ± 2.5		10
	1300°C 4h AC	20	-	577 ± 12	14 ± 3.0		10
	1200°C 17h in vac.	20	470	595	8		2
Spray[2] Joined with FeCrAlY$_2$O$_3$	1200°C 17h in vac.	20	457	636	5	Parent Sheet	3
Unwelded MA 956	1200°C 17h in vac.	800	-	136	8		2
Spray[2] Joined with FeCrAlY$_2$O$_3$	1200°C 17h in vac.	800	-	114	1	Joint	3

[1] Determined on batch of MA 956 used for joining trials.
[2] Specimens machined back to original thickness taking care to remove any excess FeCrAlY$_2$O$_3$

Good levels of tensile strength were obtained in the joined test pieces and although the ductilities appear low in comparison with those of alloy MA 956 it should be borne in mind that the joints were very small in relation to the overall size of the test piece gauge length. Further study of heat treatment is required to optimise the strength and ductility of such joints.

Discussion

A conventional plasma spray facility has been modified for use at low pressure in a chamber incorporating facilities for simultaneous peening workpiece during plasma spraying. Preliminary trials showed that high density deposits were achieved when plasma spraying was carried out without peening at low pressure, particularly when fine (<53 μm) particle size powders were employed. Peening conditions were identified that enabled 100% coverage to be achieved over an area 100 mm square on each pass of the workpiece.

Metallographic examination of plasma spray deposits showed that some improvement in density was achieved using SSP processing. With SSP there was marked evidence of mechanical working of both individual particles and of the successive layers of deposit. Also powder particles which were coarse and unmelted following LPPS became severely deformed and aligned by the peening shot and were difficult to distinguish from the overall plasma spray structure. Both types of deposit showed a pronounced layering effect after etching which was related to the number of plasma spray passes across the workpiece. In SSP material the inter-laminar layers had clearly been deformed by the shot so that the interface had the appearance of a series of flat facets in contrast with the more usual plasma sprayed 'splat' surface.

The boundaries between the layers were less than 1 m thick and showed no evidence, when examined by EPMA, of oxygen or nitrogen enrichment. The structure had no significant influence on the recrystallation and grain growth in heat treated deposits.

The use of SSP processing to reduce the levels of residual stress in plasma spray deposits was clearly demonstrated in the split tube experiments. It is recognised that residual tensile stresses in LPPS deposits are already low in comparison with deposits sprayed at atmospheric pressure enabling thick deposits to be achieved in selected materials. However, the ability to control the level of residual stress by SSP will be of particular benefit in achieving thick deposits in materials of limited ductility.

Conventional sheet alloys Nimonics 263 and 86 were both successfully processed by LPPS/SSP routes. Both SSP and LPPS materials had previously been shown to have low ductility in the as processed condition. However, subsequent heat treatments effected considerable improvements. Although the data showed some scatter, clear trends were evident. There was no marked difference between the tensile properties of LPPS and SSP processed material. However, plasma-deposited material did show remarkably good proof and ultimate strengths at room temperature and 600°C. At 800°C, better properties were obtained with conventional ingot processed materials. The amount of testing carried out to date has been severely restricted by the limited availability of good quality spray powders and it felt that conclusions drawn from this preliminary exercise may be premature.

Similar problems were encountered when evaluating the properties of the mechanically alloyed $FeCrAlY_2O_3$ powder. The test specimens were also in the form of small round bar machined from a thick plate of spray deposit. In general the tensile test data reflected the trends obtained for the more conventional superalloys. Significant improvements in proof and ultimate tensile strength were obtained at room temperature following plasma spray processing but at the expense of some ductility. It is evident that further work is required to optimise heat treatments for the structures produced via plasma spray processing. The fine grain size obtained as a result of spray deposition was usually retained after heat treatment even when simultaneous peening had been employed. The effects of the shot although significant were not sufficient to promote recrystallisation and grain growth, at least for the range of materials and heat treatments investigated.

The investigation into the spray joining of the oxide dispersion strengthened alloy MA 956 was particularly encouraging. In general a number of specimens were prepared for each condition in an attempt to reduce the effects attributable to scatter in material or specimen preparation for these relatively brittle materials. Particular care was taken when preparing joint specimens to ensure that the test section was truly representative of joined MA 956. The room temperature tensile data show that an excellent joint has been achieved with properties which compare favourably with those of the parent sheet. Failure in every case was in the MA 956 material. At 800°C although the properties of the composite specimen were disappointing the bond between the MA 956 and the $FeCrAlY_2O_3$ remained intact. The measured properties thus reflect the relatively low strength of the joint material itself. Further work to modify the grain structure by thermo-mechanical processing possibly including SSP may well enable properties equivalent to the MA 956 being achieved throughout the working temperature range of the alloy.

Conclusions

Conventional superalloys can be processed by LPPS/SSP techniques and provide an interesting combination of mechanical properties at 20°C, 600°C and 800°C.

Simultaneous Spray Peened material generally showed marginally better strength than LPPS material in the as deposited condition. However, post deposition heat treatment had a significant beneficial influence on the strength and ductility relationships.

Mechanically alloyed ODS $FeCrAlY_2O_3$ powder was processed via the LPPS/SSP route yielding significant improvements in proof strength at room temperature relative to MA 956 alloy.

Spray depositing $FeCrAlY_2O_3$ powder into profiled gaps between MA 956 strips produced joints which had strengths comparable with the MA 956 sheet at 20°C and 800°C.

REFERENCES

[1] A.R.E. Singer — "Simultaneous Spray Deposition and Peening of Materials (SSP)". Metal Technology II (1984) 99-104. UK Patent Application GB 2142858A.

[2] [3] Registered Trademarks - Haynes International and Inco Alloys International respectively see Trade Publications for Composition.

[4] A.R. Nicoll et al., "Thermal Spray Coatings for High Temperature Protection". Surface Engineering I (1985) 59-71.

ACKNOWLEDGEMENTS

The Authors would like to thank the Directors of Lucas Industries plc for permission to publish the paper and also the Ministry of Defence, Procurement Executive for their financial support. The support of colleagues within Lucas Combustion Technology and Engineering Centre, Burnley and the Royal Aerospace Establishment, Farnborough is gratefully acknowledged.

THE DEFORMATION BEHAVIOR OF P/M RENE'95
UNDER ISOTHERMAL FORGING CONDITIONS

J. M. Morra, R.R. Biederman, F.R. Tuler

Mechanical Engineering Department
Worcester Polytechnic Institute
Worcester, MA 01609

Abstract

The high temperature deformation behavior of as-HIP P/M Rene'95 as a function of powder particle size has been examined. Compression testing of +230, -270+400, and -500 mesh materials at temperatures of 1038°C, 1079°C, and 1121°C and constant true strain rates of 10^{-1}, 1, and 5 s^{-1} has shown there to be virtually no difference in their flow behavior. All three materials exhibited a rapid work-hardening to a peak flow stress, followed by a gradual flow softening which approached a steady-state of flow stress at a true strain of 0.50. Peak flow stress, rate of flow softening, steady-state flow stress, and strain at the onset of steady-state flow were all seen to increase with increased strain rate and/or decreased temperature. With lowered temperature and increased strain rate the as-HIP materials exhibited non-uniform deformation which was attributed to strain gradients created within the material at prior powder particles. Deformation maps created from compression test data predict processing instability over most of the temperature/strain rate range of this study.

Introduction

The deformation behavior of various P/M nickel-base superalloys has been studied by several researchers (1,2,3,4). The goal of this work has been to define the processing parameters which result in the optimum workability of these P/M materials. A new approach to this issue was developed (5), whereby dynamic material behavior is modeled in terms of the efficiency of energy dissipation in a workpiece during processing, as a function of temperature and strain rate. This model, known also as a deformation map, can be utilized as an aide in choosing optimum processing parameters, provided that metallurgical processes corresponding to the projected efficiencies across the map are known (6).

This paper describes a study made of the high temperature deformation behavior of as-HIP P/M Rene'95, as a function of initial powder particle size, and attempts made to generate corresponding deformation maps. Particular attention will be given to discussing some of the problems encountered with the creation and interpretation of such maps.

Experimental Procedure

Argon atomized P/M Rene'95 powder (Table I) was screened into three size lots of +230, -270 + 400, and -500 mesh. Each size lot was compacted by hot isostatic pressing (HIP) at 1121°C, 103 MPa, for three hours. Gamma prime solvus was determined by DTA to be ~1159°C.

Compression specimens measuring 1.0 cm in diameter and 1.5 cm in height were machined from the as-HIP material. Shallow spiral grooves were made in the ends of the specimens for lubricant retention at hot-working temperatures. The specimens were coated with a boron nitride lubricant. Because of the limited amount of material, the test matrix consisted of only three temperatures and three strain rates. Compression testing was conducted at temperatures of 1038°C, 1079°C, and 1121°C and strain rates of 10^{-1}, 1, and 5 s^{-1}. The compression tests were performed on a 222 kN servohydraulic Instron testing system equipped with a three-zone ATS split-type resistance furnace. Each specimen was held at temperature for at least 15 minutes prior to compression, to allow for uniform heating throughout the specimen. Specimens were compressed between silicon nitride flat dies mounted on Udimet 720 compression columns, to a total true strain of 0.50, and were immediately water quenched. Constant true strain rate was maintained by an analog exponential function generator. Load-time and stroke-time data were recorded on a digital oscilloscope and were plotted as load-stroke curves on an X-Y recorder. The data were replotted in terms of stress and strain using standard equations, assuming uniform deformation and a constant specimen volume.

Microstructures of the three starting materials and each of the compression specimens were examined for gamma prime distribution and prior particles using SEM. TEM was used to determine gamma prime size distribution and morphology.

Table I. Chemical Analysis of P/M Rene'95* (in Wt. %)

Ni	Cr	Co	Mo	W	Nb	Al	Ti	C	B	Zr	O (ppm)
Bal	13.15	8.44	3.36	3.39	3.52	3.37	2.32	0.06	-	0.05	83

*Rene'95 is a trademark of the General Electric Co.

Experimental Results

Flow Curves

The flow curves for the three materials tested are shown in Figures 1 and 2. As seen in these figures, the flow behavior appears to be independent of particle size. All three materials exhibit rapid work-hardening to a peak flow stress, followed by a gradual flow softening which approaches a steady-state of flow stress at a true strain of 0.50. Peak flow stress, rate of flow softening, steady-state flow stress, and strain at the onset of steady-state flow all increase with increased strain rate and/or decreased temperature.

Peak flow stress data were obtained from the flow curves and are plotted versus strain rate and as a function of temperature in Figure 2. These data were utilized to calculate strain rate sensitivity, m, defined as:

$$m = \frac{d\ln\sigma}{d\ln\dot{\varepsilon}} \qquad (1)$$

In order to calculate m, the logarithm of the peak flow stress versus the logarithm of strain rate was plotted based on a polynomial regression fit. The equation of the curve corresponding to

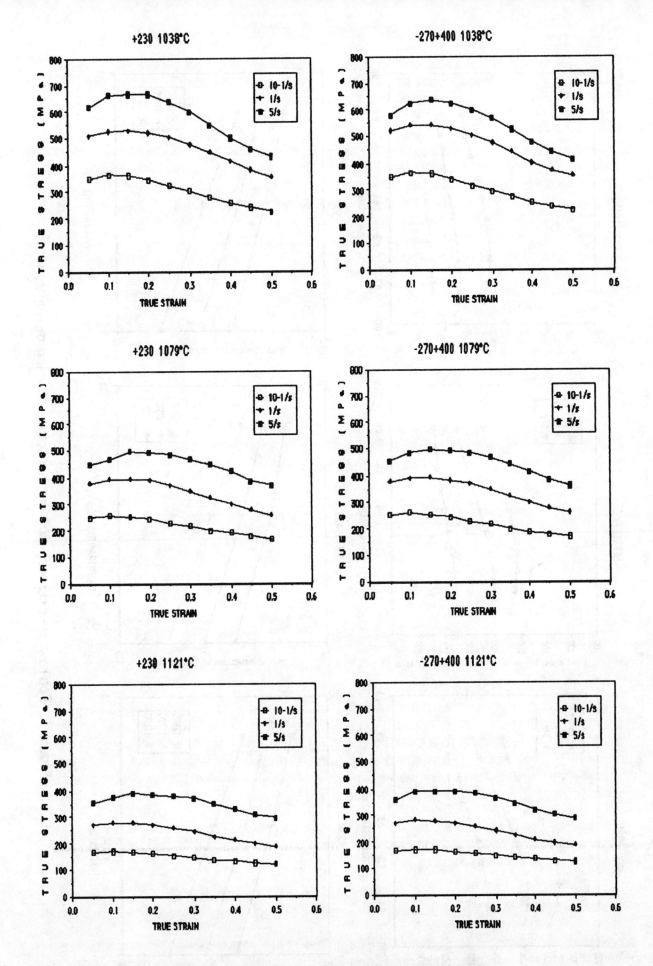

Figure 1. σ−ε Curves as a function of temperature.

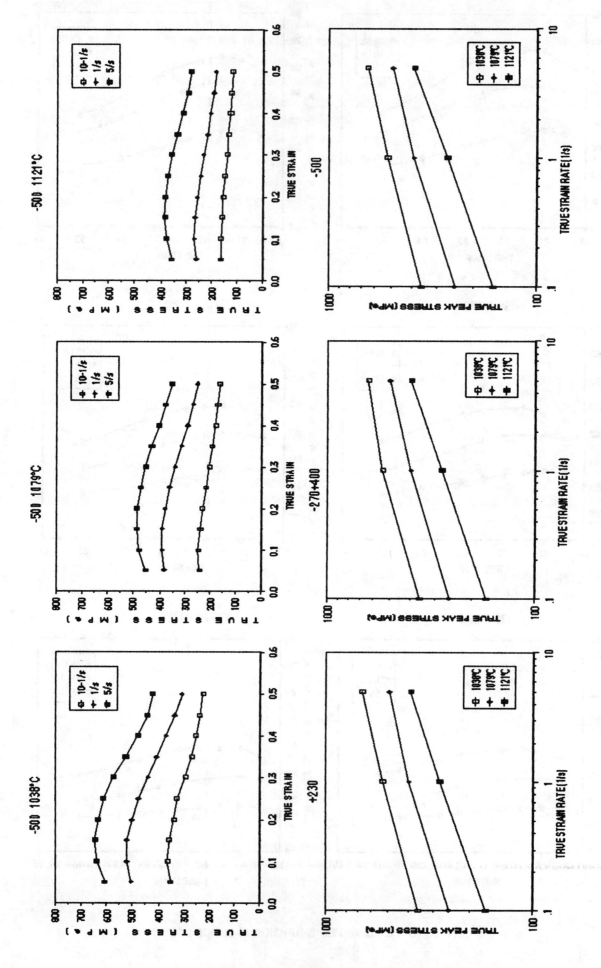

Figure 2. Top row: $\sigma - \varepsilon$ curves, -500 mesh. Bottom row: log σ_{peak} vs. log $\dot{\varepsilon}$.

the polynomial fit was differentiated to give the equation for the slope of the curve at any given point. Thus, m could be calculated at each strain rate. Table II lists the strain rate sensitivities calculated in this manner for a second order polynomial fit, all of which have values of m < 0.3. Superplastic behavior is typified by strain rate sensitivities in the range of 0.3 - 0.8, corresponding to stage II on the sigmoidal $\ln \sigma - \ln \dot{\varepsilon}$ curve, where deformation occurs by grain boundary or interphase boundary sliding. The values presented in Table II are characteristic of stage III on the $\ln \sigma - \ln \dot{\varepsilon}$ curve, where deformation occurs by intragranular flow, and conventional hot-working mechanisms such as dynamic recovery and dynamic recrystallization operate (2). As seen in Table II, there does not appear to be a transition to stage II deformation over the range of strain rates tested.

Table II Strain Rate Sensitivity at Peak Stress As a Function of Temperature and Strain Rate

Powder Size	Temperature (°C)	Strain Rate (s^{-1})		
		10^{-1}	1	5
+230	1038	0.18	0.16	0.14
	1079	0.22	0.16	0.12
	1121	0.22	0.21	0.20
-270+400	1038	0.23	0.13	0.06
	1079	0.21	0.16	0.13
	1121	0.22	0.22	0.22
-500	1038	0.18	0.14	0.12
	1079	0.23	0.16	0.12
	1121	0.21	0.22	0.23

There was a tendency for some of the test specimens to deform non-uniformly — i.e. the geometry of a right circular cylinder with parallel sides was not maintained. This tendency was seen to increase with increased strain rate or decreased temperature. Non-uniformity was in most cases accompanied by edge-cracking of the specimens.

Deformation Maps

A deformation map is a plot of the calculated efficiency of power dissipated during plastic deformation, as a function of temperature and strain rate, for a given strain. Efficiency is defined (5) as:

$$\eta = \frac{2m}{(m+1)} \qquad (2)$$

where m = strain rate sensitivity.

Deformation maps corresponding to a true strain of 0.50 were generated. Flow stress data corresponding to 0.50 strain were used to calculate m values in the same manner as cited previously. As there were only three strain rates involved, a second-order polynomial regression fit of the data was initially chosen, and was subsequently compared to a linear fit.

Values of m calculated from a linear regression curve fit of the data indicate no superplasticity (m < 0.3), and constant strain rate sensitivity. A second-order polynomial fit indicates that the materials were deforming superplastically (at true strain = 0.50) at the highest strain rate and highest temperature of this study (m >= 0.3). The possibility of superplasticity occurring at a strain rate of 5 s^{-1} appears unlikely. Curve fitting of log flow stress vs. log strain rate data based on only three data points (three strain rates) is assumed to be the basis for this discrepancy. Efficiency values calculated from these strain rate sensitivities are shown in Table III.

Deformation maps generated from the polynomial fit data are shown in Figure 3. Maps based on a second-order polynomial fit, in general, indicate an increase in efficiency with increased strain rate. A linear fit would force the maps to have a constant efficiency over the range of strain rates at each temperature. In either case, efficiency increases with increased temperature. It is obvious that the generation of deformation maps can be sensitive to the method of data manipulation chosen.

In addition to predicting conditions of high processing efficiency, deformation maps delineate regions of "stable" processing for a workpiece. Stability of a workpiece under a given set of conditions is a concept which implies that "safe" processing may result – that no unfavorable processes which produce defects, fracture, or plastic instability will occur. Stability is represented in two parts, which are defined as follows (7):

$$\text{Mechanical Stability} \quad \frac{d\eta}{d\log\dot{\varepsilon}} \quad (3)$$

$$\text{Material Stability} \quad \frac{ds}{d\log\dot{\varepsilon}} \quad (4)$$

where

$$s = \frac{1}{T}\left[\frac{d\log\sigma}{dT^{-1}}\right] \quad (5)$$

and T = absolute temperature.

In order for a process to be considered stable, equations 3 and 4 must be negative. Based on these criteria, the deformation maps shown in Figure 3 predict processing instability over most of the temperature – strain rate range used for this study.

Microstructural Analysis

Microstructural analysis of the starting materials and the compressed specimens was conducted to correlate predicted processing stability with microstructure. Microstructures of the three starting materials, shown in Figure 4, indicate the presence of prior powder particles after HIP. Prior powder particle areas are characterized by a globular, interdendritic distribution of gamma prime and are outlined by a course gamma prime precipitate. Areas surrounding prior particles contain a random distribution of gamma prime of varying size fractions. Examination of the microstructures at all test conditions after deformation to a true strain of 0.50 showed the three materials to be microstructurally similar. Each sample had undergone grain refinement via dynamic recrystallization, and exhibited a more homogeneous microduplex gamma + gamma prime structure. Gamma prime was seen to have coarsened with increased temperature, resulting in a larger recrystallized grain size at higher temperatures. Prior powder particles were noted to be present after deformation in all three materials, under all test conditions. The microstructure of the prior powder particles was noticeably different from that of the regions surrounding them, as illustrated in Figure 5. The interdendritic powder particle areas contained a high dislocation density which had not been present in the as-HIP material prior to compression, and showed no signs of having undergone recrystallization. The regions

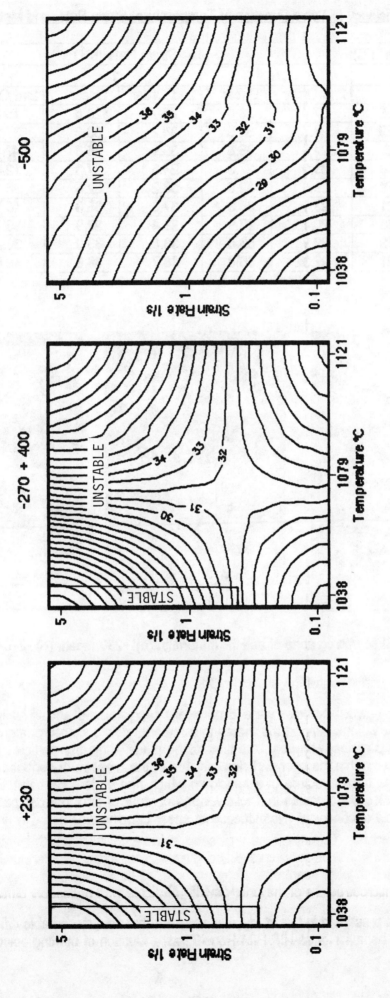

Figure 3. Deformation maps of the three materials at ε = 0.5.

Table III. Efficiency (%) as a Function of Temperature, Strain Rate, and Method of Data Fit

Size	Temp. (°C)	Strain Rate (s^{-1})					
		10^{-1}		1		5	
		2nd Order	Linear	2nd Order	Linear	2nd Order	Linear
+230	1038	37.4	28.6	26.7	28.6	18.3	28.6
	1079	28.7	33.3	34.2	33.3	37.9	33.3
	1121	24.6	36.6	38.9	36.6	47.6	36.6
-270+400	1038	39.3	27.3	24.5	27.3	12.4	27.3
	1079	29.8	32.6	33.2	32.6	35.5	32.6
	1121	23.1	35.3	37.6	35.3	46.5	35.3
-500	1038	20.4	27.9	29.4	27.9	35.2	27.9
	1079	29.9	33.0	33.6	33.0	36.2	33.0
	1121	27.3	36.8	38.7	36.8	45.8	36.8

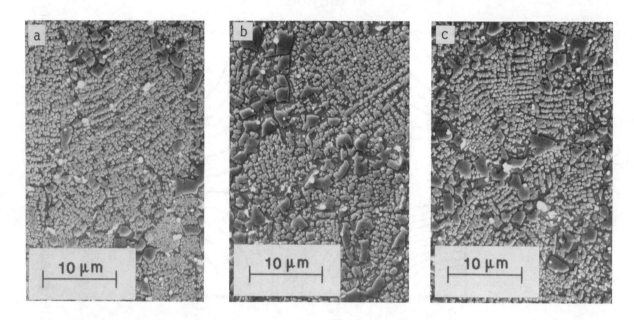

Figure 4. SEM micrographs of as-HIP materials: (a) +230 mesh, (b) -270+400 mesh, (c) -500 mesh.

surrounding prior powder particles were dynamically recrystallized and virtually dislocation-free. These observations support the theory that material flow occurred primarily in the regions surrounding prior particles, with very little flow taking place within the particles themselves. Flow that did occur within the prior particles resulted in the creation of dendritic remnants. Examination of the microstructure of a specimen which had undergone non-uniform deformation revealed edge cracks that followed the contour of prior powder particles, with voids at points of particle decohesion. No evidence of shear banding was noted in any of the specimens.

Discussion.

From the data presented in this study, it is apparent that little discernable difference exists between the flow behavior of as-HIP P/M Rene'95 as a function of starting powder particle size.

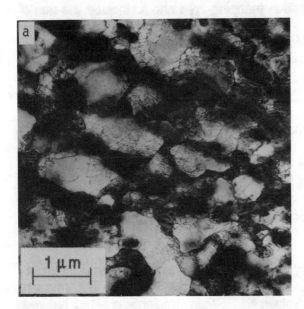
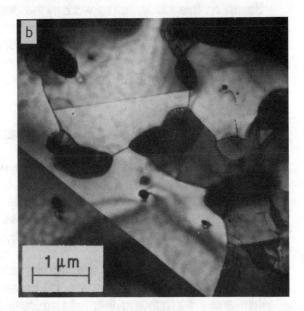

Figure 5. TEM microstructural comparison of (a) prior powder particle and (b) surrounding recrystallized area

Observed deformation behavior of the as-HIP materials is consistent with the findings of others (1,2,3,4). Correlation of strain rate sensitivities and microstructural observations leads to the conclusion that deformation over the given range of temperatures and strain rates is of the stage III type, characterized by intragranular flow, dynamic recovery and dynamic recrystallization. The tendency toward non-uniform flow behavior of all three materials with increased strain rate or decreased temperature is thought to be due to the inhomogeneous nature of the material itself. Specifically, the presence of dendritic prior powder particles introduces strain gradients in the material which cannot be accommodated at high strain rates and low temperatures, resulting in cracks and decohesion at prior particle boundaries. In general, prior particles undergo relatively little deformation, while the material surrounding them flows and dynamically recrystallizes.

Interpretation of the generated deformation maps in terms of the observations made above is unclear. Maps based on a second order polynomial fit of stress-strain data in some instances imply superplasticity, while those based on a linear fit do not. The latter case may seem to be a more realistic model, yet a polynomial fit is cited (8) as being the correct method for creating deformation maps. Maps based on more than three strain rates would provide more insight in this regard.

Both types of maps predict processing instability for the three materials over most, if not all of the temperature-strain rate range of this study. Because both parts of the stability criterion must be true in order for a map region to be deemed "unstable", the specific component that causes system instability may not be discernable. It is therefore difficult to know the nature of the instability predicted in the deformation maps. The non-uniform flow behavior observed could be assumed to be the predicted instability, yet two of the most severe cases occurred under conditions which were within stable regions of the deformation maps

Microstructural analysis showed that all of the compressed specimens had undergone substantial grain refinement, with some prior powder particle refinement, and a phase redistribution that resulted in a more homogeneous structure. Such microstructural change is generally considered favorable, yet the maps designate such processing to be unstable. It is unclear whether stability should be interpreted as being desirable for processing, especially since it does not seem to correlate with microstructure.

Based on this study, future work in the deformation mapping area should include the use of substantially more data points for the creation of deformation maps, and compression tests to total true strains which are less than that associated with the onset of steady-state for more direct observation of microstructural evolution. Such an approach may lead the development of a more realistic model of hot-working processes.

Summary

1. There is no discernable difference between the flow behavior of as-HIP P/M Rene'95 as a function of particle size. The flow curves of the three materials exhibit a rapid work hardening followed by a gradual flow softening due to dynamic recrystallization. Strain rate sensitivities calculated for peak flow stress indicate stage III deformation over the range of temperatures and strain rates in this study.

2. The three as-HIP materials are microstructurally similar after deformation to a true strain of 0.50. Prior powder particles are still present, having undergone relatively little deformation while areas surrounding them dynamically recrystallized. Strain gradients created between prior powder particles and areas surrounding them are believed to be the cause of non-uniform deformation behavior at low temperatures and high strain rates.

3. Deformation maps based on efficiency values derived from only three strain rates offer inconsistent correlation with observed flow behavior and resultant microstructure. Future work in this area should incorporate larger test matrices with a corresponding increase in microstructural analysis, so as to generate more meaningful hot-working models.

Acknowledgement

The authors gratefully acknowledge the Wyman-Gordon Co. of North Grafton, MA for their support of this study.

References

1. "High Temperature Deformation Behavior of P/M Rene'95," Superalloys 1984, eds. M. Gell et al. (Warrendale, PA: The Metallurgical Society, 1984), 275-284.
2. A.Y. Kandeil et al., "Flow Behavior of Mar M200 Powder Compacts During Isothermal Forging," Metal Science, 14(10)(1980) 493-499.
3. R.G. Menzies, J.W. Edington, G.J. Davies, "Superplastic Behavior of Powder Consolidated Nickel-Base Superalloy IN-100," Metal Science, 15(5)(1981) 210-216.
4. J.-P.A. Immarigeon, P.H. Floyd, "Microstructural Instabilities During Superplastic Forging of a Nickel-Base Superalloy Compact," Met. Trans. A,12A(7)(1981) 1177-1186.
5. Y.V.R.K. Prasad et al., "A New Systems Approach to Dynamic Modeling of Material Behavior in Metal-Working Processes" (Paper published in proceedings of the Titanium Net Shape Technologies Conference, Los Angeles, CA, 26 February - 1 March 1984), 279-289.
6. C.T. Sims, N.S. Stoloff, W.C. Hagel, ed., Superalloys II (New York, NY: John Wiley & Sons, 1987), 453.
7. H.L. Gegel, "Synthesis of Atomistic and Continuum Modeling to Describe Microstructure" (USAF Wright Aeronautical Labs, Wright-Patterson AFB).
8. J.C. Malas, "A Thermodynamic and Continuum Approach to the Design and Control of Precision Forging Processes" (M.S. thesis, Wright State University, 1985).

UTILIZATION OF COMPUTER MODELING

IN SUPERALLOY FORGING PROCESS DESIGN

T. E. Howson and H. E. Delgado

Wyman-Gordon Company
North Grafton, Massachusetts 01536

Summary

The forging of a superalloy high pressure turbine disk has been simulated using ALPID, an FEM code for analysis of large plastic deformation problems. The modeling was carried out to determine the cause of coarse grains and relatively high sonic noise at some locations in a disk. The simulation indicated a temperature increase in the forging that occurred as a result of deformation heating. Compression testing was carried out at temperatures between 1010°C and 1149°C (1950°F and 2100°F) to generate microstructures for comparison to microstructures from the forging. The comparisons of the modeling results and the microstructures support a conclusion that the coarse grains and high sonic noise in the forging occurred because the temperature in the forging rose too high in some locations, partly because of deformation heating. The changes made in the forge process as a result of the modeling are described.

Introduction

Computer simulation of superalloy metal forming is now an integral part of forging design. Process modeling is being used for tasks such as definition of preform and die geometries, prediction of forging defects such as laps or underfill, determination of stress, strain, strain rate and temperature profiles, point tracking and flow line prediction, and determination of the loading on tooling for die stress analysis. The costly cycle of physically trying out the dies, testing, and redesigning is now being replaced by an iterative process utilizing computer modeling prior to manufacturing.

A code being used in the forging industry to simulate metal flow is the finite element method code ALPID[1] (Analysis of Large Plastic Incremental Deformation.) ALPID is a 2-dimensional rigid-viscoplastic code capable of performing non-isothermal analyses of metal flow in arbitrarily shaped dies. In this paper the use of ALPID to assess the effects of forge temperature and strain rate on actual workpiece temperature and resulting microstructures in a disk forging will be described.

Experimental Procedure

Forged Microstructures

The metal flow simulations were carried out to help determine the cause of coarse grains and high sonic noise in a superalloy high pressure turbine disk forging. The shape of the forging can be seen in Figure 7 which is part of the simulation results. The alloy is a high γ' volume fraction -150 mesh extruded powder superalloy. In a disk forged at 1121°C (2050°F) with a nominal strain rate of 8.3×10^{-3} sec^{-1} (0.5 min^{-1}) the sonic inspectability was found to be nonuniform. In the ultrasonic inspection at a sensitivity of 21dB most of the forging was found to exhibit a background noise level of 10% or less of the screen. Parts of the forging, particularly at the rim, however, exhibited a significantly higher noise level.

Microstructures were obtained from both high noise and low noise areas

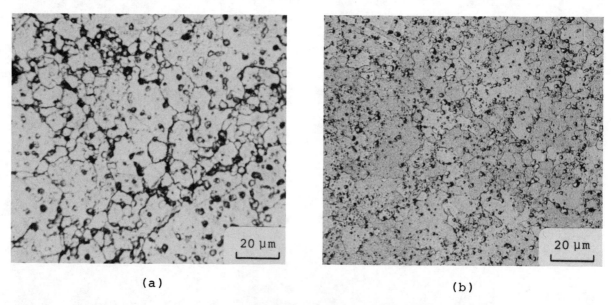

Figure 1 - Microstructures from: a) high sonic noise area at the rim of a disk forging; b) compression specimen deformed at 1149°C (2100°F) and 0.0083 sec^{-1} (0.5 min^{-1}.)

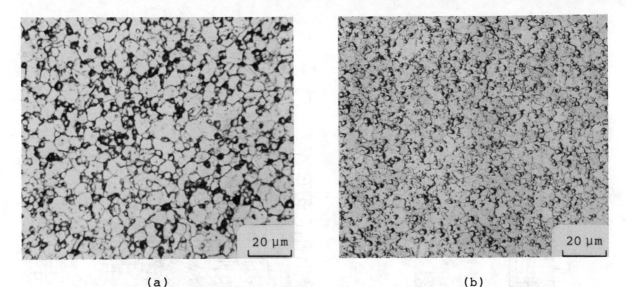

(a) (b)

Figure 2 - Microstructures from: a) low sonic noise area of a disk forging; b) compression specimen deformed at 1135°C (2075°F) and 0.0083 sec^{-1} (0.5 min^{-1}.)

of the forging. The high noise optical microstructure, Figure 1a, is characterized by a coarser grain size and a lower volume fraction of the large primary γ' than the low noise microstructure, Figure 2a. Also included for comparison in Figures 1 and 2 are microstructures from compression specimens deformed at 1149°C (2100°F) and 1135°C (2075°F), respectively. The compression testing and the comparisons in Figures 1 and 2 will be discussed in more detail below. The differences between the forged microstructures, Figure 1a and Figure 2a, suggest that the high noise material has, during forging, experienced a higher temperature and been heated closer to the γ' solvus than the low noise material. The γ' solvus for the heat of material from which the forging was made is 1152°C (2105°F.) How this nonuniform heating could happen was investigated by process modeling.

Computer Simulations

The flow data required to carry out metal flow simulations were available from compression testing. The compression specimens had been machined from extruded stock and upset at temperatures between 1010°C and 1149°C (1850°F and 2100°F) and strain rates between 10^{-3} sec^{-1} and 1 sec^{-1}. These ranges encompass the feasible hot working conditions for this alloy. Also needed were thermal property data for both the workpiece and dies: thermal conductivity and specific heat.

The first simulation was carried out for a forge temperature of 1121°C (2050°F) and a strain rate of 8.3 X 10^{-3} sec^{-1} (0.5 min^{-1}.) Both the workpiece and the die were initially at the same temperature and a constant friction factor of 0.2 was used. The initial die velocity was selected to give a strain rate of 8.3 X 10^{-3} sec^{-1} based on the initial height of the billet. The die velocity was continuously reduced throughout the simulation to maintain a constant strain rate. A two step forge operation was simulated. In this case, for lubrication and inspection, the forging was carried out part way to the finish in the first step (the prefinish step) and then forged to the final configuration in the same dies in a second step (the finish operation.)

The initial finite element meshes of the workpiece and die are shown in Figure 3. The workpiece mesh at the end of the prefinish step is shown in Figure 4. The workpiece mesh at the end of the prefinish step is not

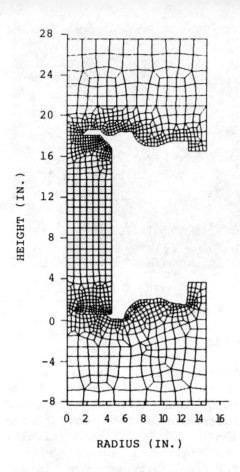

Figure 3 - Initial finite element meshes of the workpiece and dies.

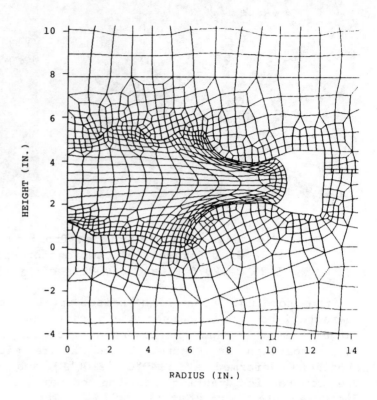

Figure 4 - Deformed workpiece mesh and die meshes at the end of the 1121°C (2050°F) prefinish forge operation.

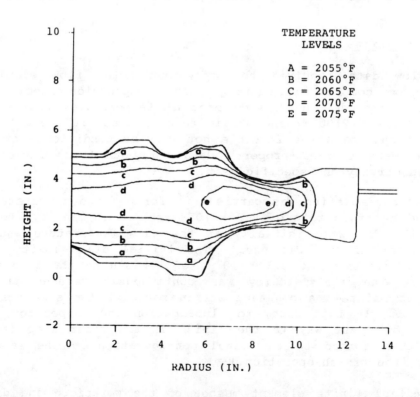

TEMPERATURE LEVELS

A = 2055°F
B = 2060°F
C = 2065°F
D = 2070°F
E = 2075°F

Figure 5 - Temperature contours in the workpiece at the end of the 1121°C (2050°F) prefinish forge operation.

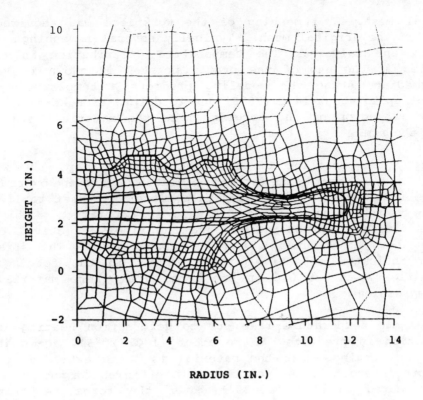

Figure 6 - Workpiece and die meshes at the end of the 1121°C (2050°F) finish forge operation. Superimposed on the workpiece mesh is the final position of material that was heated 11-14°C (20-25°F) over the nominal forge temperature during the prefinish forge operation (see Figure 5.)

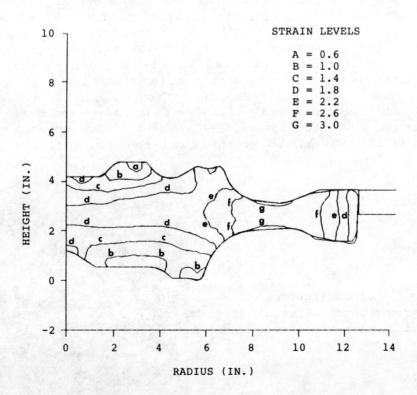

Figure 7 - Final effective strain contours in the workpiece at the end of the finish forge operation.

the original mesh. A remeshing of the workpiece was required because of distortion of the original mesh. Typically several remeshings are required to complete an analysis. The temperature distribution in the workpiece and the dies at the end of the prefinish step is shown in Figure 5. At this intermediate stage the modeling predicts a temperature rise in the workpiece of up to about 14°C (25°F) due to deformation heating, particularly in the narrow region of the forging where the material is being extruded out toward the rim.

To carry out modeling of the finish forge step, the workpiece was remeshed. In the remesh the strain field was carried over from the old mesh to the new mesh but the temperature field was reset to 1121°C (2050°F) to simulate a reheating for the finish forge operation. The mesh at the end of the finish forge operation, and the final effective strain distribution are shown in Figures 6 and 7, respectively. The strain field at the finish indicates effective strain levels in the forging as high as 3.0 and as low as about 0.6 in one local area where material is trapped in a die impression.

The maximum temperature rise due to deformation heating in the finish forge operation is predicted to be about 11°C (20°F), again in the narrow region of the forging where the material is being extruded out toward the rim. Of more interest is the position in the finish forged shape of material which was heated 11-14°C (20-25°F) over the forge temperature in the prefinish step. The final location of this material is superimposed on the workpiece mesh in Figure 6.

Compression Testing

Compression testing to obtain microstructures was performed on an MTS servohydraulic system in a controlled atmosphere (the platens were TZM.) The material was as-extruded stock with a microstructure shown in Figure 8. The γ' solvus temperature of the material was determined by DTA to be 1159°C (2119°F.) Testing was performed in the temperature range of 1010°C - 1149°C (1950°F - 2100°F) at a strain rate of 8.33×10^{-1} sec^{-1} (0.5 min^{-1}.) Test specimens, 0.5" Ø X 0.75" tall, were cut so that the axis of each cylinder was parallel to the extrusion axis. Tests were conducted under constant true strain rate control. The specimens were heated in vacuum to the test temperature for 15 minutes prior to testing. A graphite lubricant was used. All specimens were upset to a final true axial strain of 0.69 (50% upset.) At the conclusion of each test the test chamber was filled with helium gas to quench the specimens.

Figure 8 - Microstructure of as-extruded material used for compression testing.

Optical and scanning electron (SEM) microstructures are shown in Figure 9 to illustrate the microstructural changes that occur as the test temperature varies from 1093°C (2000°F) to 1149°C (2100°F.) The grain size after deformation is controlled primarily by the size and spacing of the large γ' present in the microstructure during deformation. As the deformation temperature increases, the volume fraction of γ' in the matrix at the test temperature decreases, and the resulting grain size increases. The difference in microstructures resulting from deformation at 1135°C (2075°F) and deformation at 1149°C (2100°F) is most striking. 1149°C is close to the γ' solvus temperature. The equilibrium concentration of γ' in the matrix changes rapidly with temperature near the solvus. Small changes in temperature result in big changes in the amount of γ' present and in the grain size, and the rapid increase in grain size as the deformation temperature approaches the γ' solvus is evident. A number of other observations can be made about microstructural features such as γ' morphology but that is beyond the scope of this paper. The particles that appear white in the SEM micrographs in Figure 9 are MC carbides.

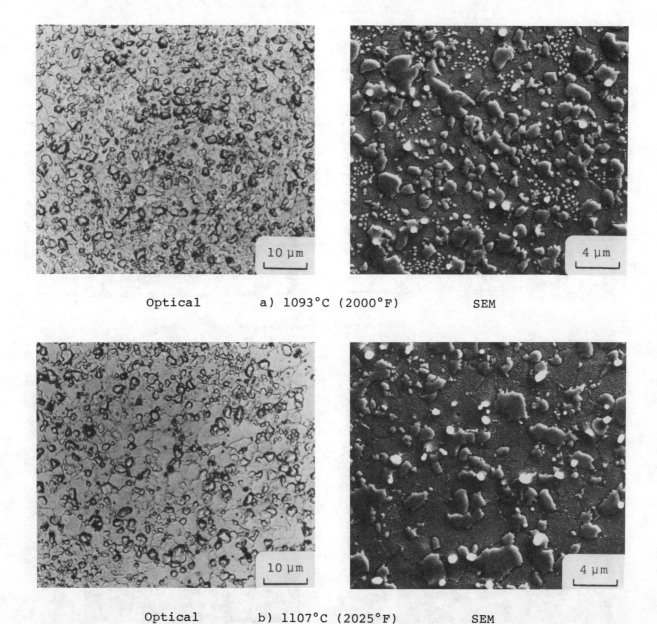

Figure 9 - Optical and SEM micrographs of compression specimens deformed at 0.0083 sec^{-1} (0.5 min^{-1}) and: a) 1093°C (2000°F), b) 1107°C (2025°F), c) 1121°C (2050°F), d) 1135°C (2075°F), and e) 1149°C (2100°F.)

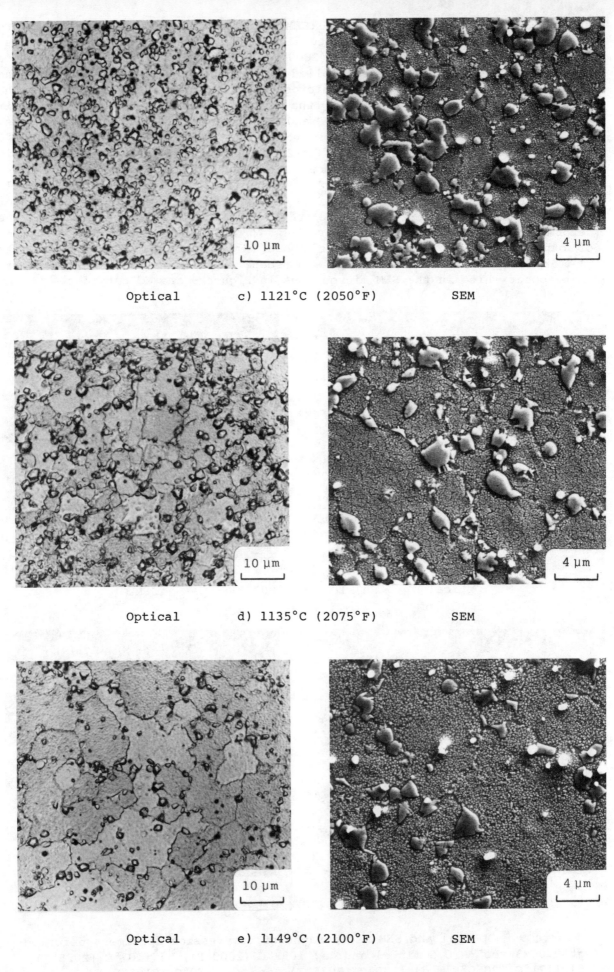

Optical   c) 1121°C (2050°F)   SEM

Optical   d) 1135°C (2075°F)   SEM

Optical   e) 1149°C (2100°F)   SEM

Figure 9 (Continued)

Discussion

Relatively high background noise in sonic inspection of the rim of a forged superalloy disk was correlated with a grain structure which is coarser than in other areas of the forging where the sonic noise is not as high. The question is whether the grain coarsening occurred as a result of the deformation heating the material experienced in the forge process. Modeling of the metal flow predicts that the workpiece temperature has been raised up to 14°C (25°F) by the heat generated by the deformation, even in this case where the nominal strain rate is only 8.3×10^{-3} sec^{-1} (0.5 min^{-1}.) However, judging from Figures 1, 2 and 9, this change in temperature caused by the deformation heating is not enough to have caused the grain coarsening evident in Figure 1a.

Another factor to consider is the variability in furnace and die temperatures. Because of equipment limitations and productivity issues, established forge practices allow furnace and die temperatures to vary within fixed, controllable limits, typically ±8-14°C (±15-25°F.) In this case the acceptable variation for the furnace and die temperatures was ±14°C (±25°F.) It is possible, then, that the allowable temperature variation added to the increment in temperature resulting from deformation induced heating could have resulted, in this forging, in temperatures about 28°C (50°F) higher than the nominal forge temperature of 1121°C (2050°F.)

In Figure 1 the 500X optical microstructure from the high noise area of the forging is compared to a 500X microstructure from a compression specimen deformed at 1149°C (2100°F.) In Figure 2 a 500X microstructure from the low noise area of the forging is compared to a 500X microstructure from a compression specimen deformed at 1135°C (2075°F.) The comparisons support the premise that the sonic noise and the forged microstructure in Figure 1a have resulted from a process in which the temperature has risen too high, up to 1149°C (2100°F), in localized areas of the forging. The forged microstructure in Figure 1a represents a position in the rim of the forging on the boundary of the region highlighted in Figure 6 of material that was heated in the prefinish forge step 11-14°C (20-25°F) above the nominal forge temperature of 1121°C (2050°F.)

Without the process modeling the deformation heating that is a part of the problem would not have been considered significant. Instead, over-heating or nonuniform heating of the workpiece by the die heating system would have been suspected. However, furnace surveys and calibration and monitoring of temperature sensors insure that process temperatures do in fact remain within prescribed limits. The process modeling indicates that in this case the workpiece temperature can locally rise too high because of a combination of factors even though the process parameters are controlled properly.

One additional observation is that the effective strains in the forging predicted by the modeling, Figure 6, are certainly higher than the effective strains in the compression specimens upset 50%. Is it appropriate to compare the forged microstructures to the compression specimen microstructures? Flow curves have been published for this superalloy in the extruded condition for deformation at 1107°C (2025°F.)[2] These curves show that the flow stresses reach a constant value at low strains (about 0.2) in the range of strain rates of interest. This suggests that at the temperatures and strain rates discussed, microstructures evolve rapidly to a steady state and the microstructural comparisons made are valid.

It is worth noting the process change made as a result of the modeling. Additional modeling carried out for a forge temperature of 1093°C (2000°F), a strain rate of 8.3 X 10^{-3} sec^{-1} (0.5 min^{-1}), and a friction factor of 0.2, showed again that in the prefinish step a temperature rise of up to about 14°C (25°F) would occur for this forging due to deformation heating. Under these conditions, taking into account the allowable process temperature variability, the maximum temperature that any part of the workpiece could experience is 1121°C (2050°F.) Another process that was modeled was forging start to finish in one step at 1093°C (2000°F) and 8.3 X 10^{-3} sec^{-1} (0.5 min^{-1}.) That simulation predicted a maximum temperature rise at the end of the forge stroke of about 22°C (40°F.) The process conditions adopted as a result of all the modeling are a forge temperature of 1093°C (2000°F) at a strain rate of 8.3 X 10^{-3} sec^{-1} (0.5 min^{-1}), in a two step operation. Areas of coarser grains and higher sonic noise have not been encountered.

Concluding Remarks

Metal flow simulation, and process modeling in general, are being utilized in forge process design and are moving from the research environment into the production environment. Several factors are responsible for this progress. The first, obviously, is that the software is available. Another is the extraordinary advances in computer technology. Computer speeds are increasing rapidly while computer costs are decreasing, and the computer-intensive finite element analyses of various processes can be done with a computer that sits on a desk. A third is that the database of material properties needed to carry out the modeling has been generated for many of the alloys of interest. Perhaps most important is that a level of confidence in the software has been reached that permits application of the modeling in production. With respect to metal flow this confidence stems from, for example, correct prediction of unfilled areas in die cavities, correct prediction of the formation of forging defects such as laps, and successful manipulation of preform and die shapes to modify strains and microstructures to enhance mechanical properties.

In this paper the application of nonisothermal metal flow simulation with ALPID to analyze suspected overheating in a forging has been described. The modeling has helped redefine the forge process parameters and overcome the problem. Other successful nonisothermal simulations have been carried out to assess the effects of die chilling (when the dies are cooler than the workpiece) on the flow behavior and strains in forgings, and to analyze titanium alloy beta forging. Process modeling is enhancing the ability of the industry to design and manufacture forgings.

References

1. S. I. Oh, "Finite Element Analysis of Metal Forming Processes with Arbitrarily Shaped Dies," International Journal of Mechanical Science, Vol. 24, 1982, pp. 479-493.

2. T. E. Howson, W. H. Couts, Jr., and J. E. Coyne, "High Temperature Deformation Behavior of P/M Rene' 95," Superalloys 1984, M. Gell et. al., eds., AIME, Warrendale, Ohio, 1984, pp. 275-284.

PROPERTY OPTIMIZATION IN SUPERALLOYS THROUGH

THE USE OF HEAT TREAT PROCESS MODELLING

R. A. Wallis and P. R. Bhowal

Cameron Forge Company
P. O. Box 1212
Houston, TX 77251-1212

Abstract

One important variable in maximizing tensile, creep, stress rupture and toughness properties in superalloys is the cooling rate from the alloy solutioning temperature. Correlations between properties and cooling rates may be conveniently developed using small blanks of material. Implementation of the data in complex geometries such as aircraft engine disks, however, is most effectively done through finite element modelling of the heat treatment process.

The application of heat treat process modelling to predict the property levels at various locations in parts having relatively complex shapes are discussed. An example of the use of models to obtain the required properties while reducing residual stress levels in the part is given together with an example where the properties are maximized against the tendency of the part to crack during quenching.

Introduction

The designers of aircraft engines are continually seeking improved properties from existing alloys so that components can be operated at higher stress levels and higher temperatures, thereby, giving increased engine performance. One important variable that determines the strength of many superalloys is the cooling rate from its solutioning temperature. Correlations between properties such as tensile, creep and stress rupture strength and cooling rate can be conveniently developed using simple pancakes of a material. The application of this data to superalloy forgings of complex shapes, however, is most effectively done through finite element modelling.

Over the past few years Cameron has developed heat transfer data and modelling techniques to enable the temperatures and stresses in components during heat treatment to be calculated. Coupling the modelling work with data correlating the properties to cooling rate has enabled the properties in relatively complex shapes to be maximized, and has eliminated the costly and time consuming trial and error methods of the past.

The Mathematical Models

The first step in the development of the mathematical models was the determination of the heat extraction rates in various quenching media. Instrumented disks were used to obtain time-temperature data upon quenching. These data were used as input into a finite difference inverse heat conduction program (1) which calculates the surface temperature of the disk together with the corresponding heat transfer coefficient. The heat transfer coefficients obtained are used in finite element models (2,3) to predict the temperatures and stresses in parts during quenching. Validation of the models was carried out with several instrumented parts, both sub-scale and full size. The comparison between the measured and calculated temperatures in a full size turbine disk made of alloy 901 is given in Figure 1 (4). In the example shown the disk was heated to 1175°C, then transferred from the furnace to a forced air cooling system in 0.83 minute, forced air cooled for 2 minutes, transferred to an oil tank in 0.75 minute and finally oil quenched. The correlation between the measured and calculated temperatures may be seen to be good even in a relatively complicated multiple step heat treatment.

Correlation of Properties with Cooling Rate

Correlations between mechanical properties and the cooling rate from the solution temperature are obtained by instrumenting small blanks of material (typically, 10 cm to 12 cm square or round and 4 cm thick). The cooling rate at different locations in the disk are thereby obtained during quenching in different media. Solid blanks of the same size and material are then similarly quenched and cut up for property testing. Testing is carried out with specimens taken from locations where the cooling rate is known. This work allows the tensile, creep and stress rupture properties of an alloy to be obtained as a function of cooling rate. The prediction of the properties that will be obtained in a forging having a complex shape is subsequently made by comparing the cooling rates in the forging, as calculated by the mathematical models, with the established property/cooling rate relationship.

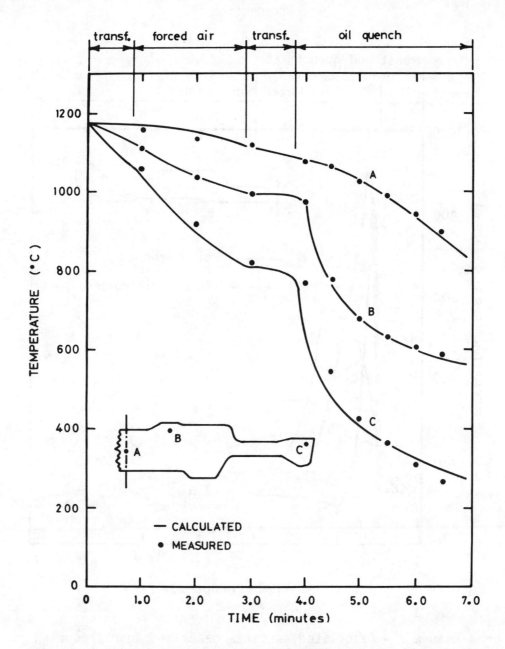

Figure 1 - Comparison between experimental and calculated cooling curves (4).

Modelling to Meet Property Requirements and Reducing Residual Stress/Distortion Problems

Forgings having relatively thin cross-sections, or large differences in cross section can present distortion problems during heat treatment. For example, a thin disk made of a nickel-base powder alloy was initially produced using oil quenching to meet the property requirements. This relatively rapid quench rate enabled the properties to be met, but resulted in residual stresses being set up in the part which made subsequent machining extremely difficult due to the part "springing" on the machine. This problem led to an investigation into the stress levels being developed in the forging during quenching. Figure 2 shows the stresses being developed at several locations during the oil quenching (solid lines). These stresses arise from the very steep temperature gradients that develop in the part shortly after it is immersed in the oil tank, as shown in

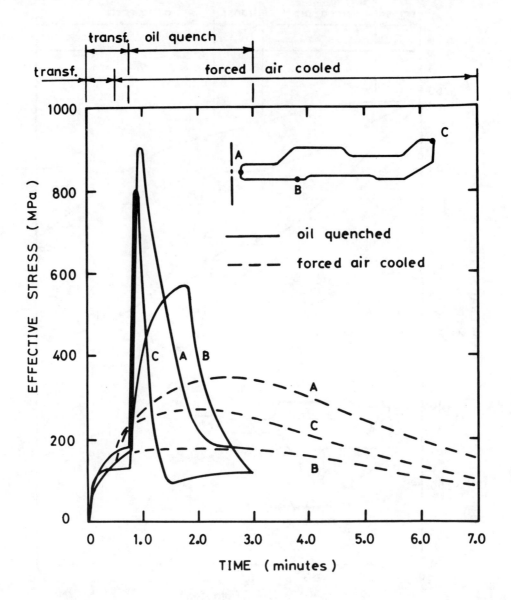

Figure 2 - Calculated stresses generated in disk during cooling.

Figure 3. The stresses developed during the quench exceeded the elastic limit of the material at high temperature, which led to plastic deformation, which, in turn, led to residual stresses being present in the part at the end of the heat treatment.

The solution to the problem, therefore, lay in reducing the cooling rate in the part such that the stresses during heat treatment were reduced. The properties of the material are a function of the cooling rate, hence, the solution automatically reduces the properties in the final part. The effect of cooling rate on the yield strength of the material (at room temperature and at 650°C) is given in Figure 4 for a solution temperature of 1120 ± 15°C. Also shown is the actual property data taken from cut up tests on oil quenched forgings (cooling rates 235-300°C/min.). These strength levels are typically 140 MPa higher than the specified requirements, leaving the possibility to reduce the cooling rate (and hence the properties) and, thereby, the residual stress levels.

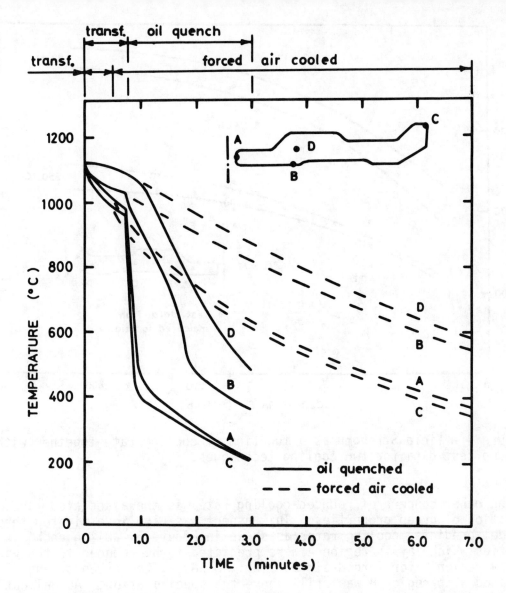

Figure 3 - Calculated temperatures in disk during cooling

Process modelling was carried out to examine alternative heat treatment processes such as static-air cooling or forced-air cooling from the solution temperature. This showed that static air cooling would give cooling rates of about 70°C/minute. This was considered inadequate to meet the property requirements. Forced-air cooling was therefore selected as the alternative to oil quenching. The resultant stresses and temperatures during forced air cooling may be seen in Figures 2 and 3. The stresses were significantly reduced (half those developed during oil quenching). The forced-air cooling also resulted in much lower cooling rates (typically 90-135°C/min.). Referring to Figure 4, the predicted cooling rates obtained with forced air cooling would be expected to give yield values between 1210 and 1300 MPa at room temperature and between 1135 and 1170 MPa at 650°C in locations A, B, C, D and E. Thus, the change from oil quenching to forced-air cooling has reduced the yield strength approximately 70-100 MPa. The properties predicted, however, remained above the specification. Following this investigation several parts were produced utilizing forced-air cooling, and cut up tests revealed properties in the range shown in Figure 4. The data median lay within the laboratory generated property bands demonstrating correlation between the predicted and the observed properties.

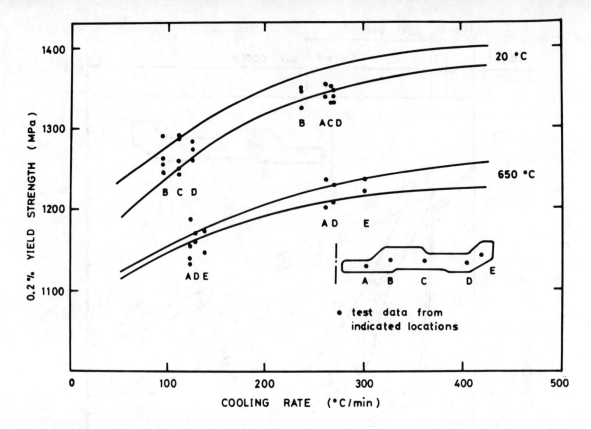

Figure 4 - Yield Strength as a function of cooling rate together with cut up test data for two cooling techniques.

The other concern of reduced cooling rate was the associated degradation of creep properties. This aspect was also investigated through the model predicted cooling rates as shown in Figure 5. Oil quench properties which lay in region A were predicted to be reduced to the values given by region B for forced-air cooling. However, the creep strength associated with region B was still above the specification. Actual cut up data were found to be consistent with the model predictions as shown by the solid scatter bars in Figure 5.

Following this investigation a series of disks were heat treated using forced-air cooling. These parts were found to meet property requirements, and did not exhibit residual stress related problems during machining.

Maximizing Properties Against Quench Cracking

In the design of components for new engines, improved properties are being called for from existing alloys. The improved properties are being obtained, in part, by the rapid quenching of forgings from their solution temperature. There is, however, a limit to how fast a particular alloy can be quenched before it will crack. This limit has been reached for several superalloys, and part geometry along with alloy structure and required cooling rates has become an important factor in the development of heat treatment practices and, hence, the property levels that can be achieved in a part.

As an example, consider the heat treatment of another powder alloy forging which required a yield strength level of 1076 MPa. Investigations were carried out to determine the effect of grain size and cooling rate on

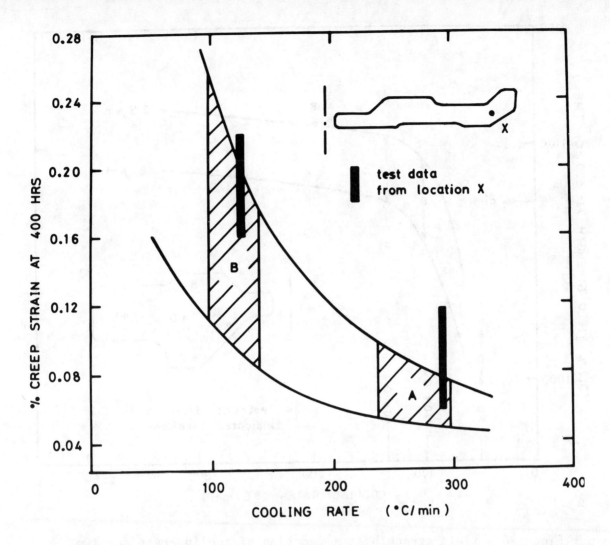

Figure 5 - Creep deformation as a function of cooling rate together with expected and actual data for two cooling techniques.

properties. Figure 6 shows the laboratory generated 500°C yield strength for two grain sizes as a function of cooling rate. To meet the strength requirement with the finer grain structure (3 μm), a cooling rate of 80°C/minute would be required at point D (the slowest cooling point in the forging). However, with a coarser grain size (5 μm) a much higher cooling rate (133°C/minute) would be required. Modelling was carried out to determine the appropriate cooling technique from the solution temperature to meet properties.

Three cases were considered: forced-air cooling and oil quenches with a 120 second and a 60 second delay from the furnace to the oil tank. The property range for the three cases are shown in Figure 6 by vertical lines at the appropriate cooling rate. It may be seen that forced-air cooling is inadequate at either grain size. The 120 second delay oil quench meets property specification with an average grain size of 3 μm, but would be inadequate for a grain size of 5 μm. A 60 second delay time appears most attractive since it meets the specification at either grain size. It should be noted that once the yield strength levels are met, it is desirable to lean toward the coarser grain size from the creep and stress rupture viewpoint. Hence, it was decided to aim at a grain size of 5 μm and a 60 second or lower delay time in oil quenching from its solution temperature.

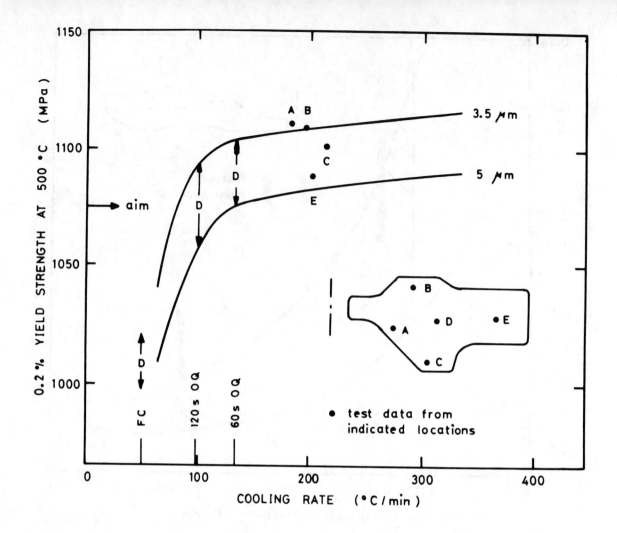

Figure 6 - Yield strength as a function of cooling rate for two different grain sizes showing correlation between expected and measured properties.

Since the cooling rate thus achieved was relatively fast, it was necessary to examine the stress levels imposed at critical locations on the disk during heat treatment. In the first series of computer simulations, a 45 second delay oil quench was used to determine the effect of disk orientation on the stresses developed. Figure 7 shows the stresses developed. The graph shows the stresses at the critical location on the disk as a function of the temperature at that location during quenching. Also given on the graph is the approximate yield strength of the material as a function of temperature. With a 45 second delay, it was found that position 2 (upside down relative to position 1) results in stress levels above the yield point. Hence, depending on the ductility of the material at the temperature, the part could crack during quenching. Also, excessive plastic yielding could cause distortion and residual stress problems. Position 1 was thus considered to be the preferred orientation. At high temperatures (980°C), even in position 1 the stresses are very close to the yield point. A further simulation was carried out with a 60 second delay oil quench which was found to reduce the stresses to safer levels.

The parts were heat treated with a 60 second delay oil quench. The properties at several locations in the disk at their calculated cooling rates are shown in Figure 6. In practice, the structure was bounded by the

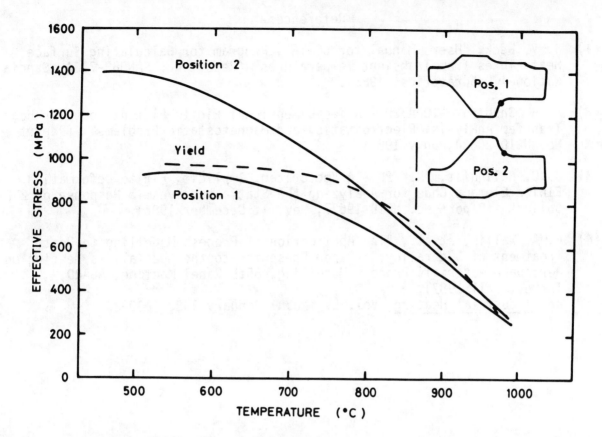

Figure 7 - Stress-temperature curves during cooling showing the effect of part orientation with respect to yielding.

grain size 5 - 3 μm. It may be seen that the properties obtained in the forging are consistent with predicted values.

The above analysis was carried out prior to the parts being made. The first part manufactured met the property specification without quench cracking and illustrated how these techniques could eliminate costly and time consuming trial and error methods.

Conclusions

Correlations between the mechanical properties of superalloys and the cooling rate from the solution temperature can be obtained using small blanks having a simple geometry. Process models are now available which allow the temperature and stress distribution within a part during quenching to be accurately predicted. Coupling the results from the models with the property/cooling rate data enable the properties in relatively complex superalloy forgings to be determined.

The techniques have proved to be very accurate, and will enable the industry to get the most out of existing superalloys. They are also capable of eliminating costly and time consuming trials, thereby, reducing the lead time from engine design to the production of usable hardware.

References

(1) J. V. Beck, "Users Manual for CONTA - Program for Calculating Surface Heat Fluxes from Transient Temperatures" (Report No. SAND83-7134, Sandia National Laboratories, 1983).

(2) A. B. Shapiro, "TOPAZ2D - A Two-Dimensional Finite Element Code for Heat Transfer Analysis, Electrostatic, and Magnetostatic Problems" (Report No. UCID-20824, July 1986).

(3) J. O. Hallquist, "NIKE2D - A Vectorized, Implicit, Finite Deformation, Finite Element Code for Analyzing the Static and Dynamic Response of 2-D Solids" (Report No. UCID-19677, Rev. 1, December 1986).

(4) R. A. Wallis, et.al., "The Application of Process Modelling to Heat Treatment of Superalloys" (Paper Presented to the Specialists Meeting on Aerospace Materials Process Modelling, 65th Panel Meeting, AGARD, Turkey, Fall 1987).
and Industrial Heating, Vol. LV, No. 1 January 1988, p30-33.

PROCESSING OF HIGH STRENGTH SUPERALLOY COMPONENTS

FROM FINE GRAIN INGOT

P. D. Genereux and D. F. Paulonis

Materials Engineering
Pratt & Whitney
400 Main Street
East Hartford, CT 06108

Abstract

The utilization of high strength superalloys, such as MERL 76, for critical gas turbine engine disks was made possible by the emergence of powder metallurgy technology. This paper clearly demonstrates that it is now possible to produce these superalloy components from fine grain ingot, without the need for powder. Potential advantages are lower cost and improved cleanliness (properties). Several alternative process methods are described, which incorporate either specialized thermal treatments and/or extrusion prior to forging. Work to date indicates that the processability and mechanical properties of parts made from fine grain ingot are at least equivalent to those made by powder metallurgy methods.

Introduction

Nickel-base superalloys are used for many critical compressor and turbine components in gas turbine engines. As engine performance and durability have increased in response to market pressures, the materials used for critical parts (e.g., disks and seals) have also had to be improved. This has been achieved by changing the basic character of the superalloys used. In the 1960's, the most common disk material was the intermediate strength alloy Waspaloy, which contains about 20 volume percent of the hardening phase, gamma prime (1). During the late 1960's and 1970's, new higher strength alloys IN100 and its derivative MERL 76 were developed for disks in the F100 and PW2037. The concentration of gamma prime was now much higher (~65%), which was achieved by large increases in alloy element content and complexity. These changes made it impossible to process alloys by the conventional VIM/VAR (vacuum induction melt/vacuum arc remelt) ingot and forging sequences used for Waspaloy due to segregation and cracking problems. The problems were solved by the development of inert gas powder atomization processes, effective consolidation (hot isostatic pressing or extrusion) procedures, and near net isothermal forging processing (2,3).

Powder products have proven to be reliable and of high quality. However, comprehensive (and frequently costly) process control measures must be implemented to assure that the amount and size of indigenous oxide inclusions (4) (originating from the melting operation) and extraneous contaminants (5), both of which can affect fatigue properties, are minimized. Recent developments in ingot technology are now making it possible to process these high strength alloys without the need for an intermediate powder making step. By eliminating powder (and simultaneously some required process controls) and through the incorporation of EBCHR (electron beam cold hearth refined) starting stock to reduce oxide content (6), the potential exists for reduced cost and improved cleanliness. Key elements of the technology needed to accomplish this are: (1) production of fine grain ingots, and (2) processing methods capable of converting these ingots into components. This paper describes the chronological development and optimization of processing methods for fabrication of hardware from cast FGI (fine grain ingot). This work was performed predominantly using an alloy with a modified MERL 76 composition (Table I).

Table I. Nominal Composition of Modified MERL 76

Ni	Cr	Co	Mo	Al	Ti	Cb	B	Zr	C
Bal.	12.0	18.5	3.2	5.0	4.3	1.4	0.02	0.06	0.025

Fine Grain Ingot Casting

Several casting approaches currently exist which are capable of producing sound, crack-free fine grain ingot (typically ASTM 1-3 grain size). Generally, these techniques involve various schemes to control superheat, mold heat extraction, and dendritic growth. The VADER (vacuum arc double electrode remelt) process (7), developed by Special Metals Corporation, has been the focus of much of this development activity, although Howmet's Microcast-X process also shows potential as a fine grain ingot casting method (8).

Most of the process development effort described in this paper used VADER material, which consists of electric arc remelting two horizontally opposing electrodes. The electrodes are fed toward each other while maintaining the desired gap and current density to produce low superheat droplets which are collected in a static or withdrawal mold (Figure 1). Various schemes are employed to distribute the molten metal and promote the desired steady-state melting and solidification mechanisms. If high purity material is required, the use of EBCHR (electron beam cold hearth refined) electrodes should be specified.

Typical macro- and microstructures of VAR and VADER cast ingots are presented in Figure 2. Grains in VAR ingots are typically 0.25 inch or larger in size (making high strength alloys virtually unworkable), while VADER grains are typically ASTM 2-3. Both microstructures exhibit similar eutectic gamma prime phases, either at interdendritic (VAR) or intergranular (VADER) locations.

Forge Process Development

Development of processing methods for FGI occurred in three stages:

1. Early Direct Forge - Simple preforge conditioning
   Tailored forge process sequence

2. Optimized Direct Forge - Improved (superoverage) preforge conditioning
   Standard forge parameters

3. Extrude Plus Forge - Extrude preforge processing
   Standard forge parameters.

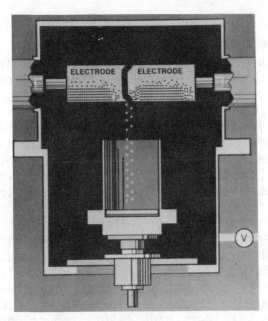 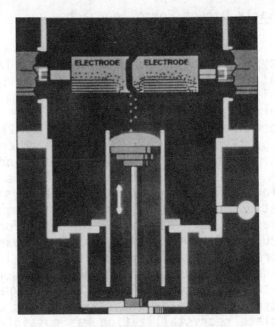

STATIC   WITHDRAWAL

Figure 1 - Schematic illustrations of static and withdrawal VADER casting modes.

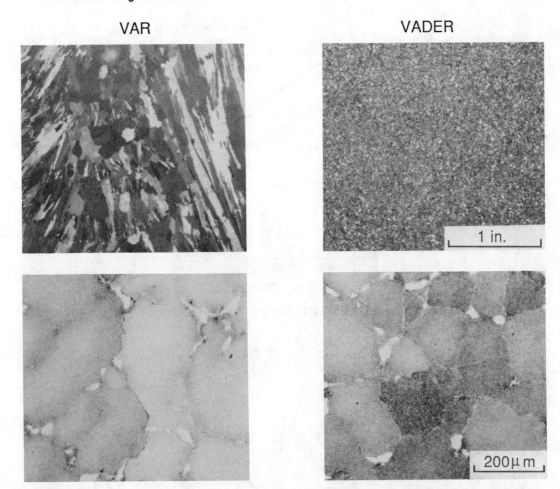

Figure 2 - Macro (top) and micro (bottom) structures of VAR and VADER cast ingots.

In most of the work described below, the fine grain ingots were hot isostatic pressed (HIP) below the gamma prime solvus to close casting porosity while maintaining the as-cast grain size.

Early Direct Forge Approach

Isothermal forging of either as-cast or cast + HIP VADER material at typical conditions (2050 to 2100°F)/$\dot{\epsilon}$ = 0.1) results in significant cracking at low reductions (<50%). The initial approach to improve forgeability consisted of an evaluation of a range of forge process parameters and simple preforge thermal treatments. Forge temperature and strain rate in this program were varied between 1900 to 2150°F and 0.005 to 0.5 in/in/minute, respectively. For the preforge thermal conditioning (overage) treatments, temperatures from 2000 to 2100°F and times from 4 to 48 hours were examined. A series of subscale forge trials were conducted and evaluated based on forge flow stress, material recrystallization, and cracking. Typical flow stress curves as a function of temperature are shown in Figure 3. Based on results such as these, the best forge process sequence (9) consisted of a 2050°F/4 hour overage and forge parameters of 2050°F/$\dot{\epsilon}$ = 0.1. To minimize cracking, forging reductions were limited to about 50%, and an intermediate anneal (2100°F/1 hour) was imposed between forging steps. Forging flow stresses were still relatively high, and the resulting grain structure was typically duplex with large grains approaching the as-cast grain size. Complete recrystallization was possible only with very high (>90%) reductions. The effect of percent reduction on recrystallization is shown in Figure 4.

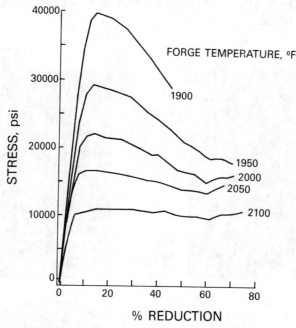

Figure 3 - Flow stress versus forge temperature for FGI given 2050°F/4 hour overage.

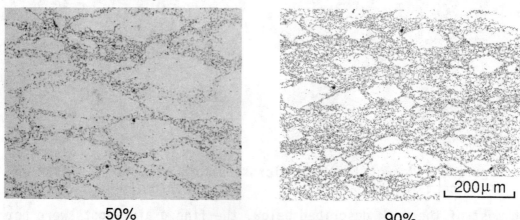

Figure 4 - Recrystallization versus % forge reduction using initial direct forge processing.

Optimized Direct Forge Approach

Although a number of experimental disks were successfully produced using the above processing approach, the forgeability improvements achieved through optimizing isothermal forging parameters and a simple overage were considered insufficient for a production process. Therefore, a more sophisticated process sequence was developed. Observing that modest improvements in forgeability could be gained from a simple overage cycle, a method was devised (10) to dramatically coarsen the gamma prime precipitates, particularly in the grain interiors. The intent was to promote, through subsequent working, dynamic recrystallization in the center of the grains resulting in an even, fine grain structure throughout a forging.

The desired superoveraged (SOA) structure is produced by taking advantage of the gradient in gamma prime solvus from grain center to grain boundary. This gamma prime solvus gradient is a result of composition gradients which exist in all castings, including fine grain ingots.

To create the SOA microstructure, the billet to be forged is heated to a temperature slightly below the solution temperature of the intergranular or eutectic gamma prime (to prevent grain coarsening). The key processing step is a slow controlled cool ($\leq 10°F$/hour) from this temperature through the gamma prime temperature formation range. A combination of enhanced high temperature diffusion and decreasing solubility for gamma prime formers on cooling are believed to be the factors responsible for the effectiveness of this gamma prime coarsening cycle, which is illustrated in Figure 5. SOA structures produced by using 2 and 10°F/hour cooling rates are shown, along with the starting microstructure and one obtained by a long (60 hour) isothermal hold at 2050°F. Compared to a gamma prime size of 1.8 microns in the isothermally held sample, the SOA samples had substantially larger sizes of 8 microns and 3.3 microns, respectively.

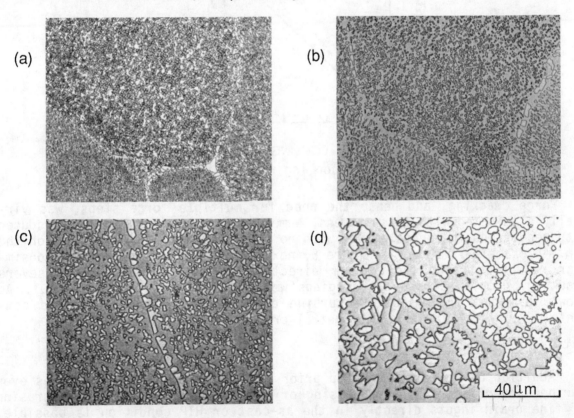

Figure 5 - Microstructures showing effectiveness of SOA in coarsening γ': (a) as-cast, (b) 2050°F/60 hour isothermal overage, (c) 10°F/hour and (d) 2°F/hour SOA cycles.

Limited forge parameter studies demonstrated that conventional isothermal forge parameters of 2050°F/$\dot{\epsilon}$ = 0.1 worked well for material given the SOA treatment. This treatment had the desired effect of resulting in a completely recrystallized structure (ASTM 7-8) at forging reductions as low as 50% (Figure 6a). Furthermore, forging flow stresses were approximately half of those experienced with the earlier direct forge process and decreased (as expected) with decreasing SOA cooling rate (Figure 7). For best results, an SOA cooling rate of ≤ 5°F/hour is preferred.

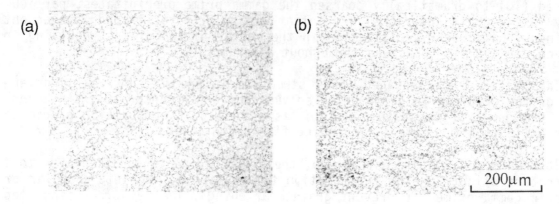

Figure 6 - Recrystallized microstructures of forgings made using the direct forge (a) and extrude plus forge (b) processes following an SOA treatment.

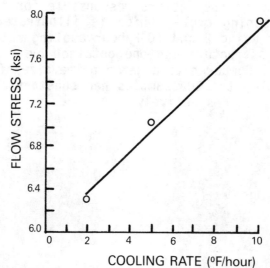

Figure 7 - Effect of SOA cooling rate on forging flow stress.

Forge cracking, and hence the need for multiple forge steps, was virtually eliminated with this process. Numerous pancake forgings were produced with reductions in excess of 90% with no sign of cracking or checking on the rim. An example is shown in Figure 8, next to another pancake forged to similar reduction using the earlier direct forge process which shows severe cracking. Complex closed die forgings were produced equally successful. As shown in Figure 9, a subscale turbine disk shape was crack-free and completely recrystallized (full diametral cross section is macroetched).

Extrude Plus Forge Approach

Addition of an extrusion step prior to isothermal forging provides even further advantages in the processing of FGI (11). Unlike forging, extrusion of fine grain ingots directly in the as-cast or HIP condition is possible, though not without some difficulty. However, an SOA preconditioning thermal treatment was found to reduce cracking and promote recrystallization, especially at low extrusion ratios.

Figure 8 - Pancake forgings with 90% reduction made using: (a) initial direct forge process, and (b) optimized SOA direct forge process.

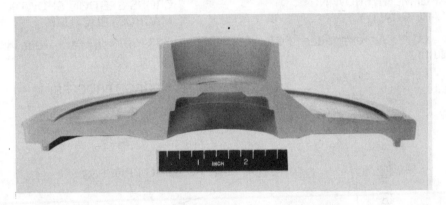

Figure 9 - Subscale turbine disk shape made using optimized SOA direct forge process. Note absence of cracking and complete recrystallization.

Because the stress induced during extrusion is primarily compressive, the propensity for cracking is less than that for forging, thus permitting use of a relatively fast cooling rate SOA cycle ($\approx 10°$F/hour). The smaller overaged gamma prime size results in a finer recrystallized grain size - from ASTM 7-8 for direct forged FGI (Figure 6a) to ASTM 10-11 for extruded plus forged material (Figure 6b).

Canning in mild steel cans and use of streamline die extrusion technology (12) were both found to be beneficial in producing a quality product. Figure 10 shows an extruded billet after decanning. The surface is crack-free and the structure completely recrystallized with only a slight texturing in the longitudinal direction.

As a result of the finer grain size, forging flow stress is also reduced compared to a direct forge process. Representative flow stress plots for all three processes described in this paper are presented in Figure 11. All curves exhibit an initially high breakthrough stress followed by a lower steady-state stress as a result of dynamic recrystallization. Ingot processed by extrusion has a flow stress approximately equal to billets prepared by powder metallurgy techniques.

Standard Gatorizing® parameters, used for powder metallurgy billet, work equally well for extruded FGI. A variety of component shapes have already been produced; an example of an 85-pound turbine disk is shown in Figure 12.

EXTRUDED LOG

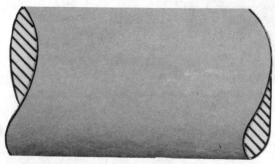

ENLARGEMENT SHOWING
SURFACE QUALITY

CROSS SECTION SHOWING
MACROSTRUCTURE

Figure 10 - As-extruded FGI showing detail of surface quality and macrostructure.

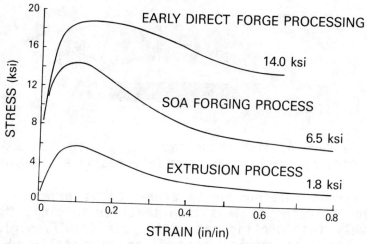

Figure 11 - Forging flow stress plots for all three FGI processing methods.

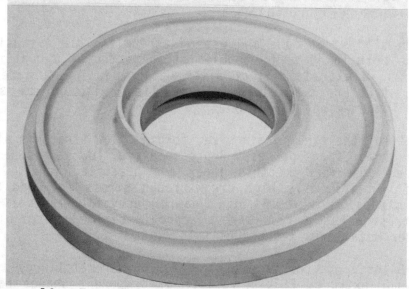

Figure 12 - Extruded and forged FGI 85-pound turbine disk.

Mechanical property evaluation has been initiated on the FGI product. Results to date have shown tensile and stress-rupture properties to be well above goal values (Figure 13). Notched rupture lives, in particular, are typically an order of magnitude better than an equivalent powder product. Other properties (low cycle fatigue, crack growth) have shown equivalence to a powder product.

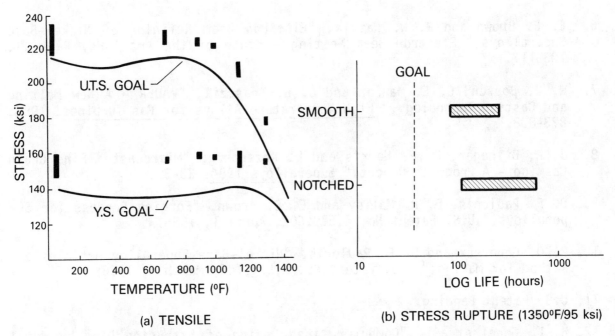

Figure 13 - Tensile and stress rupture properties of direct forge and extrude + forge FGI modified MERL 76.

Conclusions

The work described in this paper clearly demonstrates the feasibility of producing complex parts from high strength superalloy fine grain ingots. Production incorporation, however, will ultimately depend on a variety of factors such as: process economics, capital equipment availability, and additional property testing to verify that high property levels can be consistently achieved.

Acknowledgements

The authors gratefully acknowledge A. J. Nytch for his skilled technical assistance; E. E. Brown and D. R. Malley for their individual contributions to this technology; and M. J. Blackburn for his guidance and continuing support.

References

1. M. J. Donachie, Jr., "Introduction to Superalloys," Superalloys Source Book, 1984, 3-16.

2. J. B. Moore and R. L. Athey, "Fabrication Method for the High Temperature Alloys," U.S. Patent No. 3,519,503, July 7, 1970.

3. G. H. Gessinger, "Recent Development in Powder Metallurgy of Superalloys," Powder Metallurgy International, 13(1981), 93-101.

4. E. E. Brown et al., "The Influence of VIM Crucible Composition, Vacuum Arc Remelting and Electroslag Remelting of the Non-Metallic Inclusion Content of MERL 76, Superalloys 1980, 159-168.

5. D. R. Chang, D. D. Krueger and R. A. Sprague, "Superalloy Powder Processing, Properties and Turbine Disk Applications," Superalloys 1984, 245-273.

6. E. E. Brown and R. W. Hatala, "Electron Beam Refining of Nickel-Base Superalloys," Electron Beam Melting - State of the Art 1985, Part II, 103-117.

7. W. J. Boesch, G. E. Maurer and C. B. Adasczik, "VADER - A New Melting and Casting Technology," High Temperature Alloys for Gas Turbines, 1982, 823-838.

8. J. R. Brinegar, L. F. Norris and L. Rozenberg, "Microcast-X Fine Grain Casting - A Progress Report," Superalloys 1984, 23-32.

9. D. F. Paulonis, D. R. Malley and E. E. Brown, "Forging Process for Superalloys," U.S. Patent No. 4,579,602, April 1, 1986.

10. P. D. Genereux and D. F. Paulonis, "Nickel-Base Superalloy Articles and Method for Making," U.S. Patent No. 4,574,105, March 4, 1986.

11. U.S. Patent Pending.

12. H. L. Gegel et al., "Computer Aided Design of Extrusion Dies by Metal Flow Simulation," AGARD-LS-137, 8-1, 1984.

ISOCON MANUFACTURING OF WASPALOY TURBINE DISCS

Douglas P. Stewart
Senior Metallurgical Engineer
LADISH CO., INC.
Cudahy, WI 53110

Summary

Waspaloy is a vacuum melted age hardenable nickel base alloy with good strengths and elevated temperature properties for use as compressor and turbine rotor components for turbine engine applications. Advanced turbine engine requirements through defect tolerant designs require a manufacturing approach which provides increased ultrasonic inspection capabilities and cost effectiveness. By using a combination of isothermal press and conventional hammer forging techniques, the "ISOCON" manufacturing approach has been developed for Waspaloy turbine discs which meets these requirements while maintaining mechanical property and microstructural conformance.

Introduction

Today's turbine engine builders have goals to produce engines with improved life by using defect tolerant designs. These design requirements in Waspaloy turbine discs demand critical nondestructive inspection techniques while maintaining a high level of mechanical properties and cost effectiveness. A manufacturing procedure has been developed for a thirty inch diameter Waspaloy turbine disc which meets and exceeds these goals. This manufacturing approach has been entitled "ISOCON".

ISOCON is a two step manufacturing approach which utilizes isothermal press and conventional hammer forging techniques. The initial forging sequence is performed isothermally in order to produce a uniform microstructure which is required for stringent ultrasonic inspection techniques. The final forging sequence is performed on a hammer for improved tensile strengths and refined contour outline.

Background

Prior to the introduction of ISOCON, large Waspaloy turbine discs were forged using a multiple (five) step manufacturing sequence. Forging temperatures and reductions produced a duplexed or necklaced microstructure. The grain size in cross sectioned discs ranged from average ASTM 3 to 6 (rating all grains) with variations in the percentage of duplexing from the disc's bore to the rim. See Figure I.

Figure 1: Macroetched Radial Section From a Multiple Step Forged Waspaloy Turbine Disc

The mechanical properties were typical of conventionally forged Waspaloy discs solultion heat treated, stabilized, and aged. Room temperature tensile strengths were nominally 150 Ksi (1035 MPa) yield (0.2% offset) and 198 Ksi (1365 MPa) ultimate. Total plastic strain at 23 hours from creep tests were nominally 0.14% (rim, tangential) when tested at a temperature of 1240°F (670°C) and 80 Ksi (550 MPa) stress.

Ultrasonic inspection was performed to the equivalent of 100% of a 0.025 in. (0.064 mm) diameter flat bottom hole (FBH) standard. Approximately 10% of the discs were rejected due to indications which were associated with melt related defects. Also noted were discs which displayed increased noise or hash levels in the bore and rim areas. Microstructural review of the areas with increased noise noted coarse grains and variations in the percentage of duplexing. See Figure II.

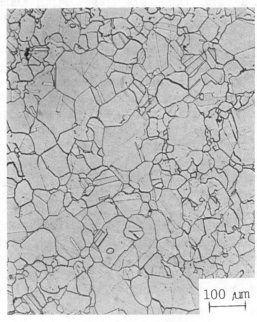

LOCATION: BORE

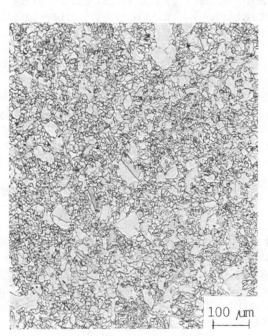

LOCATION: MID-RADIUS

LOCATION: RIM

FIGURE 2 : MICROPHOTOGRAPHS FROM A MULTIPLE STEP FORGED WASPALOY TURBINE DISC

ISOCON

Improvements necessary for upgraded turbine engines through defect tolerant designs required more sensitive ultrasonic inspection techniques while still maintaining a high level of mechanical property conformance and cost effectiveness. By combining Ladish Co.'s 10,000 Ton isothermal press and a 125,000 MKg hammer, Waspaloy high pressure turbine discs were manufactured which met the upgraded engine requirements. This manufacturing approach has been entitled "ISOCON".

The manufacturing sequence initially attempted, utilized a fourteen inch diameter Waspaloy billet (approximately 1,000 lbs.) which was isothermally press forged to a preform shape. The forging temperature was selected to dynamically recrystallize the billet to a uniform (controlled) grain size. This microstructural uniformity would therefore reduce the variation in ultrasonic noise levels from the bore to the rim of the discs.

Mechanical property evaluation after heat treatment of the preform configuration (solution, stabilize, age) however revealed low 0.2% offset yield strengths which at room temperature ranged from 122 Ksi to 131 Ksi (840 MPa to 900 MPa). Consequently the conventional hammer final forge operation was introduced. Forging temperatures and reductions were selected to increase strengths without significantly altering the uniform microstructure produced during the preform forging step.

RESULTS

The ISOCON manufacturing approach has successfully produced thirty inch diameter Waspaloy turbine discs, 3.5 in. (rim) to 5.0 in. (bore) in thickness, which have conformed to the mechanical, microstructural, and ultrasonic inspection requirements for upgraded turbine engines. ISOCON manufacturing has also reduced the number of forging sequences from multiple (five) steps to two steps thereby becoming a more cost effective process.

PHYSICAL PROPERTIES

Visual inspection of a cross section macroetched radial section revealed a uniform structure throughout the disc. Microstructural review showed a slightly wrought uniform average grain size (rating all grains) ranging from ASTM 4 to 6. There was approximately two ASTM grain size difference from the extreme outer diameter (rim to the bore. See Figures 3 & 4.

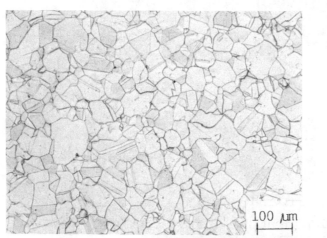

LOCATION: RIM

FIGURE 3: MICROPHOTOGRAPHS OF ISOCON TURBINE DISC

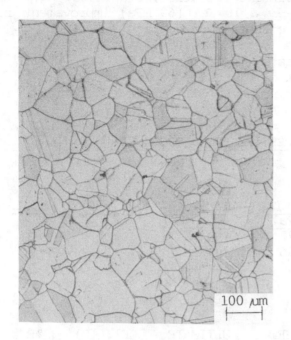 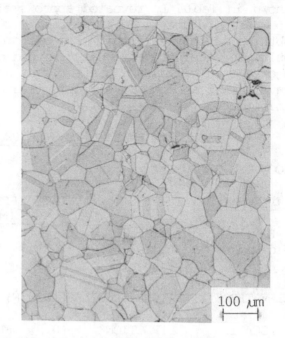

LOCATION: BORE   LOCATION: MID-RADIUS

FIGURE 3 : (Cont'd)
MICROPHOTOGRAPHS OF ISOCON TURBINE DISC

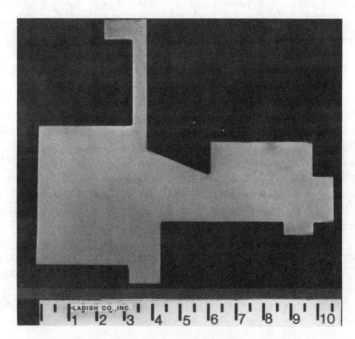

FIGURE 4 : MACROETCHED RADIAL SECTION
OF ISOCON TURBINE DISC

Tensile properties after heat treatment (solution, stabilize, age) as shown in Table I, revealed approximately a 10 Ksi (69 MPa) improvement in room temperature yield strengths (0.2% offset) when compared to values obtained from the isothermally preformed disc. However, in comparison to multiple step, conventionally forged discs, approximately a 7 Ksi (48 MPa) decrease in yield (0.2% offset) was noted. Tensile properties however, have shown excellent isotropic characteristics due to the microstructural uniformity.

TABLE I: AVERAGE TENSILE PROPERTIES FROM ISOCON TURBINE DISCS

ROOM TEMPERATURE

LOCATION	DIRECTION	(0.2% OFFSET) Ksi (MPa)	ULTIMATE Ksi (MPa)	%El.(4D)	%R.A.
Rim	Axial	141 972	194 1338	23	31
Mid-Radius	Radial	143 986	197 1358	23	30
Bore	Tangential	143 986	194 1338	24	31

ELEVATED TEMPERATURE TENSILE (1000°F)

LOCATION	DIRECTION	(0.2% OFFSET) Ksi (MPa)	ULTIMATE Ksi (MPa)	%El.(4D)	%R.A.
Rim	Tangential	127 876	178 1227	21	28
Mid-Radius	Radial	128 883	176 1214	20	27
Bore	Tangential	127 876	177 1220	20	25

Total plastic strain at 23 hours from creep tests (rim, tangential) ranged from 0.06% to 0.10% when tested at a temperature of 1240°F (670°C) and 80 Ksi (550 MPa) stress. Creep life to 0.20% extension has consistently exceeded 100 hours and has shown significant improvement when compared to multiple step, conventionally forged turbine discs.

ULTRASONIC INSPECTION

Ultrasonic inspection of the ISOCON rectilinear machined discs when inspected to the equivalent of 100% of a 0.025 inch (0.064 mm) diameter FBH standard has shown a 3 db improvement rejection limit capability utilizing a 5 MHZ transducer in comparison with the multiple step, conventionally forged turbine discs. Inspection of "critically strained" areas of the discs consistently reveal noise levels as low as -28 dbs (0.010 in. average diameter) when inspected using a 10 MHZ transducer. These ultrasonic inspection levels have been possible due to uniform grain size developed during the isothermal manufacturing sequence.

Disc rejections have been significantly reduced when compared to the multiple step, conventionally forged discs even though the ISOCON discs have been inspected to more critical inspection levels. Only one sonic indication has been associated with a melt related defect in over two hundred turbine discs manufactured to date.

CONCLUSION

ISOCON manufacturing of large Waspaloy turbine discs is a reproducible and cost effective manufacturing approach for advanced turbine engine designs. A large number of turbine discs have been manufactured and are currently being utilized in European and American commercial turbine engines. ISOCON manufacturing allows turbine discs to be critically ultrasonic inspected while maintaining mechanical property and microstructural conformance. ISOCON manufacturing has also led to other evaluations for forging age hardenable nickel base alloys which require defect tolerant designs utilizing controlled structures.

ACKNOWLEDGMENT

The author wishes to thank all the people of Ladish Co., Inc. for their help and cooperation in making ISOCON manufacturing successful.

LASER DRILLING OF A SUPERALLOY COATED WITH CERAMIC

P. FORGET*, M. JEANDIN*, P. LECHERVY** and D. VARELA**

* Ecole des Mines de Paris, Centre des Matériaux P.M. Fourt
  BP 87, 91003 EVRY CEDEX, FRANCE
** SNECMA, Centre de Villaroche, 77550 MOISSY-CRAMAYEL, FRANCE

Abstract

Laser drilling has been developed in advanced aircraft industry in particular to achieve the intricate hole network of the combustion chamber because of several advantages compared to the main competing process, that of electron beam drilling. The combustion chambers for the next generation of engines will be protected on the inner side by a thermal barrier coating (TBC), capable of working at higher temperatures 100-150°C higher, thus improving the gas turbine efficiency.

Consequently, three questions might arise :
- Which was the main phenomena involving material under irradiation during laser drilling ?
- How could these phenomena be described by models ?
- Could laser drilling be applied to multi-layered materials such as a superalloy coated with a TBC ?

An attempt, in part, to answer these questions was made by the study of laser drilling of 1.5 mm thick Hastelloy X sheet coated with conventional plasma-sprayed MCrAlY bond coat plus plasma-sprayed zirconia.

The study focused on microstructural SEM (Scanning Electron Microscopy) observation coupled with modeling. SEM observations, applied to polished (and, if necessary, etched) axial sections of the holes and to edges of the holes, allowed accurate measurements of relevant microstructural parameters which validated the models. The influences of the principal laser parameters such as pulse length, pulse rate and power density were determined. The nature (metal or ceramic) of the side of the part exposed to the beam, the locus of the beam focus and the beam entrance angle were only considered to lesser degree.

Nomenclature

V_i^k, Volume of the ith part (see figure 1) after the kth pulse; H_f^C, H_f^M, Heats of fusion respectively of the ceramic and of the metal; H_v^C, H_v^M Heats of vaporization respectively of the ceramic and of the metal; e_o, Power density; r, Cylindrical coordinate perpendicular to the Oz hole axis; l, Hole mid-depth; τ, Pulse length; ρ, Density; C_p, Specific heat; R, Reflectivity; λ, Thermal conductivity; α, Absorption coefficient; T_f, T_v, Melting and Vaporization temperatures, T_o, Room temperature, φ, ψ, Fractions of respectively melted and evaporated ceramic.

Introduction

The use of lasers is very attractive for drilling small holes at high production rates, as required in the aircraft industry for parts such as nozzle guide vanes and the combustion chamber. The prominent advantages of the process usually claimed are :
- no contact with the part being drilled;
- precise location of the holes;
- no chip problems;
- high production rates;
- large (up to 100 : 1) depth-to-diameter ratios attainable;
- applicable various materials (hard as well as soft);
- achievement of a large range of hole diameters (from about 0.2. µm to about 1.5 mm by percussion drilling, larger by trepanning).

Laser drilling is furthermore promoted by the development of a new generation of systems i.e. the so-called "face-pumped" ("slab") lasers and HSS (high power Solid State) lasers capable of delivering an average power of 1 kW or more (1). Laser drilling has been developed over the past 6 years at the SNECMA Company and is currently applied to the manufacture of air-cooled turbo-engine components such as combustor and turbine parts. A collaborative work between the "Ecole des Mines de Paris" and SNECMA was run to study the expansion of the laser drilling capability to the new ceramic-coated components. The study was conducted in the light of previous works dealing with laser drilling of monolithic materials for applications such as lubrication holes for gears, wire drawing dies, porous ceramic tiles for heat exchanger, small parts in jewellery and air-cooled parts as those involved in this work in aerospace industry (2-6).

Materials and Apparatus

Laser drilling was applied to conventional Hastelloy X sheets coated with ytrria-stabilized zirconia plasma-sprayed onto a MCrAlY bond coat also deposited by conventional air plasma spraying (compositions in table I).

A Nd:YAG rod laser drilling machine integrated into a multi-axis C.N.C. system was used. The main laser drilling parameters are given in table II.

Although the influence of energy, entrance angle and pulse rate were investigated, the work focused on drilling mechanisms occurring for given experimental conditions. All the following observations and calculations reported were made for a pulse length of 240 µs, an energy output of 3.5 J and a given pulse rate. Holes were produced without gas assist by percussion drilling and not in the trepan mode.

Table I - Chemical Compositions of the Alloys, Wt.%.

Substrate	Ni	Cr	Co	Mo	Fe
Hastelloy X	Bal.	22.0	1.5	9.0	18.5
	Mn	Si	C		
	05.	0.5	0.1		
Bond coat	Ni	Cr	Al	Y	Ta
MCRAlY	Bal.	21.0	8.5	0.6	5.7

Table II - Laser Drilling Parameters

Pulse width, µs	0 - 1000
Pulse rate, p.p.s.	⩽ 50
Lens focal length, mm	100 - 150
Beam focus	at work surface
Average output power, W	400

Observation of polished specimens required in certain cases a specific preparation technique to prevent crumbling, cracking and phase transformation (from tetragonal to monoclinic zirconia) in the ceramic layer. It consisted of impregnation under vacuum of the porous material using SHELL EPIKOTE 815 epoxy pre-polymer polymerized by addition of 50% vol. SHELL LINDRITE 15 hardener followed by sectionning and grinding from 1200 grit and then mechanically polishing to 1 µm diamond finish.

Phenomenological Description of Drilling

Energy and Matter Balance

At each pulse, so long as the hole remained blind i.e. for the first 6 shots, the volumes of heated, melted, evaporated, and re-deposited materials correlated with the amount of energy used. For more than 6 pulses, the losses of matter and energy used could not be significantly analyzed partly because some material was expelled from the entrance as well as from the backside of the sheet.

The volumes V_i^k (see the nomenclature) were determined using SEM micrographs of axial sections of the holes. The relative accuracy of the values of the volumes was estimated to be ± 10%, except thin layers (re-deposited ceramic) for which the accuracy dropped ± 20%. The main sources of error rested on the assimilation of the various actual shapes to simple shapes (cones, cylinders...) and the determination of the average thicknesses of the layers (figure 1).

The values of the previously mentioned energies were calculated assuming that :
- the relevant material, i.e. evaporated or melted at a given pulse, was solid before the pulse;
- the material was heated up to the phase transformation point, temperature remaining constant then;
- a negligible part of the energy was used to evaporate liquid metal as ascertained by previous work (7, 8) in which it has been shown that most of the material left the worked surface in the liquid state and at relatively high velocity.

The energy requirements for melting and vaporization were calculated using the values of latent heat given in table III. For example, the value of the energy used to melt the ceramic was obtained by multiplying the so-called A coefficient (figure 1) equal to $\rho \left[C_p (T_f - T_o) + H_f^c \right]$ by the volume of material involved. The energies required to evaporate the ceramic and melt the metal were similarly estimated using the so-called B and C coefficients.

Table III - Physical Properties of Hastelloy X and Zirconia

Material	ρ(kg.m^{-3})	T_f(°C)	T_v(°C)	H_f(10^4 J.kg^{-1})	H_v(10^4 J.kg^{-1})
Hastelloy X	8220	1250	2400	26.0	-
Zirconia	5200 (solid state)	2700	5000	82.0	1400

	C_p(J.kg^{-1}K^{-1})	λ(W.m^{-1}.K^{-1})	R(at 1.06μm)	α(m^{-1})
Hastelloy X	485 from 0 to 100°C 320 + 0.56 T(K)	9.1 at RT	0.72 up to T_f 1 above T_f	-
Zirconia	502	1.4 10^{-2}	0.4	10^4

Evaporated ceramic was completely removed during the pulse. Re-deposit from the vapor phase was negligible as ascertained by a proper observation of the hole wall. Evaporated metal was also neglected because calculations showed that almost all the unreflected energy (more than 90%) was used to heat Hastelloy X to T_f and to melt it.

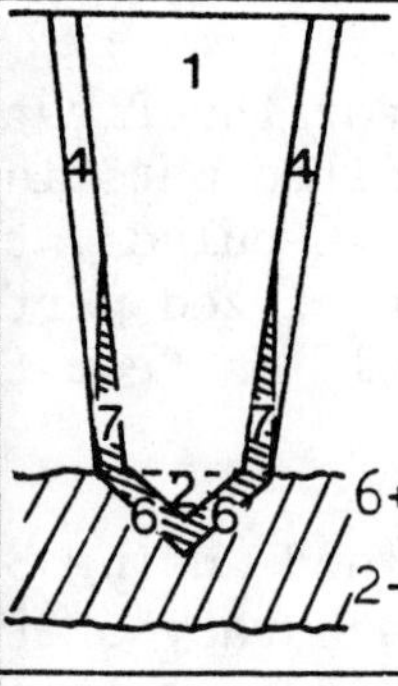

				Number of pulses					
				1	2	3	4	5	6
MATTER (10^{-12} m^3)	1:Hole in Hastelloy	V_1^k		87	190	285	390	600	705
	2:Hole in zirconia	V_2^k		0	0	0	0.32	14	45
	4:Remelted Hastelloy	V_4^k		11	10	21	40	≪ V_1^k	
	6:Remelted zirconia	V_6^k		0	0	0	0.13	1.9	6.2
	7:Re-deposited zirconia	V_7^k		0	0	0	0.22	5.6	17.5
	6+7:Melted zirconia	$V_6^k + V_7^k$		0	0	0	0.35	7.5	24
	2-7:Evaporated zirconia	$V_2^k - V_7^k$		0	0	0	0.09	8.2	27
ENERGY, E	A. $(V_6^k + V_7^k - V_7^{k-1})$, E to melt zirconia, mJ			0	0	0	4.0	83	205
	B. $(V_6^k - V_7^k - V_7^{k-1})$, E to evaporate zirconia, mJ			0	0	0	8.4	705	1200
	C. $(V_1^k + V_4^k - V_4^{k-1})$, E to melt Hastelloy, J			1.0	1.2	1.2	1.5	2.2	1.6
	Total energy used, J (for à 3.5 J output)			1.0	1.2	1.2	1.5	3.0	3.0

Fig. 1 - Energy-matter balance

With regard to the ceramic when drilling the core of the layer (for example during the 5th and the 6th pulses) half of the energy was used to melt ceramic and half to evaporate it (figure 1).

Hole Depth

The dependence of hole depth on number of pulses was practically linear in drilling Hastelloy X and zirconia. The depth per pulse was, as expected, lower in ceramic than in metal as revealed by the change at the ceramic-metal interface of the representative curve (figure 2). The total number of pulses to drill throughout was too low to show (almost parabolic) decrease of the depth per pulse versus the number of pulses, as can be found when drilling for example thicker materials (9).

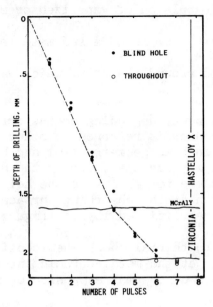

Fig.2 - Dependence of the hole depth on number of pulses.

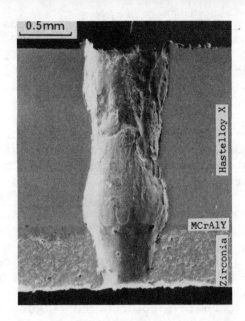

Fig.3 - SEM image of a section of a hole drilled throughout; 3.5 J, 240 μs, 9 pulses.

Hole Shape

Energy Density of the Laser Beam. Starting from the energy conservation law and assuming that radial conduction was negligible and that the profile of the hole followed an isothermal line (generally that of the melting front), the hole shape was the same as that of the energy density distribution of the beam.

The power density which could not be experimentally determined was calculated. The TEM_{00} mode used for drilling gave a Gaussian profile with a power density expressed, in cylindrical coordinates, by

$$e_0(r) = e_{00} \exp\left[-\left(\frac{r}{l}\right)^2\right] \quad (1)$$

For example for a 2 pulse hole, r at mid-depth (where $e/e_{00} = 1/2$) was measured about 245 μm thus leading to the value of 295 μm for l. Then, e_{00} was given by integrating over the whole pulse duration, i.e.

$$\frac{E}{\tau} = \int_{r=0}^{\infty} \int_{\theta=0}^{2\Pi} e_0(r) \, r \, dr \, d\theta \quad (2)$$

(1) + (2) gave

$$e_{00} = \frac{1}{\Pi} \frac{E}{\tau \, l^2} \quad (3)$$

which led to the value of $53.3 \, 10^9 \, W/cm^2$ for e_{00}.

Hole Features. The main features, typical of laser drilling, were (figure 3) :
- barrelling, due to the presence of a confined hot plasma ("plume") of evaporated material which preferentially eroded the metallic wall of the hole by thermal or pressure effects or made the focus point move within the metal;

- taper, due to erosion caused by the expulsion of vaporized or melted material;
- debris, due to resolidified ("recast") or recondensed material and ejected at the lip of the hole;
- recast, made of material, inside and around the hole, which was not completely expelled.

These features were more or less prominent, depending on the drilling conditions. To go into these aspects the reader is referred to other works centered on the effects of laser parameters on hole geometry (10) and how to improve it (11). However, qualitatively, one may say that, during the pulse, the fluidized material was submitted to pressure forces inside the hole from evaporated material and the laser radiation at the absorbing surface. The motion of the material was well revealed by resolidified layers (figure 4).

Craze cracking of ceramic recast due to thermal shock when cooling at the wall and circumferential microcracking at the metal-ceramic interface due to differences between the coefficients of thermal expansion of metal and ceramic also occurred (figure 5).

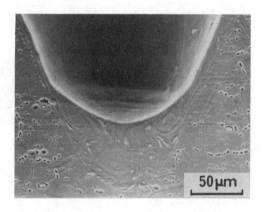

Fig.4 - SEM image of the Hastelloy resolidified layer, etched axial section.

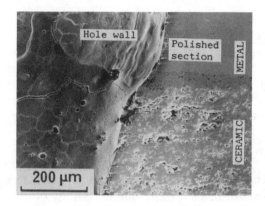

Fig.5 - Craze cracking of zirconia and cracking at the metal/ceramic interface.

Thermal Approach to Drilling

Drilling of Hastelloy X

Axial and radial heat conductions were assumed to be negligible : all the more negligible for the radial because the holes were of a small diameter and for the axial because of the high energy delivered and the short pulses. Moreover, except for the first pulse, axial conduction had no influence on the propagation of the melt front because the energy lost by axial conduction was re-used at the subsequent pulse. For points not on the axis, the previous assumptions were not valid but it did not matter since these points were much less irradiated by the beam.

The energy absorbed by expelling liquid or vaporized material was actually negligible. Lastly, the model did not consider re-focused energy due to multiple internal reflections on the hole wall and involved a reflectivity, R, independent of temperature.

If δv was the volume unit of length d and edges δx, δy at a distance r from the Oz axis and if before the ith pulse the hole surface was located at

$z = z_o(r)$ and at $z_o(r) + d(r)$ after the pulse, the energy needed to drill through a depth $d(r)$ was that used to melt the volume δv that is

$$\int_0^\tau e_o(r) \, \delta x \delta y \, dt = \rho \delta v \left[\frac{1}{1-R} \int_{T_o}^{T_f} C_p \, dT + H_f^M \right] \quad (4)$$

hence

$$d(r) = \frac{\tau(1-R)}{\rho} \cdot \frac{e_o(r)}{\int_{T_o}^{T_f} C_p \, dT + H_f^M} \quad (5)$$

with $e_o(r)$ given by the expression (1)

Thus, the depth per pulse along the axis in particular was given by

$$d(0) = \frac{\tau(1-R)}{\rho} \cdot \frac{e_{oo}}{\int_{T_o}^{T_f} C_p \, dT + H_f^M} \quad (6)$$

which corresponded numerically to a drilling depth of about 400 μm per pulse, consistent with the experimental results (figure 2) : the reflectivity for Hastelloy X, R, being taken equal to that of nickel, i.e. 0.72.

Drilling of Zirconia

The model. As in the metal, the conduction phenomena were negligible, but in this case because of the low thermal conductivity of the ceramic. Heating involved the bulk of the material mainly due to the transparency of zirconia. Assuming the reflectivity to be constant at solid and liquid states, the power density distribution within the ceramic was expressed as

$$e(r, z) = (1-R) \, e_o(r) \, \exp(-\alpha z) \quad (7)$$

with the hole surface located at z=0 at the pulse origin.
a) For $T < T_f$, the heat conduction equation applied to ceramic was :

$$\lambda \Delta T + Q = \rho C_p \frac{\partial T}{\partial t} \quad (8)$$

where Q was the heat input in the ceramic given by

$$Q(r, z) = -\frac{de}{dz}(r, z) \quad (9)$$

which, combined with (7), led to

$$Q(r, z) = \alpha(1-R) \, e_o(r) \, \exp(-\alpha z) \quad (10)$$

Although equation (8) was numerically solvable, as in a one-dimensional approach by WAGNER (2) in the case of laser heating of a finite slab, it was analytically considered, assuming the material semi-infinite and no heat losses through the bounding surface for boundary conditions, as ascertained by PAEK et al. (12). This led to a rather complex litteral expression of T.

However, when neglected conduction, relation (8) simplified to

$$Q = \rho C_p \frac{\partial T}{\partial t} \tag{11}$$

then,

$$T(r,z,t) = T_o(r,z) + \frac{1}{\rho C_p} \int_0^t Q(r,z)\, dt \tag{12}$$

therefore

$$T(r,z,t) = T_o(r,z) + \frac{\alpha(1-R)\, e_o(r)\, \exp(-\alpha z)}{\rho C_p} \cdot t \tag{13}$$

b) For $T > T_f$, the latent heats of fusion and vaporization were involved, which gave the following new expressions of the heat equation

$$Q(r,z,t) = \rho C_p [T(r,z,t) - T_o(r,z)] + \varphi \rho H_f^c + \psi \rho H_v^c \tag{14}$$

depending on the phenomena involved :
  Stage 1; heating of solid ceramic where $T < T_f$, $\varphi = \psi = 0$
  Stage 2; melting of ceramic, $T = T_f$, $0 < \varphi < 1$, $\psi = 0$
  Stage 3; heating of liquid ceramic, $T_f < T < T_v$, $\varphi = 1$, $\psi = 0$
  Stage 4; evaporation of ceramic, $T = T_v$, $\varphi = 1$, $0 < \psi < 1$
  Stage 5; heating of evaporated ceramic, $T_v < T$, $\varphi = \psi = 1$.

Then, for Stages 1, 3 and 5 :

$$T(r,z,t) = T_o(r,z) + \frac{1}{C_p}\left[\frac{\alpha(1-R)\, e_o(r)\, \exp(-\alpha z)}{\rho} \cdot t - \varphi H_f^c - \psi H_v^c\right] \tag{15}$$

for Stage 2 :

$$\varphi(r,z,t) H_f^c = \frac{\alpha(1-R)\, e_o(r)\, \exp(-\alpha z)}{\rho} \cdot t - C_p [T_f - T_o(r,z)] \tag{16}$$

and for Stage 4 :

$$\psi(r,z,t) H_v^c = \frac{\alpha(1-R)\, e_o(r)\, \exp(-\alpha z)}{\rho} \cdot t - C_p [T_v - T_o(r,z)] - H_f^c \tag{17}$$

The beam was absorbed at the melting front because liquid ceramic material was expelled during the inter-pulse time, as shown by further results of this work (13).

Consequently, assuming
- the melting front at Stages 1 and 2 and not at 2 and 3;
- room temperature at 0°C and temperature expressed in °C;
- conduction still negligible for short inter-pulse duration (less than 0.2 s), as ascertained by further calculations carried out in the frame of this program but not reported in this article (13);
- z origin at the surface of the ceramic layer for $r = 0$.

The melting front propagation given by $z = d(r,t)$ was inferred from expression (15) with 0 for r (along the axis) and e_{oo} for e_o

Then,

$$d(0,t) = \frac{1}{\alpha} \left[\ln\left(1 + \frac{\alpha(1-R)e_{oo}}{\rho C_p T_f} \cdot t\right) - \ln\left(1 + \frac{H_f^c}{C_p T_f}\right) \right] \quad (18)$$

Experimental Versus Model Results. From relation (18), in the experimental conditions used (see previous sections), the calculated hole depth per shot i.e. $d(0,\tau)$ was about 200 μm which was in keeping with experimental results (see figure 2)

Expressions (15), (16) and (17) set the temperature profile for laser-heated zirconia as function of depth (figure 6). Equivalent depths of melted and evaporated ceramic, respectively $\tilde{z}_f$ and $\tilde{z}_v$ proportional to the volume of melted and evaporated ceramic, were defined as

$$\tilde{z}_f = \int_0^{z_{m,1}} [1 - \psi(0,z,\tau)] \, dz + d(0,\tau) - z_{m,1} + \int_{z_{m,1}}^{z_{m,2}} \varphi(0,z,\tau) \, dz \quad (19)$$

and

$$\tilde{z}_v = \int_0^{z_{m,1}} \psi(0,z,\tau) \, dz \quad (20)$$

which numerically gave for $\tilde{z}_f$ and $\tilde{z}_v$ respectively, about 170 μm and 54 μm, thus a fraction of about 0.25 for $\tilde{z}_v/(\tilde{z}_f + \tilde{z}_v)$, which represented the amount of evaporated ceramic. The discrepancy between this calculated value and experimentally estimated volumes ($V_2^5 - V_7^5$ or $V_2^6 - V_7^6$ in figure 1) of about 50% of the affected ceramic was partly attributed to neglecting of solid and liquid materials expelled at the same time as the evaporated.

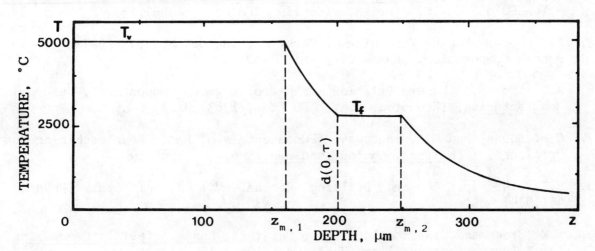

Fig.6 - Temperature profile in the ceramic along the hole axis, at the end of a pulse.

Related Studies

A fluid-mechanics analysis of the removal of melted material was developed and found in good agreement with experimental results. Solidification and cracking (especially at the ceramic-metal interface) phenomena were also investigated. However, all these aspects will be detailed in an additional contribution to be published (13).

Conclusion

The study demonstrated the capability of laser drilling of ceramic-coated superalloys. Thermal models coupled with an energy-matter balance were established. They can be used to predict the main phenomena involving material under irradiation and thus to optimize drilling parameters. They were validated through accurate SEM observations and measurements applied to axial polished section of holes.

Further testing, however like vibrational fatigue testing should be carried out to determine the influence of microcracking at the metal-ceramic interface and the influence of the ceramic layer re-deposited onto the hole wall, which might limit air cooling due to thermal insulation effects.

ACKNOWLEDGEMENTS

Prof. A. PINEAU, Dr. Y. BIENVENU and Mr. C. COLIN are gratefully acknowledged.

REFERENCES

1. Anonymous, "Face-Pumped Lasers Enter Industry", Ind. Laser R., (1988),6.

2. D.A. Belforte, "Economic Justification of Industrial Laser Applications", Applications of High Power Lasers, ed. R.R. Jacobs (Bellingham, Wa:SPIE, 1985),18.

3. H.R. Niederhäuser, "Laser Drilling of Wire Drawing Dies", Wire Ind., 53(1986),709.

4. R.W. Frye and D.M. Polk, "Laser Drilling of Ceramic for Heat Exhanger Applications", "Laser Welding Machining and Materials Processing - ICALEO'85, ed. C. Albright, (Berlin, IFS Ltd., 1985),137.

5. A. Laudel, "Laser Machining of Ceramic", (Report BDX-613-2507-UC38, The Bendix Corporation, Kansas City, Mi, 1980).

6. A.G. Corfe, "Laser Drilling of Aero Engine Components", Lasers in Manufacturing, (Krempton, U.K., IFS Ltd., 1983),31.

7. C.M. Adams and G.A. Hardway, "Fundamentals of Laser Beam Machining and Drilling", IEEE Trans. on Ind. and Gen. Appl., 1(1965),90.

8. R.E. Wagner, "Laser Drilling Mechanics", J. of Appl. Physics, 45(1974),4631.

9. W.M. Steen and J.N. Kamalu, "Laser Cutting", Laser Materials Processing, ed. M. Bass, (North-Holland Publ. Co., 1983),15.

10. B.S. Yilbas, "Study of Affecting Parameters in Laser Hole Drilling of Sheet Metals", Trans. of ASME, 109(1987),282.

11. F.R. Joslin, G.E. Palma and G.L. Whitney, "Laser Precision Drilling", U.S. Patent, Request # 875.726.

12. U.C. Paek and F.P. Gagliano, "Thermal Analysis of Laser Drilling, Processes", IEEE J. of Quantum Elec., QE-8(1982),112.

13. P. Forget et al., "Laser Drilling", Adv. Manuf.Proc., To be pub. (1988).

Microstructure and Mechanical Behavior

THE EFFECT OF MICROSTRUCTURE ON THE FATIGUE CRACK
GROWTH RESISTANCE OF NICKEL BASE SUPERALLOYS

Randy Bowman and Stephen D. Antolovich

School of Materials Engineering
Mechanical Properties Research Lab
Georgia Institute of Technology
Atlanta, Georgia 30332-0245

Abstract

The micro-mechanisms responsible for influencing fatigue crack propagation (FCP) in nickel base superalloys were investigated. Four experimental alloys were developed such that the lattice mismatch (δ), antiphase boundary energy (APBE), and volume fraction (V_f) of γ' precipitates were systematically varied. Heat treatments were also employed to obtain various grain and γ' sizes.

Constant amplitude cyclic loading revealed distinct differences in the FCP response of the four alloys. Precise load-displacement determinations indicated that crack closure was not responsible for these differences. The strength-normalized results indicate that those microstructures which can best accommodate damage are most resistant to crack growth. This is consistent with the accumulated damage model of FCP. Alloys with low V_f, low δ, and low APBE exhibited FCP rates that were approximately 50 times lower than for other treatments. FCP rates were dramatically reduced for those compositions and heat treatments that promoted planar, reversible slip. The effects of individual microstructural features on FCP rates were also determined.

Introduction

The use of superalloys as jet engine components covers a period of nearly 40 years. The early applications were as blade materials and constituted approximately 10% of the weight of the engine. Current applications now include blades, high temperature turbine and compressor discs, nozzle guide vanes, and combustion chambers [1].

Fatigue crack propagation (FCP) was chosen as the basis for this study since the introduction of the retirement-for-cause (RFC) philosophy has made FCP a design-critical property for turbine disks. Because disks constitute up to 30% of the total engine weight, improvements in FCP could result in significant savings of expensive strategic materials while increasing the margin of safety.

Not only is quantification of fatigue damage an elusive goal but defining "damage" is open to debate. The most appealing way to make progress in defining and improving FCP resistance is to work on microstructurally simple systems in which important parameters can be varied systematically. In this way the importance of slip mode, precipitate coherency, crystal structure, etc., can be established as related to FCP. The body of this paper describes the results of such a study.

Background

Deformation of Ni Base Superalloys

Cutting of the γ' by dislocations produces an antiphase boundary (APB) resulting in an overall increase of energy. Paired dislocations are very common during particle cutting where the first dislocation creates the APB and passage of the second restores the stacking sequence thus eliminating the fault. Alloys with combinations of large γ' particles, low volume fractions and high lattice mismatch promote deformation by Orowan looping. In order for shearing to occur the stress necessary for particle cutting must be less than the Orowan looping stress. Both the shearing and looping stresses are a function of the γ' properties and particle size thus allowing the deformation process to be controlled through manipulation of microstructure and chemistry. Particle shearing causes dislocation pairing, makes cross-slip difficult and thus promotes inhomogeneous planar deformation. Cutting of fine precipitates leads to softening in the active slip bands.

The parameters described above can be controlled by composition and heat treatment modifications as described elsewhere [2].

Fatigue Crack Propagation

Antolovich and co-authors [3,4] proposed modifications to an earlier theory [5] in which FCP was viewed as an LCF process occurring out to a distance ρ ahead of the main crack tip. In this model, FCP is caused by damage accumulation in small elements that undergo reversed yielding. The crack then advances by some distance when sufficient damage has accumulated in this "process zone". It was found that longer FCP lives were associated with larger process zones. The larger the process zone the smaller the average strain and the greater the number of cycles required to accumulate a critical amount of damage and advance the crack. Large grain sizes increase the process zone and should reduce FCP rates. This effect can be magnified by promoting slip reversibility (e.g. low mismatch and low APBE).

Alternatively, other authors [e.g. 6-8] attribute the lower FCP rates in coarse grain materials to increased crack closure. Closure reduces the stress intensity due to crack tip shielding of the remotely applied load. These studies found that the improvements in FCP resistance with the larger grains did not exist at high R -ratios where closure does not occur. Coarse and fine grained materials had nearly the same growth rates for R = 0.8. The explanation was that the larger grain sizes increased the fracture surface roughness resulting in more roughness induced closure. Also, the larger grains and correspondingly rougher surfaces had longer effective crack paths. Both of these effects contribute to a reduction in the FCP rates.

Research Program

Alloy Compositions

The controlled microstructural variables included APBE (Γ), mismatch (δ), and volume fraction of γ' (V_f). In addition, grain size (not reported in detail here) and γ' size were controlled by heat-treatments. These variables were chosen since they have all been shown to influence the deformation mode and, presumably, damage accumulation. The compositions and relative target levels for the control variables are shown in Table I.

Antiphase Boundary Energy, Mismatch and Volume Fraction

Control of antiphase boundary energy (APBE) was achieved through manipulation of the Ti/Al ratio. APBE was determined by measuring the spacing of dislocation pairs which are separated by the faulted region.

Table I. Alloy Compositions in Weight Percent and Associated Properties

Alloy	Ni	Al	Ti	Mo	Cr	B	Γ	δ	V_f
1	Bal	2.35	<.01	<.01	13.83	.0037	low	low	low
2	Bal	4.92	<.01	<.01	14.18	.0042	low	low	high
3	Bal	2.96	2.58	<.01	9.39	.0037	low	high	low
4	Bal	1.24	3.71	9.91	13.21	.0060	high	low	high

The effect of mismatch was investigated by producing alloys with low APBE and low V_f while varying the Ti content. It was calculated from the lattice parameter of the matrix and precipitate as determined using x-ray diffractometry.

The V_f of γ' is directly related to the amount of Al + Ti available for precipitation and was measured by phase extraction. Details of all measurement techniques as well as complete rationales are given elsewhere [9].

Mechanical Testing

Tensile specimens were tested to failure under strain control at a rate of 50%/min at room temperature to provide information on yield stress and Young's modulus. In addition, TEM examination of the tested specimens provided information concerning the deformation mode for each composition.

Two types of LCF tests were conducted. The first was a total strain controlled (0-tension-0) test to failure while the other was a interrupted constant plastic strain controlled test. All specimens were electrolytically polished prior to testing [9].

FCP tests were performed at room temperature using a closed loop servo-hydraulic test machine. Testing started in the near-threshold region and covered approximately three decades in growth rate. Crack lengths were monitored with a d-c potential drop system. In addition, crack closure was measured by compliance (load/displacement) techniques. Testing was performed at an R-ratio of 0.1 and 0.8 to examine the effect of closure. Full experimental details are given elsewhere [9].

Results and Discussion

This section describes the results obtained on the small grained alloys designated as indicated in Table I. For ease of presentation, the alloys were numbered according to their composition and gamma prime size. Therefore alloy number one with small γ' size is referred to as 1S while the large γ' material is 1L, etc. A summary of the alloys' properties is presented in Table II.

Initial Microstructures

The alloys were formed into "pancakes" by Wyman-Gordon prior to specimen fabrication. The grain size was very uniform across the entire cross section with an average grain size number of 8 for the small grained material. This uniformity of grain size in the pancake is essential to facilitate comparisons. Metallography of the ingots showed them to be free from any defects such as inclusions, porosity, etc. The γ' was unresolvable by optical microscopy for the S series of alloys whereas in the L series the γ' was clearly visible. There was some preference for γ' formation along the grain boundaries in the L series but the distribution was relatively homogeneous throughout the matrix.

To better characterize the γ' phase, transmission electron microscope (TEM) studies were undertaken on the initial structures. The γ' size in the L series alloys was seen to be approximately 0.6 μm whereas the small γ' material had precipitate sizes of around 0.08 μm. Higher magnification dark field images revealed the presence of hyperfine γ' particles in all alloys with an average size of 0.008 μm. The γ' morphology ranged from spherical in alloy 1, the low mismatch alloy, to a blocky form in the larger mismatch system of alloy 4. Such morphologies are entirely consistent with basic physical metallurgy principles. In the untested condition, the dislocation density is seen to be very low. Only isolated dislocations, which were probably generated during forging and machining, were present.

TABLE II. Measured Microstructural Properties

Alloy	grain size (μm)	γ' size (μm)	Γ (ergs/cm^2)	δ (%)	V_f (%)	σ_{ys} (ksi)	ε_f (%)
1S	52	0.08	56	.09	21	30.31	54.9
1L	55	0.50	80	.07	21	32.60	49.6
2S	51	0.09	124	.07	27	88.68	36.6
2L	52	0.62	198	.04	25	100.00	31.2
3S	36	0.07	96	.21	21	94.30	32.8
3L	42	0.54	120	.18	18	108.40	31.6
4S	23	0.07	420	.18	25	94.30	48.4
4L	34	0.68	403	.14	22	93.20	42.3

Tensile Properties

Monotonic tensile properties for all alloys are summarized in Table II. Yield stress values for each alloy are qualitatively consistent with those predicted from deformation models as discussed in detail elsewhere [9].

Fatigue Crack Propagation

In Fig. 1a, the FCP response of the small γ' alloys tested at an R-ratio of 0.1 at room temperature in air is presented. Correlation of FCP rates as a function of ΔK describes a material's response to cyclic loading. Crack extension at a given value of ΔK is a function of both the amount of damage imposed at the crack tip and the material's intrinsic ability to accommodate the damage. For a given stress level, an alloy with a low yield stress but high resistance to crack extension can exhibit the same FCP rate as an alloy with a high yield strength and low crack growth resistance. Most FCP models predict an inverse relationship between crack growth rate and yield strength. It is therefore useful to attempt to normalize the FCP results with respect to yield strength thereby eliminating the differences in strength, Fig. 1b.

Examination of the fracture surface, Fig. 2, shows that roughness of the fracture surface arises due to cracking along slip bands which formed during the test. Roughness measurements of the fracture surfaces as a function of ΔK are presented in Fig. 3. The roughness parameter, R_L, is a measure of the actual crack path divided by the projected crack path. Comparison of these results with Fig. 1 shows that the specimens with the roughest surfaces have the greatest resistance to crack growth while the specimens with a relatively smooth fracture surface have the highest growth rate. Initially this seems to imply that FCP is controlled by a roughness induced closure effect whereby the specimen with the largest surface roughness will have the most roughness induced crack closure and thus the lowest crack growth rate. This interpretation is discussed below.

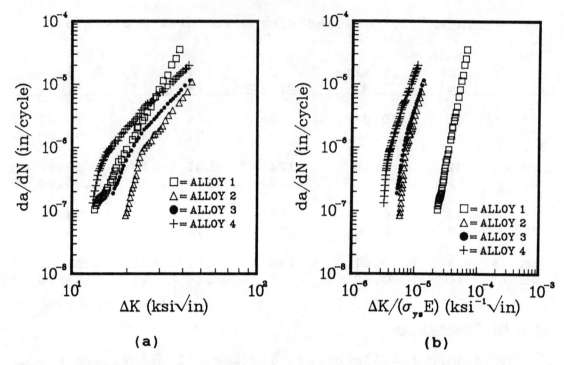

Figure 1 - FCP response of small γ' materials for R = 0.1. In (b) the data is normalized with respect to yield strength and elastic modulus.

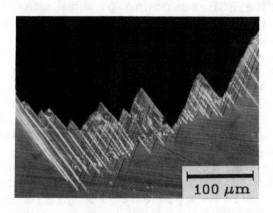

Figure 2 - Optical micrograph of alloy 4S fracture profile.

Closure

To investigate the importance of roughness, closure measurements were performed [9] at R=0.1 and at R=0.8, Fig. 4. Load-line/displacement measurements indicated a closure load of approximately $0.1 P_{max}$ for all alloys at near threshold regions which decreased to zero in the mid-Paris regime. Since the closure loads were similar and relatively low for all alloys, no significant difference in the growth rates was noted when plotted vs. ΔK_{eff} ($\Delta K_{eff} = \Delta K_{max} - \Delta K_{cl}$). The **apparent** correlation between surface roughness and FCP rates (i.e. large roughness associated with low growth rates) can not be explained on the basis of roughness induced closure, however appealing this explanation may be.

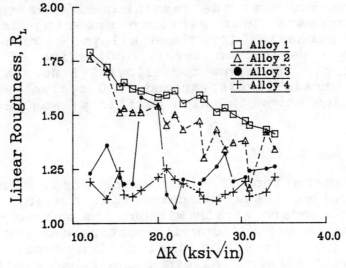

Figure 3 - Fracture surface roughness as a function of stress intensity level.

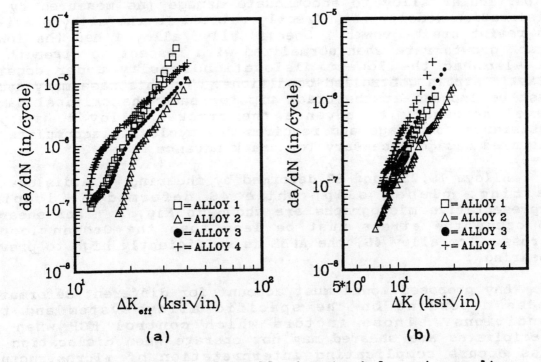

Figure 4 - FCP response of small γ' materials. a) results plotted versus ΔK_{eff} thus accounting for crack closure. b) material response at R = 0.8.

At R = 0.8 (no closure), the relative FCP rates were unchanged although the absolute values of da/dN at R = 0.8 were higher due the larger mean stress. It is clear that the observed differences in growth rates for these alloys are due to "intrinsic" differences in fundamental micromechanical processes and not to "extrinsic" effects such as closure as is often cited for other systems. With all "extrinsic" factors eliminated, the dominant features controlling FCP can be identified.

Low Cycle Fatigue

Total strain controlled tests resulted in an initial hardening response (increasing load) for alloys 1-3 followed by gradual softening over the remaining life of the tests. This behavior suggests that particle shearing is the dominant deformation mechanism for these alloys. Conversely, alloy 4 hardened to a saturation level, indicative of deformation by Orowan looping. The lives for alloys 1-3 were all very similar at the same strain levels (around 7500 cycles) whereas the life of alloy 4 (low strength, high ductility) was nearly 4 times as long.

Deformation Structures

TEM examination of the interrupted, constant plastic strain controlled tests indicated that the ability of alloys 1 and 2 to accommodate strain without the subsequent development of damage was very good (low dislocation density) whereas alloy 4 had nearly 5 times the dislocation density for the same imposed plastic strain. Alloys 1 and 2 had similar dislocation densities, any differences being hidden in the inaccuracies of the dislocation density measurement technique. The ability of a particular alloy to accommodate damage (as measured by the dislocation density) is directly related to the alloy's ability to resist crack growth. Specifically, alloy 1 had the lowest crack growth rate when normalized with respect to strength and it also had the lowest dislocation density under constant plastic strain controlled conditions. In this case many cycles must be imposed at the crack tip to reach the critical damage level necessary to advance the crack. Alloy 4 has poor resistance to damage and requires few cycles to accumulate the required damage necessary for crack advance.

Alloys 1S, 2S, and 3S deformed by shearing (i.e. dislocation pairing, planar slip) while 4S deformed by looping. Representative micrographs are shown in Fig. 5. For shearing to occur the stress must be less than the Orowan looping stress. For alloy 4S, the APBE is sufficiently high to prevent shearing.

Any proposed model must account for different deformation modes depending on the specific alloy system and test conditions. Those factors which control FCP when the precipitates are sheared may not operate when dislocation by-pass occurs complicating interpretation of microstructural influences on FCP.

FCP of Small γ' alloys

From the da/dN vs. ΔK plots (Fig. 1b), it is clear that alloy 1S (low Γ, low δ, low V_f) is most resistant to crack advance. When normalized as described above, the FCP rate of this alloy system is at least two orders of magnitude slower than the others. The low volume fraction of precipitates results in a larger mean free path between obstacles for the mobile dislocations. The imposed plastic strain is therefore more easily accommodated resulting in less damage accumulation and thus greater resistance to crack advance. The efficiency of the γ' as obstacles to dislocation motion is reduced further in this alloy by the low values of APBE and δ. Conversely, alloy 2S has the same APBE and δ but a much higher V_f of γ'. This combination results in relatively higher crack growth rates. The ability of this material to accommodate strain at the crack tip is reduced by the small mean free path of the dislocations.

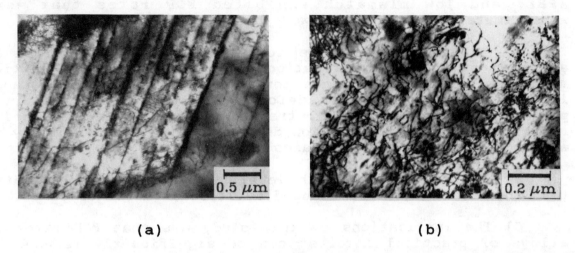

(a)   (b)

Figure 5 - Representative deformation structures illustrating a) shearing, alloy 2S and b) looping, alloy 4S.

The decreased resistance to FCP due to higher APBE (alloy 4S) is due to the difference in deformation mode caused by the APBE. In the other S series alloys, which have various combinations of APBE and V_f, particle shearing occurs, Fig. 5, whereas for alloy 4S with a high APBE particle by-pass is the dominant deformation mode and no APBE is created (i.e. no energy increase). With particle looping, the contribution of δ toward inhibiting dislocation motion becomes more pronounced since in the looping regime the CRSS is directly related to mismatch.

For small γ' precipitates, alloys 2S and 3S had differences of 3-5 times in FCP rates due only to changes in mismatch and a slight difference in V_f. In fact, from the previous argument, increased V_f is seen to increase the FCP rate when normalized with respect to σ_{ys} and E due to a lowering of the mean free path. All other factors being constant, a lower volume fraction of precipitates should reduce the FCP rate. Therefore, the differences in the FCP response of alloys 2S and 3S is even greater if alloy 2S is shifted down in the FCP plot (or 3S is shifted up) to account for the difference in mean free path. The increase in FCP rate for the higher mismatch alloy is a result of the increased resistance to dislocation motion due to the enhanced strain field around the precipitate and/or a different deformation mode.

Summary

Constant amplitude FCP tests were performed on each of four alloy compositions with two different γ' sizes to investigate the effects of microstructure and composition.

The major findings and observations were:

1) Crack closure concepts do not explain differences in the FCP rates for both near threshold and Paris regime propagation in the model Ni base alloys studied here.

2) FCP rates are dramatically low for those compositions and treatments that promote planar, reversible slip.

3) In this study, alloys having high volume fraction, low APBE, and low mismatch exhibited FCP rates that were approximately fifty times lower than other alloys.

4) Internal resistance to damage ahead of a crack is achieved by low volume fraction of precipitates, low lattice mismatch, and low anti-phase boundary energy. FCP resistance is increased by a planar deformation mode. However, in a planar slip material, on a strength/modulus normalized basis, restricting dislocation motion decreases the alloy's ability to accommodate damage and increases the FCP rate.

5) At the same strength level, it has been demonstrated that the FCP rate can be reduced by at least a factor of 50.

6) The implications of this study are that FCP rates in alloys of practical interest can be significantly reduced by heat treatment and modest compositional changes.

Acknowledgements

The authors are grateful to the AFOSR for financial support of this research (Grant # AFOSR 84-0101, Dr. Alan H. Rosenstein Program Manager). They would also like to thank Dr. Hugh Gray of the NASA-Lewis Research Center for assistance in the alloy development phase of the work and Mr. W. Couts of Wyman-Gordon for his help in processing the alloys. One of us (RB) would also like to thank the Wyman-Gordon foundation for a Wyman-Gordon Fellowship.

References

1. J.E. King, Mat. Sci. and Tech., Vol. 3, 1987, pp. 750-764.
2. The Superalloys, C.T Sims and W.L. Hagel eds., John Wiley and Sons, Inc., New York, 1972.
3. G.R. Chanani, S.D. Antolovich, and W.W. Gerberich, Met. Trans., Vol. 3, 1972, pp. 2661-2672.
4. S.D. Antolovich, A. Saxena, and G.R. Chanani, Eng. Frac. Mech., Vol. 7, 1975, pp. 649-652.
5. F.A. McClintock, Fracture of Solids, D.C. Drucker and J. Gilman Eds., John Wiley and Sons, N.Y., 1963, pp. 65-102.
6. G.T. Gray III, J.C. Williams, and A.W. Thompson, Met. Trans., Vol 14, 1983, pp. 421-433.
7. Jian Ku Shang, J.-L. Tzou, and R.O. Ritchie, Met. Trans., Vol. 18, 1987, pp. 1613-1627.
8. R.D. Carter, E.W. Lee, E.A. Starke Jr., and C.J. Beevers, Met. Trans., Vol. 15, 1984, pp. 555-563.
9. R. Bowman, Ph.D. Dissertation, Georgia Institute of Technology, Atlanta, Georgia, 1988.

ISOTHERMAL AND "BITHERMAL" THERMOMECHANICAL FATIGUE
BEHAVIOR OF A NiCoCrAlY-COATED SINGLE CRYSTAL SUPERALLOY

J. Gayda, T. P. Gabb, and R. V. Miner
NASA Lewis Research Center
21000 Brookpark Road Cleveland, Ohio 44135

Abstract

Specimens of single crystal PWA 1480 with <001> orientation, bare or with NiCoCrAlY coating PWA 276, were tested in low cycle fatigue (LCF) at 650, 870, and 1050°C, and in "bithermal" thermomechanical fatigue (TMF) tests between these temperatures. The bithermal test was examined as a bridge between isothermal LCF and general TMF. In it an inelastic strain is applied at one temperature, T_{max}, and reversed at T_{min}. The "out-of-phase" (OP) type bithermal test, which imposes tension at T_{min} and compression at T_{max}, was studied most, since it was most damaging. Specifically investigated were the effects of: inelastic strain range ($\Delta\varepsilon_{in}$), the coating, ΔT, T_{max}, time at T_{max}, T_{min}, and the environment.

On a $\Delta\varepsilon_{in}$ basis, isothermal LCF life of bare crystals exhibited classic dependence on ductility, decreasing with temperature from 1050 to 650°C. Coated crystals exhibited the same life at 1050°C, but at 650°C, cracks initiating in the coating reduced life in the low-$\Delta\varepsilon_{in}$, long-life regime. Life for various bithermal TMF tests in the high-$\Delta\varepsilon_{in}$ regime was also controlled by ductility, and approached the life exhibited for isothermal LCF at the T_{min} of the cycle. However, in the low-$\Delta\varepsilon_{in}$ regime, the OP bithermal test (which imposes tension at T_{min}) reduced lives of both bare and coated crystals, drastically so for the 650-1050°C cycle.

Damage mechanisms in the low-$\Delta\varepsilon_{in}$ regime for OP bithermal tests were different, however, for bare and coated crystals. A 650-1050°C OP vacuum test of a bare crystal was discontinued after 10000 cycles (five times the life observed in air tests) without evidence of cracking. Yet, coated crystals tested in vacuum formed cracks through the coating nearly as fast as those tested in air. The total OP bithermal lives of the coated crystals were, however, longer in vacuum than in air tests, due to slower crack entry and propagation through the crystal itself. Additional tests illustrated the effects of other OP bithermal cycle variables on the life of coated crystals in air. As expected, life decreased with increasing ΔT, increasing T_{max}, and decreasing T_{min}, since, respectively, they increase the thermal mismatch strain between crystal and coating, increase oxidation, and decrease the ductility of the crystal (in the range 650-1050°C). Time at T_{max}, however, had only a small detrimental effect.

Superalloys 1988
Edited by S. Reichman, D.N. Duhl,
G. Maurer, S. Antolovich and C. Lund
The Metallurgical Society, 1988

Introduction

Gas turbine engine components are subject to complex strain-temperature-time cycles during operation. Damage from such cyclic loading is termed thermomechanical fatigue (TMF). Advanced single crystal superalloy turbine blades are covered by coatings which are basically more Al-rich for protection from the environment. Relative to the <001> orientation of superalloys, these coatings have greater coefficients of thermal expansion; higher elastic moduli; lower flow stress, particularly at high temperature; and better ductility at high temperature. Both coatings and single crystal superalloys have rather low ductility at temperatures below about 700°C (1). Coating cracks are promoted in areas cycled into tension at low temperatures and may propagate into the superalloy below (2,3). The greater coefficient of thermal expansion for the coating adds to this problem, since it places the coating in tension simply upon cooling from high temperatures (3).

The system chosen for study, Ni-base single crystal superalloy PWA 1480 and NiCoCrAlY low-pressure-plasma-sprayed (LPPS) overlay coating PWA 276, was developed by Pratt and Whitney Aircraft. PWA 1480 crystals from the same lot studied herein have been characterized in isothermal monotonic (4,5), fatigue (5,6), and creep-fatigue tests (7), and preliminary TMF tests (8). The bulk coating alloy PWA 276, produced as thick plates by the same LPPS process, has also been studied in high temperature monotonic (9,10) and fatigue tests (11). TMF behavior in more complicated cycles than those employed herein have been performed on crystals from other heats of PWA 1480 and having different coatings (12,13).

The objective of this investigation was a greater understanding of the fatigue behavior of coated and bare crystals in isothermal low cycle fatigue (LCF) at 650, 870, and 1050°C and "bithermal" TMF cycles. The temperature 650°C represents about the upper limit of low temperature behavior in both materials and is characterized by relatively constant high strength and low ductility. Whereas, 1050°C is about the upper limit of allowable temperature in turbine engine blades and is in the regime where the coating is extremely weak and ductile. The bithermal test was proposed as a bridge between isothermal LCF and TMF (14). It is relatively simple experimentally and analytically since the mechanical and thermal strains are not superimposed. Since deformation, at least in the superalloy, occurs only at the two temperature limits some unification with isothermal LCF behavior at the two temperatures may be possible. The qualification above is presented because stresses generated during heating and cooling due only to thermal mismatch are likely to produce some inelastic deformation in the coating.

Materials and Procedures

Materials

PWA 1480 contains nominally 10Cr, 5Al, 1.5Ti, 12Ta, 4W, 5Co, in weight percent and balance Ni. It has about 65 v/o of the γ' phase, but no carbides or borides. Crystals were cast commercially as bars about 21 mm in diameter and 140 mm long. Those having <001> within 7° of the axis were selected and solution treated 4 hr at 1290°C before machining. The fatigue specimens had a 4.8 mm diameter and 15 mm long reduced section. The PWA 276 coating was applied to some specimens after machining. Both coated and uncoated specimens were given the coating cycle heat treatment of 1080°C for 4 hr and aged at 870°C for 32 hr. Fig. 1 shows the heat treated alloy and coating. Interdendritic porosity averaged about 0.3 v/o of the alloy and about 7 μm in diameter. Interdendritic areas also contained undissolved γ' eutectic nodules a few μm to tens of μm in diameter and occupying 1-2 v/o of the alloy. Elsewhere, the γ' was cuboidal averaging about 0.6 μm on edge.

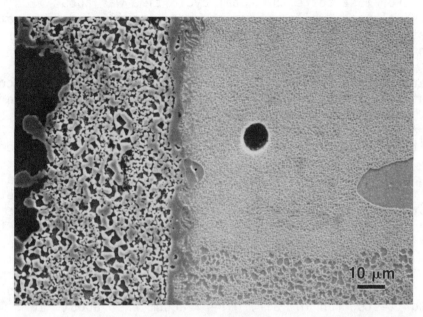

Figure 1. Microstructures of single crystal superalloy, PWA 1480, and NiCoCrAlY overlay coating, PWA 276 (left).

The PWA 276 LPPS coating composition was 20.3Co, 17.3Cr, 13.6Al, 0.5Y, in weight percent and balance Ni. It contains about 50 v/o fcc solid solution and 50 v/o NiAl-based intermetallic compound. The coating thickness was about 0.13 mm with a grain size of about 1.5 μm, and contained 1-2 v/o of pores averaging about 20 μm in diameter.

Test Procedures

All fatigue testing was done on servohydraulic, closed-loop machines, one of which was equipped with a diffusion pumped vacuum chamber. Radio frequency induction heating with closed-loop control was employed. For tests of coated specimens in air, an infrared pyrometer was used. However, for tests of bare specimens in air, and all tests in vacuum, emissivity was found to vary significantly with time, so a thermocouple wrapped against the specimen was employed. At the start of the test, the thermocouple output was calibrated against an optical pyrometer which measured the temperature of a small area of the specimen surface coated with a high temperature 'paint' of known emittance. Strain was measured using an axial extensometer with a 12.5 mm gage length. Strain/time and load/time data were recorded continually, while load/strain hysteresis loops were recorded periodically.

Isothermal LCF tests were conducted at 650, 870, and 1050°C under total mechanical strain control at a frequency of 0.1 Hz with a sinusoidal control waveform and an R ratio of -1 (minimum/maximum strain). TMF behavior was studied in simplified bithermal cycles between the above temperatures. Specimens were strained at one temperature, unloaded, changed to the other temperature, and then strained in the opposite direction to produce a completely reversed strain cycle. Fig. 2 shows the stress-strain hysteresis loop for what is termed an "out-of-phase" (OP) cycle in which the tensile and compressive strains are imposed at the lower and higher temperatures, respectively. This is reversed in the "in-phase" (IP) cycle. A 16 bit computer, equipped with dual digital/analog converters, was used to generate the control waveforms for load and temperature. For most tests the total cycle time was about 120s, of which 100 to 110s elapsed while changing and stabilizing temperature. The mechanical strain rate in the bithermal tests was within a factor of two of that used in the isothermal tests.

For large-strain bithermal tests, equal tensile and compressive inelastic strains, $\Delta\varepsilon_{in}$, were produced at the two temperatures by controlling between two fixed values of the total strain (the sum of the mechanical, $\Delta\varepsilon_{mech}$, plus thermal, $\Delta\varepsilon_{th}$, strain as shown in Fig. 2). Under this condition the material rapidly equilibrated the tensile and compressive $\Delta\varepsilon_{in}$. A constant $\Delta\varepsilon_{mech}$ test results if the temperature endpoints, and thus $\Delta\varepsilon_{th}$, are held constant. However, for small-strain, essentially elastic tests it becomes questionable whether such self-equilibration ever occurs. Thus, tests with expected $\Delta\varepsilon_{in} \leq 10^{-4}$ were run in load control with endpoints extrapolated from those of the higher strain range tests run in strain control. A sensitivity study indicated no significant effect on life for various tension/compression load ratios considered in the range of reasonable extrapolation. Out-of-phase tests in vacuum were also conducted with tensile/compressive load limits identical to tests run in air. The vacuum was about 10^{-6} Torr. The effects of various maximum temperatures, T_{max}, and times at T_{max} on OP bithermal life were investigated in a series of essentially elastic tests having a tensile stress of 620 MPa applied at 650°C, but no load at T_{max}.

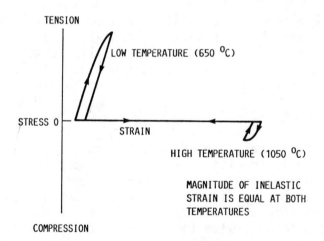

Figure 2. Stress-strain hysterisis loop for an out-of-phase (OP) bithermal test.

Results and Discussion

Specimens of single crystal superalloy PWA 1480 with <001> orientation, either bare or coated with a NiCoCrAlY alloy, PWA 276, were tested in LCF at 650, 870, and 1050°C and in 'bithermal' TMF tests between these temperatures. Fig. 3 presents previous (9) and new isothermal LCF results. It shows that isothermal LCF life of the bare crystals exhibited nearly classic dependence on inelastic strain range and monotonic tensile ductility at all temperatures investigated. Coated crystals had lives equivalent to bare specimens at the highest test temperature, 1050°C. Both failed at internal micropores (8). However, at 650°C, where the coating has little ductility (9), cracks initiating in the coating reduced life in the low-$\Delta\varepsilon_{in}$, long-life regime (8). In this and other figures comparing tests on an $\Delta\varepsilon_{in}$ basis, values of $\Delta\varepsilon_{in} \leq 10^{-4}$ are based on extrapolation of the relationship between the $\Delta\varepsilon_{in}$ and the maximum absolute value of σ at 650°C from tests with measurable $\Delta\varepsilon_{in}$. This is treated in detail elsewhere (8). For this reason, close comparisons between test types on the $\Delta\varepsilon_{in}$ basis must be avoided in the low-$\Delta\varepsilon_{in}$ regime, however the effects to be discussed are large and would also be observed on the basis of σ at 650°C.

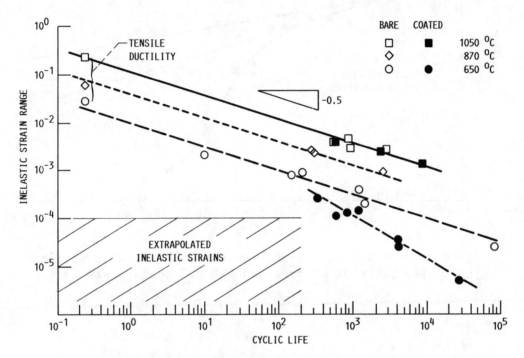

Figure 3. Isothermal inelastic strain - cyclic life behavior for bare and PWA 276 coated, PWA 1480.

The $\Delta\varepsilon_{in}$-based comparison in Fig. 4 shows the link between 650°C LCF and 650-1050°C TMF behavior in the high-$\Delta\varepsilon_{in}$ regime. On the $\Delta\varepsilon_{in}$ basis, both IP and OP bithermal life approached that for isothermal LCF at 650°C, where the single crystal has the least ductility. Fig. 4 also shows that in the low-$\Delta\varepsilon_{in}$ regime, OP bithermal lives of both bare and coated crystals decreased drastically compared with 650°C life. In both bare and coated crystals, surface cracks initiated very early, and roughly 90% of the short lives observed represented crack propagation through the single crystal (8). This was not true of the IP bithermal test, which imposed compression rather than tension at the low temperature. In the low-$\Delta\varepsilon_{in}$ regime, failure in the IP bithermal tests initiated at internal micropores just as for 1050°C isothermal tests.

Comparative tests were performed in vacuum in order to separate the effects of environment and coating on fatigue life. It may be seen in Fig. 5 that the 650°C isothermal LCF lives of coated specimens in air and vacuum are about the same, both being considerably less in the low-$\Delta\varepsilon_{in}$ regime than that for bare specimens. This small enviromental effect indicates that the decrease in isothermal LCF life at this temperature for coated crystals is largely a mechanical effect. Because the coating is weaker, large inelastic strains are produced in the coating at strains which are almost totally elastic in the single crystal. Cracks initiate rapidly in the coating and propagate into the single crystal.

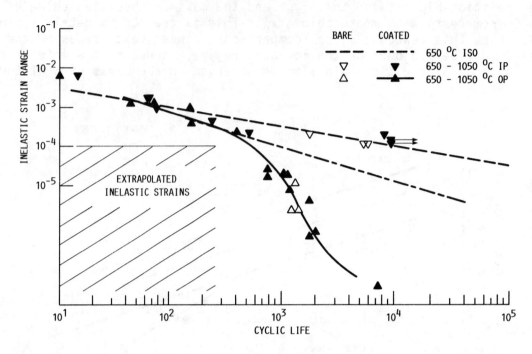

Figure 4. In- and out-of-phase bithermal inelastic strain - cyclic life behavior of bare and PWA 276 coated, PWA 1480.

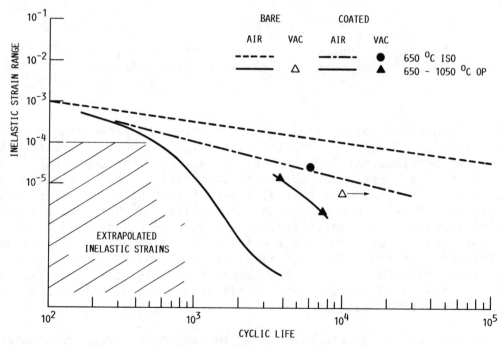

Figure 5. Comparison of cyclic life in air and vacuum for various tests.

In contrast, for the 650-1050°C OP bithermal test there was a large, detrimental effect of the air environment on bare crystals. Surface cracks initiated rapidly during the tests in air, however no cracks had yet initiated in the discontinued test in vacuum. The lives of coated crystals also showed degradation due to the air environment, however interrupted tests in vacuum showed rapid cracking through the coating, almost as soon as in air tests. Thus, the rapid crack initiation in coated crystals in both vacuum and air appears largely the result of the thermal mismatch strains in the coating. The longer total life of coated crystals in vacuum is primarily due to slower crack entry and propagation through the single crystal itself.

The effects of other T_{max}, T_{min}, and ΔT in OP bithermal tests of coated crystals with nominally reversed $\Delta\varepsilon_{in}$ are shown in Fig. 6. Again, comparison with the isothermal tests results shows that in the high-$\Delta\varepsilon_{in}$ regime, behavior was controlled by the cycle temperature with limiting ductility, which is T_{min} in this temperature range. Bithermal behavior approached isothermal LCF behavior at T_{min} in the high-$\Delta\varepsilon_{in}$ regime. In the low-$\Delta\varepsilon_{in}$ regime, the slope of the $\Delta\varepsilon_{in}$-life line became quite steep for the 870-1050°C OP bithermal tests, as for the 650-1050°C OP tests. A less rapid change in the slope of the 650-870°C OP bithermal tests was also observed. Life reductions in the low strain regime resulted from early coating failure in the OP bithermal cycle which is accelerated by the thermal expansion mismatch between the coating and single crystal. This adds to the tensile, mechanical strain in the coating[3] applied at T_{min}. Of course, increasing ΔT of the OP cycle directly increases the thermal mismatch strain and more rapidly cracks the coating. The test results in Fig. 6 reflected this in the shorter lives for the 650-1050°C OP tests in comparison with either the 650-870°C or 870-1050°C OP tests. The detrimental effect of increasing T_{max}, which increases the rate of environmental damage, is illustrated by comparison of the 870-1050°C OP and 650-870°C OP test results. Though ΔT is somewhat smaller for the 870-1050°C OP cycle than the 650-870°C OP cycle, the rate of change in the

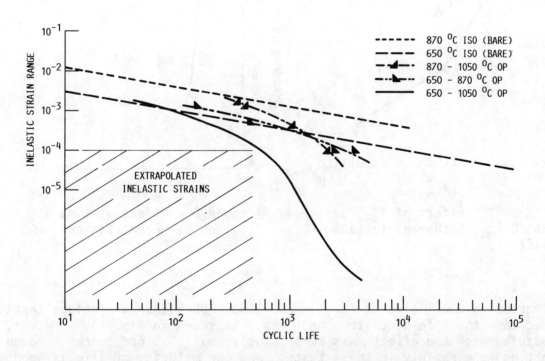

Figure 6. Effect of T_{max} and T_{min} on out-of-phase bithermal fatigue life of PWA 276 coated PWA 1480.

slope of the 870-1050°C OP cycle $\Delta\varepsilon_{in}$-life curve appears considerably more rapid.

The role of high temperature deformation on life for the 650-1050°C OP bithermal cycle was also investigated. Nominally elastic 650-1050°C OP bithermal tests without load at 1050°C were conducted and compared with the tests having compressive loads at 1050°C. In this low strain, long life regime there was little, if any, effect of high temperature deformation.

The time/temperature dependence of damage in the OP bithermal cycle was examined in more detail in tests where T_{max} or time at T_{max} was varied. These were nominally elastic tests with no compressive loading as discussed above. Fig. 7 shows the effect of T_{max} in tests having T_{min} of 650°C, and the normal cycle period of 120 seconds. As shown previously, the lives of bare and coated crystals in tests with T_{max} of 1050°C are the same due to rapid surface crack initiation in either case. The lives of both bare and coated crystals also increased as T_{max} was reduced. For bare specimens, the surface damage, which we indicated previously to be largely environmental, decreases rapidly with decreasing T_{max}. However surface damage of coated crystals, is less sensitive to T_{max} as coating cracks can arise from purely mechanical loads applied at 650°C. These coating cracks can foreshorten crack initiation in the single crystal in the absence of environmental effects, as was the case in isothermal tests at 650°C.

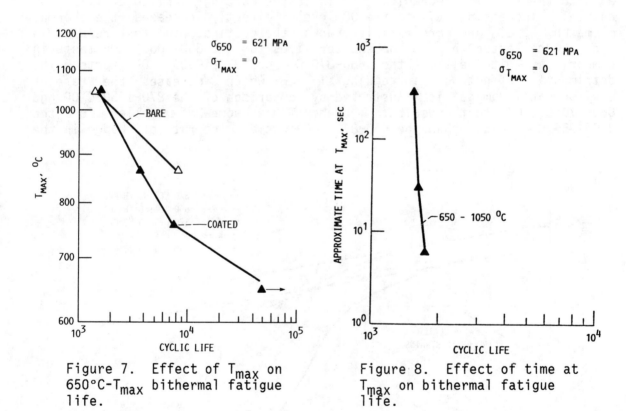

Figure 7. Effect of T_{max} on 650°C-T_{max} bithermal fatigue life.

Figure 8. Effect of time at T_{max} on bithermal fatigue life.

Increasing dwell time at T_{max} in 650-1050°C OP bithermal fatigue tests did decrease the life of coated specimens, as shown in Fig. 8. However, the magnitude of the effect was surprisingly small. Since cracks through the coating form rapidly in these tests, and the majority of life is spent in crack entry and propagation through the superalloy, a discussion of environmental effects with respect to the superalloy is necessary. It should be noted that due to the heating and cooling rates achievable, even in tests with the shortest dwell times, the exposure time above 650°C was still on the order of one minute. It therefore appears that most of the environmental damage at the crack tip occurs quite rapidly, well within the time of the fastest test, and does not increase rapidly with extended dwell time at 1050°C. A crack growth mechanism is proposed in which an environmentally damaged zone, whether actual oxides and a γ' depleted zone, or only dissolved oxygen, forms ahead of the crack tip at elevated temperatures and is completely fractured by tensile loads applied at low temperature. As virgin superalloy is exposed, the process would be repeated and the depth of environmental attack would not accumulate with cycling. Such a mechanism would also explain the steep slope of the OP bithermal life line in the low $\Delta\varepsilon_{in}$ regime, Fig. 4. The critical strain required to fracture the embrittled zone appears to be below that applied at 650°C, thus resulting in life almost independent of $\Delta\varepsilon_{in}$. The apparent turn out in life for the lowest $\Delta\varepsilon_{in}$ test, which was also observed on a stress basis, suggests that such a critical fracture strain may exist.

Conclusions

Bithermal TMF behavior was examined in the hope of providing a bridge in understanding between isothermal LCF and general TMF behavior. For the bare and NiCoCrAlY coated PWA 1480 single crystals studied, the bithermal test has shown that connection. Life in the high-$\Delta\varepsilon_{in}$ regime for isothermal LCF, IP-, and OP-type bithermal TMF was controlled by ε_{in}, and the superalloy ductility at the appropriate temperature. For TMF, this is the cycle temperature where the superalloy has least ductility, T_{min} in this case. Further, lives of coated and bare crystals were about the same, since crack initiation and propagation in the superalloy determined life.

The similarity, for example, between high-$\Delta\varepsilon_{in}$ 650°C LCF and 650-1050°C bithermal TMF is some indication that deformation at temperatures as high as 1050°C does not introduce additional damage mechanisms. We should caution, however, that one might not expect this similarity in a material in which other damage mechanisms might operate at the high temperature, say, in a polycrystalline alloy which might experience grain boundary cavitation, or in an alloy with a less stable precipitate. The similarity between IP bithermal TMF behavior and the LCF behavior at T_{min} of bare crystals continued into the low-$\Delta\varepsilon_{in}$, long-life regime. Crack initiation in the superalloy continued to control life in the IP bithermal test, and lives were similar between bare and coated specimens.

However, relative to the isothermal LCF behavior of bare specimens, cracks initiating in the coating reduced the life of coated specimens in 650°C LCF tests, and drastically so in 650-1050°C OP bithermal tests. The strictly mechanical effect of the coating was demonstrated in vacuum tests, and was, as would be expected, greater in the bithermal test due to the additional thermal mismatch strain in the coating. Additionally, for OP bithermal tests in air there was severe environmental damage. Even in bare crystals, surface cracks initiated and propagated very rapidly leading to very short lives in the low-$\Delta\varepsilon_{in}$ regime. The environmental damage was the result of the high temperature exposure to air followed by tensile loading at low temperature and was therefore limited to the OP TMF cycle.

To develop coated single crystals with improved TMF life in OP type cycles, coatings with coefficients of thermal expansion more closely matching that of the <001> single crystal direction, and greater ductility and strength are of course desirable. Yet, this work shows the importance of improved environmental resistance in single crystal superalloys themselves. Any coating defect exposing the superalloy to the environment could lead to very early failure, as exhibited in bare specimens. It also shows that in efforts to develop single crystal superalloys which can operate in gas turbine engines without coatings, OP TMF behavior may still be critical. Thus, OP TMF behavior should be assessed early on, rather than relying on simple isothermal or cyclic oxidation tests.

References

1. K. Schneider and H. W. Grunling, Thin Solid Films, **107**(1983), p. 395.

2. G. R. Leverant, T. E. Strangman, and B. S. Langer, Superalloys: Metallurgy and Manufacture (Claitor's, Baton Rouge, Louisiana, 1976), p. 285.

3. T. E. Strangman, Thin Solid Films, **45**(1977), p. 499.

4. M. G. Hebsur, and R. V. Miner, NASA TM-88950 (National Aeronautics and Space Administration), 1987.

5. T. P. Gabb, and G. E. Welsch, Scripta Met., **20**(1986), p. 1049.

6. T. P. Gabb, G. Welsch, and R. V. Miner, Scripta Met., **21**(1987), p. 987.

7. R. V. Miner, J. Gayda, and M. G. Hebsur, Low Cycle Fatigue, ASTM STP-942 (American Society for Testing and Materials, 1987), p. 371.

8. J. Gayda, T. P. Gabb, R. V. Miner, and G. R. Halford, Effects of Load and Thermal Histories on Mechanical Behavior of Materials (The Metallurgical Society, Warrendale, Pa., 1987), p. 179.

9. M. G. Hebsur, and R. V. Miner, Mat. Sci. Eng., **83**(1986), p. 239.

10. M. G. Hebsur, and R. V. Miner, Thin Solid Films, **147**(1987), p. 143.

11. J. Gayda, T. P. Gabb, and R. V. Miner, Int. J. Fatigue, **8(4)**(1986), p. 217.

12. D. P. DeLuca, and B. A. Cowles, AFWAL-TR-84-4167 (Air Force Aeronautical Laboratories, Wright-Patterson AFB, Ohio), 1984.

13. G. A. Swanson, I. Linask, D. M. Nissley, P.P. Norris, T. G. Meyer, and K. P. Walker, NASA CR-179594 (National Aeronautics and Space Administration), 1987.

14. G. R. Halford, M. A. McGaw, R. C. Bill, and P. D. Fanti, Low Cycle Fatigue, ASTM STP-942 (American Society for Testing and Materials, 1987), p. 625.

EFFECTS OF AGING ON THE LCF BEHAVIOR

OF THREE SOLID-SOLUTION-STRENGTHENED SUPERALLOYS

D. L. Klarstrom and G. Y. Lai

Haynes International, Inc., Kokomo, IN 46904-9013

Abstract

A study was conducted to examine the effects of thermal aging at 760°C (1400°F)/1000 Hrs. on the low temperature LCF behavior of HASTELLOY® alloy X, HAYNES® alloy No. 230 and HAYNES alloy No. 188. Results of LCF tests on samples in the annealed condition indicated that alloy 188 had the best fatigue resistance over the whole range of test conditions, followed by 230™ alloy and alloy X. In the aged condition, the fatigue of alloys X and 188 were significantly degraded under conditions in which the inelastic strain was greater than 0.10%. A much smaller amount of degradation was noted for 230 alloy. Taking data scatter into account, the aged fatigue resistances of alloy 188 and 230 alloy were essentially equivalent, and superior to alloy X. In all cases, the alloys were found to cyclically harden with plateaus or peak stresses not reached until near the point of crack initiation. The 230 alloy was found to harden to a greater extent than the other two alloys. The cyclic stress response of alloy 188 was unusual in the sense that observed stress amplitudes at the low end of the strain test range were higher than some tests at higher strain range levels. Microstructural observations indicated that the precipitation of carbides and brittle intermetallic compounds were responsible for the fatigue life degradations of alloys X and 188. In contrast, only carbide precipitation was observed in 230 alloy.

® HASTELLOY and HAYNES are registered trademarks, and 230™ is a trademark of Haynes International, Inc.

Introduction

Solid-solution strengthened superalloys are widely used in gas turbine engines as combustor cans, transition ducts, and other static components. In such applications, they are repeatedly exposed to cyclic thermal and mechanical stresses during the start-up, steady-state and shut-down portions of engine operation. Not surprising, therefore, is the fact that fatigue cracking is a major mode of failure in such parts. Hence, fatigue behavior must be taken into account by the designer in order to achieve satisfactory operating lives. This should include not only the behavior at maximum steady-state design temperatures, but also the behavior at lower temperatures encountered during the engine transients.

Most often the approach used to generate the required data is to run iso-thermal low cycle fatigue tests using materials in the solution annealed condition. While such data may provide a useful baseline for nominal design purposes, it cannot be considered wholly satisfactory since it does not take into account changes in mechanical properties that can occur during service exposure. Examples of such changes, especially with respect to ductility loss, have been extensively reported in the literature. The effects of these changes can be expected to be significant at low temperatures since the recovery of mechanical properties generally occurs at temperatures on the order of 0.5 Tm or higher. It was, therefore, the purpose of this investigation to examine the low temperature LCF behavior of three solid-solution strengthened superalloys in the annealed and aged conditions.

Experimental Procedures

The nominal compositions of the alloys studied in this investigation are listed in Table I. The materials consisted of 19mm (0.75-inch) thick plate of alloy X with a grain size of ASTM 5.5, 16mm (0.625-inch) thick plate of 230™ alloy with a grain size of ASTM 5.5 and 19mm (0.75-inch) diameter bar of alloy 188 with a grain size of ASTM 4 which were produced from commercial heats by Haynes International, Inc. The materials were tested in the as-received, solution annealed condition, and after aging at 760°C (1400°F) for 1000 hours. Tensile testing was conducted in accordance with ASTM standards to document mechanical properties of the alloys in the two conditions.

TABLE I

Nominal Compositions of Alloys

Alloy	Ni	Co	Cr	Mo	W	Fe	Si	Mn	C	Al	La
HASTELLOY® alloy X	Bal.	1.5	22	9	0.6	18.5	1.0*	1.0*	0.10	-	-
HAYNES® alloy No. 230	Bal.	5.0*	22	2	14	3.0*	0.40	0.50	0.10	0.30	0.02
HAYNES alloy No. 188	22	Bal.	22	-	14	3.0*	0.35	1.25*	0.10	-	0.04

* Maximum

Fully reversed, axial, strain-controlled, low cycle fatigue tests were performed by Metcut Research Associates, Inc., Cincinnati, Ohio. The LCF tests were conducted on smooth bar samples at a temperature of 427°C (800°F) and a frequency of 0.33 HZ (20 cpm). The fracture surfaces of selected samples were examined in an SEM to determine the modes of fracture initiation and propagation. Metallographic analysis was also performed on the selected samples to examine secondary fatigue cracks and general microstructural features.

Results

Tensile Properties

A summary of the 427°C (800°F) tensile properties of the three alloys in the annealed and aged conditions is presented in Table II. It can be seen that alloys X and 188 exhibit strengthening as the result of the aging process as evidenced by increases in yield and ultimate strengths. In contrast, the strength changes of 230 alloy were small and mixed, with the yield strength increasing slightly, and the ultimate strength decreasing slightly. In all cases, aging caused a significant decrease in ductility values. The largest absolute and relative decreases occurred for alloy X, while 230 alloy suffered the least such changes. In terms of absolute magnitude, alloy X exhibited the least residual ductility, followed by 230 alloy and alloy 188.

TABLE II

Summary of 427°C (800°F) Tensile Properties
For Materials In The Annealed and Aged* Conditions

Alloy	Condition	0.2% YS MPa (Ksi)	UTS MPa (Ksi)	% EL in 5D	% RA
X	Annealed	234 (34)	613 (89)	57	54
	Aged	379 (55)	841 (122)	20	24
230	Annealed	282 (41)	751 (109)	57	46
	Aged	296 (43)	730 (106)	40	35
188	Annealed	269 (39)	792 (115)	89	61
	Aged	386 (56)	841 (122)	40	40

* Aged at 760°C (1400°F)/1000 Hrs.

Low Cycle Fatigue Life Behavior

A listing of the 427°C (800°F) low cycle fatigue lives of the three alloys in the annealed and aged conditions is given in Table III. At the 0.55% total strain range level, the inelastic components were less than 0.10%, and tests ran to 10^5 cycles and beyond. In the case of alloy 188, tests were discontinued before failure. A graphical presentation of the data above 0.55% total strain range is provided in Figure 1.

TABLE III

Comparison of Annealed vs. Aged
427°C (800°F) LCF Behavior
R = -1.0, f = 0.33 HZ (20 cpm)

Nominal Total Strain Range, %	Cycles to Failure					
	Annealed			Aged @ 760°C (1400°F)/1000 Hrs.		
	Alloy X	230 Alloy	Alloy 188	Alloy X	230 Alloy	Alloy 188
1.50	2,051	2,398	3,710	1,756	2,260	2,848
1.00	7,750	8,742	12,647	4,889	7,033	6,970
0.80	14,417	16,575	21,089	10,320	15,310	12,470
0.65	28,679	46,523	59,652	21,367	34,571	38,841
0.55	100,486	115,456	>150,000*	97,325	123,200	>155,000*

* Test Discontinued

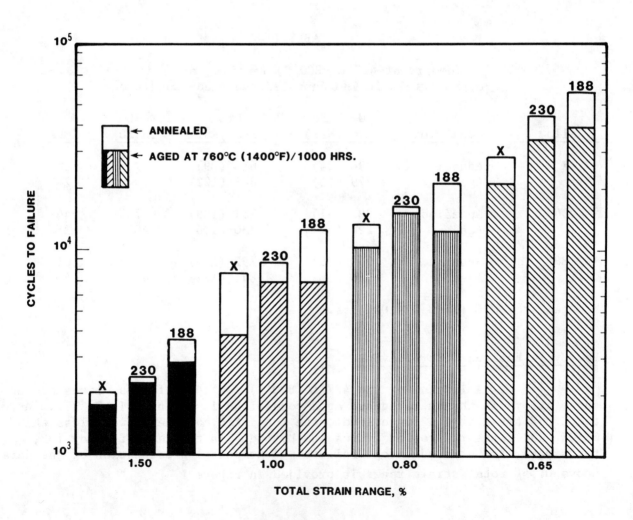

Figure 1. Fatigue life comparison of alloys in the annealed and aged condition.

In the annealed condition, the alloy ranking in terms of highest to lowest lives was in the order of alloy 188, 230 alloy and alloy X. This ranking is in good agreement with what might be anticipated on the basis of the mechanical properties of Table II. In the aged condition, a fatigue life degradation was observed for all of the alloys for total strain range levels above 0.55%. Again, alloy X exhibited the lowest fatigue lives, while near parity was reached for the fatigue lives of alloy 188 and 230 alloy taking data scatter into account. If the fatigue life degradation of each alloy is viewed on a relative basis as illustrated in Table IV, then it can be seen that overall alloy 188 was proportionately degraded the most, and 230 alloy was degraded the least.

TABLE IV

Aged/Annealed Fatigue Life Ratios

Nominal Total Strain Range, %	Ratios of Cycles to Failure Aged/Annealed		
	Alloy X	230 alloy	Alloy 188
1.50	0.86	0.94	0.77
1.00	0.63	0.80	0.55
0.80	0.72	0.92	0.59
0.65	0.75	0.74	0.65
0.55	0.97	1.07	-

Cyclic Stress Behavior

The cyclic stress behavior of the three alloys in the annealed and aged conditions is shown in Figure 2. In all cases, the alloys were found to cyclically harden, and, with few exceptions, well defined plateaus were not achieved prior to crack initiation. The greatest amount of hardening in both the annealed and aged conditions was observed for 230 alloy. Alloy 188 exhibited an unusual response in the sense that the extent of hardening observed at the lowest strain range levels was higher than that at some of the higher strain range levels. In the annealed condition, the degrees of hardening reached by alloys X and 188 were about the same.

There were some notable differences in the cyclic stress behavior of the materials in the annealed and aged conditions. Generally, the initial stress levels of the aged materials were higher as might be anticipated from the tensile data in Table II. The largest upward shifts were observed for alloy X followed by alloy 188 and 230 alloy. The extent of hardening for the aged condition tended to be about the same or slightly higher for 230 alloy and alloy X, but notably higher for alloy 188. In spite of the upward shift of the initial stress, the configurations of the curves for 230 alloy and alloy 188 were similar to the corresponding curves for the annealed condition. Alloy X, on the other hand, displayed markedly different characteristics. The curves for the aged condition exhibited more gradual and steady hardening compared to the annealed condition.

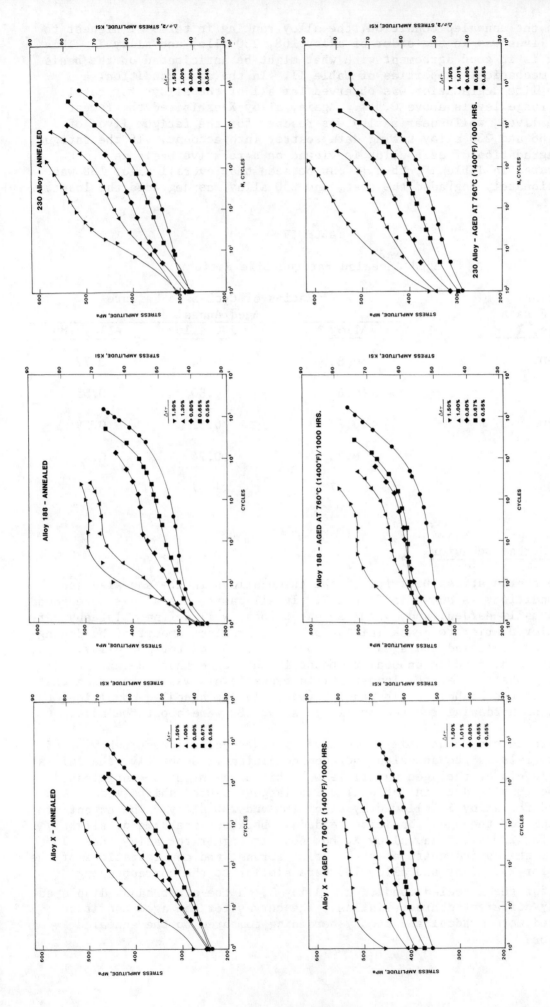

Figure 2. The cyclic stress behavior of the alloys in the annealed and aged conditions.

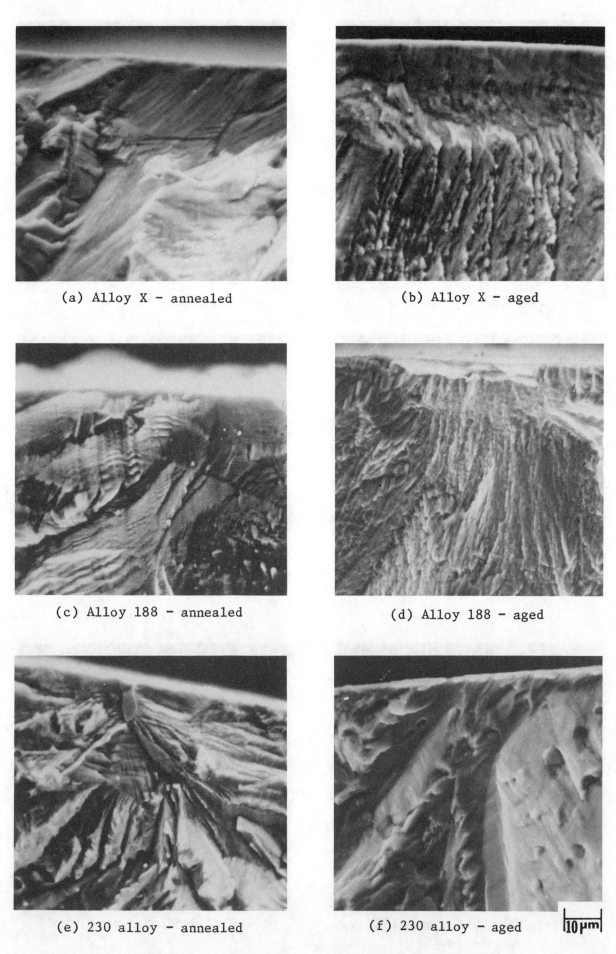

(a) Alloy X - annealed (b) Alloy X - aged

(c) Alloy 188 - annealed (d) Alloy 188 - aged

(e) 230 alloy - annealed (f) 230 alloy - aged

Figure 3. SEM photos of fracture surfaces.

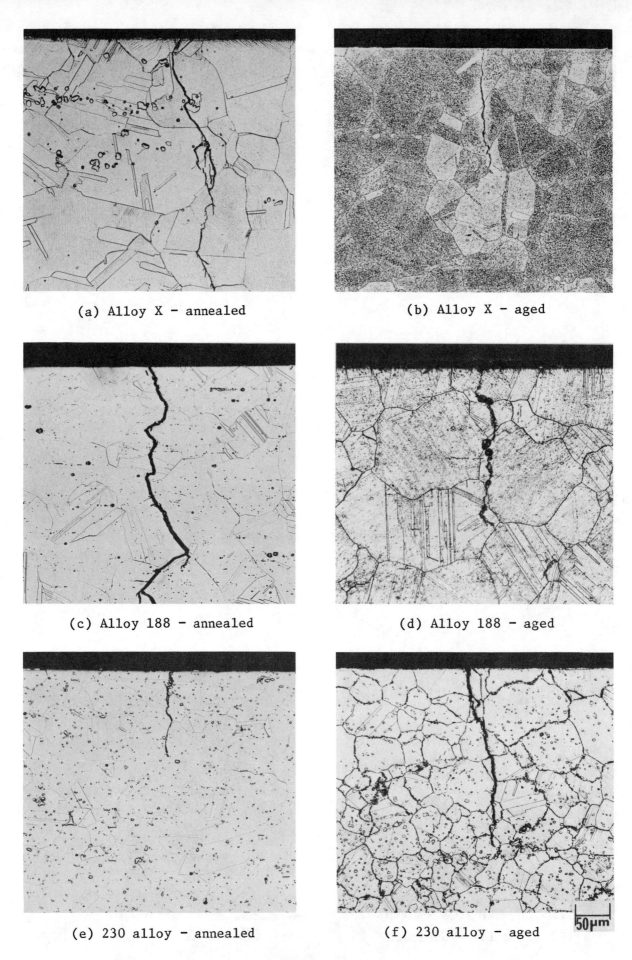

Figure 4. Secondary cracks and microstructural features.

Fracture Surface Characteristics

The fracture surfaces of the samples tested at a total strain range of 0.80% were selected for SEM examination. A summary of the fracture features observed at the fracture initiation sites is presented in Figure 3. In the annealed condition, all of the alloys exhibited Stage I type fracture initiation which is characterized by transgranular, cleavage type features. Within very small distances all of the alloys experienced a transition of Stage II crack propagation which is characterized by the familiar transgranular fatigue striations. In the aged condition, alloy 230 again displayed Stage I crack initiation and a rapid transition to Stage II crack propagation. In contrast, both alloys X and 188 in the aged condition exhibited an extensive area of transgranular crack initiation in a mode commonly described as quasi-cleavage. As will be shown later, this behavior is most likely the result of the copious precipitates that formed in these alloys during the aging treatment. Eventually, both alloys underwent a transition to Stage II crack propagation, but, in the case of alloy 188, occasional flat facets were noted in the fracture surface.

Secondary Cracking and Microstructural Features

Figure 4 summarizes the nature of the secondary cracking and the microstructural features for the samples tested at a total strain range of 0.80%. In all cases, the secondary cracks were observed to propogate in a transgranular fashion. Some branching and deviation of the cracks can be noted.

In terms of microstructure, all of the alloys in the annealed condition displayed clean grain boundaries, and there were primary M_6C-type carbides randomly distributed throughout the matrix as is typical for alloys of this class. In the aged condition, alloys X and 188 exhibited extensive grain boundary and matrix precipitation. The extent of the matrix precipitation was much greater in the case of alloy X. Previous studies have shown that the precipitates present in alloy X are $M_{23}C_6$ carbides and sigma-phase (1), and those present in alloy 188 are M_6C and $M_{23}C_6$ carbides and a Co_2 W-type Laves-phase (2). In the case of 230 alloy, extensive grain boundary and very slight matrix precipitation was in evidence for the aged condition. These precipitates have been previously identified as $M_{23}C_6$ carbides (3).

Discussion and Conclusions

The results clearly indicate that the 427°C (800°F) fatigue lives of all of the alloys were degraded by the 760°C (1400°F)/1000 Hrs. aging treatment. On a relative basis, the extent of the degradation was greatest for alloy 188 and least for 230 alloy. The cause of the decline in alloys X and 188 was the precipitation of sigma- and Laves-phases respectively in addition to the precipitation of carbides which are fully anticipated in substitutionally strengthened alloys of this class. The degradation in 230 alloy was lower since only carbide precipitation occurred.

The observed fatigue life degradation was expected based on the lower tensile ductilities recorded for each alloy in the aged condition. It is important to recognize here that this investigation covered the LCF behavior of the materials after 1000 hours of exposure time, and longer times might be expected to result in further declines in ductility. To gain some insight on this prospect, the results of a previous study of alloys X and 188 (4) were compiled along with recently developed data on 230 alloy into Table V. It can be seen from this table that additional aging would significantly reduce the ductility of alloy 188, while alloy X and 230 alloy maintain essentially constant ductility levels beyond the first 1000 hours. Accordingly, it would be expected that alloy 188 would exhibit much lower fatigue lives with additional aging than reported here, but those of alloy X and 230 alloy would be expected to remain about the same.

TABLE V

The Effect of Exposure Time at 760°C (1400°F)
On Room Temperature Tensile Elongation

Alloy	RT Tensile Elongation (%) After Indicated Exposure Time			
	None	1000 Hrs.	4000 Hrs.	8000 Hrs.
230	51	33	38	35
X	47	23	21	20
188	63	35	11	9

References

1. H. M. Tawancy, "Long-Term Aging Characteristics of HASTELLOY alloy X," J. Mater. Sci., 18 (1983), pp 2976-2986.

2. R. B. Herchenroeder, et. al., "HAYNES alloy No. 188," Cobalt, No. 54, March, 1972, pp 3-13.

3. H. M. Tawancy, D. L. Klarstrom and M. F. Rothman, "Development of a New Nickel-Base Superalloy," J. Metals, 36, No. 9, Sept., 1984, pp 58-62.

4. S. J. Matthews, "Thermal Stability of Solid Solution Strengthened Superalloys," Proc. 3rd Intl. Conf. on Superalloys, Seven Springs, Eds. B. H. Kear, et.al., Claitor's Publishing Div., Baton Rouge, 1976, pp 215-226.

OXIDE DISPERSION STRENGTHENED SUPERALLOYS: THE ROLE OF GRAIN STRUCTURE AND DISPERSION DURING HIGH TEMPERATURE LOW CYCLE FATIGUE

D. M. Elzey and E. Arzt

Max-Planck-Institut für Metallforschung
Seestrasse 92, D 7000 Stuttgart 1
Federal Republic of Germany

Abstract

The mechanisms leading to failure during high temperature, LCF and creep-fatigue are presented and discussed for two representative, commercial ODS (Oxide Dispersion-Strengthened) superalloys, Inconel MA 754 and MA 6000. The fatigue behavior of these alloys is compared with that of conventional (non-ODS) superalloys of nearly identical elemental composition. The behavioral differences may be understood on the basis of the two primary microstructural differences between ODS and non-ODS, namely, grain structure and the fine particle dispersion itself. It is found that recrystallization defects in the form of fine grains are the primary cause of crack initiation in the ODS materials. The results are discussed in light of previous creep and fatigue studies of dispersion-strengthened alloys.

Introduction

Alloys strengthened through inclusion of a fine dispersion of oxide particles have received considerable attention due to their exceptional creep properties. As a consequence, our understanding of creep-related phenomena in ODS alloys is now comparable with that of currently used, non-dispersion strengthened superalloys. However there have appeared relatively few studies in the open literature whose subject has been the behavior of ODS materials during high temperature fatigue. In the following, the results of high temperature fatigue and creep-fatigue experiments obtained for two commercially available ODS superalloys will be presented and discussed in light of previous fatigue studies and our current understanding of the creep behavior of ODS materials.

Essentially, ODS alloys are distinguished from conventional superalloys by the dispersion of fine oxide particles and by an elongated grain shape, which develops during a recrystallization heat treatment. This particular grain structure enhances the high temperature deformation behavior by inhibiting intergranular damage accumulation. This is accomplished by increased constraint in the neighborhood of cavitating, transverse grain boundaries and by increased tortuosity of intergranular crack paths (1). It has also been quite well demonstrated how an oxide dispersion improves high temperature creep strength. The movement of dislocations is impeded by the non-shearable dispersoid particles resulting in a threshold stress below which creep rates are negligible (2,3,4).

It is to be expected that the high temperature cyclic behavior of ODS alloys is also strongly influenced by the dispersion and grain morphology. Practically all investigators into the fatigue behavior of ODS materials have reported improved fatigue damage resistance, (e.g. 5-7). It has been commonly reported that the amount of slip observed is less for dispersion-strengthened material and that slip character becomes more wavy or discontinuous (8,9,10,11). TEM observations have indicated that dislocation cell structures either do not form or that their formation and diameter are determined by particle geometry (10,12,6,13,14).

Low frequency fatigue at high temperatures entails conditions which are sufficient for the operation of time dependent damage mechanisms such as those present during creep. It is clear that such processes, which normally lead to intergranular fracture, are made more difficult by an elongated grain structure during cyclic loading just as they are during monotonic loading, (e.g. 6,13). However, the creep-fatigue behavior of dispersion-strengthened alloys has not been studied in any detail to date. In the following, some results of creep-fatigue tests of MA 6000 are discussed which reveal aspects useful for the optimization of alloy behavior under high temperature creep and fatigue conditions.

Experimental

Present experimental work has been concentrated on the two yttrium-oxide dispersion-strengthened superalloys Inconel MA 754 and MA 6000. Both alloys are produced by INCO Alloys International, by the mechanical alloying process (15). The performance and behavior of each ODS alloy has been compared with that of a conventional (cast or wrought non-ODS) alloy of similar elemental composition. The compositions as determined for both ODS and non-ODS counterpart alloys are given in table I along with other microstructural data.

MA 754 is strengthened by solid solution and by the dispersion. In addition, an elongated grain structure is obtained during heat treatment (see (16) for a more detailed description of the microstructure). In addition to solution strengthening, MA 6000 is strengthened at intermediate temperatures by gamma-prime $Ni_3(Al,Ti)$ precipitates while at temperatures above 900 °C, the dispersion is the dominant source of strength. A bi-modal distribution of grain sizes exists in MA 6000; large grains with lengths on the order of 1-2 centimeters and grain aspect ratios (GAR) > 20 account for some 95% of the volume, while the remainder comprises a distribution of much smaller grains with lengths ranging from 100μm to 1 millimeter and GAR's between 1 and 20. For a more detailed description of the microstructure of MA 6000 the reader is referred to (17) and (18).

Low cycle fatigue tests have been carried out using total strain control with total or inelastic strain limits in the temperature range from 750 to 1100 °C. Symmetric and asymmetric waveforms were chosen in conjunction with test frequencies such that the prevailing damage mode would be either of fatigue-type or mixed creep-fatigue. Creep-fatigue cycle forms included tensile hold-times and slow tensile loading followed by fast compression, commonly referred to as 'slow-fast'. Samples from tests run to physical separation of the specimen as well as interrupted experiments were then subjected to optical, scanning electron and transmission electron microscopy. Further experimental details will be given elsewhere (19).

Table I. Elemental Composition of Materials and Microstructural Data

Alloy	Ni	Cr	Fe	Al	Ti	W	Mo	Co	Ta	C	B	Zr	Nb	Y_2O_3
MA 6000	Bal	15.5	–	4.5	2.5	3.8	2.0	–	1.9	.06	.01	.16	–	1.1
IN 738	Bal	15.9	–	3.5	3.5	2.5	1.6	8.3	1.6	.09	.01	.5	.7	---
MA 754	Bal	20.5	.13	.30	.35	–	–	–	–	.06	–	–	–	0.5
Nim 75	Bal	20	<5	.25	.40	–	–	–	–	.1	–	–	–	---

	Grain Size [mm]	GAR	Particle Dia [nm]	Particle Spacing [nm]	Texture
MA 6000	5–20	20	30	150	<110>
IN 738	2	1	–	–	–
MA 754	0.5–3.0	10	15	125	<100>
Nim 75	0.1	1	–	–	–

Fatigue Behavior

Slip character: ODS vs non-ODS

It has been well documented that gamma-prime-strengthened, Ni-base superalloys are subject to the development of concentrated, planar slip bands during LCF (20,21,22). The gamma-prime precipitates are sheared by such bands, causing a transition to coarse, planar slip (14). Such slip bands were found very infrequently in MA 6000 and only at high inelastic strain amplitudes (N_f < 100 cycles). In contrast with IN 738 (non-ODS counterpart for MA 6000), stage I cracks, which develop from coarse slip bands in non-ODS alloys, have not been observed in MA 6000. Additionally, TEM micrographs often show individual dislocations which are pinned at the backside of dispersoid particles (figure 1), confirming the existence of an attractive interaction in a manner analogous to that observed for creep (23). No evidence of the formation of regular dislocation networks could be found, confirming the improved slip character.

Similarly, slip bands could not be resolved in fatigued MA 754 under any conditions of temperature and strain rate tested. MA 754 deforms macroscopically along planes of maximum shear stress (figure 2). Identical behavior has been reported for TD-Ni (nickel with thorium-oxide dispersion) (24,10). This mode of deformation also results in a pronounced tendency for growing cracks to branch. By comparison, Nimonic 75 (non-ODS counterpart to MA 754), exhibits intense planar slip extending through entire grains under these conditions and initiates cracks at surface slip extrusions and at twins. Again, no stage I cracks were observed in the ODS material.

In agreement with earlier studies, these observations tend to confirm the hypothesis, e.g. (10), that planar slip deformation becomes dispersed as dislocation glide on planes oriented for slip is impeded by non-shearable particles. The subsequent activation of

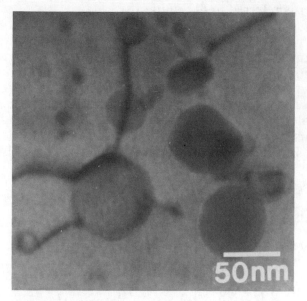

Figure 1: TEM showing dislocation configurations in fatigued MA 6000. Dispersoid-dislocation interactions similar to those which have been described for creep are observed. (850 °C, symmetric cycling, 10^{-3} [s^{-1}], $\Delta\epsilon_t/2=0.4\%$, failure after 100 cycles)

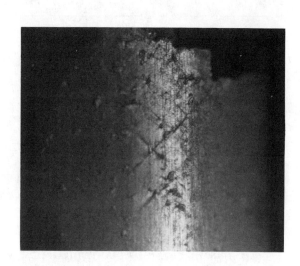

1 mm

Figure 2: Surface of a cylindrical fatigue specimen of MA 754 illustrating damage on planes of maximum shear stress. (1050 °C, symmetric cycling, 10^{-3} [s^{-1}], $\Delta\epsilon_t/2=0.4\%$, failure after 180 cycles)

alternative slip systems homogenizes the plastic glide deformation. This process has sometimes been referred to as 'slip dispersal'. The metallurgical stability of the ceramic dispersoids in ODS alloys enables this mechanism of strengthening to be effective at temperatures in excess of $0.8T_m$.

Grain boundaries as fatigue crack initiation sites

The high degree of strengthening obtained within the grain interiors of ODS superalloys enhances the susceptibility of grain boundaries to various damage processes. Countering this imbalance is the highly elongated grain structure, characteristic of these materials. Large grain aspect ratios imply less grain boundary area lying normal to the applied stress within a given volume. The influence of GAR in determining creep damage rates has been well demonstrated for MA 6000, where single crystal creep performance may be attained at GAR > 15-20 (25,18,1).

Although the average GAR may be in excess of 15-20, as is the case for the MA 6000 material tested in this study, local values of the GAR may be as low as 1-2. Such grain structure inhomogeneities, which take the form of relatively small included grains or pockets of finer grains sandwiched between elongated macro grains, might be expected to be the weakest link in the microstructural chain. That this is the case may be seen in figures 3 and 4, which show crack initiation in MA 6000 during symmetric cycle LCF at 1 Hz and 850 °C. The crack has initiated at the grain boundary of a fine grain located just beneath the sample surface. Initiation is usually observed to be on or near the sample surface but some internal initiation at fine grains has also been observed during the symmetric LCF of MA 6000. Crack propagation to failure is fully transgranular. Although macroscopically MA 6000 exhibits a very strong <110> texture, it has been shown that the incompletely recrystallized, fine grains are characterized by high misorientation (26). It is believed that the susceptibility of fine grains to fatigue damage is related to differences in the crystallographic orientation, which in some cases results in higher elastic stresses at fine grain boundaries during loading.

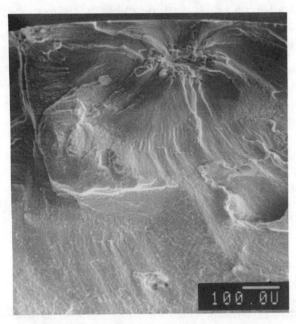 

Figure 3: MA 6000 LCF fracture surface: crack initiation has occurred at the transverse boundary of a fine grain located near the test sample surface. (850 °C, symmetric cycling, $10^{-2}[s^{-1}]$, $\Delta\epsilon_t/2=0.3\%$, failure after 5017 cycles)

Figure 4: An axial section taken through the crack initiation point shown in the figure left. The etched boundary of a fine grain is visible on the polished section surface.

Grain structure inhomogeneities are also present in MA 754 but since the average GAR is much less than for MA 6000, the disparity between fine grains (GAR $\geq$ 1) and the average GAR is also much smaller. Crack initiation is also complicated by the aforementioned shear deformation. It appears likely that the intersection of coarse shear bands at the sample surface can cause crack initiation although internal damage at fine grains has also been occasionally observed.

Creep-Fatigue

Examination of the creep fracture behavior of MA 6000 has demonstrated the extreme susceptibility of grain structure inhomogeneities to cavitation damage (18,1). This observation, and the fact that cracks are frequently seen to initiate at such relatively fine grains during symmetric-cycle LCF, indicates the likelihood that mixed creep-fatigue damage initiation should also occur at grain inhomogeneities. This hypothesis is well confirmed by the results of a number of 'slow-fast' LCF tests which were carried out on MA 6000.

Figure 5 is a plane view of a fracture surface of an MA 6000 sample tested under 'slow-fast' (S-F) conditions at 850 °C. Closer inspection of the fracture surface reveals more than 20 individual, internal crack initiation sites, each surrounded by a radially expanding crack front (figure 6). That these initiation sites are indeed fine grains, is shown in figure 7. Metallographic examination of polished axial sections revealed the nucleation and growth of pores primarily on the transverse grain boundaries of fine grains. The presence of precipitate-free zones (PFZ) between pores indicates that pores grow by the stress-directed transport of atoms from the pore surface to the grain boundary. Detailed analysis of interrupted tests has shown that approximately 60-70% of the life is spent initiating cracks intergranularly in this manner.

The extent of cavitation damage accumulated during creep-fatigue exhibits a clear dependence on grain size. Fine grains were observed to be responsible for at least 80% of all crack initiation sites during creep-fatigue. Although the average GAR of finer grains is considerably less than that of the macrograins, no clear connection between the extent of cavitation damage and GAR could be ascertained among the population of fine grains. It should be noted that this observation does not conflict with the results of Zeizinger (18) and Arzt and Singer (25), who have shown that the creep rupture life of MA 6000 increases strongly as the average GAR increases. Stephens and Nix (27) have studied the creep behavior of MA 754 and also report the deleterious effect of a duplex grain morphology on the creep resistance. The enhanced grain boundary damage associated with fine grains could be due to the greater ease with which damage can be accommodated by grain sliding. However, the observation that fine grains with aspect ratios upto 20 are still highly prone to cavitation renders this explanation suspect. It appears more likely that, as mentioned previously in connection with LCF crack initiation, the crystallographic misorientation of fine grains with respect to the fully recrystallized macrograins enhances damage formation.

The subject of creep-fatigue must include consideration of possible interactions between damage processes associated with creep (e.g., cavitation, grain sliding) and with fatigue (e.g., plastic deformation). It may then happen that for a given test temperature, the time-to-failure for creep-fatigue is much shorter than an "equivalent" stress-rupture loading. This is the case for MA 6000. To give an example, at a temperature of 850 °C, a creep-fatigue test with applied tensile and compressive strain rates of 10^{-5} and 10^{-2} [s^{-1}], respectively, and a strain amplitude of 0.3% has a time-to-failure of about 120 hours. The stress needed to cause creep rupture after 120 hrs at this temperature corresponds to 85% of the actual peak tensile stress applied during the creep-fatigue test. However, only about 20 hrs are spent at this stress or higher during the S-F test, (a stress well above the threshold for creep). Thus, only 1/6 of the creep-rupture time is required for a creep-fatigue loading to cause the equivalent damage based on the applied stress. Figure 8 illustrates the comparison of S-F lifetimes with symmetric cycle fatigue (strain rate of $2 \cdot 10^{-5}$ [s^{-1}]) at 850 °C. It may be seen from the figure that the asymmetric loading leads a reduction in cyclic life of roughly a factor of 3. From these observations, it is clear that the net damage accumulation rate is accelerated under conditions which combine slow or constant tensile loading with fully reversed, plastic compression.

One observation made while examining the cavitation damage in MA 6000 caused by creep-fatigue provides a possible clue as to the mechanism by which creep and fatigue damage interact; the maximum pore size reached prior to coalescence never exceeds approximately 2 μm. This is in contrast with pores developed under constant stress, which attain a diameter of 5-10 μm before coalescence (1,18). Although direct measurement of the cavity spacing is made difficult by the fact that regularly spaced, individual pores are rarely seen, it was clearly observed that the cavity spacing is much smaller during creep-fatigue. This is taken to be an indication that cavity nucleation is enhanced by strain reversal. A possible secondary effect could be plastic flattening of pores during compression and subsequently, the acceleration of lateral pore growth.

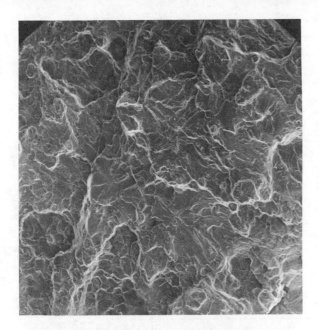

Figure 5: Fracture surface of an MA 6000 sample tested under 'slow-fast' conditions at 850 °C. Cracks initiate internally at fine, included grains, (see figs. 6 and 7). More than 20 individual crack initiation sites may be identified. ($\dot{\epsilon}_t = 10^{-5}$ [s^{-1}], $\dot{\epsilon}_c = 10^{-2}$ [s^{-1}], $\Delta\epsilon/2 = 0.3\%$, failure after 650 cycles)

1 mm

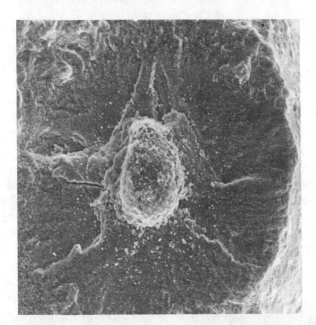

Figure 6: Detail of crack initiation site taken from the fracture surface of fig. 5. The tip of an included, fine grain is visible from which a crack has propagated transgranularly through the surrounding macrograin.

10 µm

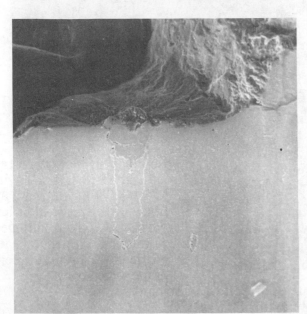

Figure 7: A section taken through a crack initiation site of a sample tested under 'slow-fast' conditions. A crack has initiated on the transverse boundary of a fine, included grain. The outlines of several small grains have been made visible by etching the polished axial section.

100 µm

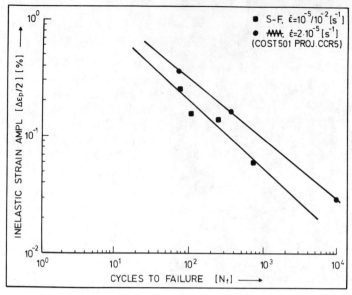

Figure 8: Fatigue resistance of MA 6000 at 850 °C: S-F cycling results in a severe degradation of the cyclic lifetime in comparison with symmetric LCF.

Creep–fatigue crack propagation

The crack propagation phase consists of the time between complete cavitation of the transverse fine grain boundary and failure. The greater part of this phase will be spent in transgranular crack extension through the macrograins with the remainder for link-up of planar cracks and subsequent failure. Metallographic investigation has revealed that transgranular cracks are blunted and that quite frequently, pores nucleate and grow ahead of the crack (see figure 9). In addition, precipitate-free zones have been observed between the main crack and individual pores. This evidence suggests that crack advance is controlled by diffusive cavity growth and the fracture of ligaments ahead of the main crack.

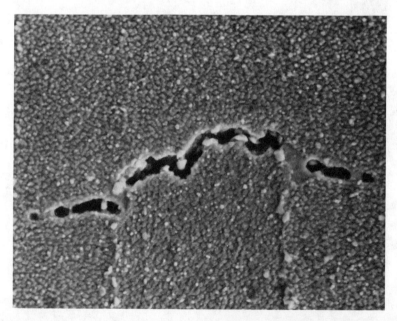

Figure 9: The transgranular growth of cracks in MA 6000 which initiate at fine grain boundaries during creep-fatigue. Note pores growing ahead of the crack tip and precipitate-free zones. White particles seen on the grain boundary are the result of an etchant reaction. ($\dot{\epsilon}_t = 10^{-5}$ [s^{-1}], $\dot{\epsilon}_c = 10^{-2}$ [s^{-1}], $\Delta\epsilon/2 = 0.4\%$, failure after 242 cycles)

The linking up of macrograin cracks occurs primarily intergranularly along grain boundaries oriented in the longitudinal direction. Thus, the coalesence of transgranular crack planes is controlled by the spatial distribution of the planes. For the purposes of modelling or lifetime prediction it is necessary to determine the probability that a given average planar crack separation will exist. Decreasing the total number of grain structure inhomogeneities will result in a corresponding improvement in the high temperature creep and creep-fatigue lifetime. Since the next largest defects (e.g. carbonitride particles) are roughly 3 orders of magnitude smaller than the grain defects which are at present responsible for crack initiation, the potential for improving the fatigue resistance by elimination of grain inhomogeneities appears to be substantial.

Summary

Commercial ODS superalloys have demonstrated improved high temperature fatigue damage resistance in comparison with non-ODS alloys of nearly identical elemental composition. The dispersion improves the fatigue behavior through the inhibition of slip band formation and the resultant suppression of crystallographic microcracking. The creep-fatigue properties are enhanced both by the dispersion and by a highly elongated grain structure. It has been shown that recrystallization defects in the form of fine grains are susceptible to the formation of damage. These defects are responsible for the initiation of cracks under LCF and creep-fatigue conditions.

Testing under creep-fatigue conditions leads to a severe degradation of lifetime in comparison with either creep or fatigue alone. The principal mechanism by which the accumulation of creep damage is accelerated by strain reversal appears to be the plastic strain-enhanced cavity nucleation rate. Creep-fatigue tests can be employed to establish the relative density of recrystallisation defects.

Our results indicate that due to the presence of fine grains, the role of the dispersoids in improving the slip character during LCF of oxide-dispersion strengthened superalloys has not yet been exploited. Further progress will depend heavily on our understanding of how recrystallization defects arise and on their eventual elimination.

Acknowledgements

The authors are especially grateful to C. Weis for assistance with SEM studies and special metallography. Parts of this work have been carried out within the frame of the European Collaborative Programme COST 501. We would like to acknowledge the financial support of the Bundesministerium für Forschung und Technologie in the Federal Republic of Germany under project number O3ZYK1228.

References

1. Zeizinger, H. and Arzt, E., to be published.
2. Arzt, E. and Wilkinson, D.S., "Threshold Stresses for Dislocation Climb Over Hard Particles: The Effect of an Attractive Interaction", Acta metall., 34(10) (1986), 1893-1898.
3. Arzt, E., "Threshold Stresses for Creep of Dispersion Strengthened Materials", Handbook of Metallic Composites, ed. T.Glasgow and J.D.Whittenberger, (New York, NY: Marcel Dekker, Inc., in press).
4. Arzt, E. and Rösler, J., "The Kinetics of Dislocation Climb Over Hard Particles - II. Effects of an Attractive Particle-Dislocation Interaction", Acta metall., 36(4) (1988), 1053-1060.
5. Blucher, J.T., Knudsen, P. and Grant, N.J.,"The Effect of Strain Rate and Temperature at High Strains on Fatigue Behavior of SAP Alloys", Trans AIME, 245 (1969), 1605-1611.
6. Weber, J.H. and Bomford, M.J., "Comparison of Fatigue Deformation and Fracture

in a Dispersion-Strengthened and a Conventional Nickel-Base Superalloy", Met Trans, 7 (1976), 435-441.
7. Hoffellner, W. and Singer, R.F., "High Cycle Fatigue Properties of the ODS Alloy MA 6000 at 850 C", Met Trans, 16A (1985), 393-399.
8. Martin, J.W. and Smith, G.C., "The Effect of Internal Oxidation on the Fatigue Properties of Copper Alloys", J. Inst. of Metals, 83 (1955), 153-165.
9. Snowden, K.U., "Some Creep and Fatigue Properties of Dispersed-Oxide-Strengthened Lead", J. Mat Sci, 2 (1967), 324-331.
10. Leverant, G.R. and Sullivan, C.P., "The Effect of Dispersed Hard Particles on the High Strain Fatigue Behavior of Nickel at Room Temperature", Trans AIME, 242 (1968), 2347-2353.
11. Starke, E.A. and Lütjering, G., "Cyclic Plastic Deformation and Microstructure", Fatigue and Microstructure, (Metals Park, OH: American Society for Metals, 1979), 205-243.
12. Leverant, G.R. and Sullivan, C.P., "The Low Cycle Fatigue of TD-Nickel at 1800 F", Trans AIME, 245 (1969), 2035-2039.
13. Kim, Y.G. and Merrick, H.F., "Fatigue Properties of MA 6000E, A Gamma-Prime Strengthened ODS Alloy", Superalloys 1980, ed., J.K.Tien et al (Metals Park, OH: American Society for Metals, 1980), 551-561.
14. Lütjering, G., "Gleitverteilung und Mechanische Eigenschaften Metallischer Werkstoffe", Ber. der Deutschen Luft- und Raumfahrt, Nr.DLR-FB 74-70.
15. Hack, G.A.J., "Fundamentals of Mechanical Alloying", Frontiers of High Temp Materials II, ed., J.S.Benjamin and R.C.Benn (USA: INCO Alloys International, 1983), 3-18.
16. Stephens, J.J. and Nix, W.D., "The Effect of Grain Morphology on Longitudinal Creep Properties of Inconel MA 754 at Elevated Temperatures", Met Trans, 16 (1985), 1307-1323.
17. Schröder, J., "Elektronenmikroskopische Untersuchung des Hochtemperatur-Härtungs-Mechanismus in einer ODS-Superlegierung", Fortschr.-Ber. VDI, Reihe 5 (1987), Nr.131.
18. Zeizinger, H., "Werkstoffschädigung in einer ODS-Superlegierung durch Hochtemperatur-Ermüdung und Kriechen", Fortschr.-Ber. VDI, Reihe 5 (1987), Nr.121.
19. Elzey, D.M., Ph.D. thesis, University of Stuttgart, 1988, to be published.
20. Merrick, H.F., "The Low Cycle Fatigue of Three Wrought Nickel-Base Alloys", Met Trans, 5 (1974), 891-897.
21. Purushothaman, S. and Tien, J.K., "Slow Crystallographic Fatigue Crack Growth in a Nickel-Base Alloy", Met Trans, 9 (1978), 351-355.
22. Lerch, B.A., Jayaraman, N. and Antolovich, S.D., "A Study of Fatigue Damage Mechanisms in Waspaloy from 25 to 800 °C", Met Sci & Eng, 66 (1984), 151-166.
23. Schröder, J. and Arzt, E., "Weak Beam Studies of Dislocation-Dispersoid Interaction in an ODS Superalloy", Scripta Metall., 19 (1985), 1129-1134.
24. Ham, R.K. and Wayman, M.L., "The Fatigue and Tensile Fracture of TD-Nickel", Trans AIME, 239 (1967), 721-725.
25. Arzt, E. and Singer, R.F., "The Effect of Grain Shape On Stress Rupture of the Oxide Dispersion Strengthened Superalloy Inconel MA 6000", Superalloys 1984, eet al (Warrendale, PA: The Metallurgical Society, 1984), 369-378.
26. Tekin, A. and Martin, J.W., "Micromechanisms of Creep/Fatigue Interactions in ODS Materials" (Final Report, European Concerted Action COST 501, Project UK2, 1987).
27. Stephens, J.J. and Nix, W.D., "Constrained Cavity Growth Models of Longitudinal Creep of ODS Alloys", Met Trans, 17 (1986), 281-293.

OBSERVATIONS OF MICROSTRUCTURAL AND GEOMETRICAL INFLUENCES ON FATIGUE CRACK

GROWTH IN SINGLE CRYSTAL AND POLYCRYSTAL NICKEL-BASE ALLOYS

David C. Wu, David W. Cameron, and David W. Hoeppner

Garrett Engine Division, Allied Signal Aerospace Company,
111 South 34th Street, Phoenix, AZ 85010

115 Clarence Street, Allegany, New York 14706

Chairman, Department of Mechanical and Industrial Engineering,
University of Utah, Salt Lake City, Utah 84112

Abstract

Understanding of single crystal crack growth can yield significant insight into the behavior of similar polycrystalline materials. Two nickel-based alloys which display planar slip under cyclic loading were examined, viz. single crystal SRR 99, and polycrystalline IN 901. The investigations were conducted at ambient temperature using a small load frame integrated with a scanning electron microscope, allowing for both 'still' photographs and dynamic videorecording. Principal goals of the experiments were to observe the nucleation and growth of fatigue cracks under sinusoidal tension-tension loading. The features associated with fracture path development and supplemental fractographic studies are shown for both materials. Orientation-dependent crack growth in the SRR 99 is presented and shown to be a manifestation of Schmid-Law controlled slip. Once this slip is activated, the crack path develops such that the strain energy release rate is maximized. Generally the behavior observed in the single crystal was found in the polycrystal, with allowance for the microstructural variables. The polycrystalline IN 901 displayed responses which reflected the constraint to free slip distance associated with multiply-oriented grains, grain boundaries, eta precipitates, etc. These influences operate to minimize crack extension by a single slip plane which can readily traverse a favorably oriented single grain/crystal. Such 'complexing and hindering' contributes substantially to the fatigue resistance of polycrystalline materials under these conditions.

Introduction

The behavior of short cracks under fatigue loading has been a topic of intense interest. One class of short cracks can be termed 'microstructurally-short', i.e. the size of the crack is small with respect to the microstructural unit sizes of the material. It is clear that in this class of short cracks, understanding of single crystal crack growth behavior can yield significant insight into the behavior of similar polycrystalline material.

The studies described in this paper were carried out on two nickel-based superalloys: cast, single-crystal SRR 99 and wrought, polycrystalline IN 901. Both materials display planar slip mode behavior under cyclic loading at room temperature. Principal goals of the studies were the observation of the externally visible circumstances of crack nucleation, the evolution of the crack path as it extended through the material, and the influences of microstructural features.

The SRR 99 alloy, a single crystal turbine blade alloy, was provided in a single bar or block sample grown in the <001> direction. Orientation of the samples was based on reference marks on the block of material whose initial orientation was established by Laue back-reflection techniques. The IN 901 material was obtained from a typical cast and wrought disk for an aeroengine application. No particular directional choices were maintained in this reasonably equiaxed material. The nominal chemistries of the two alloys investigated are given in Table I. Only the principal elements are given in the table, other control limits exist on residual and tramp elements as well.

TABLE I. Alloy Chemistries.

	C	Cr	Co	Mo	Ta	W	Al	Ti	Fe	B	Ni
SRR 99	0.015	8.5	5.0	-	2.8	9.5	5.5	2.2	-	-	bal
IN 901	0.05	13.5	-	6.0	-	-	0.25	2.7	35	0.01	bal

Despite their disparate chemistries and product forms, a substantial number of similar features and events were observed in the experiments.

Experimental Procedure

The experimental apparatus consisted of a servohydraulically-controlled 1.8 kN load frame integrated with an ISI Alpha 9 scanning electron microscope. This facilitated in-situ secondary electron observation and recording of the surface response and cracking processes. Typical 'still' scanning electron micrographs could be taken at any point of the load sequence, and dynamic operation of the system generally ranged from a maximum of 20 to 30 Hz to fractions of a Hertz. These latter frequencies were useful in obtaining dynamic video recordings of events under active loading. Details of the device have been reported previously (1).

The fatigue crack growth experiments were performed using modified single edge notch (SEN) specimens under sinusoidal tension-tension loading. The general specimen geometry is shown in Figure 1; one side of the 1mm thick by 4mm wide gage length was notched to provide the SEN configuration. The test specimen thickness approximated that of cooled airfoil sections employed in many advanced blade and stator components. In the case of the wrought IN 901, the sample was about 15 grains thick in the through-thickness direction, probably sufficient for a continuum-type response.

Notching of the specimens to obtain the SEN geometry was by mechanical means, e.g. a saw cut. The SRR 99 specimens were first mechanically polished followed by electropolishing to produce a smooth notch root and gage section surface for subsequent observation. Interdendritic porosities, carbides, and gamma prime precipitates were visible at the appropriate magnifications after such surface preparation. The IN 901 specimens were mechanically polished and chemically etched to delineate the grain boundaries and other microstructural features.

The SRR 99 specimens were tested at a load ratio of 0.2, the IN 901 specimens were tested at load ratios of 0.2 and 0.5. Although the applied load ratios were constant, constant load amplitude conditions were not necessarily maintained throughout the test. Common goals of both studies were the observation of the initial circumstances of crack nucleation and the progression of the crack through the microstructure. These two criterion were satisfied by first applying relatively high loads to nucleate crack(s) at the notch root. Following observation of cracking at the root location, the load amplitude was reduced to facilitate more extensive propagation observations, while maintaining the required load ratio.

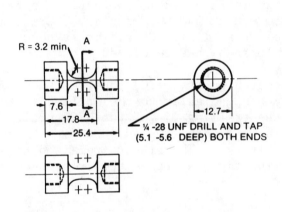

Figure 1 - Specimen geometry. All dimensions mm, except threads.

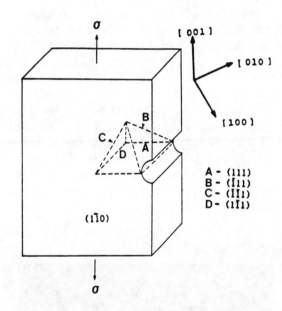

Figure 2 - Relative orientation of the $\{111\}$ tetrahedron, specimen 4.

Experimental Results

SRR 99 Specimens

Recent studies (2,3) on single crystal nickel-base alloys have shown that at temperatures below approx. 700C, crystallographic cracking on $\{111\}$ planes predominates under cyclic loading. Preliminary experiments on SRR single crystals confirmed this observation (4). Further experiments on specimens of controlled primary and secondary slip orientations were carried out to define the necessary and sufficient conditions for such crystallographic crack growth. A summary of the specimen orientations tested is shown in Table II.

Table II. SRR 99 Specimen Orientations.

Specimen I.D.	Loading Axis	Specimen Face
1,2	[001]	(100)
3,4	[001]	(1$\bar{1}$0)
5,6	[.12 .12 .986]	"
7,8	[3$\bar{1}$0]	(131)

In all cases, mixed mode I, II, and III cracking was observed. The geometry of the resulting cracks could be visualized based on the relative orientation of the {111} tetrahedron with respect to the specimen. The crack nucleation and growth process in all specimens were similar. In all the cases the crack(s) nucleated at material discontinuities in the notch root. Crack evolution in a particularly illustrative specimen will be presented in detail.

The orientation of the {111} tetrahedron in specimen 4 is shown in Figure 2. The initial maximum applied net section stress was 475 MPa. At 176,000 cycles, cracking at a pore was observed at the notch root. The maximum applied stress was reduced to 400 MPa and held there until 340,000 cycles, the crack at this point is shown in Figure 3. The applied stress was further reduced in two steps to 250 MPa at 540,000 cycles. Cracking continued at the notch root, a view of it at 1,110,000 cycles is shown in Figure 4. It can be seen that a crack 'branch' has developed.

At 1,160,000 cycles, a fissure appeared on the (1$\bar{1}$0) face, as shown in Figure 5. This opening was nominally perpendicular to the applied stress and an uncracked ligament remained at the trailing tip of the fissure. A higher magnification view of the opening's leading tip is shown in Figure 6, note the manifestations of slip activity. The fissure remained tight until 1,200,000 cycles, when the trailing tip penetrated the free edge of the specimen.

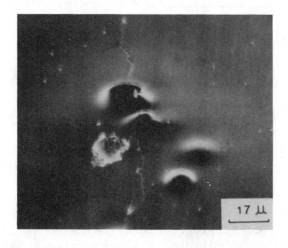

Figure 3 - Crack nucleation at pore in notch root, N = 340,000 cycles.

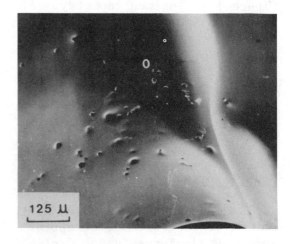

Figure 4 - View of crack at notch root, N = 1,110,000 cycles. Crack branching at nucleating pore indicated by 'O'.

Figure 5 - Notch root and fissure on the free surface, N = 1,160,000 cycles (stress axis horizontal).

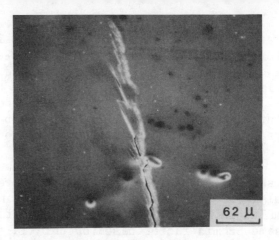

Figure 6 - Leading tip of fissure, N = 1,190,000 cycles.

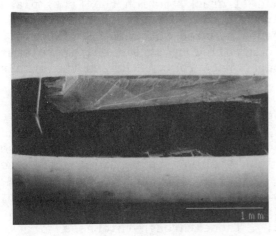

Figure 7 - Plan view of the failed specimen; compare crack geometry to Fig. 2.

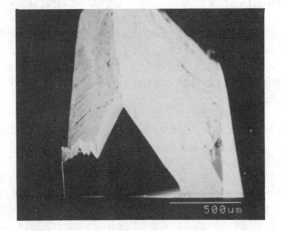

Figure 8 - End-on view from the notch of the failed specimen; compare geometry to Fig. 2.

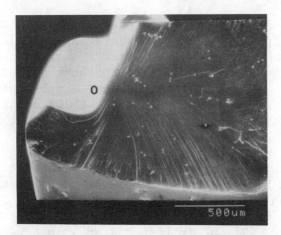

Figure 9 - View of the (111) plane adjacent to notch root. Crack nucleation site shown in Figures 3 to 5 marked by 'O'.

Specimen failure occurred at 1,225,580 cycles. Two views of the separated specimen are shown in Figures 7 and 8. It can be seen that adjacent to the notch, the crack was a three-sided pyramid, as the crack progressed, two sides of the pyramid continued to form a roof-top fracture; all the planes that made up the pyramid and the roof-top were $\{111\}$ planes. Figure 9 shows the fracture surface on the (111) plane associated with the fissure described earlier. It can be seen that part of the crack had to propagate 'backwards' to complete this particular plane.

Beachmarks and radial ridge lines on Figures 7, 8, and 9 showed simultaneous subsurface cracking on three different planes. This can be explained by the fact that this specimen, with a [001] stress axis, has a multiple slip orientation (eight different octahedral slip systems have the same Schmid factor based on uniaxial loading). Similar behavior in specimens 1 through 3 with the same stress axis had been observed. However, an additional criterion was required to explain the observed crack path.

The crack geometry seen in specimen 4 involved three slip planes initially, but switched to two planes that made up a roof-top. This roof-top approximates a mode I crack. As such, its strain energy release rate is greater than that of a mixed mode I and II crack, which the (111) plane would have formed had it continued to propagate. Using this observation, it was postulated that once slip had been activated according to Schmid's Law, the selection of the crack path is controlled by strain energy release rate considerations: the crack grows along a path that maximizes the strain energy release rate (4).

Specimens 5 through 8 were designed to reduce the number of active slip systems (four in specimens 5 and 6, two in specimens 7 and 8), based on Schmid factor comparison. In both orientations, at least one active slip system was associated with a slip plane that was perpendicular to the broad face of the specimen. If cracking were to continue on that plane, the resulting crack geometry would have been greatly simplified to that of a mixed-mode I & II crack at about 45 degrees to the stress axis, producing unambiguous da/dN crack growth rate data. These through-thickness cracks were not observed. Nevertheless, the actual crack path met the Schmid factor and the maximum strain energy release rate requirements. A generalized version of the energy release rate requirement can be found in reference 5.

IN 901 Specimens

The IN 901 experiments were conducted along with two other materials having markedly different response characteristics (6). The anticipated general planar slip character of the polycrystalline material was manifested early in the tests and continued throughout the area of cyclic crack extension. Initial cracking in all of the samples was controlled by a taper cut notch to force nucleation in the thinned extreme of the notch root, near, if not at, the observed surface of the specimen.

A selection of photos from three tests is included here so that only the most illustrative are present. Control parameters of maximum net section stress and R ratio on these specimens were: sample 1, constant 350 MPa at R = 0.2; sample 2, 400 to 600 MPa at R = 0.5; and sample 5, 400 to 500 MPa at R = 0.2. For reference purposes, total lives of these specimens are: 980,423 cycles, 3,902,662 cycles, and 485,516 cycles, respectively.

Planar slip developed almost immediately after beginning cycling and was reflected in the real-time observations and subsequent fractography (Figure 10). Faceted fracture, very similar to that of the single crystal

 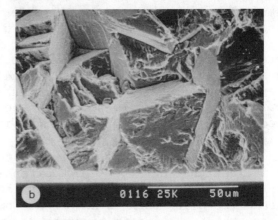

Figure 10 - Fractographs of a) sample 2 and b) sample 1 illustrating the distinctly faceted surfaces.

reported above, including 'roof top' features were found. Very few 'classic' striations were observed. The general character of the facets was similar to that of the single crystal, but the readily observed radiating lines in the SRR 99 were only infrequently seen in the IN 901.

The planar slip character of this polycrystalline material provided for quite varied crack extension responses. In some cases two to four grains would be coupled through mutually compatible, closely oriented slip planes as in sample 2 at 3,250,000 cycles (Figure 11). Within an individual grain these cases frequently gave little evidence of slip on identical, neighboring slip planes of the same orientation, or other related planes. Crack growth was relatively rapid under these circumstances. The other extreme was where multiple octahedral systems were active within an individual grain or where other incompatibilities were present. In such locations relative, local crack extension slowed, reflecting the greater difficulty in extending under these conditions (e.g. Sample 1 at 175,000 cycles, Figure 12).

In the single crystal work described above the extension of cracks 'backwards' was noted. No specific evidence was gathered in the polycrystalline IN 901 to support an identical mechanism, i.e. a crack extending 'backwards' to complete the fracture path to the free surface of the notch root. This is probably due to constraint associated with mean-free-slip path and perhaps notch geometry details. Many times during the real-time

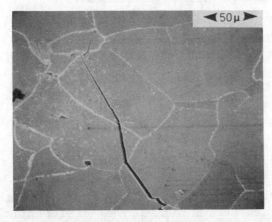 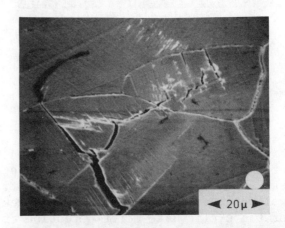

Figure 11 - Multiple grains traversed by a single crack on complementary slip planes (Sample 2, N = 3,250,000 cycles).

Figure 12 - Diffuse, multiple slip at the crack tip of sample 1, N = 175,000 cycles.

studies of IN 901 cracking was observed 'ahead' of the leading crack tip in locations apparently susceptible to nucleation within the crack tip strain field, e.g. planar slip band crack in a grain, a favorable grain boundary, etc. The general effect is seen in Figure 13 (Sample 5 at 450,000 cycles). The leading cracks would then extend in both directions, forward and backward, until they linked up with the main crack or were by passed by the principal crack front, becoming secondary cracking. Link up did not always occur along slip planes, but might involve grain boundaries, etc.

The eta colonies present in the IN 901 also provided for additional complications of the crack extension response. The presence of the oriented platelets serves to obstruct slip, further shortening the possible mean-free-slip path to an extent sometimes perhaps an order of magnitude less than the grain boundary restriction. This further confounds the cracking and quite tortuous crack paths can develop. One of a series of micrographs reflecting the extension of a crack through an eta colony in the vicintiy of the notch root of specimen number 2 is provided in Figure 14. The platelets were frequently observed to deflect the crack path even where the eta concentration was quite diffuse.

Many other interactions were noted, including: the development of interfering features that correspond to 'roughness' effects, emergence of film-type features from slip bands, rejection of detrital material from between the approximating surfaces, fracture and separation of interfering pieces from along the crack face, etc. These last two items indicate the possibility of crack plane orientation or attitude dependent closure effects. If the movement of debris is 'toward' the crack tip due to potential or other forces, crack tip closure may be inhibited.

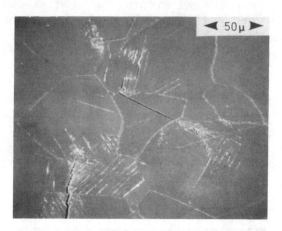
Figure 13 - A fissure 'ahead' of the main crack (Sample 5, N = 450,000 cycles).

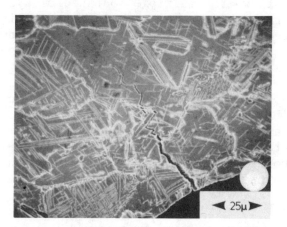
Figure 14 - Crack extending through an eta colony in the notch root of sample 2 at 100,000 cycles.

Discussion

At the stress levels employed in this study, most of the initial cracking activities in the single crystal occurred below the observed surface, adjacent to the notch root. Typically, the cyclic life after a well-defined through-thickness crack had developed was less than one-tenth of the cyclic life after detection of crack nucleation. The differential crack growth rate along different portions of the crack front led to uncracked ligaments in the specimen that played an important role in determining crack behavior.

The complexity of the crack front, especially at the early stages of crack growth, gives a direct illustration of the difficulty in unambiguously measuring and characterizing crack growth rate. Testing at higher stress levels would tend to homogenize the crack front at the notch root, however this brings into question whether crack growth rate can be uniquely characterized by K, since the initial crack morphology changes with stress level. Although the measurements of extension were based on surface observations, this difficulty would be encountered no matter which mensuration technique were employed; the specific direction of growth is not part of the information available.

The polycrystalline results illustrate a number of surface and fractographic features similar to the single crystal. The faceted nature of the polycrystal's fractography is a direct reflection of the planar slip response of the material. There was one notable difference between the crystallographic fracture characteristics of the materials. The radiating lines, which were so characteristic of the single crystal SRR 99, were rather uncommon on the IN 901 samples examined; this is remains a puzzle.

The typical nucleation-propagation life fraction observed on the polycrystalline specimens was not consistent with those of the single crystal. This could be due to both the particular method of notching and the more general, inherent 'hinderance and complexing effects' of a polycrystalline microstructure (7).

The original observations and review of the real time recordings revealed local displacements in both materials that were consistent with modes I, II and III. This multiplicity of modes also reflects on the difficulty of K_I as a correlative parameter under some conditions. In cases where local crack displacements are not substantially mode I for general anisotropic reasons (e.g. single crystals) or due to immediate structual alterations (e.g. local grain response in polycrystals), K_I does not apply very well.

The formation of ligaments in both alloys is interesting, even if they both seem to have different origins. This is another area which provides a distinct modelling challenge.

Comparing the fractographic studies to the specimen surface observations revealed the external views to be reasonably reflective of internal specimen response, especially in the IN 901. In addition, there was not a noticeable variation due to the free surface.

Conclusions

Orientation and microstructure-sensitive crack growth in a single crystal and a polycrystalline nickel-base superalloys have been presented. In the case of the single crystal SRR 99, the onset of slip and crystallographic cracking can be predicted by Schmid's law of maximum resolved shear stress. Once cracks are nucleated, the crack path can be shown to be controlled by the strain energy release rate: the crack propagates along a path that maximizes strain energy release rate.

Planar slip was also very obvious in the IN 901 and the predominant form of cracking was by related crystallographic cracking. The polycrystal also showed the influence of other microstructural variables such as grain boundaries, definite indication of slip system dependent grain-to-grain interactions, and the ability of individual precipitates to alter crack path within an eta colony. An individual grains' crystallographic cracking in IN 901 should be predictable by arguments similar to the preceding paragraph, but is confounded by the polycrystalline structure.

A variety of similarities were observed in the two materials. In a very general sense, the principal difference was polycrystal's constraint of mean free slip path. This behavior is directly related to mixed-orientation grains, their boundaries, and the influence of secondary precipitates.

The applicability of da/dN - (delta) K_I LEFM-type approach is questioned because of the multiplicity of modes present. Further studies are necessary to quantify the effects of various crystallographic constraints that affect crack growth.

Observation of geometrical and microstructural influences on the cracking process highlight the necessity to further define the relativistic issues associated with varying life-prediction approaches for both materials. An additional 'scale' effect is present in the polycrystal alloy: the level of prediction could be the entire specimen/structure or a specific grain at a given location.

Acknowledgements

The experimental efforts reported here were conducted as part of the the Ph.D. dissertations of authors Wu and Cameron at the University of Toronto, Canada. The sponsorship of Rolls-Royce, Plc and the University of Toronto is acknowledged. The assistance of Professor D. McCammond during the latter part of Dr. Wu's studies is also acknowledged.

References

1. D.W. Cameron and D.W. Hoeppner, "A Servohydraulic-Controlled Load Frame for SEM Fatigue Studies," International Journal of Fatigue, 5 (1983) 225-229.

2. C. Howland, "The Growth of Fatigue Cracks in a Nickel-base Single Crystal", The Behaviour of Short Fatigue Cracks, ed K. Miller and E. de los Rios (London, UK: Mechanical Engineering Publications, 1986), 229-239.

3. K. Chan, "Effect of Cross Slip on Crystallographic Cracking In Anisotropic Single Crystals," Acta Metallurgica, 35 (1987), 981-987.

4. D.C. Wu, "An Investigation into the Fatigue Crack Growth Characteristics of a Single Crystal Superalloy" (Ph.D. Thesis, University of Toronto, 1986).

5. J.S. Short, D.W. Hoeppner, and D.C. Wu, "The Maximal Dissipation Rate Criterion II: Analysis of Fatigue Crack Propagation In FCC Single Crystals", Submitted to Engineering Fracture Mechanics, 1988.

6. D.W. Cameron, " Perspectives and Insights on the Cyclic Load Response of Metals" (Ph.D. Thesis, Univeristy of Toronto, 1984).

7. D. McLean, Grain Boundaries in Metals (Oxford, England: Clarendon Press, 1957) 151 ff.

ACCELERATED FATIGUE CRACK GROWTH BEHAVIOR OF PWA 1480

SINGLE CRYSTAL ALLOY AND ITS DEPENDENCE ON THE DEFORMATION MODE

Jack Telesman and Louis J. Ghosn*

NASA Lewis Research Center
Cleveland Ohio, USA 44135

Abstract

An investigation of the fatigue crack growth (FCG) behavior of PWA 1480 single crystal nickel base superalloy was conducted. Typical Paris region behavior was observed above a ΔK of 8 MPa$\sqrt{m}$. However, below that stress intensity range, the alloy exhibited highly unusual behavior. This behavior consisted of a region where the crack growth rate became essentially independent of the applied stress intensity. The transition in the FCG behavior was related to a change in the observed crack growth mechanisms. In the Paris region, fatigue failure occurred along {111} facets, however at the lower stress intensities, (001) fatigue failure was observed. A mechanism was proposed, based on barriers to dislocation motion, to explain the changes in the observed FCG behavior. The FCG data were also evaluated in terms of a recently proposed stress intensity parameter, K_{rss}. This parameter, based on the resolved shear stresses on the slip planes, quantified the crack driving force as well as the mode I ΔK, and at the same time was also able to predict the microscopic crack path under different stress states.

Introduction

The relatively recent advent of the directionally solidified and single crystal nickel based superalloys for aerospace applications has focused attention on the ability to understand and predict the fatigue behavior of these alloys. In particular, with the recent emphasis on damage tolerant design for turbine engine components, the understanding and modeling of fatigue crack growth (FCG) behavior has become increasingly important.

Stress intensity range (ΔK) has been used for the past 25 years as the correlating parameter for fatigue crack growth. It has shown to be a good empirical parameter for correlating the FCG data of polycrystalline materials. While this crack driving force parameter is well suited for polycrystalline alloys, it may not be the best parameter to use for very large grain or single crystal alloy FCG data correlation. Studies in the recent years have shown that short cracks exhibit lower threshold stress intensities and accelerated FCG rates in comparison to long cracks when compared at similar values of ΔK (1-5). The short crack behavior is usually limited to a situation where the crack size is of the same order of magnitude or smaller than the grain size (3-5). This points to the weakness of the ΔK parameter to correlate the FCG data when the grain orientation and microstructure become important factors in controlling FCG behavior.

---

* Resident Research Associate

In addition to pointing out the weaknesses of the use of the ΔK parameter, the above discussion underscores the importance of microstructure and the associated deformation mechanisms in controlling FCG behavior. A single crystal alloy offers the best opportunity to study in detail the effect of microstructure and the deformation mechanisms on the FCG behavior. For a single crystal alloy, the deformation mechanisms are active on a substantially larger scale, making their observation and identification considerably easier. In addition, the grain orientation of a single crystal can be easily determined allowing for detailed calculations of the stresses on the active slip systems.

Chen and Liu (6) recently proposed a crack driving force parameter for correlating FCG data. The new parameter, also with its roots in linear elastic fracture mechanics, is based on the resolved shear stresses on the active slip plane. This parameter may be a better candidate than ΔK for the correlation of FCG data since it takes into account the deformation mechanisms and the actual crack path.

A test program was undertaken to determine the FCG behavior of a single crystal alloy in the near-threshold and intermediate ΔK region. The alloy chosen was a single crystal nickel based superalloy PWA 1480. The emphasis was placed in relating the fatigue damage mechanisms to the observed crack growth behavior. The shear stress intensity parameter proposed by Chen and Liu (6) was evaluated for correlation with the FCG data.

Experimental Procedure

Single crystal PWA 1480 slabs were obtained with the following composition in weight percent: 4.7 Al, 0.005 C, 4.8 Co, 9.4 Cr, 0.9 Si, 11 Ta, 1. Ti, 5.2 W and the balance in Ni. The slabs were solution treated for 4 hr at 1290°C, followed by the usual coating cycle diffusion treatment of 1080°C for 4 hr and aging at 870°C for 32 hr. The diffusion treatment was used to simulate the typically performed heat treatment, even though the specimens were uncoated. The alloy contained 60 to 65 vol % of the γ' phase, which has a cuboidal morphology and a cube size range between 0.4-0.6μm (7).

Five compact tension specimens were machined with the loading axis being 7° from the (001) orientation and the side faces of the specimens in the near (010) orientation as shown schematically in Fig. 1. In order to avoid crack closure effects, the testing was performed at a load ratio R (min load/max load) of 0.5. Three specimens were tested to achieve the near-threshold region by using a load shedding procedure recommended by ASTM (8). Crack length and crack closure were measured through the use of the compliance method and the crack length was also verified through occasional optical measurements. After the near threshold region was achieved, the tests were restarted using a constant load range mode, also at R=0.5, to obtain a ΔK increasing data base. This was done to assure that load shedding had no effect on FCG data. Two other specimens were tested only at a constant load range mode to achieve a FCG data base in the intermediate ΔK region. All the tests, with one exception, were conducted in laboratory environment and at room temperature. One test in the near threshold region was conducted in a nitrogen atmosphere to analyze any possible environmental effects on FCG behavior. All the testing was done at a frequency of 20 Hz. For comparison purposes, three additional tests were performed on a readily available polycrystalline nickel base superalloy. These tests were performed on specimens machined from a polycrystalline Waspaloy disk forging under test conditions identical to those of PWA 1480, with the exception that no nitrogen testing was performed.

The stress and displacement fields, along a given crack path, were determined for FCG data correlation using the two dimensional boundary integral equation (BIE) method (9). The two surfaces of the crack are modeled in separate subregions with appropriate continuity along the interface. A typical two dimensional multidomain BIE mesh for the compact tension specimen is shown in Fig. 2. The number of elements and the rigid body constraints are also shown. Quadratic variations of the displacements and tractions are assumed. An isotropic solution is used since it has been shown by Chan and Cruse (10) that the difference between anisotropic and isotropic solutions are neglegible for the single crystal nickel based superalloy. The shear stress intensity factor parameter was determined from the stress field solution near the crack tip by the projection of the traction on a slip plane in the direction of slip, as proposed by Chen and Liu (6). This procedure is explained in more detail in a later section.

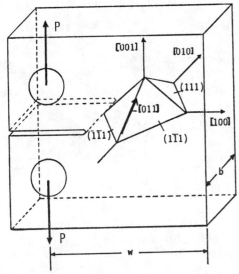

Fig. 1 Approximate orientation of {111} planes with respect to the compact tension specimens tested.

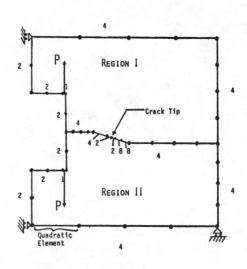

Fig. 2 Two-dimensional multidomain boundary integral mesh.

Results

Macroscopic Observations of Failed Specimens

For all the PWA 1480 specimens tested, the macroscopic failure planes were approximately 7° to the plane of the starter notch. Thus the macroscopic failure occurred on the (001) plane. Small amounts of secondary cracking were observed on the {111} type planes at higher stress intensity ranges. The Waspaloy specimens failed in the plane of the starter notch.

Fatigue Crack Growth Results

Chan and Cruse (10) have shown that the ASTM stress intensity solution for a Mode I crack is valid for single crystal compact tension specimens having an inclined crack, provided that the crack angle is less than 30° from the starter notch. The PWA 1480 specimens tested exhibited only a 7° crack angle, thus Mode I solutions were used to correlate the FCG data.

The PWA 1480 fatigue crack growth data is shown in Fig. 3. At stress intensities above a ΔK of approximately 8 MPa$\sqrt{m}$, the FCG behavior is rather normal, exhibiting typical region II (Paris region) characteristics where the crack growth rate is directly proportional to the applied ΔK, on a log-log basis. However, below that value, the observed behavior is very different in comparison to the typical long crack behavior. Below a ΔK of 8 MPa$\sqrt{m}$, the FCG rate becomes essentially independent of the applied ΔK. This region continues until a ΔK of approximately 2.5 MPa$\sqrt{m}$ is reached, after which the FCG rate again starts decreasing with decreasing ΔK. This behavior was identical under both the load shedding and constant load portions of the test, indicating that the test procedure was not a factor in causing this behavior. Crack closure was monitored throughout the test and was shown to be below that of K_{min}, thus it also had no effect on the test results. The results of the test performed in nitrogen, shown in Fig. 4, reveal identical behavior indicating that the environment is probably not the cause of this unusual behavior.

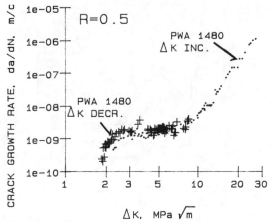

Fig. 3 Fatigue crack growth rate of PWA 1480 as a function of ΔK.

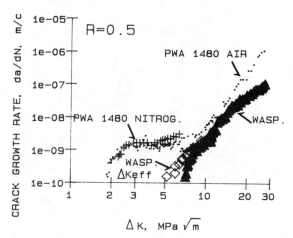

Fig. 4 Comparison of FCG rates of PWA 1480 and Waspaloy.

A comparison of the FCG behavior of PWA 1480 and Waspaloy is also shown in Fig. 4. Waspaloy FCG behavior is quite different and resembles that of typically observed long crack growth curves. For Waspaloy, in the near threshold regime, the crack closure stress intensity factor, K_{cl}, was reached at K values somewhat above the K_{min}. The data was corrected for crack closure and is also shown in Fig. 4 based on the effective ΔK, (ΔK_{eff}). The corrected data still exhibits the typically observed FCG trends.

Review of the literature failed to reveal any previous observations in a polycrystalline alloy under prevailing linear elastic conditions, which are similar to the observed PWA 1480 behavior. Nor was any near-threshold single crystal data found in the literature to which the current results could be compared.

The region of unusual FCG behavior of PWA 1480 is similar in appearance to the accelerated crack growth behavior reported for short cracks (1-5). Whether this similarity is coincidental or whether it is an indication of a single phenomenon is a topic for future studies.

Fractography of PWA Specimens

A detailed fractographic evaluation was performed on the PWA 1480 specimens to determine the microscopic deformation modes and their relationship to the observed FCG behavior.

Outside Surface Fractography: Even though on the macroscopic level the crack propagated on the (001) plane inclined 7° to the starter notch, the surface observations on the microscopic level revealed a presence of slip traces ±45° to the macroscopic crack (or 52° and -38° to the starter notch). Thus crack propagation was a result of slip on at least two different planes as shown in Fig. 5.

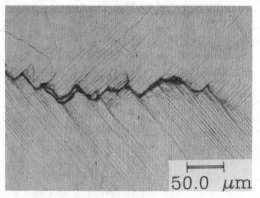 

a) General behavior b) Slip offsets (etched)

Fig. 5 Outside surface observations of the crack growth process.

Through-Thickness Fractography: Fig. 6 reveals the progressive change in the fatigue failure mechanism as a function of the applied stress intensity in the mid-thickness of the specimens. At the lowest stress intensities (Fig. 6a), cuboidal facets of the strengthening precipitates are seen throughout the mid-thickness sections resulting in a (001) fatigue failure appearance. As the stress intensity was increased, areas containing facets on {111} planes also became apparent (Fig. 6b). With a further increase in the stress intensity, the (001) fatigue failure completely disappeared, and was replaced by the {111} fatigue failure (Fig. 6c-d). As seen in these figures, the increase in the ΔK resulted in an increase in the size of the {111} failure facets. Also, in the mid-thickness areas the failure occurred on all four {111} planes.

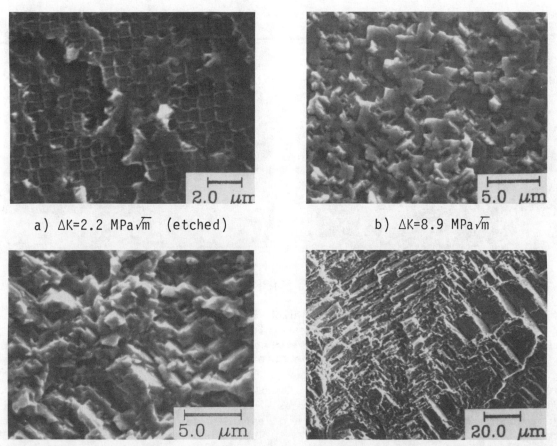

a) $\Delta K = 2.2$ MPa$\sqrt{m}$ (etched)   b) $\Delta K = 8.9$ MPa$\sqrt{m}$

c) $\Delta K = 18.5$ MPa$\sqrt{m}$ d) Final failure $\Delta K > 40$ MPa$\sqrt{m}$

Fig. 6 Mid-thickness failure appearance at various ΔK.

An interesting phenomenon was observed with regard to the (001) fatigue failure. Examination of the small ridges (or steps) on the etched failure surface, as viewed in Fig. 6a and at a higher magnification in Fig. 7, suggests that the failure was confined only to the matrix phase. The results obtained by Miner et al (11) can be used to support the above suggestion. They have shown that only octahedral {111} slip is active for a (001) oriented single crystal at room temperature. Since no {111} facets were visible on the failure surface at low ΔK, and [001] cube slip is unlikely, the only plausible mechanism which could explain the presence of (001) γ' cuboidal facets on the failure surface is the confinement of the {111} slip deformation to the matrix network. The {111} matrix slip deformation cannot be resolved without transmission electron microscopy (TEM). A hypothesis explaining why such mechanism is occurring and its influence on the FCG behavior is described later on in the paper.

Fig. 7 Fatigue failure at ΔK of 2.2 MPa$\sqrt{m}$ in the mid-thickness. Arrows point to matrix failure.

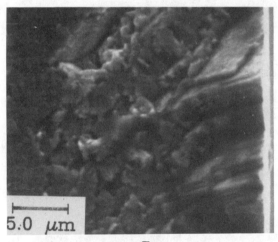

a) ΔK= 4 MPa$\sqrt{m}$

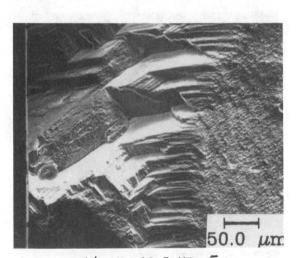

b) ΔK= 18.5 MPa$\sqrt{m}$

Fig. 8 {111} failure facets near outside surface.

The crystallographic planes on which fatigue failure occurred were not only dependent on the applied stress intensity but also on the through-thickness location. Near the outside surface, in the lower ΔK region, there was still an area approximately 5-10 μm thick of {111} failure (Fig. 8a). At somewhat higher stress intensities the regions of {111} failure extended deeper into the thickness, with the (001) regions being confined to small mid-thickness areas. The size of {111} facets was largest near the outside surfaces and rapidly decreased with increasing distance from the outside surfaces (Fig. 8b). While the failure occurred on all four {111} planes in the mid-thickness, failure occurred on only two {111} planes in the near surface. The summary of the fractographic findings is shown in Table I.

Table I Observed failure planes as a function of the applied ΔK.

ΔK MPa√m	Outside Surface (Plane Stress)	1/4 Thickness	Mid-thickness (Plane Strain)
2	(111)	(001)	(001)
2.3			(001)
3.1			(001)
4.3		(001)	(001)
4.9		(001)/(111)	(001)
6.4			(001)
6.8		(001)/(111)	
7.4	(111)		
7.7		(111)/(001)	
8.9			(001)/(111)
11.3	(111)	(111)	(111)/(001)
13	(111)	(111)	(111)/(001)
16	(111)		(111)
18.5			(111)
23.1			(111)
25.3			(111)
27.5			(111)
29.7			(111)

The changeover from a predominantly (001) fatigue failure to {111} occurs at a ΔK of 7 to 10 MPa√m, as seen in Table I. This corresponds closely to the stress intensity at which the transition from accelerated to a more typical FCG behavior occurs (Fig. 3). Thus the unusual FCG behavior is associated with the (001) fatigue failure mechanism.

Resolved Shear Stress Intensity Parameter

The newly proposed K_{rss} parameter was used to explain some of the aspects of the observed microscopic failure mechanisms and the associated FCG behavior.

Definition: The resolved shear stress intensity parameter, K_{rss}, is defined, following Chen and Liu formulation (6), as the limiting value of the resolved shear stress, τ_{rss}, multiplied by $\sqrt{2\pi r}$, as r approaches zero:

$$K_{rss} = \lim_{r \to 0} \tau_{rss} \sqrt{2\pi r} \qquad (1)$$

where r is the distance to the crack tip and τ_{rss} is defined as the projection of the stress tensor $[\sigma]$ on a plane whose outward normal is $\vec{n}$ in the direction of slip $\vec{b}$.

$$\tau_{rss} = \vec{b} \cdot [\sigma] \vec{n} \qquad (2)$$

Fatigue cracking is postulated to occur along slip systems which have the highest K_{rss} value.

In this study, the three dimensional stress state $[\sigma]$ is approximated from the two dimensional boundary integral solution, assuming the asymptotic singularity of the stress field, under plane stress and plane strain conditions. Values of K_{rss} for PWA 1480 are determined by projecting the stresses near the crack on the {111} planes in a particular slip direction. The calculated maximum ΔK_{rss} values as a function of the normalized crack length for constant load range testing are shown in Fig. 9. The results indicate that the magnitude of the plane stress ΔK_{rss} is always approximately twice that of plane strain.

Application to FCG: To check the validity of the resolved shear stress intensity factor, K_{rss}, for correlating FCG data, ΔK_{rss} was calculated and plotted versus the da/dN data and is shown in Fig. 10, together with the da/dN versus the mode I ΔK plot. The shape of the two curves is virtually identical. A linear relationship exists (on the log-log basis) between ΔK_{rss} and da/dN which is similar in nature to the classical Paris region. This shows that the ΔK_{rss} parameter is as good as the mode I ΔK for FCG data correlation. But the advantage of ΔK_{rss} lies in the ability to predict the actual microscopic fatigue fracture mechanisms, as described next.

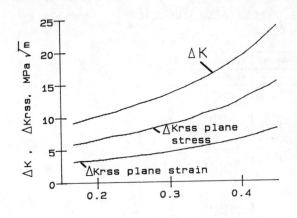
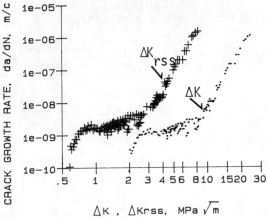

Fig. 9 Calculated values of ΔK_{rss} and ΔK as a function of crack length.

Fig. 10 Fatigue crack growth rate as a function of ΔK_{rss} and ΔK.

Prediction of the Microscopic Crack Behavior: An attempt was made to predict the microscopic crack path by the use of K_{rss} in the region where failure occurred on the {111} facets. It was assumed that the facets, through the entire thickness of the specimen, made either a 52° or -38° crack angle to the starter notch (taking into consideration the measured 7° macroscopic crack angle and in agreement with the observed slip surface traces). Table II shows the values of K_{rss} for different slip systems after a small amount of crack growth has occurred on either one of the crack angles, for both plane stress and plane strain conditions. Values in Table II can be used to predict whether the crack growth will continue in a self similar manner or change to a different slip system. For reference, the calculated Schmid factors are also given in the table.

Table II Normalized values of K_{rss} for different slip systems under plane stress and plane strain conditions.

Slip Plane	Slip Direc.	Schmid Factor	$K_{rss}b\sqrt{W}/P$ Plane Stress -38°	52°	Plane Strain -38°	52°
(111)	[$\bar{1}$10]	.08	-1.77	-2.58	0.28	0.29
(111)	[$\bar{1}$01]	.34	2.24	0.10	2.56	0.55
(111)	[01$\bar{1}$]	.32	-3.97	-2.64	-2.28	-0.26
(1$\bar{1}$1)	[$\bar{1}\bar{1}$0]	.08	0.94	1.87	-0.69	-0.58
(1$\bar{1}$1)	[$\bar{1}$01]	.43	3.06	1.68	2.72	1.18
(1$\bar{1}$1)	[011]	.34	3.93	3.52	1.97	0.56
(11$\bar{1}$)	[1$\bar{1}$0]	.01	1.69	1.55	-0.63	-0.48
(11$\bar{1}$)	[101]	.42	-0.66	-2.32	-0.74	-2.38
(11$\bar{1}$)	[011]	.42	-2.30	-3.82	-0.13	-1.92
($\bar{1}$11)	[$\bar{1}\bar{1}$0]	.08	1.74	1.73	-0.86	-0.46
($\bar{1}$11)	[101]	.36	1.24	2.42	1.13	2.33
($\bar{1}$11)	[01$\bar{1}$]	.44	-3.05	4.25	-0.27	-1.91

◯ First pairing systems of (1$\bar{1}$1) and ($\bar{1}$11)
☐ Second pairing systems of (111) and (11$\bar{1}$)
__ Observed Failure Planes under Plane Stress Condition

For the plane strain case, the fractography of the mid-thickness region at intermediate ΔK_{rss} or (ΔK), showed that all four slip planes were active (Fig. 6b-d). A closer look at the fractography shows that the crack path zigzagged on two distinct pairs of planes (either ($\bar{1}$11) and (1$\bar{1}$1) or (111) and (11$\bar{1}$) as shown in Fig. 6c and d). The K_{rss} parameter can predict this type of crack growth behavior. For instance, after a small amount of crack growth at a 52° crack angle on the ($\bar{1}$11) plane in the [101] direction, the maximum K_{rss} changes to the (1$\bar{1}$1) plane in the [$\bar{1}$01] direction. Slip on this new system results in crack growth at a -38° crack angle. However after a small extent of crack growth at the -38° crack angle, the maximum K_{rss} switches back to the original slip systems as seen in Table II. The same type of behavior occurs for the other pair of slip systems thus creating the observed fatigue crack behavior. For plane stress, after a small amount of crack growth at -38° on the (111) [01$\bar{1}$] system, the maximum normalized K_{rss} changes to (111) [01$\bar{1}$] slip system, which results in the crack switching to the 52° inclination. The near surface failure was observed to occur on these two predicted planes. The surface slip offsets, as shown in Fig. 5b, suggest <011> slip direction in agreement with K_{rss} predictions. Presence of these two active slip systems suggests that cross slip might have been activated.

From the successful description of the FCG single crystal behavior, it can be concluded that the K_{rss} is a microscopic parameter that can quantify the crack driving force and at the same time be used to predict the microscopic crack propagation path under different stress states. K_{rss} can also be used to explain other aspects of the PWA 1480 single crystal behavior, as is described later on.

Relationship Between Slip Mechanisms and FCG Behavior

One of the main findings in this study was the identification of a relationship between the fatigue failure mode and the FCG behavior. Fatigue failure along {111} planes was associated with Paris region crack growth behavior, and along (001) planes was associated with the accelerated FCG behavior at low ΔK_{rss} (or ΔK). Accelerated FCG behavior refers to the comparison of the actually measured FCG rates versus the FCG rates obtained through the extrapolation of the Paris region to the low ΔK_{rss} region (Fig. 10). Examination of the effect of shear stresses on the dislocation motion in the matrix and in the γ' precipitates may explain this behavior.

It requires a certain critical value of the resolved shear stress, τ_o, to move a dislocation. The τ_o for dislocations to cut through the γ' precipitate is considerably higher than τ_o for dislocation motion through the matrix, as was shown by Copley and Kear (12). In the Paris region of Fig. 10 (intermediate ΔK_{rss} or ΔK), the shear stresses are high enough to allow for dislocations to slip through both the matrix and the precipitates (Fig. 5b). This process results in a {111} fatigue failure and the corresponding FCG behavior. However as ΔK_{rss} (or ΔK) is decreased, the accompanying resolved shear stresses also decrease. When the resolved shear stress falls below the critical value needed for a dislocation to cut through the precipitates, the slip becomes confined to the matrix. As mentioned earlier, the localization of the damage to the {111} matrix planes results in a preferential failure in that area, exposing the cuboidal facets of the precipitates and creating a (001) failure appearance. This scenario is supported by the previously mentioned fractographic evidence which indicated that the (001) failure appearance was associated with the observed failure in the matrix (Fig. 7). TEM will be performed to confirm this hypothesis.

The above hypothesis can be used to explain the accelerated FCG behavior encountered in the PWA 1480 at low stress intensities. The localization of the {111} slip to only the matrix will result in a higher local dislocation densities and increased dislocation interactions when compared to a situation where the dislocations can spread over a much longer active slip plane, as would be the case when both the matrix and the precipitates are sheared. For a given ΔK_{rss} (or ΔK), the localization of the slip exclusively to the matrix is a more damaging process than the one through which shearing of both the matrix and precipitates occurs. Therefore, at the lower ΔK_{rss} (or ΔK) the crack growth rates are higher than the extrapolated Paris region would predict.

The dependence of the failure mode on the state of stress, as shown previously in Table I, can be explained by reviewing the effect of stress state on the value of K_{rss}. As shown earlier, the plane stress K_{rss} is approximately twice that of plane strain (Fig. 9). The resolved shear stresses on the slip planes are directly proportional to the K_{rss} (eq. 1). Thus the resolved shear stresses on the dislocations are smaller in the mid-thickness (plane strain) than near the surface (plane stress) for a given crack length. The difference in the K_{rss} values, together with the previously postulated mechanism, explains the prevalence of (001) fatigue failure in the mid-thickness as compared to the more extensive {111} failure closer to the surface. The presence of considerably larger {111} surface facets which diminish rapidly with the distance from the outside surface (Fig. 8b), is due to a decrease in the K_{rss} caused by the change from plane stress near the surface to a mixed stress state away from the surface.

Conclusions

1. At ΔK greater than 8 MPa $\sqrt{m}$, the FCG rate exhibited the classical linear relation on a log-log basis between da/dN and ΔK (Paris region). In this region the microscopic crack propagation was along {111} slip planes. In the mid-thickness, all four planes were activated, however near the surface only two {111} slip planes were active.

2. At ΔK below 8 MPa$\sqrt{m}$, the FCG rate became almost independent of the applied ΔK. The change in the FCG behavior was related to a change in the fatigue failure mechanism. With the decrease in the applied ΔK, the (001) fatigue failure appearance became progressively more dominant.

3. The state of stress had a substantial influence on the fatigue failure mechanisms. At lower ΔK, while the (001) failure predominated in the mid-thickness regions, the near surface regions still exhibited {111} fatigue failure.

4. A mechanism was proposed, based on the barriers to dislocation motion, to explain the changes in the observed FCG behavior.

5. The FCG data was also evaluated in terms of a resolved shear stress intensity parameter, K_{rss}. This parameter, based on the shear stresses resolved into the slip planes, quantified the crack driving force as well as mode I ΔK, and at the same time predicted the crack path. The differences in the K_{rss} values were used to explain the observed dependence of the fatigue failure mode on the state of stress.

References

1. S. Pearson, Eng. Fract. Mech., vol 7, pp 235-247, 1975.
2. W.L. Morris, Met. Trans. A, vol 11, pp 1117-1123, July 1980.
3. J. Lankford, Fat. of Eng. Matl. and Struct., vol 5, no. 3, pp 233-248, 1982.
4. P. Newman and C.J. Beevers, R.O. Ritchie and J. Lankford eds., AIME, pp 97-116, 1986.
5. J. Telesman, D.M. Fisher and D. Holka, NASA TM 87208, 1985.
6. Q. Chen and H.W. Liu, NASA CR-182137, 1988.
7. R.V. Miner, J.Gayda and M.G. Hebsur, Low Cycle Fatigue, ASTM STP 942, pp. 371-384, 1988.
8. Standard Test Method for Measurement of Fatigue Crack Growth Rates, E 647, ASTM, 1987.
9. L.J. Ghosn, To appear in the ASME Journal of Tribology, vol 111, July 1988.
10. K.S. Chan and T.A. Cruse, Eng. Fract. Mech.,vol 23, no. 5, pp 863-874, 1986.
11. R.V. Miner, R.C. Voigt, J.Gayda and T.P. Gabb, Met. Trans., vol 17A, pp 491-496, 1986.
12. S.M. Copley and B.H. Kear, Trans. AIME, vol 239, pp 984-992, July 1967.

CREEP BEHAVIOR OF MAGNESIUM

MICROALLOYED WROUGHT SUPERALLOYS

Peili Ma, Ying Yuan, Zengyong Zhong

Central Iron and Steel Research Institute
Beijing, PR China

Abstract

By adding minor amount of magnesium to the wrought superalloys, the rupture life and ductility are considerably improved, the steady stage of creep, and especially the tertiary stage of creep are prolonged, and the steady creep rate is decreased at low strain rates. It is found that magnesium segregates to grain boundaries and interfaces, and also exists in matrix, γ' phase and carbides. The magnesium causes the change in interfacial energy and enhances the cohesion of the grain boundaries and of the interfaces. Magnesium can spheroidize the carbides in the grain boundaries, purify the grain boundaries and decreases the vacancy concentration in the alloys. In order to clarify the mechanism of these effects, detailed studies have been made.

Introduction

Magnesium had been seldom utilized as a microalloying element in commercial alloys although the studies of the effects of it on the superalloys could be found in the literatures. Over the years, a series of studies have been carried out by Chinese materials scientists and the application of magnesium has being introduced into commercial wrought superalloys. The results of the studies indicated that minor amount of magnesium would favor the properties of superalloys, especially considerably improve the rupture life, rupture ductility, high-temperature tensile ductility, notch sensivity and hot workability. Some studies of the mechanisms of these effects have also been done.

In this work, a systematic investigation about the effects of magnesium on the creep and creep rupture behaviors in wrought superalloys was carried out and the mechanism was investigated.

Experimental Procedures

In the present work, two differently alloyed wrought superalloys were used for study. Alloy A (GH33), simply alloyed, has the similar composition to Nimonic 80A and approximately 8wt-% γ' phase. Alloy B (GH220), complexly alloyed, contains molybdenum and tungsten, etc, as strengtheners and approximately 45 wt-% γ' phase. It can operate at 900°C. The compositions of these two alloys are listed in table 1.

Table I. Compositions of the Alloys (wt,%)

Alloy	C	Cr	Co	W	Mo	Al	Ti	V	B
A (GH33)	0.02	20	--	—	—	0.7	2.5	—	0.007
B (GH220)	0.06	10	15	5.5	5.5	4.3	2.5	0.3	0.008

The contents of magnesium were investigated to be 0.0003%, 0.006%, 0.013% and 0.030% for alloy A, and 0.0003%, 0.007%, 0.0125% and 0.028% for alloy B.

Rupture life, elongation and creep curves under various stresses were measured. Fracture characteristics and the microstructures of the tested specimens were studied by metalloscope, SEM and TEM, etc.

The concentration of magnesium in the phases was determined by means of electroextraction and chemical analysis. The lattice constant of γ' phase and carbides were determined by Nonus II Guinier camera. Tests of grain boundary sliding during creep was taken in a vacuum creep-testing equipment. Vacancy activation energy of the alloys was calculated from the values of electrical-resistivity.

Experimental Results

1. Stress Rupture Properties

The effect of magnesium on the rupture life is shown in fig.1. It shows clearly that minor amount of magnesium can increase the rupture life, particularly for alloy A. Excessive amount of magnesium, however, decreases the rupture life. The elongation is also significantly improved with enhancing of the rupture life. Similar results have come out for some other superalloys. It should be noted that the optimal content of magnesium varies in different alloys.

2. Creep Behavior

Fig. 2 shows the effect of magnesium on creep behavior. Proper amount of magnesium lengthens the duration of second stage of creep and especially the tertiary stage, thereby the creep rupture life is enhanced and the total creep strain increases.

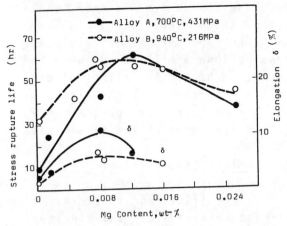

Fig. 1 Effect of Mg-Content on the Stress-Rupture Properties

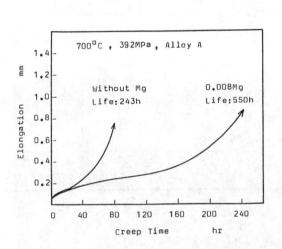
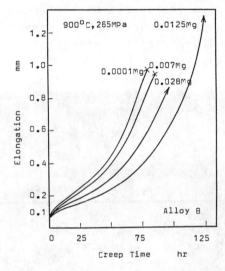

Fig. 2 Creep Curves of Alloy A and B with various Mg-Addition

The experimental results of the effect of magnesium on the steady creep rate (fig. 3, 4 and 5) show that minor amount of magnesium could

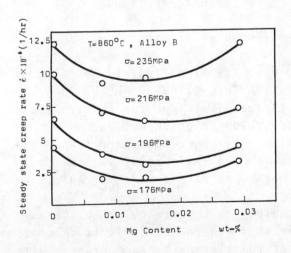
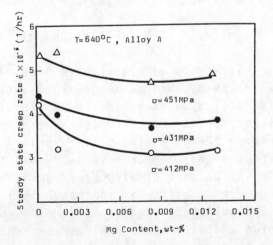

Fig. 3, 4 Influence of Mg on the Steady-State Creep Rate

decrease the rate of steady state creep at low creep rate($\dot{\varepsilon} < 10^{-5}$ mm·mm^{-1}·hr^{-1}) At the strain rate greater than a critical value($\dot{\varepsilon} > 10^{-5}$ mm·mm^{-1}·hr^{-1}), the effect of magnesium vanishes, i.e. the effect on the steady creep is present only within the range of low strain rate.

3. Fracture Characteristics

The fractured specimens of alloy A were studied. The specimens without Mg-addition show smooth surface, no necking and insignificant local deformation near fracture place. The fracture surface exhibites intergranular brittle rupture (fig. 6a). The specimens that contained magnesium, however, show many micro-cracks on the surface and obvious necking down. Ductile intergranular rupture is yielded at the fracture surface (fig. 6b).

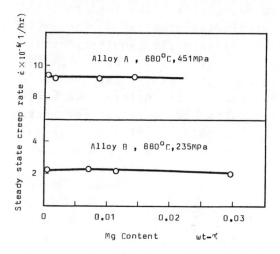

Fig.5 Influence of Mg on the Steady-State Creep Rate

a) without Mg-addition   b) with Mg-addition

Fig.6   SEM-Fractographs of Stress-Rupture Specimens

Two types of cracks have been found in the longitudinal section of the specimens. One is wedge-like cracks which form owing to the grain boundary sliding, as shown in fig.7a. In fig.7b, the other type cracks are shown which arises from the coalescence of micro-voids in grain boundaries, verticle to the direction of the applied stress. As magnesium increases, the percentage of wedge cracks decreases from 70% in the case of without Mg to 40% in the case of with Mg. Excessive magnesium raises the percentage of wedge cracks again, as indicated in fig.8.

Significant changes in carbide morphology take place owing to the magnesium addition. The carbides are considerably spheroidized. It is shown in fig. 9, for alloy B, that the percentage of rod-like carbides decreases with

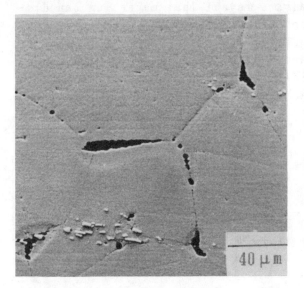
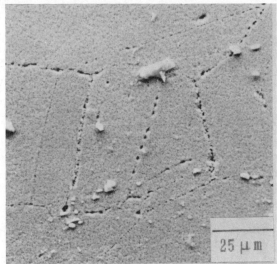

a) Wedge-crack   b) Cavitation-crack

Fig. 7 SEM - Micrographs of Creep Specimens

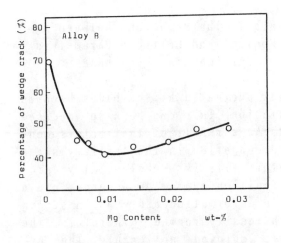

Fig.8 Relation between the Percentage of Wedge-Cracks and Mg-Content

Fig.9 Relation between the Percentage of Rod-like Carbides and Mg-Content

the addition of magnesium.

Discussion

In publications, it is an opinion of some workers that the role of magnesium is owing to its desoxydation and desulphidation. In fact, however, magnesium is also a microalloying element. This problem will be discussed subsequently.

1. The Mg-distribution is an essential factor for the study on the mechanism of its effects. The results of the studies by the authors of this papar [1,2,3,4] have indicated that magnesium is a surface-active element inclined to segregate to grain boundaries and carbide/matrix interfaces. Magnesium causes the change in the energy of grain boundaries and interfaces and thereby results in the spheroidization of carbides. On the

other hand, the results of phase analyses reveal that magnesium can dissolve in γ matrix in small amount, and enter into γ' phase and carbides. For alloy B with 0.005% Mg, the magnesium concentration in γ, γ' and carbide phases is 0.0024 %, 0.0076% and 0.0238% respectively. The magnesium dissolved in γ' phase and carbides, leads to the compositional changes, i.e. increases the contents of W, Mo and Ti in the phases and therefore changes their lattice constants (fig. 10). The presence of Mg in grain boundaries and matrix improves the properties of the alloys. Besides, very small amount of inclusion formed by magnesium acting with P, S and Si can usually be found in the alloys.

2. Mg located at the grain boundaries and carbide/matrix interfaces lowers the grain boundary energy and the interfacial energy, increases the interface cohesion and reduces the rates of crack initiation and propogation [5] Furthermore, the spheroidized carbides improve the stress distribution state at grain boundaries and thus decreases the possibility for wedge crack formation. Hence the

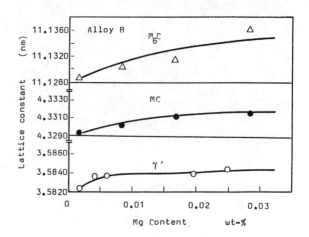

Fig. 10 The Relation between Mg-Content and Lattice Parameters of Carbides and γ'-Phase

coalescence of grain boundary microvoids nucleated by carbides becomes the prominent mechanism of carck initiation. This can be also deduced from the similarity of figures 8 and 9. As a result of the factors mentioned above, the crack initiation and propogation in the tertiary stage of creep are retarded and thus the rupture life is substantially prolonged. On the other hand, magnesium addition can allow a large grain boundary sliding to coordinate the grain deformation, and even in strongly deformed specimen, no cracks would be easily formed. Therefore the stress rupture and creep elongation are improved noticeably. The experiments conducted by the authors [5] indicated that magnesium could purify grain boundaries through binding with S and P, etc. This also contributes to the improvement of stress rupture property.

Magnesium addition is proved to be favorable to the grain boundary sliding by vacuum creep test. In the test, line markes were set on the polished surfaces of specimens so that the grain boundary sliding could be measured. For the specimen containing magnesium, small sliding occured at the time of 113 hour during the test. Great sliding happened at 293 hour, but no cracks appeared at that time. For the specimen without Mg-addition, the grain boundary sliding was smaller compared with that containing magnesium. No visible sliding occured at 130 hour on the test, but noticeable grain boundary sliding happened abruptly at 170 hour and wedge cracks were found at the triple grain boundary (fig. 11).

3. The fact that Mg reduces steady creep rate within the range of low strain rate is related to the presence of magnesium in matrix. It is well known that creep mechanism is very complicated and varies under different conditions. Lagneborg et al [6] studied the creep behavior of

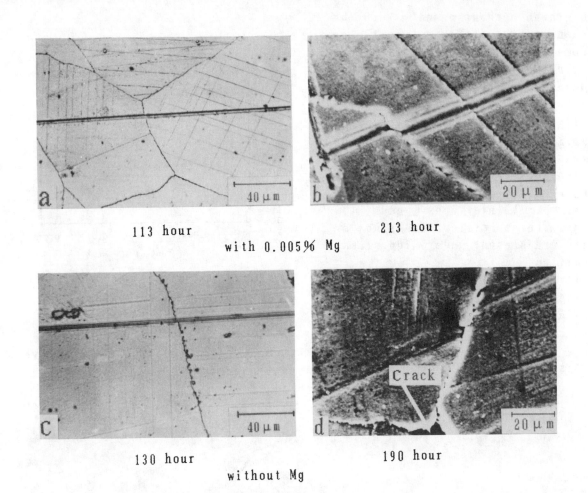

113 hour　　　　　　　　213 hour
with 0.005% Mg

130 hour　　　　　　　　190 hour
without Mg

Fig.11 Grain Boundary Slip in Creep Test

precipitation-strengthened alloys under various stresses at intermediate temperatures and pointed out that under low stress condition, i.e. at low strain rate, dislocation movement could not be realized by Orowan mechanism or cutting γ' particles, but otherwise by dislocation climbing over γ' particles. Only when the stress increased to a certain value, the Orowan mechanism or that of cutting γ' particles would play a dominent role.

Within the range of low strain rate, the climbing of dislocation over γ' particles depends upon the vacancy concentration in matrix. The experimental result indicates that magnesium has an influence on the vacancy activation energy of the alloys. Fig.12 demonstrates that activation energy of vacancy increases as concentration of magnesium increases in the alloys. Compared with that containing no magnesium, the alloy with magnesium has a faster increase in the activation energy. The vacancy activation energy, Q_v, is calculated from the increment of electrical-resistivity based on the following equation, where $\Delta \rho$ is the in-

$$\Delta \rho = D \cdot \exp(-Q_v / KT)$$

crement of electrical-resistivity due to vacancy and D is a constant.

The increase in vacancy activation energy of Mg-contained alloy causes the decrease in vacancy concentration, therefore the climbing velocity of dislocation over γ' particles reduces and a low steady creep rate proceeds. When the stress is raised to a certain value and the Orowan mechanism and / or the mechanism of cutting γ'-particles become to be prominent in creep strain, the influence of the vacancy concentration arising from magnesium will disappeare. The dislocation configurations are demonstrated in fig. 13 for alloy A under both low stress and high stress creep conditions. It can be seen that only single dislocations exist at the matrix / γ' interfaces under low stress, whereas dislocation rings or pairs are present under high stress condition.

It has been found that magnesium influences the steady creep rate in the range of $\dot{\varepsilon} < 10^{-5} mm \cdot mm^{-1} \cdot hr^{-1}$, this result agrees with that proposed by Lagneborg et al introducing the mechanism of dislocation climbing over γ' particles.

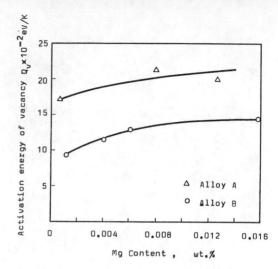

Fig. 12 The Relation between Mg-Content and the Activation Energy of Vacancy

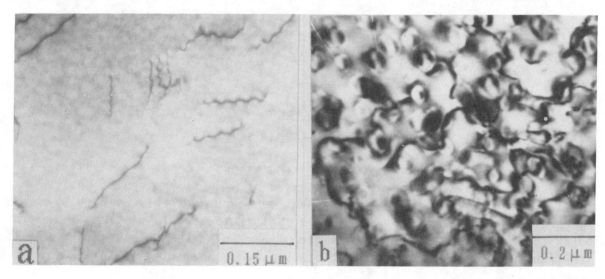

Fig. 13 Dislocations in the Creep Specimens at various Strain Rate
a) low strain rate   b) higher strain rate

Conclusions

By adding minor amount of magnesium to the wrought superalloys, rupture life and ductility can be significantly improved, the percentage of wedge cracks in the fractured specimens decreases and part of ductile intergranular fracture on the fracture surface increases. These are mainly related to that magnesium segregating to the grain boundaries could decrease the grain boundary enery, enhance the cohesion of the grain boundaries and the interphase interfaces, promote the spheroidization of carbides in grain boundaries and accordingly change the stress distribution around carbides, etc.

In the range of low strain rate ($\dot{\varepsilon} < 10^{-5} mm \cdot mm^{-1} \cdot hr^{-1}$), minor amount of magnesium would decrease the steady creep rate. At high strain rates, however, the effect of magnesium on creep rate would vanish. These are caused by the magnesium dissolved in the matrix. A mechanism accounting for it is proposed in this paper.

References

1. Danian Ke, Zhengyong Zhong, Acta Metall, 19(1983), No.5, A377(in Chinese).
2. Peili Ma et al, Acta Metall, 23(1987), No.3, A195 (in Chinese).
3. Peili Ma et al, Acta Metall, 23(1987), No.6, A484 (in Chinese).
4. Danian Ke, Zhengyong Zhong, Central Iron and Steel Research Institute Technical Bulletin, 3(1983) No.1, 73(in Chinese).
5. Danian Ke, Thesis for master's degree, (supervisor: Zhengyong Zhong), Central Iron and Steel Research Institute, Beijing, CHINA, 1981.
6. Lagneborg, R. and Bergmen, B., Met. Sci., (1976), No.10, Jan. 20.

THE ROLE OF Mg ON STRUCTURE AND MECHANICAL PROPERTIES IN ALLOY 718

Xishan Xie, Zhichao Xu, Bo Qu and Guoliang Chen
University of Science and Technology, Beijing 100083, China

John F. Radavich, School of Materials Engineering
Purdue University, W. Lafayette, In 47906, USA

Abstract

The role of Mg in alloy 718 has been systematically investigated. Mg raises not only high temperature tensile and stress-rupture ductilities but also increases considerably smooth and notch stress-rupture life. Mg containing alloy 718M is free of stress-rupture notch sensitivity. Mg improves creep and fatigue interaction properties (LCF or cyclic stress rupture) at any grain size. The basic role of Mg is equilibrium segregation at grain boundaries which helps to change continuous grain boundary δ-Ni_3Nb morphology to discrete globular form which has a retardation effect on intergranular fracture. Mg promotes the change from intergranular to transgranular fracture mode.

INTRODUCTION

For the past several decades, alloy 718 continues to be used in gas turbines in greater volume and for many applications. High performance demands and high quality requirements expecially in disk application, have required material homogenity, grain size control, and high mechanical properties (such as LCF or cyclic stress rupture) at operating conditions.

In the early 70's Couts et al. [1] studied the effect of Mg (from 1-350 ppm) on mechanical properties of alloy 718 and showed stress rupture ductility improvement in the range of Mg content from 30 to 200 ppm, but little data was presented in the lower content range (up to 100 ppm) of Mg. In 1971 Muzyka et al. [2] showed beneficial stress rupture ductility improvement at 30 ppm Mg in alloy 718. Recently, Moyer [3] in his extra low carbon alloy 718 study showed a remarkable stress rupture ductility and life improvement with a small addition of Mg (13-19 ppm). However, the true effects of Mg have not been fully understood. A systematic research study of Mg effect in nickel- and iron-base superalloys has been conducted in China for a long time [4]. Our previous studies [5,6] show optimum small addition of Mg (less than 100 ppm) not only can increase stress rupture ducitlity but also prolong stress rupture life. The beneficial effect of Mg in alloy 718 can be still maintained even after 5000 hrs long time exposure at 650°C. For further understanding the role of Mg in wrought alloy 718, especially for disk application, an investigation of Mg and grain size effects on structure and mechanical properties, especially on stress rupture notch sensitivity and cyclic stress rupture or LCF was undertaken.

MATERIALS AND EXPERIMENTAL PROCEDURE

Two 79 Kg heats of alloy 718 containing 4 ppm (mg free) and 59 ppm (Mg containing) were VIM melted. Chemical composition and alloy designation are listed in Table I.

Table I. Alloy Chemical Compositions (wt %)

Alloy	C	Mn	Si	P	S	Cr	Fe	Mo	Al	Ti	Nb	B	Mg
718	0.057	0.04	0.23	0.006	0.004	19.10	18.24	2.95	0.68	1.01	4.98	0.0054	0.0004
718M	0.052	0.04	0.23	0.006	0.004	19.04	18.10	2.95	0.67	1.00	4.98	0.0058	0.0059

Alloy ingots were partially homogenized at 1150°C for 6 hours and then forged to produce different grain size experimental disks (ϕ 200 x 45 mm). Typical structure of coarse, fine and mixed grains are shown in Fig. 1.

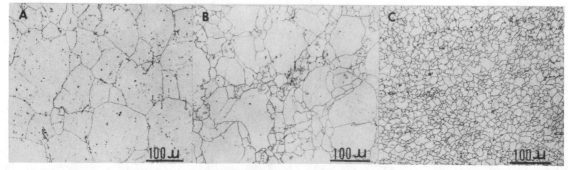

Fig. 1. Typical grain structure of experimental disks with different ASTM grain sizes (after heat treatment); (A) coarse grain, (b) duplex grain, (C) fine grain.

Samples were cut from the rim of disks with different grains sizes and given the ASM 5596C heat treatment, i.e. 950°C/1h/AC +720°C/8h/FC 50°C/h→620°C/8h/AC Mechanical property samples were tensile tested at 650°C, smooth and notched bar stress rupture tested at 650°C/686 MPa, cyclic stress rupture tested at 650°C/686 MPa with different holding times or LCF tested.

Structural characterization techniques included optical, SEM and TEM microscopy, fractography, Auger analysis and X-ray analysis of extracted residues.

EXPERIMENTAL RESULTS

Mechanical Properties

In order to study systematically the grain size and Mg effects on mechanical properties, different grain size disks of Mg free (718) and Mg containing (718M) alloys were made. Different forging procedures were used which produced various amounts of recrystallized and unrecrystalized grains especially in the mixed grain disks. After the ASM 5596°C heat treatment, grain sizes of the experimental disks of the two alloys varied from ASTM 3 to ASTM 10. The mixed grain disks of alloy 718 and 718 M displayed a necklace structure of ASTM 7-8 fine grains surrounded by ASTM 3-4 coarse grains (see Fig. 1).

Results of 650°C tensile tests on all grain size disks showed that Mg can greatly increase ductility but had little effect on ultimate strength, which is only increased by grain refining (see Fig. 2).

Similar to that seen in tensile tests, Mg can remarkably increase the 650°C stress rupture ductility as shown in Fig. 3. It should be noted that Mg not only can increase smooth S/R life but also increase notch S/R life considerably even in mixed grain samples of alloy 718M. Smooth bar S/R tests of mixed grain samples from Mg free alloy 718 disk show only 109 hrs/4.7% elongation, but 176 hrs/20.2% elongation from Mg containing 718M disk. Smooth bar S/R life decreases where notch S/R life increases with finer grain size in Mg free alloy 718. It is clear from Fig. 2 that Mg free alloy 718 will be susceptible to S/R notch sensitivity when grain size is coarser than ASTM 5. A positive advantage of alloy 718M is that Mg increases the notch S/R life remarkably; consequently, Mg containing alloy 718 M is not susceptible to S/R notch sensitivity even at coarse grain and mixed grain conditions. This should be of great benefit for forging of disks.

High temperature LCF or cyclic stress rupture characters are the most important mechanical properties for gas turbine disk application. A study of the grain size and Mg effects on cyclic stress rupture life with different holding times (5, 180, 1800 sec) at maximum stress of 686 MPa/650°C showed that Mg really improved cyclic stress rupture (namely stress controlled LCF with dwelling time) properties at fatigue and creep interaction conditions, representative of disk service conditions.

Microstructure Analyses

Microchemical phase analysis results show that the amount of main strengthening phase of γ' and γ phases is not affected by Mg addition or grain size in alloy 718 and 718M as shown in Fig. 5. Mg free alloy 718 or Mg containing alloy 718M both contains approximately 14% $\gamma' + \gamma''$ strengthening phase, independent of grain size. However, δ–Ni_3Nb phase precipitated at grain boundaries increases with grain refinement and amount of Mg. Consequently, the amount of δ–Ni_3Nb is much higher in fine grain alloy 718M as compared to Mg free alloy 718.

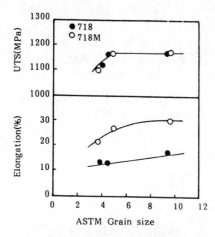

Fig. 2. Grain size and Mg effect on 650°C tensile properties.

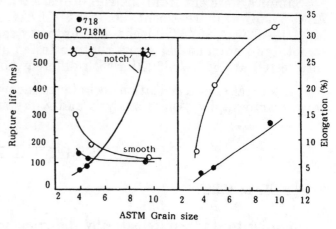

Fig. 3. Grain size and Mg effect on stress rupture life and elongation at 650°C, 686 MPa.

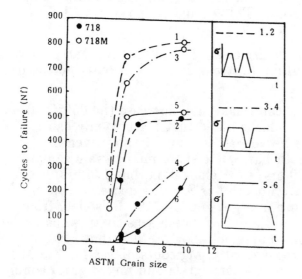

Fig. 4. Grain size and Mg effect on cyclic stress rupture life with different holding times at maximum stress of 686 MPa, 650°C.
1,2-5sec, 3,4-180sec, 5,6-1800sec

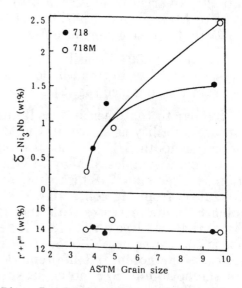

Fig. 5. Grain size and Mg effect on the amount of r'' and σ-Ni_3Nb.

Mg addition to alloy 718 increases not only δ-Ni_3Nb amount but also changes its morphology from plate-like form to globular and discrete form as shown in Fig. 7 (A and D). Quantitative analysis on the amount of grain boundary δ-Ni_3Nb shows that concentration coefficient of δ-Ni_3Nb at grain boundaries (number of δ particles/wt%δ in certain area) increases with grain refinement because of the increment of total grain boundaries. Mg addition can raise δ-Ni_3Nb concentration coefficient at grain boundaries to a higher level in alloy 718M than in alloy 718 (see Fig. 6). Thus, much smaller and more particles of δ-Ni_3Nb phase appear in Mg containing alloy 718M with fine grain structure.

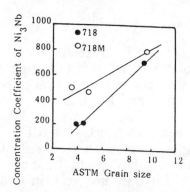

Fig. 6. Grain size and Mg effect on concentration coefficient of N Ni₃Nb at grain boundaries.

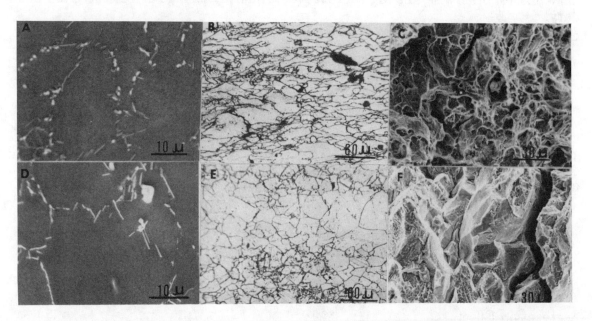

Fig. 7. Mg effect on grain boundary behavior of alloy 718M (A,B,C) and 718 (D,E,F).   A, D - grain boundary Ni₃Nb behavior,
B, E - grain boundary crack behavior,
C, F - intergranular fracture behavior at 650°C, 686MPa.

Fractography Observation

Optical microscopy observation on longitudinal sections of stress rupture samples shows extended elongated grain structure of Mg containing alloy 718M with grain boundary cavities because of high stress rupture ductility (see Fig. 7). In contrast, very small elongation of grains and scarce grain boundary cracks appear in Mg free alloy 718 (see Fig. 7E)

SEM fractographic study of the various grain size stress rupture samples shows that when Mg is present the coarse and mixed grain samples have many more dimples on the intergranular fracture surfaces than in Mg free samples (compare Fig. 7C and F). As the grain size decreases, the intergranular fracture mode changes into a partially transgranular mode. The change from intergranular to transgranular fracture occurs in the mixed grain stress rupture samples of Mg containing alloy 718M while in Mg free alloy 718 transgranular fracture is never totally achieved even in fine grain stress rupture samples.

Auger Analysis

Semi-quantitative Auger analysis on intergranular fracture surface of Mg containing alloy 718M samples shows the profile of Mg content distribution at the grain boundary regions. It can be seen from Fig. 8 that the concentration of Mg at grain boundaries characterizes an equilibrium segregation and Mg has been further concentrated at grain boundaries during long time stress aging time, i.e. Mg content at grain boundaries increases after 526 hrs stress aging at 650°C/686 MPa. After AMS 5596C heat treatment and 526 hrs stress aging conditions, the concentration of Mg decreases gradually away from the grain boundary. The gradual change of Mg content in the region of grain boundary shows that Mg does not exist in grain boundary phases; otherwise, the Mg content would sharply change across the grain boundary.

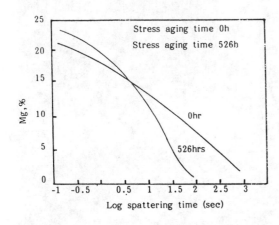

Fig. 8. Grain boundary segregation behavior of Mg in alloy 718M before and after stress aging at 650°C, 686 MPa.

Segregation Effect

Because the degree of Nb segregation remaining after any conversion process results in non-uniform grain sizes and affects the δ–Ni_3Nb solvus temperature, the heat treated samples were given the "TAG" heat treatment [7] to detect residual segregation of Nb. It was found that the Mg containing alloy 718M samples appear to have less segregation than Mg free alloy 718.

DISCUSSION

The great advantage of adding Mg to alloy 718 is that Mg can greatly increase 650°C tensile and stress-rupture ductility and also increase smooth and notch stress-rupture lives remarkably. Mg containing alloy 718M is free from stress-rupture notch sensitivity, which is important for material used for disk application. In addition, another benefit of Mg addition in alloy 718 is that Mg improves creep and fatigue interaction properties (LCF or cycle stress-rupture), so necessary for turbine disk applications.

The diffusion to and segregation of Mg at grain boundaries changes grain boundary behavior. Magnesium addition can change grain boundary precipitation of δ–Ni_3Nb from continuous plate-like form to discrete globular shapes, and retards intergranular crack growth as schematically shown in Fig. 9. This grain boundary precipitation behavior was also confirmed in nickel-base [5] and iron-base superalloys [8]. The amount of δNi_3Nb precipitation at the grain boundaries depends on grain size, amount of Mg, and heat treatment.

Because Mg is concentrated at the grain boundaries, Mg cannot severely affect precipitation behavior in bulk grains. As a result, the amount of strengthening phase ($\gamma' + \gamma''$) in grains is not affected by Mg addition in alloy 718 and is nearly constant ($\sim 14\%$ $\gamma' + \gamma''$) in both alloy 718 and 718M at all grain sizes.

Concentration of Mg at grain boundaries plays a strengthening role on grain boundaries. It allows more deformation in bulk grains before intergranular fracture occurs in stress-rupture tests. From the viewpoint of creep, Mg prolongs the secondary creep stage and develops a tertiary creep stage in nickel-base and iron-base superalloys [8] which should raise both stress rupture ductility and failure life. Ductile stress-rupture fracture surfaces with much more dimples should appear in Mg containing alloy 718M samples.

It appears Mg can reduce the Nb segregation in cast alloy 718 ingots which allows for material homogeneity improvement during conversion practice. Detail study of Mg effect on segregation behavior will be discussed in other papers [9].

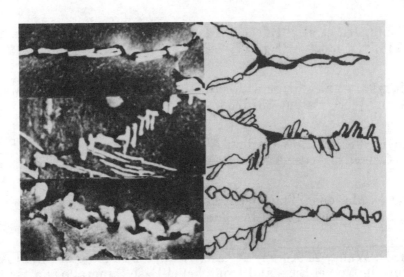

Fig. 9. Grain boundary σ-Ni$_3$Nb behavior and intergranular crack propagation mode suggested.

CONCLUSIONS

1. Mg increases 650°F tensile and stress-rupture ductilities remarkably but has little effect on tensile strength; however, the smooth and notch stress-rupture lives both increase considerably and Mg containing alloy 718 M is free of stress-rupture notch sensitivity at any grain size.

2. Mg improves 650°C creep and fatigue interaction properties (LCF or cyclic stress-rupture) at any grain size.

3. Mg does not appear to have effect on the wt.% of strengthening phase ($\gamma' + \gamma''$). However, the wt.% of δ–Ni$_3$Nb is much greater in fine grain alloy 718M indicating that Mg affects the precipitation of δ–Ni$_3$Nb at grain boundaries.

4. Mg plays a role of equilibrium segregation at grain boundaries and changes grain boundary δ–Ni$_3$Nb morphology from continuous plate-like form to discrete globular shapes, producing a retardation effect on intergranular fracture which simultaneously increases stress-rupture ductility and prolongs failure life.

5. Mg can produce ductile stress rupture fracture and hasten the change from intergranular fracture mode into partially transgranular mode.

6. Mg may appear to be beneficial to improve Nb segregation in alloy 718.

ACKNOWLEDGEMENTS

The authors are grateful to the Daye Steel Works for melting and forging alloys in plant and conducting time consuming stress-rupture tests. Special thanks go to Shimu Zhou and Ziufeng Cheng for experiment arrangement and Yingzhi Zhu for conducting microchemical analyses of samples in the R&D of Daye Steel Works.

REFERENCES

1. W. H. Couts, Jr., et al., "Effect of Magnesium as an Alloying Element in Inconel 718," (Report AEML-TR-7s-76, 1971).

2. D. R. Muzyka, et al., "Process for Making Nickel Base Precipitation Hardenable Alloys," (U.S. Patent 3575734, April 20, 1971.

3. J. M. Moyer, "Extra Low Carbon Alloy 718," in Proceedings of Superalloys, 1984, M. Gell et al., eds. AIME (1984) 443-454.

4. Z. Xr and P. Ma, eds., The effect and Control of Trace Elements in Superalloys, (Beijing, Metallurgical Press, In Chinese, 1987.

5. G. Chen, D. Wang, Z. Xu and X. Xie, et al., "The Role of Small Amounts of Magnesium in Nickel-Base and Iron-Nickel-Base Superalloys after High Temperature Long Time Exposures," in Proceedings of Superalloys 1984, M. Gell et al., eds. AIME (1984) 611-620.

6. Z. Xu, X. Xie and G. Chen, et al., "Mg Microalloyed CrH (6°)" in Proceedings of the Effect and Control of Trace Elements in Superalloys, Z. Xu and P. Ma, eds., Metallurgical Press, Beijing (1987) 147-153.

7. J. A. Corrado, W. H. Couts, Jr. and J. F. Radavich, "A Microstructural Test for Chemical Homogeneity in Inconel 718 Billet," TMS Technical Paper No. A86-34, (1986).

8. X. Xie, et al., "Effect of Small Amount of Magnesium on High Temperature LCF Behavior in Iron-Base and Nickel-Base Superalloys," in Proceedings of Low Cycle Fatigue and Elasto-Plastic Behavior of Materials, K. T. Rie, ed., Elsevier Applied Science (1987) 719-723.

9. X. Xie and Z. Xu, et al., "Magnesium Effect on Segregation Behavior in Alloy 718," to be published.

Torsional creep of

Alloy 617 tubes at high temperature

H.J. Penkalla, F. Schubert, H. Nickel

Nuclear Research Centre, KFA-Jülich
Institute for Reactor Materials,
P.O. Box 19 13, 5170 Jülich, Germany

Abstract:
The multiaxial creep of Alloy 617 tubes at temperatures above 900 °C is evaluated by theoretical calculations and by experimental tests with the main emphasis to torsion loading. The stress-strain-time behaviour can be discribed satisfactorally by the v. Mises theory and the use of Norton's typ creep law.

I. Introduction

With the increase of application temperatures for components in power plants up to 1000 °C the main concern for the design and analysis of the component behaviour is the time dependent materials behaviour. To describe the stress-strain-time relationship data from uniaxial creep tests are used. From these test results structural design values such as creep strain limits and creep rupture strength are derived. The mathematical description of the strain-time behaviour is based on experimentally obtained creep curves. The inelastic analysis of the component gives the base for the evaluation of component behaviour under complex loading conditions. The mathematical description of the strain-time relationship, obtained from the uniaxial test, is transfered to a three-dimensional formulation /1/. The aim of this presentation is to investigate and verify or modify the above mentioned formalism for multiaxial creep.

2. Principles

The investigation of creep in tubes is part of a general evaluation programme of materials development for high temperature components of an intermediate heat exchanger (IHX) in a nuclear process heat plant. The loading conditions for the test reflect mainly upset and emergency conditions. For the theoretical description of multiaxial creep the theory of invariances, in which the v. Mises hypothesis and the Norton's creep law are integrated, is applied.

2.1 Test pieces

The present candidates for the high temperature components in a nuclear process heat plant are Alloy 800 (X10NiCrAlTi 32 20) and Alloy 617 (NiCr22Co12Mo). Alloy 617 is a commonly used material for combustion chambers, hot ducting and piping in gas turbines as well as for tubing of heat exchanging components and shows the highest creep rupture strength of the two candidates.

tube dimensions		spec. form (tube test)
IHX – tube:	22 x 2.2 mm	
reformer tube:	120 x 10 mm	
buckling test:	40 x 3,5 mm	

mean chemical composition (wt–%)								
alloy	Fe	Ni	Cr	Al	Ti	Mn	Co	Mo	C
Alloy 800	bal.	32	22	0.4	0.4	1.0	–	–	0.06
Alloy 617	–	bal.	22	0.9	0.4	0.2	12	9	0.06

Table 1: Table of specimens form and investigated materials.

2.2 Loading and temperature ranges and material behaviour

The material behaviour under mechanical loading is principally divided into elastic, plastic (time independent), creep (time dependent).
Fig. 1 shows schematically the separation of the material behaviour dependent on the loading and temperature. The different ranges overlap partly. The loading conditions discussed here lead to a time dependent creep, so only the range III must be considered.

An analysis of the operation and emergency conditions of heat exchanging components provides the following conditions: temperature 900 °C; stress intensity 30 MPA; strain rates 3×10^{-2} % min^{-1}. Fig. 2 shows the stationary creep rate as a function of stress, obtained in the uniaxial creep tests, and illustrates the range of the considered loading condition.

2.3 Uniaxial creep

A scatterband evaluation of test results from more than 300 creep tests showed that Norton's creep law (power law creep) /3/ is a good fit to describe the stationary creep. Fig. 2 shows this result for one melt. In Norton's creep law the stationary creep rate $\dot{\varepsilon}$ is a function of σ as follows

$$\dot{\varepsilon} = k'(\sigma/E)^n = k\sigma^n \text{ with E as Youngs Modulus.} \qquad (1)$$

2.4 Multiaxial creep

The transfer of the uniaxial creep law to multiaxial exposure leads to a static indefinite problem. The solution of this problem requires additional postulates and a materials law. For the further considerations, the following postulates are important:
- Constancy of volume:
  This is equivalent with the application of the stress deviator for the calculation of the deformation
- Compatibility:
  This means the continuity of the strain rate distribution
- Invariance of coordinates:
  The result of the calculation must be independent of the selection of coordinates
- No hardening rules:
  There should be always an equilibrium between strain rate and the true stresses in the component

- Isotropy:
  The materials behaviour is the same in all dimensions of the component
- Materials law:
  For stationary creep Norton's creep law is assumed to apply.

For a given loading σ_{ij} is the stess tensor and σ^*_{ij} the related deviator. Reflecting the assumed postulate and using Norton's creep law, the tensor of stationary creep rate is given by

$$\dot{\varepsilon}_{ij} = \frac{3}{2} k \sigma_v^{n-1} \cdot \sigma^*_{ij} \qquad (2)$$

with σ_v as the deviatoric stress according the v. Mises hypothesis. The tube geometry suggests cylindrical coordinates, therefor, the compatibility leads to

$$\frac{d \sigma_r}{d r} = \frac{1}{r} (\sigma_u - \sigma_r) \qquad (3)$$

with σ_r as radial stress component, σ_u the circumferential stress component and r the radius. In thin walled tubes, the stress distribution across the wall can be replaced by membrane stresses, so that the calculation is static definite without the application of compatibility /4, 5, 6/.

Under additional estimation of the elastic strain proportion, the complete description of multiaxial strain rate follow the equation:

$$\dot{\varepsilon}_{ij} = \frac{1+\nu}{E} \dot{\sigma}^*_{ij} + \frac{1-2\nu}{E} S \cdot \delta_{ij} + \frac{3}{2} k \sigma_v^{n-1} \sigma^*_{ij} \qquad (4)$$

with ν as Poisson's ratio, S the trace of the stress tensor, δ_{ij} the Kronecker symbol.

2.5 Discussion of typical tube loadings

In power plants tubes are stressed by primary loads such as internal pressure, tension, bending and torsion. Except for the shear stresses caused by a torque all stresses are directed in the cylindrical main axis of the tube; therefore, an analytical description is possible and quite simple. Secondary stresses due to temperature transients are mostly in the axial direction. In helix tubes, e.g. in helix heat exchanger component, secondary stresses can be torsional shear stresses.

The simplest case of loading is the unique tensile stress in axial direction. The result of the calculation is identical with the normal creep curve. The stress tensor and stress deviator are

$$\sigma_{ij} = \begin{pmatrix} 0 & 0 & 0 \\ 0 & 0 & 0 \\ 0 & 0 & \sigma \end{pmatrix} \quad \text{and} \quad \sigma^*_{ij} = \begin{pmatrix} -1/3\sigma & 0 & 0 \\ 0 & -1/3\sigma & 0 \\ 0 & 0 & 2/3\sigma \end{pmatrix} \qquad (5)$$

and the deviatoric stress is given by $\sigma_v = \sigma$.

The strain rates are:

$$\dot{\varepsilon}_z = k \sigma^n; \quad \dot{\varepsilon}_r = \dot{\varepsilon}_u = -1/2 k \sigma^n \qquad (6)$$

For tubes under internal pressure p the stress tensor and deviator is given by

$$\sigma_{ij} = \begin{pmatrix} \sigma_r & 0 & 0 \\ 0 & \sigma_u & 0 \\ 0 & 0 & \sigma_z \end{pmatrix} \quad \text{and} \quad \sigma^*_{ij} = \begin{pmatrix} \sigma_r - 1/3 S & 0 & 0 \\ 0 & \sigma_u - 1/3 S & 0 \\ 0 & 0 & \sigma_z - 1/3 S \end{pmatrix} (7)$$

with $\sigma_r = 0$, $\sigma_u = \Delta p r/d$, $\sigma_z = 1/2 \Delta p r/d$, $S = \sigma_r + \sigma_u + \sigma_z$ and $\sigma_v = \sqrt{3}/2 \ \Delta p \ r/d$

Under unique internal pressure tubes show no axial strain rate, this means the plane strain condition. This leads to the strain rates:
$$\dot{\varepsilon}_z = 0; \text{ and } \dot{\varepsilon}_u = -\dot{\varepsilon}_r = (\sqrt{3}/2)^{n+1} k (\Delta p r/d)^n \tag{8}$$

The loading of a tube by a torque represents a plane stress condition. The stress tensor and the deviator are indentically is given by
$$\sigma_{ij} = \sigma_{ij}^* = \begin{pmatrix} 0 & 0 & 0 \\ 0 & 0 & \tau \\ 0 & \tau & 0 \end{pmatrix} \tag{9}$$

The shear stresses are eigenvalues of the tensor and provide after principle axis transformation the tensor
$$\sigma_{ij}^t = \begin{pmatrix} 0 & 0 & 0 \\ 0 & -\tau & 0 \\ 0 & 0 & \tau \end{pmatrix} \tag{10}$$

The deviatoric stress is then give by $\sigma_v = \sqrt{3}\,\tau$, and the calculated strain rates are:
$$\dot{\varepsilon}_{uz} = \dot{\varepsilon}_{zu} = \frac{\sqrt{3}}{2}^{n-1} k \tau^n \tag{11}$$

Combinations of these types of loading are of special interest for the evaluation of these constitutive equations. In all cases in which the stress components are parts of the trace of the stress tensor, e. g. under tension load and internal pressure, a combination of these loads lead to a redistribution and to an alteration of the stresses and strain rates.

Superpositions with shear stress components, caused by an additional torque, show no redistribution but only a simultaneous increase of all strain rate components caused by the increase of the deviatoric stress.

As an example, the circumferential strainrate of a tube under constant internal pressure is given by
$$\dot{\varepsilon}_u = \left(\frac{\sqrt{3}}{2}\right)^{\frac{n-1}{2}} k \left(\Delta p \frac{r}{d}\right)^n \tag{12}$$

Under an additional tensile load the three main strain rates show a behaviour as shown in Fig. 3. While the radial and axial strain rates $\dot{\varepsilon}_r$ and $\dot{\varepsilon}_z$ increase, the circumferential strain rate $\dot{\varepsilon}_u$ increases in first stage and decreases at higher axial stresses.

A tube specimen under internal pressure and a superimposed torque show another behaviour of the strain rates. The axial strain rate remains zero and the circumferential strain rate increases directly dependent on the increase of the deviatoric stress. Additionally the shear strain rate $\dot{\varepsilon}_{uz}$ occurs (Fig. 3b).

Furthermore such loading conditions are of interest, in which one portion is a primary loading and another portion is caused by secondary stresses. A combination of constant tensile stress in axial direction and a constant shear strain cause a relaxation of the torsional shear stress, where

- the axial strain rate decreases in the same way as the deviatoric stress decreases
- the relaxation rate of the shear stress increases with an increase of the axial tensile stress.

In Fig. 4a the behaviour of the axial strain rate dependent on time under constant tensile stress and relaxing shear stress is shown. With the time dependent decrease of the shear stress the axial strain rate decreases to the strain rate under unique constant tensile stress. Fig. 4b shows the relaxation behaviour of the shear stress superimposed with different constant tensile stresses.

These theoretical investigations are the basis for the experimental work.

3. Experimental Investigations

3.1 Test programme

The investigations of multiaxial creep on tubes are based on an extended materials evaluation program for HTR components in which a certain number of high temperature creep resistant alloys are subjected to standardised tests /7, 8/. This programme provided the design values. The experimental examination concerning multiaxial creep should help to answer questions related to inelastic analysis and simplified approximations. This results in a test matrix given in Table 2. With the exception of internal pressure all loadings were applied either stationally (stat), cyclically (cycl.) or relaxing (rel). Some of the test combinations were carried out with the same tube specimen in order to avoid the effects of scatter in the materials parameters.

number of tests	internal pressure	tension			torsion		
		stat	cycl	rel	stat	cycl	rel
6	x						
5		x					
12		x	x				
4	x	x					
3	x	x	x				
2	x			x			
2					x		
6					x	x	
3	x				x		
2	x				x	x	
2	x						x
3		x			x		
3		x			x	x	
2		x					x
2		x	x				x
2	x	x			x		

Table 2: Test program to investigate the multiaxial creep behavior

3.2 Experimental Realisation

The experiments for multiaxial creep were performed on typical tubes as used for construction of the prototype heat exchanger and methane steam reformer. The dimensions of the tube specimens are listed in Table 1. The mechanical loadings were applied using servo hydraulic Instron test machines with load capacity of 100 kN for tensile and 500 Nm for torsion for heat exchanger tubes and 500 kN tensile load for reformer tubes. The tubes were heated by multizone resistance furnaces. The internal pressure was supplied by a gas inlet through the flanges at the ends of the tube specimens and regulated by a pressure valve.

The strains in axial direction were measured by the movement of the piston and calibrated by specific tests. The angle of torsion of the specimen was determined by the twist angle of piston and by calibration tests. A continuous measurement of the radial deformation was not carried out because of the distortion of the continuous temperature profile. The radial deformation was measured after periodical interruptions of the tests.

4. Results and discussion

The numerous test results /9, 10, 11/ of this multiaxial creep programme are now discussed with the main emphasis on torsional creep.

4.1 Constant tensile and torsion loading

At loading the shear strain rate is described by equation 11. The results in Fig. 5 represent the shear creep strain for the shear stresses of 20 MPa and 13.3 MPa at a test temperature of 950 °C The analysis of this curves provide a value of the Norton's exponent n of 5.7; typical values for Alloy 617 obtained in uniaxial creep tests are between 4.5 to 6.5.
Under combined tension and torsion the axial strain rate and the shear strain rate increases in direct proportion to the increase of the v. Mises deviatoric stress

$$\sigma_v = \sqrt{\sigma_z^2 + 3\tau^2} \qquad (13)$$

and the relation ship between tensile and shear stress

$$\beta = \sigma_z / \sqrt{3}\tau \qquad (14)$$

results in an increase of shear strain rate proportional to $(1 + \beta^2)^{\frac{n-1}{2}}$ and an increase of axial strain rate proportional to $(1 + \beta^{-2})^{\frac{n-1}{2}}$.

Fig. 6 illustrates the increase of the axial strain rate dependent on β. The measured points (dots) fit well in the range of values, compared with the calculated curves.

4.2 Relaxing of torsion under constant tension

A constant twist angle on the tube results in the temperature range of creep to a relaxation of the shear stress combined with the constant tensile stress with application of equation the relaxation of the shear stress behaves according to

$$\dot{\tau} = -(3/2)^2 \, (1 + \beta^2)^{\frac{n-1}{2}} \quad k \, E \, \tau^n \qquad (15)$$

This means with continuing constant tensile stress the relaxation accelerate.
The experimental obtain shear stress relaxation (Fig. 7a) corresponds with an decrease in the axial strain rate (Fig. 7b).

4.3 Superimposed cyclic load

The superimposition of a cyclic stress in axial direction of the form

$$\sigma = \bar{\sigma} + \tilde{\sigma} \sin\omega t \qquad (16)$$

results in a increase of axial strain rate represented by the factor:

$$f = 1/t_c \int_{t}^{t+t_c} (1 + \alpha \sin\omega t)^n \, dt \qquad (17)$$

whereby is defined as $\alpha = \tilde{\sigma}/\bar{\sigma}$.

Fig. 8 summarizes the results of a test under tensile pulsation stress combined with relaxing shear stress. On the left hand the relaxation curve of the shear stress once combined with cyclic tensile stress and once combined with constant tensile stress are given. The right hand graph shows the related axial strain behaviour. The postulation of volume constancy requires that under accelerated strain rate, caused by the superimposed cyclic tensile stress, the shear strain rate and the relaxation rate of shear stress decreases.

4.4 Test with load reversals

The calculations of the test were made by application of the v. Mises theory and by Norton's creep law and can describe only multiaxial stationary creep. In order to calculate loading conditions containing changes in load level and inversions of load a consideration of hardening rules is necessary. For the verification of well known rules or to create new rules which fit the component behaviour better, some tests with stepwise chances and inversions of stresses strain rates under different combinations of loadings were carried out.

As an example Fig. 9 shows the torsional shear strain in dependence time for a test with periodical inversions of torque. The calculation based on Norton's creep law without hardening terms results in creep strain after the 2nd. cycle of zero. As a matter of fact the material shows a softening of the torque. After further inversions the amount of softening becomes less and comes to a steady value.

Fig. 10 shows the results of a test with periodical inversion of torque combined with a constant tensile load. As expected the shear strain is higher than in the test before. Meantime the softening increases with each load inversion. The diagram below shows the axial strain of the tube, which is influenced marginally by the behaviour of the shear strain.

5. Final remarks

Multiaxial creep of Alloy 617 tubes can be described mathematically by the v. Mises' theory using the Norton's creep law as the constitutive equation. The theoretically derived formulas for the stress-strain-time behaviour give an acceptable approximation to the obsered deformation behaviour of tubes under multiaxial loading conditions.

Combinations of tension and torsion loadings results depending on the kind of combination, in an acceleration either of the shear stress relaxation or the axial creep strain rate.

Acknowledgment

This investigation was carried out for the German HTGR project PNP under the sponsorship of The Federal Ministry for Science and Technology, The Federal Ministry for Reactor Safety and environment, and The Ministry of Economics, Small Business and Technology of North Rhine Westphalia.

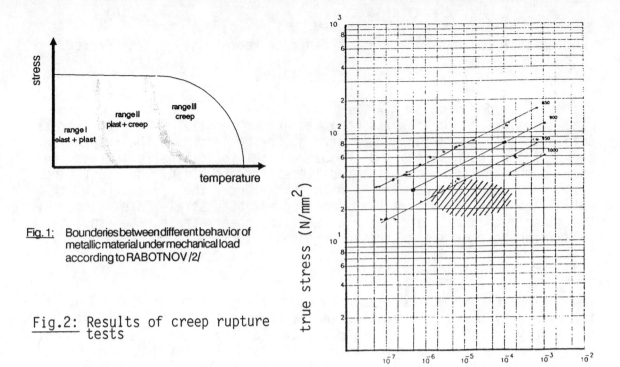

Fig. 1: Bounderies between different behavior of metallic material under mechanical load according to RABOTNOV /2/

Fig. 2: Results of creep rupture tests

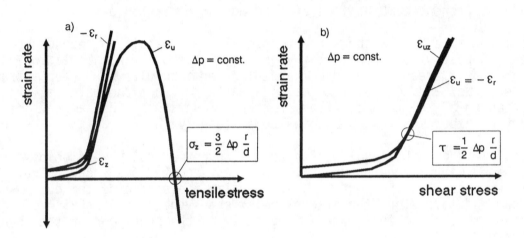

Fig. 3: Redistribution of strain rates under different loading conditions:
— a) internal pressure and additional axial tension
— b) internal pressure and additional torque

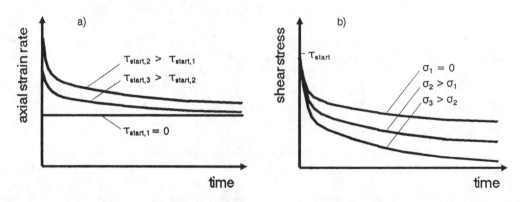

Fig. 4: Interaction between the constant tensile stress and a relaxing torsional shear stress:
— a) Influence of the relaxing shear stress on the axial strain rate
— b) Accelleration of the relaxation rate of the shear stress under konstant tensile stress

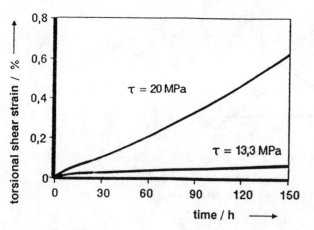

Fig. 5: Creep strain of INCONEL 617 tubes loaded by torsion

Fig 6: Increase of the axial strain rate for IHX-tubes under superimposed torsional loading

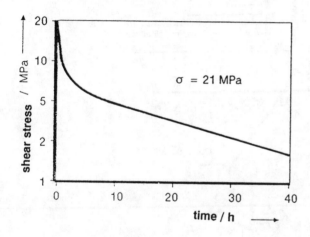
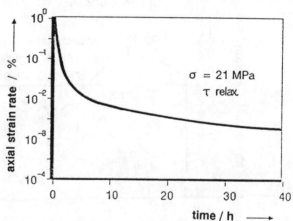

Fig. 7: Relaxing shear stress under superimposed constant axial stress (a) and corresponding axial strain rate (b), INCONEL 617, IHX-tubes

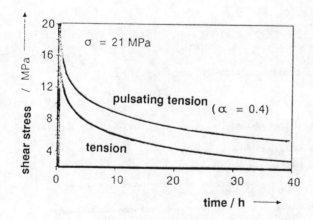
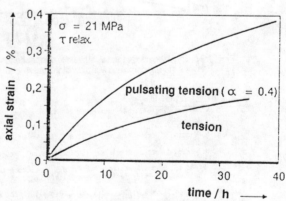

Fig. 8: Relaxing shear stress under superimposed tension and pulsating tension respectively (a) and related strain rate (b), INCONEL 617, IHX-tubes

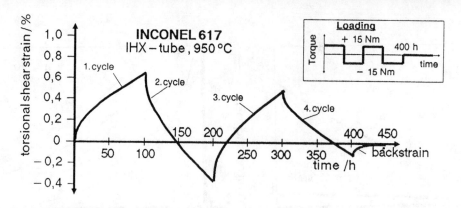

Fig. 9: Torsional shear strain in denpendence on time as result of a test with periodicaly inversed torque

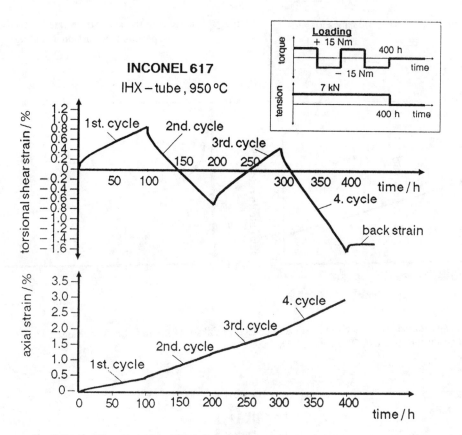

Fig. 10: Torsional shear strain and axial strain in denpendence on time as result of a test with periodicaly inversed inversed torque combined with a constant tensile load.

6. Literature

/1/ R. v. Mises: "Mechanik der festen und flüssigen Körper im plastisch deformablen Zu- stand", Königl. Ges. der Wiss., Göttingen, 1913

/2/ Y. N. Rabotnov: "Creep Problems in Structural Members", North Holland Publishing Com- pany, Amsterdam 1969

/3/ F. H. Norton "Creep of Steels at High Temperature", McGraw Hill, New York, 1929

/4/ K. Franzke, H.-J. Penkalla, M. Rödig, F. Schubert, H. Nickel: "Untersuchungen zum Kriechverhalten von Rohren aus X10NiCrAlTi 32 20 (Nicrofer 32 20) im Anlieferungszustand bei mehrachsiger Belastung", Jül-2127, 1987

/5/ G. Breitbach, S. Kragel, M. Rödig, H.-J. Penkalla: "Zur Berechnung von Kriechverformungen und Spannungen in dickwandigen Rohren", Jül.-Spez. 373, 1986

/6/ F. K. G. Odqvist, J. Hult: "Mathematical Theory of Creep and Creep Rupture", Oxford, 1966

/7/ M. Rödig, H.-J. Penkalla, W. Hannen, H. Hellwig: "Versuche zur komplexen Beanspruchung von Rohrabschnitten bei Tempera- turen oberhalb 800 °C" VDM-Tagung "Werkstoffprüfung 87", Dez. 1987, Bad Nauheim

/8/ H. Nickel, F. Schubert: "Hochtemperaturwerkstoffe für Rohrleitungen in Kernkraftwerken", 3R international 20, 8, 1981, S. 396-404

/9/ H.-J. Penkalla, M. Rödig, M. Hoffmann: "Übertragbarkeit von zeitabhängigen Kennwerten und von Stoffgesetzen auf mehrachsige Belastungsfälle", 10. MPA-Seminar, 1984, Bd. 2

/10/ M. Rödig, H.-J. Penkalla, K. Franzke, F. Schubert, H. Nickel: "Untersuchungen an Rohrproben aus INCOLOY 800 H bei einachsiger und mehrachsiger Beanspruchung", Jül.-Bericht 1975, 1985

/11/ H.-J. Penkalla, F. Schubert, H. Nickel: Kriechverhalten von Rohren aus X10NiCrAlTi 32 20 und NiCr 22 Co 12 Mo unter mehrachsiger statischer und zyklischer Belastung; 13. MPA-Semi- nar, Oktober 1987

IDENTIFICATION OF MECHANISMS RESPONSIBLE FOR

DEGRADATION IN THIN-WALL STRESS-RUPTURE PROPERTIES

Mehmet Doner and Josephine A. Heckler

Allison Gas Turbine Division, General Motors Corporation

P. O. Box 420, Speed Code T-27

Indianapolis, IN 46206-0420

Abstract

A series of critical experiments was performed on single crystal CMSX-3 and equiaxed grain Mar-M246 specimens to identify the mechanisms responsible for the degradation in stress-rupture properties of 0.020 in. thick mini-flats machined from airfoils compared with 0.250 in. diameter standard test bars. Both materials exhibited approximately a factor of 3X life degradation at a stress level of 20,000 $lb/in.^2$, when uncoated airfoil mini-flats were tested in air.

To determine the influence of specimen geometry on test results, a series of tests was conducted on CMSX-3 alloy using 0.020 in. thick mini-flats and 0.020 in. wall-thickness cylindrical hollow specimens. Both types of specimens were machined from 5/8 in. diameter bars. The results obtained indicated no noticeable difference in stress-rupture lives, suggesting that, in this material, specimen geometry does not influence the test results.

The next series of tests was conducted using aluminide coated CMSX-3 airfoil mini-flats tested in air and uncoated airfoil mini-flats tested in high purity argon. The stress-rupture lives obtained were equivalent to those obtained on 0.250 in. diameter baseline specimens, suggesting that, in this material, the life degradation observed in airfoil mini-flats is primarily due to environmental effects.

The last series of tests was conducted using aluminide coated equiaxed grain Mar-M246 airfoil mini-flats. No improvement in stress-rupture lives was obtained compared with uncoated airfoil mini-flats tested in air, suggesting that the primary life degradation mechanism in this material is related to the behavior of grain boundaries in thin sections.

Introduction

An important consideration in gas turbine airfoil design is the effect of section thickness on creep and stress-rupture properties of the turbine blade and vane alloys. Past experience with equiaxed grain nickel- and cobalt-base superalloys indicates a reduction of a factor of 3 to 5 in stress-rupture properties of thin-wall castings (0.020-0.025 in. section thickness) compared with standard 0.250 in. diameter test bars. A recent study by the authors (1) showed a similar reduction in stress-rupture properties of uncoated single crystal mini-specimens machined from thin-wall hollow airfoil castings and tested in air.

The objective of the present study, therefore, was to ascertain which specific mechanisms were responsible for the observed thin-wall effects in stress-rupture properties of the equiaxed and single crystal castings. A series of critical experiments was conducted with the aim of separating various contributions due to microstructural, environmental and geometric factors. The present study is an extension of the work previously reported by the authors (1).

Background and Experimental Approach

Figure 1 provides a comparison of the stress-rupture properties of 0.020 in. thick single crystal CMSX-3 mini-flat specimens machined from thin-wall hollow airfoils with those of 0.250 in. diameter standard specimens machined from 5/8 in. diameter bars (1). It is seen that at stress levels less than 30,000 $lb/in.^2$, the stress-rupture properties of thin-wall specimens show considerable degradation. At 20,000 $lb/in.^2$, for example, the stress-rupture lives of thin-wall specimens are reduced by a factor of approximately 3. A similar behavior is displayed in Figure 2 by the equiaxed grain Mar-M246 alloy. At a stress level of 20,000 $lb/in.^2$, the stress-rupture lives of 0.020 in. thick mini-flat specimens are reduced, in this case, by a factor of approximately 2.5.

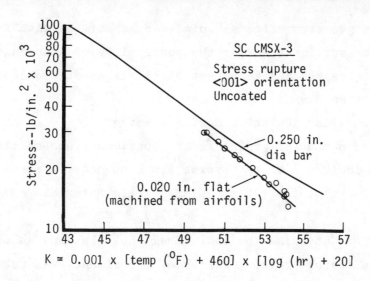

Figure 1: Larson-Miller representation of the stress-rupture properties of 0.020 in. thick mini-flat specimens machined from hollow CMSX-3 airfoils. The solid line represents the data obtained on 0.250 in. diameter bars (1).

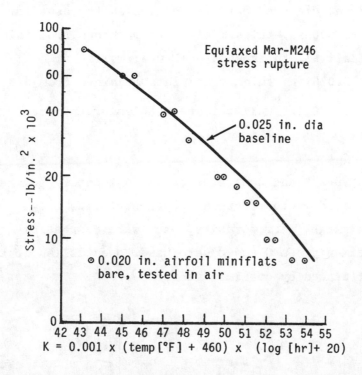

Figure 2: Larson-Miller representation of the stress-rupture properties of 0.020 in. thick mini-flat specimens machined from equiaxed grain Mar-M246 airfoils. The solid line represents the data obtained on 0.250 in. diameter bars.

A number of mechanisms are expected to contribute to the observed thin-wall degradation phenomenon in single crystal and equiaxed grain alloys:

(1) differences in the degree of microsegregation and/or microshrinkage/microporosity between thin and thick section castings

(2) specimen geometry effects: mini-flat specimens with free edges versus round specimens and/or the possibility of inducing greater bending stresses due to potential misalignment problems with mini-flat specimens

(3) oxidation/alloy depletion during testing

(4) decreased number of grains in the specimen cross-section and the increased incidence of transverse grain boundaries across the specimen thickness (mini-flats for equiaxed grain materials only)

In order to determine the individual contributions associated with the mechanisms listed above, a series of experiments was performed using four different specimen geometries illustrated in Figure 3. These specimens were fabricated as follows:

(a) Type A: 0.250 in. diameter test section standard specimens machined from 5/8 in. diameter bars (to establish the baseline behavior)

(b) Type B: 0.020 in. thick mini-flats machined from thin-wall (0.025 in. nominal wall thickness) hollow airfoils

(c) Type C: 0.020 in. thick mini-flats machined from 5/8 in. diameter

(d) Type D: 0.020 in. wall thickness hollow round standard size specimens

Specimen types B and C were selected to identify to what extent solidification related microstructural characteristics (e.g., microsegregation, microshrinkage, microporosity, etc) were responsible for thin-wall effects. Specimen types C and D were used to establish if the test results were influenced by specimen geometry.

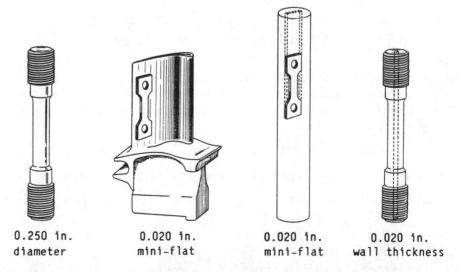

Figure 3: Specimen geometries used to investigate the section thickness effects on stress-rupture properties.

The first series of tests was conducted in air on uncoated specimens Hence, environmental effects, if present, were included in the test results. In another series of experiments, the environmental effects were eliminated while the specimen geometry was held constant. This was accomplished by conducting the following tests:

o   coated mini-flat (Type B) specimens tested in air
o   uncoated mini-flat (Type B) specimens tested in high purity argon

The test temperature range was 1650 to 2000°F for the CMSX-3 alloy and 1520 to 1950°F for the Mar-M246 alloy. The stresses used were in the range of 13 to 50 ksi for the CMSX-3 and 8 to 80 ksi for the Mar-M246 alloy. The results are presented in terms of stress versus the Larson-Miller parameter K.

Results and Discussion

Figure 4 provides a comparison of the stress-rupture lives of uncoated 0.020 in. thick mini-flats (Type C) with those of uncoated 0.020 in. wall cylindrical specimens (Type D). Both types of specimens in these tests were machined from 5/8 in. diameter bars and tests were conducted in

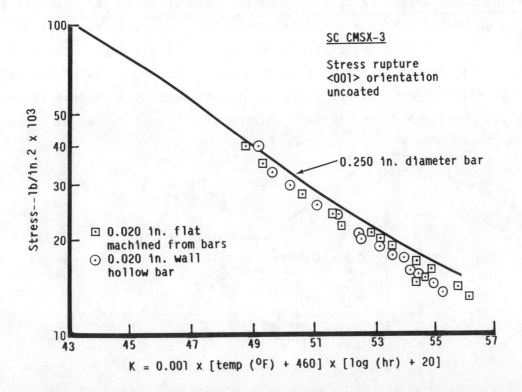

Figure 4: Stress-rupture properties of 0.020 in. thick mini-flat and 0.020 in. wall thickness hollow cylindrical specimens machined from 5/8 in. diameter bars. The solid line represents the data observed in 0.250 in. diameter bars (1).

air. No appreciable difference exists between the behavior of these specimens, implying that the specimen geometry is not responsible for the observed thin-wall degradation in the CMSX-3 alloy.

A comparison of the Type C and D specimen results described above with the results obtained in Type B specimens (i.e., mini-flats machined from airfoils) is presented in Figure 5. It is seen that at stresses below 20 ksi (and at test temperatures above 1875°F), the reduction in stress-rupture lives of Type C and D specimens appears to be less than that observed for the mini-flats machined from airfoils (Type B specimens). Implication here is that differences may exist between thin-flats machined from airfoils (Type B) and those machined from 5/8 in. diameter bars (Type C), in terms of either microstructural characteristics or surface characteristics (i.e., as-cast versus ground surface). More will be said about this observation in the following paragraphs.

The next series of tests on CMSX-3 alloy was conducted using aluminide coated mini-flat specimens machined from airfoils (Type B). The results are plotted in Figure 6. The stresses for the coated mini-flat specimens in Figure 6 were calculated assuming the load was carried by the unaffected area only, i.e., the coating thickness was not included in the cross-sectional area. It is seen that when presented in this manner, the results on coated mini-flats become essentially identical to those obtained on 0.250 in. diameter standard specimens. This implies that the observed degradation in stress-rupture lives of the mini-flats machined from airfoils (Type B) is due primarily to environmental effects (Mechanism 3). To further substantiate this observation, a series of tests

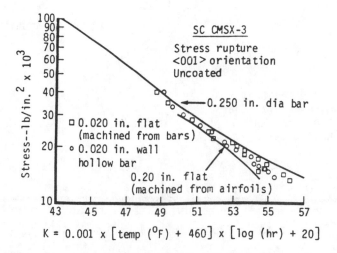

Figure 5: Comparison of stress-rupture properties of 0.020 in. thick CMSX-3 specimens of various geometries. Solid lines represent data obtained on 0.020 in. thick mini-flats machined from airfoils and on 0.250 in. diameter bars (1).

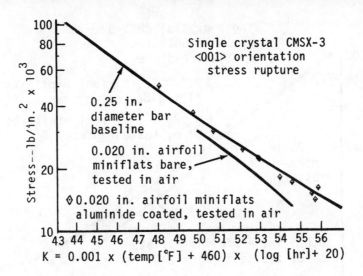

Figure 6: Stress-rupture lives of aluminide coated 0.020 in. thick mini-flats machined from hollow CMSX-3 airfoils. Solid lines represent data obtained on uncoated 0.020 in. thick mini-flats and 0.250 in. diameter bars. All tests are conducted in air.

were, then, conducted in high purity argon using uncoated mini-flat specimens (Type B). The results are presented in Figure 7 along with the data displayed in Figure 6. These results indicate essentially no difference between the stress-rupture lives of the uncoated mini-flats tested in high purity argon and the aluminide coated mini-flats tested in air. Further, the stress-rupture lives of both types of specimens are about the same as those of 0.250 in. diameter baseline specimens. These observations clearly demonstrate that for the CMSX-3 alloy it is the environmental effects (Mechanism 3) that are primarily responsible for the observed degradation in the stress-rupture lives of 0.020 in. mini-flats machined from airfoils. The observation that the degree of degradation in stress-rupture lives of Type B specimens increases with decreasing stress (Figure 1) is also in accord with this mechanism since, almost in all cases, the decrease in stress was accompanied with an increase in test temperature. The higher the test temperature, the greater will be the effect of oxidation on uncoated specimens.

In view of the observations made in the preceding paragraph, the differences between the stress-rupture lives of Type B and Type C specimens (Figure 5), however, appear to be puzzling. As noted previously, the data presented in Figure 5 implies that differences exist between thin-flats machined from airfoils (Type B) and those machined from 5/8 in. diameter bars (Type C) in terms of either microstructural or surface characteristics. The results presented in Figure 6 rule out the possibility that

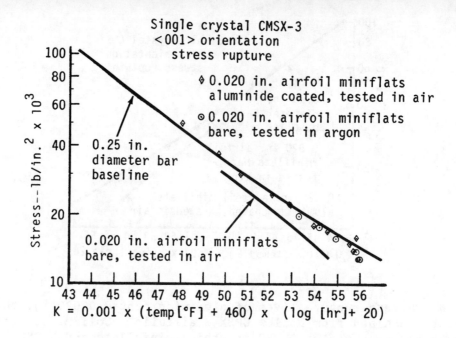

Figure 7: Comparison of the stress-rupture lives of aluminide coated 0.020 in. thick CMSX-3 mini-flats tested in air with those of uncoated mini-flats tested in high purity argon. Solid lines represent data on uncoated 0.020 in. thick mini-flats and 0.250 in. diameter bars tested in air.

microstructural differences (Mechanism 1) may be responsible for the observed behavior. The only other plausible explanation at this time, therefore, appears to be the possible differences in the response of these specimens to environmental degradation due to differences in their surface conditions. Additional work is needed to fully understand this phenomenon.

In tests conducted on equiaxed grain Mar-M246 alloy, however, the observed degradation in stress-rupture lives of thin-wall specimens cannot be attributed to the environmental effects (Mechanism 3). As noted previously in Figure 2, the stress-rupture lives of uncoated mini-flats machined from equiaxed grain Mar-M246 airfoils (Type B specimens) are degraded compared with baseline (Type A) specimens, with the amount of degradation being comparable to that observed for the single crystal CMSX-3 alloy. In contrast to CMSX-3, however, the application of an aluminide coating to the Mar-M246 mini-flats did not result in any improvements in the stress-rupture lives; see Figure 8. These results suggest that in the case of the equiaxed Mar-M246 mini-flats, the primary mechanism responsible for the stress-rupture life degradation is related to the behavior of grain boundaries in thin sections (Mechanism 4). This mechanism apparently overrides the environmental effects in this material within the range of test temperatures employed.

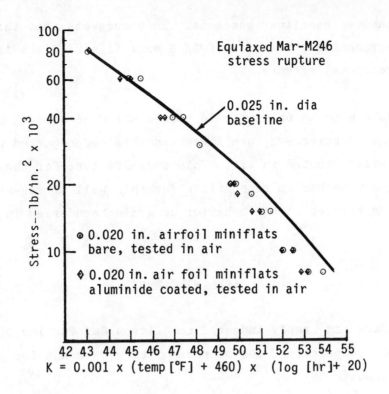

Figure 8: Comparison of the stress-rupture lives of aluminide coated 0.020 in. thick Mar-M246 airfoil mini-flats with those of uncoated mini-flats tested in air. The solid line represents the data obtained on 0.250 in. diameter bars.

Summary and Conclusions

Stress-rupture tests conducted on 0.020 in. thick mini-flat specimens machined from hollow airfoil castings indicated degradation in stress-rupture lives of both CMSX-3 and Mar-M246 alloys at stress levels below 40,000 lb/in.^{2}. The life degradation is approximately 3X at 20,000 lb/in.^{2}.

Tests conducted on CMSX-3 alloy on 0.020 in. thick mini-flats machined from 5/8 in. diameter bars and 0.020 in. wall-thickness cylindrical hollow specimens (also machined from bars) indicated no noticeable difference in stress-rupture lives. This implies that in this material the specimen geometry does not influence the thin-wall stress-rupture test results.

Experiments conducted using aluminide coated CMSX-3 airfoil mini-flats tested in air and uncoated airfoil mini-flats tested in high purity argon, resulted in stress-rupture lives equivalent to those obtained on

0.250 in. diameter baseline specimens. This suggests that the stress-rupture life degradation observed in CMSX-3 airfoil mini-flats is primarily due to environmental effects.

In tests conducted on aluminide coated Mar-M246 airfoil mini-flats, no improvement in stress-rupture lives was obtained compared with uncoated airfoil mini-flats tested in air. This suggests that for the Mar-M246 alloy, the primary mechanism responsible for thin-wall stress-rupture life degradation is related to the behavior of grain boundaries in thin sections.

Acknowledgment

The authors are deeply indebted to Allison Gas Turbine Division, General Motors Corporation, for support of this work and for permission to publish the results.

References

1. M. Doner and J. A. Heckler, "Effects of Section Thickness and Orientation on Creep-Rupture Properties of Two Advanced Single Crystal Alloys", SAE Paper 851785.

MICROSTRUCTURAL DEVELOPMENT UNDER THE INFLUENCE OF

ELASTIC ENERGY IN Ni-BASE ALLOYS CONTAINING γ' PRECIPITATES

Minoru Doi and Toru Miyazaki

Department of Materials Science and Engineering,
Metals Section, Nagoya Institute of Technology,
Gokiso-cho, Showa-ku, Nagoya 466, Japan

Abstract

The changes in the size and the distribution of γ' precipitate particles during ageing of some Ni-base alloys were investigated by means of transmission electron microscopy (TEM).

In the alloy system which has smaller lattice misfit and hence smaller elastic energy (e.g. Ni-Cr-Al or Ni-Si-Al), the γ' particles coarsen steadily and the mean particle size $\bar{r}$ at an ageing time t is proportional to $t^{1/3}$. The size distribution of γ' particles does not change during the ageing, and the standard deviation σ of size distribution is practically constant.

In the alloy system which has larger elastic energy (e.g. Ni-Cu-Si), the deceleration of the coarsening of γ' particles occurs during ageing. At the same time, the size distribution of γ' particles becomes sharper gradually, that is, the σ decreases in the course of coarsening.

When the elastic energy is small (e.g. Ni-Cr-Al or Ni-Si-Al), the γ' particles are uniformly (i.e. homogeneously) distributed in the γ matrix. However, when the elastic energy is large (e.g. Ni-Al or Ni-Al-Ti), the γ' particles have a tendency to exhibit non-uniform (i.e. inhomogeneous) distribution in the γ matrix. When the volume fraction of γ' is higher, this tendency is less obvious, that is, the γ' distribution appears uniform, even if the elastic energy is large.

The deceleration of coarsening, the decrease in and the formation of non-uniform distribution are the results of elastic interaction energy. When understanding the microstructure of Ni-base superalloys strengthened by γ' particles, we should always take account of the important role of elastic interaction energy.

Introduction

It is well known that the high temperature strength of Ni-base super-alloys is a result of a particular microstructure consisting of finely and regularly distributed γ' precipitate particles. Such a microstructure accompanied by desirable properties is almost always in a thermodynamically metastable state because it is usually obtained by interrupting the phase transformation in the course of heat-treatment. Therefore, during further heat-treatment, the favourable microstructure is very likely to develop into a thermodynamically stabler microstructure which has no longer favourable properties in most cases: the individual precipitate particles change their size, shape and distribution to minimize their energy state.

The energy state of the individual coherent particles can be expressed by three energies: surface energy of the particle; elastic strain energy due to the lattice misfit between the particle and the matrix; elastic interaction energy between particles which originates from the overlap of the elastic strain fields around the individual particles. The latter two are known as elastic energy. We have often pointed out the important effects of elastic energy on the morphology of coherent precipitates (1-4). A typical example is the case where the elastic interaction energy plays an essential role in forming various types of γ' precipitate morphology: a single γ' cuboid splits into a pair of parallel small plates or into eight small cuboids in the course of coarsening (1-3). Furthermore, a number of important things which cannot be understood without considering the effects of elastic energy have been reported so far: one example is the structural and/or stability bifurcations (5-7); another is the inhomogeneous (i.e. non-uniform) distribution of precipitate particles (8).

Regarding the change in the size of precipitate particles, the well-known process is the coarsening due to the surface energy of the particle, i.e. "Ostwald ripening". Almost all the conventional theories of precipitate coarsening are based on the theoretical treatment of Ostwald ripening which is widely known as "LSW (Lifschtz, Slyozov and Wagner) theory" (9,10). In such theories, the larger particles coarsen by absorbing the smaller particles to release their excess surface energy, and hence the total energy of the microstructure decreases. Contrary to the conventional theories, our new theory of microstructural stability named "bifurcation" theory (6,7) predicts that in elastically constrained systems, sometimes the smaller particles can grow at the expense of the larger particles to bring a uniform distribution in particle size. Furthermore, it has widely been recognized so far that coherent precipitates are distributed uniformly in the matrix owing to the elastic interaction (11). However, our theoretical calculations predict that coherent precipitates tend to exhibit non-uniform distribution (8). Although a number of attempts are now being made to verify the above two predictions, they leave something to be desired.

The aims of the present studies are (I) to investigate the changes in the size and the distribution of γ' precipitate particles during ageing of Ni-base alloys by means of transmission electron microscopy (TEM), and (II) to discuss the effects of elastic energy, especially the effects of elastic interaction energy, on the microstructural developments.

Experimental Procedures

The larger is the lattice misfit δ between the particle and the matrix, the larger is the elastic energy. Therefore, the following Ni-base alloys which have different misfits were used in the present studies: Ni-18.2%Cr-6.2%Al (δ=0.008 %), Ni-7.0%Si-6.0%Al (δ=0.10 %), Ni-12.5%Al (δ=0.56 %), Ni-8.0at.%Al-5.0at.%Ti (δ=0.65 %), Ni-36.1%Cu-9.8%Si (δ=-1.3 %) and Ni-47.4%Cu-

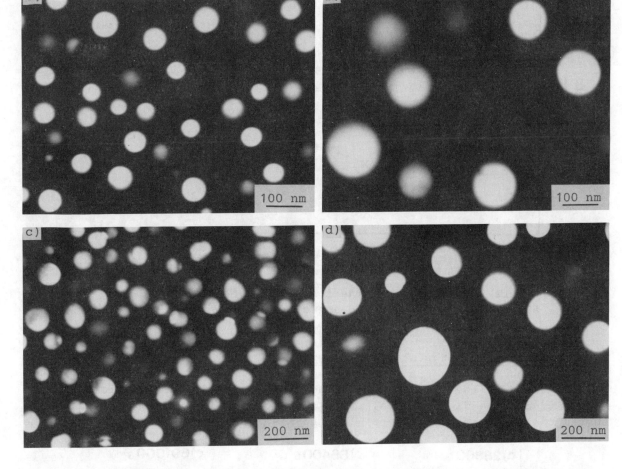

Figure 1 - TEM images of γ' precipitate particles in Ni-base alloys with small elastic energy: Ni-18.2%Cr-6.2%Al aged at 1073 K for 86400 s (a), and 691000 s (b); Ni-7.0%Si-6.0%Al aged at 1073 K for 7200 s (c), and for 57600 s (d).

5.0%Si (δ=-1.3 %). All the compositions are given in atomic %. Each alloy was quenched into iced water after homogenizing at a high temperature (i.e. solid solution treatment), and then was aged at a temperature lower than the γ' solvus line. Foil specimens for TEM observations were prepared by electropolishing the aged samples. The changes in the size and the size distribution of γ' particles during ageing were calculated from the TEM images.

Experimental Results

Alloy Systems with Small Elastic Energy

Figure 1 illustrates the TEM images of γ' precipitate particles in the Ni-Cr-Al and the Ni-Si-Al alloys. The shape of the individual particles is spherical and the particles are uniformly and randomly distributed in the γ matrix. Figure 2 illustrates the coarsening kinetics of γ' particles in the Ni-Cr-Al and the Ni-Si-Al alloys aged at 1073 K. It can be seen from this figure that the mean particle size (radius) $\bar{r}$ at an ageing time t is proportional to $t^{1/m}$, and the exponent 1/m is 0.33 for the former and 0.32 for the latter. Figure 3 illustrates the changes in the size distribution of γ' particles during ageing of the Ni-Cr-Al and the Ni-Si-Al alloys at 1073 K. The size distribution does not change essentially during coarsening, and the standard deviation σ remains practically constant: the σ is about 0.25 for the former and about 0.27 for the latter.

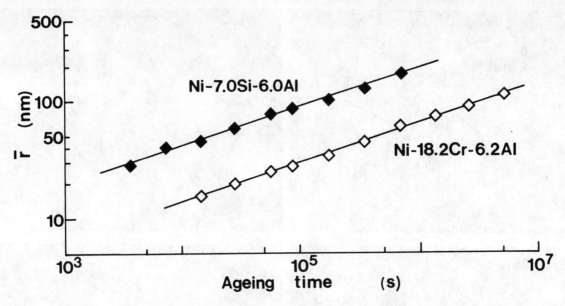

Figure 2 - Coarsening kinetics of γ' particles in Ni-base alloys with small elastic energy.

Alloy Systems with Large Elastic Energy

Figure 4 illustrates the TEM images of γ' precipitate particles in the Ni-Cu-Si and the Ni-Al alloys. The shape of the individual particles is substancially cuboidal except for the Ni-Al alloy aged for a long time

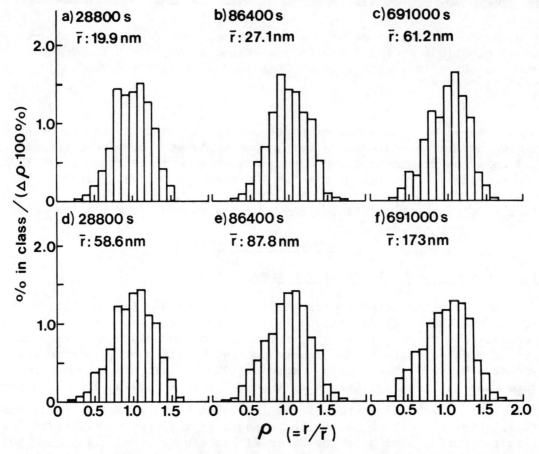

Figure 3 - Size distribution of γ' particles in Ni-base alloys with small elastic energy: Ni-18.2%Cr-6.2%Al aged at 1073 K (a-c); Ni-7.0%Si-6.0%Al aged at 1073 K (d-f).

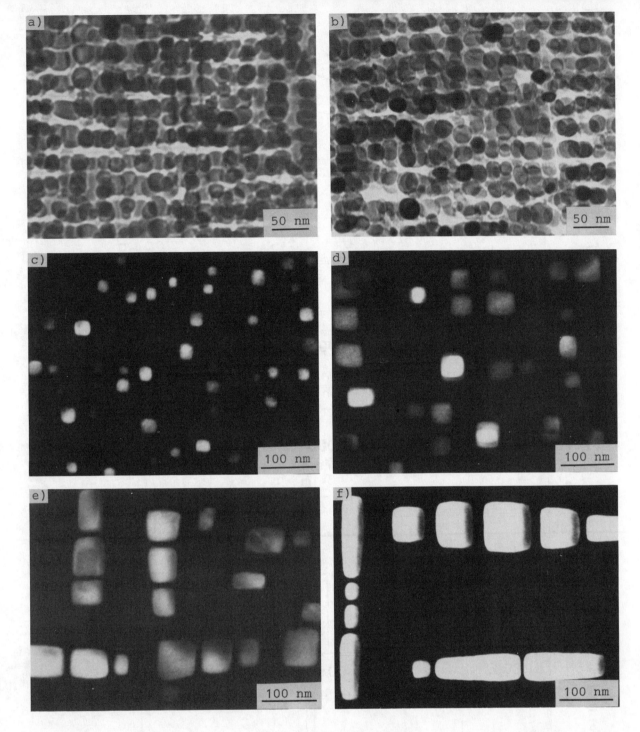

Figure 4 - TEM images of γ' precipitate particles in Ni-base alloys with large elastic energy: Ni-36.1%Cu-9.8%Si aged at 823 K for 690000 s (a), and 1200000 s (b); Ni-12.5%Al aged at 973 K for 43200 s (c), 173000 s (d), 346000 s (e), and 1210000 s (f).

($t > 3 \times 10^5$ s). In the Ni-Al alloy, the shape change from cuboid to plate occurs during ageing. Therefore, we must discuss the coarsening behaviour during only the shorter ageings which cause no such serious shape-changes, because we cannot deal with plates on the same basis as cuboids. The half length of the edge of a cuboid is regarded as r value for the cuboid. Furthermore, a prolonged ageing of the Ni-Al alloy brings an extremely non-uniform (i.e. inhomogeneous) distribution of γ' particles, forming clusters. Figures 5 and 6 illustrate the coarsening kinetics and the changes in the size distribution of γ' particles during ageings of the Ni-Cu-Si at 823 K

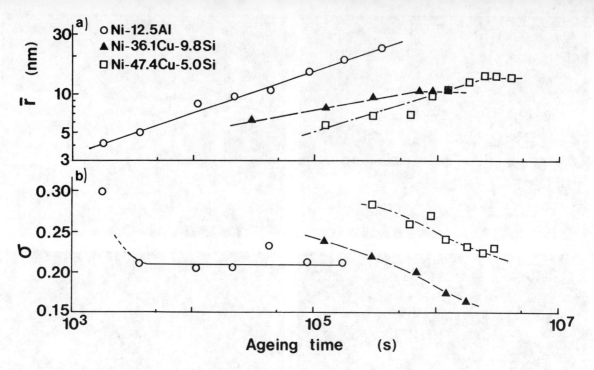

Figure 5 - Coarsening kinetics (a) and standard deviation σ of size distributions (b) of γ' particles in Ni-base alloys with large elastic energy.

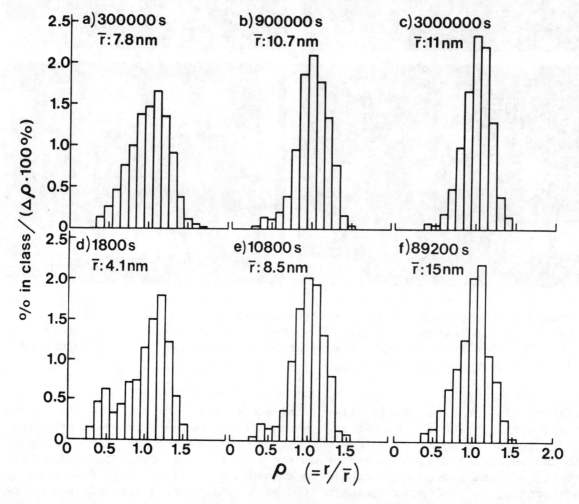

Figure 6 - Size distribution of γ' particles in Ni-base alloys with large elastic energy: Ni-36.1%Cu-9.9%Si aged at 823 K (a-c); Ni-12.5%Al aged at 973 K (d-f).

and the Ni-Al alloy at 973 K. It can be seen from these figures that the deceleration of coarsening and the simultaneous sharpening of size distribution (i.e. the decrease in σ) occur during ageing of the Ni-Cu-Si alloys. In the Ni-Al alloy, however, the σ value remains constant ($\simeq 0.21$), and the $\bar{r}$ is proportional to $t^{1/3}$: i.e. the deceleration of coarsening does not occur.

Discussions

Coarsening Kinetics and Size Distributions

In the alloy system which has small elastic energy, the coarsening of γ' particles obeys the $t^{1/m}$ law and the $1/m$ value (0.32 or 0.33 in Fig. 2) is practically equal to the value 1/3 predicted by the LSW theory of Ostwald ripening. However, the size distributions observed in the actual alloy systems with small elastic energy are significantly different from that predicted by the LSW theory. The former is always wider, more symmetric and less peaked than the latter: the observed σ value (0.25 or 0.27 in Fig. 3) is larger than the value 0.215 predicted by the LSW theory. The LSW theory has been modified so far by many investigators to explain the difference between the observed size distribution and the predicted one; many attempts aimed at modifying the LSW theory with respect to the volume fraction of particles (12-17). The results obtained here are in accord with either the LSW theory or the LSW theories modified with respect to the volume fraction. In the alloy systems having smaller elastic energy, the driving force for the coarsening of γ' precipitate particles is their excess surface energy.

A noticeable result obtained in the present studies is that the coarsening of γ' particles is decelerated in the Ni-Cu-Si alloys which have large elastic energy (see Fig. 5-a). This tendency toward deceleration is more obvious when the volume fraction of γ' is higher. Another noticeable result is that the size distribution of γ' particles becomes sharper gradually, i.e. the σ decreases gradually, in the course of coarsening. It should be noted that the deceleration of coarsening and the decrease in σ occur simultaneously. These results suggest the important thing that the sizes of the individual particles become less scattered (i.e. uniform) and the microstructure converges to a particular state if the elastic energy is large and the volume fraction is high: i.e. the unification of particles and the stabilization of a particular microstructure occur.

Stability Bifurcation in γ'-Particle Coarsening

The already well-known theories of particle coarsening, i.e. the LSW or the modified LSW theories, indicate that larger particles continue to coarsen by absorbing smaller particles (hence the microstructure continues coarsening) because the driving force for coarsening is considered to be only the excess surface energy of the particles. As long as we accept the already recognized theories, the unification of particles and the stabilization of a particular microstructure seem incredible. However, if we take account of the new idea named "stability bifurcation" (6,7) in which the elastic energy, and especially the elastic interaction energy plays an essential role, we can clearly explain such unbelievable phenomena.

Figure 7 illustrates the variation in the energy state of a pair of γ' particles as a function of $\bar{r}$ and R (7). The parameter R ($\equiv (r_\alpha - r_\beta)/(r_\alpha + r_\beta)$; $|R| \leq 1$) is used to describe the relative sizes of the paired particles where r_α and r_β are the radii of the paired particles: when R=0, the paired particles are identical and take the average size $\bar{r}$; when R=±1, only one of the pair exists and takes the maximum size $\sqrt[3]{2} \cdot \bar{r}$. When the $\bar{r}$ is small, e.g. $\bar{r}_1$ in Fig. 7, the state at R=0 has the highest energy (indicated by the open circle O) and the states at R=±1 have the lowest energy (indicated by the solid circles ●); the total energy decreases from O to ●, as indicated by the bold arrows (➔). In the Region I, the energy state always decreases toward R=±1, and one of the paired particles can coarsen by absorbing the other just like the case predicted by the LSW or the modified LSW theories: the microstructure coarsens to decrease its surface energy and hence its total energy.

However, when the $\bar{r}$ is large, e.g. $\bar{r}_2$ in Fig. 7, the energy state takes a minimum at R=0 (indicated by the rectangular □): the total energy decreases toward R=0, as indicated by the open arrows (⇨). As the $\bar{r}$ becomes large, the elastic interaction energy becomes dominant as compared with the surface energy. The appearance of the energy minimum at R=0 is a result of the elastic interaction energy between the paired particles. It is clear from Fig. 7 that in the shadowed Region II, smaller particles can coarsen by absorbing larger particles to decrease the energy state, which is opposite to the LSW or the modified LSW theories. According to the above bifurcation theory, no wonder the deceleration of coarsening and the sharpening of size distribution (the decrease in σ), i.e. the unification and the stabilization of microstructure, occur.

It is also clear from the present discussions that the deceleration of coarsening is a result of the change in driving force for coarsening, i.e. the change from the mechanism controlled by surface energy to that by elastic interaction energy. The equation $\bar{r}=K \cdot t^{1/m}$ is usually used to describe the coarsening kinetics. When the coarsening rate changes, we usually take account of only the change in 1/m. However, we had better conclude that not the exponent 1/m but the K should change during ageing, which results in the deceleration of coarsening and the decrease in σ. Here we should remember the coarsening behaviour of γ' particles in the Ni-Al alloy (see Fig. 5). Although the elastic interaction in the Ni-Al alloy

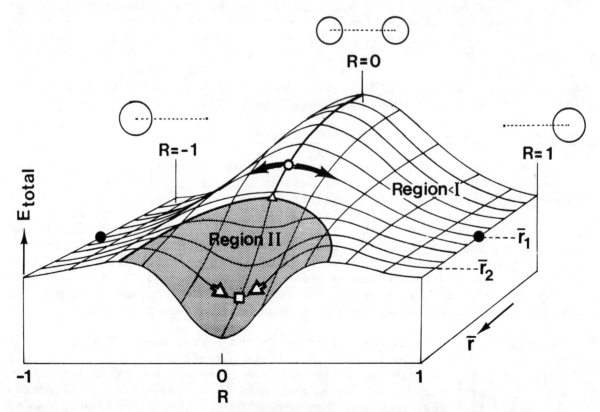

Figure 7 - Variation in the energy state of a pair of γ' particles as a function of the average particle size $\bar{r}$ and the parameter R.

system is strong, the coarsening of γ' particles is not decelerated but obeys the $t^{1/3}$ law when their size distribution and hence the σ do not change. This fact clearly supports the above idea that the exponent $1/m$ is always constant (=1/3) and a constantly varying K brings a constantly varying coarsening-rate, i.e. the deceleration of coarsening.

Inhomogeneous Distribution of γ' Particles

The third noticeable result is that the γ' particles in the Ni-Al alloy have a tendency to exhibit an extremely non-uniform distribution in the matrix, as shown in Figs. 4-e and -f. This tendency is more obvious when the volume fraction of γ' is lower: e.g. in the Ni-Al-Ti alloy during reversion, the γ' particles are locally distributed forming clusters in the γ matrix. The present results agree with our result obtained by theoretical calculations (8) that such a non-uniform distribution is energetically stabler than the uniform distribution. Even if the elastic interaction is strong, however, the higher is the volume fraction, the less obvious is the tendency toward non-uniform distribution. The distribution of γ' particles appears uniform when their volume fraction is high.

Summary and Conclusions

We have been believing for very long that the precipitate coarsening proceeds steadily and never slows down to converge to a particular state. We have also been believing that the coherent precipitates are uniformly distributed in the matrix owing to the effect of elastic interaction energy and never have the tendency toward clustering. However, the present results are just the opposite and clearly urge us to change the already well-known ideas on the development of microstructure containing coherent precipitates. It is an effect of elastic interaction energy that the γ' precipitates become uniform in their size to converge to a particular microstructure in a stable state (desirable effect). It is another effect of elastic interaction energy that the γ' precipitates become non-uniform in their distribution to form clusters (undesirable effect).

The present studies clearly show that the elastic interaction energy has remarkable effects on the microstructural developments, in other words, on the instability of the desirable microstructure accompanied by good properties. Of course, it is very important for the practical use whether such a desirable microstructure is stable or unstable during further heating. When understanding the microstructures and hence the properties of Ni-base superalloys strengthened by γ' precipitate particles, we should not neglect the important role of elastic energies.

Acknowledgments

The authors are pleased to acknowledge considerable assistance from Dr. T. Kozakai, Mr. D. Konagaya and Mr. H. Yukawa of Nagoya Institute of Technology. A part of this work was financially supported by a Grant-in-Aid for Scientific Research from the Ministry of Education, Science and Culture, Japan.

References

1. M. Doi, and T. Miyazaki, "The Effect of Elastic Interaction Energy on the Shape of γ'-Precipitate in Ni-Based Alloys," Superalloys 1984, ed. M. Gell, C. S. Kortovich, R. H. Bricknell, W. B. Kent and J. F. Radavich (Warrendale, PA: The Metallurgical Society, 1984), 543-552.
2. M. Doi, T. Miyazaki, and T. Wakatsuki, "The Effect of Elastic Interac-

tion Energy on the Morphology of γ' Precipitates in Nickel-based Alloys," <u>Materials Science and Engineering</u>, 67 (1984) 247-253.

3. M. Doi, T. Miyazaki, and T. Wakatsuki, "The Effects of Elastic Interaction Energy on the γ' Precipitate Morphology of Continuously Cooled Nickel-base Alloys," <u>Materials Science and Engineering</u>, 74 (1985) 139-145.

4. M. Doi, and T. Miyazaki, "γ' Precipitate Morphology Formed under the Influence of Elastic Interaction Energies in Nickel-base Alloys," <u>Materials Science and Engineering</u>, 78 (1986) 87-94.

5. W. C. Johnson, and J. W. Cahn, "Elastically Induced Shape Bifurcations of Inclusions," <u>Acta Metallurgica</u>, 32 (1984) 1925-1933.

6. T. Miyazaki, K. Seki, M. Doi, and T. Kozakai, "Stability-bifurcations in the Coarsening of Precipitates in Elastically Constrained Systems," <u>Materials Science and Engineering</u>, 77 (1986) 125-132.

7. T. Miyazaki, M. Doi, and T. Kozakai, "Shape Bifurcations in the Coarsening of Precipitates in Elastically Constrained Systems," <u>Proc. Int. Symp. on Non Linear Phenomena in Materials Science</u>, Aussoia, September 1987 [in printing].

8. M. Doi, M. Fukaya, and T. Miyazaki, "Inhomogeneity in a Coherent Precipitate Distribution Arising from the Effects of Elastic Interaction Energies," <u>Philosophical Magazine</u>, (1988) [in printing].

9. I. M. Lifshitz, and V. V. Slyozov, "The Kinetics of Precipitation from Supersaturated Solid Solutions," <u>Journal of Physics and Chemistry of Solids</u>, 19 (1961) 35-50.

10. C. Wagner, "Theorie der Alterung von Niederschlägen durch Umlösen (Ostwald-Reifung)," <u>Zeitschrift für Elektrochemie</u>, 65 (1961) 581-591.

11. A. J. Ardell, R. B. Nicholson, and J. D. Eshelby, "On the Modulated Structure of Aged Ni-Al Alloys," with an Appendix "On the Elastic Interaction between Inclusions," <u>Acta Metallurgica</u>, 14 (1966) 1295-1309.

12. A. J. Ardell, "The Effect of Volume Fraction on Particle Coarsening: Theoretical Considerations," <u>Acta Metallurgica</u>, 20 (1972) 61-71.

13. A. D. Brailsford, and P. Winblatt, "The Dependence of Ostwald Ripening Kinetics on Particle Volume Fraction," <u>Acta Metallurgica</u>, 27 (1979) 489-497.

14. C. K. Davies, P. Nash, and R. N. Stevens, "The Effect of Volume Fraction of Precipitate on Ostwald Ripening," <u>Acta Metallurgica</u>, 28 (1980) 179-189.

15. J. A. Marqusee, and J. Ross, "Theory of Ostwald Ripening: Competitive Growth and Its Dependence on Volume Fraction," <u>Journal of Chemical Physics</u>, 80 (1984) 536-543.

16. P. W. Voorhees, and M. E. Glicksman, "Solution to the Multi-Particle Diffusion Problem with Applications to Ostwald Ripening — I. Theory," <u>Acta Metallurgica</u>, 32 (1984) 2001-2011.

17. Y. Enomoto, M. Tokuyama, and K. Kawasaki, "Finite Volume Fraction Effects on Ostwald Ripening," <u>Acta Metallurgica</u>, 34 (1986) 2119-2128.

NEW INTERPRETATION OF RUPTURE STRENGTH

USING THE POTENTIAL DROP TECHNIQUE

Ioannis Vasatis
GE Corporate Research and Development
Materials Research Laboratory
P.O. Box 8
Schenectady, NY 12301

Summary

The effects of zirconium and carbon on the rupture behavior of an experimental blade alloy were examined and analysed by using the notched rupture test and the dc potential drop technique. Differences in the rupture behavior, not detectable by simply comparing rupture times, revealed by using the experimental technique described in this paper, provided more in depth understanding of the material's behavior and useful guidance to the alloy developer. The results of this work clearly demonstrate that minor chemistry modifications in nickel base alloys do not necessarily affect the resistance to crack initiation and crack propagation in a similar manner.

Introduction

The performance and reliability of industrial gas turbines can be reduced by the premature degradation of critical hot gas components. The principal degradation mechanisms are creep, hot corrosion and thermal fatigue (1,2). The first two presently seem to be limiting in some of the designs, and as a result the alloy developer aims to improve creep and hot corrosion resistance.

Recent advances in understanding high temperature behavior of nickel base superalloys, indicated that improved creep performance implies resistance to high temperature deformation, crack initiation, crack propagation, and equally important resistance to environmental degradation during exposure to high temperatures (3-5). All these aspects of materials behavior should be taken into account when designing a new alloy for high temperature applications.

For this purpose, new technologies combined with conventional practices, have been developed to provide more in-depth understanding of materials behavior. One such procedure based on the dc potential drop technique and the notch stress rupture test is described in this paper. The dc potential drop technique, commonly used to measure crack length (6,7), has also been used to monitor deformation, detect crack initiation and measure crack propagation characteristics in axisymmetric notched bars (8-10).

The necessity to look at initiation and propagation separately, results from the fact that the relative importance of degradation mechanisms is not necessarily the same during the initiation and propagation stages. Chemistry modifications designed to improve rupture strength can adversely affect the material's resistance to crack initiation and propagation. The work in this paper demonstrates these concepts and offers a new way of evaluating rupture strength.

Materials

The material used in this work was an experimental high strength cast bucket type nickel base superalloy. Two modified chemistries, the first with lower zirconium than normal and the second with higher carbon concentration, were compared with the baseline chemistry. The exact chemical composition of these experimental alloys, are absolutely unimportant to the arguments presented here.

Specimen Geometry and Testing Procedures

Notched stress rupture tests were performed on a dead-load lever arm testing machine on axisymmetrically notched bars. Two types of notched geometries were used, the "U" notch geometry, shown in Figure 1, and the "V" notch, shown in Figure 2. In the "U" notch a significant volume of material is subjected to almost the same stress and state of stress (11-13) and fracture usually initiates internally, thus minimizing environmental interactions (14,15). These features in addition to the fact that deformation and fracture occur in a confined volume of material which can be easily monitored, make the "U" notch specimen the ideal tool for studying rupture behavior. On the other hand the "V" notch geometry is designed, mainly to measure the material's resistance to crack initiation. Two identical notches 1/2" apart were ground accurately into the gage length of the notched specimens in order to facilitate metallographic examination of creep damage distribution at fracture.

The tests were monitored in real time using the dc potential drop technique. A constant current of 4A was passed through the specimen and the change in potential across the two notches was measured during testing. For this purpose, two pairs of platinum wires were attached by spot welding at points where the notch intersects the straight section of the specimen. A third pair of probes, serving as the "reference probe", attached on the straight section of the specimen, 1/2" away from the notch. The reference probe is used to compensate for current and temperature fluctuations (9). Extensometry was also used for measuring the total axial displacement. A diagram with the location of the potential probes and the extensometer is shown in Figure 3.

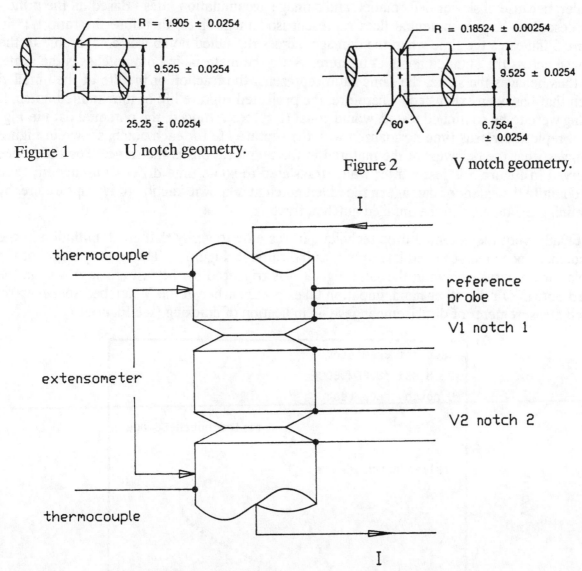

Figure 1 U notch geometry.

Figure 2 V notch geometry.

Figure 3 Schematic of notched specimen indicating the location of potential probes and extensometer.

Basics - Fundamentals

The potential change across the notch, measured during a notch rupture test is due to creep deformation and damage accumulation (i.e., cavities and cracks). Prior to the nucleation of damage the increase in potential is caused only by creep deformation. By establishing a calibration relationship between strain and potential for the specific specimen type, creep strain can be measured by using the specimen as a strain gage (5,9). With the nucleation and growth of cracks, the potential increase measured across the notch, becomes greater than the value expected from creep deformation.

Experimental support of the above arguments is given in Figure 4, where the potential across the failed notch of duplicate tests is plotted versus time. Despite comparable initial deformation rates among the two specimens, early initiation, marked with the departure of the potential curve from linearity, resulted in shorter rupture life. For the same reason, in Figure 5 the potential curve across the failed notch, in a double notched specimen, deviated first from linearity. The potential increase across the unfailed notch was from the beginning smaller, because of slower deformation and damage accumulation rates. Based on the principles described here, isopotential lines represent isodamage lines, therefore the ratio t/t^* in Figure 5 indicates the time when the damage across the failed notch was comparable to the damage across the unfailed notch at fracture. Since the material in both notches is the same, and thus behaves the same, the same ratio represents the fraction of the life of the unfailed notch that has been exhausted. Therefore, the projected rupture life of the unfailed notch, if testing were to be continued tested, would equal $(t/t^*) \times t_r$. Indeed, the potential data in Figure 5, replotted versus time normalized with the rupture life for each notch, shown in Figure 6, suggest that the behavior of the material in the two locations is the same. The difference observed in Figure 5 is just scatter which translated to some time drift. Consequently, one can quantify the extent of damage in the failed notch at a known fraction of its rupture life, by examining the damage in the unfailed notch at fracture.

Finally, with the potential drop technique it was easy to verify that crack initiation in the U notched specimen occurred late in life, as indicated in Figure 7. The tertiary stage of the displacement curve, shown in the same figure, was triggered by the fast crack growth in the failed notch. On the other hand, initiation takes place earlier in the V notched specimen, in which measurements of displacement gave no indication of cracking (see Figure 8).

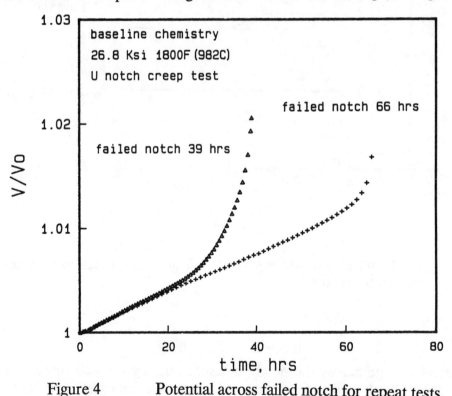

Figure 4   Potential across failed notch for repeat tests.

Results and Discussion

The approach described in the previous section was used to analyze the effects of zirconium and carbon on the rupture behavior of a cast bucket alloy. The evaluation of the three alloys was limited to a specific test condition. The rupture times for both type of notched tests, shown in Figures 9 and 10, were comparable, within scatter, for all three alloys.

The potential drop measurements across the failed notch of the U notched specimens also indicated similarity in behavior (see Figure 11). The initial deformation rate and the onset of crack propagation, detected between 75% - 80% of the rupture life for all the three alloys, were very close. Significant differences were observed in the potential drop across the failed notch of the V notched specimens (see Figure 12). Crack initiation occurred around 20%,

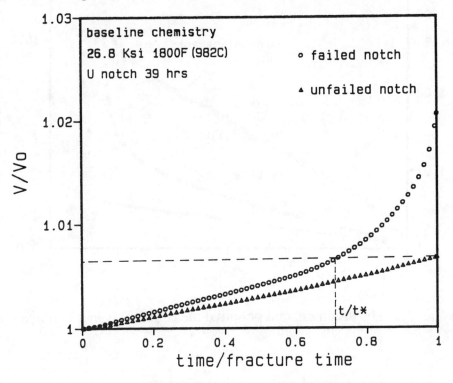

Figure 5   Potential curves for failed and unfailed notch.

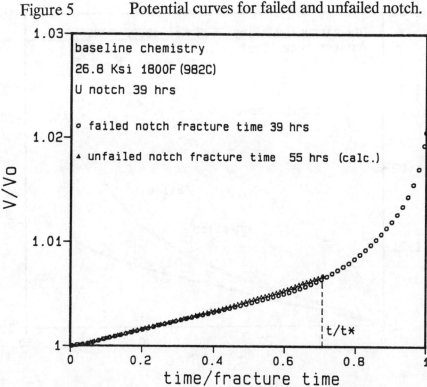

Figure 6   Potential changes versus time normalized with the fracture time for the failed notch and a calculated fracture time for the unfailed notch.

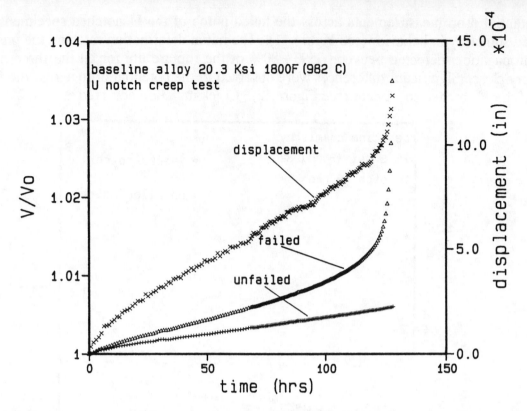

Figure 7   Displacement and potential measurements in a U notch creep test.

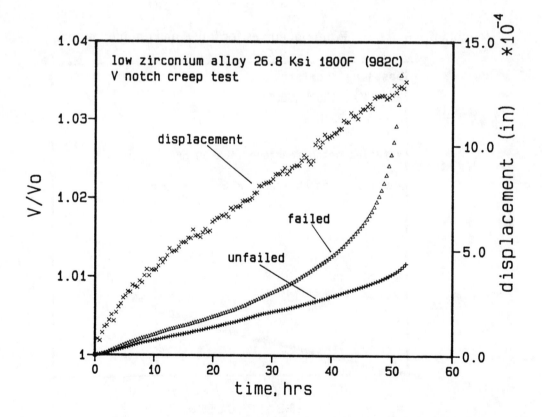

Figure 8   Displacement and potential measurements in a V notch creep test.

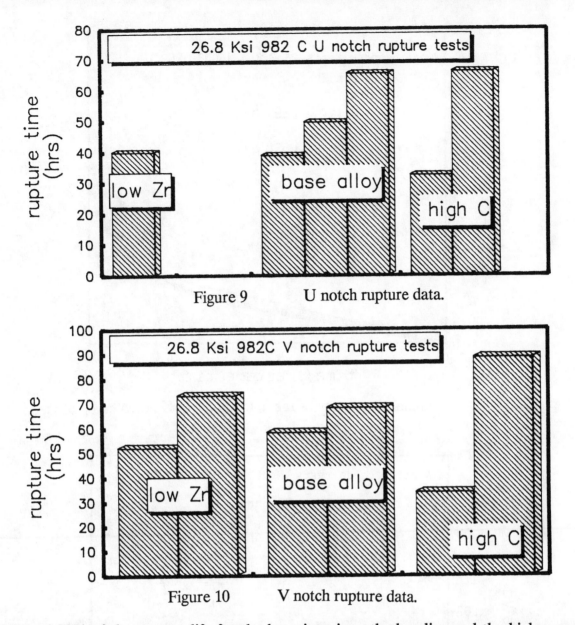

Figure 9 U notch rupture data.

Figure 10 V notch rupture data.

40% and 80% of the rupture life for the low zirconium, the baseline and the higher carbon alloys respectively. The reduction in the zirconium level made the alloy less resistant to crack initiation but better in sustaining cracking. The opposite effects were observed with the addition of carbon in the alloy. Thus, the specific modifications changed both the crack initiation and propagation characteristics of the alloy. The effects on crack propagation are expected to be more pronounced, since the U notched creep data, which measure the resistance to creep deformation and crack initiation, were not significantly different. Indeed, crack growth rates for the high carbon alloy increased by an order of magnitude with respect to the baseline alloy (see Figure 13). Unfortunately crack propagation data for the low zirconium alloy were not available. It is important to point out that these results are pertinent to the specific test condition and no generalizations should be made without additional testing.

The enhancement of creep properties of nickel base superalloys with zirconium and carbon has been often reported in the literature (16,17). In contrast to their unambiguous effects, the mechanisms have resisted clarification. Both change the chemistry and microstructure of the grain boundaries and consequently affect time dependent properties. The addition of carbon to the alloy under consideration caused the precipitation of fine grain boundary carbides between the larger carbides observed in the baseline alloy. Fine carbides at the grain boundaries can further inhibit grain boundary sliding and make crack initiation more difficult. At the same time they reduce the alloy's tolerance to sustain damage by

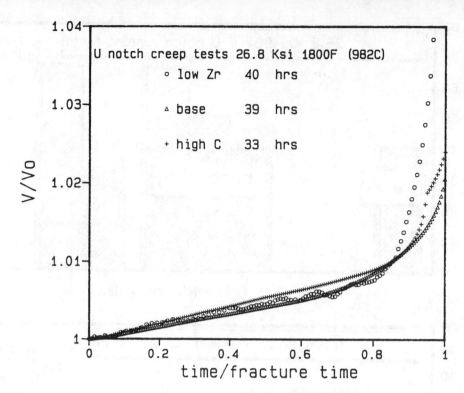

Figure 11   Potential change across the failed notch in U notch creep tests.

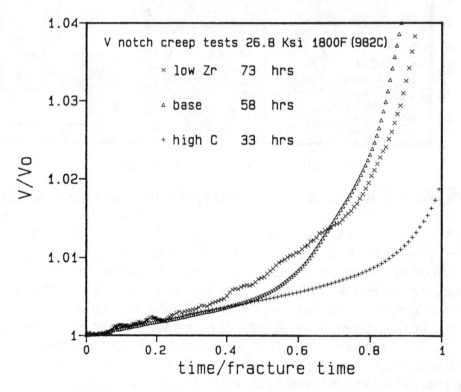

Figure 12   Potential change across the failed notch in V notch creep tests.

weakening its resistance to crack growth. Carbides have been reported to be detrimental to the resistance of the material to environmentally assisted cracking (18,19). The potential drop technique and the analysis described in this work can certainly provide additional information necessary for understanding adverse effects and develop a knowledge base for improving the performance of superalloys.

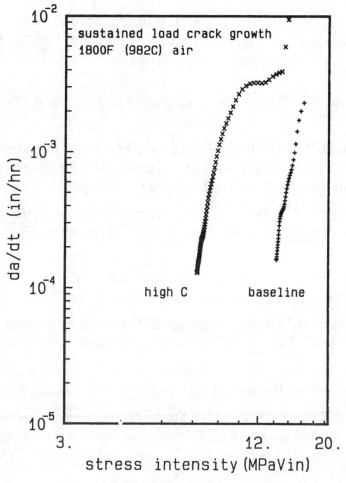

Figure 13   Sustained load crack propagation data.

Concluding Remarks

The resistance of a material to high temperature deformation, crack initiation and crack propagation constitutes its rupture strength. The work presented in this paper demonstrates that chemistry modifications can adversely affect resistance to crack initiation and propagation and as a result these processes ought to be looked at separately. The dc potential drop technique and notched rupture test, combined with the experimental methodology described here, helped interpret and understand rupture strength by analyzing life to crack initiation and crack propagation.

Acknowledgements

This work was partly funded by GE/TBO and numerous technical discussions with Lance Peterson, Jack Wood of GE/TBO and Michael Henry of GE/CRD were greatly appreciated. The work could have not been accomplished without the dedicated effort on the part of Mark Benz and Harvey Schadler of GE/CRD to provide the resources.

References

1. A.S. Radcliff, "Influencing Gas Turbine Use and Performance", <u>Materials Science and Technology</u>, Vol. 3, July 1987, p. 554

2. A.G. Dodd, "Mechanical Design of Gas Turbine Blading in Cast Superalloys", <u>Materials Science and Technology</u>, Vol. 2, May 1986, p. 476.

3. S. Taira and R. Ohtani, "Creep Crack Propagation and Creep Rupture of Notched Specimens", Paper C213, Proc. Int. Conf. on <u>Creep and Fatigue in Elevated Temperature Application</u>, Philadelphia, 1973, published by J. Mech. E, 1975.

4. M.F. Ashby and B.F. Dyson, "Creep Damage Mechanics and Micromechanisms", <u>Proc. of ICF6</u>, 1, 3, 1984.

5. I.P. Vasatis and R.M. Pelloux, "Application of the dc Potential Drop Technique in Investigating Crack Initiation and Propagation Under Sustained Load in Notched Rupture Tests", <u>Met. Trans. A</u>, Volume 19A, April 1988, p. 863.

6. M.D. Halliday and C.J. Beevers, Symposium on <u>The Measurement of Crack Length and Shape During Fracture and Fatigue</u>, ed. C.J. Beevers, EMAS, Birmingham, UK, 1979, p. 85.

7. B.L. Freeman and G.J. Neate, "The Measurement of Crack Length During Fracture at Elevated Temperatures Using the dc Potential Drop Technique", <u>Ibid</u>, p. 435.

8. G.B. Thomas and H.R. Tipler, "Changes in Electrical Resistance During Creep of Cr-Mo-V Steels", <u>INPL Reports</u> DMA74, NPk 7/05, September, 1981.

9. I.P. Vasatis and R.M. Pelloux, "Application of the dc Potential Drop technique to Creep Stress Rupture Testing", <u>Journal of Metals</u>, 10, 1985, p. 44.

10. I.P. Vasatis and R.M. Pelloux, "A Study of the Creep Deformation and Rupture Behavior of Notched Bars of IN-X750 with the dc Potential Drop", Proc. Conf. on <u>Life Prediction for High Temperature Gas Turbine Materials</u>, eds. V. Weiss and W. Bakker, EPRI, Syracuse, NY, August 1985, p. 12-1.

11. D.R. Hayhurst, J.A. Lechie and J.T. Henderson, "Creep Rupture of Notched Bars", <u>Proc. R. Soc.</u>, London A360, 1978, p. 243.

12. D.R. Hayhurst and J.T. Henderson, "Creep Stress Redistribution in Notched Bars", <u>Int. J. Mech. Sci.</u>, 19, 1977, p. 133.

13. D.R. Hayhurst, in <u>Engineering Approaches to High Temperature Design</u>, eds. B. Wilshire and D. Owen, Pirendge Press, Swansea UK, 1983, p. 85.

14. B.F. Dyson and M.J. Loveday, in <u>Creep of Structures</u>, I.U.T.A.M. Symposium, Ceicester, UK, eds. A. Porter and D. Hayhurst, Sprinyes Verlay, Berlin, 1981, p. 406.

15. H.S. Loveday and B.S. Dyson, <u>ICM3</u>, Cambridge, UK, 2, 1979, p. 213.

16. R.F. Decker, "Strengthening Mechanisms in Nickel-Base Superalloys", Aimax Molybdenum Company Symposium, Zurich, May, 1969.

17. C.T. Sims, N.S. Stolloff and W.C. Hagel, eds. <u>Superalloys II</u>, John Wiley & Sons, 1987.

18. D.A. Woodford and R.H. Bricknell, <u>Treatise on Materials Science and Technology</u>, eds. C.L. Briant and S.K. Banerji, Vol. 25, 1983, p. 157.

19. I.P. Vasatis and R.M. Pelloux, "The Effect of Environment on the Sustained Load Crack Growth Rates of Forged Waspaloy", <u>Met. Trans.</u>, Vol. 16A, 1985, p. 1515.

A MODEL BASED COMPUTER ANALYSIS OF CREEP DATA (CRISPEN): APPLICATIONS TO NICKEL-BASE SUPERALLOYS

A Barbosa[1,2], N G Taylor[3], M F Ashby[3], B F Dyson[1] and M McLean[1]

1. Division of Materials Applications, National Physical Laboratory, Teddington, Middlesex TW11 OLW, UK

2. Imperial College of Science and Technology, Prince Consort Road, London, SW7 2BP, UK

3. University Engineering Department, University of Cambridge, Trumpington Street, Cambridge CB2 1PZ, UK

Abstract

The physical mechanisms of deformation and fracture that control the shapes of creep curves in superalloys are reviewed and creep by these mechanisms is modelled using the approach of continuum damage mechanics. The resulting constitutive and damage evolution laws for creep are thus fully compatible with the underlying physics of the process. The equation-sets are implemented in a computer system (CRISPEN) that operates on an IBM AT or SYSTEM 2 personal computer to allow: (a) analysis of creep curves; (b) compilation of a database representing full creep curve shapes; (c) simulation of creep curves for abitrary test conditions and (d) comparison of simulated curves with available data. CRISPEN has been applied successfully to a range of nickel-base superalloys (wrought, conventionally cast, directionally solidified, single crystal). Examples are given of the agreement between creep simulations and experimental data for both representations of isolated creep curves and extrapolations to different stress and/or temperature and to variable stress/temperature conditions.

Introduction

Although considerable effort and expense have been directed to characterising the highly non-linear creep behaviour of engineering alloys, such as nickel-base superalloys, most current compilations of creep data for engineering design represent only simple measures of creep performance. Indicators such as rupture life, time to 1% extension, minimum creep rate and ductility are adequate for traditional design-to-code methods. However, advanced computer-aided design procedures will require an increasingly sophisticated description of the full strain-time evolution of the material during complex loading cycles. There have been many empirical analyses of the shapes of creep curves that have met with varying degrees of success, but these have had no clear physical basis and, consequently, cannot be extrapolated with confidence beyond the field of the database to predict either longer lives or the effects of typical service cycles.

The understanding of high temperature deformation and fracture of superalloys has developed to the point where the important mechanisms are believed to have been identified. Detailed models of the time dependent movement of dislocations around γ'-particles by cutting, bowing and climb have been proposed that have the potential of accounting for the principal features of the creep behaviour of these materials[1]. Similarly, theories of the development of damage that leads to tertiary creep and fracture in these materials have been advanced[2,3]. The purpose of the present study has been to translate this knowledge of the underlying physics of deformation and fracture into a constitutive description of the creep behaviour that can be readily used in engineering calculations. The principal aim of the programme has been to assess the viability of developing a computer-system, incorporating these physically-based constitutive laws, and a base of reliable data to provide a capability for describing the full shapes of creep curves and to predict creep behaviour under conditions for which data are not available. The approach adopted has been to use the formalism of continuum damage mechanics expressed in terms of state variables that have clear physical significance.

Mechanisms and Models

Creep curves for nickel-base superalloys do not exhibit a long steady state behaviour. Rather, after a short primary transient of progressively decreasing creep rate until a minimum value $\dot{\epsilon}_{min}$ is achieved, most of the life is characterised by an extensive period of tertiary-creep where the creep rate progressively increases. Consequently, we shall restrict the following discussion to the causes of primary and tertiary creep.

Primary creep

In simple single-phase alloys, the strength of the material is a <u>consequence</u> of the dislocation network within the material. Primary creep is the result of hardening due to an increase in dislocation density with increasing strain until a steady state is reached when recovery exactly balances the rate of hardening. The situation is rather different in multi-phase engineering alloys where primary creep can be caused by stress redistributions between the various heterogeneities in the material (eg between differently oriented grains, between soft matrix and hard particles). Ion et al[4] have proposed a general model that describes the latter situation and is compatible with the successful empirical formulation of Webster et al[5] which has similarities to the primary component of the θ-projection description of creep curves[6].

$$\epsilon = \frac{\sigma}{E} + \epsilon_p \left\{ 1 - \exp(-D \dot{\epsilon}_{min} t) \right\} + \dot{\epsilon}_{min} t \qquad (1)$$

where E is Youngs modulus, ϵ_p is the total primary strain and D is a constant characterising the rate of hardening.

Tertiary creep

Ashby and Dyson[7] have reviewed the range of phenomena that can lead to an acceleration in creep rate in the tertiary stage. The acceleration is caused by changes in external geometry and internal structure, with clear physical meaning, that can be defined as damage. For superalloys there appear to be three important types of damage that can contribute to tertiary creep and these are shown schematically in Figure 1.

a) Intrinsic softening of superalloys is associated with changes in the dislocation substructure. Dyson and McLean[2] have analysed a wide range of creep data for superalloys and shown that creep softening results from accumulated plastic strain, rather than from a time dependent coarsening of the microstructure as had been previously advocated.[8] They proposed that the accelerating creep rate was due to increasing dislocation densities and support for this view has been obtained from electron microscope studies.[9,10] Ion et al[4] have developed these ideas into strain softening models of tertiary creep. A linear strain softening model gives precisely the equation that accounts for the tertiary creep component of the θ-projection approach[6].

$$\epsilon = \frac{1}{C_\ell} \left[\exp\left(C_\ell \dot{\epsilon}_i t\right) - 1 \right] \tag{2}$$

where $\dot{\epsilon}_i$ is the initial creep rate and C_ℓ is a constant. An alternative exponential softening model[4], which is used in the subsequent analysis, leads to the following equation.

$$\epsilon = -\frac{1}{C} \ln \left[1 - C \dot{\epsilon}_i t \right] \tag{3}$$

where C is also a constant.

b) Development of grain boundary cavitation causes a loss of internal load bearing section that leads to an increased stress on the remaining sound ligaments. Dyson and Gibbons[3] have examined the shapes of creep curves of nickel-based superalloys with different ductilities resulting from, for example, increased cavitation associated with detrimental trace element concentrations. They show that Equation 3 adequately describes the tertiary parts of the creep curves, but that C increases with decreasing fracture strain ϵ_f. They modelled this aspect of the behaviour showing that cavitation leads to an addition to the intrinsic strain softening constant of $\sim n/3\epsilon_f$ where n is the stress exponent associated with the minimum creep rate.

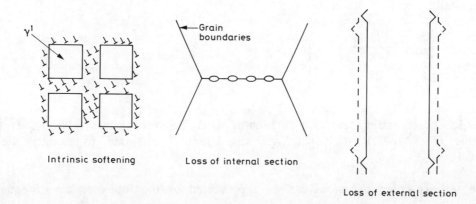

Figure 1   Schematic illustration of the three categories of micro- and macro-structural changes that cause tertiary creep in superalloys.

c) Loss of external section through the reduction in area of cross-section associated with tensile strain during constant load tests also leads to a progressive increase in stress and this causes strain rate acceleration. Assuming constant volume, the reduction in area dA scales with the change in strain $d\epsilon$ which, for constant load, simply relates to the change in stress. Thus:

$$d\epsilon = -\frac{dA}{A} = \frac{d\sigma}{\sigma} \qquad (4)$$

Integrating Equation 4 and combining it with a power-law representation of the initial creep rate, $\dot{\epsilon}_i = A\sigma^n$, leads to the following expression

$$\epsilon = -\frac{1}{n} \ln\left[1 - n\dot{\epsilon}_i t\right] \qquad (5)$$

Equations 3 and 5 are clearly of the same form but with different constants.

Constitutive Laws and CRISPEN

In the approach of continuum damage mechanics, pioneered by Kachanov[11], and extended by several authors[12-14], the creep rate at any instant is expressed as a function of both the operating conditions (σ,T) and of one or more state variables or damage parameters. Equations describing the evolution of the state variables (ω_1, ω_2) are also expressed in differential form and the creep behaviour is determined by integrating the coupled set of differential equations.

$$\left.\begin{array}{rl} \dot{\epsilon} &= f(\sigma, T, \omega_1, \omega_2 \ldots\ldots\ldots) \\ \dot{\omega}_1 &= g(\sigma, T, \omega_1, \omega_2, \ldots\ldots\ldots) \\ \dot{\omega}_2 &= h(\sigma, T, \omega_1, \omega_2, \ldots\ldots\ldots) \end{array}\right\} \qquad (6)$$

In previous published work[11-14], the explicit forms of Equation 6 used to represent creep behaviour have been derived empirically. The important feature of the present study has been to derive explicit forms of Equation 6 that are consistent with the current understanding of deformation and fracture outlined in the previous section. By retaining the differential formulation and integrating numerically, it is relatively simple to consider the effect of variable loading conditions.

The primary creep behaviour can be considered to be due to the development of an internal stress σ_i, due to stress redistribution in the heterogeneous material. Taking as the state variable $S = \sigma_i/\sigma$ where σ is the applied stress we can write for the linearised form of the model[4]:

$$\left.\begin{array}{rl} \dot{\epsilon} &= \dot{\epsilon}_{min} \frac{(1-S)}{(1-S_{ss})} \\ \dot{S} &= H\dot{\epsilon} - RS \end{array}\right\} \qquad (7)$$

where H, R are constants describing hardening and recovery respectively, S_{ss} is the steady state value of S and $R = H\dot{\epsilon}_{min}/S_{ss}$. Equations 7 integrate to exactly the form of Equation 1.

The tertiary creep behaviour can be represented by the following set of equations

$$\left.\begin{array}{rl} \dot{\epsilon} &= \dot{\epsilon}_i e^{\omega} \\ \dot{\omega} &= C\dot{\epsilon} \end{array}\right\} \qquad (8)$$

where $\dot{\epsilon}_i$ is the initial creep rate and the constant C is the sum of three terms representing intrinsic softening, loss of internal section and loss of external section[3]. Thus:

$$C = C_{int} + \frac{1}{3}\frac{n}{\epsilon_f} + n \qquad (9)$$

Integration of Equations 8 leads to Equation 3.

When both primary and tertiary creep occur, a two state variable description is required. The two expressions describing $\dot{\epsilon}$ are combined in product (rather than addition) form to account for interactions between the two mechanisms; addition would imply that primary and tertiary creep were independent and co-existed throughout the entire creep life. Thus:

$$\begin{aligned}\dot{\epsilon} &= \dot{\epsilon}_i\,(1-S)\,e^{\omega} \\ \dot{S} &= H\dot{\epsilon} - \frac{H\dot{\epsilon}_i}{S_{ss}}S \\ \dot{\omega} &= C\dot{\epsilon}\end{aligned} \qquad (10)$$

The four-parameter set ($\dot{\epsilon}_i$, S_{ss}, H, C) completely describes changes in the variables at a given loading condition and each may vary with stress and temperature.

A software package, designated CRISPEN, has been developed that utilises sets of equations such as Equation 10 to represent the full creep curves of engineering alloys. By analysing raw creep curves, a database of model parameters ($\dot{\epsilon}_i$, S_{ss}, H, C) can be constructed. Using this database, the strain/time evolution during arbitrary stress and temperature conditions can be simulated and, when available, compared with experimental data. The system is designed to allow interpolation and limited extrapolation to unknown conditions and to cope with variable stresses and temperatures. It has been written to operate on an IBM AT or SYSTEM 2 personal computer in order to facilitate dissemination. Consequently the analysis involves acceptable approximation consistent with limited memory available.

Data Analysis

Equation Sets 7 or 8 taken alone can be integrated analytically to describe primary and tertiary creep, but it has not proved possible to derive an analytical solution to the complete three-equation set (Equation 10). Consequently, derivation of the model parameters ($\dot{\epsilon}_i$, S_{ss}, H, C) from an arbitrary creep curve is not straight forward. Two different methods have been used with varying success depending on the nature of the creep curve.

a) Method 1 depends on the analysis of the creep curve where strain ϵ is displayed as a function of time. As discussed by Ion et al[4] the general shape of the creep curve can be represented by the four operational parameters (ϵ_p, t_p, $\dot{\epsilon}_{min}$, t_t) defined in Figure 2a. If primary creep is largely exhausted before significant strain rate acceleration occurs, then the model parameters required for subsequent calculation are simply related to these operational measures of the creep curve. This approach requires that $\dot{\epsilon}_{min} \sim \dot{\epsilon}_i$: where the inequality is not satisfied due to, for example, a dominant primary behaviour or distortion of these minimum creep rate by microstructural changes, then an alternative analysis is required.

b) Method 2 analyses the differential form of the creep curve where strain rate $\dot{\epsilon}$ is displayed as a function of strain ϵ. The strain softening parameter $C = d\ln\dot{\epsilon}/d\epsilon$ when $S \to S_{ss}$. Although $\dot{\epsilon}_i \neq \dot{\epsilon}_{min}$, and indeed is not directly measurable from the creep curve, it can be estimated by the procedure indicated graphically in Figure 2b. The primary parameters can similarly be derived as described more fully by Taylor and Ashby[15].

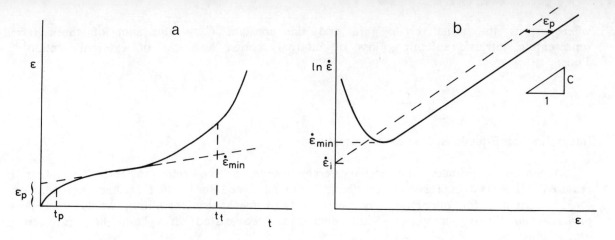

Figure 2 Schematic creep curves showing analysis procedures used in determining model parameters.
a) strain versus time – Method 1.
b) log (strain rate) versus strain – Method 2.

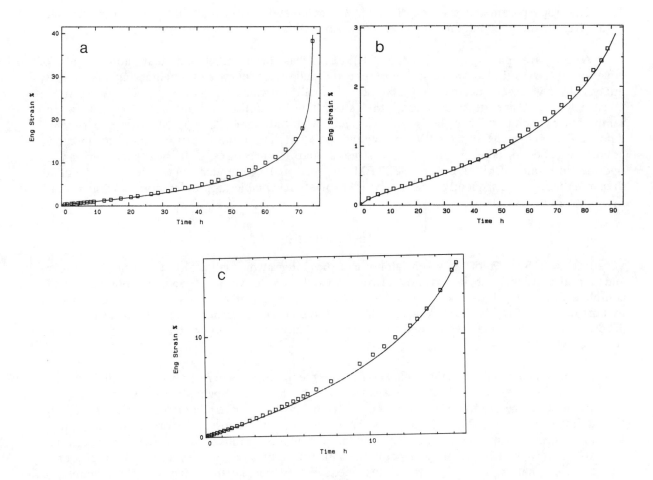

Figure 3 Examples of the agreement between experimental data (points) and the CRISPEN representation (line)
a) IN738LC, directionally solidified, constant load
    375 MPa/850°C
b) MarM002, conventionally cast, constant stress
    240 MPa/950°C
c) SRR99, single crystal, constant stress
    600 MPa/900°C

Figure 3 shows the agreement between the creep curves calculated using Equations 10 together with the model parameters derived by these procedures and the original creep data. Examples are shown for (i) constant load creep tests for directionally solidified IN738LC where the contribution of loss of internal section is insignificant, (ii) constant stress tests on MarM002 where low ductility is due to extensive grain boundary cavitation and (iii) single crystal SRR99 in constant stress conditions where crystal orientation is an important factor. Equation set 10 is clearly capable of representing the creep behaviour in all of these cases.

The parameter sets derived in this manner constitute a database representing the full shapes of the creep curves from which further calculations can be made. Inspection of Figure 3 shows that primary creep is relatively unimportant for most long term tests. The tertiary behaviour and the life are largely controlled by the model parameters $\dot{\epsilon}_i \sim \dot{\epsilon}_{min}$ and C. The procedures adopted for representing $\dot{\epsilon}_{min}$ as a function of stress and temperature in terms of power or exponential laws are well known and will not be discussed further here. Figure 4a shows that for directionally solidified IN738LC $C (\approx C_{int} + n)$ is relatively constant over a wide range of stresses and temperature. However, for MarM002 a variable C can be associated with differences in creep ductility, as shown in Figure 4b, indicating the dominance of loss of internal section by cavitation: from Equation 9, $C \sim n/3\epsilon_f$.

Predictive Calculations

The procedure used by CRISPEN is to base predictions of creep curves for arbitrary stress/temperature conditions on the known model parameters for a reference creep curve at a neighbouring condition. It is assumed that the values of H, S_{ss} and C for the arbitrary condition are identical to those for the reference curve, but an appropriate value of $\dot{\epsilon}_i$ is calculated from the representation of $\dot{\epsilon}(\sigma,T)$. In practice, the reference curve would be chosen to have a stress and temperature as close to the unknown condition as possible.

The following illustrations indicate extrapolations in time and to cyclic test conditions and compare the predictions with available experimental data.

a) Figure 5 shows a series of calculated creep curves for directionally solidified IN738LC using the data for a test at 200 MPa/1223K, which had a life of $\sim$ 50h, as a reference. The predictions are displayed as bands of ± 10% in life. Extrapolations across stress and/or temperature to times of up to $\sim$ 20,000h give agreement within this range.

b) The use of a single reference curve to predict the strain/time behaviour in conditions of changing stress and temperature are shown in Figures 6a and 6b respectively for directionally solidified IN738LC. Good agreement is obtained for these relatively small changes in test conditions. However, this should not be taken to indicate that CRISPEN is appropriate to calculate fatigue behaviour.

c) It is difficult to evaluate the overall self-consistency of such predictions from isolated comparisons. Figure 7 shows a Larson-Miller plot constructed for conventionally cast MarM002 by using CRISPEN to calculate lives for arbitrarily chosen conditions of stress and temperature using the available database. Experimental points for the tests constituting the database are shown for reference. The scatter in this predicted Larson-Miller plot of $\sim$ ± 10% in stress is no greater than the inherent scatter of the experimental data or of the normal scatter found in such parametric representations used for design purposes. The ability of CRISPEN to generate acceptable Larson-Miller curves provides confirmation of the reliability of the strain/time predictions.

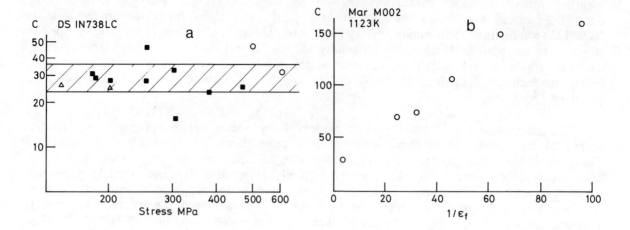

Figure 4 a) Tertiary creep parameter C for directionally solidified IN738LC as a function of stress and temperature.
b) Tertiary creep parameter C for conventionally cast MarM002 shown as a function of $1/\epsilon_f$.

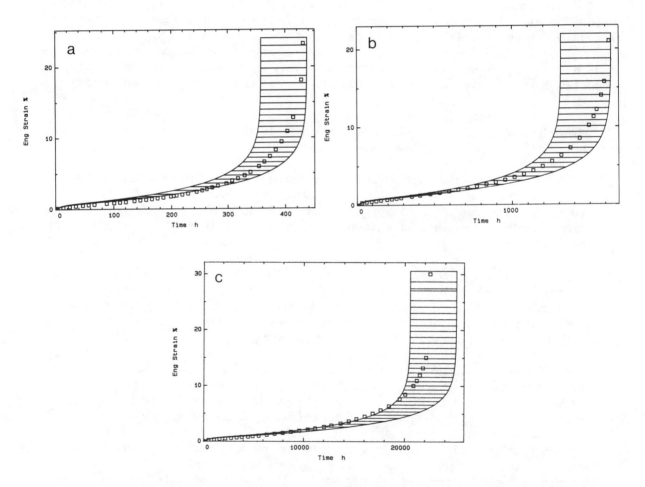

Figure 5 Comparison of experimental creep data with CRISPEN generated creep curves calculated using short term reference data (viz 200 MPa, 950 °C, t_f=40h).
a) 150 MPa, 950 °C   b) 250 MPa, 850 °C
c) 170 MPa, 850 °C.

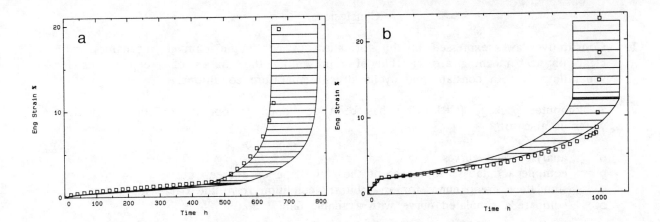

Figure 6 As Figure 5 but for conditions of changing stress and temperature
a) σ_1 = 250 MPa, 0 - 497 h; σ_2 = 300 MPa, 497 - 672.5 h;
T = 850 °C
b) T_1 = 900 °C, 0 - 49.5 h; T_2 = 850 °C, 495 - 9815 h;
T_3 = 900 °C, 981.5 - 999.2 h; σ = 250 MPa.

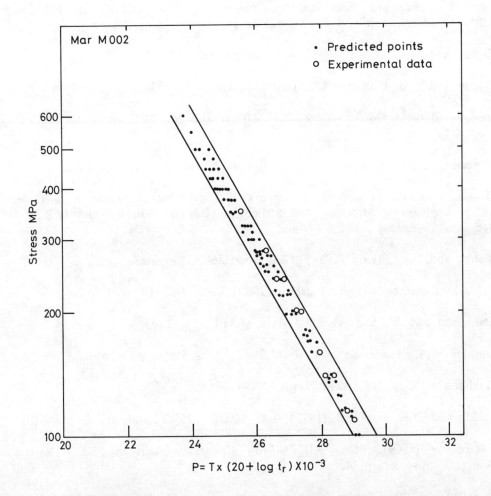

Figure 7 Larson–Miller plot for conventionally cast MarM002 calculated from the CRISPEN database for random stress/temperature conditions. Experimental data are included for comparison.

Conclusions

1. Constitutive laws, expressed in the formalism of continuum damage mechanics, with clear physical meaning are capable of representing the shapes of creep curves for superalloys in both constant and cyclic stress/temperature conditions.

2. A computer package (CRISPEN) has been developed to operate on these equations which is able to:

    o   analyse creep curves
    o   compile a database to represent the full shape of a creep curve
    o   simulate a creep curve for an arbitrary condition from the database
    o   compare a simulated curve with available data

3. The system has been evaluated on data for several nickel-base superalloys.

Acknowledgements

The authors thank Drs J C Ion and M Maldini for contributions at various stages of this NPL/Cambridge University collaborative programme.

References

1. L.M. Brown (1982) 3rd Risø Int. Symposium. "Fatigue and Creep of Composite Materials" (Eds H. Lilholt and R. Talrega) p.1.

2. B.F. Dyson and M. McLean (1983) Acta Metall., $\underline{31}$, 17.

3. B.F. Dyson and T.B. Gibbons (1987) Acta Matall., $\underline{35}$, 2355.

4. J.C. Ion, A. Barbosa, M.F. Ashby, B.F. Dyson and M. McLean. NPL Report DMA A115, April 1986.

5. G.A. Webster, A.P.D. Cox and J.E. Dorn (1969) Met. Sci. Jnl. $\underline{3}$, 221.

6. R.W. Evans, J.D. Parker and B. Wilshire (1982) in "Recent Advances in Creep and Fracture of Engineering Materials and Structures (Eds B. Wilshire and D.R.J. Owen) Pineridge Press Swansea, p.135.

7. M.F. Ashby and B.F. Dyson (1984) ICF6 New Delhi, Pergamon, p.3.

8. H. Burt, J.P. Dennison and B. Wilshire (1979) Metal Sci. $\underline{13}$, 295.

9. P.J. Henderson and M. McLean 1983 Acta Metall., $\underline{31}$, 1203.

10. P.N. Quested, P.J. Henderson and M. McLean. Acta Metall., (in press).

11. L.M. Kachanov (1958). Izv. Akad. Nank. SSSR No.8, 26.

12. F.A. Leckie and D.R. Hayhurst (1977) Acta Metall., $\underline{25}$, 1059.

13. Yu.N. Rabotnov (1969). Proc. XII IUTAM Congress, Stanford, Eds. Hetenyi and W.G. Vincenti, Springer, p.342.

14. J. Hult (1974) In "On Topics in Applied Continuum Mechanics" (Eds. J.L. Zeman and F. Ziegler) Springer p.137.

15. N.G. Taylor and M.F. Ashby work to be published.

EFFECT OF MINOR ELEMENTS

ON THE DEFORMATION BEHAVIOR OF NICKEL-BASE SUPERALLOYS

D. M. Shah and D. N. Duhl

Materials Engineering
Pratt & Whitney
400 Main Street
East Hartford, CT 06108

Abstract

Alloying with the minor elements carbon, boron, zirconium, and hafnium has played a very critical role in the development and application of nickel-base superalloys. The elements are known to improve strength and grain boundary cohesion, but a clear understanding of the underlying mechanisms has been lacking. To systematically address this, a series of 15 polycrystalline alloys were prepared with various minor element additions using Alloy 454 (alloy chemistry of PWA 1480) as the base alloy. Intermediate temperature (1400°F) creep testing was used as a means for evaluating the effectiveness of the minor elements. Typically, a fifty-fold improvement in creep-rupture life was attained with ~0.5 atom percent minor element additions. Although the increase in elongation to failure, in an engineering sense, was insignificant, the synergistic effect of adding combinations of the elements resulted in even greater enhancements. It was not possible to rationalize the results with any consistent grain boundary microstructural changes. To further understand the effects of minor element additions, eight single crystal alloys were cast with optimum additions of minor elements using Mar-M200 as the base alloy. Intermediate temperature (1400°F) creep behavior was evaluated in the three major orientations <001>, <011>, and <111>. In general, the addition of the minor elements did not alter the behavior of the <001> or the <011> orientations, but decreased the creep resistance of the <111> orientation. Again, the synergistic effect of combinations of the minor elements was very significant. The decrease in the creep rate of the polycrystalline material is believed to be a manifestation of the mechanism that leads to the decrease in the creep rate of <111> oriented single crystals. This in turn is attributed to an intrinsic strengthening of the cube slip system which is geometrically favored near the <111> orientation. It is proposed that the suppression of cube slip, with lower multiplicity than octahedral slip, enhances strain compatibility in the grain boundary regions. These areas are enriched in the minor elements, due to preferential segregation of minor elements to the grain boundaries.

Introduction

Alloying with the minor elements carbon, boron, zirconium, and hafnium has played a very critical role in the development and application of nickel-base superalloys (1). Ductile polycrystalline superalloys would not be a practical reality without the presence of these elements within a narrow composition range well defined by experience. Similarly, hafnium additions to directionally solidified superalloys has been found to be a significant milestone in the application of columnar grained hollow turbine airfoil castings (2). While the underlying mechanisms are not well understood, the benefits of adding minor elements to nickel-base superalloys are generally believed to be related to improvements in grain boundary strength and cohesion. With the advent of single crystal nickel-base superalloys, these elements were deemed unnecessary and their concentration reduced to increase alloy melting point, widen the solution heat treatment range, and broaden alloying flexibility (3). While interest in minor element additions to superalloys has matured, the ductilizing effect of boron additions to the intermetallic compound Ni_3Al is generating renewed enthusiasm for the application of the concept to other high temperature materials and in particular intermetallics (4).

Preferential segregation of minor elements to grain boundaries is easily rationalized on the basis of the atomic size difference of over 25 percent between the minor elements and the average size of the nickel atom. Consequently, it is natural to attribute and limit the role of minor elements to grain boundaries and a host of mechanisms related to second phases, microstructural changes, electronic structure, scavenging effects, and other processing effects (5,6). To systematically understand the effect of minor elements, an intermediate temperature creep study of a series of minor element modified polycrystalline superalloys was undertaken. The experimental details and results of this study are presented in Part I below.

Besides these obviously grain boundary related mechanisms, the intrinsic effect of these elements on the deformation behavior of the grain cannot be ignored. A comparison of the yield strength anisotropy of single crystal Mar-M200 with minor elements (7) and PWA 1480 with no minor elements (8) first hinted at such a possibility. To establish the validity of any intrinsic effect of the minor elements, the effect of these elements on single crystal superalloys was also evaluated. The experimental details and results of this effort are presented in Part II of this paper. This is followed by a discussion of a consistent reconciliation of the results of both Parts I and II in terms of the intrinsic effects of minor elements on the slip behavior of superalloys.

Part I - Polycrystalline Alloys

Experimental Procedure

A series of 15 polycrystalline alloys were prepared with varying concentrations of carbon, boron, zirconium, and hafnium individually, with four of these alloys having combinations of these elements as presented in Table I. Alloy 454 (alloy chemistry of PWA 1480) was chosen as the base alloy, and minor elements were added as late additions using a conventional casting process. The liquidus and solidus temperatures were determined using differential thermal analysis (DTA) as listed in Table I, and subsequently casting parameters were adjusted to avoid unusually coarse grain material. All the cast material was hot isostatically pressed (HIP) and heat treated at temperatures listed in Table I. Intermediate temperature creep testing (1400°F/ 45 ksi) was used as a sensitive means for evaluating the effectiveness of the minor elements.

Table I. Composition, DTA Results and Heat Treatment Temperatures of Conventionally Cast Polycrystalline Alloys

Alloy	Weight % Minor Elements				DTA Results		HIP Temp. °F	Solution Heat Treat Temp. °F
	C	B	Zr	Hf	Liquidis °F	Solidus °F		
454	-	-	-	-	2485	2377	2250	2350
455	0.018	-	-	-	2477	2372	2250	2350
456	0.062	-	-	-	2477	2346	2250	2350
457	0.12	-	-	-	2477	2379	2250	2350
458	0.15	-	-	-	2483	2379	2250	2350
459	-	-	-	0.38	2467	2307	2150	2225
460	-	-	-	1.05	2468	2205	2150	2175
461	-	0.015	-	-	2464	2288	2150	2225
462	-	0.10	-	-	2451	2249	2150	2175
463	-	-	0.05	-	2477	2366	2250	2325
464	-	-	0.1	-	2464	2359	2250	2325
465	-	0.015	0.05	-	2470	2359	2250	2275
466	0.1	0.015	0.05	-	2468	2373	2250	2275
467	0.1	0.015	0.05	1.0	2462	2225	2150	2175
468	0.05	0.1	0.05	-	2462	2221	2150	2175

Base Composition: Ni-5 Co-10 Cr-4 W-12 Ta-5 Al-1.5 Ti (weight %)
Heat Treatment: HIP 4 hours/15 ksi + Sol. heat treat. 4 hours/air cool
 + 1975°F/4 hours + 1600°F/32 hours

Results

The creep curves for alloys with individual additions of carbon, boron, zirconium, or hafnium along with a creep curve for the minor element free baseline Alloy 454 are presented in Figures 1(a) through 1(d), respectively. Results for alloys with combinations of the minor elements are presented in Figure 2. It is clear from Figure 1 that among the individual elements, zirconium is the least and hafnium the most effective in reducing the creep rate. The synergistic effect of combining the elements was very great as evident in Figure 2. In all cases, improvement in rupture life is a result of a significant decrease in the minimum creep rate. The increase in the elongation to failure from 0.1% to 0.5%, in an engineering sense, is not very significant. These results could not be rationalized with any consistent microstructural changes. For example, a fivefold decrease in the grain size between hafnium modified alloys and alloys modified with a combination of minor elements does not have any significant effect on creep behavior. In all cases, except carbon, the rupture life continually improves with increasing concentrations of the minor elements. In the case of individual additions of carbon, 0.05 weight percent seems to be optimum. Typically, a fifty-fold improvement is attained with ~0.5 atom percent additions as shown in Figure 3, where the rupture life is plotted against total analyzed atom percent of the minor elements. Beyond that concentration level, the gain is marginal or declines for the case of carbon.

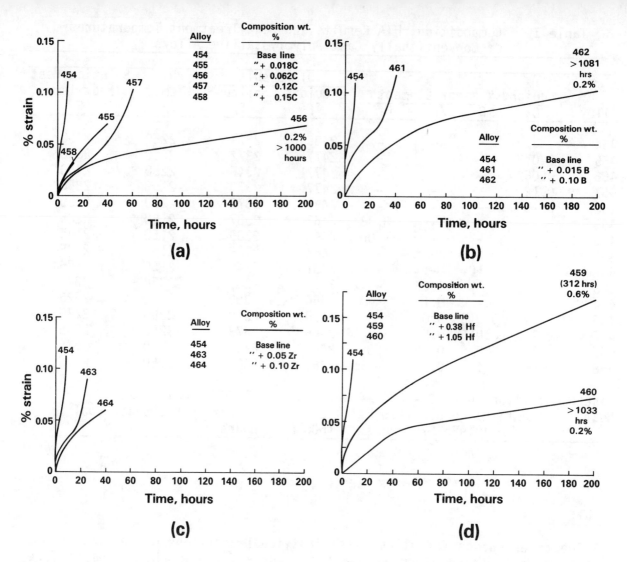

Figure 1 - Effect of individual additions of minor elements (a) carbon, (b) boron, (c) zirconium, and (d) hafnium, on 1400°F/45 ksi tensile creep behavior of conventionally cast polycrystalline alloys.

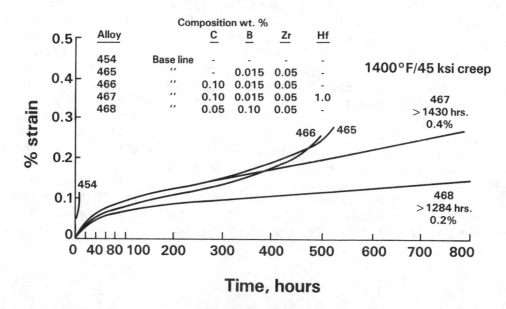

Figure 2 - Effect of combined additions of minor elements on 1400°F/45 ksi creep behavior of conventionally cast polycrystalline alloys.

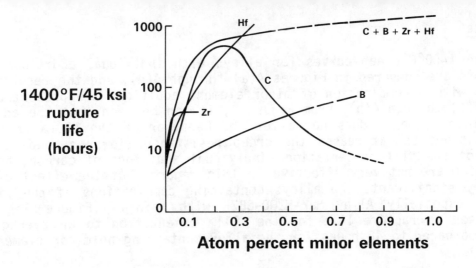

Figure 3 - Rupture life at 1400°F/45 ksi versus total atom percent of minor elements in conventionally cast polycrystalline Alloy 454.

Part II - Single Crystal Alloys

Experimental Procedure

In the second part of this study, single crystals of a series of eight alloys were cast using a Mar-M200 baseline. These included one alloy devoid of minor elements; four alloys containing optimum concentrations of carbon, boron, zirconium, and hafnium; and three alloys having combinations of the elements. Nominal compositions of the alloys are presented in Table II along with the adjusted solution heat treatment temperature for each alloy. Standard creep specimens within 10° of the three major orientations <001>, <011>, and <111> were machined and tested at 1400°F/110 ksi for each alloy. In a few cases, compression specimens were also machined and tested at the same condition. In addition, the temperature dependence of the yield strength was determined for several alloys.

Table II. Composition and Heat Treatment Temperatures of Single Crystal Alloys

Alloy	Weight Percent Minor Elements				Solution Temperature °F
	C	B	Zr	Hf	
Mar-M200-NME	-	-	-	-	2300
Mar-M200-C	0.10	-	-	-	2300
Mar-M200-B	-	0.012	-	-	2300
Mar-M200-ZR	-	-	0.018	-	2300
Mar-M200-HF	-	-	-	1.82	2300
Mar-M200-BZ	-	0.095	0.015	-	2300
Mar-M200-CBZ	0.089	0.013	0.032	-	2300
Mar-M200-CBZH	0.14	0.018	0.064	2.07	2300

Base Alloy: Ni-10 Co-9 Cr-12.5 W-1.0 Cb-5 Al-2 Ti (weight %)
Heat Treatment: Sol. Heat Treat 4 hours/air cool + 1975°F/4 hours + 1600°F/32 hours.

Results

The 1400°F creep curves for alloys with individual additions of minor elements are compared in Figures 4(a) through 4(e), and the results for the alloys with a combination of minor element additions are presented in Figures 4(f) through 4(h). From Figure 4, it can be seen that the addition of the minor elements does not alter the behavior of the <001> or the <011> orientations but increases the creep resistance (lower second-stage creep rate) of the <111> orientation. Individual additions of carbon, boron, and zirconium are not very effective in this regard, but the effect of hafnium is very significant. The alloys containing combinations of the minor elements, especially Alloy Mar-M200-CBZH with hafnium (Figure 4h), exhibit increases in rupture life for the <111> orientation to an average of 240 hours compared to 48 hours for the alloy containing no minor elements (Figure 4a).

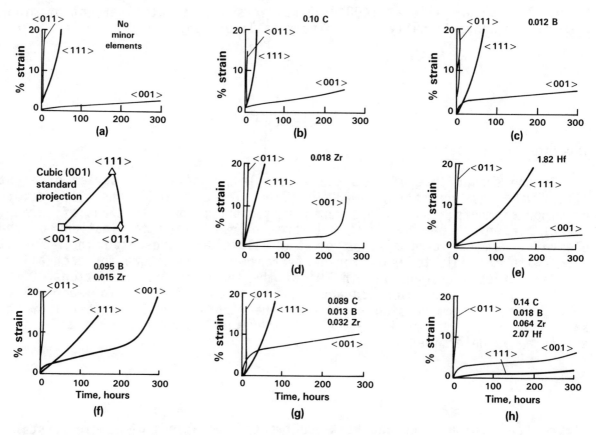

Figure 4 - Effect of minor element additions on 1400°F/110 ksi creep anisotropy of single crystal Mar-M200; (a) no minor elements, (b) to (e) individual additions, and (f) to (h) combination of elements.

In all cases, <001> is observed to be the most creep resistant and <011> the least creep resistant orientation in tension. Differences in rupture life are observed to be very significant, being less than 10 hours for the <011> oriented specimens in contrast to over 300 hours for the <001> oriented specimens in most cases. However, the tensile creep behavior for the two orientations is completely reversed in compression, as shown in Figure 5 for the alloy with no minor elements and Alloy Mar-M200-CBZH containing a combination of minor elements. For the alloy with no minor elements, the <011> orientation with the highest creep rate in tension becomes the one with the lowest creep rate in compression and vice versa for the <001> orientation. For Alloy Mar-M200-CBZH, the effect is not as dramatic. However, neither the creep rate, the creep anisotropy, nor the tension compression asymmetry of the <001> and the <110> orientation are altered by the addition of minor elements. The effect of minor elements is limited to the deformation behavior of the <111> orientation at intermediate temperatures.

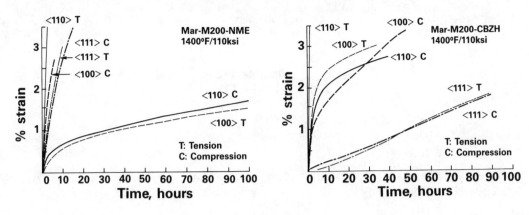

Figure 5 - Observation of tension compression assymetry in 1400°F/110 ksi creep behavior of <100> and <111> orientations for Mar-M200 with and without minor elements.

Discussion of Results

The observed decrease in the creep rate of polycrystalline superalloys with minor element additions, described in Part I, is consistent with earlier work (9). The general validity of the effect for different superalloys and for various combinations of minor element additions indicates the intrinsic difficulty in correlating the effect of minor element additions with microstructural details. In contrast, the parallel between the decrease in the creep rate of polycrystalline alloys (Part I) and the reduction in the creep rate of <111> oriented single crystals (Part II) with the addition of minor elements is very revealing. It points to a common mechanism at least not completely related to grain boundaries. The fact that the base alloy in the polycrystalline study was Alloy 454, and for the single crystal study was Mar-M200, is not relevant. Published results of a 1400°F creep anisotropy study of single crystal alloys CMSX-2 and 454 are comparable to the results presented here for single crystal Mar-M200 containing no minor elements (10).

To understand the effect of minor elements, it is necessary to focus on the deformation behavior of <111> oriented single crystals versus <001> and <011> oriented single crystals. This is well understood for superalloys in terms of octahedral and cube slip behavior as in the case of $L1_2$ compounds (11). Octahedral slip is favored near the <001>-<011> orientations, while cube slip is favored near the <111> orientation as shown geometrically in Figure 6. To what extent cube slip participates in the deformation process depends on alloy composition and temperature, but at higher temperatures, a greater participation of cube slip is well recognized (11). Without going into the details of the cross slip and dislocation core-constriction mechanisms, it is apparent that as the primary effect of the minor elements is on the deformation behavior of the <111> orientation, the minor elements must affect cube slip more than they affect octahedral slip. The deformation behavior of the <001> and <011> orientations, which are predominantly controlled by octahedral slip, are not affected by minor element additions. The observed tension/compression asymmetry in creep, irrespective of minor element additions, is similar to that observed with respect to the yield strength and can be attributed to cross slip and dislocation core-constriction mechanisms associated with octahedral slip which are not available for cube slip (8).

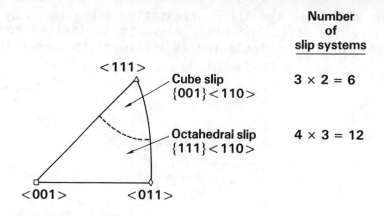

Figure 6 - Relationship between single crystal orientation and operative slip systems in superalloys.

A comparison of the plot of yield strength versus temperature for Mar-M200 with and without minor elements is presented in Figure 7 along with the 1400°F/110 ksi creep curves for the <001> and <111> orientations. For the alloy with no minor elements, the <111> orientation has a lower yield strength and poorer creep resistance. For the alloy with minor elements, both orientations have comparable yield strength and creep resistance.

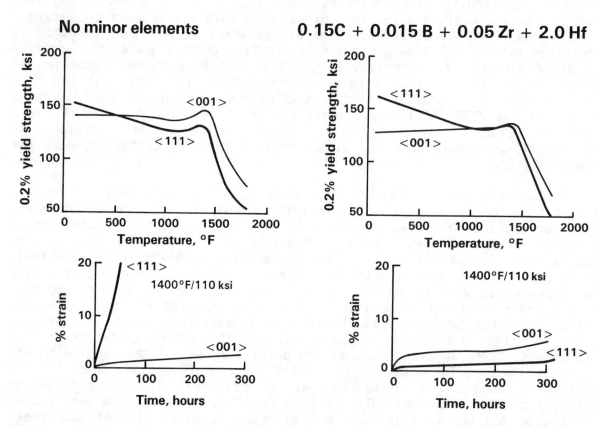

Figure 7 - Comparison of temperature dependence of yield strength and 1400°F/110 ksi creep behavior of Mar-M200 single crystal in <001> and <111> orientations with and without minor elements.

The observation that the minor elements make deformation occurring by cube slip more difficult than that occurring by octahedral slip may be rationalized in terms of an intrinsic dislocation core effect. One does not expect the minor elements to affect the antiphase boundary energy or the stacking fault energy and hence slip on closely packed octahedral planes. However, with the significant difference in atomic size between the minor elements and nickel, the lattice friction is expected to increase and hence slip on more widely spaced cube planes should be affected. Addition of minor elements can change the morphology and size of the gamma prime precipitates which can affect alloy properties. Gamma prime rafting has been demonstrated to affect creep anisotropy (12), but such a mechanism is difficult to isolate from an intrinsic effect without further controlled experiments.

To reconcile the results obtained for the polycrystalline and single crystal material, let us approximate a randomly oriented polycrystal as a collection of single crystals. A polycrystalline material is more like a group of <123> oriented grains because this orientation has the highest (i.e., 24) crystallographically equivalent variants. In this orientation, both octahedral and cube slip systems are equally stressed. If cube slip is favored, however, then qualitatively the <123> direction is more like the <111> direction than the <001> direction. In this simplistic model, the decrease in the creep rate for polycrystalline material with minor element additions parallels the decrease in the creep rate for the <111> oriented single crystals. The mechanics of deformation within a grain is not as simple as in the case of an unconstrained single crystal, but the argument seems consistent with the observations. Consistent with the argument, it can be further conjectured that the addition of minor elements improves the intergranular strain compatibility of polycrystalline superalloys by suppressing cube slip with lower multiplicity, thereby promoting octahedral slip with higher multiplicity. With preferential segregation of the minor elements at the grain boundaries, accommodation of localized deformation in the grain boundary is enhanced with increased multiplicity of the slip systems in the regions near the boundaries.

Summary

1. A fifty-fold improvement in intermediate temperature creep rupture life of polycrystalline superalloys is achieved with about a 0.5 atom percent addition of a combination of minor elements, especially with hafnium as one of the minor elements. In an engineering sense, the elongation to failure is not enhanced significantly.

2. Minor element additions significantly enhance the intermediate temperature creep resistance of <111> oriented single crystals which are favorably oriented to deform by cube slip, but have no effect on the creep behavior of <001> or <011> oriented single crystals which are expected to deform by octahedral slip. The <001> and <011> orientations also show a tension/compression asymmetry in creep similar to the observed asymmetry in yield strength.

3. The beneficial effect of minor elements on polycrystalline superalloys and their effect on single crystal superalloys is reconciled in terms of suppression of cube slip with lower multiplicity relative to octahedral slip with greater multiplicity.

Acknowledgements

The authors gratefully acknowledge the technical assistance of Mr. C. L. Calverley and Mr. R. E. Doiron as well as other support groups within Pratt & Whitney; and Ms. D. L. Avery and Mr. E. T. Brown for their assistance in the preparation of the final manuscript.

References

1. R. F. Decker, "Strengthening Mechanisms in Nickel-Base Superalloys," Steel-Strengthening Mechanisms (Climax Molybdenum Company, 1969), 147-183.

2. D. N. Duhl and C. P. Sullivan, "Some Effects of Hafnium Additions on the Mechanical Properties of a Columnar-Grained Nickel-Base Superalloy," J. of Metals, 7(1971), 38-40.

3. M. Gell, D. N. Duhl and A. F. Giamei, "The Development of Single Crystal Superalloy Turbine Blades," Superalloys 1980, (American Society for Metals, 1980), 205-214.

4. K. Aoki and O. Izumi, "Improvement in Room Temperature Ductility of the $L1_2$ Type Intermetallic Compound by Boron Addition," J. Japan Inst. Met., 43(1979), 1190-1196.

5. A. K. Jena and M. C. Chaturvedi, "Review the Role of Alloying Elements in the Design of Nickel-Base Superalloys," J. of Mat. Science, 19(1984), 3121-3139.

6. R. F. Smart, "Effects of Foundry Variables on Cast Nickel-Base Superalloys," Metallurgia and Metal Forming, 7(1977), 286-294.

7. B. H. Kear and B. J. Piearcey, "Tensile and Creep Properties of Single Crystals of the Nickel Base Superalloy Mar-M200," Trans. TMS-AIME, 239(8)(1967), 1209-1215.

8. D. M. Shah and D. N. Duhl, "The Effect of Orientation, Temperature and Gamma Prime Size on the Yield Strength of a Single Crystal Nickel Base Superalloy," Superalloys 1984, (American Society for Metals 1984), 105-114.

9. D. R. Wood and R. M. Cook, "Effects of Trace Contents of Impurity Elements on the Creep-Rupture Properties of Nickel-Base Alloys," Metallurgia, 3(1963), 109-117.

10. P. Caron, T. Khan and Y. G. Nakagawa, "Effect of Orientation on the Intermediate Temperature Creep Behavior of Ni-Base Single Crystal Superalloys," Scripta Metall., 20(1980), 499-502.

11. D. P. Pope and S. S. Ezz, "Mechanical Properties of Ni_3Al and Nickel Base Alloys with High Volume Fraction of Gamma Prime," Int. Met. Rev., 29(1984), 136-167.

12. D. D. Pearson, F. D. Lemkey and B. H. Kear, "Stress Coarsening of Gamma Prime and its Influence on Creep Properties of Single Crystal Superalloy," Superalloys 1980, (American Society for Metals, 1980), 513-520.

SUPERALLOYS WITH LOW SEGREGATION

Zhu Yaoxiao, Zhang Shunnan, Xu Leying
Bi Jing, Hu Zhuangqi and Shi Changxu

Institute of Metal Research, Academia Sinica,
Shenyang, P.R.China

Abstract

Segregation due to solidification of highly alloyed superalloy is a very complex process. The sequence of phase formation, change of composition and solute distribution during solidification have been determined. Dendritic segregation and ($\gamma+\gamma'$) eutectic reaction segregation have been discussed, that results in finding out a way to suppress the precipitation of harmful phase. A series of low segregation superalloys have been successfully developed with excellent mechanical properties and prolonged stability by careful control of alloy composition and solidification parameters.

Introduction

The history of superalloy application to gas turbine lasts for more than forty years, and the developed superalloys possess promising mechanical properties to a very high level. The service temperature of cast nickel-base superalloy used as blade material climbs up to more than 80% of its absolute melting point. The yield stress of turbine disk materials exceeds 1000MPa, even without drastic drop at 700°C. However, the progress seems to be dragged in this decade, and it requires much more effort to have a step forward. It is primarily due to the ever increasing of alloying level, that makes serious segregation, lowers phase stability and impairs workability and castability, that requires strict control in the processing. Therefore, decreasing the solidification segregation of superalloy is a main powerful method to improve the known superalloys and develop new alloys with better combined properties.

Because there exists serious segregation in cast superalloy and easy formation of ($\gamma+\gamma'$) eutectic in the interdendritic region, that accelerates the precipitation of harmful phases (σ, μ, etc.). For this reason, it restricts further alloying and improvement of the mechanical properties of superalloy (1-5).

In this paper, the segregation mechanism and formation of ($\gamma+\gamma'$) eutectic during solidification has been studied in order to seek for a method to control the precipitation of harmful phases and develop new superalloys with low segregation.

Experimental Procedure

Samples of 15x15x15 mm were cut from the alloy bars. The sample placed in a graphite boat was remelted and resolidified in a silicon carbide tube furnace (6). The fluctuation of the furnace temperature was controlled within ±3°C. There is no serious oxidation, but a little carburization, usually less than 0.03%, that will not exceed the maximum carbon level according to the alloy specification. Table 1 is an example of the composition change before and after remelting.

Table 1. Composition Change before and after Sample Remelting (wt%)

Element	C	Cr	Mo	Co	Al	Ti	V	B	Zr
Befor Remelting	0.18	9.0	2.80	14.0	5.41	4.76	0.73	0.010	0.055
After Remelting	0.21	8.9	2.81	13.8	5.35	4.71	0.73	0.012	0.063

The sample was heated to and kept at 1673K for 5 min to guarantee complete dissolution of TiC and composition homogenization. It was slowly cooled to a certain temperature, maintained at this temperature for 10 min. and rapidly quenched in a saline water solution. The segregation and phase compositions were analyzed by electron microprobe. In the above resolidification condition, it is considered that there is no change in the solid portion, but complete uniformity in the liquid portion. Sometimes the residual liquid after quenching still showed inhomogeneity microscopically. Particularly when residual liquid portion is much, there appeared dendritic structure after resolidification. Therefore it was suitable to use electron beam with large size to determine the liquid composition. The average composition was calculated by repeated determinations at various places. Because there is a difference between the supposed condition and real solidification process, in addition to the analytical error of electron microprobe, the discussion is based on a semiquantitative scale.

Experimental Results

Order of Phase Formation During Solidification

Table 2 summarizes the sequence of formation of various phases during solidification and their rough quantitative relationship. The precipitation of γ from the liquid phase is a main process during solidification, i.e. the reaction L→γ+L' goes nearly from the beginning to the end, and most part is taken place between 1613K to 1563K. Fig.1 shows the microstructures quenched at various temperatures. When nearly one half of the liquid solidifies into γ, TiC starts to precipitate as shown in Fig.2. Most of the TiC is formed above 1566K.

Table 2 Order of Phase Formation During Solidification

T K(°C)	Liquid Phase (%)	γ (%)	(γ+γ') Eutectic (%)	TiC (%)	Y	High Cr Phase
1623(1350)	100					
1613(1340)	87	13				
1603(1330)	57	42.8		0.2		
1583(1310)	21	77.7		1.3		
1566(1293)	10	88		2		
1533(1260)	2.2	95.8		2	Trace	
1523(1250)	1.6	95.1	1.3	2	Minor	
1503(1230)	0.7	95	2.3	2	Minor	
1483(1210)	Minor	95	3	2	Minor	Trace

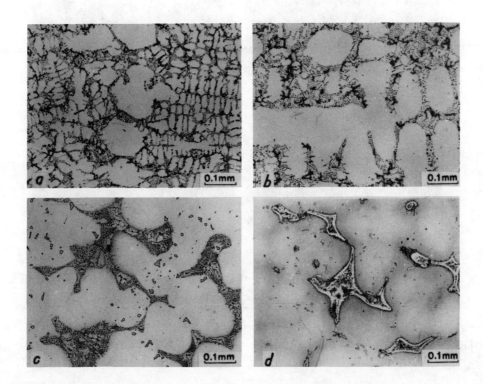

Fig.1 Microstructures quenched at various temperatures
(a) quenched at 1613K (13%γ+87%L) (b) quenched at 1603K (43%γ+57%L)
(c) quenched at 1583K (79%γ+21%L) (d) quenched at 1566K (90%γ+10%L)

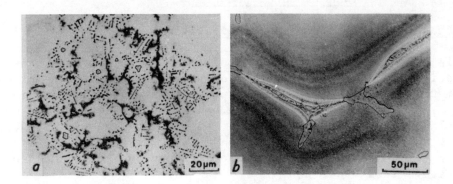

Fig.2 Morphology of TiC precipitated from residual liquid
(a) quenched at 1603K, (b) quenched at 1533K.

Beginning from 1523K, L → (γ+γ')+L' eutectic reaction takes place, and nearly completes in a narrow temperature range of 1523K - 1503K (Fig.3-a,b). In (γ+γ') eutectic formed at higher temperature, the structure is finer (Fig.3-c), while that formed at lower temperature, it is much thicker and the majority is γ' phase (Fig.3-d).

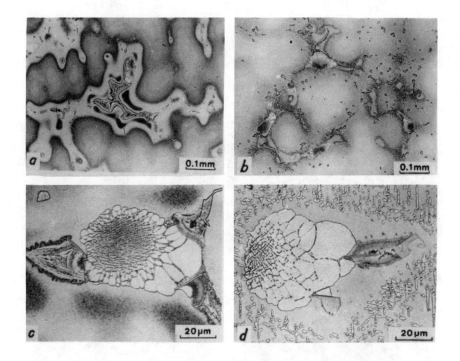

Fig.3 Morphology of (γ+γ') eutectic (a) and (c) quenched at 1523K,
(b) quenched at 1503K, (d) quenched at 1483K.

Between 1533K and 1483K, Y phase (Ti_2CS) forms. Below 1483K, a phase having high Cr, Mo contents was found, roughly expressed as 25Ni-9Co-34Cr-15Mo-13Ti-3Al-1V.

Change of Phase Compositions and Distribution of Alloying Elements

Table 3 collects the compositions of phases formed at various temperatures and the distribution coefficients (K) of alloying elements. In the temperature range when L→γ+L' reaction is dominated, it is possible to use $K(C_r/C_L)$ to express the redistribution of alloying elements. Among them, C_L and C_r represent the concentrations in liquid phase and γ phase respectively. K_γ of Ni and Co are greater than 1, and those of Cr, Mo and Ti smaller than 1. Particularly K_γ of Ti is far less than 1, that means the Ti content in γ phase is far less than that in the residual liquid, or as we say it is positive dendritic segregation. K_γ of Al and V are nearly at 1.

Table 3. Chemical Composition (wt%) and Distribution Coefficient (K) of Various Phases Formed at Different Temperatures

T (K)	phase	compositions						
		Ni	Co	Cr	Mo	Ti	Al	V
1613	γ	65.1	14.8	8.7	2.4	2.8	5.0	0.74
	L	63.1	12.9	8.8	2.9	4.8	5.4	0.76
	K_γ	1.03	1.15	0.99	0.83	0.58	0.92	0.96
1603	γ	64.9	14.0	8.5	2.3	3.1	4.7	0.76
	L	62.1	13.3	9.0	3.1	5.4	4.6	0.75
	K_γ	1.04	1.05	0.94	0.73	0.59	1.02	1.01
1583	γ	65.6	14.0	9.0	2.6	4.1	5.1	0.75
	L	59.9	13.2	9.2	3.7	7.1	4.8	0.75
	K_γ	1.10	1.06	0.92	0.72	0.57	1.04	1.00
1566	γ	65.4	14.1	8.9	2.5	4.7	5.2	0.66
	L	59.2	12.9	9.7	3.8	7.8	4.3	0.67
	K_γ	1.10	1.10	0.92	0.66	0.60	1.21	0.99
1533	γ	64.0	12.6	8.5	2.4	5.1	5.2	0.55
	L	56.0	12.9	10.8	3.7	7.2	3.9	0.55
	K_γ	1.14	0.98	0.79	0.65	0.73	1.33	1.00
1523	γ+γ'	67.1	12.4	6.1	6.1	7.3	5.2	0.56
	γ'	69.6	10.1	3.4	3.4	8.7	6.4	0.49
	γ	63.2	14.2	9.3	9.3	5.5	4.8	0.59
	L	49.9	14.4	12.7	5.3	7.2	3.2	0.48
	K_γ	1.27	0.99	0.76	0.47	0.77	1.50	1.21
	$K_{(\gamma+\gamma')}$	1.34	0.88	0.48	0.34	1.01	1.60	1.20
1503	γ+γ'	69.9	10.1	3.8	0.94	9.3	5.7	0.47
	γ'	70.5	10.0	3.4	0.82	9.5	5.8	0.46
	γ	62.8	13.8	11.6	2.0	5.6	4.8	0.57
	L	46.9	13.0	14.6	6.8	6.6	2.6	0.53
	K_γ	1.34	1.06	0.79	0.30	0.85	1.81	1.07
	$K_{(\gamma+\gamma')}$	1.50	0.78	0.26	0.14	1.40	2.20	0.90

In the range of 1523K-1503K, besides L→γ+L' reaction, an additional reaction L→(γ+γ')+L' takes place. Therefore, two distribution ceofficients, K_γ and $K_{(\gamma+\gamma')}$ can be obtained. K_γ and $K_{(\gamma+\gamma')}$ of Cr, Mo and Co are less than 1, that means they are enriched in liquid phase to a great extent. On the contrary, K_γ and $K_{(\gamma+\gamma')}$ of Al, Ni and V are greater than 1, that means the residual phase is depleted of Al, Ni and V. Below 1483K, a phase with high Cr and Mo contents is precipitated.

Discussion

Formation of (γ+γ') Eutectic

A well grown eutectic shown in Fig.4 can be divided into three parts: I is eutectic nucleation region, II is eutectic core and III is eutectic cap region. The lower frame of Fig.5-a represents the interdendritic residual liquid, and solidified γ phase is outside the frame. Suppose the residual liquid is homogeneous, in which Al+Ti content is a, less than the necessary Al+Ti content in eutectic composition L_1, so there has no eutectic reaction.

The Al+Ti content in γ formed in equilibrium with liquid is b. γ beyond the interface has a lower (Al+Ti) content than b.

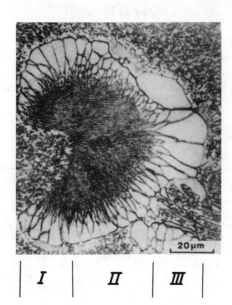

Fig.4 Typical microstructure of (γ+γ') eutectic
I-eutectic nucleation region
II-eutectic core
III-eutectic cap region

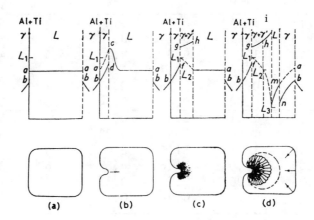

Fig.5 Schematic representation of (γ+γ') eutectic formation
(a) liquid, (b) incubation stage, (c) eutectic core formation
(d) formation of eutectic cap and depleted γ' zone

Fig.5-b represents the incubation stage of eutectic solidification. Usually a local place will grow a little bit faster, extruded into the liquid region. Due to strong positive dendritic segregation of Ti, the liquid adjacent to γ phase is progressively enriched with Ti and depleted with Ni, as expressed from a to c. At the same time, the (Al+Ti) content in γ phase also rises from b to d. When the Al+Ti content reaches the required amount, this stage ends.

Fig.5-c represents the stage of eutectic core formation. At the beginning, the reaction does not need any help for long range diffusion. Solidification proceeds faster and the lamellar structure is finer. Cr and Mo contents in liquid phase increase, and Al+Ti contents drops from L_1 to L_2. If L_2 is equal to a, this stage stops. In this stage, the Al+Ti content in eutectic rises from g to h.

Fig.5-d explains the formation mechanism of the eutectic cap. On the one side, the eutectic reaction exhausts Al and Ti, and on the other side, the L→γ+L' reaction liberates Al and Ti. They promote mutually. When the eutectic reaction goes on, the Al+Ti content in eutectic increases from h to i, while in liquid phase decreases from L_2 to L_3. The difference between i and L_3 becomes larger, but the amount of Al+Ti provided from the surrounding is not enough, only about m-n, that means the short supply of Al+Ti from the liquid phase in front of the eutectic, and retards the eutectic reaction. The lamellar tends to thicken in the eutectic cap region. The composition of the liquid phase decreases from a to m, and γ composition from b to n. If the Cr and Mo solutes accumulating in the residual liquid is not high enough to form a new phase, the above-mentioned reactions will continue until all the residual liquid solidifies.

Control of σ Phase Formation

There are two methods to control the σ phase formation, adequate design of alloy composition and strict control of solidification process. The main strengthening alloying elements in superalloy are Al and Ti, but the total amount of Al and Ti cannot be too high, otherwise σ phase will be formed. Because Ti is a solute having strong positive dendritic segregation, that promotes the eutectic reaction and accumulates Cr and Mo in the residual liquid. It is advisable to control the ratio of Al and Ti in order to suppress the σ phase formation.

To increase the solidification rate in order to refine the eutectic structure and lower the segregation is another useful way to suppress the σ phase precipitation. The reasons are: First, the dendritic arm spacing is smaller or the room for the eutectic reaction is smaller, which restricts the segregation. Second, it is shown in Fig.9-b that the concentration gradient of Al+Ti on the liquid side at the solid-liquid interface becomes steeper, so the eutectic core and cap are smaller.

Development of Superalloys with Low Segregation

Under the guidance of the above view point, we systematically studied the solidification behavior of various superalloys and the segregation characteristics of various elements. A series of low segregation superalloys have been developed. By means of the new idea, the ($\gamma+\gamma'$) eutectic usually appears in commercial cast nickel-base superalloys is now not observed (Fig.6). We can raise the alloying level of superalloys without appearance of the brittle σ phase under the service condition. It means that the stress rupture strength of the superalloys can increase by 30/40 MPa. In addition, it should be pointed out that the new way specially favors the directional solidification process. No Hf is necessary for directionally solidified superalloys in order to enhance the transverse performance. Table 4 shows the compositions of the low segregation superalloy and their stress rupture properties.

Iron-nickel-base superalloys are main disk materials for gas turbine. But, they usually contain many big block-shaped brittle phases. For example, Laves phase (AB_2) in electroslag ingot of IN718 which cannot completely be eliminated by long-time homogeneous annealing and repeated forging, and so it is difficultly forged into a big disk because of both lower starting forging temperature (below 1390K) and higher finishing forging temperature (above 1220K). However, no big brittle phase precipitates in the iron-nickel-base superalloys with low segregation (Fig.7). Their hot workability is remarkably improved, and the superalloys can be forged at a higher starting forging temperature, about 1453K-1493K. Hence, the cost of production of the parts can be reduced and their performance can also be improved greatly.

The low segregation superalloys can be easily produced by use of ordinary equipment without need of further investment. One point to be emphasized on is that the profit offered by means of the new way is far more than its slight cost increase of 5%.

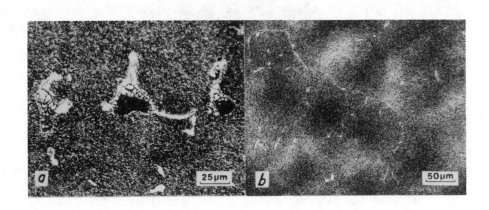

Fig.6 Comparison of structures of cast supralloy K38(IN738) and M38G.
(a) Commercial superalloy K38(IN738) (b) Superalloy M38G with low segregation

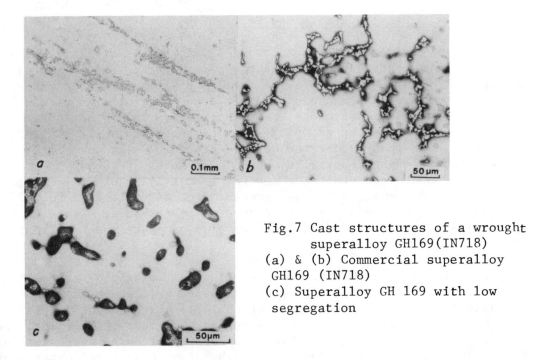

Fig.7 Cast structures of a wrought superalloy GH169(IN718)
(a) & (b) Commercial superalloy GH169 (IN718)
(c) Superalloy GH 169 with low segregation

Table Comparison of rupture stress (100 hrs/MPa) and resistance to hot corrosion

Superalloy	Composition (wt%)									Temperature K(°C)				Resistance to Hot Corrosion
	C	Cr	Co	W	Mo	Nb	Ta	Al	Ti	1023 (750)	1073 (800)	1123 (850)	1173 (900)	1223 (950)	
K17(IN100)	0.17	9.0	15.0		3.0			5.3	4.8	686	540	421	312	206	poor
M17F	0.17	9.0	9.0	3.9	2.0		3.9	5.0	4.5	745	608	480	349	255	poor
IN792	0.18	12.7	9.0	3.9	2.0		3.9	3.2	4.2	686	540	421	312	206	good
M40	0.18	12.7	9.0	3.9	2.0		3.9	3.8	4.8	696	573	451	343	245	good
K38(IN738)	0.17	16.0	8.5	2.6	1.7	0.7	1.7	3.5	3.3	598	451	363	255	176	very good
M38G	0.17	16.0	8.5	2.6	1.7	0.7	1.7	4.0	3.8	666	529	402	299	206	very good
DZ38G	0.10	16.0	8.5	2.6	1.7	0.7	1.7	4.0	3.8	L 706 / T 670	569 540	456 415	333	235	very good
M36	0.17	20.0	8.5	2.6	1.7	0.7		3.8	3.8	627	500	363	260	186	excellent

Notes: (1) M17F, M40, M38G and M36 are superalloys with low segregation and good phase stability.
(2) DZ38G is a directionally solidified M38G alloy.
(3) L-longitudinal T-transversal

Conclusions

(1) During solidification, Ti, Cr and Mo are alloying elements of positive dendritic segregation, and Ti is the most serious one.

(2) Ti promotes the eutectic reaction. A great deal of Cr, Mo and Co enriched in the residual liquid in front of the grown eutectic provides the condition for σ phase precipitation.

(3) By careful control of alloy composition and the solidification parameters, acicular σ phase, big blocky Laves phase or other harmful phases can be successfully suppressed. A series of new superalloys with low segregation have been developed.

Acknowledgement

The authors wish to thank the Chinese National Science Foundation Committee for supporting this research program.

References

1. King Chu and Ma Shih-chi, "The Phase and Structure of a Cast Nickel-Base Superalloy with High Aluminum and Titanium Contents", Acta Metallurgica Sinica, 10(1974)12-26.

2. J. R. Mihalisin, C. G. Bieber, and R. T. Grant, "Sigma-Its Occurrence, Effect, and Control in Nickel-Base Superalloys", Trans. AIME, 242(1968) 2399-2414.

3. C. S. Barrett, "Some Industrial Alloying Practice and Its Basis", J. Inst. Metals, 100(1972)65-73.

4. C. T. Sims, "A Contemporary View of Nickel-Base Superalloys", J.Metals, 18(1966)1119-1130.

5. L. R. Woodyatt, C. T. Sims, and H. J. Beattie, "Prediction of Sigma-Type Phase Occurrence from Compositions in Austenitic Superalloys", Trans. AIME, 236(1966)519-527.

6. Zhu Yaoxiao et al., "The Influence of Boron on Porosity of Cast Ni-Base Superalloys", Acta Metallurgica Sinica, 21(1985)A1-8.

RELATION BETWEEN CHEMISTRY, SOLIDIFICATION BEHAVIOUR, MICROSTRUCTURE AND MICROPOROSITY IN NICKEL-BASE SUPERALLOYS

J. Lecomte-Beckers

Department of Metallurgy and Materials Science
Faculty of Applied Science, University of Liège
Hall de Métallurgie, Sart Tilman, B17
Liège, Belgium

Abstract

Microporosity formation in nickel-base superalloys is studied experimentally and theoretically. A microporosity index, ΔP^*, which depends on solidification parameters and alloy properties has been deduced. This index can be determined from parameters obtained by quantitative differential thermal analysis. The effect of mean alloying elements on the formation of microporosity is evaluated. Thus, aluminum, titanium and cobalt are found to increase and chromium to decrease microporosity. The effect of carbon depends on aluminum content and can be beneficial or detrimental.

Introduction

Gas turbine components of complex geometry such as blades are currently produced by the technique of investment casting. Thus the microstructure and associated mechanical properties depend on solidification sequence although modifications can be introduced by the subsequent heat treatments. One of the solidification variables influencing mechanical properties is microporosity. This microporosity may result from shinkage, dissolved gases or a combination of both. However it must be noted that the gas content of modern superalloys is generally kept low and that most of it is either chemically bounded or kept into solution (1).

Microporosity is mostly caused by solidification shrinkage. It is well known that microporosity formation in nickel-base superalloys can be greatly affected by the casting conditions and by the chemical compositions of the melt. These parameters are not independent and influence deeply the solidification sequence which is the determinent factor in microporosity formation.

The aim of this work was to study superalloy proneness to microporosity in relation with solidification behavior and to point out the influence of the six major alloying elements : carbon, chromium, cobalt, molybdenum, titanium and aluminum.

Basic considerations on the mechanism of microporosity formation have led to a theoretical model. This one is based on pressure drop evaluation in the interdendritic liquid and introduce the so called ΔP^* coefficient which makes use of several solidification features (2)(3). Some parameters contained in this model are related to solidification sequence and have been extensively studied (4). This ΔP^* coefficient can only be computed if the evolution of the liquid fraction with time or with temperature is accurately known. Therefore we have developed a quantitative analysis of DTA based on the fact that peak intensity is proportional to transformation rate (2) (3). It is then possible to derive formulas giving solid and liquid fractions and solidification enthalpy (3).

Basic considerations on the mechanisms of microporosity formation have been summarized in the next section to introduce our experimental analysis of solidification behaviour which is presented in the following sections along with our main results and conclusions.

Basic considerations on microporosity formation

Nickel-base superalloys contract on solidifying. Most of this contraction takes place when efficient feeding mechanisms are still operative : liquid and mass feeding (6). However mass feeding lose efficacity when about 70% of the alloy is solidified ; capillarity (interdendritic or intergranular) feeding then becomes the only operative feeding mode. Microporosity formation occurs during the last stages of solidification when capillarity feeding becomes insufficient. Micropores form mainly because of the pressure drop resulting from flow of fluid through the liquid-solid mushy zone, which counterbalances the atmospheric pressure acting on the dendrite top (7). Microporosity occurs when the local pressure $p(x)$ drops below some critical value which may be predicted from nucleation theory (8).

The local pressure in the mushy zone during dendritic solidification have been calculated using the fluid flow calculations available for porous media, i.e. the interdendritic flow velocity obeys Darcy law and is linearly related to gravity and pressure variation (7)(8)(9).

Using the Darcy law :

$$\underline{V}_L = -\frac{K}{\mu f_L}(-\underline{\nabla} p + \rho_L \underline{g})$$

where : $\underline{V}_L$ is the interdendritic liquid velocity
K is the permeability of the porous material
μ is the viscosity of the liquid moving through it
ρ_L is the interdendritic liquid viscosity
$\underline{g}$ is the gravity acceleration
f_L is the liquid fraction.

the law of continuity and the generally accepted relation

$$K(x) = \gamma f^2_L(x)$$

where γ is a constant depending on dendrite structure

$$\gamma = \frac{1}{24} \pi n \tau^3$$

where n is the dendrite number density, and
τ the tortuosity of the interdendritic channels (9),

lead to an analytic expression of the local pressure in interdendritic channels (2)(3).

This local pressure results from atmospheric effect and pressure drop in interdendritic channels. The pressure drop is related to the evolution of the liquid fraction along the interdendritic channels of the mushy zone. Experimental data obtained using quantitative analysis of DTA charts show that liquid fraction can be fairly approximated by a linear law over a substantial portion of the mushy zone (2)(3). This analysis leads to the definition of the microporosity index ΔP^*(2)(3). A high ΔP^* value correspond to an important tendency for defects formation while a low ΔP^* is related to less defects formation. This microporosity index ΔP^* may be expressed by :

$$\Delta P^* = \frac{24\mu \beta' n \tau^3}{\rho_L g} \left(\frac{\Delta T}{G^2}\right) \left(\frac{df_S}{dt}\right)$$

The typical meaning of this parameter is straightforward. It suggests that internal soundness is favoured by a short solidification range ΔT, low dendrite number density n and "tortuosity" τ^3, high interdendritic liquid density ρ_L and fluidity, high thermal gradient G, low solidification rate df_S/dt and small contraction β' ($\beta' = \rho_S - \rho_L/\rho_L$).

Experimental procedure

Choice and Preparation of the Alloys.

Eight experimental alloy compositions were chosen following factorial design of 2^3 experiments to study the influence of the six major elements, each used at only two nominal contents (table I). The elements carbon, chromium and cobalt have been chosen as main factors, and the elements molybdenum, titanium and aluminum have been identified with the two-factor interactions.

The three-factor interaction was used to estimate σ, the experimental error. Statistical significance of the element effects was tested by an F-test (analysis of variance) and the level of significance of the effects was fixed at 10%. The significant effects are framed in the table presented.

The experimental alloys were elaborated under vacuum in an induction furnace. After homogenization, the liquid alloys were cast into preheated cylindrical sillimanite molds. These cylindrical rods were used for microstructural characterizations and DTA measurements. They were also subsequently turned to adequate specimen dimensions for QDS experiments.

Directionaly Solidified and Quenched Samples

The major part of this investigation was conducted on directionally solidified and quenched samples (Q.D.S.). This technique of unidirectional growth interrupted by quenching of the remaining liquid at a given moment offers a convenient way of studying the microstructure of the solid during solidification, as well as after its completion. It allows a fairly easy control of the two important parameters : thermal gradient and solidification rate.

The turned sample was contained in an alumina crucible (0.006m diam x 0.015m long) which is placed on a bottom water-cooled chill. This assembly was withdrawn at a constant rate R (0.060 mh^{-1}) from an induction furnace equiped with a graphite suceptor. The temperature of the sample was recorded by a second thermocouple placed at the central axis of the sample, in the solidified part which will be melted and unirectionally solidified.

The sample was melted under vacuum and superheated by 120°C above the equilibrium temperature, which was determined by DTA (4). After the temperature equilibrium was attained the crucible was withdrawn.

When steady state was established the width of the mushy zone as well as the temperature profile in the sample was stationnary with respect to the furnace and the grow rate of the solid in the heat flow was equal to the rate at which the crucible was withdrawn. The average thermal gradient in the solid-liquid region was determined from the temperature-distance charts and was $1 \times 10^{4} °Km^{-1}$. Under these steady-state growth conditions, solidification was dendritic. Growth of the dendritic specimen was interrupted at a given moment by quenching the remaining liquid achieved by pneumatically pulling the crucible from the furnace at very high speed (1m/s) and simultaneously cooling it with the helium jet.

Table I - Factorial design

EXPERIMENTS	A % C	B % Cr	C % Co	C(=AB) % Mo	E(=AC) % Ti	F(=BC) % Al	ALLOY NUMBER	ELEMENT TESTED
1 (def)	0.06	7.9	0	4	4.6	5.4	I	-
a (f)	0.14	7.5	0	0.5	1	5.5	II	carbon
b (e)	0.05	12	0	0.5	4.7	2.2	III	chromium
ab (d)	0.14	11.7	0	4	1	2.2	IV	molybdenum
c (d)	0.05	7.7	11.7	4	1	2.0	V	cobalt
ac (e)	0.15	8	10.2	0.5	4.6	2	VI	titanium
bc (f)	0.04	11.7	10.3	0.6	1.1	4.8	VII	aluminum
abc (def)	0.16	12	10.7	3.9	4.6	5.1	VIII	-

Metallographic Preparation

Each directionally solidified sample was sectioned longitudinally, polished and etched to study morphological features of the mushy zone. Transverse sections taken within the mushy zone, in the zone which was completely solidified at the moment of quench, were used to measure microporosity.

Morphological analyses were performed on transverse sections after appropriate etching on a (Quantimet 720) computerized image analyser. Forty-nine fields were analyzed on each sections. At each field (7 x 5 mm^2) corresponded 500.000 pixels.

Differential Thermal Analysis

Small cubic samples, approximately 0.3g in weight, were taken from each master rods and subjected to Differential Thermal Analysis (DTA) against a Pt reference. The tests were carried out under the protection of a constant argon flow (300ml/min) at a constant heating and cooling rate λ (600°C/h, peak temperature of 1450°C), corresponding to QDS experiments cooling rate, to observe the solidification sequence.

Results

Microporosity

Microporosity was measured in sectioned QDS samples. Quantitative measurements have been performed in the zone which was completely solidified at the moment of quench and the surface fraction was measured (total micropore area divided by total area examined). This parameter was choosen because it combines number and size of micropores present.

Comparison with the model

The predictions suggested by ΔP^* have been compared to the experimental results. This ΔP^* calculated is in good correlation with the measured surface fraction of micropores (fig.1) : for most of the alloys the level of the ΔP^* corresponds well to the surface fraction of the micropores.

Discussion on the effects of the alloying elements on the microporosity index parameters

This microporosity index depends on several parameters :

1) μ : viscosity of interdendritic liquid.
   This viscosity is strongly influenced by the effect of segregation and precipitation during solidification. It is difficult to calculate accurately the value of this term but it can be evaluated in relation with the total enthalpy which is determined by quantitative DTA. It can be assumed that for an identical solid fraction, the liquid temperature is higher in the alloy with the highest enthalpy and its liquid viscosity is less.

2) β' : solidification shrinkage $\beta' = \dfrac{\rho_S - \rho_L}{\rho_L}$

   ρ_S and ρ_L are the densities of the solid and liquid, respectively.

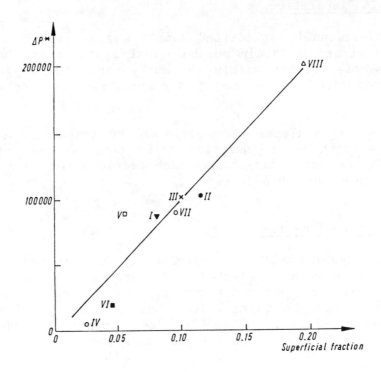

Figure 1 - Correlation between ΔP^* and micropore surface fraction
(to appear in Met. Trans.)

In the interdendritic channels the remaining liquid near the end of the solidification is always enriched in titanium and molybdenum (important for alloys containing cobalt), chromium and aluminum (for alloys rich in aluminum) and impoverished in cobalt when compared with the dendritic solid (4).
It is assume that the shrinkage coefficient obeys linear law of mixtures and is the combination of the shrinkage of each element remaining in the liquid. Moreover β' depends on the coefficient of linear thermal expansion.
Knowing the linear thermal expansion coefficient it is assumed that a nickel solid solution exhibits a high β' when poor in titanium, chromium and molybdenum and mostly rich in aluminum.

3) n : number of interdendritic channels per unit area.
In first approximation this can be assimilated to the number of dendrites per unit area which can be measured metallographically on QDS transverse sections.

4) τ : "tortuosity" of the dendrites.

It is an estimation of the importance of secondary dendrite arms of the dendrites. This factor can be evaluated metallographically on transverse sections. It can also be determined by quantitative DTA tests which gives the solidification curve (i.e. the fraction of solid in relation with time or temperature). The approach used has been described in details elsewhere (2)(3).

It has been shown that the shape and tortuosity of the dendrite structure obtained on QDS samples were strongly depending on the evolution of the parameter $1/\lambda \; df_s/dt$ (λ is the DTA cooling rate) during solidification. The initial increase of this parameter relates to dendrite growth and its subsequent shape after the maximum, to dendrite coarsening. More precisely, a high peak value at a high solid fraction corresponds to smooth dendrites, whereas a low peak values at a lower solid fraction favours growh rathes than coarsening and is indicative of a more tortuous dendrite structure (2) (3).

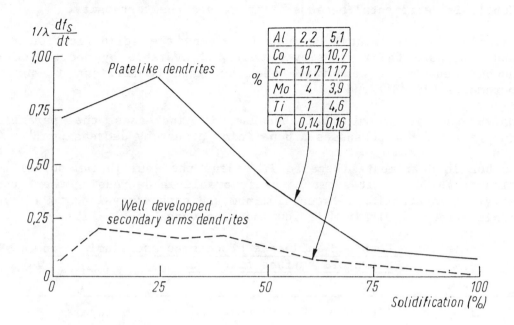

Figure 2 - Evaluation of normalised solidification rate

5) g : gravity acceleration considered as a constant.

6) ΔT : solidification range, determined by DTA.

7) G : thermal gradient in the mushy zone, constant in QDS experiments.

8) $(\frac{df_s}{dt})$: average solidification rate at the end of solidification calculated by quantitative DTA.

Table II shows the results of the factorial analysis of the various parameters n, τ^3, df_S/dt ΔT, ΔH. It must be kept in mind that the influence of ΔT and τ^3 (which have a high value) is more important than that of n, and df_S/dt ones.

Table II - Effects of alloying elements on the parameters used in ΔP^*

	n	τ	$(\frac{df_S}{dt})_{end}$	ΔT	ΔH	β'
Carbon	↓	⬇	↓	⬆	↑	−
Chromium	↓	↓	↑	↑	↑	⬇
Molybdenum	↑	↓	⬆	↓	↓	⬇
Cobalt	↑	⬆	⬆	↑	↓	−
Titanium	⬇	⬆	⬆	⬆	↓	⬇
Aluminum	↓	⬆	↓	⬆	⬇	⬆

It can be seen that :

1) Aluminum is detrimental because it widens the equilibrium solidification range ΔT, raises the contraction β' and decreases the viscosity

μ of the liquid metal by decreasing ΔH, the total enthalpy.

2) Chromium possesses some beneficial effect by decreasing β'.

3) Cobalt is detrimental because it increases the tortuosity.

4) Titanium is detrimental because it widens the solidification range and increases the tortuosity but it is favorable by decreasing the number density of dendrites, the shrinkage coefficient β' and the parameter (df_S/dt) end.

5) Molybdenum is detrimental because it increases the parameters (df_S/dt) end but possesses a beneficial effect by decreasing β'.

6) Carbon is detrimental because it widens the equilibrium solidification range ΔT but it seems to be favorable in decreasing the tortuosity. An interaction between aluminum and carbon may lead to a favourable effect if aluminum is low and vice versa (table III).

Table III - Combined influence of carbone and aluminum on micropore surface fraction

Alloy	Aluminum	Carbon	Micropore surface fraction
I	5.5	0.05	0.081
II	5.5	0.15	0.131
III	2	0.05	0.101
IV	2	0.15	0.025
V	2	0.05	0.055
VI	2	0.15	0.046
VII	5.5	0.05	0.095
VIII	5.5	0.15	0.188

Acknowledgements

The author expresses her deep gratitude to CRM (Centre de Recherches Métallurgiques - Belgique) and IRSIA (Institut pour la Recherche Scientifique dans l'Industrie et l'Agriculture - Belgique) for support of this work. A sincere appreciation is extended to Mr. D. Coutsouradis and particularly to Dr. M. Lamberigts of CRM for providing support and evaluable comments.

Thanks are also due to Professor T.Z. Kattamis, University of Connecticut for helpful discussions.

References

1. S. Rupp, Y. Bienvenu, J. Massol. Rapport final-actions concertées Cost-50/III, Novembre 84, Ecole des Mines de Paris, France.

2. J. Lecomte-Beckers, M. Lamberigts. Proceedings of "High temperature alloys for gas turbine and other applications 1986", vol.I, pp.745-757. Ed. D. Reidel Publishing Co, Dordrecht, The Netherlands, 1986.

3. J. Lecomte-Beckers. "Study of microporosity formation in nickel-base superalloys". Metall. Trans. (to be published).

4. J. Lecomte-Beckers. "Study of solidification features of nickel-base superalloys in relation with composition". Metall. Trans. (to be published).

5. H. Fredriksson, G. Rogberg. Metal Science, December 1979, 685-690.

6. J. Campbell. The British Foundryman, April 1969, p.147.

7. R. Mehrabian, M. Keane, M.C. Flemings. Metall.Trans., 1970, vol.1, 1209.

8. D. Apelian, M.C. Flemings, R. Mehrabian. Metall.Trans., 1974, vol.5, 2533.

9. M.C. Flemings. Solidification processing, Materials Science and Engineering Series. Mc Graw Hill, USA, 1974.

PHASE EQUILIBRIA IN MULTICOMPONENT ALLOY SYSTEMS

P. WILLEMIN*, M. DURAND-CHARRE

Institut National Polytechnique de Grenoble, L.T.P.C.M.-E.N.S.E.E.G,
BP 75, 38402 SAINT MARTIN D'HERES Cédex, France
*presently Centre de Recherches, IMPHY S.A., 58160 IMPHY France.

Abstract

Model alloy systems were chosen for their ability to illustrate competing elements involved in the formation of A3B phases, i.e., aluminium and titanium, aluminium and tantalum. A solid solution was substituted for nickel in the pseudoternary systems "S"-Al-Ti and "S"-Al-Ta. The range of compositions of the primary solidification phases was determined in order to observe the changes due to the elements present in the solid solution.

The different solubility ranges were studied on the basis of our experimental results on specimens annealed at 1250°C and analyzed in the light of the numerous data concerning ternary systems obtained from the literature.

Particular attention was focused on high temperature reactions and transformations involving the liquid phase. A correlation was established between certain microstructures obtained and the peritectic transformations occuring between the γ, γ' and liquid phases.

Introduction

The continually increasing demands on the creep resistance of turbine blade materials has led to a marked evolution in their composition in recent years, corresponding in particular to larger volume fractions of the γ' (Ni3Al) strengthening phase. Alloy chemistries have moved gradually closer to eutectic compositions, with a tendency toward higher solvus and lower solidus temperatures.

In order to overcome this disadvantage, there has been a "shift" in the latest single-crystal superalloy compositions, in which tantalum is used in preference to titanium and niobium. Even if the total range of elements can now be considered to be relatively stable, there remains considerable scope for obtaining improved creep properties by the optimization of compositions. This involves the close control of phase equilibria, both in the solid state, in particular the precipitation of γ' in γ, and between liquid and solid, in ordre to facilitate homogenization and solutioning of γ' without incipient melting and to enhance the aptitude for single crystal solidification. However, currently available phase diagrams are not sufficiently accurate or complete to achieve this goal.

Since modern industrial single-crystal blade alloys such as CMSX 2 or PWA 1480 contain at least 8 elements, it is essential in such studies to adopt a clearly defined strategy. For this reason, the pseudoternary model systems "S"-Al-Ti or "S"-Al-Ta were chosen to represent the real complex alloys. The symbol "S" here represents the matrix solid solution, which was taken as Ni-8Cr-5Co-8W (weight %), which is typical of the composition of numerous commercial alloys such as CMSX 2. The compositions ("S", 30%Al,30%Ti or Ta) were chosen to ensure the presence of all the main phases and to compare the contribution of aluminium and titanium/tantalum to their formation. The approach adopted was to compare compositions in these complex materials with alloys of similar Al, Ti and Ta contents in the Ni-Al-Ti and Ni-Al-Ta previously determined nickel rich corner (1,2).

The main basic ternary systems Ni-Cr-Ti (3), Ni-Ti-Al (4,5), Ni-Al-X (6) were investigated in the fifties. More recently, further developments or new investigations have concerned both these systems, Al-Cr-Ni (7), Cr-Ni-Ti (8), Ni-Cr-Ta (9), and pseudoquaternary systems Ni3Al-Ni3Cr-Ni3Ta (10), Ni3Al-Ni3Cr-Ni3Mo (11) or even more complex systems (12,13), but in this case limited to phase equilibria between γ and γ'. Our investigation differs from these studies, aimed chiefly at determining solid phase equilibria, in that it focuses on equilibria involving a liquid phase.

Experimental method

The experimental method (14) consists in a characterization of the structures of alloys solidified under controlled conditions. Freezing was monitored in a differential thermal analysis (DTA) furnace and a rapid quench was operated in order to interrupt the solidification process at different stages, particularly when the first crystal was formed at the liquidus temperature. This method, used in conjunction with electron beam analysis, enables the determination of the liquid-solid tie-lines. In addition, the observation of the so-quenched structure also allows the specification of the crystallization sequences.

The liquidus surfaces are determined from the transformation temperatures. The liquidus temperature range is found to be rather narrow for alloys in the γ' field and in the γ/γ' border line case. Temperatures alone are not sufficient to discriminate the different fields for each primary phase.

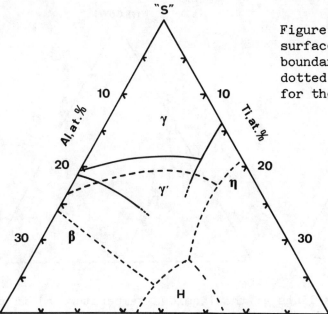

Figure 1 – Projection of the liquidus surface "S"-Al-Ti. Full lines, the boundaries of the primary fields, dotted lines, the monovariant lines for the system Ni-Al-Ti.

Consequently the investigation has to be completed by morphological examination, careful phase identification and microanalysis. All the usual characterization methods are employed : TEM, STEM, electron probe and X-ray diffraction.

Some sixty compositions were prepared for each "S"-Al-Ti or Ta system, as well as a few compositions with 0 and 4 wt% tungsten content.

The "S"-Al-Ti system

Phases in the system

Figure 1 shows for a pseudoternary system "S"-Al-Ti, the projection of the liquidus surfaces superimposed on the corresponding projection in the Ni-Al-Ti ternary system. The primary phase within these limits will be examined with respect to their ability to generate precipitations at lower temperatures. To better illustrate chemical interactions, all the results are given in atoms per cent or in atomic concentration.

The γ phase field is shown to be reduced, shifted towards the high nickel contents. This result reflects the marked tendency of Cr and Co, and to a lesser degree of W, to become part of the γ' phase. The γ liquidus hypersurface can be represented by a polynomial, after smoothing the experimental values, and expressed as a function of the atomic concentration in Al, Ti and also W. Overall precision is of the order of 4°C.

$$T+C = 1447 - 606\, x_{Ti} - 151\, x_{Al} + 193\, x_W - 3256\, x^2_{Ti} - 5051\, x_{Ti} x_{Al} + 74 x_{Ti} x_W - 1774\, x_{Ti} x_W - 1774\, x^2_{Al} + 2021\, x_{Ti} x_W + 8945\, x^2_W$$

Tungsten raises the liquidus temperature and this influence is enchanced for low aluminium contents and even more so for low titanium contents.

The η(Ni3Ti) primary phase has significant solubility for alloying elements : cobalt (6%), tungsten (1,3%), aluminium (4%) or chromium (3,4%). The η field is not reduced to a point in the compositional space as could be previously assumed (6). However at 1250°C tungsten becomes totally insoluble

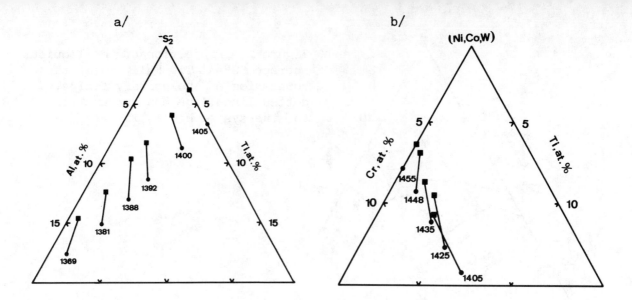

Figure 2 - Projection of the tie-lines at the liquidus temperature a) in the "S"-Al-Ti plane b) in the (Ni, Co, W)-Cr-Ti plane.

while the solubility of chromium is greatly reduced. Small very rounded precipitates are formed, composed almost exclusively of tungsten (5%Ni, 5%Cr, 1%Ti).

The β(NiAl) primary phase has also a significant solubility for alloying elements : cobalt (3,5 %), tungsten (0,7 %), titanium (7,7 %) or chromium (7%). Various precipitates are formed at 1250°C : γ', αW and αCr, depending on the respective Al, Cr and Ti contents. These γ, γ' and αCr precipitations can be interpreted by referring to the Ni-Al-Cr phase diagram (7). Tungsten is totally eliminated in the form of αW precipitates. Similar precipitations have been observed in the Ni-Al-Ta-Cr-W system (15).

The γ' phase covers a wide solid solution range, in particular in the case of titanium which can be contained in contents five times greater than that of aluminium. Comparing the systems Ni-Al-Ti and "S"-Al-Ti, the solubility of Al and Ti in the pseudoternary system is found to be reduced in a proportion of up to 30 %. Interactions between Cr and Ti, and also between Cr and W, are observed. In the solid phase, chromium diminishes the solubility of W in the γ' phase (16) and also that of titanium.

Tie-lines and crystallizations paths

The experimental method can be used to determine the tie-lines of complex alloys; this is of practical interest in that at the liquidus temperature the tie-line indicates both the direction of the crystallization path, to which it is tangent, and the extent of segregations. Figure 2a shows that the respective Al and Ti contents of the liquid and solid phases are equivalent for low titanium contents, as is the case for similar Ni-Al-Ti ternary compostions. Tungsten is concentrated in the initial solid, as shown on the electronic image of Figure 3a where the centre of the dendrite appears quite white because a heavy element is present. For titanium contents exceeding 5% a strong interaction between Cr and Ti is noticeable. There is much less segregation of Cr when the Ti content increases (Fig. 2b). All in all, for alloys pertaining to the γ liquidus surface, the crystallization paths can be interpreted by referring to the ternary system and result in the formation of γ'.

On the contrary, when the primary phases are η and β, tungsten is strongly segregated in the liquid. For titanium rich alloys with primary phase η, chromium is segregated in the interdendritic liquid. After solidification, different binary or ternary eutectic structures are observed, implying the presence of η, γ', γ, αCr and H (Ni2AlTi) phases. The γ phase has the most extensive solubility range, accommodating up to 30 %Cr. The αW phase is also found next to the eutectic zones, in the form of coarse blocks formed from the tungsten-enriched liquid where they have been trapped in the interdendritic groove. Such structures can be detrimental since they comprise zones with elements that diffuse poorly, intermetallic compounds not readily resoluble in the solid soultion, eutectics of relatively low melting point, and, last but not least, contain all the elements capable of forming the σ phase.

Morphologies and nature of three-phase equilibria

In a earlier publication (2) we showed that the nature of the equilibria between γ-γ'-liquid changes along the monovariant line. For Ti contents of over 5 % the γ-γ' liquid equilibria are peritectic in nature whereas γ'-β liquid equilibria become eutectic. The same effect is observed with tantalum (1). In the case of our complex alloys, such changes are practically impossible to prove by means of the insufficiently precise conodes and it is necessary to rely on morphology similarities with the ternary system. The electron micrograph of Figure 3 show γ phase dendrites completely surrounded by a γ' phase fringe of peritectic origin. A eutectic zone is observed in the interdendritic groove. Temperature variations along a section such as "S"-Al (Fig. 4) confirm this observation. On the other hand the alloys on the border of the β-γ' regions have a typically eutectic morphology. This means that solidification of alloys with compositions lying in the primary γ field can result in the formation of β phase since the solidification path can cross the peritectic region.

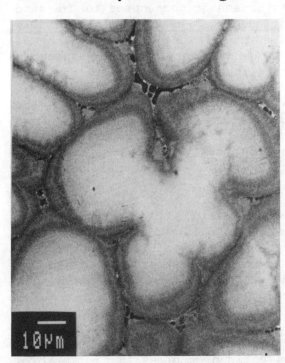
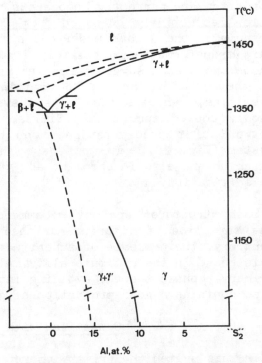

Figure 3 - (Left hand) Electron micrograph of alloy "S"-10Al-7,5Ti showing γ dendrites surrounded by a γ' fringe and a eutectic structure in the interdendritic groove with Ni2AlTi (dark phase).

Figure 4 - (Right hand) Vertical section for the system "S"-Al and Ni-Al.

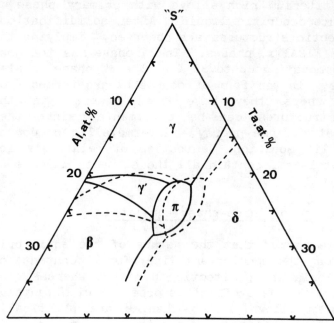

Figure 7 - Projection of the liquidus surface "S"-Al-Ta. Full lines, the boundaries of the primary fields, dotted lines, the monovariant lines for the system Ni-Al-Ta.

In the Ni-Al-Ti system, for titanium-rich alloys with compositions close to the monovariant line, the microstructure change markedly during high temperature exposure, the variations being more complex than a simple precipitation reaction. This change is illustrated by the example of an alloy containing 8 at%Al and 12 at%Ti. Samples were quenched at different stages of 300°C/h programmed cooling cycle. Figure 5a shows that, when quenching is carried out at the liquidus temperature, the primary γ' dendrites are perfectly homogenous. Figures 5b and 5c correspond to the same alloy respectively quenched from 20°C solubility of nickel and especially titanium, the γ' has undergone a transformation with rejection of γ phase. This transformation ultimately leads to the formation of spherical pockets whose appearance suggests partial remelting. These pockets appear (Fig. 5b) in the magnified area, to contain a dense precipitation of hyperfine γ' particles, whereas the latter have a coarser dendritic morphology in the slowly cooled specimens. Similar remelted pockets surrounded by a fine network of γ precipitation can be observed in alloys of the "S"-Al-Ti system (Fig. 6). In this case the remelted pocket, enriched in nickel, titanium and also in chromium and tungsten results in more complex phases on solidification.

Such structures are not recommended for industrial alloys. Consequently the formation of titanium-enriched primary γ' should be avoided. On the contrary, the presence of a peritectic γ' phase ensures a high proportion of this phase in the ultimate alloy. Alloy AM1, for instance, solidifies in the primary γ phase thereby avoiding interdendritic segregation of tungsten, but also contains a good proportion of peritectic γ'.

Comparison with the "S"-Al-Ta system

In this system the primary phase π(Ni6AlTa) is present as in the Ni-Al-Ta system (1). It also takes the same type of morphology i.e. relatively coarse primary phases and platelike precipitations along preferred planes. Figure 8 shows the projection of the liquidus surfaces in a "S"-Al-Ti pseudoternary

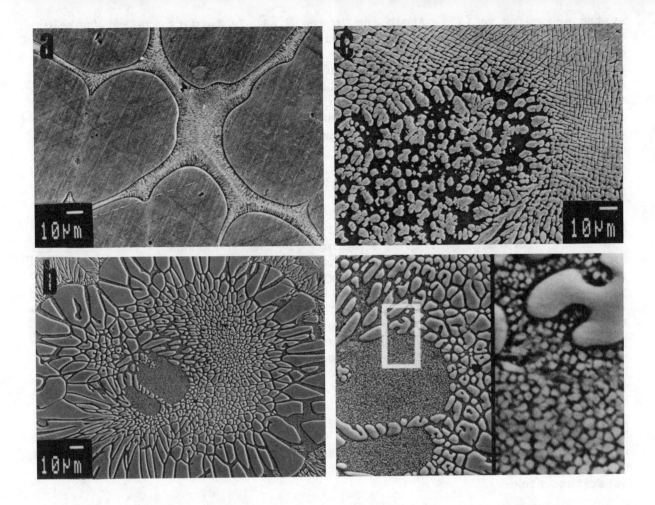

Figure 5 - Electron micrograph for alloy Ni-Al8-Ti12 a) specimen quenched at the liquidus temperature; b) specimen quenched 20°C below the liquidus temperature ; a magnified area shows the remelted zone ; c) specimen cooled at a rate of 5°C/mn.

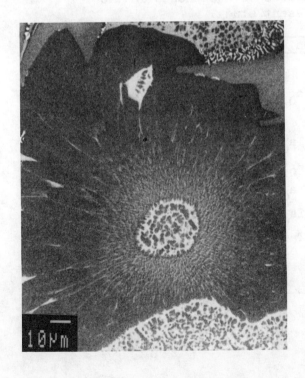

Figure 6 - Electron micrograph of alloy "S"-7,5Al-11,5Ti cooled at a rate of 5°C/mn.

system according to (17). Again, as in the Ni-Al-Ta ternary system, tantalum diminishes the solubility of W, Cr and Co in the γ' phase. In the presence of tantalum instead of titanium, the liquidus temperature of the alloy is lowered, but the drop is much less marked, more than halved, than in the case of titanium. As for the crystallization paths of the primary γ alloy, they are characterized by less segregation of tantalum in the interdendritic liquid and the formation of the γ' phase. For Ta contents exceeding 10 %, annealing of the γ matrix at 1250°C precipitated intermingled γ', δ (Ni_3Ta) and π phases.

Stability of the Ni3X type phases

Various approaches can be adopted to study the stability of A3B compounds. A thermodynamics approach at macroscopic level is not easy to realize due to the small differences in free enthalpy of the compounds that make it difficult to predict the formation of one or other of these compounds. This remains true even when using models extended to the ternary system such as the sub-lattice model which takes into account the distribution of atoms in the lattice (18). In terms of crystal structure, Ni3Al can be described as two cubic sublattices containing respectively nickel atoms and aluminium atoms. However, the location of the other elements is not so clearly established; they are disposed on either sublattice according to their bonding characteristics. Assumptions can be made considering the extent of the γ' field illustrated in Figure 8. The nickel content appears to be limited, thus indicating that nickel cannot be replaced on both lattices. This limitation no longers applies in the case of the Al-Cr-Ni system (7). Time-of-flight atom probe experiments (19) give more precise information : aluminium can be easily replaced by titanium and tantalum and also nickel, whereas chromium can substitute on either sublattice and tungsten can only substitute nickel.

These A3B compounds are also similar in their crystallographic structure. They are composed of stacks of compact planes called T or R planes depending on whether they have 3-fold or 4-fold symmetry. These planes are closely stacked with spacings of between 0.2059 and 0.2126 nm. The differences in crystallographic structure correspond to stacking sequences involving several tiers of planes.

Table I : Compositions and crystal structures of A3B phases in the Ni3Al-Ni3Ti-Ni3Ta system. * present study.

Spatial group	Phases	Composition	Type of planes	References
P m3m	γ'	Ni_3Al	T	(21)*
$P6_3/mmc$	π	$Ni_3(Al_{0.5}Ta_{0.5})$	T	(22)*
P mmm	δ	Ni_3Ta	R	(23)*
undefined		$Ni_3(Ti_{0.33}Ta_{0.67})$	R	(24)*
		$Ni_3(Ti_{0.5}Ta_{0.5})$		*
$P 6_3/mmc$		$Ni_3(Ti_{0.67}Ta_{0.33})$	T	(24)*
R 3m		$Ni_3(Ti_{0.83}Ta_{0.17})$	T	(24)*
undefined		$Ni_3(Ti_{0.96}Ta_{0.04})$	T	(24)
$P 6_3/mmc$	η	Ni_3Ti	T	(21)*

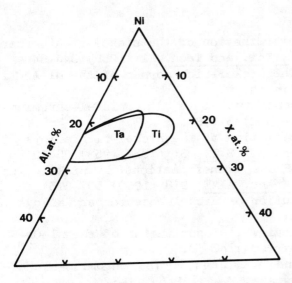

Figure 8 - γ' phase field at 1250°C for the system Ni-Al-Ti and Ni-Al-Ta

Electron entropy is a valuable means of characterizing these transition metal compounds due to the diversity of hybridized orbitals associated with the transition metals. Sinha (20) showed that, depending on the e/a ratio i.e. average number per atom of electrons of the outer shell participating in the bond, it was possible to predict the formation of phases with R planes ($8.65 \leqslant e/a \leqslant 8.75$) or phases with T planes ($7.75 \leqslant e/a \leqslant 8.65$). Unfortunately, this criterion does not help to predict the stacking sequence and therefore the crystallographic structure and resulting differences in energy.

We studied the Ni3Ti-Ni3Ta system experimentally. In addition to the η and δ extreme phases, four known intermediate phases and one new phase were identified in specimens annealed at 1250°C (Table I) (2,21,23,24). This section therefore comprises seven phases with A3B stoichiometry - a fair measure of the complexity of the problem.

Conclusion

The strategy for this investigation was based on the model systems method. The model systems chosen clearly illustrated the competition between, first, titanium and aluminium, secondly, tantalum and aluminium, in forming A3B type phases with nickel. The great similarity of "S"-Al-Ti and "S"-Al-Ta systems with their corresponding ternary system is adequate evidence that additional cobalt, chromium and tungsten elements have little influence on the outcome of this competition. Chromium and tungsten principally affect the crystallization paths since the η, β, π, δ primary phases, and to a lesser degree the γ' phase, segregate these elements in the interdendritic liquid. By a more thorough investigation of these reference systems we established that the specific morphologies where linked with the peritectic or even metatectic nature of the γ-γ' liquid reactions. Finally we point out that thermodynamic models can hardly be used to represent the competing reactions in these A3B systems because of the small differences in their energies due to their overall similarity.

Acknowledgements
This work was performed in the scope of the French scientific group "Superalloys for single crystal turbine blade materials" with the financial support of the "Direction des Recherches, Etudes et Techniques".

References

1. P. Willemin et al, "Experimental determination of the nickel-rich corner of the Ni-Al-Ta phase diagram", Mat. Sci. and Tech., 2 (1986) 344-348.
2. P. Willemin and M. Durand-Charre, The nickel-rich corner of the Ni-Al-Ti system, J. of Materials, to be published.
3. A. Taylor and R.W. Floyd, The constitution of the nickel-chromium-titanium system, J.I.S.I., 80 (1951-1952) 577-87.
4. A. Taylor and R.W. Floyd, The constitution of the nickel-rich alloys of the nickel-titanium-aluminium system, J.I.S.I., 81 (1952-1953) 25-32.
5. J.R. Milahisin and R.F. Decker, "Phase transformations in nickel-rich nickel-titanium-aluminium alloys", Trans. AIME, 218 (1960) 507-515.
6. R.W. Guard and E.A. Smith, "Constitution of nickel-base ternary alloys", J.I.S.I., 88 (1959-1960) 2883-2887.
7. S.M. Merchant and M.R. Notis, "A Review : Constitution of the Al-Cr-Ni system", Mat. Scien. and Engin., 66 (1984) 47-60.
8. K.P. Gupta, S.B. Rajendraprasad and A.K. Jena, "The chromium-nickel-tantalum system, J. Alloy Phase Diagrams, 2, 1 (1986) 31-37.
9. S.U. Schittny, E. Lugscheider, O. Knotek, "Melting behaviour and phase equilibria in the system nickel-chromium-tantalum", Thermochimica Acta, 85 (1985) 167-170.
10. S. Chakravoty and D.R. West, Ni_3Al-Ni_3Cr-Ni_3Ta section of Ni-Cr-Al-Ta system", Mat. Scien. and Techn., (1985) 978-985.
11. S. Chakravoty and D.R. West, "Ni_3Al-Ni_3Cr-Ni_3Mo section of the Ni-Cr-Al-Mo system", Journal of Mat. Science, 19 (1984) 3574-3587.
12. A. Havalda, "Les changements structuraux dans le système Ni-Cr-Ti-Al-W à une température de 850°C", Métaux Corrosion Industrie 940 (1966) 225-234.
13. R.L. Dreshfield and J.F. Wallace, "The Gamma-Gamma prime region of the Ni-Al-Cr-Ti-W-Mo system at 850°C", met. Trans. 5 (1974) 71-78.
14. M. Durand-Charre, N. Valignat et F. Durand, "Méthode expérimentale pour évaluer l'influence de chaque élément sur les microségrégations de solidification dans les alliages complexes : application à une gamme d'alliages (Fe, Ni, Co, Al, Ti)", Mem. Rev. Scien., (jan. 1979), 51-61.
15. O. Faral et J. Davidson, "Etude du système Ni-Al-Ta-(Cr)-(W) comme base de superalliages pour aubes monocristallines", Colloque National "Superalliages monocristallins", 26-28 fev. 1986, Villard de Lans (France).
16. P. Gustavson, "A thermodynamic evaluation of the Cr-Ni-W system", TRITA-MAC-0320, oct. 1986, Royal Institute of Technology, STOCKHOLM.
17. S. Boutarfaia, "Equilibres de phase liquide-solide du système multiconstitué (Ni, Cr, Co, W)-Ta-Al", Thèse de Magistère (1986), Université Harry Boumédiène, ALGER (Algérie).
18. P. Willemin, M. Durand-Charre et I. Ansara, "Liquid-solid equilibria in the system Ni_3Al-Ni_3Ta and Ni_3Al-Ni_3Ti", High Temperature Alloys for Gas Turbines and Other Applications 1986, ed. W. Betz et al., D. Reidel Publishing Company, DORDRECHT (Holland).
19. D. Blavette, A. Bostel and J.M. Sarrau, "Atom-Probe Microanalysis of a Nickel-base superalloy", Met. Trans. 16A (1985) 1703-1711.
20. A.K. Sinha, "Close-packed ordered AB_3 structures in ternary alloys of certain transition metals", Trans. AIME (1969) 911-917.
21. W.B. Pearson, A handbook of lattice spacings and structures of metals and alloys, Pergamon Press, ed. G.V. RAYNOR (1964).
22. B.C. Giessen and N.J. Grant, "New intermediate phases in transition metal systems", Acta Cryst., 18 (1965) 1080-1081.
23. B.C. Giessen and N.J. Grant, "The crystal structure of $TaNi_3$ and its change on cold working", Acta Met., 15 (1967) 871-877.
24. Van Vutch, "Influence of radius ratio on the structure of intermetallic compounds of the A_3B type", J. of Less Common Metals, 11 (1966) 308-322.

PHASE CALCULATION AND ITS USE

IN ALLOY DESIGN PROGRAM FOR NICKEL-BASE SUPERALLOYS

H. Harada, K. Ohno, T. Yamagata,
T. Yokokawa, and M. Yamazaki

National Research Institute for Metals
2-3-12 Nakameguro, Meguro-ku, Tokyo 153, Japan

Abstract

A series of phase calculation equations for γ and γ' phases in nickel-base superalloys was established by regression analysis on the EPMA data obtained from 30 experimental or commercial alloys together with data from a Ni-Al binary phase diagram. By using these equations, an accurate calculation of γ/γ' equilibrium, at 900°C, was made possible for various compositions of alloys, from Ni-Al binary to Ni-Co-Cr-Mo-W-Al-Ti-Nb-Ta-Hf-Re multi-component alloys. These phase calculation equations were substituted for those having been used in our alloy design program to extend the applicable composition range and to improve the accuracy of phase and property calculations of the program. Furthermore, a calculation equation for creep rupture life of SC alloys was established and adapted to the program. Using this new version of alloy design, the phase and property were successfully calculated for various types of γ' precipitation hardening nickel-base superalloys.

Introduction

In the design of γ' precipitation hardening nickel-base superalloys, it is of prime importance to calculate the compositions and fractions of γ and γ' phases at high temperatures with a high accuracy, not only for the prediction of the formation of undesirable phases but also for the further calculations of high temperature properties of the alloys.

Some of the authors previously developed an alloy design program for γ' precipitation hardening nickel-base superalloys [1,2]. By this program, calculations were made possible for γ/γ' phase equilibrium as well as other structural parameters (e.g., lattice misfit values). Furthermore, high temperature properties, such as creep rupture life and high corrosion resistance, could also be calculated by this program. This program has been successfully applied to the development of conventionally cast (CC), directionally solidified (DS), single crystal (SC), powder metallurgical (PM), and oxide dispersion strengthened (ODS) alloys in national projects of Japan [2-6]. However, this program still had a limitation in the applicable composition range. For instance, the phase calculation was not possible for some existing alloys, such as Alloy 454. This could be attributed to the data for γ and γ' compositions [7,8] on which our phase calculation equations were established; the compositions were obtained from alloys with various heat treatments and, in addition, the alloy compositions were not widely distributed in the γ/γ' two phase region. Both of these prevented us from establishing very accurate phase calculation equations between the two phases at high temperature.

This study was performed to reconstruct our alloy design program using our own careful EPMA analysis on experimental and commercial alloy specimens having various compositions which were aged for long times at temperature.

Experimental

Specimens of 20 experimental and 10 commercial alloys, including CC, DS, and SC alloys, were heated at 1100 to 1175°C for 30 hrs after cold work so that γ/γ' two phase recrystallization occurred to form a coarse γ/γ' structure. The specimens were then slowly cooled from these temperatures to 900°C and aged for 1500 hrs in order to achieve the equilibrium state at 900°C followed by water quenching.

The specimens thus heat-treated had γ/γ' lamellar or γ' rod/γ structures of 3 to 5 microns thickness in both phases. These structures were coarse enough to be analyzed by EPMA with a high accuracy. On the 30 pairs of thus analyzed γ and γ' compositions together with a pair from Ni-Al binary phase diagram, regression analysis was carried out to establish the phase calculation equations. A series of equations obtained were substituted for those having been used in our alloy design program to improve accuracy and extend applicable composition range of the program. Furthermore, the calculation of creep rupture life of properly heat-treated SC alloys was made possible on the basis of this phase calculation.

Phase Calculation

The $\gamma + \gamma'$ region in multi-component nickel-base superalloys is presented schematically in Fig. 1. The γ' hypersurface in this figure, on which compositions of γ' equilibrated to γ should be lying, could be expre-

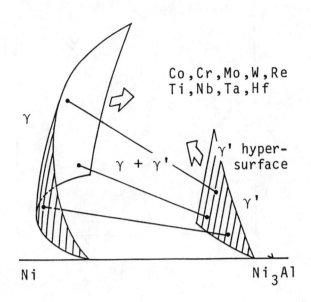

Fig. 1. Schematic phase diagram of multi-component nickel-base super-alloys.

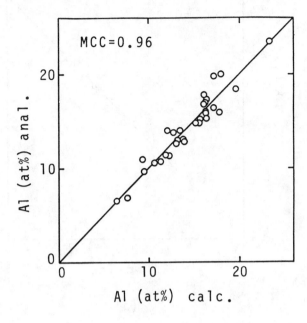

Fig. 2. Relationship between analyzed and calculated Al concentrations in γ' phase.

ssed by a regression equation, eq.1, where X'_i is a concentration of i-th element in γ' phase.

$$X'_{Al} = 23.4 - 0.03 \cdot X'_{Co} - 0.55 \cdot X'_{Cr} - 0.71 \cdot X'_{Mo} - 0.74 \cdot X'_{W}$$
$$-0.86 \cdot X'_{Ti} - 0.96 \cdot X'_{Nb} - 0.48 \cdot X'_{Ta} - 1.08 \cdot X'_{Hf} (-0.74 \cdot X'_{Re}) \quad \text{-------- (1)}$$

For this regression analysis, the 30 analyzed γ' compositions were used together with a γ' composition from Ni-Al binary phase diagram at 900 °C[9]. Since the number of alloys containing Re was too small to give a confident coefficient value, the same value as W was used for Re. MCC(Multiple Correlation Coefficient) of the equation was 0.96. Although this value was slightly smaller than that of the equation in our previous paper, the value being 0.98, the applicable composition range was found to be remarkably extended. Fig. 2 shows a good agreement between calculated(by eq.1) and analyzed Al concentrations in the γ' phase. The agreement holds in a wide range, from 23.4 at% (in Ni-Al binary alloy) to 6.5 at% (in highly solid solutioned multi-component alloy), whereas the previous equation derived from the data in literature[7,8] tended to give significant amount of error when it was used in a composition range of Ni-Al binary and near to it.

In order to calculate the composition of γ which is equilibrated to a γ' on the γ' hypersurface, partitioning ratios defined as $R_i = X_i/X'_i$, where X_i is a concentration of i-th element in γ phase, are needed. Those of Co, Cr, W, and Al could be well expressed by regression analysis on the 31 pairs of γ and γ' compositions, as functions of the γ' composition, eqs. 2, 3, 4, and 5. The MCCs were 0.98, 0.92, 0.89, and 0.83, in order. They were increased by 0.07 to 0.13 compared with those in our previous work[2], except W, a

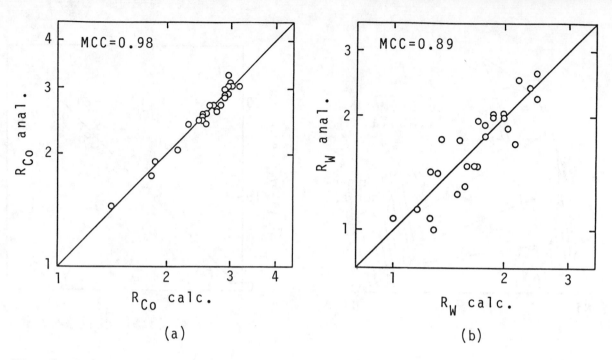

Fig. 3. Relationship between analyzed and calculated partitioning ratios for Co (a) and W (b).

constant value having been used for it. Partitioning ratios of other elements were found to be well expressed as constants, $R_{Mo}=3.93$, $R_{Ti}=0.18$, $R_{Nb}=0.30$, $R_{Ta}=0.22$, $R_{Hf}=0.15$, and $R_{Re}=10.6$.

$$\log_{10}(R_{Co}) = 0.529 - 0.012 \cdot X'_{Co} - 0.012 \cdot X'_{Cr} - 0.008 \cdot X'_{W}$$
$$+ 0.005 \, X'_{Ti} \quad\quad\quad\quad\quad\quad\quad (2)$$

$$\log_{10}(R_{Cr}) = 1.429 - 0.009 \cdot X'_{Co} - 0.061 \cdot X'_{Cr} - 0.103 \cdot X'_{Mo}$$
$$- 0.033 \cdot X'_{W} - 0.016 \cdot X'_{Al} \quad\quad\quad\quad\quad\quad (3)$$

$$\log_{10}(R_{W}) = -0.141 - 0.015 \cdot X'_{Co} + 0.052 \cdot X'_{W} + 0.037 \cdot X'_{Ti}$$
$$+ 0.085 \cdot X'_{Nb} + 0.077 \cdot X'_{Ta} \quad\quad\quad\quad\quad\quad (4)$$

$$\log_{10}(R_{Al}) = -0.274 + 0.013 \cdot X'_{Co} - 0.058 \cdot X'_{Mo} - 0.073 \cdot X'_{W}$$
$$- 0.023 \cdot X'_{Ti} - 0.087 \cdot X'_{Nb} - 0.057 \cdot X'_{Ta} \quad\quad\quad\quad (5)$$

Fig. 3(a) shows a good agreement between analyzed and calculated (by eq. 2) partitioning ratios of Co; analyzed ones, distributed from 1.5 to 3, are perfectly expressed by the calculation. In case of W, as shown in Fig. 3(b), the agreement is not as good as with Co, but the equation still expresses the partitioning behavior, the values being distributed from 1 to 2.5. Obviously, using these four equations together with the constant values, the error in phase calculation is expected to be minimized. This is in contrast to PHACOMP [10] where using constant values for all the partitioning ratios to calculate the γ and γ' compositions could give a significant amount of error in the result.

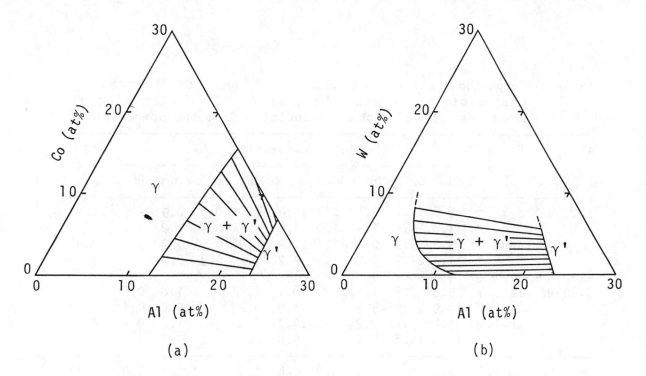

Fig. 4. Calculated γ+γ' region with tie-lines, at 900°C, for Ni-Al-Co (a) and Ni-Al-W (b) ternary systems.

By substituting the phase calculation equations thus obtained for those having been used in our alloy design program, the effective range of the program was remarkably extended; this new version could be applied to almost all the γ/γ' two phase composition range, from Ni-Al binary to Ni-Co-Cr-Mo-W-Al-Ti-Nb-Ta-Hf-Re multi-component systems.

The γ/γ' equilibrium in Ni-Al-X ternary systems, X= Co, Cr, Mo, W, Ti, Nb, Ta, and Hf, at 900°C, could be successfully calculated by the phase calculation equations. Fig.4 (a) shows the calculation for Ni-Al-Co ternary system. The γ+γ' region calculated was very consistent with phase diagrams at 800, 1000, and 1200°C available from literature[11]. In Ni-Al-W ternary system, the γ+γ' region calculated and shown in Fig. 4 (b) was again very consistent with an available phase diagram at 1250°C[12], tie-lines being not shown in it.

In multi-component alloys, a very good agreement was observed between calculated and analyzed chemical compositions of γ and γ' phases, examples being shown in Table I. The first four alloys in the table are alloys included in the 30 alloys used for EPMA analysis and so the agreement in these alloys indicates a good 'interpolation' of the phase calculation. It is to be noted that the calculation for Alloy 454, which was not possible by the old version, becomes possible here. The agreement observed in the last four alloys, whose analyzed data were not used to establish phase calculation equations, indicates a very good 'extrapolation' of the phase calculation in multi-component systems.

From all of these examinations, it was concluded that the γ/γ' phase equilibrium at 900°C could be calculated by our phase calculation equations with a very high accuracy, for almost all the γ/γ' two phase composition range, from Ni-Al binary to multi-component alloys.

Table I. Comparison between calculated and analyzed chemical compositions of γ and γ' phases, at 900 °C, with atomic fractions (f) of γ' phase calculated from the compositions.

Alloy	Phase		Chemical composition (at%)									f
			Co	Cr	Mo	W	Al	Ti	Nb	Ta	Hf	
IN738LC	calc.	γ'	3.8	3.1	0.4	0.5	13.0	7.7	0.9	0.9	-	0.41
		γ	11.4	26.8	1.4	1.0	3.4	1.3	0.3	0.2	-	
	anal.	γ'	3.8	3.3	0.3	0.5	12.9	7.4	0.9	0.9	-	0.42
		γ	11.7	27.2	1.5	0.9	3.2	1.4	0.3	0.2	-	
B1900Hf	calc.	γ'	5.5	2.7	1.4	-	18.1	1.3	-	1.8	0.6	0.54
		γ	14.9	15.9	5.6	-	7.0	0.2	-	0.4	0.1	
	anal.	γ'	5.4	3.4	2.2	-	19.7	1.2	-	1.8	0.6	0.52
		γ	14.8	14.8	4.7	-	5.5	0.3	-	0.4	0.1	
MM247DS	calc.	γ'	6.8	3.2	0.2	2.9	16.6	1.6	-	1.1	0.6	0.64
		γ	16.8	19.7	0.6	3.2	5.2	0.3	-	0.3	0.1	
	anal.	γ'	6.4	3.3	0.2	2.8	17.3	1.7	-	1.2	0.6	0.59
		γ	16.2	17.5	0.5	3.2	5.5	0.4	-	0.3	0.1	
ALLOY 454	calc.	γ'	3.1	3.2	-	0.9	15.8	2.7	-	5.6	-	0.64
		γ	8.8	26.5	-	2.1	3.3	0.5	-	1.3	-	
	anal.	γ'	2.9	3.1	-	0.8	16.8	2.7	-	5.3	-	0.59
		γ	8.2	23.6	-	2.0	3.3	0.7	-	1.4	-	
CMSX-2	calc.	γ'	2.7	2.7	0.2	2.1	17.4	1.7	-	2.7	-	0.61
		γ	7.8	19.6	0.7	3.3	4.4	0.3	-	0.6	-	
*	anal.	γ'	3.2	2.4	0.2	2.4	16.7	1.6	-	3.0	-	***0.68
		γ				n.d.						
MXON	calc.	γ'	3.4	4.0	0.7	2.4	17.6	-	-	2.6	-	0.69
		γ	8.9	20.9	2.6	3.2	4.5	-	-	0.6	-	
**	anal.	γ'	3.0	2.7	0.7	2.7	17.9	-	-	2.5	-	***0.69
		γ				n.d.						
TM-53	calc.	γ'	5.1	2.9	-	2.0	13.0	7.7	-	1.1	-	0.67
		γ	14.5	24.3	-	3.7	3.3	1.4	-	0.3	-	
	anal.	γ'	5.6	3.5	-	1.6	11.5	7.6	-	1.2	-	0.66
		γ	13.6	26.5	-	4.2	3.1	1.5	-	0.2	-	
TMS-1	calc.	γ'	5.2	2.3	-	4.7	17.3	-	-	2.5	-	0.64
		γ	13.2	14.6	-	7.7	3.6	-	-	0.6	-	
	anal.	γ'	5.5	2.6	-	5.0	16.2	-	-	2.4	-	0.68
		γ	13.6	15.5	-	7.4	3.9	-	-	0.6	-	

* at 850°C [13], ** at 850°C [14], *** Wt. fraction

Table II. Some of the structural parameters and properties calculated by the new version of alloy design program.

Alloy	γ' atomic fraction	Lattice misfit (%)	Density (g/cm^3)	Creep rup.life(h) *CC	**SC	***Hot corrosion rate
CC alloys						
Inconel713C	0.55	-0.21	7.96	51	-	8.3
IN 738LC	0.41	0.23	8.21	24	-	1
B1900Hf	0.54	-0.07	8.21	189	-	2.4
MarM200	0.61	0.08	8.50	298	-	2.3
TM-321	0.61	0.21	8.82	767	-	1.5
DS alloys						
MarM247DS	0.64	0.17	8.57	(688)	-	0.7
Rene 125	0.63	0.14	8.51	(533)	-	0.4
TMD-5	0.59	0.19	8.96	(1094)	-	1.4
SC alloys						
Alloy 454	0.64	0.58	8.68	(591)	209	(3.6)
NASAIR100	0.72	-0.13	8.59	(551)	420	(4.1)
SRR 99	0.67	0.12	8.50	(505)	367	(1.9)
RR2000	0.57	-0.02	7.90	(99)	181	(0.8)
Rene N4	0.56	0.39	8.51	(249)	52	(1.2)
CMSX-2	0.61	0.29	8.61	(496)	763	(4.6)
MXON	0.69	0.01	8.62	(626)	728	(9.1)
MMT143	0.60	-0.38	8.61	(930)	>10000	(188.5)
TMS- 1	0.64	0.16	9.16	(1492)	1747	(8.9)
TMS-12	0.60	0.33	9.06	(894)	3486	(10.5)
SC 83	0.55	0.18	8.84	(498)	2168	(21.2)
PM alloys						
MERL 76	0.49	0.20	8.05	(50)	-	0.6
Rene 95	0.48	0.04	8.26	(31)	-	5.3
AF 115	0.60	0.01	8.34	(158)	-	0.8
TMP-3	0.57	0.16	8.26	(75)	-	6.3
ODS alloys(Base metal)						
MA 6000	0.49	-0.03	8.21	(63)	-	1.6
TMO-2	0.47	0.30	8.98	(281)	-	8.3

* at 1000°C-118MPa, assuming CC structure, grain size about 3 mm.
** at 1040°C-137MPa, assuming solutioned and aged SC structure, tensile axis within 10° of <100>.
***Penetration depth in burner rig test at 850°C, normalized by IN738LC(=1).

Structural Parameters and Properties

Calculations of other structural parameters, as well as high temperature properties, were made possible on the basis of the phase calculation mentioned above. The calculation equations established in our previous paper could be successfully used for this purpose after a little modification. The structural parameters and properties thus calculated were / density / solubility index(our criterion of phase stability) / PHACOMP parameters / lattice parameters and misfit(at room temperature) / liquidus-solidus temperatures / creep rupture strength(for CC) / tensile properties(for CC, at 900°C) / hot corrosion resistance(in crucible test at 900°C and burner rig test at 850°C) / and so on.

In addition to these calculations, a calculation of creep rupture life of SC alloys, at 1040°C and at 137MPa, became possible by a regression equation, as a function of γ' composition, γ' at. fraction, and γ/γ' lattice misfit. The MCC of the equation was 0.91 when the analysis was carried out using data from our experimental alloys which were cast, heat-treated, and tested in a unified condition. When the data from 6 commercial or experimental alloys in literature were used together with ours, the MCC reduced to 0.84. This was probably because of the difference in heat treatment conditions, which will be taken into account in the equation.

It was remarked in both equations that the effect of the lattice misfit, as well as other factors, was found to be very large. The coefficients of the lattice misfit in the equations suggested that the rupture life became longer by factor of 2 or more per every 0.1% change of lattice misfit toward negative($a_{\gamma'} < a_{\gamma}$), e.g., from +0.3% to +0.2%. This strongly supported the effect of large negative lattice misfit on the formation of rafted structure which increased the creep resistance. Because almost all the alloys used for the regression analysis were expected to have negative lattice misfit values at high temperatures such as 1040°C, the change toward negative at room temperature was very likely to correspond to a change toward larger negative lattice misfit in the negative range at 1040°C.

```
ALLOY    TMS-12 **** PHASE(AT   900.C) & PROPERTY CALCULATION *************
         NI      CO      CR      MO      W       AL      TI      NB      TA      HF      RE
GP      72.89   0.00    2.09    0.00    3.04   18.06    0.00    0.00    3.92    0.00    0.00
G       72.56   0.00   16.74    0.00    6.38    3.45    0.00    0.00    0.87    0.00    0.00

ATPCT   72.76   0.00    8.01    0.00    4.39   12.16    0.00    0.00    2.68    0.00    0.00
WTPCT   67.70   0.00    6.60    0.00   12.80    5.20    0.00    0.00    7.70    0.00    0.00

  F.GP     LAT.GP(A)   NV.GP    LIQ(C)    SOL2(C)    H.COR.C    YS(MPA)    LIFE.CC(H)
  0.596     3.594      2.325    1411.4    1354.4     5863.47    477.4      893.6

 DENSITY   LAT.G(A)    NV.G     SOL1(C)   SOLV(C)    H.COR.B    UTS(MPA)   LIFE.SC(H)
  9.062     3.582      1.870    1396.3    1309.3     10.48      556.3      3485.8

  SI        LM(%)     NV.G-NVC  RANGE(C)  WDW (C)               EL(%)      SPC.STRGTH
  1.163     0.327      0.197    15.1      45.1                  6.8        15.689
*****************************************************************************
TRY AGAIN ?  < YES(0)  NO(1) >
```

Fig. 5. Output of 'Phase & Property' calculation, for alloy TMS-12, by the new version of alloy design program.

The calculation equation for creep rupture life of SC alloys having MCC of 0.84 was adapted to the alloy design program. Calculation equations for solidus and solvus temperatures were also established on microstructure observation data as functions of the alloy composition, and this was also adapted to the program. An example of the phase & property calculation by this new version of alloy design program is shown in Fig. 5, for Alloy TMS-12, a single crystal alloy developed by the authors using the previous alloy design program.

The calculation was successfully made for various currently used and recently developed alloys, including CC, DS, SC, PM, and ODS alloys. Some of the results are shown in Table II. In the table, TM-321, TMD-5, TMS-1,12, TMP-3, and TMO-2 are alloys designed by authors using previous alloy design program. Alloy SC 83 was designed by Ohno and Watanabe[15], the actual creep rupture life being 3100 hrs at this testing condition. In some SC alloys, the actual creep rupture life was longer than the calculation, probably because of the optimum heat treatments performed for the alloys; MXON, for example, had a life of 1500hrs. It is to be noted here that alloy MMT-143 is calculated to have very long rupture life mainly because of the large negative lattice misfit, although the calculation is effective so far as the misfit dislocation is not formed before testing.

Application of the Alloy Design Program

This new version of alloy design program will be used for alloy developments of various types of γ' precipitation hardening nickel-base superalloys, especially the SC alloys, using the calculation equation of creep rupture life as well as all the other calculation equations.

This new version can also be used for quality control of various types of alloys during processing. For example, it is possible to calculate the 'Phase and Property' from the check analysis before tapping the molten metal to make remelt bars or castings and, if needed, one could adjust the chemical composition so that the alloy could exhibit the needed high temperature properties.

Conclusions

1. A series of phase calculation equations for γ and γ' phases in nickel-base superalloys was established by regression analysis on our EPMA data from 30 experimental or commercial alloys and a data from Ni-Al binary phase diagram.
2. An accurate calculation of γ/γ' equilibrium, at 900°C, was made possible by using these equations for various compositions of alloys, from Ni-Al binary to Ni-Co-Cr-Mo-W-Re-Al-Ti-Nb-Ta-Hf multi-component alloys.
3. A calculation equation for creep rupture life of SC alloys, revealing the strong effect of lattice misfit, could be obtained on the phase calculation and adapted to our alloy design program.
4. Using this new version of alloy design program, phase and property calculation was successfully carried out for various types of γ' precipitation hardening nickel-base superalloys.

Acknowledgment

This work was performed as a part of the R&D project of Basic Technology for Future Industries sponsored by Agency of Industrial Science and Technology, MITI.

References

[1] H.Harada and M.Yamazaki;Tetsu-to-Hagane,65(7)(1979),p.1059(written in Japanese).
[2] H.Harada, M.Yamazaki, Y.Koizumi, N.Sakuma, N.Furuya, and H.Kamiya; Proc. of Conference, High Temperature Alloys for Gas Turbines 1982, held in Liege, Belgium, 4-6 Oct. 1982, D.Reidel Publishing Co, p.721.
[3] T.Yamagata, H.Harada, S.Nakazawa, M.Yamazaki, and Y.G.Nakagawa; presented at Conference, "5th international symposium on Superalloys", held at Seven Springs, Pa, U.S.A., 7-11 Oct. 1984(Proceedings: Superalloys 1984, published by the Metallurgical Society of AIME,p157)
[4] M.Yamazaki; Progress in Powder Metallurgy, vol.41(1986),p.531.
[5] T.Yamagata, H.Harada, S.Nakazawa, and M.Yamazaki; Trans. Iron and Steel Inst.of Japan, vol.26,(1986),p.638.
[6] T.Yamagata, H.Harada, and M.Yamazaki: Proc. of Conference "The 1987 Tokyo International Gas Turbine Congress", held in Tokyo, Japan, 26-31 Oct. 1987, sponsored by Gas Turbine Society of Japan, vol 3, p.239.
[7] O.H.Kriege and J.M.Baris: Trans. ASM, vol.62(1967),p.195.
[8] J.E.Restall and E.C.Toulson: Metals and Materials, March(1973),p.134, April,p.187.(Part 1 and 2).
[9] Constitution of Binary Alloys, 2nd Ed. by M.Hansen and K.Anderko, McGrow-Hill, New York, 1958. p.119.
[10] R.F.Decker: Symposium on Steel Strengthening Mechanisms, Climax Molybdenum Company, May 1969,Greenwich, Connecticut, U.S.A.,p.147.
[11] J.Schramm, Z.Metallk. vol.33(1941) p.403.
[12] P.Nash, S.Fielding, and D.R.F.West: Met. Sci.,vol.17(1983),p.194.
[13] P.Caron and T.Khan: Mat. Sci. and Engineering, vol.61(1983),p.173.
[14] T.Khan, P.Caron and C.Duret: presented at Conference, "5th international symposium on Superalloys", held at Seven Springs, Pa, U.S.A., 7-11 Oct. 1984(Proceedings: Superalloys 1984, published by the Metallurgical Society of AIME,p.145).
[15] T.Ohno and R.Watanabe: Tetsu to Hagane, vol.72, No.13(1986), S1509(written in Japanese).

Repair, Post-Service Evaluation and Environmental Behavior

AIRCRAFT GAS TURBINE BLADE AND VANE REPAIR

K.C. Antony and G.W. Goward

Turbine Components Corporation
Branford, Connecticut 06405

Abstract

The high replacement costs of aircraft gas turbine blades and vanes have created a fast-growing, highly-specialized segment of the aircraft repair industry. The economic rationale for repair in lieu of replacement is prima facie. The metallurgical rationale is less certain, but satisfactory flight performance over the past twenty-five years at least attests to the technical adequacy of blade and vane repairs. Blade repairs, other than re-coating, generally consist of weld overlays on low-stress blade tip or tip shroud locations. On average, high-pressure turbine blades undergo two repair cycles before replacement. Low-pressure turbine blades similarly require repair but are replaced less frequently. Dimensional restoration of a subtly distorted component to demanding original equipment standards is a real engineering challenge. Coating technology is an equally sophisticated aspect of blade repair. Vane repairs are more avant-garde ranging from platform dimensional restoration to complete airfoil replacement. Vane airfoil crack repair is commonplace and involves a multiplicity of metallurgical processes many of which are still maturing. The future of blade and vane repair, like the past, will be determined by replacement part pricing. The technical considerations will remain as they are but repair will probably become less labor-intensive as automation becomes more feasible.

Introduction

Gas turbine blades experience dimensional and metallurgical degradation during engine operation. Dimensional degradation derives from wear, nicks, dents, hot corrosion and, in the case of coated blades, stripping and re-coating as in repair. Metallurgical degradation derives from fatigue and high-temperature creep. Some degradation, depending on location and extent, is ammenable to "repair"; the definition of repair being rather broad. That is, only very rarely does a repair provide a part that is equivalent in all respects to a new part. This is not to say that the repair is not safe and cost effective. For example, dimensional restoration of a blade tip by welding provides a part which is functionally satisfactory even though the weld area does not have properties equivalent to the parent material.

Repair procedures and limits are established by the engine manufacturer and are interpreted and applied by the engine operator and/or repair facility. This paper will review some statistics relating the needs for repair or replacement, coating technology related to blade and vane repair, and the types of repairs that are commonly applied to contemporary hardware. Finally, some predictions on the future of repair technology will be attempted.

Some Statistics on Blade and Vane Repairs

Approximately twelve percent of the blades in a typical, moderately aged commercial engine are classified as non-repairable during routine overhaul. Typical blade rejection rates, by stage, are summarized in Table I.

Table I. Typical Blade Repair Rejection Rates

Stage	Rejection Rate
1	30%
2	7%
3	6%
4	3%

The average service lives of solid and air-cooled first-stage blades seem to correspond to three and four repair cycles respectively. Inspection criteria vary depending on blade configuration, but the major causes for retiring first-stage blades from service are generally related to cracks, dimensional discrepancies, hot corrosion, or creep as summarized in Table II.

Table II. Typical Causes for First-Stage Blade Rejection

Cause	Solid	Air-Cooled
1. Cracks	37%	71%
2. Dimensional	32%	20%
3. Hot Corrosion	13%	8%
4. Creep	18%	1%

Hot isostatic pressing (HIP) is being investigated as a means to reduce the number of blades that are being retired because of creep but it is doubtful that blade retirement rates will otherwise change significantly (1).

It is more difficult to ascertain the percentage of non-repairable vanes in a typical, moderately aged commercial engine inasmuch as many, otherwise non-repairable vanes are currently salvaged by airfoil replacement. Approximately forty-five percent of the high pressure turbine (HPT) vanes corresponding to the blades referred to in Table II require airfoil replacement

(2). The average service life of a HPT vane airfoil, if not the entire vane, is therefore only two repair cycles. Airfoil retirement causes vary; but approximately three quarters of the HPT vane airfoils are replaced because of dimensional discrepancies whereas only one-quarter are replaced because of excessive airfoil cracking. Replacement percentages in low pressure turbine (LPT) vanes are similarly difficult to ascertain since many LPT vanes are cast or assembled into multiple vane segments. One discrepant vane, in effect, necessitates replacement of four or more vanes thereby skewing replacement percentages. Dimensional discrepancies caused in many instances by multiple repair cycles, are the principle causes of LPT vane replacement.

Repair-Related Coating Technology

All HPT and some LPT blades and vanes are protectively coated to maintain airfoil geometry, with the intent that the coating endure at least until the parts must be otherwise repaired. Coatings are applied principally to protect against Types I and II hot corrosion (3) and simple oxidation. It is frequently stated that coatings also protect against particle erosion. What little data exist in the open literature indicates that the matter is quite complex and if coatings do provide some protection, the improvement is relatively small (4). There is some evidence that coatings improve the thermal fatigue resistance of equiaxed cast blades by minimizing crack initiation at grain boundaries (5). Conversely coatings generally degrade the thermal fatigue resistance of directionally solidified and single crystal castings (6). In these cases tradeoffs must be made between shortened lives due to thermal fatigue and extended lives afforded by coatings.

Each engine manufacturer develops and/or specifies the types of coatings used on blades and vanes in its engines. The most widely used coatings are the simple aluminides applied by pack cementation (7) or by gas phase aluminizing (8, 9). Aluminide coatings applied by gas phase processes are particularly useful for coating the interior surfaces of complex air-cooled HPT blades. Simple aluminide coatings are often modified with chromium and/or platinum applied in separate gas phase or electroplating processes to further enhance hot corrosion resistance.

At least one engine manufacturer specifies aluminide coatings formed by electrophoretic deposition and subsequent diffusion of aluminum alloy powders. Still another engine manufacturer specifies MCrAlY (where M represents nickel and/or cobalt) coatings applied by overlay processes. The simpler types of MCrAlY overlay coatings are applied by electron beam evaporation, whereas the more complex types, for example, those containing additions of silicon and hafnium, are currently applied exclusively by low pressure plasma spraying (10). Other methods in various stages of development include sputtering and suspension electroplating (11).

Considerable effort has been devoted over the past decade to adaptation of thermal barrier coatings to turbine airfoils. The most advanced of these in production consist of a low pressure plasma sprayed MCrAlY type bond coat followed by an air plasma sprayed partially (yttria) stabilized topcoat. The coating is so far used only on non-rotating airfoils. Zirconia coatings applied by electron beam evaporation show promise of superior thermal stress resistance compared to those applied by plasma spraying but are not yet at the production stage (12).

Masking to preclude deposition of coatings on mechanically critical areas of parts, for example, blade roots is an integral and usually complex part of all coating processes. High temperature diffusion processes usually require some type of slurry mask which acts as a barrier to, or getter of, the coatings species (13). Mechanical masks can sometimes be used for diffusion coating processes but may not always be completely efficient (9). Mechanical masks are satisfactory for low temperature coating processes, such as electrophoresis and high temperature overlay coating processes.

The repair of coated blades and vanes is generally preceded by localized or full removal of existing coatings. Some manufacturers require chemical cleaning with strong acid or alkali mixtures to remove field service debris and/or hot corrosion products prior to coating stripping. Others allow grit blasting to accomplish the same ends. Complex cooling passages in blades can accumulate dust or other debris in service and this may have to be removed with hot caustic at elevated pressure in an autoclave (14).

Physical methods, such as grit blasting or belt grinding can, for example, be used to locally remove coating prior to repair welding of unshrouded blades. Full removal of coatings is universally accomplished by selective dissolution of the coating phase(s) by various simple or complex mixtures of acids. Again, each engine manufacturer, and some repair houses have more or less independently developed their own acid formulations. Most procedures depend on selective attack of beta (NiAl or CoAl) phases. If the coatings are depleted of the beta phases, selective coating dissolution can become difficult or impossible and residual coatings must then be removed by physical means such as belt grinding. Great care must be exercised in the development and control of stripping solutions in order to avoid localized or general attack of the superalloy base. For example, dilute nitric acid is a commonly used, safe coating strippant. Contamination with a few percent of hydrochloric acid can, however, cause selective attack of the gamma prime (Ni_3Al) strengthening phase as illustrated in Figure 1 (15).

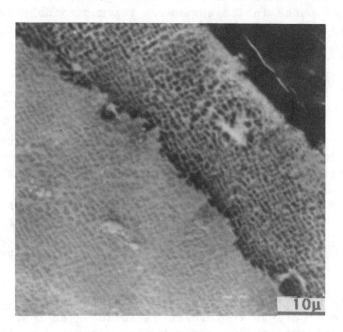

Figure 1. Gamma prime attack caused by coating stripping with nitric acid contaminated by chloride.

Inspection for completion removal of coating is usually accomplished by heat tinting at about 1000°F in air. Gold colors indicate residual coating while the fully stripped alloys are blue.

Since the diffusion aluminide coatings are formed, in part, by nickel (or cobalt) from the substrate alloys it follows that removal of the coatings will cause finite dimensional losses of the substrates. This usually amounts to 0.001 to 0.002 inches for commonly used coating thicknesses of about 0.003 inches.

To preclude the possibility of even minute amounts of attack on uncoated surfaces, such as blade roots and internal cooling passages, wax masking is usually employed. Coated cooling passages present special problems; because of the difficulty of inspecting for complete removal of coating, the passages are usually waxed to preclude any coating stripping.

Clearly gas turbine blade and vane repair involves the acquisition and implementation of a wide variety of coating technologies. Thus, access to coating technology and engine manufacturer coating approvals, in a very real sense, limit complete aircraft gas turbine airfoil repair to relatively few, technically sophisticated repair facilities.

Types of Blade Damage and Repairs

Blade repairs vary depending on blade configuration but are largely matters of restoring blade tips or tip shrouds to demanding dimensional criteria using weld overlay processes. Tip shroud repairs may also include hardfacing replacement and knife-edge seal restoration. Manual gas tungsten arc welding (GTAW) processes are most commonly used for blade repair, but plasma, laser, and electron-beam welding processes are finding increased usage as the trend to automation continues.

Blade repair is complicated by weld cracking problems inherent to blade alloys and by subtle dimensional changes which occur during engine service. Virtually all aircraft gas turbine blades are cast out of nickel-base superalloys with aluminum/titanium contents well above those empirically associated with moderately acceptable weldability, i.e. $Al + Ti/2 \leq 3\%$. Weld overlaying is not particularly difficult provided adequate precautions are taken; but some amount of microcracking is inevitable.

Weld filler alloys are usually selected to minimize microcracking but coating compatibility and strength should also be considered. Weld filler alloys containing even a small amount of aluminum are more compatible with diffused aluminide coatings than are weld filler alloys totally void of aluminum. Aluminum oxide and/or continuous alpha chromium (16) tend to form at the coating/substrate interface in the absence of aluminum in the weld repaired area. As illustrated in Figure 2, AMS 5837 (Inconel 625, $\sim 0.2\%$ aluminum) weld overlays are more compatible with diffused aluminide coatings than are AMS 5679 (Fm 62, nil aluminum) weld overlays.

Further, weld overlay filler alloys generally derive from solid solution or gamma-prime lean nickel-base superalloys considerably lower in strength than the blade alloys they replace. It follows, therefore, that weld repairs are generally not made on highly stresses airfoil surfaces. Some small-engine manufacturing are however considering weld-overlay repairs over the upper two-thirds of the leading and trailing edges of the blade airfoils as well as the blade tips.

Types of Vane Damage and Repair

Some HPT vanes are cast out of nickel-base superalloys similar to those used for blade applications; but the majority of the HPT vanes currently circulating through overhaul facilities are cast out of cobalt-base superalloys. The cobalt-base superalloys, as a rule, are more amenable to repair than are the nickel-base superalloys.

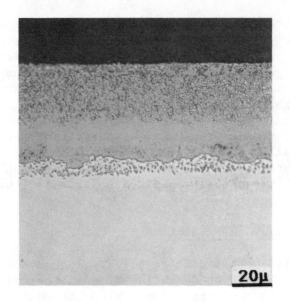 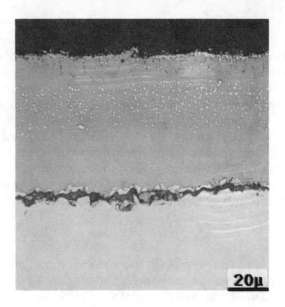

Figure 2. Comparison of compatibility of diffusion aluminide coating with weld filler materials. Left, diffusion coating on AMS 5837 (Inconel TM 625, ~0.2% aluminum). Right, same coating on AMS 5679 (Fm 62, nil aluminum).

HPT vanes encounter rather severe thermal gradients in engine service, distort, and frequently crack by thermal fatigue. HPT vane repair is largely a matter of straightening distorted airfoils in order to restore gas path flow area and re-joining airfoil cracks.

Straightening is accomplished by bending. A limited amount of bending can be accomplished at ambient temperature. However, straightening is best accomplished at elevated temperatures using hot dies, i.e., isothermal forging. Consideration of tensile ductility data suggests it is actually counterproductive to straighten heated nickel-base superalloys using ambient temperature dies or straightening tools. Straightening, if improperly accomplished, can, in and of itself cause airfoil cracking or result in subtle changes in gas path flow area due to relaxation at elevated temperature during subsequent repair operations or engine service.

Cracks in cobalt-base HPT vane airfoils, depending on location and severity, are usually repaired by manual GTAW using cobalt-base filler alloys. The metallurgical rationale for such weld repairs is prima-facie and substantiated by extensive laboratory testing, test stand validation, and field experience. Bonafide weld repairs are usually achieved in cobalt-base HPT vane components provided the crack surfaces are clean, the vane is fully annealed prior to weld repair, and reasonable care is taken while making weld repairs.

Cracks in nickel-base HPT vane airfoils are considerably more difficult to repair by GTAW. Microcracking is virtually certain in nickel-base superalloy crack weld repair; and, macrocracking is likely during post weld heat treatment. The consequences of such weld-induced defects on vane engine performance is uncertain; but it is a misnomer to refer to GTAW as a method of crack "repair" in nickel-base superalloy vanes.

The metallurgical problems associated with GTAW repairs have stimulated interest in braze welding as an alternate repair process for cracks in nickel-base superalloy HPT vane airfoils (17). Braze welding which, by definition, involves distribution of a high-temperature filler material by some mechanism other than capillary action seems well-suited to the repair of wide-gap (0.002 to 0.060 inches in width) cracks such as occur in HPT vane airfoils during engine service (18). However, the metallurgical rationale for braze welding is unclear and substantiating laboratory data are meager. Nevertheless, virtually every vane repair facility currently offers a proprietary three-initial acronym braze weld repair process for cracked HPT vanes. There is no question that cracks can be at least partially filled by braze welding and made acceptable to visual acceptance standards. However, true crack repair is another matter, requiring that the filler material wet and bond to all crack surfaces and approach vane base alloy mechanical/physical properties.

Complete oxide removal from all interior crack surfaces is essential to crack repair by braze welding. Those alloying elements that complicate crack repair by fusion welding in nickel-base superalloys also complicate crack repair by braze welding in that they contribute to the formation of extremely stable oxides that are difficult to reduce or otherwise remove in situ. Various fluoride-ion "cleaning" processes have been developed for the purpose of reducing or removing oxides that can form on nickel-base superalloy crack surfaces; but the efficacy of these processes remains uncertain (19, 20). The ability of these processes to reduce or remove oxides in deep, high-aspect-ratio cracks or as thick layers without detrimentally affecting other superalloy surfaces is suspect. Gamma prime solute depletion and spongy surfaces are natural consequences of fluoride ion treatment of unoxidized nickel-base superalloy surfaces. In addition intergranular attack is often a side-effect.

Braze welding filler materials, as used in vane crack repair, are generally mixtures of vane base alloy and high-temperature braze alloy powders. Typical base to braze alloy ratios range between 50-50 to 75-25. The braze alloys invariably contain boron and/or silicon as melting range depressants. In theory, interdiffusion between the base and braze alloys at temperatures above the solidus of the braze alloy dynamically increases the solidus temperature of the diffused filler material. In practice, the remelt or solidus temperature of weld brazing mixtures containing substantial amounts of boron seldom exceeds 1950°F.

In theory, interdiffusion retards flow without affecting wetting thereby accommodating much wider crack gaps than can be accommodated with braze alloys alone. In practice, flow is controlled by the base to braze alloy ratio and the braze welding temperature in relation to the liquidus of the diffused filler. It is extremely difficult to fill a wide-gap crack or crevice without leaving voids.

Published data concerning the mechanical properties of braze weldments are meager. High-temperature tensile and stress-rupture strengths of braze weldments approaching, and in some instances exceeding, corresponding base alloy strength levels have been measured (21). However, ductility is invariably impaired and thermal fatigue resistance is suspect. Braze welding, never-the-less, is "in vogue" as a method of repairing cracks in nickel-base superalloy HPT vane airfoils.

Thermal fatigue cracking is less frequent in LPT vanes but still occurs. Since virtually all LPT vanes are cast out of nickel-base superalloys, the problems associated with crack repair in nickel-base HPT vanes apply equally to LPT vanes; and complete airfoil replacement is not a repair option as it is with cobalt-base HPT vanes. LPT vanes are generally replaced for dimensional causes such as short chord, wall or airfoil thickness, etc. LPT vanes suffer dimensional degradation due to wear and corrosion. However, the collective effects of prior repairs such as blending and/or coating replacement eventually take their toll and can become the primary causes for LPT replacement. Repair cleaning operations such as grit blasting reduce airfoil thickness by approximately 0.00006 inches. However, chemical cleaning and/or chemical removal of diffused aluminide coatings reduce airfoil thickness by as much as 0.002 to 0.004 inches. Such thickness reductions, though seemingly small, can ultimately limit the useful life of a LPT vane or necessitate some form of dimensional restoration.

Undersized LPT vane platforms are dimensionally restored by weld overlay, thermal spray, or braze welding overlay processes depending on base alloy, location, service conditions, etc. LPT vane airfoil dimensions can be similarly restored. GTAW overlays are impractical for airfoil dimensional restoration because of distortion. However, low heat-input weld overlay processes, such as "Rapid-Arc", hold promise for LPT vane airfoil dimensional restoration applications but are limited to overlay materials that are available in continuous lengths of wire (22). Low pressure plasma spray, without alloy restrictions, holds even more promise for such applications. Weld braze overlays are being actively investigated for airfoil dimensional restoration purposes but are limited by base alloy embrittlement and coating compatibility considerations. The more avant-grade in LPT vane repair are evaluating methods of restoring airfoil dimensions while simultaneously changing gas path flow area to changing engine requirements (23). The "recast" process would seem to have the most potential in this regard although other processes/repair schemes are already being engine tested (24).

Coating Replacement

Subsequent to mechanical repair, coatings are replaced, usually in accord with the original requirements of engine manufacturers. In most cases, coating processing is identical to that used to coat new parts. In some cases, masking may be more complicated, if for example, parts were originally coated before machining to avoid masking costs which would be incurred, for example, in excluding coating from blade roots.

Recovery of alloy properties by heat treatment, if necessary, is accomplished in conjunction with or following the heat cycle of the coating process.

Future Trends

The majority of the processes used to repair aircraft gas turbine blades and vanes are labor intensive. Some processes such as brazing, coating, and heat treatment are batch processes and benefit from volume. However, even these processes require some amount of manual materials handling. There is, of course, considerable interest in automating blade and vane repair processes; but some processes such as crack weld repair must be addressed on an individual piece-to-piece basis. Weld overlay repair processes are, in theory, amenable to automation. However, weld overlay automation is proving difficult in practice due to dimensional inconsistencies intrinsic to engine-run components. Original equipment reference surfaces simply do not apply in repair and local component thickness or mass vary considerably on blades and vanes from the same engine. Automated weld overlay repair systems must be capable of adjusting the weld torch position over variable surfaces as well as adjusting the weld process heat input and deposition rate to accommodate a variable heat sink. The primary focus of automation in aircraft gas turbine blade and vane repair must be toward the development and implementation of adaptive control schemes that will permit automatic repair devices to alter their pre-programmed directions or torch conditions to suit the specific requirements of each individual blade or vane.

The future of blade and vane repair will be determined by replacement part pricing, "break-out" sales activity, and labor cost. The technical issues will probably remain as they are, but repair will become less labor-intensive as automation becomes more feasible.

References

1. J. Wortmann, "Improving Reliability and Lifetime of Rejuvenated Turbine Blades," Materials Science and Technology, 1 (1985), 644-650.

2. J. Riccitelli, private communication with authors, Turbine Components Corporation, Branford, CT.

3. G. W. Goward, "Low-Temperature Hot Corrosion: a Review of Causes and Coatings Therefor," Journal of Engineering for Gas Turbines and Power, 108 (1986), 421-425.

4. R. H. Barkalow, J. A. Goebel, and F. S. Pettit, "Erosion - Corrosion of Coatings and Superalloys in High Velocity Hot Gases," Erosion: Prevention and Useful Applications, ASTM STP 664, (American Society for Testing and Materials, 1979), 163-192.

5. M. Gell, G. R. Leverant, and C. H. Wells, "The Fatigue Strength of Nickel-Base Superalloys," Achievement of High Fatigue Resistance in Metals and Alloys, ASTM STP 467, (American Society for Testing and Materials, 1970), 113-153.

6. K. Schneider and H. W. Grünling, "Mechanical Aspects of High Temperature Coatings," Thin Solid Films, 107 (1983), 395-416.

7. G. W. Goward and L. W. Cannon, "Pack Cementation Coatings for Superalloys: a Review of History, Theory, and Practice," Journal of Engineering for Gas Turbines and Power, 110 (1988), 150-154.

8. R. S. Parzuchowski, "Gas Phase Deposition of Aluminum on Nickel Alloys," Thin Solid Films, 45 (1977), 349-355.

9. G. Gaujé and R. Morbioli, "Vapor Phase Aluminizing to Protect Turbine Airfoils," High Temperature Protective Coatings, ed. S. C. Singhal (Warrendale, PA: The Metallurgical Society of AIME, 1983), 13-26.

10. D. K. Gupta and D. S. Duvall, "A Silicon and Hafnium Modified Plasma Sprayed MCrAlY Coating for Single Crystal Superalloys," Superalloys 1984, ed. M. Gell et al. (Warrendale, PA: The Metallurgical Society of AIME, 1984), 711-720.

11. J. E. Restall, "Development of Coatings for the Protection of Gas Turbine Blades Against High Temperature Oxidation and Corrosion," ibid., 721-730.

12. R. Shankar, "Electron Beam Physical Vapor Deposition Development of Zirconia Coatings" (Paper presented at Coatings for Advanced Heat Engines Conference, U. S. Department of Energy, Castine, Maine, July 27-30, 1987).

13. A. L. Baldi, "Diffusion Treatment of Metal," U.S. Patent 3,958,047 (1976).

14. E. A. Ault and C. E. Bevan," Cleaning Process for Internal Passages of Superalloy Airfoils," U.S. Patent 4,439,241, (1984).

15. G. W. Goward, "Effects of Superalloy and Coating Structures on Hot Corrosion," J. Vac. Sci. Technol., A4 (1986), 2905-2906.

16. G. W. Goward and D. H. Boone, "Mechanisms of Formation of Diffusion Aluminide Coatings on Nickel-Base Superalloys," Oxid. of Metals, 3 (1971), 475-495.

17. J. W. Lee, J. H. McMurray, and J. A. Miller, "Development of a New Brazing Technique for Repair of Turbine Engine Components," Welding Journal, 64 (1985), 18-21.

18. Metals Handbook. Ninth Edition, (American Society for Metals, 1983) Glossary 3.

19. D. L. Keller, D. L. Resor, "Superalloy Article Cleaning and Repair Method," U.S. Patent 4,098,540, (1978).

20. J. W. Chasteen, "Method for Cleaning Metal Parts," U.S. Patent 4,405,379, (1983).

21. N. Czech, W. Esser, and R. Wolters, "Hot Corrosion and Air Creep Rupture Properties of Superalloy Weldings and Brazings," (Final Report, Cost 50-111 Dig. 1983).

22. E. Craig, "A Unique Mode of GMAW Transfer," Welding Journal, 66 (1987), 51-55.

23. P. J. Draghi, J. P. Arrigoni, "Turbine Vane Nozzle Reclassification," U.S. Patent 4,726,101, (1988).

24. S. J. Cretella, M. Bernardo, and R. T. DeMusis, "Refurbished Turbine Vanes and Method of Refurbishment Thereof," U.S. Patent 4,028,787, (1977).

REJUVENATION OF SERVICE-EXPOSED

IN 738 TURBINE BLADES

A.K. Koul*, J-P. Immarigeon*, R. Castillo**, P. Lowden*** and J. Liburdi***

*National Aeronautical Establishment
National Research Council
Ottawa, Ontario
Canada, K1A 0R6

**Turbine and Generator Division
Westinghouse Canada
Hamilton, Ontario
Canada, L8N 3K2

***Liburdi Engineering
Burlington, Ontario
Canada, L9J 1E7

Abstract

A HIP rejuvenation study for an aluminide coated, internally cooled IN 738, turbine blade from an aero engine is described. The study, which involved assessing changes in blade shape, microstructure and creep properties of the airfoil portions of the blades as a function of service time, has shown that airfoils have a tendency to lengthen along their longitudinal axis during service and also that the rate of lengthening increases with increasing service life. The paper discusses on that basis, when to apply HIP rejuvenation for cost-effective blade life extension. It is also shown that the microstructure of the blades is substantially modified by service. The γ' particles coarsen through agglomeration, continuous networks of $M_{23}C_6$ carbides form along the grain boundaries and the blade surface along the internal cooling passages oxidizes. The design of a HIP rejuvenation/heat treatment cycle to recover microstructures and creep properties is discussed. Finally, a rejuvenation cycle that incorporates a diffusion treatment for recoating the blade is proposed that is shown to restore the loss of creep ductility induced by service, while improving time to rupture by a factor of 3 relative to new blades.

Introduction

Ni-base superalloys are used as investment cast turbine blades in many aircraft gas turbine engines. In their conventionally cast polycrystalline equiaxed forms, these alloys derive their high temperature strength from the precipitation of γ' within the grain interiors, as well as from the precipitation of γ' and $M_{23}C_6/M_6C$ carbides at the grain boundaries, the formation of a serrated grain boundary structure and the segregation of trace elements such as B and Zr at the grain boundaries[1]. At service temperatures, microstructural changes occur in blade airfoils which decrease their creep strength and promote airfoil distortion[2]. These time dependent microstructural changes include coarsening of the γ' phase, changes in the grain boundary and grain boundary carbide morphologies and precipitation of brittle intermetallic phases such as σ-phase[2,3]. The loss of creep strength eventually promotes temperature and stress assisted creep cavitation which leads to internal cracking and ultimately to failure. While the majority of microstructural changes can be reversed by reheat treatments, the elimination of creep cavities is only possible through hot isostatic pressing, (HIPing)[4,5]. During HIPing, the blades are subjected to the simultaneous application of high temperature and high inert gaseous pressure, such that the cavities collapse and the cavity walls diffusion bond together. HIPing must be followed by heat treatment for the optimization of microstructure and creep properties[5,6].

In this paper we discuss the results of a rejuvenation case study conducted on an aluminide coated and internally cooled IN 738 aero engine blades with uncoated internal cooling passages. The objectives of the paper are threefold. Firstly, the paper discusses the basis for establishing the need for rejuvenation for a given blade set and for deciding at what stage of service life to apply the rejuvenation for safe and cost effective blade life extension. Secondly, the paper presents a strategy for designing a HIPing rejuvenation cycle for conventionally cast IN 738 turbine blades and thirdly, it discusses problems associated with applying rejuvenation technology to coated blades.

Experimental Materials and Methods

Three blade sets with service times since new (TSN) of 0h, 4200h and 8400h were evaluated in terms of blade lengthening and microstructural damage in the top, middle and bottom sections of the airfoil and in terms of airfoil creep or stress rupture properties. The creep tests were conducted over a range of stresses (70 to 315 MPa) and temperatures (800 to 927°C), using miniature flat specimens machined from airfoils near the trailing edge (concave side) of the blades. The specimen geometry and dimensions are shown in Fig. 1. The specimen thickness was limited by the airfoil wall thickness (1.14mm) of the hollow blades. The gauge length of the miniature specimen corresponded to the mid-airfoil section of the blades. Extensometers were attached to the specimen grips for monitoring creep elongation with the help of a linear voltage digital transducer.

Results and Discussion

Blade Lengthening

A survey of user experience has indicated that, as service time accumulates on these IN 738 blades, their twisted airfoil sections have a tendency to untwist, while the blades lengthen along their longitudinal axis. A plot of reported increase in blade length as a function of blade

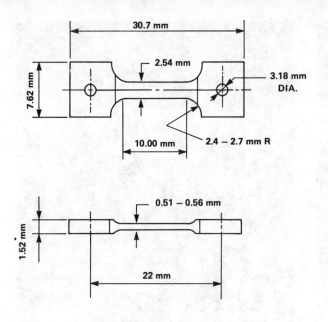

Figure 1. Drawing of the miniature creep-rupture specimen machined from the trailing edge of the blade airfoils.

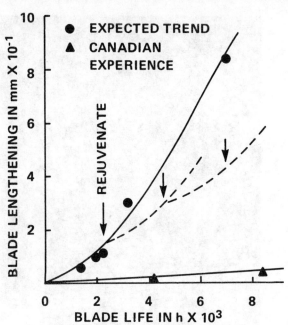

Figure 2. Diagram of blade lengthening versus blade service life illustrating how to decide when to HIP rejuvenate.

life is shown in Fig. 2. Rejuvenation would be expected to decrease the rate of airfoil growth and increase the life expectancy of the blades through the recovery of microstructure and creep properties. For example, assuming that the rejuvenation treatment is applied every 2300 hours, and that the creep properties are fully recovered each time, the blades would lengthen by only 0.43 mm over 7000 hours of service as compared to 0.84 mm for a blade with the same amount of service and no rejuvenation applied, Fig. 2. For a given lengthening allowable, blade life could be substantially increased in this manner. In order to decide when to apply the treatment for cost effective rejuvenation, airfoil growth rates and growth limits should be known. In the present case, less frequent than every 2300 hour rejuvenation would not be attractive because the rate of airfoil growth beyond that time becomes excessively large, as shown in Fig. 2. Furthermore, if the blade growth limits are too small, the rejuvenation may not be cost effective because the blades may be nearing their allowable growth limits after the first engine overhaul. Finally, since rejuvenation will not restore the original dimensions of the blades, the potential impact of dimensional changes for instance on engine performance or vibration characteristics should also be addressed.

The service exposed blades in the present study exhibited much lower lengthening values than those reported by other users, Fig. 2, which indicates that their operating conditions must have been less severe. The rejuvenation study was nevertheless conducted to establish whether any service-induced reduction in creep ductility of the blades, due for instance to grain boundary embrittlement, could be circumvented.

Microstructural damage

Considerable primary and secondary γ' coarsening and agglomeration was observed in the mid-airfoil sections of the service-exposed blades, Fig. 3, but no creep cavities were detected during optical and SEM examination. However, this does not rule out the possibility that

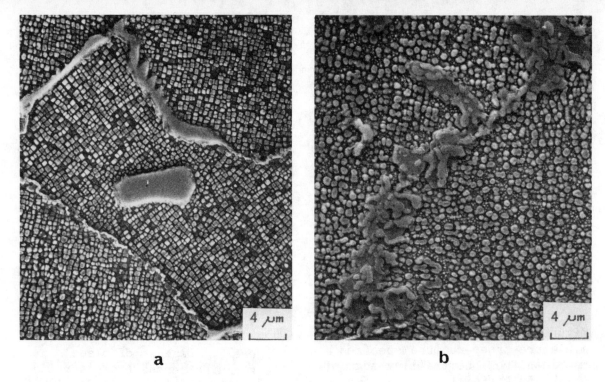

Figure 3. Microstructures of the IN 738 blades in a) the new, fully heat-treated condition and b) after 8400 hours of service in an aircraft engine, illustrating that service induces coarsening of the γ' precipitate phase and modifies the morphology of the grain boundaries. (SEM micrographs).

ultrafine creep cavities were present that were not resolved by the optical or SEM techniques employed(7). There was also some evidence of continuous networks of grain boundary $M_{23}C_6$ carbide formation in the service exposed blades, Figs. 3.

Both batches of service exposed blades had formed oxides along the uncoated walls of the internal cooling passages within the blade airfoils, where oxides were observed to have formed over a depth of approximately 25 μm, with some intergranular oxide spikes penetrating as deep as 125 μm below the surface, Fig. 4. None of these microstructural changes are unusual for blades after long time service in the oxidizing atmosphere of a gas turbine. Finally, shrinkage cavities, an inherent feature of investment cast components, were observed in both new and service exposed blades, Fig. 5.

Design of HIP Rejuvenation Cycles

The HIP plus reheat treatment conditions were selected on the basis of past experience with IN 738(8), IN 738LC(2), Inconel 700(9), Inconel X-750(1), Alloy 713C, Nimonic 105(5), Nimonic 115(9) and Nimonic 80A(17) turbine blades, which has shown that:

(a) The HIPing temperature should lie above the γ' and $M_{23}C_6/M_6C$ solvus temperature but preferably below the MC solvus temperature and should be selected to avoid incipient melting. HIPing above the γ' and $M_{23}C_6/M_6C$ solvus temperatures reduces resistance to plastic flow and ensures complete closure of shrinkage and/or creep cavities, while keeping the HIP temperature below the MC solvus temperature prevents rapid grain growth that otherwise occurs in these materials.

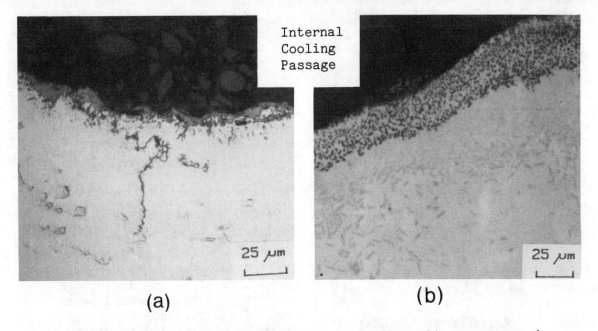

(a) (b)

Figure 4. Evidence of oxidation of the internal cooling passages in service-exposed blades a) oxide spike penetrating to a depth of approximately 125 μm and b) oxidized and alloy depleted surface layer roughly 25 μm thick lining the walls of the cooling passages.

Rapid grain growth in IN 738 commences at 1225°C. Grain growth reduces the grain boundary area available for carbide precipitation which in turn can lead to continuous carbide film formation along the grain boundaries during post HIP ageing treatments, thereby embrittling the grain boundaries[6,9,11]. In addition, HIPing in the MC solvus temperature range can dissolve sulphocarbides and the free S can segregate at the grain boundaries[5], which can decrease the grain boundary cohesive strength.

(b) HIPing above the γ' solvus temperature destroys the original serrated grain boundary structure and, a controlled cool from the HIPing or post HIPing solution treatment temperature through the γ' precipitation range is necessary to reproduce the serrations[8,12]. Serrated grain boundaries suppress grain boundary sliding[13], the deformation mechanism that predominates at service stresses and temperatures[14]. Therefore, reproducing the serrated grain boundaries during rejuvenation is extremely important, Fig. 6.

(c) The post-HIPing aging treatments for controlled precipitation of primary and secondary γ' precipitates are usually (but not necessarily) the same as those designed for the virgin alloy. In a number of older Ni-base alloys, γ' precipitate sizes and distributions have been optimized on the basis of short term creep testing where the dominant deformation mechanism may not be the same as that operative at service stresses and temperatures[14].

Based on this past experience, three HIPing rejuvenation cycles were considered for eliminating the shrinkage cavities and rejuvenating the damaged microstructure of the service exposed blades. The three cycles consisted of:

Cycle 1: HIPing[8] + the standard high temperature-low activity aluminide coating cycle.

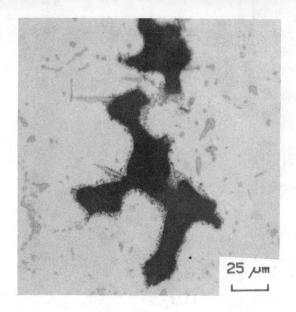
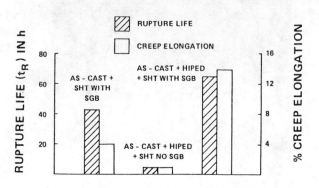

Figure 5. Evidence of shrinkage microporosity present in the airfoil sections of the blades.

Figure 6. Effects of serrated grain boundaries on stress-rupture properties of IN 738 at 586 MPa and 760°C in the standard heat treated conditions. Serrated grain boundaries are formed by controlling cooling rate from the solution treatment temperature.(8)

Cycle 2: HIPing + a silicon modified low temperature-high activity aluminide coating cycle.

Cycle 3: HIPing at 1200°C/2h/105MPa/F.C. + Solution treat at 1200°C/2h $\xrightarrow{\text{controlled cool}}$ 1130°C/AC + the slurry coating heat treatment cycle.

All HIPing rejuvenation cycles were designed to avoid primary MC carbide dissolution, suppress rapid grain growth and regenerate the original serrated grain boundary structures. Cycle no. 1 produced a uniform distribution of primary and secondary γ' precipitates, Fig. 7(a), whereas cycle no. 3 produced a bimodal distribution of secondary spherical γ' with a mixture of $M_{23}C_6$ and MC carbides at the grain boundaries, Fig. 7(c). Cycle no. 2 which incorporated the slurry coating diffusion treatment of cycle no. 3 also incorporated a hold time at 1120°C and this produced some primary γ' precipitates in addition to the biomodal distribution of secondary spherical γ' precipitates, Fig. 7(b).

In some rejuvenated blades, the severity of oxide penetration along the walls of the internal cooling passages appeared to have increased during HIPing or post-HIPing heat treatments, Fig. 8. This increase could be due to contaminants present within the atmospheres of the HIP vessel or heat treatment furnaces used. Oxide penetration and spalling can reduce the load bearing capabilities of blades and intergranular oxide spike formation can locally increase the stress concentrations, thus increasing the risks of in-service stress-rupture failures. The influence of oxide formation was not studied in detail because, while machining the creep specimens, the oxides were removed from the gauge length.

Creep properties

The rupture life (t_R) of the new, service exposed and rejuvenated blades are presented in the form of a Larson-Miller (L-M) plot in Fig. 9. The t_R data for all materials generally fall within the scatterband for

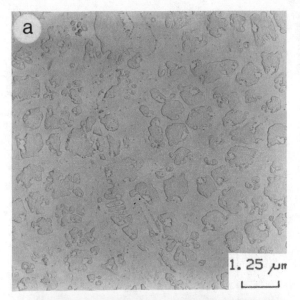
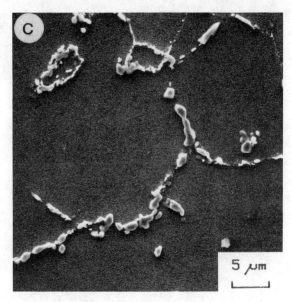
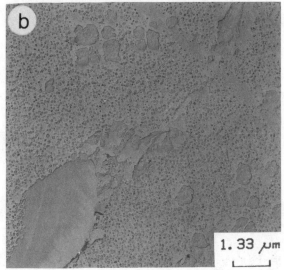

Figure 7. Microstructures of the HIP rejuvenated blades a) after Cycle 1, b) after Cycle 2 and c) after Cycle 3 (see text for HIP cycle details).

the new blades although some data points from blades subjected to rejuvenation cycle no. 1 fall below the lower bound at lower stresses, Fig. 9. Recent investigations have shown that the use of parametric functions is not the best method for assessing creep damage in turbine blades(2,5). Instead, it has been suggested that creep testing parameters should be selected in a manner such that a single deformation mechanism remains dominant under a given set of creep stresses and temperature. Information from these creep tests can then be used to decide whether resistance to a specific creep deformation mechanism has been restored in the rejuvenated blades(2). Following this suggestion, several creep tests (6 tests per material condition) were conducted at a low temperature and high stress (800°C/315MPa) where intragranular deformation is likely to predominate and at high temperature and low stress (927°C/76MPa) where grain boundary sliding is expected to predominate.

Relative to new blades, the service exposed blades showed a marginal reduction in t_R and a slight increase in rupture strain (ϵ_R) at 800°C/315MPa, Fig. 10a. The rejuvenation cycles nos. 1 and 2 improved t_R with a marginal decrease in ϵ_R, Fig. 10a. Assuming intragranular deformation prevailed under these testing conditions, these trends in t_R and ϵ_R are to be expected because service exposed blades contained coarse γ' precipitates whereas rejuvenation cycle nos. 1 and 2 considerably refined the γ' relative to both new and service exposed blades, thereby increasing grain strength. The minimum creep rates ($\dot{\epsilon}_m$)

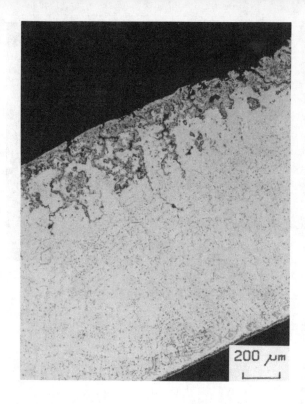

Figure 8. Evidence of increased oxide penetration along the walls of the internal cooling passages after HIPing.

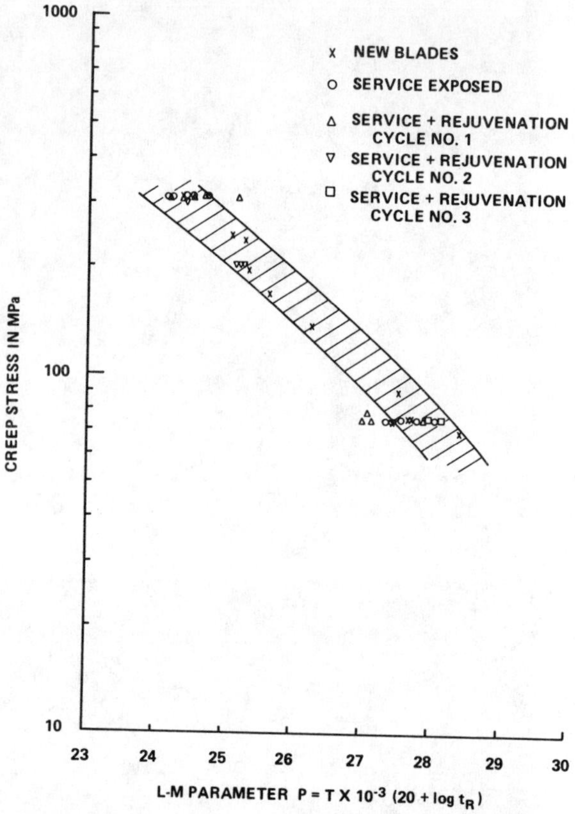

Figure 9. Larson-Miller parameter plots for new, service-exposed and HIP rejuvenated IN 738 blades.

for rejuvenated blades were noted to decrease by a factor of 5 relative to the new and service exposed blades which lends support to this argument.

At 927°C/76MPa, where grain boundary sliding was expected to predominate, the service exposed blades revealed a dramatic reduction in ϵ_R with a marginal increase in t_R relative to new blades, Fig. 10b. Since service exposed blades contained continuous networks of grain boundary $M_{23}C_6$ carbides, the reduction in ϵ_R under grain boundary sliding condition ought to be expected. It is however surprising to note that rejuvenation cycle no. 1 did not restore ϵ_R and t_R to as new condition whereas rejuvenation cycle no. 2 not only restored ϵ_R but improved t_R by a factor of 3 relative to new blades, Fig. 10b. It is suggested that these differences are related to differences in substrate microstructure and composition arising from differences in processing conditions during recoating of the blades.

The improved creep properties of blades subjected to rejuvenation cycle no. 2, where a high aluminum activity Si modified aluminide coating was applied, can be attributed to the precipitation of discrete grain boundary carbides and a uniform refinement of γ'. During grain boundary sliding, deformation is not confined to the grain boundary plane alone but is accommodated within a finite zone adjacent to the grain boundaries by intragranular flow(14). Thus, γ' refinement would be expected to decrease creep rate and increase t_R, whereas discrete grain boundary carbides would be expected to delay fracture and improve ϵ_R.

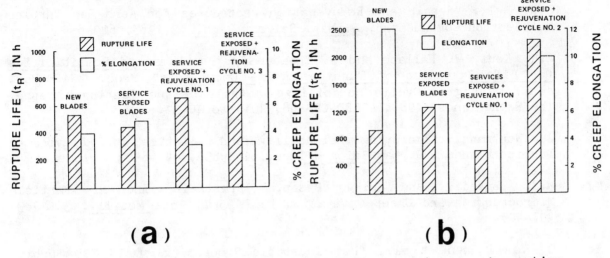

Figure 10. Effects of service exposure (8400 hours) and rejuvenation treatments on the stress-rupture properties of IN 738 at a) 315 MPa and 800°C and b) 76 MPa and 926°C.

In contrast, blades subjected to rejuvenation cycle no. 1, where a low activity aluminide coating was applied, Ni and Cr are known to diffuse into the coating(15). Since the blades were recoated as part of the rejuvenation treatment, the Cr content of the substrate may have been reduced considerably to cause premature creep failure during testing in air due to accelerated oxygen attack along the grain boundaries. Investigations are currently underway to validate this hypothesis.

Conclusions

The intrinsic loss of creep ductility induced by service in an IN 738 blade material can be fully recovered by HIP rejuvenation while time to rupture relative to new blades can actually be improved, providing post-HIP heat treatments are carefully designed. Internal surface degradation along blade cooling passages during HIP or post-HIP processing is however a concern that needs to be addressed.

Acknowledgements

Financial assistance from the Department of National Defence of Canada in the conduct of this work is gratefully acknowledged.

References

1. P.H. Floyd, W. Wallace and J-P. Immarigeon, "Rejuvenation of Properties in Turbine Engine Hot Section Components by Hot Isostatic Pressing", Proc.Int. Conf. Heat Treatment 81, International Congres on Metals Engineering, The Metals Society, 1983, 97-102.

2. A.K. Koul and R. Castillo, "Assessment of Service Induced Microstructural Damage and its Rejuvenation in Turbine Blades", Met. Trans. A., 1988, in press.

3. A.K. Koul and W. Wallace, "Microstructural Changes During Long Term Service Exposure of Udimet 500 and Nimonic 115", Met. Trans A., 14(1983), 183-189.

4. S.R. Bell, "Repair and Rejuvenation Procedures for Aero Gas Turbine Hot Section Components", J. Mat. Sci and Tech., 2(1985) 629-634.

5. A.K. Koul, W. Wallace and R. Thamburaj, "Problems and Possibilities for Life Extension of Turbine Engine Components" Proc. AGARD-PEP, conf. on Engine Cyclic Durability by Analysis and Testing, Liesse, The Netherlands, 1984, AGARD-CP-368, 10-1 to 10-31.

6. J. Wortmann, "Improving Reliability and Lifetime of Rejuvenated Turbine Blades", J. Mat.Sci and Tech., 1(1985), 644-650.

7. M.S. Loveday and B.F. Dyson, "Prestrain Induced Particle Microcracking and Creep Cavitation in IN 597, Acta Metall., 31(1983) 397-405.

8. G. Van Drunen et al., "Hot Isostatic Processing of IN 738 Turbine Blades", Proc. AGARD-SMP conf. on Advanced Fabrication Processes, Florence, Italy, 1978, AGARD-CP-256, 13-1 to 13-12.

9. A.K. Koul and W. Wallace, "Microstructural Dependence of Creep Strength in Inconel 700", Met. Trans A, 13(1982) 673-675.

10. A.K. Koul, "Rejuvenation of Nimonic 115 Turbine Blades - Microstructural Considerations" (Report NRC/NAE LTR-ST-1284, National Research Council of Canada, 1981.)

11. P. Lowden and J. Liburdi, "Rejuvenation of Nimonic 80A Blades" (Liburdi Eng. Report No. 84-12-87B, 1986).

12. A.K. Koul and R. Thamburaj, "Serrated Grain Boundary Formation Potential of Ni-Base Superalloys and its Implication", Met. Trans A, 16 (1985) 17-26.

13. M.F. Ashby, R. Raj and R.C. Gifkins, "Diffusion Controlled Sliding at a Serrated Grain Boundary", Scripta Met., 4(1970) 737-742.

14. R Castillo, A.K. Koul and J-P. Immarigeon, "Creep Behaviour of IN 738 at Low Stresses", Present conference proceedings, 1988.

15. P. Patnaik and J-P. Immarigeon, "High Temperature Protective Coatings for Aero Engine Gas Turbine Components", (Report NRC/NAE NAE-AN-42, National Research Council of Canada, 1986).

APPLICATION OF MELT-SPUN SUPERALLOY RIBBONS

TO SOLID PHASE DIFFUSION WELDING FOR NI-BASE SUPERALLOY

K.Yasuda, M.Kobayashi, A.Okayama, H.Kodama, T.Funamoto, M.Suwa

Hitachi Research Laboratory Hitachi, Ltd.
Hitachi-shi, Ibaraki-ken, 317 JAPAN

Abstract

Solid phase diffusion welding using an insert metal having almost the same compositions as the base metal is investigated in order to realize high welding strength for Ni-base superalloys. For the purpose of this study, melt-spun superalloy ribbons, which are noticed due to its dimensions in thickness and fine grain structures, are examined about superplastic property and application as insert metals. Welding is carried out for various welding pressures, times, temperatures, atmospheres, cleaning treatment for welded surface, and heat treatment after welding. Base metal used is René80, and insert metals are ribbons of IN738LC, IN738FC and Ni-20Al-5Cr produced by a twin-roll rapid solidification process. In this investigation, welding strength, evaluated by tensile tests at 980°C using ribbons aged at 1200°C for only 5h, exhibits high strength almost equal to that of the base metal.

Introduction

Liquid phase diffusion welding using a thin interlayer having a melting point lower than that of base metal is widely used for Ni-base superalloy welding.(1)(2) This process has an advantage of low welding pressure, but is has several disadvantages as outlined below.
(1) A long homogenizing treatment time.
(2) A decrease in the welding strength because of the existence of a heterogeneous layer composed of boride, low melting point elements, etc.
(3) Aggregation of the dispersed oxide because of joint melting in case of an oxide dispersion strength (ODS) alloy.

On the other hand, solid phase diffusion welding using an insert metal having almost the same composition as the base metal eliminates these problems. Here, effects such as less deformation of the base metal and improvement of contact are realized by deformation of the insert metal itself. However in solid phase diffusion welding, higher welding pressure is usually needed and it is hard to fabricate the thin interlayer needed for welding γ' precipitate-hardened Ni-base superalloy.

Therefore, we looked at grain refining and making a thin interlayer by rapid solidification (twin-roll process). If fine grain Ni-base superalloy ribbons are made by rapid solidification, it is expected that they deform with low stress by superplastic deformation and the welding pressure for solid phase diffusion welding can be decreased due to their easy deformability. We found that superalloy ribbons have fine grain structures and superplastic potential.(3)

In this paper, we investigate further the superplastic propriety of the superalloy ribbons and apply them as an insert metal to solid phase diffusion welding for cast γ' precipitate-hardened Ni-base superalloys, in order to realize high welding strength.

Experimental procedure

The welding base metal studied was γ' precipitate-hardened Ni-base superalloy Rene80, which was cast in 10mm diameter rods, and machined to 8mm in diameter, 10mm in height and 1μm in average roughness on the welding surfaces. The insert metals (ribbons) used were: IN738LC which was expected to have a driving force for diffusion due to the difference in composition; IN738FC which contained less C than IN738LC; and Ni-20Al-5Cr had a composition like that of the main elements of Ni-base superalloy and was expected to increase the volume of the γ' phase in order to improve the strength. Table 1 lists the chemical compositions of alloys used.

Table 1 Chemical compositions of alloys used (wt%).

Alloy	Co	Cr	Mo	W	Al	Ti	Nb+Ta	C	Zr	B
Rene80	9.51	13.94	3.90	4.06	2.96	5.02	—	0.16	0.04	0.01
IN738LC	8.63	15.99	1.71	2.52	3.32	3.38	0.48	0.1	0.08	0.01
IN738FC	8.20	15.60	1.65	2.59	3.23	3.33	2.65	0.007	0.10	—

Melt-spun ribbons were produced in Ar gas on two Cu-2%Be rolls (150mm in diameter). The two roll surface velocities were 10m/s, roll compression force was 6900N, and initial roll gap was 0mm. The molten metal was ejected through an orifice (1mm in diameter), using an Ar gas flow, after the alloy was melted. Thickness and widths of ribbons were about 50-100μm and 8-15mm, respectively.

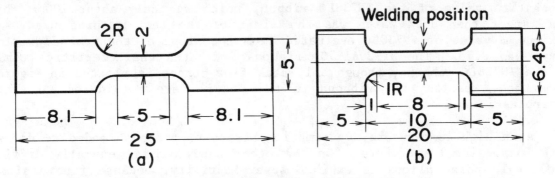

Figure 1 Dimensions of tensile specimens:
(a) for ribbons, (b) for welding strength.

Superplasticity of the ribbons was evaluated by tensile tests carried out at various strain rates at 1100°C in 10^{-3}Pa vacuum atmosphere. In this test, the strain rate was fixed by controlling the crosshead speed correspondingly to elongation of the ribbon through the test. The dimensions of a tensile specimen for the ribbon are shown in Figure 1(a).

Solid phase diffusion welding was performed in 10^{-3}Pa vacuum atmosphere using a tungsten mesh heater. Heat treatment after welding was conducted by aging at 1200°C for 5-10h followed by air cooling(A.C.) in Ar gas, and then 1100°C for 1h followed by furnace cooling (F.C.) in the welding furnace. Welding strength was estimated by tensile tests at a 1.7μm/s in tensile rate. (Initial strain rate was 3.3×10^{-4}/s.) The dimensions of a tensile specimen for welding strength are shown in Figure 1(b).

Results and discussion

Superplasticity of melt-spun superalloy ribbons

Figure 2 shows relationships between peak flow stress or elongation and strain rate for IN738LC and Ni-20Al-5Cr ribbons in tensile tests at 1100°C. The peak flow stress decreased monotonously with deceresing strain rate, and the minimum peak flow stress for IN738LC was 16MPa. The maximum

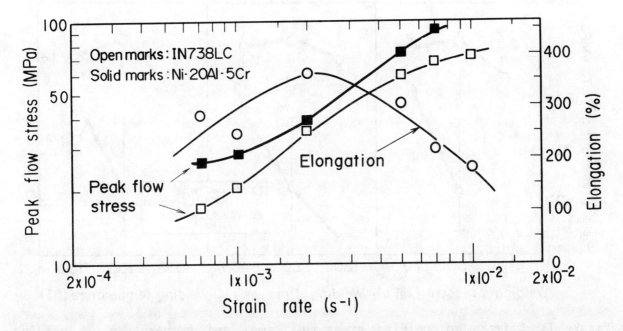

Figure 2 Relationships between peak flow stress or elongation and strain rate for IN738LC and Ni-20Al-5Cr ribbons at 1100°C.

elongation value of the IN738LC ribbon, which was only about 100μm thick, was larger than 360% which was the limit of the test machine used. The maximum m value of IN738LC evaluated from the slope of the peak flow stress curve was 0.73. The Ni-20Al-5Cr ribbons indicated characteristics similar to the IN738LC ribbons, though only peak flow stress is plotted in Figure 2. From these results, the ribbons in this study were considered to exhibit superplasticity.

At a high temperature, deformation stress of ribbons tended to decrease with increasing temperature. On the other hand, it is generally difficult for single phase alloys to exhibit superplasticity, because fine grains of single phase alloys are easily coarsened by aging at high temperature. Ni-base superalloy is usually a single phase at a higher temperatures than 1200°C because of dissolution of the γ' phase. Then in this study, welding were carried out at 1050-1150°C, considering the low deformation stress and ductility of ribbons.

Solid phase diffusion welding for Ni-base superalloy

Solid phase diffusion welding was carried out using Renē80 base metal with and without IN738LC ribbons while changing the welding pressure, time, and temperature. Standard welding conditions were set as 49MPa for pressure, 3.6ks for time and 1100°C for temperature. Figure 3(4) shows the effects of welding pressure, time, and temperature on welding strength at room temperature with and without IN738LC ribbons. The welding strength using ribbons was higher than that without ribbons under all conditions. This meant that the ribbons contributed were considered to the improvement of welding strength. These effects of ribbons were considered to be caused by easier deformability with small grains. The base metal was hardly deformed during welding because cast Renē80 had large grains and high strength at 1050-1150°C. For example, 0.2% yield stress at 1100°C of the base metal was 126 MPa, when the initial strain rate was 2×10^{-4}/s. On the other hand, the fine grained ribbon, which showed superplasticity in the casting direction, exhibited about a 20% reduction in thickness after welding. As a result, an active green surface appeared on the ribbons, and

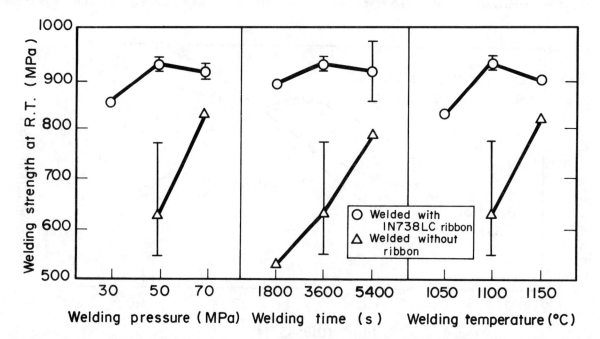

Figure 3 Effects of welding pressure, time, and temperature on welding strength at room temperature of joints with or without IN738LC ribbons.

the joint was securely welded. However, the fractured positions of the joints in Figure 3 were all at the welded interface.

Figure 4 shows the results of electron probe microanalyzer (EPMA) analysis of the welded interface using the IN738LC ribbon under the standard welding conditions. Peaks of C and O were observed at the welded interface which suggested that carbides and oxides caused the decrease in welding strength. The O source for oxides at the welded interface was considered to be the welding atmosphere and an oxide film covered the welded surfaces before welding. An oxide film, about 10-20nm in thickness, was detected by Auger electron spectroscopy (AES) analysis sputtered in the depth direction by Ar ion.

Next, effects of aging at 1200°C on the welding strength were investigated. The following welding was performed at the standard welding conditions, which provided the highest welding sterngth in Figure 3. Figure 5 shows the relationships between welding strength (a) at room temperature, or (b) at 980°C and aging time at 1200°C. At room temperature, joints using Ni-20Al-5Cr ribbons exhibited the highest welding strength, and these fractured at the base metals. However, they had low welding strength and fractured in the insert metals at 980°C.

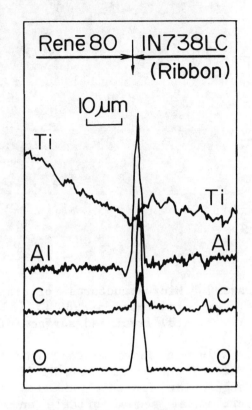

Figure 4 Results of EPMA analysis of the welded interface of the joint using the IN738LC ribbon.

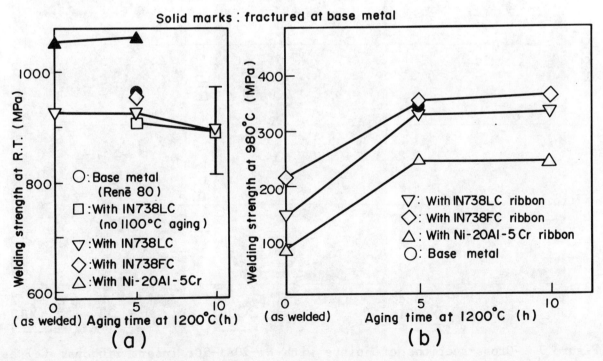

Figure 5 Relationships between welding strength (a) at room temperature or (b) at 980°C and aging time at 1200°C.

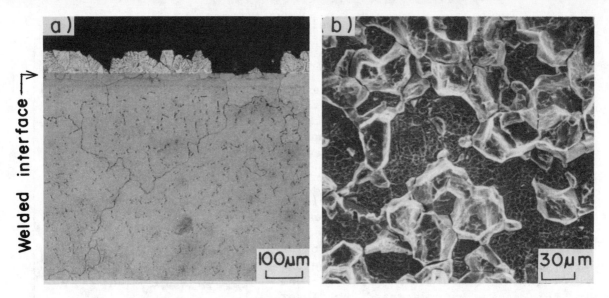

Figure 6 Microstructures of the joint using Ni-20Al-5Cr ribbon which fractured at 980°C: (a) cross-section of the fractured point and (b) fractured surface of the joint.

Figure 6 shows microstructures of the cross-section and the fractured surface of the joint using Ni-20Al-5Cr ribbon which fractured at 980°C. The intergranular fracture occurred in the insert metal, and grains in the insert metal seemed brittle and coarse, being about 50μm. In order to understand these results, microstructures of the cross-sections of the joints using Ni-20Al-5Cr ribbons were observed. Figures 7(a), (b) and (c) show microstructures of the joints as welded (no etching), aged at 1200°C for 5h→A.C., and aged at 1100°C for 1h→F.C. after the heat treatment of (b), respectively. In Figure 7(a), small carbide precipitates were observed on the base metal near the welded interfaces. On the other hand, a precipitation-free zone (PFZ) with no carbide existed near the welded

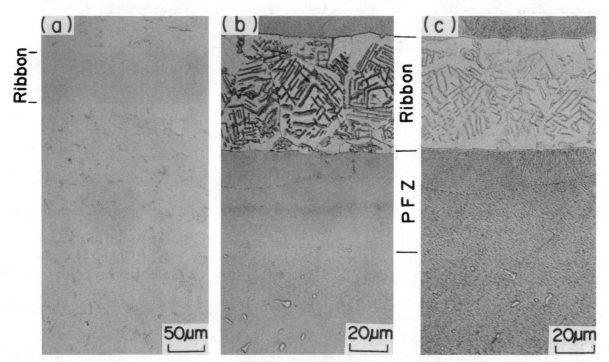

Figure 7 Cross-sections of joints with Ni-20Al-5Cr insert ribbons: (a) as welded at 1100°C (no etching), (b) aged at 1200°C for 5h→A.C. and (c) aged at 1200°C for 5h→A.C. and 1100°C for 1h→F.C. after welding.

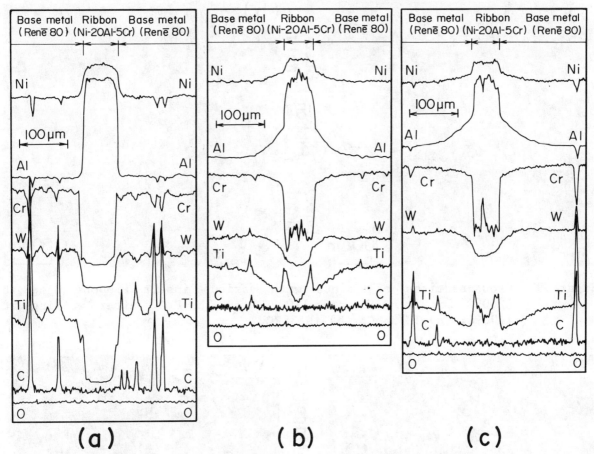

Figure 8   Results of EPMA analysis of the welded interfaces of the joints using Ni-20Al-5Cr ribbons: (a) as welded at 1100°C, (b) aged at 1200°C for 5h→A.C. and (c) aged at 1200°C for 10h→A.C. after welding.

interfaces in Figure 7(b). In Figure 7(c), a fine γ' phases precipitated, but no carbides existed in the PFZ. Figure 8 shows the resutlts of the EPMA analysis of the welded interfaces of the joints using the Ni-20Al-5Cr ribbons as welded and aged at 1200°C for 5 or 10h. Here, the remarkable distributions of Ti were observed. On aged joints, Ti content in an area of the base metals which corresponded to the PFZ decreased to the welded interfaces, and increased abruptly at the welded interfaces. This suggested that the insert metal became a brittle compound due to Ti movement from the base metal to the insert metal. Consequence, it was considered that the low welding strength at 980°C using the Ni-20Al-5Cr ribbon was caused by the brittleness at 980°C of the compound formed in the ribbon. A similar phenomenon was observed in other Ni-Al-X system ribbons, so the Ni-Al-X system ribbons were considered to be unsuitable as insert metals for welding of Ni-base superalloys.

On the other hand, in Figure 5, the joints using the IN738FC and IN738LC ribbons exhibited high welding strength at 980°C, almost equal to the base metal strength when aged for only 5h. Figure 9 shows microstructures of cross-sections of the joints with the IN738LC and IN738FC ribbon aged at 1200°C for 5h followed by A.C. and 1100°C for 1h followed by F.C. after welded. On the joint with the IN738LC ribbon, white lines were observed at the welded interfaces. Figure 10 shows microstructures of cross-sections of the joints with the IN738LC ribbons aged at 1200°C for 5-10h followed by A.C. (γ' phase was prevented from precipitating). Here, fine carbides were found at the welded interfaces, and the white lines observed in Figure 9 were such carbides. From the above results, the

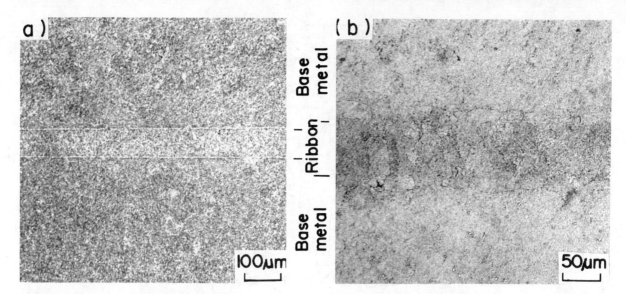

Figure 9 Cross-sections of joints with various insert ribbons aged at 1200°C for 5h→A.C. and 1100°C for 1h→F.C. after welding at 1100°C: (a) IN738LC, (b) IN738FC.

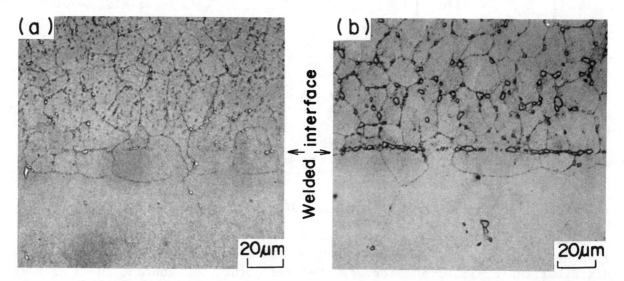

Figure 10 Cross-sections of joints with IN738LC insert ribbons: (a) aged at 1200°C for 5h→A.C. and (b) aged at 1200°C for 10h→A.C. after welding at 1100°C.

difference in the welding strength between IN738FC and IN738LC ribbons was due to carbide formation the source of which was the insert metal, at the welded interfaces. In Figure 5, the effect of the aging treatment at 1200°C on the welding strength at 980°C was remarkable, though that was little effect on the room temperature strength. This was caused by the grain sizes. That is, a large grain was more favorable to high temperature strength than a small grain was. In Figure 10, the coarse grains of the ribbon were observed on aging at 1200°C.

Next, a new apparatus was developed and the welding strength was examined in order to realize higher welding strength. In this apparatus, the surfaces of the base metals and ribbons were cleaned by Ar ion beam bombardment before welding in order to remove the oxide film, and all processes in solid phase diffusion welding could be carried out in a better vacuum atmosphere (10^{-5}Pa) without exposing the base metals and ribbons to air. Figure 11 shows the effects of welding atmosphere and cleaning of welded surfaces on the welding strength at room temperature, with or without

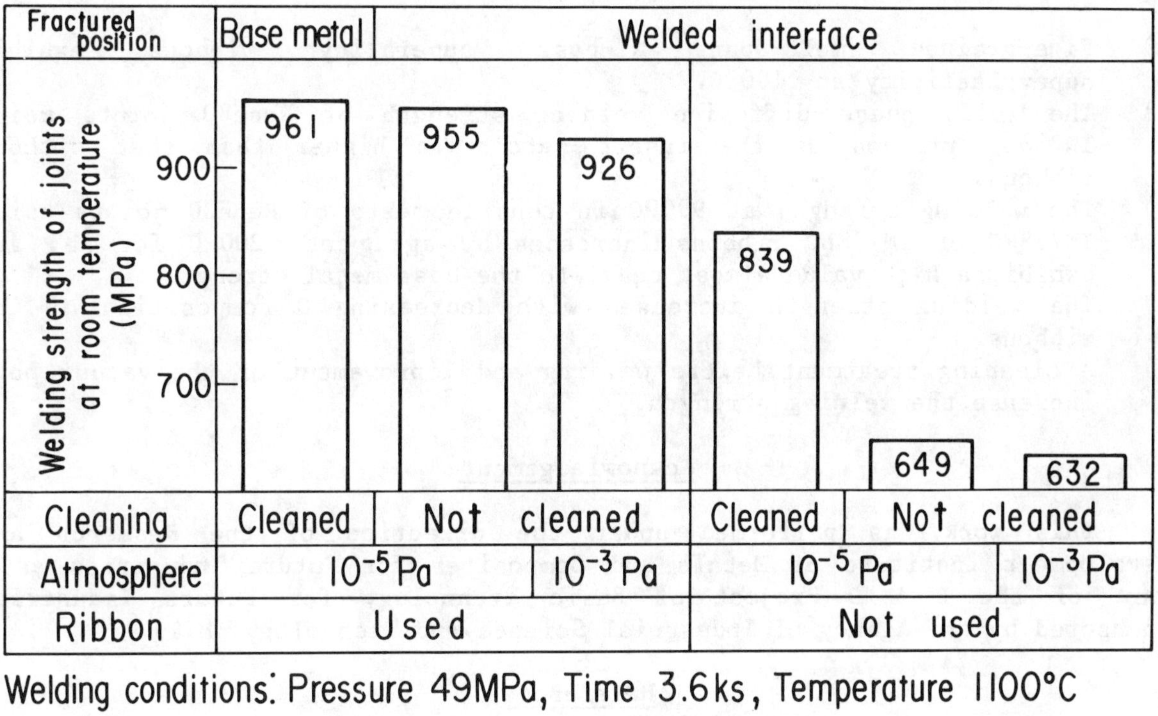

Fractured position	Base metal	Welded interface			
Welding strength of joints at room temperature (MPa)	961	955	926	839	649	632
Cleaning	Cleaned	Not cleaned		Cleaned	Not cleaned	
Atmosphere	10^{-5} Pa		10^{-3} Pa	10^{-5} Pa		10^{-3} Pa
Ribbon	Used			Not used		

Welding conditions: Pressure 49MPa, Time 3.6ks, Temperature 1100°C

Figure 11 Effects of welding atmosphere and cleaning of the welded surfaces on welding strength of Reneˉ80 joints with or without IN738LC ribbons.

the IN738LC ribbons, using the developed apparatus and the standard welding conditions. Only the joint welded using the ribbon with the cleaning treatment at 10^{-5}Pa fractured in the base metal, the other joints fractured at the welded interfaces. Moreover in Figure 11, the welding strength of all joints increased when the welded surfaces were cleaned or welding was done at 10^{-5}Pa. From these results, the cleaning treatment and the improvement of vacuum atmosphere were considered to increase the welding strength.

Figure 12 shows the microstructure of a cross-section of a MA6000 (ODS alloy) joint aged at 1200°C after welding using the IN738FC ribbon using the developed apparatus and the standard welding conditions. This joint fractured in the base metal at room temperature, so this welding method was judged as also effective for ODS alloys.

Figure 12 A cross-section of MA6000 joint with IN738FC insert ribbons aged at 1200°C for 5h→A.C. and 1100°C for 1h→F.C. after welding at 1100°C.

Conclusions

1) Fine-grained melt-spun Ni-base superalloy ribbons exhibit superplasticity at 1100°C.
2) The solid phase diffusion welding strength of Renè80 joints using IN738LC ribbons as the insert metals is higher than that without ribbons.
3) The welding strength at 980°C in tensile tests of Renè80 joints using IN738FC or IN738LC ribbons increases by aging at 1200°C for 5h, and exhibit a high value almost equal to the base metal strength.
4) The welding strength increases with decreasing C composition of the ribbons.
5) A cleaning treatment before welding and improvement of the vacuum both increase the welding strength.

Acknowledgement

This work was performed under the direction of the Research and Development Institute of Metals and Composites for Future Industries as a part of the R & D Project of Basic Technology for Future Industries sponsored by the Agency of Industrial Science and Technology, MITI.

References

1) D.Duvall, W.Owczarski and D.Paulonis, "TLP Bonding; A New Method for Joining Heat Resistance Alloys," Weld.J., 53-4(1974),203-214.
2) G.Hoppin and T.Bery, "Activated Diffusion Bonding," Weld.J., 49-11(1970),505-509.
3) K.Yasuda M.Tsuchiya, T.Kuroda and M.Suwa, "Mechanical Properties and Microstructure of Melt-spun Superalloy Ribbons," Proc. 5th Int. Symp. on Superalloys, Seven Springs Mountain Resort, Champion, Pa, USA,(1984) 477-486.
4) K.Yasuda et al., "Effect of Welding Condition and Carbon Content of an Insert Metal on Joint Strength for Ni-base Superalloys," to be published in Quarterly Journal of the Japan Welding Society.

THEORETICAL RESEARCH ON TRANSIENT LIQUID INSERT METAL

DIFFUSION BONDING OF NICKEL BASE ALLOYS

Yoshikuni NAKAO, Kazutoshi NISHIMOTO, Kenji SHINOZAKI
and Chungyun KANG

Department of Welding and Production Engineering
Faculty of Engineering, Osaka University
Osaka, Japan

Abstract

This paper is concerned with the theoretical consideration on the transient liquid insert metal diffusion bonding(TLIM bonding) process.
Main results obtained in this study are as follows;
(1) The dissolution phenomenon of base metal into liquid insert metal can be explained by Nernst-Brunner theory.
(2) The isothermal solidification process can be explained theoretically on Ni-B and Ni-P binary systems and the complete time of the process can be estimated by eq.(13).
The isothermal solidification process of Mar-M247, MM007, Alloy 713C, Inconel 600 joints with using Ni-15.5Cr-3.7B insert metal can be also interpreted by the same theoretical equation applied to the process of Ni-B and Ni-P binary systems.

Introduction

Ni-base cast superalloys are a class of materials that are generally used hottest parts in aircraft or land-base gas turbines. As fair amount of γ' forming elements such as Al, Ti, Nb, Ta is contained, these alloys are very sensitive to hot cracking during fusion welding[1].

TLIM bonding has been recently developed to join hot cracking susceptible Ni-base cast superalloys and called 'TLP bonding' or 'Activated diffusion bonding'[2]. A liquid film temporarily forms at bonding interlayer during TLIM bonding and solidifies isothermally during interdiffusion of elements for depressing melting point of filler metal. Consequently, the excellent joint quality can be expected without any fusion welding defects. Therefore, TLIM bonding has been applied for bonding of Hastelloy X[2], Inconel X[3], Udimet 500[3], 700[2], Alloy 713C[2], Mar-M247[4], René 80[5], IN-100[6], TD-NiCr[7] and MA 754[8].

The bonding process of TLIM bonding is composed of the following three processes. Namely, the first one is the melting process of base metal adjacent to the liquefied insert metal. The second one is the isothermal solidification process and the last one is the homogenizing process of alloying elements. It is essentially important to make clear these processes theoretically in order to understand the mechanism of TLIM bonding.

In this research, the melting process of base metal and the isothermal solidification process were studied theoretically.

Experimental Materials and Procedures

Experimental Materials

In this study, commercially pure Ni and Ni-base superalloy, MM007, Mar-M247, Alloy 713C, Inconel 600 were used for base metals. Amorphous alloys commercially available and electroless Ni-B plating were adopted as insert metals. MM007 and IM-1 were used to study the melting process of base metal. Commercially pure Ni, MM007, Mar-M247, Alloy713C, Inconel 600 were used as base metals and IM-1, IM-2, Ni-B plating were used as insert metals to clarify the isothermal solidification process. Chemical compositions of these materials are presented in Table I.

Table I Chemical compositions of materials used(wt%)

Materials		Melting range(K)	Thickness(μm)	Chemical composition (wt%)													
				Ni	B	Co	Cr	Al	Ti	Mo	W	Hf	Ta	C	Si	Fe	S	P
Base metal	Ni			Bal.	-	-	-	-	-	-	-	-	-	0.02	0.19	0.04	0.01	-
	MM 007			Bal.	0.016	10.2	7.98	6.01	0.96	6.17	0.05	1.34	4.16	0.09	-	-	-	-
	MarM 247			Bal.	0.015	9.87	8.27	5.53	0.99	0.69	9.92	1.51	3.06	0.15	-	-	-	-
	Inconel 713C			Bal.	0.011	0.4	13.4	5.8	0.9	4.2	-	2.1	2.1	0.11	-	-	-	-
	Inconel 600			Bal.	-	-	15.5	-	-	-	-	-	-	0.03	-	7.7	-	-
Insert metal	IM-1	1293-1338	38	Bal.	3.7	-	15.5	-	-	-	-	-	-	-	-	-	-	-
	IM-2	1153	44	Bal.	-	-	-	-	-	-	-	-	-	-	-	-	-	11.5
	Ni-B	1353	20	Electroless plating														

Experimental Procedures

Bonding Procedures. A bonding equipment is illustrated in Fig.1. Specimens for bonding were held together with the characterized insert metal under slight compressive pressure as shown in Fig.1 and heated to the each bonding temperature in vacuum(13.3mPa) by the high frequency induction heating source and held for predetermined time. Prebond surface finishing of each specimen was achieved by #1500 polishing paper and its prebond cleaning was conducted in acetone by an ultrasonic cleaner. Each specimen size of MM007, Mar-M247, Alloy713C and Inconel 600 is φ10 X 10mm and that of commercially pure Ni is 5 X 10X 12mm.

Figure 2 shows commercially pure Ni joints for examining the isothermal solidification process. Ni stoppers were spot welded in order to keep the width of the bonding interlayer constant. These specimens were also heated to predetermined temperatures in vacuum by a high frequency induction heating source.

<u>Procedure for Measuring Dissolution Width of Base Metal.</u> Prebond alignment of the piece being joined and bonded test piece are schematically illustrated in Fig.3. Dissolution width of base metal, X was calculated by the following equation.

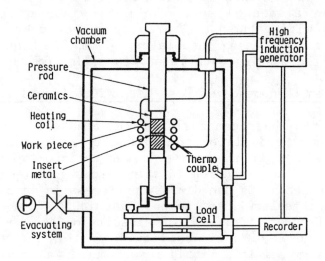

Fig.1 Schematic diagram of bonding apparatus

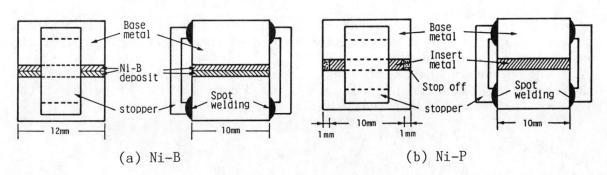

Fig.2 Schematic illustration of test specimen

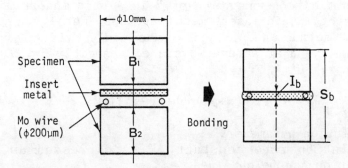

Fig.3 Prebond alignment of the piece being joined and bonded test piece

777

$$X=\{(B_1+B_2)-(S_b-I_b)\}/2 \tag{1}$$

where B_1, B_2 ;length of base metal before bonding, S_b ;length of bonded test piece and I_b ;width of interlayer. Mo wire(ϕ 200 μm) was used to maintain I_b constant and IM-1 was used as an insert metal. B_1, B_2 and S_b were measured by a dial gauge-strain amplifier system(accuracy of the measurement;0.5 μm). I_b was measured by a microscopy.

Procedure for Measuring the Mean Width of Retained Liquid Phase in the Bonding Interlayer. Metallurgical observation(X 300) of the bonding interlayer was performed and the contour line of eutectic was traced at the central region(4.5mm length) of the bonding interlayer on each specimen heat-treated at predetermined conditions. Total area of the eutectic in the central region was examined by measuring area of the outlined eutectic figure using an area-analyser. The mean width of the eutectic was determined by dividing the total area of eutectic by the length of the central region of the bonding interlayer, 4.5mm. The mean width of retained liquid phase in the bonding interlayer was estimated by eq.(14).

Dissolution Phenomenon of Base Metal into Liquid Insert Metal

A kinetic study has been carried out to make clear the dissolution phenomenon of base metal into liquid insert metal. The isothermal phase transfer from solid to liquid can be broken up into two consecutive steps. The first is the surface reaction in which atoms go from the solid into the liquid phase. The second is the diffusion process in which the solute atoms diffuse from the interface into the bulk liquid.

The Nernst-Brunner theory [9] of dissolution has been applied to the dissolution phenomenon of solid to liquid. In this study, the dissolution phenomenon of base metal into liquid insert metal was analysed based on the Nernst-Brunner theory which was formulated as follows.

$$n=n_s[1-\exp\{-(KA/V)t\}] \tag{2}$$

where n_s, n ;solute concentrations at saturated state and at spontaneous time, t respectively, K ;dissolution rate constant, A ;surface area of the solid and V ;volume of the liquid.

When n and n_s are expressed in weight %, C_t and C_s respectively, eq.(2) can be given by eq.(3). C_t and C_s of B can be given by eq.(4) in the case of the joint illustrated in Fig.3.

$$dc/dt=K(A/V)(C_s-C_t) \tag{3}$$

$$C_t=\{ph/(x+ph)\}C_i$$
$$C_s=\{ph/(x_s+ph)\}C_i \tag{4}$$

where C_i ;initial concentration of B in liquid insert metal, p ;(ρ_l/ρ_s), ρ_l, ρ_s ;density of liquid insert metal and base metal, 2h ;initial width of liquid insert metal and x, x_s ;dissolution width of base metal at spontaneous time, t and at saturated state, respectively. Eq.(5) is given by eq.(3) and (4).

$$Kt=h[\ln\{x_s(x+ph)/ph(x_s-x)\}]=P \tag{5}$$

where P;dissolution parameter.

If dissolution rate constant, K is expressed as $K=K_o\exp(-Q/RT)$, following equation can be obtained.

$$\ln K=-Q/RT + \ln K_o \tag{6}$$

where Q ;apparent activation energy of dissolution of base metal into liquid insert metal, T ;temperature ,K_o ; frequency factor and R ; gas constant.

The linear relationship between holding time,t and dissolution parameter,P can be expected during dissolution process of base metal from eq.(5) and Q can be estimated from correlation between ln K and 1/T. Figure 4 indicates the relationship between holding time,t and P. As predicted from eq.(5), there is a good linear correlation between holding time,t and P. Figure 5 also shows a good linear relationship between ln K and 1/T which is predicted from eq.(6). The apparent activation energy of dissolution, Q was estimated 163 kcal/mol from the result in Fig.5. It is concluded that the dissolution phenomenon of base metal into liquid insert metal can be explained by Nernst-Brunner theory.

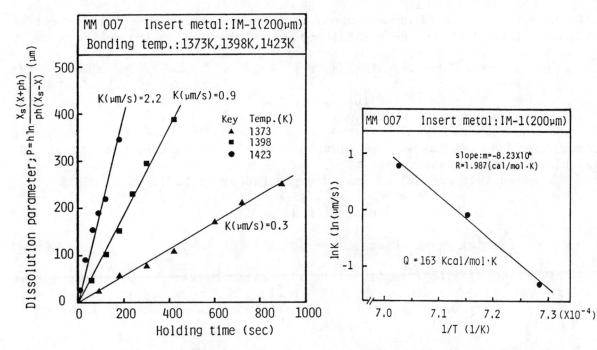

Fig.4 Dissolution parameter,P/ time,t relationships at various temperature

Fig.5 Temperature dependence of dissolution rate constant

Isothermal Solidification Process on Transient Liquid Insert Metal Diffusion Bonding

As shown in a schematical Ni-X binary phase diagram(Fig.6(a)), when the bonding pair is heated to the bonding temperature, T_B, a liquid film is formed at the bonding region. Shaded area of Fig.6(b) indicates such a liquid film and a schematic X content profile at the bonding region is shown in Fig.6(c). At the bonding temperature, T_B, a liquid phase of which X content is C_l is in equilibrium with a solid phase of which X content is C_s. Consequently, at the interface between the liquid phase and the solid phase, both X contents of the liquid phase and the solid phase should be kept constant. If X atoms diffuse into the solid phase from the liquid phase, the solid-liquid interface must move to the minus direction of x axis to make X content in the liquid phase constant, C_l.

At the solid-liquid interface, the concentration gradient of X in the solid phase is assumed to be constant even if the solid-liquid interface travels. And as the liquid film was very thin comparing with the length of base metal, one dimensional diffusion model was adopted. X flux, J, per unit area of the solid-liquid interface and unit time is given by eq.(7).

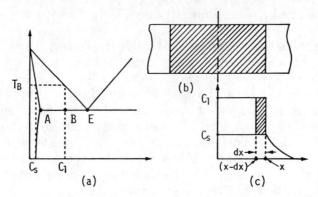

Fig.6 Schematic diagrams of Ni-X binary alloy and concentration profile of X at bonding region

If the solid-liquid interface moves dx for time interval, dt on account of X atoms diffusing into the solid phase, J is given by eq.(8).

$$J = -(D/V_s)(dc/dx) \tag{7}$$

$$J = -(dx/dt)(C_1/V_1 - C_s/V_s) \tag{8}$$

X concentration at arbitrary position x in the solid phase is described by eq.(9) under the condition of the solid phase at the solid-liquid interface being constant, C_s.
From eq.(9), (dc/dx) at x=h is expressed by eq.(10).

$$(C/V_s) = (C/V_s)[1 - \mathrm{erf}\{(x-h)/2\sqrt{Dt}\}] \tag{9}$$

$$(dc/dx)_{x=h} = -C_s/\sqrt{\pi Dt} \tag{10}$$

From eqs. (7),(8) and (10), the thickness of the liquid film, 2x at arbitrary time, t is given by eq.(11). If the gradient of linear relationship expressed by eq.(11) is represented as m, natural logarithm of m is given by eq.(12).

$$2x = 2h - (4C_s/\sqrt{\pi}V_s)(C_1/V_1 - C_s/V_s)^{-1}\sqrt{Dt} \tag{11}$$

$$\ln m = \ln\{(4C_s/\sqrt{\pi}V_s)(C_1/V_1 - C_s/V_s)^{-1}\} + (\ln D_0)/2 - (Q/2RT) \tag{12}$$

From eq.(12), activation energy of X diffusion in the solid phase can be estimated approximately. C_1 is mol fraction of X in the liquid phase at T_B, C_s is mol fraction of X in the solid phase at the solid-liquid interface at T_B, V_1 is molar volume of the liquid phase at T_B, V_s is molar volume of the solid phase at T_B, 2h is initial thickness of the liquid phase at arbitrary time, t, D is diffusion coefficient, R is gas constant T is absolute temperature and D_0 is frequency factor.
The complete time(t_f) of the isothermal process can be estimated by eq.(13).

$$\sqrt{t_f} = (4C_s/\sqrt{\pi}V_s)^{-1}(C_1/V_1 - C_s/V_s)2h/\sqrt{D} \tag{13}$$

As it was a little difficult to determine the solid-liquid interface at room temperature, the thickness of Ni-X eutectic was measured in this experiment and the eutectic thickness, W_e was converted into the thickness of the liquid film, W_1 using eq.(14).

$$W_1 = W_e\{1 + (BE/AB)(\rho_e/\rho_s)\} \tag{14}$$

where ρ_e; density of Ni-P eutectic and ρ_s; density of Ni solution.

As expected from eq.(11), there are good linear correlations between reduced width of eutectic and square root of holding time at each holding temperature, in Ni-B and Ni-P systems, as shown in Fig.7.

Figure 8 shows the relation between ln m and 1/T. As presumed from eq.(12), there are good linear correlations between ln m and inverse of absolute temperature, T in Ni-B and Ni-P systems. Each activation energy of B and P diffusion in Ni, obtained from Fig. 8 is 226 KJ/mol and 284 KJ/mol, respectively. As the term, $(4C_s/\sqrt{\pi}V_s)^{-1}(C_1/V_1-C_s/V_s)$ in eq.(13) is nearly independent of the temperature in the temperature range from 1373K to 1523K on both Ni-B and Ni-P systems, the square root of t_f is directly proportional to $2h/\sqrt{D}$ according to eq.(13).

As indicated in Fig.9, there are good proportional relations between $\sqrt{t_f}$ and $2h/\sqrt{D}$ in both Ni-B and Ni-P systems. The disappearance time of liquid phase at bonding interlayer can be estimated by eq.(13) in both Ni-B and Ni-P systems. The isothermal solidification process on Ni-base superalloy joints were also examined.

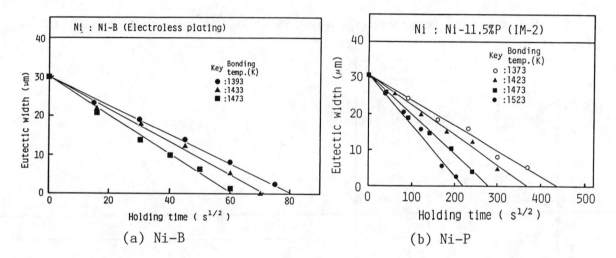

Fig.7 Effect of holding time at each temperature on eutectic width

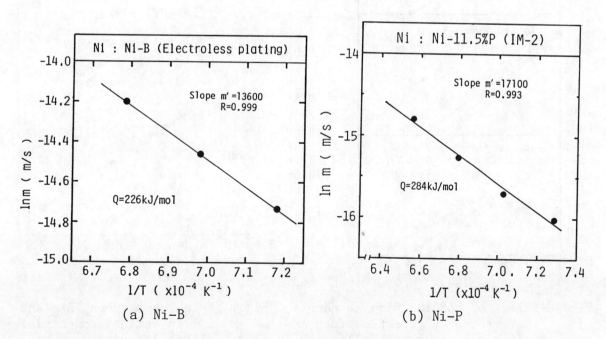

Fig.8 Relation between ln m and 1/T

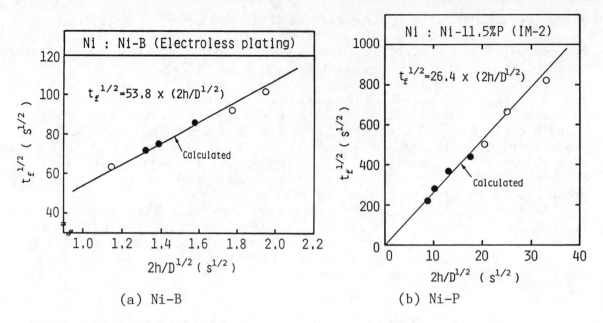

(a) Ni-B

(b) Ni-P

Fig.9 Comparison between calculated value of $\sqrt{t_f}$ and experimental ones

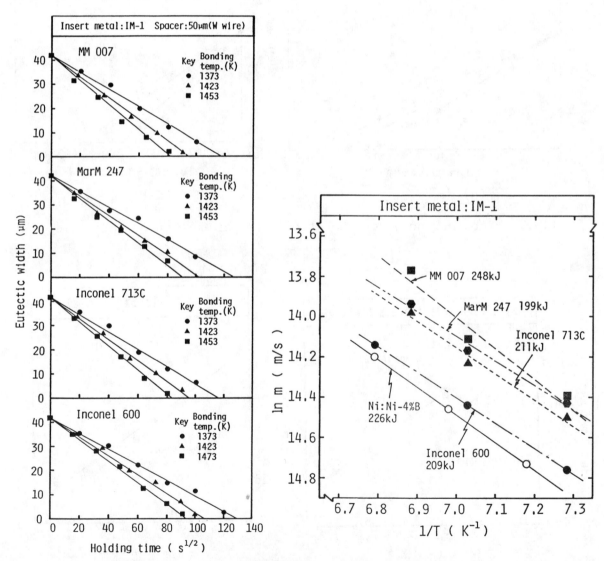

Fig.10 Effect of holding time at each temperature on eutectic width in various commercial alloys

Fig.11 Relation between ln m and 1/T in various commercial alloys

Figure 10 indicates that there are also good linear relationships between reduced width of Ni solid solution- B eutectic in MM007, Mar-M247, Alloy713C and Inconel 600.

In these Ni-base superalloy joints, linear relationships between ln m and 1/T are also obtained as shown in Fig.11.

Based on the experimental results mentioned already, it was made clear that the solidification process of liquid film during TLIM bonding was controlled by the diffusional process of dipressant element in the base metal.

Conclusions

In this study, the bonding process on the transient liquid insert metal diffusion bonding(TLIM bonding) of Ni-base alloys was considered theoretically.

Experimental results obtained in this study are as follows;
(1) The dissolution phenomenon of base metal (MM007) into liquid insert metal(Ni-15.5Cr-3.7B) can be explained by Nernst-Brunner theory.
(2) The isothermal solidification process can be explained theoretically on Ni-B and Ni-P binary systems and the complete time of the process can be estimated by eq.(13).

The isothermal solidification process of Mar-M247, MM007, Alloy713C, Inconel 600 joints with using Ni-15.5Cr-3.7B insert metal can be also interpreted by the same theoretical equation applied to the process of Ni-B and Ni-P binary systems.
(3) It is clarified that the isothermal solidification process of TLIM bonding is controlled by the diffusional process of dipressant element in the base metal.

Acknowledgment

The authors appreciate the assistances of Messrs. Y.Hori, K.Kuriyama and K. Matsuhiro and also wish to express much appreciation to Mitsubishi Metal Co. for their supports to this study.

References

[1] R.Thamburaj, W.Wallace and J.A.Goldak;Inter. Metals Reviews,28-1 (1983),1
[2] D.S.Duvall, W.A.Owczarski and D.F.Paulonis;Weld. J.,53-4(1974),203
[3] H.Tamura et al.;Journal of Japan Welding Society,49-7(1980),462
[4] M.Nakahashi, H.Takeda et al.;TLP Bonding for Heat Resistant Alloy MarM-247,Proceedings of Japan-US Semminor on SUPERALLOYS(1984),239
[5] G.S.Hoppin et al.;Weld. J.,49-11(1970),505s
[6] R.Sellers et al.;New Approaches to Turbine Airfoil Cooling and Manufacturing, AIAA/SAE,13th Propulsion Conference(1977),77-948
[7] A.Kaufman et al.;ASME publication,71-GT-32(1971)
[8] T.Hirane et al.;Journal of Tetsu-to-Hagane,70-13(1984),184
[9] E.A.Moelwyn-Hughes;The Kinetics of Reaction in Solution, Clarendon Press,Oxford(1947),374-377

REPAIR WELDABILITY STUDIES OF

ALLOY 718 USING VERSATILE VARESTRAINT TEST

C. P. Chou and C. H. Chao
Department of Mechanical Engineering
National Chiao Tung University
Hsinchu, Taiwan, R.O.C.

Abstract

The effect of multiple thermal cycles on the weldability of Alloy 718 was investigated using the Versatile Varestraint test. Metallography, SEM fractography, and EDAX analysis were also utilized to study the nature and cause of the fissures occurred during the weldability testing. The results of the Spot-On-Bead (SOB) test, conducted using the Versatile Varestraint test device, showed an adverse effect of multiple thermal cycles on hot cracking susceptibility of Alloy 718. The metallographic and fractographic observations revealed that the Heat Affected Zone (HAZ) cracking mechanisms are associated with the constitutional liquation of (Nb,Ti)C, remelting of Laves phase, and niobium segregation.

Introduction

Alloy 718 is a high strength precipitation-hardened Ni-base alloy suitable for service in the temperature range of -250°C to 705°C (1). This alloy was developed primarily for fabricability and, in particular, for weldability. Alloy 718 possesses excellent corrosion and oxidation resistance as well as good tensile, fatigue, and creep properties at elevated temperatures (2). As a result, Alloy 718 weldments have been widely employed in structural components for heat resistance purposes. Although the welding characteristics of Alloy 718 are very attractive, some welding problems may occur, such as poor penetration, microfissuring in the HAZ, and inferior impact toughness and ductility in the weld metal (3). Large shrinkage contraction that occurs after welding may be a practical problem too (4).

It is well known that several high temperature alloys are easily reheated to temperatures in the vicinity of the solidus by multiple thermal cycling during multipass welding, repair welding after fabrication, and repair welding after service (5). With the increased usage of heavy section weldments and expensive welded structures, much concern has been concentrated on the effect of multiple thermal cycling on the weldability of these high temperature alloys, for example, Alloy 718.

In actual welding, Alloy 718 is susceptible to HAZ cracking (6-13). Therefore, reheat cracking (including liquation cracking and ductility dip cracking) may be a problem when weldments are reheated during multipass welding or repair welding. In order to simulate the actual welding cracks that can occur during multiple thermal cycling, a new hot cracking test method, called Spot-On-Bead (SOB) test, was proposed and employed to evaluate the hot cracking susceptibility of Alloy 718.

Experimental Procedures

The Versatile Varestraint Test

The Versatile Varestraint test device was proposed and employed to evaluate the hot cracking susceptibility of alloys. It is a modification of the Moving Torch Tigamajig Varestraint test adopted by Lundin et al (14). The Versatile Varestraint test device is shown in Figure 1. The main apparatus is the Varestraint Moving Torch Unit which consists of three major components: the torch moving assembly, the air operated strain assembly, and the electrical control unit. The welding power is applied by the Miller Syncrowave 500. The welding arc time control panel, with a precise delay timer, can control the spot welding time for reproducibility and regulate the time at which the strain is operated. This modification allows the test device to have a multiple capability to permit testing to be accomplished in the normal longitudinal varestraint mode, the trans-varestraint mode, or the spot varestraint mode. With this versatile capability, a new test concept called Spot-On-Bead (SOB) test was developed in this study.

Spot-On-Bead (SOB) Test

In order to reach the maximum utility of the Versatile Varestraint test, a new testing concept combining the longitudinal varestraint mode and the spot varestraint mode was conducted. This new methodology can simultaneously evaluate the hot cracking susceptibility of the fusion zone, the weld metal HAZ, and the base metal HAZ in a single test sample. It is also economical, effective, and time-saving for studying the weldability of multiple pass welding and repair welding.

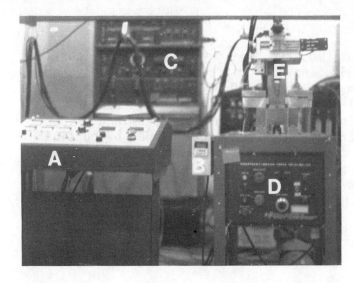

A - Welding Parameter Controller
B - Delay Time Controller
C - Power Supply
D - Varestraint Tester Controller
E - GTA Torch and Test Fixture

Figure 1 - The apparatus of the Versatile Varestraint test device.

The specimen used for the Versatile Varestraint test is 5" x 1" x 0.125". After finishing by power-driven grinding wheel with 320 grit and cleaning by acetone, the specimen is positioned in the test device and welded by a longitudinal autogenous GTA pass. The position of weld bead should be carefully controlled such that one of the fusion lines lies along the center line of the specimen. The width of the weld bead is wide enough to accommodate the subsequent spot weld puddles. After depositing the longitudinal bead, the specimen is then refinished and recleaned.

The actual testing is accomplished by initiating a stationary GTA weld at the center of the specimen. Sufficient arc time is required to ensure that the specimen reached approximately steady-state thermal conditions. The weld puddle produces a base metal HAZ and a HAZ in the previously deposited weld metal with same area. A second or third spot weld can be superimposed directly on the first spot weld puddle after using a stainless steel brush and acetone to clean the previous spot weld puddle.

As the arc current of the last spot weld is interrupted, a predetermined delay time is counted to allow some solidification of the weld pool to occur prior to bending. Then the air cylinder is extended, forcing the sample to conform to the radius of a preselected die block. A schematic drawing and five SOB tested specimens are shown in Figure 2.

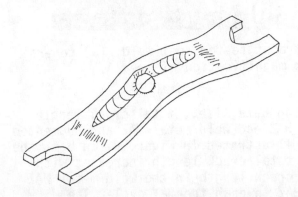
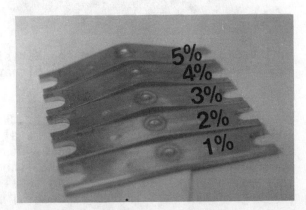

Figure 2 - "SOB" tested specimen used in the Versatile Varestraint test.

Metallographic and SEM Examination

Metallographic samples were cut, mounted, wet ground with a silicon carbide abrasive using 120, 240, 360, and 600 grit paper successively, and polished using alumina powder from 1 μm through 0.05 μm. The etching solution contains 40 ml HCl, 40 ml CH_3OH, and 2 gm $CuCl_2$. The fracture surface of the Versatile Varestraint tested weldments was examined using SEM (Hitachi model 200). The hot cracking surfaces were opened by cutting through the weldment to the crack tips with a diamond wafering saw and then bending the section apart. EDAX analyses were made on special features of the fracture surface and particles of interest in the microstructure of the metallographic samples.

Results

Spot-On-Bead Test

In this study, the Versatile Varestraint test was first introduced to investigate the incidence of hot cracking of alloys. Figure 3(a) shows the general appearance of the SOB tested weld puddles and the induced cracks of Alloy 718. The GTA spot weld was made on the longitudinal weld bead. Therefore, a spot which covers equal parts of weld metal and base metal was produced. Figure 3(b) shows the fusion zone cracking and HAZ cracking in the base metal which occurs in the lower region of the spot puddle. The phenomena of backfilling on the crack tip and liquation on the crack edge are clearly observed. The crack length was measured from the HAZs of the as-welded specimen surface.

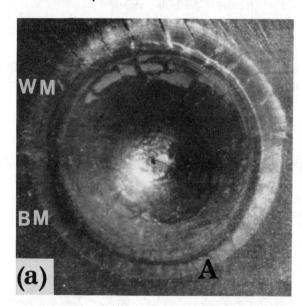
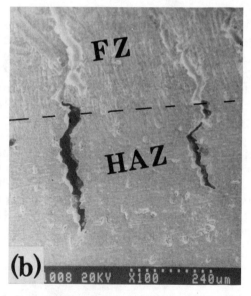

Figure 3 - Hot cracking in the Varestraint tested spot weld: (a) general appearance, and (b) higher magnification of area A.

The relationship between the cracking data, i.e., total crack length and maximum crack length of base metal HAZ and weld metal HAZ and augmented strain, after three thermal cycles, is illustrated in Figure 4. In both the weld metal HAZ and base metal HAZ, the total crack length increases with increasing augmented strain. The total crack length in the weld metal HAZ is larger than that in the base metal HAZ in each thermal cycle. The magnitude of augmented strain has a slight influence on the maximum crack length of both the weld metal HAZ and base metal HAZ. A similar trend can be seen from the results obtained for the "SOB" tested specimens after one and two thermal cycles.

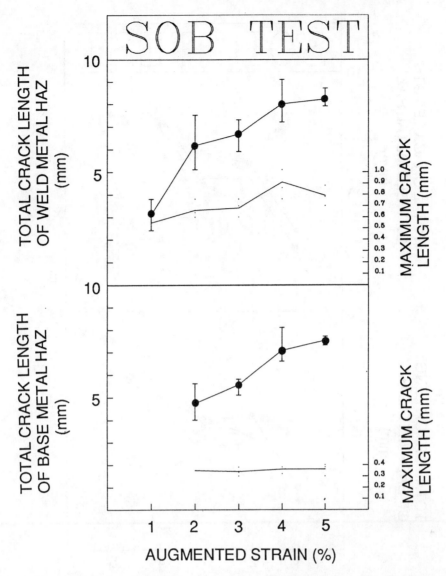

Figure 4 - Cracking lengths versus augmented strain of SOB tested specimens with three thermal cycles.

Figure 5 compares both the total crack length in the weld metal HAZ and base metal HAZ with different numbers of thermal cycles. The total crack length of weld metal HAZ increases with increases in the number of thermal cycles, which indicates that multiple thermal cycling reduces the hot cracking resistance of weld metal HAZ. The increment of total crack length reached 40% when 2% augmented strain was used, and 30% increment with 3% strain as the number of thermal cycles increases from one to three.

Metallographic Observations

After SOB testing, selected sections of tested samples were cut and mounted for metallographic observation. Figure 6(a) illustrates the general appearance of the SOB tested sample, revealing a fusion zone with epitaxial solidification, and both the HAZ region of base metal (BM) and of the weld metal (WM) with different structures. The base metal HAZ cracking is presented in Figure 6(b) at higher magnification. The cracking is intergranular in nature and propagates along the grain boundaries of the base metal HAZ. Some healing reaction may be initiated from the fusion zone (FZ), across the mixed zone (MZ), backfilling the HAZ cracking.

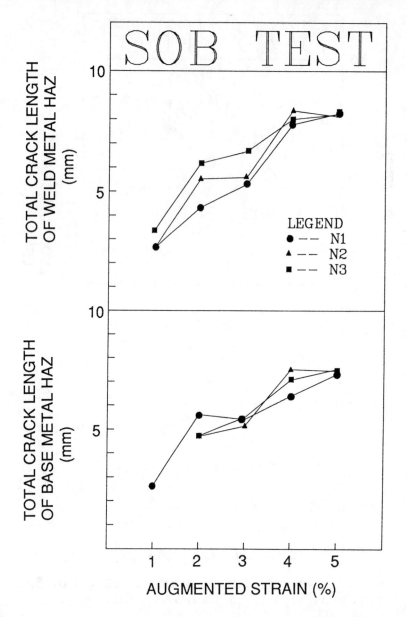

Figure 5 - Comparison of crack lengths of SOB tested specimens with different numbers of thermal cycles.

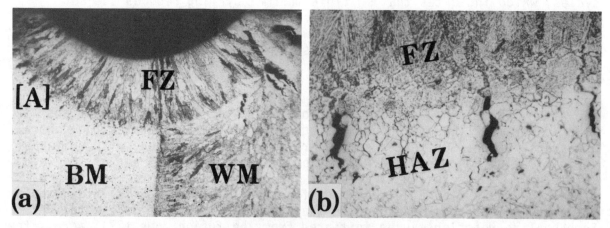

Figure 6 - Microstructures of SOB tested specimen: (a) general appearance, and (b) higher magnification of area A.

In order to clarify the microstructure of HAZs, a polished and etched transverse specimen bent by 3% augmented strain with one thermal cycle was investigated with the scanning electron microscope. Figure 7 shows clustering of niobium carbides in a HAZ crack. The wide crack region reveals that cracking was opened in a semi-liquated condition to accommodate the clustering carbides. In the weld metal HAZ, the predominant constituent is Laves phase, as shown in Figure 7(b). The Laves phase is the terminal solidified constituent of predeposited weld metal, which is remelted by subsequent thermal cycles, and then resolidified as lamellar, eutectic Laves/gamma constituent or blocky Laves phase.

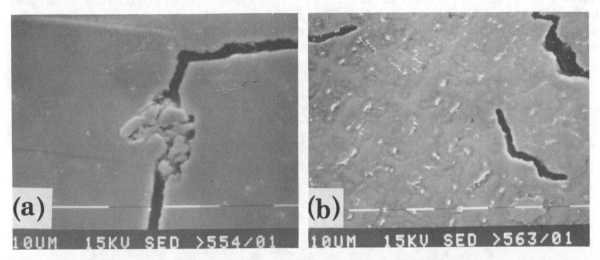

Figure 7 - Microstructures of SOB tested Alloy 718 weld:
(a) base metal HAZ, and (b) weld metal HAZ.

Utilizing the Versatile Varestraint test, the specimen is bent under strain which initiates hot cracks at the solid-liquid interface, which then propagate radially outward, along the grain boundaries of the predeposited weld metal and base metal. When the artificial cracks open, liquid metal from the molten weld pool, in advance of the solid-liquid interface, may be immediately drawn into the cracks by capillary action. This process, which has been termed "backfilling" or "healing", is under investigation using SOB test specimens. Of primary concern is the region near the intersection of the spot weld puddle and the longitudinal weld bead. In this region, backfilling of both HAZ cracks is easier to see, owing to the relative lower stress level. Figure 8(a) shows a typical backfilled crack observed in the base metal HAZ. The constituent in the "backfilled" grain boundaries is lamellar, eutectic-like Laves phase as shown in Figure 8(b).

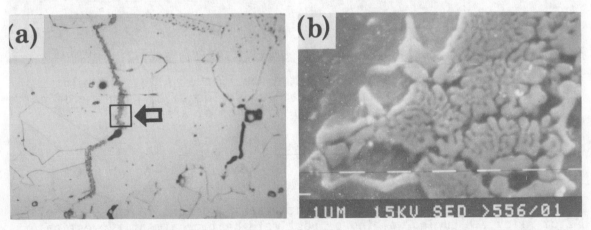

Figure 8 - (a) Backfilled crack in the base metal HAZ of Alloy 718, and
(b) high magnification of constituent in the grain boundary.

Fractographic Examination

In order to reveal the crack morphology, several specimens with different conditions were sectioned and cracks were opened up. Figure 9(a) shows the cracks occurred in both the base metal and fusion zone which experienced one thermal cycle. Three kinds of hot cracking were observed: solidification cracking, liquation cracking, and ductility dip cracking. In the fusion zone the liquated dendritic structure, as shown in area A, was clearly verified to be solidification cracking. In the HAZ region the cracking near the fusion boundary, revealing the liquated intergranular nature of fracture, is liquation cracking. Figure 9(b) illustrates at higher magnification liquation cracking. The fracture surfaces show wavy, rounded shapes coupled with many distinct speckles (15). Figure 9(c), a higher magnification of region C, shows numerous thermal facets which may be slip bands on the fracture surface.

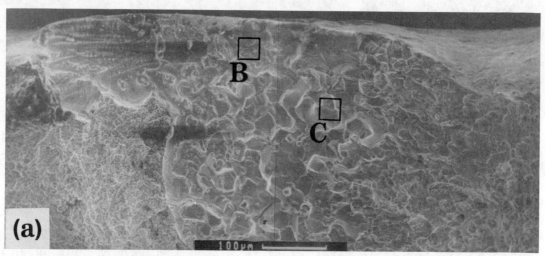

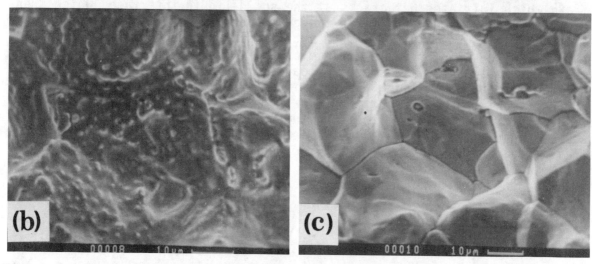

Figure 9 - Fracture surface of SOB tested specimen of Alloy 718 weld: (a) low magnification of cracking surface, (b) high magnification of area B, and (c) high magnification of area C.

Discussion

Nature of the HAZ Cracking

Studies have been reported that hot cracking occurs in the base metal HAZ of Alloy 718 during welding or post weld heat treatment. Investigators have concluded that microfissuring susceptibility is intimately related to the formation of liquid films at grain boundaries during welding. This phenomenon is similar to the liquation cracking proposed for other alloys. But the source of the formation of the liquid films is not clear. Thompson et al (12) found that the formation of intergranular eutectic-type liquid was caused by constitutional liquation of NbC precipitates. Cieslak et al (13) observed that the cause of hot cracking was the presence of eutectic gamma/Laves phase as the low melting interdendritic species.

The present results indicate the preferred cracking site is in the HAZs of the weld metal and base metal. For the case of the base metal HAZ, the primary constituent of the base metal is NbC, and the constituents NbC and Laves phase have been reported to impair the hot cracking susceptibility of Alloy 718. In the weld metal, the terminal solidification constituents are Laves phase and niobium carbides (13). Because the weld metal HAZs are easily reheated to temperatures in the vicinity of the solidus, Laves phase and niobium carbides are no doubt able to remelt and leave NbC back in the matrix or grain boundaries. These partially melted grain boundaries are easily opened into cracks with adequate welding strain.

Effect of Multiple Thermal Cycling

Other investigators (5, 16) have found that multiple thermal cycles imposed on the weld metal HAZ can significantly increase the hot cracking tendency of austenitic stainless steel welds. The degradation of HAZ cracking resistance could be due to decreases in the ferrite level, enhanced segregation, thermally-induced and/or strain-induced precipitation, or accumulation of thermal and restraint strains. The results in Figure 5 indicate a similar tendency, although the alloy involved, Alloy 718, is different than austenitic stainless steel. It was observed that the distribution of Laves phase became loose and small in the weld metal HAZ with thermal cycling (17). The EDAX results showed that the matrix is depleted of niobium after thermal cycling. Therefore, it is proposed that Laves phase is remelted by subsequent thermal cycling, leaving niobium enriched in the matrix or grain boundaries. Grain boundary migration in the weld metal HAZ then increases the degree of segregation by "sweeping up" niobium solute into preferred interdendritic subgrain boundaries as grain coarsening occurs. This sweeping effect plus the constitutional liquation of carbides increase the HAZ cracking tendency of Alloy 718 weld metal.

Conclusions

1. Multiple thermal cycling increases the weld metal HAZ cracking tendency of Alloy 718.

2. Hot cracking occurs primarily along dendritic boundaries in the weld metal HAZ.

3. The constitutional liquation of (Nb,Ti)C, remelting of Laves phase, and niobium segregation are probable causes of hot cracking.

References

1. J.F. Barker, "A Superalloy for Medium Temperatures," Metal Progress, 80(5)(1962), 72-76.

2. R.C. Hall, "The Metallurgy of Alloy 718," Journal of Basic Engineering, 89(9)(1967), 511-516.

3. J. Gordine, "Some Problems in Welding Inconel 718," Welding Journal, 50(11)(1971), 480s-484s.

4. J. Gordine, "Welding of Inconel 718," Welding Journal, 49(11)(1970), 531s-537s.

5. C.D. Lundin, C.P.D. Chou and C.J. Sullivan, "Hot Cracking Resistance of Austenitic Stainless Steel Weld Metals," Welding Journal, 59(8)(1980), 226s-232s.

6. R. Vincent, "Precipitation Around Welds in the Nickel-Base Superalloy, Inconel 718," Acta Metal., 33(7)(1985), 1205-1216.

7. M.J. Lucas and C.E. Jackson, "The Welded Heat-Affected Zone in Nickel Base Alloy 718," Welding Journal, 49(2)(1970), 46s-54s.

8. D.S. Duvall and W.A. Owczarski, "Further Heat-Affected-Zone Studies in Heat-Resistant Nickel Alloys," Welding Journal, 46(9)(1967), 423s-432s.

9. E.G. Thompson, "Hot Cracking Studies of Alloy 718 Weld Heat Affected Zones," Welding Journal, 48(2)(1969), 70s-79s.

10. R.G. Thompson and S. Genculu, "Microstructural Evaluation in the HAZ of Inconel 718 and Correlation with the Hot Ductility Test," Welding Journal, 62(12)(1983), 337s-345s.

11. W.A. Baeslack III and D.E. Nelson, "Morphology of Weld Heat-Affected Zone Liquation in Cast Alloy 718," Metallography, 19(1986), 371-379.

12. R.G. Thompson, J.R. Dobbs and E.D. Mayo, "The Effect of Heat Treatment on Microfissuring in Alloy 718," Welding Journal, 65(11)(1966), 299s-304.

13. M.J. Cieslak et al., "The Use of New PHACOMP in Understanding the Solidification Microstructure of Nickel Base Alloy Weld Metal," Metall. Trans., 17A(12)(1986), 2107-2116.

14. C.D. Lundin et al., "New Concepts in Varestraint Testing for Hot Cracking," Proceeding of the JDC University Research Symposium, 1985 International Welding Congress, 33-42.

15. J.C. Lippold, "An Investigation of Heat-Affected Zone Hot Cracking in Alloy 800," Welding Journal, 62(1)(1983), 1s-11s.

16. C.D. Lundin and C.P.D. Chou, "Hot Cracking Susceptibility of Austenitic Stainless Steel Weld Metals," WRC Bulletin, 289(1983).

17. C.H. Chao, "Effect of Multiple Thermal Cycles on the Weldability of Alloy 718," (M.S. Thesis, National Chiao Tung University, R.O.C., 1987).

INTERGRANULAR SULFUR ATTACK IN NICKEL AND NICKEL-BASE ALLOYS

J.P. Beckman and D.A. Woodford

Department of Materials Engineering
Rensselaer Polytechnic Institute
Troy, New York

Abstract

The phenomenology of intergranular sulfur attack in nickel and nickel-base alloys and the resulting embrittlement on post-exposure tensile testing was investigated. Using Ni 270 as a model material, specimens were exposed at temperatures between 450°C and 800°C to gaseous sulfur environments with sulfur partial pressures ranging from 10^{-8} atm. to 10^{-4} atm. Increasing exposure temperature and increasing sulfur partial pressure increased the severity of the sulfur attack as measured by post-exposure tensile testing in air. Measurements of the maximum depth of intergranular fracture on room temperature tensile specimens were used to calculate the activation energy for the sulfur penetration as 74 KJ/mol. Auger analyses indicate that the level of sulfur on the grain boundaries increases with increasing temperature and increasing sulfur partial pressure. The implications of these data for nickel-base alloys are discussed.

Introduction

Many high temperature alloys will be operating in increasingly corrosive environments in the future. Mechanical property degradation resulting from oxygen attack as well as attack by more aggressive gaseous species such as sulfur and chlorine has become a critical issue. There is now a growing realization that assessment of the full extent of environmental damage to these alloys requires looking beyond the surface scale formation. The diffusion of oxygen from the environment down grain boundaries, resulting in embrittlement and enhanced crack propagation, has been well-documented in nickel and nickel-base alloys (1). Limited preliminary work on the effects of grain boundary penetration of other aggressive environmental gaseous species on nickel and nickel-base superalloys implies that tensile ductility, stress rupture life, and crack propagation resistance are all more seriously impaired by sulfur than by oxygen (2). For example, Figure 1 compares the elongations obtained in Ni 270 after 50 hours exposure at 800°C in air, sulfur, and chlorine atmospheres and readily demonstrates the profound effect of gaseous sulfur on the ductility of nickel. Figure 2 shows that the stress rupture life of IN738, a cast nickel-base alloy, is drastically reduced by two orders of magnitude after exposure in a sulfur environment at 800°C. Clearly then, the intergranular embrittlement of nickel and nickel-base alloys by sulfur vapor warrants serious consideration.

The embrittlement of nickel and nickel-base alloys due to the intergranular segregation of sulfur from the nickel-base matrix is widely reported in the literature (3-7). The presence of sulfur reduces grain boundary cohesion and results in decreased tensile ductility, enhanced crack propagation, and lower impact resistance. The current investigation was undertaken to verify as well as characterize the embrittlement caused by sulfur which is present in the operating environment, diffusing down grain boundaries and causing a degradation of mechanical properties. This study is based on the premise that the intergranular penetration of sulfur from the environment leads to the same reduction in grain boundary cohesion as that caused by intergranular sulfur segregation from the matrix. Therefore, pre-exposures of sufficient time and temperature in sulfur-containing environments may embrittle the grain boundaries

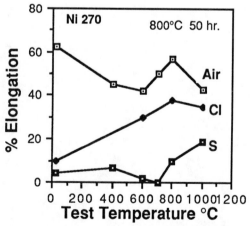

Fig.1 Tensile elongation of Ni270 after 50 hr. exposure in air, Cl, and S at 800°C.(2)

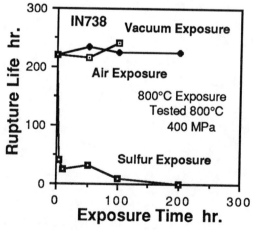

Fig. 2 Effect of prior 800°C exposure in air, vacuum and sulfur on the 800°C stress rupture properties of IN738.(2)

throughout the cross-section and result in a general reduction of ductility on subsequent tensile testing in air. The depth of intergranular mode on the tensile fracture surface could be used to delineate the extent of this embrittlement. On the other hand, fracturing specimens in a sulfur-containing environment would result in sulfur penetration localized at the crack tip. Such localized penetration is likely to be diffusion-controlled and stress-accelerated and may be the mechanism for reported instances of enhanced crack propagation in sulfur-containing environments (8,9).

The objectives of the current investigation are:
1) To confirm that elemental sulfur from a gaseous environment may enter nickel and nickel-base alloys via the grain boundaries and degrade their fracture resistance.
2) To establish the basic parameters of such a sulfur attack, describing the phenomenology of embrittlement, including its dependence on the sulfur activity, the time and temperature of exposure, as well as the alloy composition sensitivity.

Experimental Procedure

Metal/Metal Sulfide Mixtures

The initial experimental approach was similar to that used successfully to study oxygen embrittlement (1), including the use of nickel as a model material. Pre-exposures at elevated temperatures in sulfur-containing environments were followed by post-exposure mechanical testing in air. Powder mixtures of metal/metal sulfides were used to establish thermodynamically predetermined sulfur partial pressures at elevated temperatures. The tensile specimens of 2.5 mm gauge diameter and 9.5 mm gauge length, together with a quartz boat of the metal/metal sulfide mixture, were sealed in evacuated quartz tubes and heated in a furnace for 24 hours at temperatures between 400°C and 800°C. Elevated temperature tensile testing over the range of 250°C to 700°C was done at a crosshead speed of 2.5 mm per minute, particularly looking for an effect on mid-range ductility minima.

Controlled Sulfur Partial Pressure

A limitation of the initial experimental approach was that for a given metal/metal sulfide mixture, the exposure temperature of the specimens could not be varied independently of the established sulfur partial pressure. Also, the sulfides are notoriously non-stoichiometric, making it difficult to define precisely the sulfur partial pressure by thermodynamic calculations. Therefore, a second experimental procedure for exposure was designed. Elemental sulfur, for which there is well-established vapor pressure data (10), was used as the source of the sulfur partial pressure. The sulfur and the specimens were placed in a two-zone furnace which allowed for independent temperature control: a lower temperature for the sulfur powder to establish the desired sulfur partial pressure and a higher temperature for the specimens. The two zones are controlled in such a manner to develop a monotonic thermal gradient to insure that the specimens will see the partial pressure established by the temperature in the locality of the sulfur. A schematic of the actual equipment is shown in Figure 3. The specimens, along with a magnetically-movable quartz container for the sulfur, are sealed under

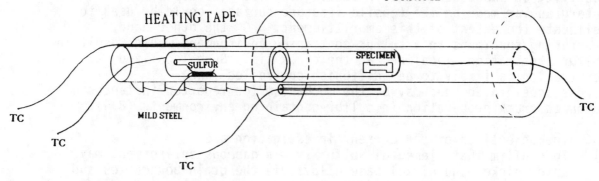

Fig. 3 Schematic of two-zone furnace for controlled sulfur partial pressure using elemental sulfur.

vacuum in a 50 cm long evacuated quartz tube. During evacuation, the vacuum is achieved by a Hg diffusion pump and is routinely better than 10^{-5} torr (10^{-8} atm). Prior to placement in the furnace, the sealed reaction tube is carefully positioned horizontally and the sulfur container is magnetically moved to the opposite end from the specimens. The detail of the schematic shows the unique design of the sulfur container. A small piece of mild steel, encased in quartz and attached to the bottom of the sulfur well, allows the container to be moved magnetically. The narrow neck is lightly stuffed with quartz wool prior to encapsulation to keep the sulfur powder contained during evacuation and handling. This simple gadget has worked well and has eliminated the need for a more elaborate system.

In order to establish the kinetics of the sulfur penetration, it became necessary to use specimens large enough to result in only partially intergranular fracture at higher temperatures and longer exposure times. Larger size specimens with 6.35 mm gauge diameter and 25 mm gauge length were exposed to the gaseous sulfur. Early in the investigation it was found that when sulfur attack occurred to the extent that it could be identified by a decrease in tensile ductility, it occurred with nearly equal severity across the range of test temperatures from 250°C to 700°C. Since the embrittlement could be measured at room temperature just as well as at elevated test temperatures, the subsequent tensile testing was done conveniently at room temperature.

Auger Studies

Pins for Auger analysis were machined to 2.5 mm diameter and were exposed to the various sulfur environments created by either the sulfide mixtures or by elemental sulfur in the two-zone rig, as described earlier. For sulfur analysis of the grain boundaries, the pins were fractured in situ in the high vacuum (10^{-8} torr) of the Auger and immediately analyzed. Additional pins were cut perpendicular to the growth direction of a directionally solidified ingot of Ni 270. These had a bamboo-type grain structure which could be fractured to expose only a single grain boundary in order to eliminate grain to grain differences in S/Ni peak ratios and possibly reveal a surface to center gradient in the amount of sulfur present on the grain boundary. The particular DS ingots we had were extremely gassy and only a few pins could be used. Nevertheless, Auger results were obtained in both polycrystalline and DS pins for one of the exposure conditions.

Results and Discussion

In the experiments involving exposures to the sulfide mixtures, it was found that at a constant temperature, increasing sulfur partial pressures led to decreased tensile ductility in Ni 270 as shown in Figure 4. At 800°C the partial pressure of sulfur in equilibrium with the Ni/NiS mixture is approximately 4×10^{-4} atm. (11), and in equilibrium with the Cr/Cr_2S_3 mixture is approximately 1×10^{-8} atm. (12). Also, as shown in Figure 5, for a given sulfide mixture, increasing exposure temperature, and concomitant increased sulfur partial pressure, results in decreased ductility. The sulfur partial pressures in equilibrium with the Ni/NiS mixture are: 4×10^{-4} atm. at 800°C, 9×10^{-6} atm. at 600°C, and 2×10^{-8} atm. at 400°C (11). Because of the imprecision of final elongation measurements when there is prevalent surface cracking, the nonelastic elongation to maximum load is reported here. As mentioned previously, these figures indicate that when sulfur embrittlement did occur in Ni 270, it appeared to be equally severe over the full test temperature range, revealing no obvious mid-range ductility minimum.

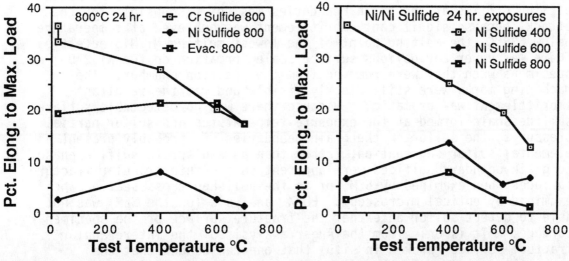

Fig. 4. Effect of various exposures at 800°C on tensile elongation of Ni 270

Fig. 5. Effect of Ni/Ni Sulfide exposures on post-exposure tensile elongation of Ni 270

Room temperature test results from exposures of Ni 270 to sulfur vapor in the two-zone rig are shown in Figures 6 and 7. As can be seen in Figure 6, at a constant sulfur partial pressure of 1.45×10^{-6} atm., equivalent to sulfur at 80 °C (10), increasing specimen temperature results in a systematic decrease in subsequent room temperature tensile ductility. At this sulfur partial pressure, little sulfur attack was noted at 450°C, but at 500°C the dramatic effect of the sulfur became evident, and increased in severity with increasing exposure temperature. Figure 7 shows that at a constant specimen temperature of 600°C, increasing sulfur partial pressure tends to decrease the room temperature tensile ductility. Sulfur partial pressures of 2×10^{-8} atm., equivalent to sulfur at 40°C (10), had no effect on the tensile properties. However at 6×10^{-8} atm., equivalent to sulfur at 50°C (10), there is a dramatic effect.

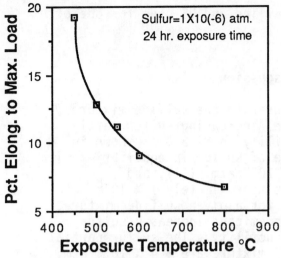

Fig. 6 Effect of exposure temperature at a constant sulfur partial pressure on room temperature tensile ductility of Ni 270

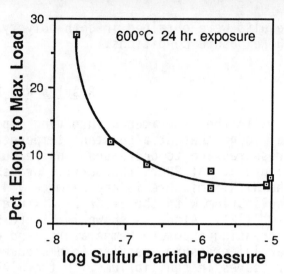

Fig. 7 Effect of sulfur partial pressure at 600°C on room temperature tensile ductility of Ni 270

It is the grain boundary diffusion of sulfur from the environment which causes the intergranular embrittlement. The possible formation of liquid films at the grain boundaries was not responsible for the attack, because significant embrittlement was achieved at temperatures below 635°C, the melting point of the lowest melting Ni/NiS eutectic. There was never any obvious surface scale formation on the Ni 270 specimens when they were removed from the reaction chamber. The machining marks were still clearly visible and yet the resultant embrittlement was dramatic. Because there was no detectable surface sulfide scale formed at the exposure temperatures and sulfur partial pressures, the sulfur on the grain boundaries is probably present as elemental sulfur segregation, rather than as a discrete sulfide phase. No grain boundary sulfides were apparent on the intergranular fracture surfaces when examined by SEM, or in the polished cross section when examined by optical microscopy. EDAX analysis (on the SEM) was not able to detect sulfur on either the fracture surface or the specimen surface. It is only from the Auger analysis of the intergranular fracture surface (broken in situ) that one can deduce that the embrittling species is the sulfur which has penetrated from the exposure environment. Brief ion milling readily depletes the sulfur on the grain boundary surface indicating that it is present in only the first few atom layers.

Some results of AES analyses on pre-exposed Ni 270 pins fractured in situ in the Auger, are shown as S/Ni peak height ratios in Table I. The analyses were made using a 10 KV beam with 2.5 nA beam current.

TABLE I

Auger Results for Various Exposures

Exposure	Sulfur Partial Pressure	Structure	Ave.S/Ni ± 2σ
600°C/Ni/NiS: 24 hr.	9×10^{-6} atm.	Poly.	.512 ± .30
600°C/S at 80°C: 24 hr.	1.45×10^{-6} atm.	Poly.	.440 ± .23
600°C/S at 80°C: 24 hr.	1.45×10^{-6} atm.	D.S.	.566 ± .08
800°C/S at 80°C: 24 hr.	1.45×10^{-6} atm.	Poly.	.624 ± .13

Ten spots were analyzed moving across the diameter on each fracture surface. No obvious sulfur gradient from surface to center was found on either the polycrystalline or DS fracture surfaces. Due to the variability in the measurements from grain to grain, it was assumed that the data were normally distributed and the mean ± 2σ is indicated for each. As can be seen from the data, the average sulfur level detected on the intergranular fracture surface increased with increasing sulfur partial pressure and temperature of exposure. However, the level of sulfur was greater for the DS than for the polycrystalline pin of the same condition, and as was expected, variation among individual measurements was less for the DS pin than for the polycrystalline pin.

Measurements of the maximum radius of intergranular fracture were made on SEM fractographs of the larger size 6.35 mm gauge diameter tensile specimens as well as on those smaller 2.5 mm gauge diameter specimens where only partial intergranular fracture occurred. These measurements relate to the depth of sulfur penetration and were used to calculate an activation energy for the process. The typical diffusion relationship of $x \propto t^{1/2}$ gives a good description of the data, where x is the penetration distance in centimeters and t is the exposure time in seconds; for data at one temperature, x^2/t plots are not ordered with t. Therefore, one can assume an equation of the form:

$$x = K\sqrt{D_{gb}\,t} \text{ where } D_{gb} = D_o \exp(-Q/RT)$$

Although K and D_o cannot be obtained directly, the activation energy, Q, for the process can be determined by plotting $\ln x^2/t$ vs. $1/T$, and the slope of the best fit line through the data will be $-Q/R$. In this case the activation energy for intergranular sulfur penetration in Ni 270 was found to be 74 KJ/mol.

It is interesting to compare the kinetics of this sulfur penetration with those of oxygen penetration as shown in Figure 8. The present data for sulfur in Ni 270 are plotted alongside results for oxygen penetration made by measuring the depth of gas bubble formation (CO_2) in Ni 270 (13). Also shown on the graph are some results for a nickel-base superalloy, Rene 80, where the depth of oxygen penetration was determined by sequentially machining away embrittled material until ductility was restored (14). The kinetics of the sulfur penetration are much faster (the line lies higher on the graph) and the activation energy is much lower (the slope is less steep) at 74 KJ/mol for sulfur than for oxygen in Ni 270 at 274 KJ/mol or oxygen in Rene 80 at 280 KJ/mol.

The activation energy for S^{35} volume diffusion in pure nickel has been reported as 376 KJ/mol in polycrystalline specimens at temperatures above 1000°C (15) and as 218.6 KJ/mol for single crystals at temperatures between 800°C and 1225°C (16). For polycrystalline specimens at temperatures between 800°C and 1000°C, it has been measured as 192 KJ/mol (17). Investigators have noted that sulfur prefers to diffuse along grain boundaries at lower temperatures and that the diffusivity becomes non-linear with inverse temperature (18). For example, data for the diffusion of S^{35} into fine grained TD nickel (18) indicates an activation energy of 159 KJ/mol at

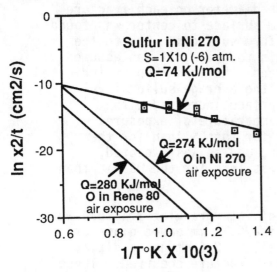 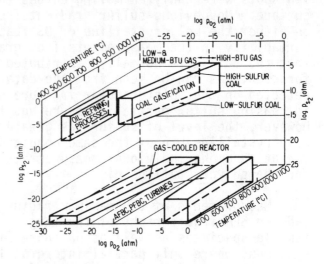

Fig. 8 Plot of ln x^2/t vs. 1 1/T comparing sulfur penetration in Ni 270 to oxygen penetration in Ni 270 (11) and Rene 80 (12).

Fig. 9 Sulfur-Oxygen-Temperature diagram indicating operating conditions in various commercial processes. (17)

temperatures above 900°C. However, data were reported for 850°C and 775°C, although these lower temperature data were purposefully disregarded in the calculation of the activation energy because of the non-linearity with the higher temperature data. If a separate calculation is done for only these lower temperature data, an activation energy of 51 KJ/mol is found. Therefore, our value of 74 KJ/mol for the activation energy of sulfur penetration in Ni 270 measured at temperatures between 450°C and 800°C, although much lower than any found in the literature, seems reasonable in light of the TD nickel data. Typically the activation energies for grain boundary diffusion are significantly lower than for volume diffusion, which corresponds to our findings here.

Commentary

To predict the implications of these data relative to the performance of nickel-base alloys in sulfur-containing environments is not a simple task. Certainly, the kinds of sulfur partial pressures used in this study are typically found in many industrial environments as shown in Figure 9 (19). For instance, in both coal gasification and oil refining processes, the sulfur partial pressures are in the range 10^{-5} atm. to 10^{-10} atm. at temperatures from 500°C to 1000°C. Even in environments where sulfur partial pressures are much lower, localized concentrations are always a possibility. For example, gas turbines often encounter severe sulfidation even though the overall conditions would indicate such reactions should not occur. Although elemental sulfur penetration and embrittlement in Ni 270 has been demonstrated in the absence of visible corrosion products, the effects of external scale formation and internal alloying elements on such sulfur penetration in specific nickel-base alloys must be addressed on an individual basis. For certain alloys such as IN738 the embrittlement by sulfur penetration has been clearly shown (2). In general, nickel-base alloys do show an acceleration of fatigue crack propagation in sulfur environments indicating their general susceptibility to the detrimental effects of sulfur (8,9). Having demonstrated the drastic intergranular embrittlement which can occur in

Ni 270, it is hoped that by identification of such a phenomenon, a broad awareness of the possibility of intergranular sulfur embrittlement in nickel-base alloys will be developed.

Conclusions

1. Sulfur penetrates from the environment down grain boundaries in Ni 270 during pre-exposures at temperatures above 450°C and sulfur partial pressures above 6×10^{-8} atm., resulting in a degradation of tensile ductility on subsequent room temperature tensile testing in air, by inducing intergranular fracture.
2. Increasing exposure temperature in the range 450°C to 800°C while maintaining a constant sulfur partial pressure 1.45×10^{-6} atm., equivalent to the partial pressure of elemental sulfur at 80°C (10), greatly increases the severity of the embrittlement.
3. Increasing sulfur partial pressures above 6×10^{-8} atm. at a constant exposure temperature of 600°C increases the severity of the embrittlement.
4. Measurements of the depth of intergranular fracture indicative of the extent of the sulfur penetration were used to calculate an activation energy of 74 KJ/mol for sulfur penetration in Ni 270 in the range 450°C to 800°C. This is lower than the activation energy typically given for volume diffusion of sulfur in nickel, which is consistent with the fact that the sulfur penetration is a grain boundary phenomenon. Intergranular sulfur penetration is faster than intergranular oxygen penetration in Ni 270.
5. Auger analyses indicate that both increasing sulfur partial pressure and increasing temperature increases the level of sulfur found on the grain boundaries. After the same exposure, the level of sulfur detected on a single grain fracture was greater than on a corresponding polycrystalline fracture surface. No gradient in sulfur level from surface to center was detected by Auger.
6. Results similar to these found for sulfur attack in Ni 270 are indicated for some nickel-base alloys. However, extension of this phenomenon to individual nickel-base superalloys will require more specific experiments.

References

1. D.A. Woodford and R.H. Bricknell,"Environmental Embrittlement of High Temperature Alloys," Treatise on Materials Science and Technology Vol. 25, ed. C.L.Briant and S.K. Banerji (New York, NY: Academic Press, 1983), p. 157-199.

2. D.A. Woodford and R.H. Bricknell, "Penetration and Embrittlement of Grain Boundaries by Sulphur and Chlorine -- Preliminary Observations in Nickel and a Nickel-base Superalloy," Scripta Met., 17 (1983) 1341-1344.

3. R.A. Mulford, "Grain Boundary Segregation in Ni and Binary Ni Alloys Doped with Sulfur," Met. Trans., 14A (1983) 865-870.

4. C. Loier and J. Boos, "Influence of Grain Boundary Sulfur Concentration on the Intergranular Brittleness of Nickel of Different Purities," Met. Trans., 12A (1981) 1223-1233.

5. M.G. Lozinskiy, G.M. Volkogon, and N.Z. Pertsovskiy, "Investigation of the Influence of Zirconium Additions on the Ductility and Deformation Structure of Nickel Over a Wide Temperature Range," Russian Metallurgy, 5 (1967) 65-72.

6. C.L. White, J.H. Schneibel, and R.A. Padgett, "High-Temperature Embrittlement of Ni and Ni-Cr Alloys by Trace Elements," Met. Trans., 14A (1983) 595-610.

7. J.H. Westbrook and S. Floreen, "Kinetics of Sulphur Segregation at Grain Boundaries and the Mechanical Properties of Nickel," Canadian Metallurgical Quarterly, 13 (1974) 181-186.

8. R.H. Kane and S. Floreen, "The Effect of Environment on High Temperature Fatigue Crack Growth," Met. Trans., 10A (1979) 1745-1751.

9. M.Y. Nazmy, "The Effect of Sulfur Containing Environment on the High Temperature Low Cycle Fatigue of A cast Ni-Base Alloy," Scripta Met., 16 (1982) 1329-1332.

10. K.C. Mills, Thermodynamic Data for Inorganic Sulphides, Selenides and Tellurides, (London: The Butterworth Group, 1974) 63-64.

11. R.Y. Lin, D.C. Hu and Y.A. Chang, "Thermodynamics and Phase Relationships of Transition Metal-Sulfur Systems: II. The Nickel-Sulfur System," Met. Trans., 9B (1978) 531-538.

12. W.D. Halstead, "A Review of Saturated Vapor Pressures and Allied Data for the Principal Corrosion Products of Iron, Chromium, Nickel and Cobalt in Flue Gases," Corr. Sci., 15 (1975) 603-625.

13. R.G. Iacocca and D.A. Woodford, "The Kinetics of Intergranular Oxygen Penetration in Nickel and its Relevance to Weldment Cracking," Met. Trans., in press (1988).

14. W.H. Chang, "Tensile Embrittlement of Turbine Blade Alloys After High-Temperature Exposure" (Report R72AEG219, General Electric Company, Cincinnati, Ohio, 1972) 6.

15. I. Pfeiffer, "Uber das Verhalten und die Diffusion von Schwefel in Nickel," Zeitschrift fur Metallkunde, 46 (1955) 516-520.

16. A.B. Vladimirov, et. al., "Diffusion of Sulphur in Nickel," Fiz. Metal. Metalloved, 39 (1975) 319-323.

17. S.J. Wang and H.J. Grabke, "Untersuchungen der Diffusion des Schwefels in Metallen bei der Reaktion in H_2S-H_2-Gemischen," Zeitschrift fur Metallkunde, 61 (1970) 597-603.

18. R.K. Hotzler and L.S. Castlemen, "Diffusion of Sulfur into Thoriated Nickel," Met. Trans., 3 (1972) 2561-2564.

19. K. Natesan, "High-Temperature Corrosion in Coal Gasification Systems," Corrosion, 44 (1985) 646.

THE EFFECT OF SERVICE EXPOSURE ON THE CREEP PROPERTIES OF CAST IN-738LC SUBJECTED TO LOW STRESS HIGH TEMPERATURE CREEP CONDITIONS

R. Castillo,[*] A.K. Koul[**] and J-P.A. Immarigeon[**]

[*]Product Reliability Department
Turbine and Generator Division
Westinghouse Canada Inc.
Hamilton, Ontario, Canada

[**]Structures and Materials Laboratory
National Aeronautical Establishment
National Research Council of Canada
Ottawa, Ontario, Canada

SUMMARY

Constant load (90 MPa) creep properties of specimens machined from new and service exposed IN-738LC turbine blades are reported for testing temperatures in the range of 899 to 996°C. The rupture lives in these tests varied between 250 to 10,000 hours.

There appears to be a transition temperature ($\sim$ 960°C) above and below which intragranular and grain boundary sliding deformation mechanisms predominate in IN-738LC at 90 MPa. Under intragranular deformation conditions, service exposed blades exhibit an increase in $\dot{\epsilon}_m$ and ϵ_r relative to new blades because the coarse γ' precipitates in service exposed blades facilitate flow. Under grain boundary sliding deformation conditions, however, the service exposed blades exhibit a decrease in $\dot{\epsilon}_m$ and ϵ_r relative to new blades because service induced break down of MC carbides produces continuous networks of grain boundary $M_{23}C_6$ carbides which suppress the sliding more effectively during creep testing.

The rupture life in both new and service exposed blades appears to be governed by stress assisted environmental cracking rather than any deformation mechanisms 'per se.' Fracture in both materials occurs through the link-up of environmentally induced surface cracks with the creep induced internal cracks. Final fracture occurs by transgranular shear. The service exposed blades contain slightly lower Cr content in the grain boundary regions because of heavy $M_{23}C_6$ precipitation along the grain boundaries. It is suggested that the reduced oxidation resistance of the grain boundary regions in service exposed blades increases the severity of oxidation and results in marginally lower rupture lives during creep testing.

Introduction

The effects of extended service times upon the mechanical properties of cast nickel base superalloy components in hot sections of gas turbine engines are not well understood. A series of time and stress dependent solid state precipitation reactions occur at service temperatures in these components that can adversely affect their structural integrity.[1]

Accelerated creep testing has generally been employed to demonstrate the effects of service exposure on mechanical properties, usually by conducting tests at stresses and temperatures higher than those experienced during service. The problem with this approach is that the deformation and fracture mechanisms during testing may be quite different from those prevailing under service conditions. In particular, the fracture mode can change from transgranular to intergranular as the deformation temperature is increased.[2] Such transition has also been observed in IN-738LC at constant creep testing temperature with changes in grain boundary microstructure.[3] Therefore, in order to analyze the effects of service-exposure on the mechanical properties of hot parts, it is important to understand the microstructure property relationships during creep testing at near service stresses and temperatures.

This paper reports the results of a study on the relationship between microstructure and creep behaviour of conventionally cast IN 738LC, a nickel base superalloy, tested at a low stress and high temperatures representative of approximate service conditions. The main purpose of the study was to provide information leading to the identification of the failure mechanism, which is the subject of some controversy for alloys of this type.[4] The investigation involved an evaluation of the microstructural changes caused by service and their influence on post-exposure properties when compared to the material in the unexposed condition.

Experimental Procedure

The material selected for the study was obtained as precision cast turbine blades with nominal compositions as shown in Table I. Creep data were generated for one new blade and two used blades which had been in service for 14,000 and 31,000 hours respectively. The blade temperature in mid-airfoil section during service is known to be in the range of 830°C to 845°C.[3,5]

All as-cast blades had been hot isostatically pressed (HIPed) at 1200°C/2 hours, ramp-cooled to 1120°C and then furnace cooled to room temperature. Post-HIP heat treatments included solutioning at 1120°C/2 hours followed by an air cool and aging at 845°C/24 hours followed by air cooling to room temperature. Microstructures of blades were evaluated by optical and scanning electron microscopy. The details of this evaluation are reported elsewhere.[3,5] Creep tests were conducted in air, on

TABLE I

NOMINAL COMPOSITION IN WEIGHT % OF COMMERCIAL CAST IN-738 LC ALLOY

Material Condition	C	Mo	Cr	Ni	W	Fe	Co	Al	Ti	Nb	Ta	Zr	B	Mn	Si	P	Cu
Unexposed	.10	1.74	16.01	Bal.	2.66	.12	8.35	3.44	3.40	.81	1.66	.03	.01	<.10	<.10	<.10	<.10
14,000 hr Service	.09	1.77	16.03	Bal.	2.51	.12	8.36	3.44	3.49	.83	1.66	.03	.01	<.10	.03	.005	<.05
31,000 hr Service	.09	1.70	15.9	Bal.	2.59	.10	8.29	3.37	3.36	.70	1.64	.04	.01	<.10	<.10	.004	<.10

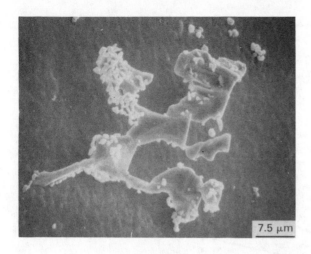

Fig. 1. Intragranular MC carbide and $M_{23}C_6$ carbides attached to its surface as a result of its decomposition.

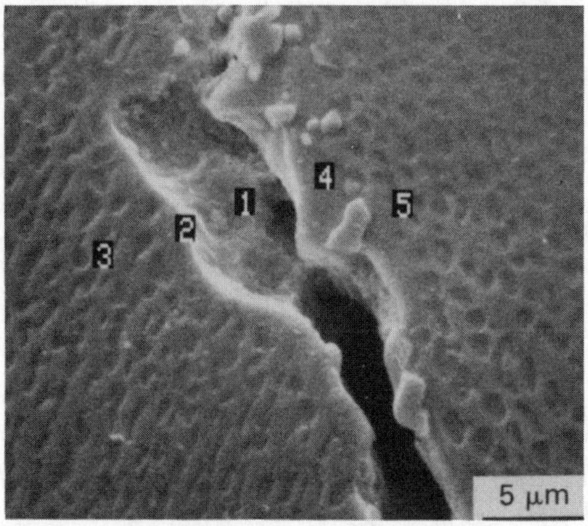

Fig. 2. Fractured grain boundary area of a crept specimen. The Cr Concentration in spot nos. 1 = 9.63%, 2 = 14.86%, 3 = 15.12%, 4 = 7.83% and 5 = 15.96% in wt. %.

specimens machined from mid-airfoil sections of blades, in constant load creep machines at 90 MPa and at temperatures between 899 and 996°C. The creep specimens had a gage diameter of 4.0 mm and a gage length of 19.0 mm. Creep strain was measured with an accuracy of greater than ± 0.01 pct by means of a linear variable displacement transducer (LVDT).

Results

Previous studies have indicated that γ' precipitates coarsen and MC carbides degenerate in these blades during service.(3,5) Primary MC carbides break down into $M_{23}C_6$ carbides which precipitate preferentially along grain boundaries, thus modifying the original grain boundary microstructure. Evidence of MC carbide degeneration is shown in Fig. 1. Free carbon released by MC carbide decomposition diffuses rapidly to the grain boundaries, reacting with surrounding Cr to form $M_{23}C_6$ carbides, with the result that the grain boundary regions are depleted in Cr, Fig. 2. Since Cr is the key element responsible for the corrosion resistance of the alloy, the reduction in Cr content close to the grain boundaries increases the susceptibility of the material to intergranular oxidation.

In both new and service exposed materials, several types of MC carbides were detected that were rich in the usual carbide forming elements, namely, Ti, Nb and Ta. Approximate compositions of the four types of MC carbides that were identified are given in Table II.

TABLE II

METALLIC ELEMENT COMPOSITION (in wt. pct.) OF INTRAGRANULAR MC CARBIDES

	Ta	Nb	Ti	Cr	Co	Ni
MC_I	47.0	19.0	30.0	1.4	1.6	3.0
MC_{II}	28.0	31.0	30.0	2.0	1.3	7.9
MC_{III}	15.0	6.5	75.0	1.2		2.0
MC_{IV}	26.0	11.0	58.0	1.6		3.0

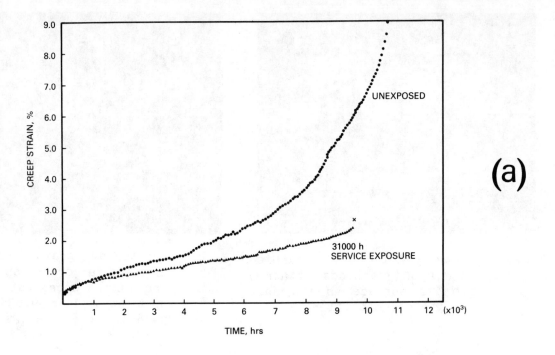

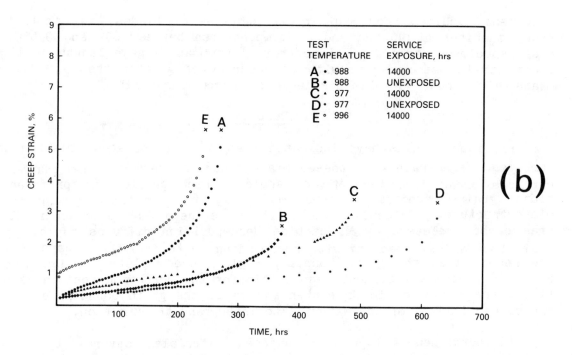

Fig. 3   Effect of service exposure on the creep behaviour of new and service exposed IN-738LC blades tested at 90 MPa and (a) 899°C and (b) 977, 988 and 996°C.

Creep testing Results

Creep curves for long term (low temperature) and short term (high temperature) testing conditions are shown in Figs. 3a and 3b respectively. The data indicate that compared to new blades higher creep rates prevail during short term tests in service-exposed material whereas the reverse is observed during long term tests of the order of 2,000 to 10,000 hours.

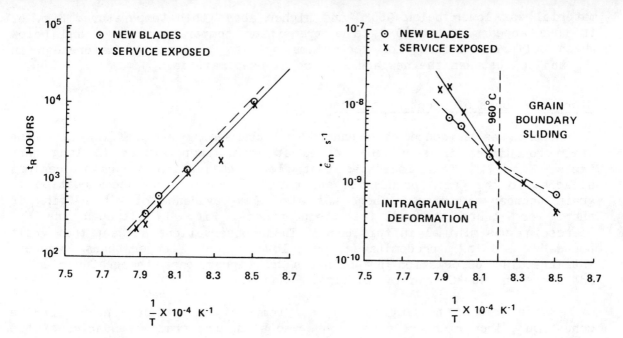

Fig. 4 Effect of creep testing temperature on t_r of new and service exposed blades at 90 MPa.

Fig. 5 Effect of creep testing temperature on $\dot{\epsilon}_m$ of new and service exposed blades at 90 MPa.

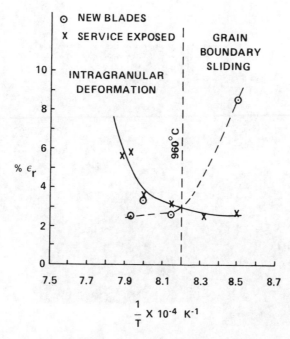

Fig. 6 Effect of creep testing temperature on ϵ_r of new and service exposed blades at 90 MPa.

The effects of service-exposure on rupture life, t_r, minimum creep rate, $\dot{\epsilon}_m$, and creep ductility, ϵ_r, for all test conditions examined can be seen in Figures 4, 5 and 6, respectively. Under comparable testing conditions, the service-exposed material exhibited rupture lives that were marginally lower than those for the new material, Figure 4. However, there were noticeable differences in $\dot{\epsilon}_m$ and ϵ_r that varied in magnitude with test temperature. Below approximately 960°C, the service-exposed material exhibited lower $\dot{\epsilon}_m$ values whereas above this temperature, its $\dot{\epsilon}_m$ was higher, Fig. 5. Similarly, compared to new blades ϵ_r for service exposed material was lower below 960°C and higher above this temperature, Fig. 6. It thus appears that 960°C is a transition temperature above and below which different deformation mechanisms operate leading to differences in $\dot{\epsilon}_m$ and ϵ_r between the new and service-exposed material.

Microstructures of Crept Specimens

There was evidence of enhanced γ' coarsening and rafting near the grain boundaries in specimens tested at lower temperatures in long term tests, Fig. 7. This indicates that flow localization occurs in regions adjacent to the grain boundaries and may be viewed as an accommodation to grain boundary sliding. There was also some evidence of γ' rafting in short term tests at high test temperatures, Fig. 8, although the γ' coarsening was minimal in this case. These observations suggest that grain boundary sliding predominates at lower test temperatures whereas intragranular deformation predominates at higher test temperatures below and above the 960°C transition, Figs. 5 and 6.

After creep testing at lower temperatures, under near service conditions, iron rich areas were observed along the grain boundaries of the new material Fig. 9. It is not clear what role these Fe-rich areas play in the creep behaviour of Ni-base superalloys.

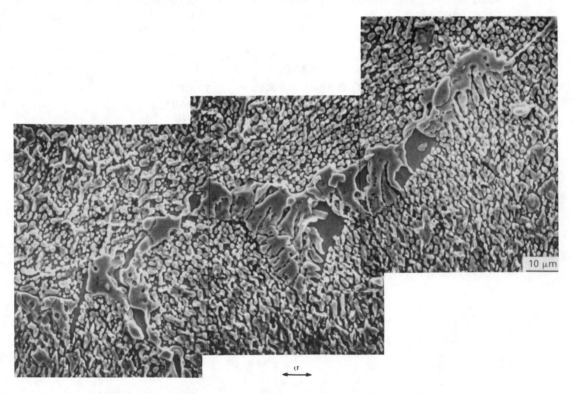

Fig. 7   Typical grain boundary structure of the new blade specimen creep tested at a lower temperature (899°C) at 90 MPa showing grain boundary γ' coarsening and rafting.

Fracture Behaviour of Crept Specimens

All specimens tested to rupture contained cavities and cracks at grain boundaries and oxidized surface cracks at intergranular sites, Fig. 10. In all cases, surface grain boundary cracks were larger than internal cracks and were mostly normal to the tensile stress axis. Intragranular /interdendritic fracture features were observed in both new and service exposed materials up to approximately one-half of the specimen gage section, Fig. 10. The fast fracture areas showed the appearance of typical transgranular shear inclined at about 45° to the specimen axis, Fig. 10.

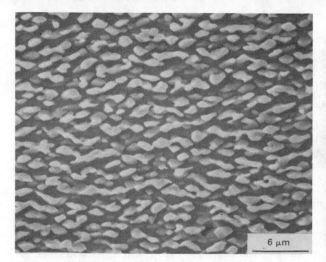

Fig. 8 Some γ' rafting in new blade specimen creep tested at 977°C and 90 MPa.

Fig. 9 Example of an Fe rich region in a new blade sample creep tested at 899°C and 90 MPa. In spot No. 1 Fe = 4.18 wt.% and in the matrix Fe = 0.12 wt.%.

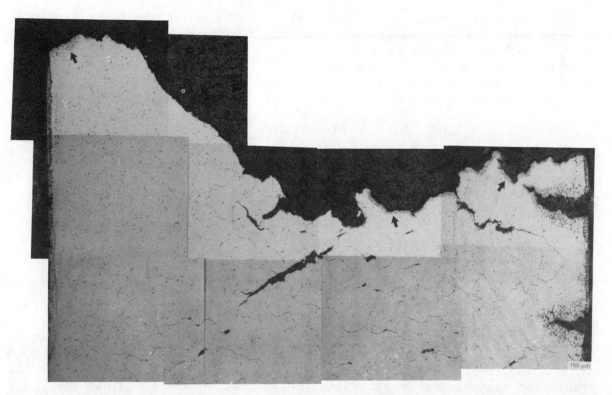

Fig. 10 Optical micrograph showing evidence of intergranular oxide penetration (arrows), micro-cracking and transgranular shear.

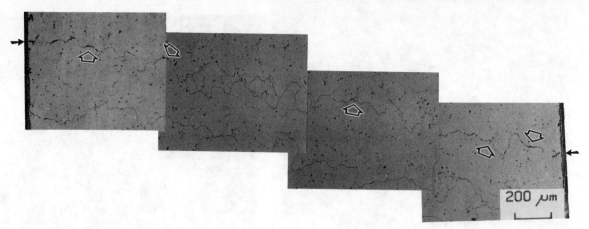

Fig. 11 Grain boundary surface cracks (arrows) and early stages of internal crack link-up (open arrows) in a service exposed specimen creep tested at 927°C and interrupted during early stages of secondary creep.

Damage in the gage length of the crept specimens showed important differences between the new and service exposed specimens that were creep tested at temperatures below 960°C. The 899°C creep tests indicated that the new material contained relatively more creep cavities than the service exposed material under identical creep loading conditions. Internal cracking was also more extensive in the new material than the service exposed material. These observations suggest that below 960°C the continuous networks of grain boundary $M_{23}C_6$ carbides in the service exposed material suppress deformation in the grain boundary regions. At creep testing temperatures above 960°C, however, the differences in the fracture behaviour of the new and service exposed specimens were less obvious.

Interrupted creep tests indicated that environmentally assisted intergranular cracking occurs early during secondary creep in both new and service exposed materials, Fig. 11, and concurrently the internal grain boundary cracking also occurs within the bulk of the material. Final fracture appears to take place through the link up of the surface crack with the internal cracks in all cases, Figs. 10 and 11.

Discussion

The results indicate that there is a transition in the predominant deformation mechanism as the creep testing temperature decreases. At 90 MPa, the transition occurs around 960°C in the alloy investigated, Figs. 5 and 6. The activation energies (Q) for the deformation processes, below and above the transition temperature, were calculated based on power law fits through the data of the form:

$$\dot{\epsilon}_m = A\ e^{-Q/RT} \sigma^n \qquad \qquad \qquad \ldots \ldots \ldots \ldots \ldots \ldots (1)$$

where σ is the applied stress, Q is the activation energy, T is the test temperature, R is the gas constant and A and n are material constants. Below 960°C the Q values were of the order of 16-22 KCal/mole whereas above the transition temperature they were approximately 30-40 KCal/mole for both new and service exposed materials. While the Q values below and above the transition temperature do not match the activation energy values for grain boundary or volume diffusion of various alloying additions in Ni, it is significant that below 960°C the Q values are roughly one half the value observed above 960°C. The deviation in the actual magnitude of Q from that reported for intragranular and grain boundary sliding deformation mechanisms is not totally unexpected. The true Q values for a given deformation mechanism can only be obtained providing the structure is kept constant, a condition that cannot be satisfied in metastable Ni-base alloy systems. However, in view of the evidence for flow localization adjacent to the grain boundaries, (Fig. 7), the crossover in $\dot{\epsilon}_m$, (Fig. 5) and the smaller Q values below the transition temperature it can be argued that grain boundary sliding predominates below 960°C whereas intragranular flow predominates above 960°C.

The differences in the overall creep behaviour of new and service exposed blades can be rationalized in terms of the differences in the intragranular and grain boundary microstructural features of the two materials. Above 960°C, the intragranular deformation mechanism predominates and the service exposed blades are expected to show an increase in $\dot{\epsilon}_m$, and ϵ_r, Figs. 5 and 6, because they contain coarse γ' precipitates which lower the grain strength. In contrast, below 960°C, the grain boundary sliding mechanism predominates and the service exposed blades would be expected to show lower $\dot{\epsilon}_m$ and ϵ_r values, Figs. 5 and 6, because continuous carbide networks in service exposed blades will suppress the grain boundary sliding more effectively.

The similar t_r values of the new and service exposed material at comparable testing temperatures can be attributed to the fact that stress and environment assisted crack nucleation and propagation controls t_r rather than a specific deformation mechanism 'per se'. For both materials final fracture occurs in air as a consequence of the propagation of one or several of the surface initiated cracks nucleated early during the secondary stage of creep. Fig. 10 shows the profile of a crack, which was able to follow the transverse grain boundary to a greater depth causing final fracture. Oxides and nitrides were observed around the crack surface. Similar surface nucleated cracks, located in the specimen shoulder section, showed minimal grain boundary oxidation. Clearly, the rate of oxygen penetration along the grain boundaries is greatly enhanced by applied stress. The manner in which oxygen prompts grain boundary embrittlement leading to intergranular failures could be explained by a mechanism originally proposed by Briknell and Woodford.(6) Initially, the boundary is embrittled by oxygen penetration in the near surface region. This embrittled boundary fails afterwards in tension and the free surfaces thus produced are oxidized while oxygen diffuses down the boundary ahead of the crack. It appears that oxygen penetration along grain boundaries indeed precedes actual crack formation. The metallographic evidence suggests that the extremely low Cr content in the boundary regions plays some role during fracture because of the reduced oxidation resistance of the grain boundaries. The marginally lower t_r values of the service exposed blades are perhaps related to this effect, Fig. 4.

Conclusions

The results demonstrate that short term creep testing does not reveal loss of creep ductility due to grain boundary embrittlement in service exposed blades. This is because intragranular deformation mechanisms are dominant during short term testing whereas grain boundary sliding predominates under service conditions.

References

1. J.W. Martin and R.D. Doherty, "Stability of Microstructure in Metallic Systems," (University Press: Cambridge, 1976).

2. H.R. Tipler et. al, Proc. Conf. "High Temperature Alloys for Gas Turbines," Liege, Belgium, September 25-27, 1978, Applied Science Pub., 359-407
3. R. Castillo and A.K. Koul, Proc. Conf. "High Temperature Alloys for Gas Turbines and Other Applications," Liege, Belgium, October 6-9, 1986, p. 1395-1410.
4. S. Ogersby and T.B. Gibbons, "Creep Cavitation in Cast Ni-Cr Base Alloy," J. Mat. Sci and Eng., 59 (1983), L11-L14
5. A.K. Koul and R. Castillo, "Assessment of Microstructural Damage and Its Rejuvenation in Turbine Blades," Met. Trans. A, In Press, 1988.
6. R.H. Briknell and D.A. Woodford, "Grain Boundary Embrittlement of the Fe-Base Superalloy IN-903," Met. Trans A., 12 (1981) 1673-1680.

DEGRADATION OF ALUMINIDE COATED DIRECTIONALLY SOLIDIFIED SUPERALLOY TURBINE BLADES IN AN AERO GAS TURBINE ENGINE

P.C. Patnaik, J.E. Elder and R. Thamburaj

Hawker Siddeley Canada Inc.
Orenda Division
Box 6001
Toronto AMF, Ontario, Canada
L5P 1B3

Abstract

The conventional polycrystalline nickel based superalloys for gas turbine blade application are gradually replaced by the newer and more advanced directionally solidified (DS) superalloys due to their improved mechanical properties. In spite of this development, cheaper aluminide coatings are still widely used on the DS superalloys to protect them from high temperature oxidation and corrosion attack. This paper discusses the modes of service induced degradation such as oxidation, thermal fatigue (TF) and interdiffusion in a DS nickel based superalloy with a diffusion aluminide coating which is currently being used in the high pressure turbine (HPT) of an advanced aero engine. The presence of complex internal cooling passages in the blade along with poor TF properties of the aluminide coating give rise to the initiation of a large number of cracks in the coating. Ingress of oxidants through these cracks leads to the failure of the component.

Introduction

Directionally solidified (DS) superalloys are generally superior to their polycrystalline counterparts because of their improved high temperature mechanical properties. This has been achieved as a result of better control on grain morphology and crystallographic texture, lower levels of porosity and a significant difference in the nature and distribution of precipitating phases during directional solidification (1). As a means of protecting these advanced superalloys from the aggressive turbine environment, coatings are applied by a variety of processes (2). In advanced aero engines, NiCrAlY or NiCoCrAlY type overlay coatings are most commonly used although ceramic type thermal barrier coatings are finding increasing application in the turbine hot section particularly in stationary components. However, aluminide coatings obtained by pack cementation process, at a much lower cost are still widely used by aero engine manufacturers.

This paper describes the modes of service induced degradation in several prematurely retired directionally solidified high pressure turbine (HPT) blades. The blades were air cooled DS René 80H with a diffused aluminide coating on the airfoil and platform surfaces applied by a pack cementation process. The modes of coating degradation in these blades were examined and the influence of coating degradation on the substrate alloy was investigated.

Experimental

Several prematurely retired DSRené 80H HPT blades of chemical composition given in Table 1 were retrieved from the service engines. The blades were electroless nickel plated to preserve the corrosion products on the surface. Blade sections were cut and metallographic examination was carried out using a Novascan 30 scanning electron microscope. Energy dispersive X-ray analyses were done using a Tracor Northern energy dispersive X-ray analyzer attached to a scanning electron microscope.

Table 1

Approximate Composition in Weight % of Alloy René 80H* (3)

Ni	Cr	Co	Mo	W	Ti	Al	B	Zr	Hf
Base	13	9.6	4.0	4.9	4.5	3.0	0.015	0.01	0.74

* Trade Mark of General Electric Company

Results and Discussion

Figure 1(a) and (b) respectively show the front and top view of a service exposed HPT blade illustrating A) tip oxidation, B) tip cracking as well as C) thermal fatigue cracking in the leading edge area. The aluminide coating showed signs of oxidation, thermal fatigue and interdiffusion between the coating and substrate alloy. These degradation modes are discussed individually in the following.

Oxidation and Plugging of Tip Cooling Holes

The DSRené 80H blade has several cooling holes in the blade tip region as shown in Figure 1(b). These cooling holes were found in the partially or fully plugged state after service exposure (Figure 2). As high as 6 to 9 tip cooling holes were plugged in a single blade. The debris was identified as nickel and aluminum rich oxides. Minor quantities of other elements such as titanium, chromium and cobalt were also found in these oxides.

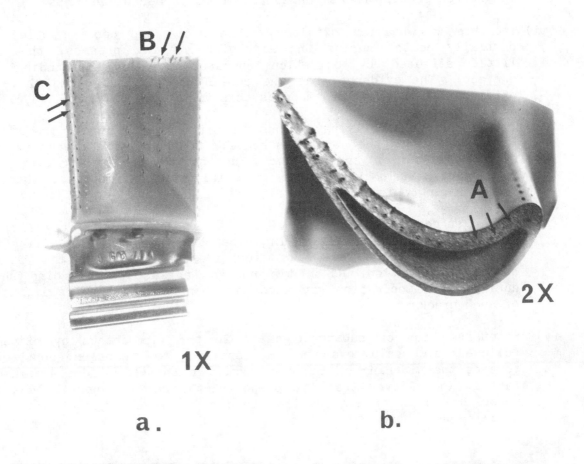

Figure 1. a) Front and (b) top view of a service exposed HPT blade showing regions of: A. Tip oxidation. B. Tip cracking. C. TF cracking in the leading edge.

The implications of partially blocked cooling holes are that these oxides have a lower thermal conductivity than the blade material and act as thermal insulators between the internal cooling air and the hot internal blade surface, while fully blocked holes limit the flow of cooling air into the blade. Consequently, both conditions would act to increase the blade temperature which in turn can lead to a higher oxidation rate of the substrate alloy if exposed directly to the gas turbine atmosphere. This condition was actually observed in the service exposed turbine blades. Since the new blades were "tip ground" prior to engine installation and/or were allowed to cut through the high pressure turbine shroud during testing, the aluminide coating in both cases was removed thus exposing the bare tip surface. As a result, severe oxidation of the DSRené 80H substrate alloy was observed at the tip in the vicinity of cooling holes as depicted in Figure 2.

Tip Oxidation and Cracking

The high pressure turbine blade had undergone catastrophic oxidation and radial cracking at the blade tip, specifically between the mid-chord and trailing edge regions. The damage to the blade trailing edge was found to be much more severe than in other regions. This damage is illustrated in Figure 3. The tip degradation found by examination of a number of service exposed blades is summarized as follows:

(i) The number of tip cracks ranged from 6 to 12 per blade.

(ii) The tip was heavily oxidized. Since the tip cooling holes were partially coated during the aluminide coating process, they were little affected by oxidation compared to the surrounding base surface. The evidence that the cooling hole walls stand above the oxidized substrate alloy supports this view point, (Figure 1(a)).

(iii) The tip cracks usually have a broad mouth (1-3 mm) towards the trailing edge of the concave surface. Some of these cracks extended up to the convex surface i.e., across the entire trailing edge thickness. Such cracks were in fairly advanced stages, therefore their origin could not be established.

(iv) It is suggested that these cracks were of thermal fatigue nature resulting from the variation in temperature across the trailing edge region of the blade. Loss of aluminide coating due to tip grinding before installation and/or tip rubbing against the turbine shroud during engine operation may also have contributed towards initiation of such cracks.

(v) Extensive loss of substrate alloy in the tip cracks by oxidation followed by scale evaporation, spalling or erosion contributed towards the widening of these cracks. In the HPT blades examined so far, it is believed that the predominent mode of material loss from these cracks is by evaporation since no visible signs of erosion or spalling were observed.

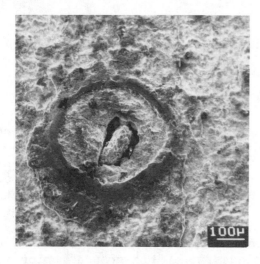

Figure 2. SEM micrograph showing tip oxidation and plugged cooling hole.

Figure 3. SEM micrograph showing the oxide morphology on the blade tip surface.

(vi) The tip cracks ranged from 0.1 to 6 mm in length and propagated into the substrate alloy downwards in the radial direction. Some of these radial cracks have been observed to turn to become axial. This could have a serious consequence such as blade tip separation.

(vii) The radial cracks proceeding downwards gave rise to a number of fine axial cracks.

The oxidation of substrate alloy DSRené 80H was not confined to the tip regions, but also extended into the interior of the blade through the tip cracks. Figures 4 (a) and (b) show longitudinal cross-sections of a blade in which cracks extended from the tip and propagated in the radial direction. Some of these cracks originated at the grain boundaries near the blade tip and some far away from it, suggesting that the grain boundary channels of the DS alloy did not have any major impact on the material degradation process.

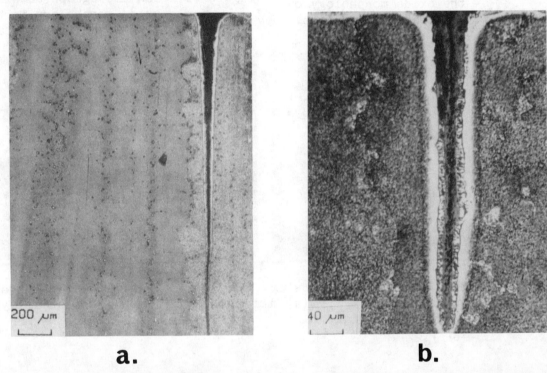

Figure 4. Scanning electron micrographs showing a crack originating from the blade tip (a) near a grain boundary, b) in the bulk grain.

Figure 5 shows a general morphology of the oxidized layer at the blade tip in a longitudinal cross section. No trace of the aluminide coating was found near the tip. The outer scale layer contained Ti and Cr rich oxides and the inner layer contained small elongated internal oxide particles rich in Al and Ti. The morphology of oxidation products in a growing crack was quite similar to the general tip oxidation morphology except in certain regions where the internal oxides had formed a complete layer (Figure 6).

Elemental X-ray analyses near the internal oxidation zone revealed a depletion of Al, Ti and Cr in the alloy matrix. Absence of gamma prime (γ') was noticed in this depleted zone. The severity of the depletion was best judged by comparing the morphology, size and volume fraction of the γ' phase in the depleted zone and the alloy underneath. Figure 7 shows the size and distribution of γ' precipitates in the substrate alloy just below the depleted zone. Loss of Al and Ti in this region resulted in a loss of gamma prime phase in the alloy.

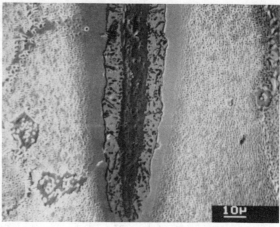

Figure 5. SEM micrograph showing the oxide morphology on the blade tip surface.

Figure 6. SEM micrograph showing oxide morphology in the vicinity of a tip crack.

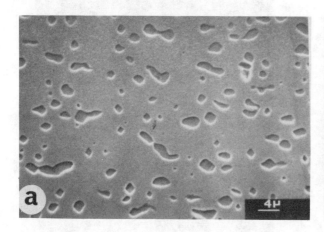
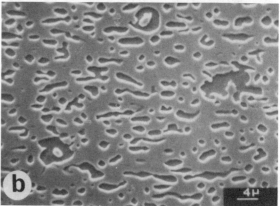

Figure 7. SEM micrograph showing variation in morphology and distribution of γ' (a) immediately below the depleted layer as shown in Fig. 5. (b) well below the region shown in Fig. 7(a).

Degradation of Aluminide Coating

The aluminide coating on the HPT blade consisted of a coarse grain β-NiAl structure in the outer layer and a diffusion zone containing TiC, $M_{23}C_6$ carbides and sigma phases. Three modes of coating degradation have been observed in the service exposed turbine blades. They are respectively oxidation, thermal fatigue and interdiffusion.

Oxidation

Oxidation of the aluminide coating leads to the formation of Al_2O_3 which if mechanically stable can provide very good oxidation resistance. In actual service operation, cyclic variation in turbine blade temperature results in spalling of the Al_2O_3 from the coating. In the HPT blades investigated, the aluminide coating provided moderate oxidation resistance in the regions where the coating was undamaged. Localized oxidation attack of the coating occurred near the diffusion zone which resulted in coating delamination (Figure 8). In some instances, the coating had spalled from the leading edge region of the blade due to oxidation attack at its diffusion zone (Figure 9).

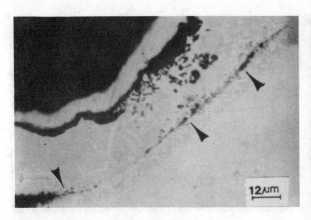
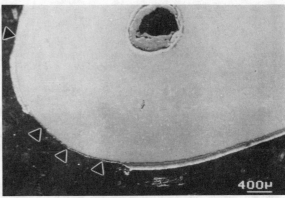

Figure 8.

SEM micrograph illustrating preferential oxidation attack at the coating/substrate interface.

Figure 9.

SEM micrograph illustrating coating loss at the leading edge due to oxidation attack at the diffusion zone.

The role of hafnium in providing oxide scale adhesion on MCrAl alloys is well known (4). Hafnium in a directionally solidified MM-200 alloy has been observed to diffuse into the EB-PVD NiCoCrAlY coating and provide better Al_2O_3/coating adhesion than on a NiCoCrAlY coated Hf free alloy (5). The theory is that hafnium diffuses into the overlay coating and forms a higher density of hafnium rich oxide "pegs". Similar behaviour could be expected from the DSRené 80H/aluminide coated blades which contains an equivalent amount of Hf. However, in this investigation service exposed blades did not show any such pegs. Recent Auger and SIMS analyses on these blades also confirmed the absence of Hf enriched oxide pegs (6). This observed difference could be due to the limitations imposed on diffusion of Hf through various carbide and intermetallic phases present in the coating diffusion zone.

Thermal Fatigue

The HPT blade experiences a complex thermal and mechanical history during a typical mission of the aircraft. Temperature gradients and mechanical constraints during such complex cycles give rise to cyclic thermal stresses and therefore induce thermal fatigue (TF) damage in the HPT blade. Since the aluminide coating on the HPT blade has a fairly high brittle to ductile transition temperature, the coating can be somewhat brittle at near service temperatures. Any thermal fatigue crack formed in the coating can then be propagated into the substrate alloy.

Figures 10(a) to (d) illustrate the thermal fatigue cracks present in various regions of an HPT blade. Depending on the location of the crack, minor to extensive substrate alloy attack has been observed. Figure 10(a) shows the initiation of a TF crack in the coating whereas Figures 10(b) and (c) show attacked regions near the blade leading edge. It should be noted in Figure 10(d) that near the leading edge of the blade, the diffusion zones of the external airfoil coating and the internal coating on the cooling hole surface have completely overlapped each other. In other words, no substrate alloy was found to be present in between these two coating interfaces. As a result of this and poor TF properties of the coating, a crack that had initiated in this region of the coating propagated through the entire blade wall thickness. Gamma prime depletion zones beneath the interior oxidized surfaces were also observed in these cracks.

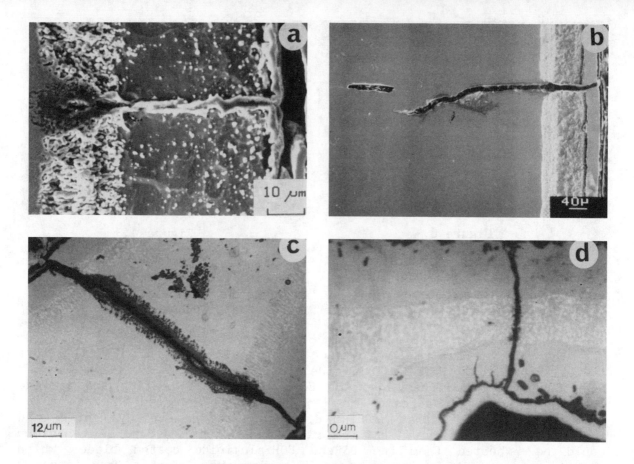

Figure 10. a) SEM micrograph showing TF cracking of the coating and b) subsequent substrate alloy attack, c) optical photomicrograph showing complete attack on a thin blade wall d) TF cracking at the region where coating diffusion zones overlapped.

Interdiffusion

The difference in chemistry of the aluminide coating and the substrate DS René 80H alloy gives rise to chemical potential gradients of Al, Ni and other alloying elements. Predominant processes are the diffusion of Al from the coating into the substrate alloy and diffusion of Ni, Co, Ti and Cr in the opposite direction.

Compositional variation of different alloying elements such as Ni, Al, Co, Ti and Cr were determined across the coating on a qualitative basis by a line scan technique using an Energy Dispersive X-ray analyzer attached to a scanning electron microscope. The results are shown in Figures 11(a) and (b). Figure 11(a) shows a gradual depletion of Al from the coating due to its diffusion zone into the substrate alloy. Sigma phase formation beneath the coating diffusion has been observed quite extensively, particularly towards the trailing edge of the blade. This is shown in Figures 12(a) and (b) which represent respectively, a longitudinal and a transverse section of the blade at the substrate/coating interface region. Sigma phase is formed as a result of Al diffusion from the coating into the substrate alloy and enrichment of elements like Cr, Mo, W and Ti at the coating/substrate alloy interface. Formation of sigma phase in thin wall sections of the blade can offer an easy path for crack propagation from the coating into the substrate alloy, although no such phenomenon was observed in the present investigation (7).

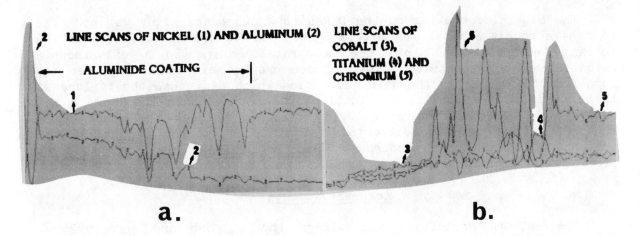

Figure 11. Elemental X-ray intensity profiles of a) Ni, Al and b) Co, Ti and Cr obtained by an energy dispersive X-ray analyzer.

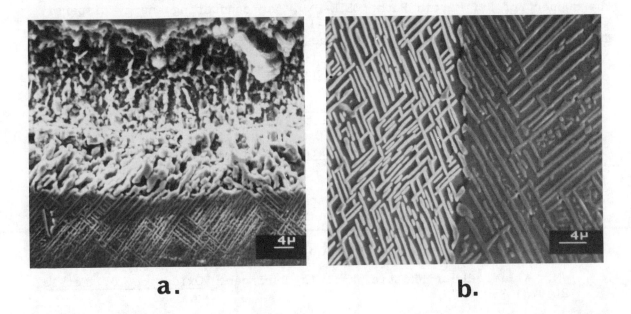

Figure 12. SEM micrographs illustrating the sigma phase formation below the coating diffusion zone, a) longitudinal section and b) transverse section through the region containing the sigma phase.

Summary and Conclusions

Investigation of several prematurely retired HPT blades showed extensive tip damage such as cracking and oxidation particularly towards the trailing edge region. The initiation of this damage is believed to have started from the tip surface due to the loss of the protective aluminide coating by tip grinding and/or rubbing against the HPT shroud. Initiation and propagation of TF cracks at the trailing edge subsequently accelerated the substrate alloy attack by oxidation. Resistance of the DSRené 80H substrate alloy to oxidation at the turbine operating temperature was observed to be poor. The plugging and oxidation of tip cooling holes also were attributed to tip grinding and/or rubbing against the turbine shroud. The oxide in the cooling hole can act as an insulator and limit heat transfer from the blade to the internal cooling air.

The aluminide coating on the DSRené 80H substrate alloy was observed to behave poorly against thermal fatigue. The TF cracks initiated in the coating and propagated into the substrate leading to a catastrophic oxidation of the substrate alloy. These cracks have often been seen to turn from the radial to the axial direction which can cause the failure of the entire blade. Localized oxidation attack underneath the coating was observed. Interdiffusion of alloying elements between the coating and substrate alloy formed a deleterious "sigma" phase at the interface. The presence of these phases underneath the coating can cause embrittlement particularly in thin sections of the blade and therefore induce cracking.

Acknowledgements

The authors gratefully acknowledge the support and encouragement given by Mr. J.A. Barber, Director of Engineering, Orenda Division during this investigation. Support from the Structures and Materials Laboratory of the National Aeronautical Establishment, National Research Council Canada is greatly appreciated. The authors would also like to acknowledge the support of Dr. Martin Roth, QETE-DND, for conducting energy dispersive X-ray analyses of specimens. Financial support for this work was obtained from the Department of National Defence Canada.

References

1. M. McLean, *Directionally Solidified Materials for High Temperature Service*, (London, The Metals Society, 1983) 11 and 151.

2. E. Lang, ed., *Coatings for High Temperature Applications*, (New York, Applied Science Publishers, 1983), 341, F.S. Pettit and G.W. Goward.

3. S.D. Antolovich, "Elevated Temperature Low Cycle Fatigue of Nickel Base Superalloys in the Conventionally Cast, Directionally Solidified and Single Crystal Forms" (Report AFSOR-TR-83-0827, Air Force Office of Scientific Research, Washington, D.C., 1983).

4. I.M. Allam, D.P. Whittle and J. Stringer, *Oxidation of Metals*, 12(1978) 35.

5. D.K. Gupta, "Effects of coating-substrate interdiffusion on the performance of plasma sprayed MCrAlY coatings", (Proc. International Conference on Metallurgical Coatings, Elsevier Sequoia S.A., Lausanne and New York, 1980) 477.

6. P.C. Patnaik, I. Sproule and D.F. Mitchell. Unpublished work.

7. E. Lang and L. Tottle, "Some Observations Concerning Structural Stability of Aluminide Coatings on Alloy IN 738LC" (JRC Report on Special Activities P/COST/II 1.3, March 1981).

HIGH TEMPERATURE CORROSION FATIGUE AND GRAIN SIZE CONTROL

IN NICKEL-BASE AND NICKEL-IRON-BASE SUPERALLOYS

M. Yoshiba and O. Miyagawa*

Department of Mechanical Engineering, Faculty of Technology
Tokyo Metropolitan University
Fukazawa, Setagaya-ku, Tokyo 158, Japan

* Tokyo Metropolitan University, Professor Emeritus

Abstract

In order to clarify an effect of hot corrosive environment both on the fatigue strength and the fatigue fracture behavior of superalloys, the rotating-bending fatigue tests were conducted for nickel-base Inconel 751, nickel-iron-base Inconel 718 and Fe-42Ni-15Cr-3Mo alloy, at 800°C in the Na_2SO_4-NaCl molten salt environment as well as in air. A grain size was controlled in a wide range from the viewpoint of its practical importance. It was revealed that hot corrosive environment brings about a significant fatigue strength degradation for all the alloys, in particular with reduced grain size, by affecting the initiation and propagation processes of the fatigue cracks. A grain size was confirmed to concern strongly in the corrosion fatigue crack propagation behavior. In particular, the detrimental effect of reducing a grain size was emphasized on the enhanced intergranular crack propagation. A fundamental criterion of the grain size control for the improved corrosion fatigue strength properties of superalloys was presented in connection with a Ni content.

Introduction

Superalloys used predominantly for hot section components of various heat engines such as jet engines, gas turbines, diesel engines and so on, are inevitably subjected to the simultaneous effects of both the mechanical damages such as creep and/or fatigue and the corrosive damage due to a hot corrosion. For the practical applications of superalloys to be successful, it should be important for the mechanical performances to be evaluated in such an aggressive environment. For the creep rupture properties, from this standpoint, a lot of useful knowledge about the corrosion-environmental effect has been accumulated recently through a variety of worldwide studies containing authors' study (1,2). However, the environmental effect on the fatigue strength and fatigue fracture behavior remains unclear, although it is also a serious problem. The fatigue fracture behavior, and hence the fatigue strength, should be more affected by the corrosive environment than the case of a creep, because a fatigue failure is apt to initiate from the alloy surface. Such a corrosion-environmental effect on the fatigue properties is predicted to be most significant at the temperature range between approximately 800 and 900°C in which both hot corrosion and mechanical fatigue damages are dominant (3).

On the other hand, it has been well known that grain size is one of the most important material factors affecting the fatigue properties of many kinds of wrought alloys. However, the grain size dependence of high temperature corrosion fatigue properties has been hardly clarified.

In the present study, then, the high cycle fatigue tests were conducted at 800°C for some superalloys with a range of grain sizes controlled, in air and in the Na_2SO_4-NaCl molten salt environment, and the corrosion-environmental effect on the fatigue strength as well as its grain size dependence were investigated in connection with the initiation and propagation behavior of the corrosion fatigue cracks.

Materials and Experimental Procedures

Materials

Three kinds of wrought superalloys were used in this study: γ'-precipitation-hardened nickel-base Inconel 751 alloy, γ''- and/or γ'-precipitation-hardened nickel-iron-base Inconel 718 alloy, and γ'-hardened Fe-42Ni-15Cr-3Mo alloy. These chemical compositions are given in Table I. The latter one is a new alloy developed by authors group for a purpose of high performances

Table I. Chemical Compositions of Superalloys Used (Mass %)

Alloys	C	Si	Mn	P	S	Ni	Cr	Ti	Al	Mo	Fe	Others
Inconel 751	0.09	0.20	0.53		0.007	Bal.	16.07	2.05	1.10	-	5.74	Nb+Ta 1.18
Inconel 718	0.03	0.11	0.22	0.011	0.005	Bal.	17.54	0.92	0.42	3.02	18.54	Nb+Ta 4.99
Fe-42Ni-15Cr-3Mo	0.05	0.26	0.52	0.010	0.002	41.91	15.02	2.82	0.87	3.05	Bal.	B 0.012

Table II. Typical Heat Treatment Conditions and the Corresponding Grain Size and Vickers Hardness Number

Alloys	Heat Treatment Conditions	Grain Size (μm)	HV (196N)
Inconel 751	1100°C×2h→WQ + 800°C×24h→AC.	44	313
	1200°C×2h→WQ + 800°C×24h→AC.	125	306
Inconel 718	1000°C×1h→AC + 720°C×8h→FC(600°C)→AC.	15	441
	1100°C×2h→WQ + 800°C×24h→AC.	145	341
Fe-42Ni-15Cr-3Mo	1000°C×2h→WQ + 800°C×24h→AC.	30	352
	1200°C×2h→WQ + 800°C×24h→AC.	130	353

in aggressive environments (4).

Different kinds of heat treatments were adopted for these three alloys in order to control mainly a grain size in a wide range. The typical heat treatment conditions adopted are listed in Table II, along with the corresponding average grain size and Vickers hardness number. For Inconel 751, a grain size has been controlled most widely between 10 and 290 μm in grain diameter, which approach is already described elsewhere in detail (5,6).

The smooth bar specimens with 8 mm in diameter and 15 mm in gage length were machined from the heat-treated rods. Furthermore, in order to examine the fatigue crack behavior in detail, the notched bar specimens also were adopted with 5 mm in notch radius and 8 mm in specimen diameter at the notch root. They were emery-polished through 500 grit, and were cleaned ultrasonically in acetone.

Experimental Procedures

The fatigue tests were conducted at 800°C using the rotating-bending fatigue testing machines (1500 rpm). In order to simulate an actual hot corrosive environment, the specimens were coated with synthetic salt mixture composed of 90% Na_2SO_4 plus 10% NaCl (m.p.=785°C). The amount of salt mixture precoated was 0.4 kg/m^2. For the prolonged fatigue tests, the coating of the same amount of salt mixture was repeated at every 3×10^5 cycles to refresh the corrosive media. The tests in static air also were carried out using the specimens without coating the salt mixture. These two types of tests will be termed as "in hot corrosive environment" and "in air", respectively.

For both the fatigue-failure and the interrupted specimens, the metallographic examinations were made mainly on the longitudinal sections by means of an optical microscope and an electron probe X-ray microanalyzer (EPMA). The observation of the fracture surface also was conducted by a scanning electron microscopy (SEM).

Results and Discussion

Fatigue Strength in Air and in Hot Corrosive Environment

Figures 1, 2 and 3 show the S-N curves (Wöhler curves) obtained from the fatigue tests at 800°C both in air and in hot corrosive environment for Inconel 751, Inconel 718 and Fe-42Ni-15Cr-3Mo alloy, respectively. In air, the fine-grained specimens exhibited longer fatigue life at the higher stress levels. However, such a superiority of fine-grained specimens disappeared as a stress level lowers. In hot corrosive environment, on the other hand, a significant fatigue life reduction occurred in all the alloys, although the degree of degradation is different dependingly on the alloy chemistry and the grain size as mentioned below. High temperature corrosion fatigue is also characterized by the disappearance of the fatigue limit (the endurance limit) along with the increased data scatter. In such an aggressive environment, the fine-grained specimens are found to yield rather lowered fatigue life or at best the same one as compared with the coarse-grained specimens.

The fatigue strengths at the representative cycles in both environments are summarized in Fig. 4, along with the corrosion fatigue strength ratio as a measure of the corrosion sensitivity of the fatigue strength. It is noted again that in air a grain size favorable for the increased fatigue strength is reversed dependingly on the cycle life: a fine grain is favored for the short life term, and a coarse one is for the long life term. As regards the alloy comparison, two nickel-iron-base superalloys were found to have the

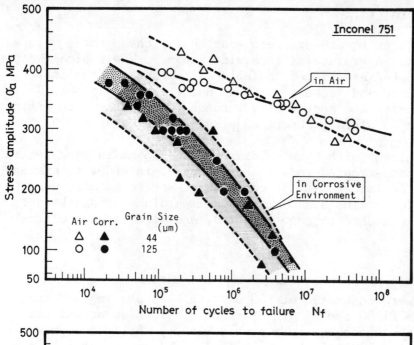

Figure 1 –
S-N curves of Inconel 751 specimens with different grain sizes at 800°C in air and in hot corrosive environment.

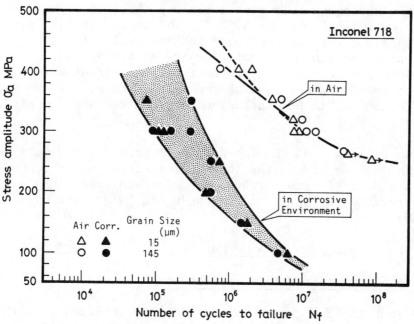

Figure 2 –
S-N curves of Inconel 718 specimens with different grain sizes at 800°C in air and in hot corrosive environment.

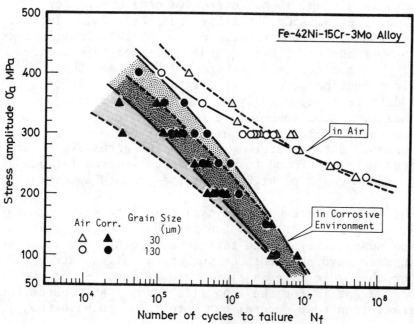

Figure 3 –
S-N curves of Fe-42Ni-15Cr-3Mo alloy specimens with different grain sizes at 800°C in air and in hot corrosive environment.

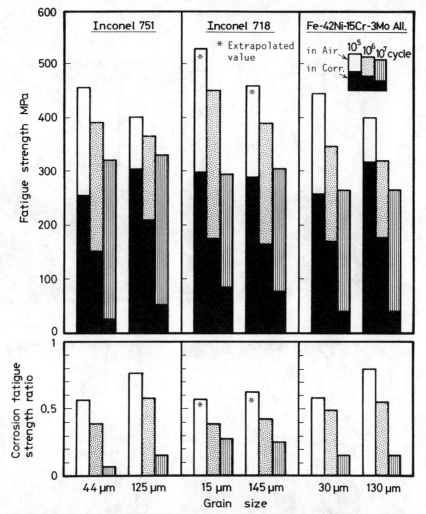

Figure 4 - Summary of the fatigue strengths in air and in hot corrosive environment along with the corrosion fatigue strength ratio.

fatigue strength comparable to Inconel 751 with the similar grain size in the short life regime, although in the long life regime their fatigue strengths were fairly lowered as compared with Inconel 751.

In hot corrosive environment, both the fatigue strength and its corrosion sensitivity were found to show somewhat different grain size dependence between nickel-base Inconel 751 and two nickel-iron-base alloys. For Inconel 751, an advantage of grain size coarsening is pronounced for the improved corrosion fatigue strength, along with the minimized corrosion sensitivity. For two nickel-iron-base alloys, on the contrary, the corrosion fatigue strength is hardly dependent on a grain size except for the short life term of Fe-42Ni-15Cr-3Mo alloy, so that an advantage of coarse grain is rather diminished.

Fatigue Fracture Behavior in Air and in Hot Corrosive Environment

From the metallographic examinations about the longitudinal sections of fatigue-failured specimens, along with fatigue-interrupted specimens in air and in hot corrosive environment, a fracture mode was found to change from a transgranular mode to an intergranular mode as a grain size decreases. Figure 5 shows the schematic illustrations of the fatigue fracture modes observed in three alloys as functions of a grain size, a stress level and an environment, along with the fracture surface morphologies of Fe-42Ni-15Cr-3Mo alloy, as a typical example. The present results about the grain size dependence of fatigue fracture mode are consistent with the previous result for Inconel 751 (5,6): an intergranular fracture becomes dominant as a grain size decreases less that about 50 μm, while a transgranular one is dominant for

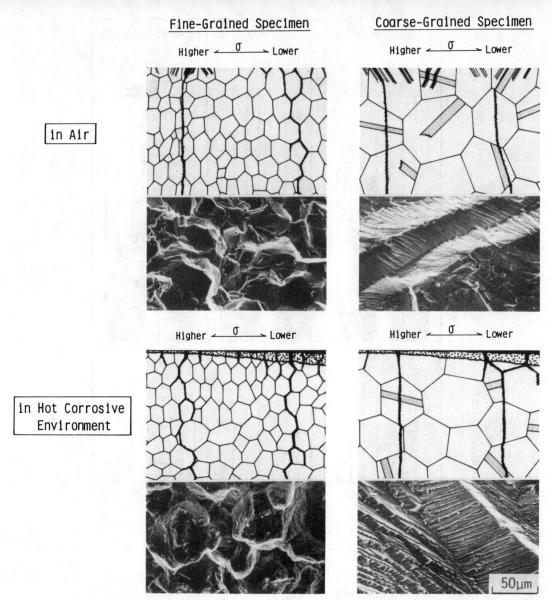

Figure 5 – Schematic illustrations showing different types of fatigue fracture behavior and the corresponding fractgraphs of Fe-42Ni-15Cr-3Mo alloy specimens in connection with a grain size, a stress level and an environment.

the coarser grain size more than about 100 μm, and an intermediate grain size between 50 and 100 μm results in a mixed fracture mode. Hot corrosive environment appears to bring about little change in the fracture mode. However, it should be noted that in the aggressive environment the fatigue cracks always initiate at the grain boundary subjected to the intergranular penetration of sulfides followed by oxides and so on, regardless of an alloy chemistry and a grain size. Figure 6 shows a typical intergranular attack by sulfides etc. occurred in the Inconel 751 corrosion fatigue specimen (5). Such a very small corrosion pit, with a depth of approximately one grain diameter, is possible to develop already in the early fatigue stage in which the mechanical fatigue damage is hardly accumulated, so that it provides itself the crack nucleation site. Therefore, the intergranular-attack-stimulated premature crack initiation should be one of the principal causes for the marked fatigue strength degradation, as schematically shown in Fig. 7. It should be noted that the development of intergranular attack is affected strongly by an alloy chemistry, i.e. the alloy composition, and is rather insensitive to a stress level and a grain size. Then it is suggested that the alloy with the minimized intergranular attack sensitivity is favored to inhibit the corrosion-induced strength degradation. In this respect, nickel-iron-base superalloys

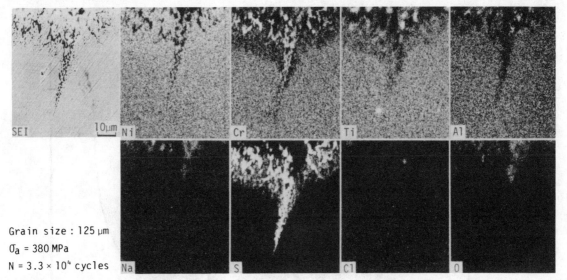

Figure 6 – Characteristic X-ray images at the early stage of intergranular attack in a coarse-grained Inconel 751 specimen fatigue-tested in hot corrosive environment.

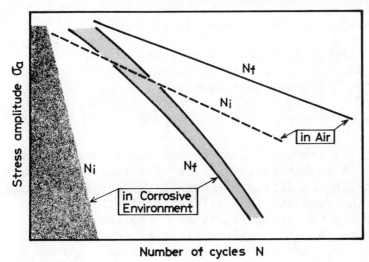

Figure 7 – Schematic representation showing the N_f and N_i in air and in hot corrosive environment. N_f and N_i denote the number of cycles to failure and the number of cycles to crack initiation, respectively.

seem to be more beneficial because of their inherent good intergranular attack resistance (7).

It has been known that the fatigue crack propagation process also is affected by the corrosive environments (8,9). Figure 8 shows the initiation and propagation behavior of fatigue cracks for Inconel 751 specimens in air and in hot corrosive environment, which was obtained from the microstructural measurement of the interrupted fatigue specimens. It is evident again that hot corrosive environment brings about a premature fatigue crack initiation. Furthermore, in the fine-grained specimens the corrosion fatigue cracks tend to propagate so rapidly almost along the grain boundary as to cause a premature intergranular fracture. This should be attributed to the combined effect of the mechanical fatigue and chemical corrosive damages concentrated to the grain boundary region (5). On the contrary, the coarse-grained specimens are apt to be subjected to the isolated damage: the fatigue damage points into the grain, while the corrosive damage prefers the grain boundary.

Grain Size Dependence of Corrosion Fatigue Strength

It has been already discussed for a nickel-base Inconel 751 alloy that in air the grain size dependence of high temperature fatigue strength is determined by two competitive factors: one is a tendency of the intergranular fracture and the other is an ability to disperse the cyclic slip deformation

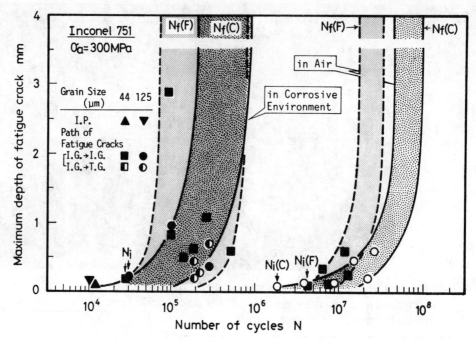

I.P.: Intergranular Penetration, I.G.: Intergranular, T.G.: Transgranular

Figure 8 — Initiation and propagation behavior of fatigue cracks in the Inconel 751 specimens with different grain sizes in air and in hot corrosive environment.

(6). Figure 9 shows schematically the grain size dependence of high temperature fatigue strength in air. A reduction in a grain size tends to promote the slip dispersion associated with the diffusion-controlled recovery process such as a dislocation climb. This results in a difficulty in the fatigue crack initiation and propagation in the grain interior. Whereas, this in return tends to cause the fracture rather along the grain boundary with relatively high strain concentration, so that the intergranular-cracking-induced strength degradation becomes more significant as a grain size decreases (10). Then, it should be reasonable to consider that for the increased fatigue strength in air a grain size should be controlled rather small as far as a transgranular fracture is dominant.

On the contrary, the grain size dependence of the high temperature corrosion fatigue strength seems to be different among the alloy systems. For a nickel-base Inconel 751 alloy, the corrosion fatigue strength has been found to lower monotonically with reduction in grain size, as schematically shown in Fig. 10 (5). Again, a reduction in a grain size in this alloy system is very harmful to the corrosion sensitivity as well as the corrosion fatigue strength itself, because of the corrosion-enhanced intergranular crack propagation rate due to the combined mechanical and chemical damages with the low melting nickel sulfide formation such as Ni_3S_2-Ni eutectic (m.p. =637°C). In this respect, nickel-iron-base alloys seem to be more useful than nickel-base alloys for suppressing the nickel sulfide formation, which should be reflected in the mitigated grain size dependence of the corrosion fatigue strength for nickel-iron-base alloys as well as in their restrained crack initiation and propagation behavior. Then, it can be concluded that for the improved corrosion fatigue strength properties of superalloys, in particular with higher Ni content, a grain size has to be controlled appropriately large enough not to induce an intergranular fracture.

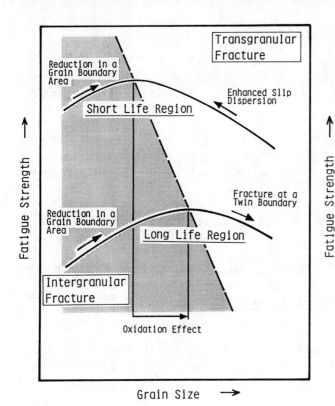

Figure 9 –
Schematic representation showing the grain size dependence of high temperature fatigue strength of superalloys in air in connection with various affecting factors.

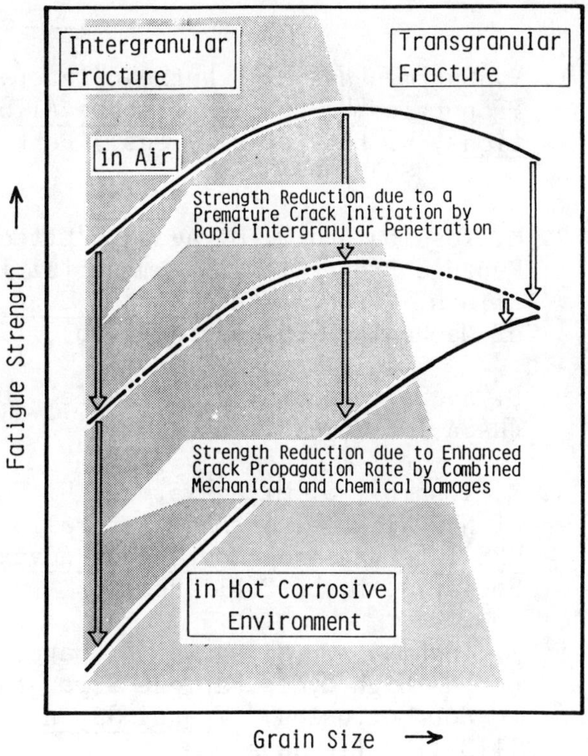

Figure 10 –
Schematic representation showing the grain size dependence of the corrosion fatigue strength of nickel-base superalloys in connection with various affecting factors.

Concluding Remarks

1) Hot corrosive environment was found to bring about a significant fatigue strength degradation for the nickel-base and nickel-iron-base superalloys. Such a marked strength degradation was confirmed to result partly from the intergranular-attack-stimulated premature crack initiation, depending mainly on a corrosion resistance, particularly on the intergranular attack resistance, of the alloys.

2) Corrosion fatigue crack propagation behavior was found to depend strongly on a grain size. In particular, a reduction in grain size results in the enhanced intergranular crack propagation rate because of the combined effect of the mechanical fatigue and chemical corrosive damages concentrated to the grain boundary, so that the corrosion-induced strength degradation is more significant than the case of a coarse grain size.

3) Hot corrosive environment caused a different grain size dependence of the fatigue strength in air. In order to improve the corrosion fatigue strength properties of superalloys, a grain size has to be controlled large enough not to induce an intergranular fracture. Such a demand becomes more serious in nickel-base superalloys with higher Ni content.

Acknowledgments

The authors wish to acknowledge Mr. T. Masaki of Shimadzu Corp. for the valuable analyses by EPMA. The experimental work in this study was competently carried out by Messrs. K. Hosoi, M. Maruyama, Y. Iwasaki, T. Ishii and J. Izumi, who were formerly undergraduates at Tokyo Metropolitan University.

References

1. V. Guttman and M. Schutze, "Interactions of Corrosion and Mechanical Properties," <u>High Temperature Alloys for Gas Turbines and Other applications</u>, W. Betz et al., eds., Part I (Dordrecht, Holland: D. Reidel Publ. Co., 1986), 293-326.

2. M. Yoshiba and O. Miyagawa, "Effect of Hot Corrosive Environment on Creep Rupture Properties of Commercial Iron-Nickel-Chromium Heat Resisting Alloys," Proc. <u>International Conference on Creep</u>, (Tokyo: Japan Society of Mechanical Engineers, 1986), 193-198.

3. D. A. Spera and S. J. Grisaffe, "Life Prediction of Turbine Components," (NASA TMX 2664, 1973).

4. M. Yoshiba, O. Miyagawa, and H. Fujishiro, "High Temperature Performances of New Nickel-Iron-Base Superalloys in Aggressive Environments," Proc. <u>MRS International Meeting on Advanced Materials</u>, (Materials Research Society), to be Published.

5. M. Yoshiba, O. Miyagawa, K. Sato, and H. Fujishiro, "Effect of Grain Size on the High Cycle Fatigue Strength of a Nickel-Base Superalloy Subjected to Hot Corrosion," <u>Trans. Japan Society of Mechanical Engineers</u>, 50(1984) 1113-1122 (in Japanese).

6. M. Yoshiba, O. Miyagawa, and H. Fujishiro, "Effects of Grain Size and Notch on the High Cycle Fatigue Strength of a Nickel-Base Superalloy at Elevated Temperature," <u>Trans. Japan Society of Mechanical Engineers</u>, 50(1984), 1443-1452 (in Japanese).

7. M. Yoshiba, O. Miyagawa, and H. Fujishiro, "Hot Corrosion Behavior of Heat Resisting Alloys," <u>J. Iron and Steel Inst. Japan</u>, 67(1981), 996-1005 (in Japanese).

8. W. Hoffelner and M. O. Speidel, "The Influence of the Environment on the Fatigue Crack Growth of the Nickel-Base Superalloy IN 738LC and IN 939 at 850°C," <u>Behavior of High Temperature Alloys in Aggressive Environments</u>, I. Kirman et al., eds., (London: The Metals Society, 1979), 993-1004.

9. S. Floreen and R. H. Kane, "Effects of Environment on High-Temperature Fatigue Crack Growth in a Superalloy," <u>Metall. Trans.</u>, 10A(1979), 1745-1751.

10. M. Gell and G. R. Leverant, "Mechanisms of High-Temperature Fatigue," <u>Fatigue at Elevated Temperatures (ASTM STP 520)</u>, (1973), 37-67.

QUANTITATIVE MICROSTRUCTURE ANALYSIS TO DETERMINE

OVERHEATING TEMPERATURES IN IN100 TURBINE BLADES

H. Huff and H. Pillhöfer

Department of Materials Technology
MTU Motoren- und Turbinen-Union München GmbH
Munich, Federal Republic of Germany

Abstract

The paper presents a method for the determination of overheating temperatures in nickel alloys by the evaluation of the gamma prime microstructure.

In the temperature range 1080 to 1200 °C, a stable fine structure with a close relationship between the gamma prime volume fraction and the temperature forms within 30 seconds.

Accordingly, overheating of turbine blades can be determined qualitatively above 1040 °C, and quantitatively above 1080 °C.

Furthermore, these relationships provide a better understanding of the formation of the fine structure of precipitation-hardening nickel alloys by showing how heat treatment in the partial solution range of the gamma prime phase and ageing influence the high-temperature strength-determining gamma prime structure.

Superalloys 1988
Edited by S. Reichman, D.N. Duhl,
G. Maurer, S. Antolovich and C. Lund
The Metallurgical Society, 1988

Introduction

In nickel-alloy turbine blades, overheating can produce certain, at times pronounced changes in the gamma prime phase. In the assessment of overheated parts for reusability, the criteria applied have been a combination of gamma prime particle size and shape (1). This approach, however, is relatively unsafe. It was thus appropriate to find a quantitative, practically exclusively temperature-dependent relationship between the formation of gamma prime particles and overheating conditions (T > 1050 °C). An attempt also had to be made to establish a component-related overheat ceiling.

Procedure

Principle of Measurement and Nomenclature

The volume fraction of gamma prime phase that does not go into solution in the gamma matrix under overheat conditions was selected and stereologically measured as a characteristic, temperature-dependent structural feature (2).

If cooling from a temperature condition in the partial solution range of the gamma prime phase (approximately 1000 to 1200 °C) is relatively fast, two different gamma prime particle dispersions occur:

1) <u>Primary gamma prime particles</u>; these include the usually larger, cubic to globular gamma prime particles stemming from the original condition

2) <u>Secondary gamma prime particles</u>; these include the fraction of small to very small, finely dispersed gamma prime particles that go into solution and reprecipitate at high temperature

Test Parameters and Microstructure Specimens

Airfoil sections in IN100 were heat treated and fan-cooled (cooling rate 120 °C/ min to 800 °C) in a lab-type vacuum furnace following long-term ageing (1000 °C/100 h + 900 °C/20 h + 800 °C/20 h)

Temp. °C	1040	1080	1100	1120	1140	1160	1180	1200	1210
Time	2 h	2 h	5 min	5 min	5 min	5 min	5 min	5 min	2 h
			2 h	2 h	2 h	2 h	2 h	2 h	
			24 h		24 h				

To simulate very short overheat periods, small, flat IN100 specimens (thickness < 2.5 mm) were fitted with thermocouples and immersed for short, specific holding times in a hot salt bath (Semper Neutral 950) at a preheat temperature of 900 °C. Preheating and overheating in the salt bath were both carried out in a furnace with high temperature gradient. The heating period lasted less than 20 seconds. The salt-bath treatment was carried out following the procedure of Böhm (3). Other heat-treatment parameters were additional temperature cycles following overheating, ageing isotherms at 800 to 950 °C, and different cooling rates after complete gamma prime dissolution.

Mechanical Properties

Separately-cast stress-rupture specimens (Ø 9.5 mm) in HIP + aged condition were overheated for 10 minutes at 1050 to 1200 °C on a hot gas test rig, and cooled at over 240 °C/min. One set of specimens was additionally overheated under a stress of 200 MPa (5 min at 1050 °C and 1075 °C, 30 sec at 1100 °C and 10 sec at 1125 °C). Other specimens were additionally aged for 8 h at 900 °C after overheating only.

Microstructure

For quantitative evaluation of the microstructure, the gamma prime phase was etched selectively, using an etchant according to Ref. (4). SEM micrographs (5000X) were analysed with the aid of a Hewlett Packard HP 1000 process computer using the linear intercept method and special stereological programs (5, 6).

Results

Influence of Temperature on Gamma Prime Microstructure

The SEM micrographs in figures 1 and 2 show the gamma prime microstructure of IN100 following exposure to several different temperature levels from 1040 to 1210 °C. The starting condition is also shown. The gamma prime particles appear as a dark phase.

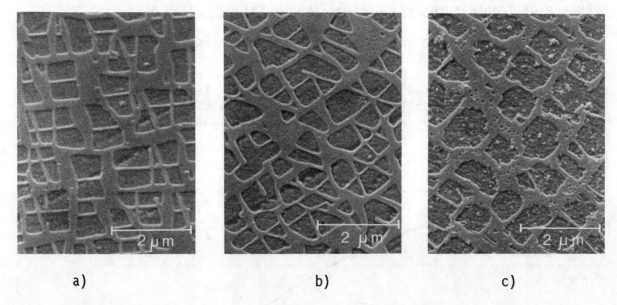

a)   b)   c)

Figure 1 - Influence of temperature on the gamma prime phase in IN100; a) Starting condition, showing a uniform dispersion of the primary gamma prime particles   b) 1040 °C   c) 1080 °C

A second dispersion of small, secondary gamma prime particles appears in addition to the approx. 0.8 µm gamma prime particles of the starting condition as early as at 1040 °C, but even more clearly at 1080 °C. As can be seen in figure 2, the area fraction of the primary gamma prime particles reduces visibly from a temperature of 1100 °C on, whereas the gamma matrix fraction, appearing as a light area on the micrograph, increases. Secondary gamma phase particles are embedded in these light matrix areas. At 1200 °C, the primary gamma prime particles have been completely dissolved and re-precipitated.

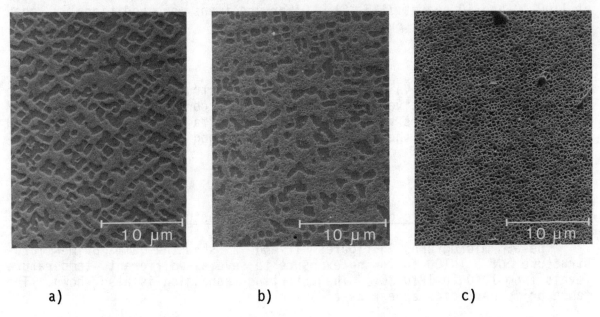

a) b) c)

Figure 2 - Influence of temperature on the gamma prime phase
a) 1100 °C, b) 1140 °C, c) 1210 °C

Influence of Temperature and Holding Time on Volume Fraction

The results of the quantitative microstructure analysis of the primary gamma prime volume fraction are shown in figure 3. Regardless of the holding time (5 minutes, 2 hours, 24 hours), the volume fraction decreases continuously from an initial 55% with increasing temperature. The scatter in the values for each temperature is less than 5% by volume.

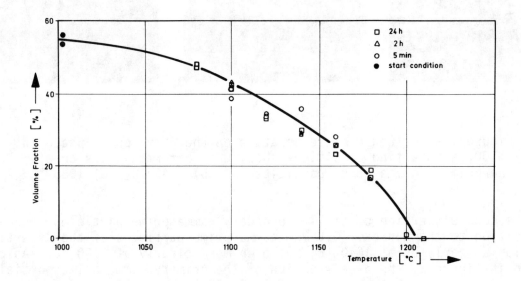

Figure 3 - Influence of temperature and holding time on the primary gamma prime particle volume fraction in IN100

Minimum Time Required for Change in Volume Fraction

The primary gamma prime volume fractions from the tests with short overheat times in the salt bath are shown in figure 4. The last two values shown in each curve are the results of overheat tests in a chamber furnace and represent the temperature-related equilibrium volume fraction for holding times of 5 minutes and 2 hours.

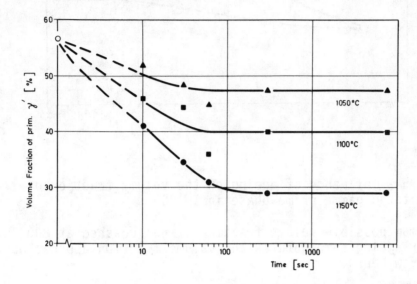

Figure 4 - Primary gamma prime particle volume fraction in relation to short holding times at different temperatures. In each case the last two values are equilibrium values (acc. to Böhm (3))

Influence of Cooling Rate

The influence of the cooling rate on the occurrence of secondary gamma prime particles was investigated optically, because it was also desirable to ascertain the secondary gamma prime volume fraction. After cooling from 1160 °C at a rate of 70 °C per minute, secondary gamma prime particles were still clearly present as a second dispersion of separate particles in addition to the primary gamma prime particles. At a cooling rate of 30 °C per minute most of the secondary particles have merged with the primary particles, and at the even slower rate of 8 °C per minute there is no secondary dispersion distinguishable in the SEM micrograph. Only the irregular shape of the primary particles still betrays partial redissolution.

Cooling rate to 800 °C (°C /min):	2	8	72	240	800
Gamma prime volume fraction (%):	54	52	50	43	21

The maximum fraction of precipitated gamma prime phase after long-term ageing is 55 to 57% by volume. The precipitation of the gamma prime phase grows less and less with increasing cooling rate. For comparison, the cooling rate of the airfoil area of a blade in still air is roughly 200 °C per minute in the upper temperature range.

Volume Fraction after Ageing

IN100 specimens with a starting volume fraction of 46% uniform gamma prime phase were aged at temperatures ranging between 800 and 950 °C for periods of between 30 minutes and 125 hours. The results can be seen in figure 5.

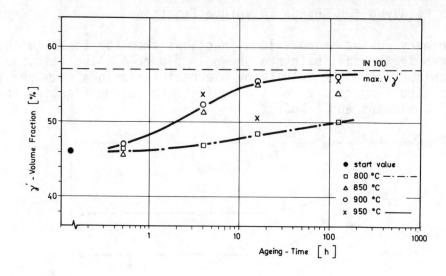

Figure 5 - Influence of ageing on the volume fraction of precipitated gamma prime phase in IN100

The maximum possible volume fraction is not reached at 800 °C. But, at the two higher ageing temperatures the volume fraction approaches its maximum within a few hours.

Stress-rupture Strength

To ascertain the length of time a blade will withstand overheat temperatures when subjected to centrifugal force, and to establish reasonable periods of stress for simulation of overheat conditions, stress-rupture tests were carried out on IN 100 specimens at elevated temperatures at a stress of 200 MPa. The results are summarised below:

Test temperature (°C):	1050	1075	1100	1125	1150
Time to rupture (minutes):	18	9.5	2.7	1.0	0.5
(2 specimens)	163	7.2	3.2	1.5	0.8

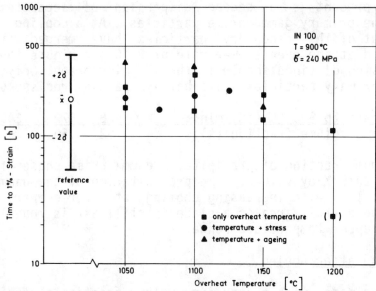

Figure 6 - Influence of overheating at various temperatures on the stress-rupture strength of IN100 at a test temperature of 900 °C and 240 MPa stress

Figure 6 shows the stress-rupture strength of IN100 specimens overheated deliberately at various temperatures. One set of specimens was overheated only, whilst a second set was both overheated and subjected to a load, and a third was aged at 900 °C for 8 hours after overheating. For comparison, a typical mean value with standard deviation for non-overheated material is also plotted for the test conditions 900 °C/240 MPa. The value at 1200 °C is shown in brackets because a fault was suspected in the simulation conditions.

The 1% strain values of the specimens up to an overheat temperature of 1150 °C are in the middle of the range for normal, non-overheated material. A decrease in the values sets in only after this temperature has been reached. Remarkable are the clearly higher strain values at 1050 and 1100 °C of the material aged after overheating and before the stress-rupture test.

Discussion

The noted progressive solutioning of the primary gamma prime phase at temperatures in excess of 1000 °C, as verified stereologically at 1080 °C and above, is caused by the temperature-dependent solubility of the gamma prime phase or of the gamma prime producers Al and Ti in the solid nickel solution. This relationship is illustrated in the Ni-Al and Ni-Ti binary alloy diagrams of Ref. (7). The equilibrium of the primary gamma prime phase and solid solution fractions is frozen in by rapid cooling and is preserved as an optically assessable change in the original gamma prime distribution.

A similar relationship between the solution or precipitation temperature and the primary gamma prime fraction has been found in other nickel-based alloys with high gamma prime-content, such as U700 (4, 8), SRR99 (3) and DS MAR-M 200 (9). Holmes (10) also observed differing gamma prime dispersion after heat treatment below the gamma prime solvus in IN100.

Minimum Time Required to Dissolution of the Gamma Prime Phase

It was demonstrated experimentally that approximately 60 seconds are sufficient for the temperature-related phase equilibrium to set in, and that just 30 seconds are enough for 90% of the equilibrium value to be approached (Figure 4).

Estimation of the diffusion rates from an investigation into the high-temperature brazing of IN100 (1) and experience with grinding of blade roots (12) confirm that the gamma prime phase goes rapidly into solution.

Evaluation of Other Factors Influencing the Gamma Prime Fraction

Other factors having a critical influence on the gamma prime distribution and its evaluation are:
a) The rate of cooling after overheating, which must be high enough to preserve the gamma prime fraction typical for the overheat temperature in question. However, secondary gamma prime particles can no longer form if the rate is too high, but they will form following brief ageing.

b) Treatment after overheating; long holding times after overheating cancel out overheat-related differences in the gamma prime structure.

c) The accuracy of the quantitative evaluation is determined essentially by the preparation of the microsection and the stereological evaluation, where care must be taken to ensure that stereologically correct preconditions for area or volume analysis are created by avoiding excessive selective etching of the gamma prime particles.

Other factors affecting the gamma prime structure, such as local and absolute fluctuations in the alloy composition, caused for example by dendritic segregation of the gamma prime producers, effect of stress and life-related changes to the gamma prime particles, do not seem to be of great significance with regard to the structure - temperature evaluation.

Conclusions

A method has been developed that uses microstructural analysis to clearly identify and quantify past overheat temperatures on IN100 rotor blades.

The relationship noted to exist between the volume fraction of primary gamma prime phase and temperature appears to permit

1) Safe and quantitative (± 20 °C scatter) identification of past overheating in the 1080 to 1200 °C range

2) At least qualitative indication of overheating in the lower temperature range of 1040 to 1080 °C

The method operates on and is thus also limited by the premise that a metallographically apparent structural differentiation of the gamma prime phase is allowed to form and survive. This premise again involves the following constraints:

a) The material needs at least 30 seconds or so at overtemperature

b) The rate of cooling from overtemperature should be faster than about 100 °C per minute

c) Overheating must not be followed by an extended holding time (of an hour or longer) at a lower temperture level (above 1000 °C)

The method of measurement can generally be transferred to all blade materials with high gamma prime fraction.

From data obtained in stress-rupture tests with simulated overheating, an overheat ceiling has been derived for IN100.

Taking degradation in stress-rupture performance as a criterion and allowing a safety margin of 20 °C, the overheat ceiling is 1130 °C for IN 100.

Quantitative stereological analysis of the gamma prime microstructure appears to provide a key to a better understanding of the effects of heat treatment on nickel-based superalloys. New insights are gained into the effects, for example, of homogenizing and ageing.

References

(1) A. Rossmann, "Overheating and Reusability of HPT and IPT Rotor Blades in IN100" (Technical Report 84/254, MTU München Materials Laboratory, 1984).

(2) H. Pillhöfer, "Quantitative Structural Analysis for Determination of Overheat Temperatures, Part 1, Test Method" (Technical Report 86/298, MTU München Materials Laboratory, 1986); German Patent 37 31 558.

(3) U. Böhm, "Influence of Overheating on the Structure of Blade Alloys IN100 and SRR 99" (Thesis 1987, Institut für Wissenschaften, University of Erlangen).

(4) H. Flöge, "Structure Variations in U700 PM and Their Influence on the Static RT Strength" (Thesis, Fachhochschule Aalen, MTU München, 1986), published in "4th World Conference on PM, Oak Ridge, Chicago, June 1988".

(5) F. Pschenitzka, private communication (1985-1987, Institut für Werkstoffwissenschaften, Lehrstuhl Prof. Dr. Mughrabi, University of Erlangen).

(6) G. Schuhmann, "Creep Mechanisms in PE16" (Dissertation, Institut für Werkstoffwissenschaften, University of Erlangen).

(7) M. Haasen, "Constitution of Binary Alloys" (McGraw Hill, 1958).

(8) N.G. Ingesten, D. Jacobson, R. Warren, "The Microstructure of PM Astroloy" (COST 501 Final Report, Project S-2, 1987).

(9) J. Jackson et al., "Effect of Volume Percent of Fine Gamma Prime on Creep in DS-Mar M200" _Metallurgical Transactions_, A Vol 8A, 10/1977, 1615-1620.

(10) D. Holmes, "Some Microstructural Aspects of HIPed Cast Nickel-Based Superalloys" (Paper presented at the 2nd Int. Conf. on Isostatic Pressing, Stratford-upon-Avon, 1982).

(11) D. Schneefeld, W. Eichmann, "Study of Brazeability of Guide Vanes" (Technical Report 86/257, MTU München Process Technology Laboratory, 1986).

(12) H. Bronn, "Machining Damage during CD Grinding" (Technical Report 86/290, MTU München Materials Laboratory, 1986).

NON-DESTRUCTIVE ANALYSIS BY SMALL ANGLE NEUTRON SCATTERING

P. Bianchi°, F. Carsughi*, D. D'Angelo°, M. Magnani˟, A. Olchini°,

M. Stefanon˟, and F. Rustichelli*

°C.R.T.N. ENEL, Milano, Italy

˟E.N.E.A., Bologna Italy

*Istituto di Fisca Medica, Universita di Ancona, Italy

Abstract

An investigation of service effects on the Ni-based superalloy UDIMET® 720 was performed by using Small Angle Neutron Scattering (SANS). In particular, gamma prime precipitate evolution was considered. The investigated sample was obtained from a turbine blade used in a turbogas power plant where the working temperature was about 850°C. Transmission Electron Microscopy (TEM) observations were compared to SANS data, which were obtained at D11 Diffractometer at ILL in Grenoble (France). The SANS theory allowed the determination of useful microstructural information, such as the volume fraction, the correlation length (which gives an idea of the packing of the system), and the Particle Size Distribution (PSD) of the precipitates. A comparison with as-heat-treated turbine blades showed the evolution of the microstructure of the superalloy during its operating life. In conclusion, SANS is a powerful non-destructive technique, which could be routinely employed to appraise the condition of materials used in technological applications.

®UDIMET is a registered trademark of Special Metals Corporation

Introduction

The main aims of research and development related to advanced materials for elevated temperature applications in electrical energy production are the optimization of power plant efficiency and component availability.

These aims could generally be pursued by means of comprehensive control of material behavior during component operation life. Such action should be split up into a wide characterization of the mechanical properties of the material to be considered at the design stage, as well as a reasonable description of the material response to the in-service damage phenomena.

To pursue the double purpose of better design and reliable operation, it is advisable to extend the microstructural assessment of new and service-exposed materials and to introduce advanced non-destructive diagnostic techniques, respectively.

With respect to this approach, attention has been focused on UDIMET 720, a Ni-based superalloy for power plant gas turbine blade applications. An investigation of thermal mechanical treatment effects was performed by using Small Angle Neutron Scattering (SANS) to analyze the precipitation evolution during the operation life.

SANS technique, which is non-destructive on small components, allows the quantitative determination of relative microstructural parameters such as precipitation, dislocation and cavitation, in a precise, reproducible and efficient manner as discussed in Ref. 1, where some of these investigations on superalloys are reported.

To calibrate the figures obtained by the SANS method, a metallographical survey was performed by scanning (SEM) and transmission (TEM) electron microscopes.

Materials

Nickel-based superalloys are widely used in high temperature applications. UDIMET 720 has been developed for this purpose and, in particular, is used for turbine blades of turbogas power plants, where the material working temperature is about 850°C.

The nominal composition (wt%) of the alloy is:

Ni	56.10
Co	14.00
Mo	3.00
W	1.30
Cr	17.20
Al	2.50
Ti	4.80
C	0.035
B	0.035
Fe	0.16
Si	0.2

and gamma prime phase precipitation hardening occurs to increase heat resistance.

The material consists of a gamma phase matrix with a bimodal distribution of gamma prime precipitates. In particular, the primary gamma prime phase is larger in size and cuboidal in shape, while the secondary gamma prime is smaller in size and spherical in form.

Three different UDIMET 720 samples have been investigated in order to reveal thermomechanical effects on the microstructure. First, an un-heat-treated sample was studied for the presence of precipitates due to solidification. Also, an as-heat-treated sample, taken from a virgin turbine blade, was investigated to show the wide precipitation induced by the heat treatment. Finally, a post-service turbine blade was considered in order to investigate the damage produced during service.

The operation time was 8000 hours at a maximum metal temperature of 850°C with thermal gradients and thermal fatigue. In combination with the presence of combustion gases, it produced a high stress field inside the turbine blade accompanied by corrosive attack.

Microstructural Characterization

The as-heat-treated sample's microstructure has been investigated by light, electron scanning, and transmission microscopy techniques; the results are presented in Ref. 2 and 3.

By the same method, the post-service sample has been studied in order to show microstructural changes in comparison with other samples.

Such material shows an equiaxed grain structure, where the grain dimensions are about doubled compared with those of the as-heat-treated turbine blades.

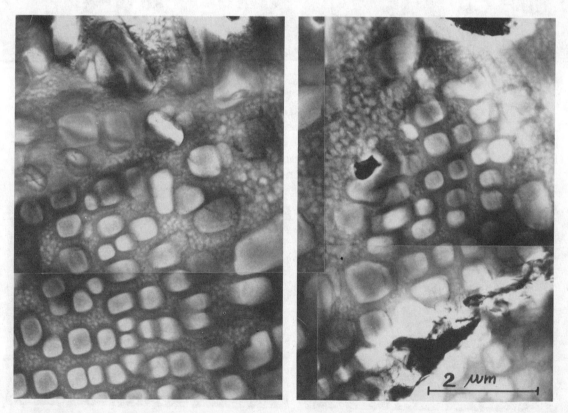

Figure 1 - TEM observations of primary and secondary gamma prime precipitation of the as-heat-treated turbine blade.

Both gamma and gamma prime phases have an FCC lattice with cell dimensions of 0.3524 nm and 0.3600 nm, respectively. Gamma prime particles show a cuboidal and globular shape for primary and secondary precipitates, respectively (Fig. 1).

Secondary gamma prime precipitates, with a dimension of the order of magnitude between 10^1 and 10^2 nm, have been detected.

Grain boundaries are subjected to recrystallization and subgrain nucleation occurs, especially when the gamma prime primary precipitates are drawn out into a radial crown morphology (Fig. 2).

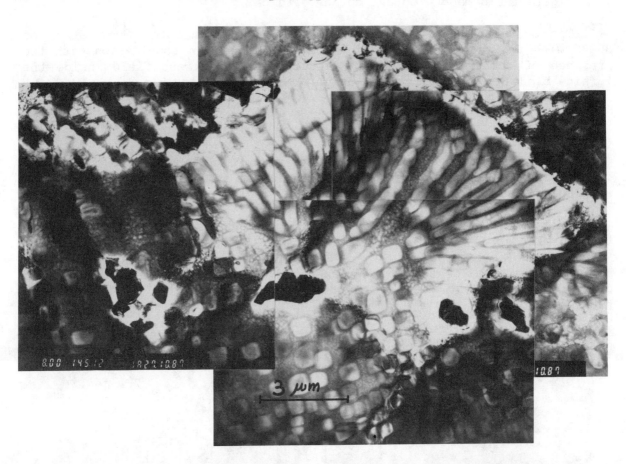

Figure 2 - Gamma prime precipitation close to grain boundaries where recrystallization processes occur in the post-service turbine blade.

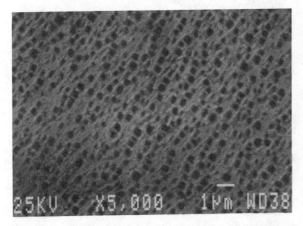

Figure 3 - Gamma prime precipitation inside the "normal" matrix of the post-service turbine blade

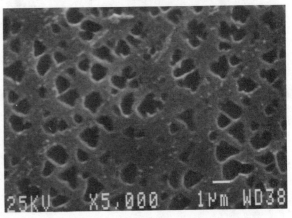

Figure 4 - Gamma prime precipitation inside the "islands" in the matrix of the post-service turbine blade.

The most evident change induced in the microstructure of UDIMET 720 during operation is the presence of wide "islands" where the gamma prime primary precipitates have larger dimensions, up to 0.8 μm, while inside the "normal" matrix, the gamma prime dimensions are at a maximum of about 0.3 μm (Figs. 3-4).

The mechanism of the gamma prime dimensional increase seems to be due to single particle growing related to multiple coalescence.

SANS Theoretical Introduction

When incident radiation strikes material containing inhomogeneities with dimensions up to two or three orders of magnitude larger than the radiation wavelength, the so-called Small Angle Scattering phenomenon occurs. The outcoming wave presents a maximum at the same direction of the incident wave and decreases very quickly when larger angles are observed.

With K the wave vector, and i, o indices of the Incident and Outcoming wave front, we can define an EXCHANGED WAVE VECTOR, Q, by:

$$Q = K_i - K_o \qquad (1)$$

Where the wave vector, K, is related to the wavelength by:

$$K = 2\pi/\lambda \qquad (2)$$

As in the small angle scattering technique, elastic scattering is considered (one that has $K_o = K_i$). Under these conditions, by simple geometrical considerations, it is possible to equate Q with the scattering angle, 2Θ, and the wavelength of incident radiation by:

$$Q = \frac{4\pi}{\lambda} \sin \Theta \qquad (3)$$

The outcoming wavefront drops to zero when:

$$2\Theta = \frac{\lambda}{D} \qquad (4)$$

where D is the dimension of the scattering inhomogeneities.

It is clear that the precipitated phases inside a matrix play the role of the previous scattering inhomogeneities. Small Angle Neutron Scattering (SANS) offers, in metallurgical application, a powerful method which can be used to investigate important elements of the microstructure such as magnetic domain, precipitated phases, and microvoids produced by creep, fatigue and irradiation. The dimension of the scattering particles are connected to the Q range to be investigated by:

$$Q \cdot D \approx 2\pi \qquad (5)$$

The formal theory of SANS is beyond the aim of this work but, if we consider the simplified model of a two-phase material, it is possible to develop some important relationships between the experimental data and the scattering particle parameters.

During an experiment, we can measure the coherent macroscopic cross section $d\Sigma/d\Omega(Q)$ that we call the scattering curve or function, whose properties will be discussed in the next section.

The use of a neutron beam allows the investigation of samples whose thickness is of the order of a cm, because:

$$\Sigma_t = \frac{1}{L} \qquad (6)$$

where L is the free mean path of neutrons inside the sample and Σ_t is the total neutron cross section taking into account both absorption and scattering phenomena.

Another kind of radiation could be used for the Small Angle Scattering experiment but, due to the high interaction of electrons or X-ray, the thickness of samples in these cases must be very thin to avoid total absorption by the sample.

As a consequence, it is clear that a nondestructive test is possible only if a neutron beam is used.

Scattering Function and Its Properties

The interaction between the neutrons and the investigated sample depends upon the difference of the scattering length density, ρ_b, between the two phases: the precipitates and the matrix.

Since p and m indices are related to the precipitates and matrix, the SANS theory shows that, if we consider a set of N_p identical particles, the scattering function is:

$$\frac{d\Sigma}{d\Omega}(Q) = (\Delta\rho)^2 \cdot N_p V_p^2 \, |F_p(Q,R)|^2 \qquad (7)$$

when

$$(\Delta\rho) = \rho_{bp} - \rho_{bm} \qquad (8)$$

where $F_p(Q,R)$ is the form factor of a single particle whose volume is V_p, depending upon its form and dimension, R.

Guinier approximation

It has been shown that it is possible to connect the scattering curve to a Giration Radius, R_g, of the particle, depending upon the form and dimension, D.

In particular, for small values of the product Q·D (for example, for spherical shape, this approximation holds for Q·R<1.2), is:

$$|F_p(Q,R)|^2 \alpha \exp(-Q^2 R_g^2/3) \qquad (9)$$

the value of R_g is the slope of the straight line fitting the experimental data in the $\text{Ln}(d\Sigma/d\Omega)$ vs Q^2 plane; if a sphere has a radius R_s, the giration radius is:

$$R_g = \sqrt{(3/5)} R_s \qquad (10)$$

Porod approximation

In the case of well defined outlines of the scattering particles, Porod has shown that the asymptotic value of the scattering function is:

$$\frac{d\Sigma}{d\Omega} = (\Delta\rho)^2 \cdot \frac{2\pi}{Q^4} \cdot S_P \qquad (11)$$

where S_P is the total surface of the particles per unit volume.

To verify if the asymptotic value has been reached, a check of this law is necessary. The plot of the scattering curve in the $Log(d\Sigma/d\Omega)$ vs the $LogQ$ plane must be approximated, at high values, by a straight line whose slope is -4.

Volume fraction

In the case of isotropic scattering, it is possible to demonstrate that [1]:

$$\int_0^\infty Q^2 \frac{d\Sigma}{d\Omega} dQ = 2\pi^2 (\Delta\rho)^2 \cdot C_p \cdot (1-C_p) \qquad (12)$$

where, V being the investigated volume of the sample, the volume fraction C_p is:

$$C_p = \frac{N_p \cdot V_p}{V} \qquad (13)$$

Correlation length

When a material shows a wide precipitation ($C_p > 10\%$), the system is said to be not diluted, and a correlation length, L_o, is necessary to express the packing of the particles.

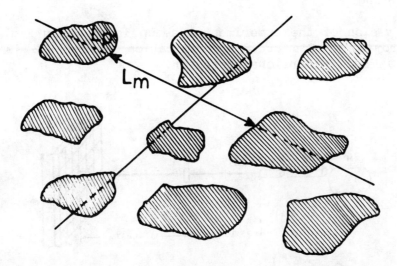

Figure 5 - The correlation length, L_o, depends upon the average particle dimensions, L_p, and on the average interparticle distances, L_m

L_m and L_p are the average chords for all directions between and through particles, respectively; L_o is defined by Fig. 5:

$$\frac{1}{L_o} = \frac{1}{L_p} + \frac{1}{L_m} \qquad (14)$$

and is connected to the scattering curve by [1]:

$$\int_o^\infty Q \frac{d\Sigma}{d\Omega} dQ = 2\pi \cdot (\Delta\rho)^2 \cdot L_o \cdot C_p \cdot (1-C_p) \tag{15}$$

Particle size distribution

When the hypothesis of identical particles is not valid, the scattering curve is the superposition of scattering curves from each of the different particles.

In this case, it is very helpful to determine the particle size distribution, $N(R)$, to characterize the polydisperse system.

It can be shown that:

$$\frac{d\Sigma}{d\Omega}(Q) = (\Delta\rho)^2 \cdot \int N(R) \cdot V^2_p(R) \cdot |F_p(Q,R)|^2 \, dR \tag{16}$$

This relation holds when the different contributions add incoherently and the particles have a known shape.

In our analysis, we have fitted the scattering curve by a procedure that uses a constrained least-square calculation [6].

Experimental Technique

Our experiment was performed on D11 equipment at ILL in Grenoble, connected to the cold neutron source of the High Flux Reactor (HFR).

A neutron wavelength of 1 nm was used to avoid double Bragg and multiple scattering.

A simple vision of the experimental setup is shown in Fig. 6, where M is the monochromator; G's are the collimation guides; S is the sample position, and D is the dimensional detector.

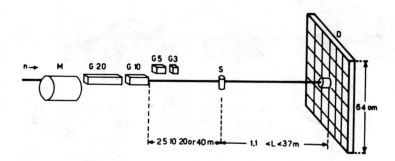

Figure 6 - Schematic view of the D11 Diffractometer

Different sample-detector distances were used to determine the investigated Q range by:

$$1.3 \cdot 10^{-2} < Q < 9 \cdot 10^{-1} \, nm^{-1} \tag{17}$$

which allows the investigation of particles whose dimensions, R, are included from 10 to 10^3 nm.

To determine the absolute value of the scattering curve, a calibration with a water sample was performed.

This calibration is necessary to bypass the measurements of important quantities like the neutron flux on the sample, the solid angle subtended by the detector and, at least, the detector efficiency [8].

Once we know the theoretical value of the incoherent scattering cross section of water, we can calculate the scattering curve by:

$$\frac{d\Sigma}{d\Omega}(Q) = \frac{d_w \cdot T_w}{d_s \cdot T_s} \cdot \frac{(I_s-b)-(T_s/T_h)\cdot(I_h-b)}{(I_w-b)-(T_w/T_q)\cdot(I_q-b)} \cdot (d\Sigma/d\Omega)w \qquad (18)$$

where d and T represent the thickness and the transmission factor, respectively; I and B the measured experimental intensities and the background, respectively; and w, s, h, and g are indices related to water, sample, sample holder and quartz cell, respectively.

Results and Discussion

The results concerning the evolution of the material microstructure relating to the thermal treatment have already been discussed in Ref. 2,3. In this work, only the microstructural evolution due to the operation is considered.

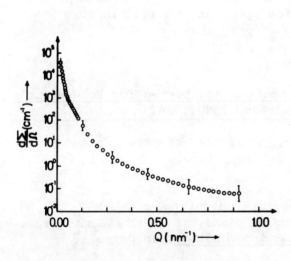

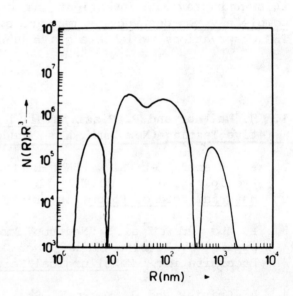

Figure 7 - Scattering pattern of the post-service turbine blade

Figure 8 - Logarithmic volume size distribution of the post-service turbine blade

Fig. 7 shows the scattering pattern of the post-service turbine blade. The asymptotic behavior of the scattering curve does not have a perfect Q^{-4} dependence because the slope of the approximating straight line is about -3.2. This could be associated with scattering contribution due to very small particles (1 to 3 nm) or to a high dislocation density [1].

However, by the extrapolation by the Guinier law to Q=0, and by the Porod law to high Q values, the volume fraction and the correlation length have been obtained. A volume fraction of about 30% was found by equation 12, which is about the same as the value found for the new turbine, which was 35%. Moreover, the packing of the system is defined by the correlation length for which a value of 85 nm was found by equation 15, compared to 172 and 129 nm for the untreated and as-heat-treated samples, respectively.

These values give an idea of the different interparticle distance in the three samples. This is strictly connected to those islands found inside the matrix, where gamma prime precipitates are bigger and more packed than in other zones.

The particle size distribution obtained by a fitting procedure of the experimental data shows an increase in the dimension of the primary gamma prime particles and the presence of a new family of small particles (Fig. 8), whose average dimension is around 4 nm.

These particles could be related to microvoids produced by high stress fields inside the blade during operation, but this is not verified by TEM observations. In this case, the two-phase model would not be completely appropriate, and only a comparison with an unstressed, but thermally treated sample would allow the determination of the stress contribution to the damage of the material.

In conclusion, the importance of the Neutron Scattering technique in the field of non-destructive testing of materials is evident. For analyzing very thick samples, other phenomena, such as multiple scattering, have to be taken into account and refraction effects have to be considered.

An approximated elaboration of the SANS data, as it was performed in this and previous investigations on UDIMET 720, seems to provide, in a completely non-destructive manner, structural information in agreement with TEM observations, which are quite useful for operating applications.

References

1. H. Walther and P. Pizzi, Small Angle Neutron Scattering for Nondestructive Testing (New York, NY: Academic Press, 1980), 341.

2. P. Bianchi et al., "Non-Destructive Analysis of the Gamma Prime Precipitation in UDIMET 720 Ni Superalloy Turbine Blade by SANS," Material Science Forum, in print.

3. P. Bianchi et al., "Neutron Small Angle Scattering on UDIMET 720 Ni Superalloy Turbine Blade. Non-Destructive Analysis of the Gamma Prime Precipitation," Metallurgical Science and Technology, in print.

4. A. Guinier and J. Fournet, Small Angle Scattering of X-Rays (New York, NY: John Wiley, 1955), 5.

5. G. Kostorz, Neutron Scattering (New York, NY: Academic Press, 1979), Volume 15, p.227.

6. M. Magnani, P. Pulit, and M. Stefanon, to be published.

7. G. D. Wignall and F. S. Bates, "Absolute Calibration of Small-Angle Neutron Scattering Data," J. Appl. Cryst., 10 (1987), 28-40.

INFLUENCE OF THERMAL FATIGUE ON HOT CORROSION

OF AN INTERMETALLIC NI-ALUMINIDE COATING

John W. Holmes[1] and Frank A. McClintock[2]

[1]Department of Materials Science and Engineering
[2]Department of Mechanical Engineering
Massachusetts Institute of Technology
Cambridge, Massachusetts 02139

Abstract

The influence of thermal fatigue strain history on the hot corrosion attack of a Ni-aluminide coating was examined. Coatings were applied by pack aluminization to stepped-disk fatigue specimens machined from a monocrystalline Ni-base superalloy (Rene N4). Induction heating of the stepped-disk specimens was used to simulate the severe thermal and strain transients experienced by gas turbine airfoils. Hot corrosion was studied by applying Na_2SO_4 to the specimen, and controlling the partial pressures of O_2, SO_2 and SO_3 in the test atmosphere.

Hot corrosion of the Ni-aluminide coatings was found to be strongly influenced by strain history. After 6000 fatigue cycles, between peak strains of -0.26% at 925°C and 0.03% at 650°C, extensive hot corrosion attack occurred, with Al and Ni sulfides found throughout the coating. By contrast, only minor surface oxidation of the coating was observed after 6000 cycles between lower peak strains of -0.16% at 925°C and 0.01% at 650°C.

The pronounced dependence of hot corrosion attack on strain history is attributed to cracking of protective surface oxide scales during thermal fatigue cycling, allowing direct interaction between molten Na_2SO_4 and the Ni-aluminide coating. This interaction results in an increase in the oxygen ion (O^{2-}) activity in the vicinity of the coating/oxide interface, preventing reformation of a protective oxide scale.

Superalloys 1988
Edited by S. Reichman, D.N. Duhl,
G. Maurer, S. Antolovich and C. Lund
The Metallurgical Society, 1988

Introduction

Hot corrosion can seriously limit the service life of the Ni- and Co-base substrates and coatings used for gas turbine hot-section components (1,2). Although a vast amount of information is available regarding the susceptibility of various alloy and coating systems to hot corrosion attack, much of this work has been performed isothermally (3,4) or with cyclic strain histories that are mild in comparison to those encountered by gas turbine components (5,6). Many of the alloys and coatings which show adequate resistance to hot corrosion under isothermal or low-strain cyclic conditions, may prove inadequate for use in high-strain cyclic applications.

To make life predictions for turbine components operated in marine environments, it is necessary to know the influence of strain history on the rate of hot corrosion attack. Moreover, to develop future corrosion resistant coatings, it is important to determine how strain history influences the mechanisms of hot corrosion observed in current coating systems; these mechanisms may differ considerably from those responsible for hot corrosion attack under isothermal conditions.

The present investigation examines the influence of thermal fatigue strain history on the hot corrosion of an intermetallic Ni-aluminide coating, a coating commonly used to protect gas turbine airfoils and shrouds from environmental attack.

Experimental Procedure

Specimen geometry, substrate and Ni-aluminide coating. Stepped-disk fatigue specimens (Fig. 1a) were machined from a single crystal rod of the nickel-base superalloy Rene N4. The disks were removed from the rod such that their faces were normal to the primary [001] growth direction of the crystal (Fig. 1b). The composition and heat treatment of the Rene N4 substrate are given in Holmes et al (7).

Ni-aluminide coatings were applied to the stepped-disk specimens by pack aluminization (8). The coating was approximately 70 μm thick along the specimen periphery. A secondary electron micrograph showing the coating microstructure is given in Fig. 2, along with results of a microprobe analysis for initial coating composition. Coating composition and structure were independent of crystallographic orientation of the substrate (9).

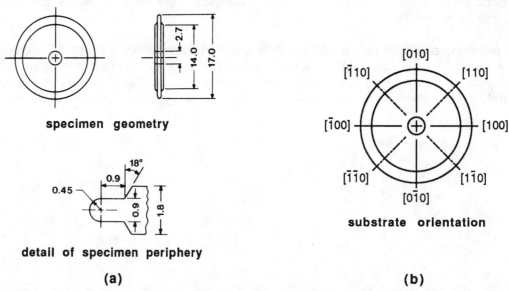

Figure 1a,b - **(a)** Stepped-disk specimen (10) used in thermal-fatigue/hot-corrosion experiments. Dimensions given in millimeters. **(b)** Machining specimens with their faces normal to the [001] growth direction of the single crystal substrate gives four <100> and four <110> orientations along the periphery of the disk. As discussed in the text, the elastic anisotropy of the substrate produces different strain histories around the specimen periphery; the circumferential strain range is a maximum at low modulus <100> orientations and a minimum at higher modulus <110> orientations.

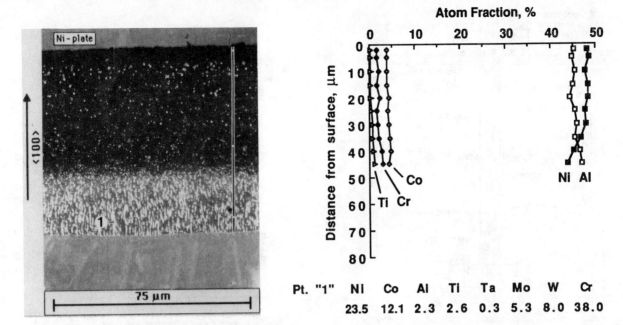

Figure 2 - Secondary electron micrograph of Ni-aluminide coating formed on a monocrystalline Rene N4 substrate. Electron microprobe analysis for coating composition was performed along the white trace line. The outer coating layer (black region of micrograph), which forms above the substrate surface, has a composition within the β phase field of the Ni-Cr-Al phase diagram (11). Outward diffusion of Ni during coating formation produces a Ni-depleted zone within the superalloy substrate; this depleted zone contains a high density of Cr-rich precipitates (Point "1" in micrograph) (9).

Test apparatus and control of strain history. High-frequency induction heating of the specimen periphery was used to produce cyclic thermal strains in the substrate and coating. This technique, developed earlier for studying the influence of strain history on oxidation of superalloys (7,10), allows close control of specimen temperature, environment *and* strain history. Detailed schematics of the test apparatus and induction coil are given in Figs. 3a,b.

The induction heating unit had an output rating of 2.5 kW and a frequency of 450 kHz. Containment of the corrosive test atmosphere (discussed below) was accomplished by inserting a quartz tube between the induction coil and test specimen (Fig. 3a). Specimens were positioned within the induction coil using a 6 mm diameter stainless steel support-rod attached to the exhaust manifold of the apparatus. This support-rod also served as a heat-sink to aid in maintaining a radial temperature gradient between the specimen periphery and core. Specimen temperature was monitored and controlled by use of two 0.20 mm diameter Pt/Pt-10%Rh thermocouples embedded in the specimen surface at a radius of 6.5 mm. To calibrate the control system for the particular temperature history used in the experiments (Fig. 4a), and to obtain detailed temperature profiles needed for finite element analysis of strain history, a specimen with 0.13 mm diameter chromel-alumel thermocouples positioned at specimen radii of 3.5, 4.0, 5.0, 6.0, 6.5 and 8.5 mm was used (7).

At a frequency of 450 kHz, the depth of induction heating in Ni-base superalloys is approximately 1 mm (12). With this shallow heating depth, the specimen periphery heats before the core, generating compressive circumferential strains in the periphery of the disk. The magnitude of the transient compression developed in the specimen periphery is controlled by varying the periphery heating rate (the low thermal mass of the stepped-disk specimens allows periphery heating rates in excess of 300°C/s). Transient tensile strains, developed during specimen cooldown, can be controlled by the flow-rate of gas passing through the quartz tube containing the specimen. Since the Ni-aluminide coating was applied over an elastically anisotropic substrate, the strain history of the coating is a function of angular position around the specimen periphery. The strain range developed in the coating (discussed below) varies from a maximum at low modulus <100> substrate orientations, to a minimum at the higher modulus <110> orientations (Fig. 1b). This variation of coating strain with substrate orientation is considered a key advantage of this specimen design, since it allows one to simultaneously determine the influence of different strain ranges on hot corrosion attack (10).

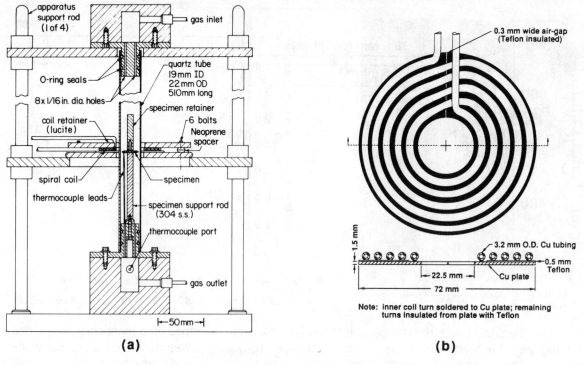

Figures 3a,b - **(a)** Schematic of test apparatus. A resistance heater (not shown), which encircled the upper portion of the quartz tube, was used to heat the gas-mixture used in the experiments. **(b)** Detailed sketch of spiral induction coil. With this coil design, induced currents concentrate along the inner radius of the thin Cu plate; the magnetic flux associated with this current induces eddy currents in the specimen periphery. Further details of the test apparatus are given in Holmes (10).

<u>Test atmosphere</u>. Hot corrosion was studied by coating the specimen periphery with 1.0 mg/cm^2 Na_2SO_4, and flowing an O_2-SO_2-SO_3-argon gas-mixture through the quartz tube containing the specimen. Prior to flowing past the specimen, the initial gas-mixture (20.00% O_2, 0.010% SO_2, 79.99% argon) passed through a 250 mm long furnace hot-zone, where gas equilibrium was established at 930°C. A platinum catalyst (6 grams of 80 mesh Pt screen) was used to enhance establishment of equilibrium. Assuming complete equilibrium, the volumetric composition of the gas exiting the hot-zone was: 20.00% O_2, 0.0089% SO_2, 0.0011% SO_3 and 79.99% argon. A gas flow rate of 100 cm^3/min was used in all experiments.

<u>Temperature and strain history of test specimens</u>. Two sets of experiments were performed in the hot corrosion environment:

(1) isothermal exposure for 100 and 200 hr at 930°C and,

(2) thermal fatigue cycling between 450 and 930°C, for 3000 and 6000 cycles (total test time at 930°C was 50 and 100 hr, respectively).

The temperature cycle used in the thermal fatigue tests (Fig. 4a) incorporated a rapid 9s heating transient from 450 to 930°C, followed by a 60s hold, and cooling to 450°C in 30s. As shown in Fig. 4b, the substrate (and coating)[1] strain range along low modulus <100> substrate orientations was approximately 60% higher than that for the stiffer <110> orientations (from the strain limits shown in Fig. 4b, $\Delta\varepsilon_{coat}^{mech}$ = 0.29% at <100> orientations *vs.* $\Delta\varepsilon_{coat}^{mech}$ = 0.17% at <110> orientations).

---

[1] Note that since the thin coating provides negligible constraint to the much stiffer substrate, the total in-plane strain history of the coating equals the total in-plane strain history of the substrate periphery. Furthermore, since the Ni-aluminide coating studied has a thermal expansion coefficient close to that of the Rene N4 substrate (7), the mechanical strain range developed in the coating is approximately equal to that of the substrate (where, $\varepsilon^{mech} \equiv \varepsilon^{total} - \varepsilon^{thermal}$ (13)).

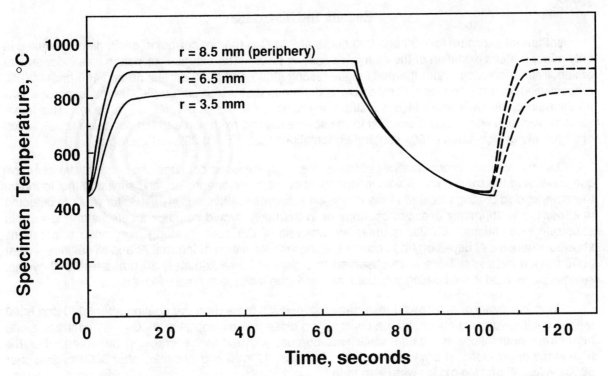

Figure 4a - Temperature history of stepped-disk specimen used in thermal-fatigue/hot-corrosion experiments. Data was obtained from thermocouples located at specimen radii of 3.5, 6.5 and 8.5 mm.

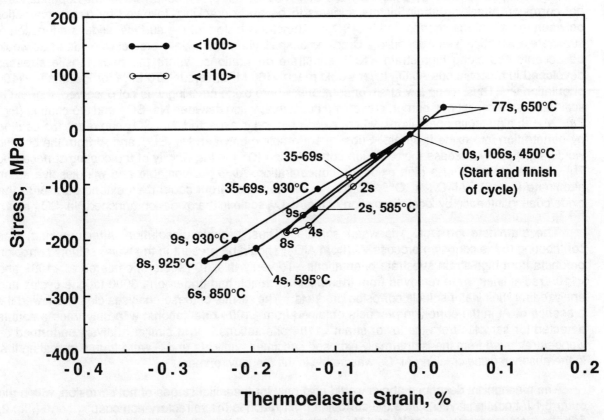

Figure 4b - Circumferential stress-strain history of <u>substrate</u> periphery at <100> and <110> orientations (obtained from a thermoelastic finite element analysis (7,10)). The thermoelastic strains shown represent the "mechanical" part of the total fatigue strain, where $\varepsilon^{mech} \equiv \varepsilon^{total} - \varepsilon^{thermal}$. As discussed in the text, the mechanical strain range of the substrate and coating are approximately equal. Recent finite element analysis of stepped-disk specimens has shown that substrate creep produces a shift in mean stress towards zero, but does not significantly alter the strain range from that predicted by a thermoelastic analysis (13).

Results and Discussion

Isothermal exposure for 100 and 200 hours at 930°C in the hot corrosion environment resulted in only minor surface oxidation of the aluminide coating (10). The extent of oxidation was independent of substrate orientation. After thermal fatigue testing (3000 and 6000 cycles, 50 and 100 hr at 930°C, respectively), coating oxidation, comparable in degree to that observed after isothermal exposure, was present along the low-strain <110> substrate orientations ($\Delta\varepsilon^{mech}$ = 0.17%) (Figs. 5a,c). This minor surface oxidation was in sharp contrast to the severe coating oxidation and hot corrosion attack which occurred along high-strain <100> substrate orientations ($\Delta\varepsilon^{mech}$ = 0.29%) (Figs. 5b,d).

Quantitative microprobe analysis of the isothermally exposed coatings showed that the chemical composition of the coating matrix was independent of substrate orientation, and similar to that found for thermally cycled coatings located along low-strain substrate orientations (10). Sulfur was not detected in either the isothermally exposed coatings or in thermally cycled coatings located along low-strain substrate orientations. Similar quantitative analysis of the coating along high-strain orientations showed extensive Al depletion had occurred, along with formation of internal Al and Ni sulfides. After 3000 fatigue cycles, sulfides were observed to a depth of approximately 30 μm; after 6000 cycles, sulfides penetrated to the coating/substrate interface (see sulfur X-ray map, Fig. 6).

Along low-strain orientations, the surface oxide which formed on the coating after 3000 and 6000 fatigue cycles was continuous, with a composition close to stoichiometric Al_2O_3. By contrast, along high-strain orientations the oxide scale was porous, with extensive cracking observed. For the high-strain orientations, analysis showed the presence of Cr, Ni and Al oxides after 3000 cycles; after 6000 cycles, W and Mo oxides were also found.

Discussion

Following earlier work by Elliott (14) and Steinmetz et al (15) a mechanism for the dependence of hot corrosion attack on strain history appears to be oxide cracking, followed by direct interaction between the aluminide coating and Na_2SO_4. The stress history of the surface oxide is a function of substrate orientation (see Appendix). Cracking or spallation of an initially protective oxide scale would be accentuated along high-strain <100> substrate orientations, where the peak tensile stresses developed in a continuous Al_2O_3 layer would reach ≈180 MPa, versus ≈70 MPa for low strain <110> orientations.[2] Thus, along low-strain orientations, where oxide cracking was not observed, the Al_2O_3 scale acts as a continuous barrier, preventing direct interaction between Na_2SO_4 and the coating (Fig. 7a). Along high-strain orientations, where oxide cracking is expected, Na_2SO_4 can reach the coating by penetration into oxide fissures. Subsequent reaction between Na_2SO_4 and Al from the coating would significantly increase the oxide ion concentration (O^{2-}) in the vicinity of the coating surface (Fig 7b). In the presence of a high oxide ion concentration, Al_2O_3 is unstable and will dissolve as an aluminate (14,16): $Al_2O_3 + O^{2-} = 2AlO_2$. The corrosion front could then extend parallel to the oxide/coating interface by local dissolution of the Al_2O_3 scale in the oxide-ion enriched Na_2SO_4 melt.

The aluminate ion (AlO_2) is water soluble. Thus, if oxide dissolution (after cracking) was contributing to the corrosion process, Al (from AlO_2) should be present in the water soluble corrosion products from high-strain substrate orientations. To verify this, 10° sectors, centered at <100> and <110> radial lines, were removed from the specimen which had undergone 3000 fatigue cycles and analyzed for their water soluble corrosion products. The results of these analyses clearly showed the presence of Al in the corrosion products obtained from <100> orientations, with only trace amounts detected for samples from the lower strain <110> orientations. In a similar analysis performed on wedges removed from the isothermally exposed specimens only Na and S were found, indicating that in the absence oxide cracking Al_2O_3 was stable in the test environment.

The mechanism described above would hold only for the initial stages of hot corrosion, within the outer β-NiAl coating layer. Once the corrosion front reaches the refractory-rich coating zone (Fig. 2, 5d), the high concentrations of W and Mo in this region would significantly alter the salt chemistry, as described in detail by Goebel et al (16) and Elliott (14). Moreover, rapid oxidation of the refractory-rich precipitates would quickly consume this coating layer.

---

[2]For a similar temperature and strain history (Fig. 4a,b), oxide cracking at high-strain substrate orientations was verified by testing in air. After 500 cycles, extensive oxide cracking was observed along the periphery at <100> orientations, with only a few random cracks observed at <110> orientations. Oxide spalling was not observed.

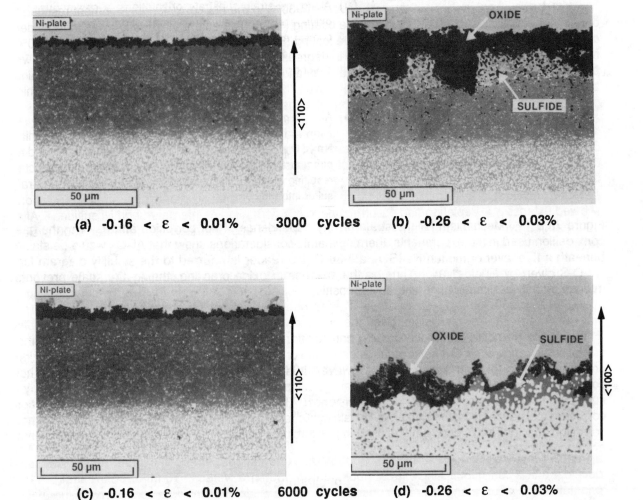

Figures 5a-d - Secondary electron micrographs showing typical Ni-aluminide coating degradation along <110> and <100> substrate orientations after 3000 and 6000 fatigue cycles. The micrographs were taken at the specimen periphery, on a cross-section parallel to the disk face and through the specimen mid-plane. The limits of substrate strain (Fig. 4b) are given beneath each micrograph. These fatigue strains act from left to right in the plane of the micrographs.

Figure 6 - X-ray image showing evidence of sulfur diffusion into coating along high strain <100> orientation after 6000 fatigue cycles (companion secondary electron micrograph is shown in Fig. 5d).

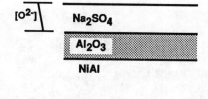

(a) Along low-strain substrate orientations, a continuous oxide scale prevents direct interaction between the coating and molten Na_2SO_4. Under these conditions, the oxide ion concentration is not sufficient to cause direct dissolution of Al_2O_3.

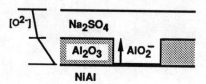

(b) Along high-strain substrate orientations, oxide cracking allows direct interaction between the coating and Na_2SO_4. This interaction increases the oxide ion concentration in the vicinity of the coating surface, resulting in local dissolution of Al_2O_3 and diffusion of sulfur into the coating.

Figure 7a,b - Mechanism for the strain history dependence of hot corrosion attack. For the gas composition used in the experiments, thermodynamic considerations show that Al_2O_3 would be stable beneath a thin layer of molten Na_2SO_4 at 930°C (the reader is referred to the stability diagram for Na-O-S given by Elliott (14)). Thus, in the absence of oxide cracking, the Al_2O_3 scale prevents interaction between the coating and environment.

It should be noted that since coating composition can influence hot corrosion attack (17), the critical strain range observed here for the acceleration of hot corrosion may be different for Ni-aluminide coatings applied to other substrates. However, it is expected that the strain dependence of hot corrosion observed with thin Ni-aluminide coatings will parallel that obtained for structural β-NiAl alloys subjected to similar temperature and strain histories. The results obtained indicate that evaluating hot corrosion resistance by isothermal or low-strain cyclic tests could greatly underestimate the rate of hot corrosion attack of alloys and coatings used in gas turbines, where severe thermal and strain transients are encountered.[3]

Conclusions

1. Hot corrosion of Ni-aluminide coatings depends critically on strain history. Only minor surface oxidation was observed after 6000 cycles between peak strains of -0.16% at 925°C and 0.01% at 650°C. However, after 6000 cycles, between peak strains of -0.26% at 925°C and 0.03% at 650°C, the coating was completely penetrated by sulfides. These results clearly show the importance of including strain history as a variable when determining the hot corrosion resistance of coatings and alloys.

2. The dependence of hot corrosion on thermal fatigue strain history can be attributed to breakdown of initially protective oxide scales <u>by cracking</u>, followed by direct interaction between Na_2SO_4 and the coating. This interaction results in internal sulfide formation and rapidly accelerated coating oxidation. This mechanism qualitatively explains the acceleration in hot corrosion attack observed along highly strained regions of gas-turbine airfoils.

3. Induction heating of stepped-disk fatigue specimens allows close control over specimen temperature, environment and strain history. Applying coatings to a single crystal substrate allows several coating strain histories to be studied simultaneously. However, the experimental technique is equally well suited for use with polycrystalline substrates.

---

[3]High-velocity gas burner rigs (Mach 0.3-0.8) are commonly used to determine the cyclic hot corrosion resistance of alloys and coatings used in gas turbine applications (17). However, due to the low heat transfer coefficient associated with *atmospheric pressure* burner rig testing, specimen heating rates may not be sufficient to reproduce the severe *strain* histories experienced by turbine airfoils (5,6), which operate at pressures as high as 10 to 20 atmospheres. In ASTM sponsored evaluations of various burner rig test facilities (18), significant differences in hot corrosion rates were found for rigs burning similar fuel. Results obtained from the present work suggest that, in addition to salt deposition kinetics, strain history may play an important role in explaining the variation in test results observed (note that strain history is a function of gas velocity *and* specimen geometry).

Acknowledgements

The authors would like to thank Bill Connor of General Electric Aircraft Engines for his critique of the manuscript. This work was supported by the National Science Foundation through Grant DMR 84-18718 to the Center for Materials Science and Engineering at M.I.T.

Appendix: Stress-Strain History of Surface Oxide

Although the surface oxide is in a state of biaxial stress, the peak stresses will be approximated here by a simpler 1-D analysis, which allows a more intuitive description of the important parameters controlling oxide stress. Modulus and thermal expansion data for polycrystalline Al_2O_3 were obtained from Samsonov (19) and Touloukian et al (20), respectively. Thermal expansion data for the Rene N4 substrate was obtained from Holmes et al (7).

Since the thin oxide offers negligible constraint to the substrate, the change in total oxide strain must equal the change in total substrate strain (note that the thin coating does not enter into the analysis, since it only acts as a vehicle to transfer substrate strain to the oxide). The increment in total oxide strain $\delta\epsilon_{ox}$ is equal to the sum of the elastic strain $\delta\epsilon_{ox}^e$ and the strain due to thermal expansion of the oxide $\alpha_{ox}\delta T$:

$$\delta\epsilon_{ox} = \delta\epsilon_{ox}^e + \alpha_{ox}\delta T. \tag{A1}$$

Equating $\delta\epsilon_{ox}$ from Eq. A1 to the increment in total substrate strain $\delta\epsilon_s$ which, ignoring inelastic deformation*, also has only elastic ($\delta\epsilon_s^e$) and thermal ($\alpha_s\delta T$) components, gives

$$\delta\epsilon_{ox} = \delta\epsilon_s^e + (\alpha_s - \alpha_{ox})\delta T. \tag{A2}$$

For a change in oxide strain given by Eq. A2 the change in oxide stress is

$$\delta\sigma_{ox} = E_{ox}\delta\epsilon_{ox}^e = E_{ox}[\delta\epsilon_s^e + (\alpha_s - \alpha_{ox})\delta T]. \tag{A3}$$

Due to the orientation dependence of elastic substrate strain $\delta\epsilon_s^e$ (Fig. 4b), the oxide stress $\delta\sigma_{ox}$ is a function of substrate orientation.

Application of Eq. A3. From Eq. A3, the stress increment in the oxide $\delta\sigma_{ox}$ depends upon the relative magnitudes of the elastic $\delta\epsilon_s^e$ and thermal $(\alpha_s - \alpha_{ox})\delta T$ strain components. Since the surface oxide forms primarily during the 60s hold at 930°C, the oxide will be stress free prior to specimen cooldown (ignoring growth strains) (10).** In the **early** stages of specimen cooldown from 930°C, where $\delta\epsilon_s^e$ is initially positive and of greater magnitude than the negative thermal strain $(\alpha_s - \alpha_{ox})\delta T$, **tension** is produced in the oxide (note that $\alpha_s > \alpha_{ox}$). At a later stage of cooldown the situation reverses, and the negative contribution from $(\alpha_s - \alpha_{ox})\delta T$ dominates, driving the oxide into compression. The substrate strain increment is largest along <100> orientations (Fig. 4b), whereas the thermal strain increment is independent of substrate orientation.*** Therefore, from Eq. A3, the oxide tension produced during cooldown will be highest at <100> orientations. Peak tension develops in the oxide at approximately 800°C (≈180 MPa for <100> orientations and ≈70 MPa for <110> orientations) and decreases in magnitude with further cooling, becoming compressive at approximately 620°C for <100> orientations and 700°C for <110> orientations.

During **heating**, $\delta\epsilon_s^e$ is negative (up to 925°C, Fig. 4b), resulting in a negative contribution to oxide stress, whereas the thermal strain increment produces a positive contribution to the oxide stress $((\alpha_s - \alpha_{ox})\delta T > 0)$ which is independent of orientation. The peak compressive stress developed in the oxide during heating would be approximately 400 MPa (720°C) at <100> orientations and 350 MPa (600°C) at <110> orientations. Since oxide growth stresses and inelastic oxide behavior are negligible, the oxide stress and strain must return to zero at the completion of a full cycle (930°C to 930°C).

It is important to note that had the oxide formed on an unstrained substrate, such as might occur with cyclic corrosion testing in an electric furnace, the oxide stress would remain compressive during the entire temperature cycle (this can be seen by setting $\delta\epsilon_s^e = 0$ in Eq. A3). Furthermore, the magnitude of the compressive oxide stress would be independent of substrate orientation.

---

*Note that, for this case, $\delta\epsilon_s^{mech} = \delta\epsilon_s^e$. Recent finite element analysis has shown that substrate creep leaves the strain limits largely unaffected (13); therefore, its omission does not alter the results of the present analysis.
**Increments are calculated from the end of the steady state hold at 930°C (Fig. 4b).
***For cubic materials, thermal expansion is isotropic.

References

1. R. E. Fryxell and G. E. Leese, "Effects of Surface Chemistry on Hot Corrosion Life," NASA CR-179471 (1986).

2. J. J. Grisik, R. G. Miner and D. J. Wortman, "Performance of Second Generation Airfoil Coatings in Marine Service," Thin Solid Films, 73 (1980) 397-406.

3. Y. Bourhis and C. St. John, "Na_2SO_4 and NaCl-Induced Hot Corrosion of Six Nickel-Base Superalloys," Oxidation of Metals, 9 (6) (1975) 507-526.

4. M. N. Richards and J. Stringer, "Some Aspects of the Hot Corrosion of Cobalt-Base Alloys," British Corrosion Journal, 8 (1973) 167-173.

5. A. Kaufman and R. H. Halford, "Engine Cyclic Durability Analysis and Materials Testing, NASA Technical Memorandum 83577 (1984).

6. E. D. Thulin, D. C. Howe and I. D. Singer, "Energy Efficient Engine: High Performance Turbine Detailed Design Report," NASA Report CR-165608 (1982).

7. J. W. Holmes, F. A. McClintock, K. S. O'Hara and M. E. Conners, "Thermal Fatigue Testing of Coated Monocrystalline Superalloys," Low Cycle Fatigue, ASTM STP 942, H. D. Solomon et al., eds. (Philadelphia, PA: American Society for Testing and Materials, 1988), 672-691.

8. G. W. Goward and D. H. Boone, "Mechanisms of Formation of Diffusion Aluminide Coatings on Nickel-Base Superalloys," Oxidation of Metals, 3 (5) (1971) 475-495.

9. J. W. Holmes and F. A. McClintock, "The Chemical and Mechanical Processes of Thermal Fatigue Degradation of an Aluminide Coating," accepted for publication in Metallurgical Transactions A.

10. J. W. Holmes, "Thermal Fatigue Oxidation and SO_2 Corrosion of an Aluminide Coated Superalloy," (Doctoral thesis, Department of Materials Science and Engineering, Massachusetts Institute of Technology, Cambridge, Massachusetts, September 1986). Available through: Micro-reproduction Laboratory, MIT, 77 Massachusetts Avenue, Cambridge, MA 02139.

11. A. Taylor and R. W. Floyd, "The Constitution of Nickel-Rich Alloys of the Nickel-Chromium-Aluminum System," Journal of the Institute of Metals, 81 (1952) 451-464.

12. J. Davies and P. Simpson, Induction Heating Handbook, (London: McGraw Hill, 1979) Chapter 12.

13. E. S. Busso and F. A. McClintock, "Stress-Strain Histories in Coatings on Single Crystal Specimens of a Turbine Blade Alloy," accepted for publication in International Journal of Solids and Structures.

14. J. F. Elliott, "Chemistry of Hot Corrosion," Solid State Chemistry of Energy Conversion and Storage, eds. John B. Goodenough and M. Stanley Whittingham (Columbus Ohio, American Ceramic Society, 1977), 225-239.

15. P. Steinmetz. at al, "Hot Corrosion of Aluminide Coatings on Nickel-Base Superalloys," High Temperature Protective Coatings, ed. S.C. Singhal (Warrendale, PA: The Metallurgical Society, 1982) 135-157.

16. J. A. Goebel, F. S. Pettit and G. W. Goward, "Mechanisms for the Hot Corrosion of Nickel Base Superalloys," Metallurgical Transactions, 4 (1973) 261-278.

17. R. E. Fryxell and G. E. Leese, "Role of Diffusion Zone Structure in the Hot Corrosion of Aluminide Coatings on Nickel Based Superalloys," Surface and Coatings Technology, 32 (1987) 97-110.

18. Hot Corrosion Task Force for ASTM Gas Turbine Panel: Round Robin Test, (Philadelphia, PA: American Society for Testing and Materials, 1970).

19. G. V. Samsonov, The Oxide Handbook, (New York: Plenum, 1973), pg. 183.

20. Y. S. Touloukian et al, Thermophysical Properties of Matter, Vol. 13, (New York: Plenum, 1979).

AN INVESTIGATION ON MAGNETRON SPUTTER DEPOSITED ALLOY-OXIDE COATING

Ye Ruizeng, Zhou Lang, Chang Shouhua, Gao Lian and Lu Fanxiu

Division of Superalloys,
Faculty of Materials Science and Engineering,
Beijing University of Science and Technology
Beijing 100083, P. R. China

Abstract

Dispersive rare earth oxide in MCrAlY alloy, compared with elemental rare earth additions, can be more effective in increasing the alloy's resistance to oxidation and hot corrosion. In order to make the alloy-oxide composite coating, experiments on sputter deposition were conducted using D.C. magnetron sputtering devices with the targets specially made for co-sputtering of both alloy and oxide. The deposited coatings were analyzed by TEM, XPS, XFS and EMPA methods. The TEM shows that there are very fine (5-20nm) dispersive particles in the annealed composite coating, which inhibit the grain coarsening in annealing. The XPS shows that there is measurable Y_2O_3 or SiO_2 in the composite coating, and the EMPA shows a homogeneous Y distribution along the cross section of the coating. The results show that a sputter deposited oxide dispersed alloy coating has been developed.

Introduction

MCrAlY-type coatings have been extensively developed and applied since their creation in the 1970's, with chemistries being continually adjusted and improved. But, up to now, the addition of the active element yttrium, which plays a key role in resistance to oxidation and corrosion, has been restricted by its limited solubility in MCrAlY. It is prone to forming needle- or plate-like intermetallic compounds; for example, a Co_3Y phase has been formed in CoCrAlY with Y additions of more than 0.1wt% (1,2). Therefore, it is difficult to obtain a homogeneously distributed high Y content in MCrAlY. Also, the plasticity of the coating is decreased if the amount and size of the yttrium compounds are increased to a certain extent. So the potential of further developing MCrAlY coatings in this respect is restricted.

From research on mechanisms of the role of Y in the oxidation resistance of alloys, a new way to develop the MCrAlY system can be expected. J. K. Tien (3) found that, in the oxidation of FeCrAlY, below the inward growing oxide scale, yttrium is oxidized, forming a dispersion of Y_2O_3 particles. This means that all of the beneficial effects of Y on decreasing susceptibility to scale spallation and reducing the oxidation rate are exerted in the form of Y_2O_3 particles. Tien proved one of the main mechanisms is "vacancy sinks": the boundaries between the oxide particles and matrix can absorb the Kirkendall vacancies produced by the external oxidation, thus reducing the scale spallation induced by voids formed by vacancy coalescence at the interface of the scale and the alloy. Ramanarayanan et al (4,5) showed that the microstructure of the oxide scale formed on FeCrAlY is the same as that on the ODS alloy MA-956 ($FeCrAlTi-Y_2O_3$) and the authors proposed that the effects of dispersive Y_2O_3 on oxidation behavior is similar to that of elemental Y, with the former being more effective because of its homogeneity. The experimental results support this argument (5).

The mechanisms of rare earth elements and their oxides on hot corrosion behavior are seldom studied, but from the fact that hot corrosion resistance is mainly based on the protectivity and continuity of the surface oxide scale, it can be expected that the dispersed rare earth oxide can enhance the hot corrosion resistance of coated alloys exactly as the elemental yttrium has done. Hot corrosion tests for ODS alloys might support this (6).

Furthermore, according to the particle strengthening effect, dispersive oxide particles in the coatings increase their erosion resistance, which has become important as industrial turbines develop. The successful practice of introducing oxide particles into sprayed wear-resistant and erosion-resistant coatings is evidence of this.

So, it can be expected to obtain better protective properties than those of the currently used MCrAlY coatings by substituting rare earth oxides for the more expensive elemental rare earth additives.

Although the spraying technique can be a feasible way to obtain the alloy-oxide composite coating, the serious porosity of the sprayed coating makes it less reliable for service in gas turbines. Ion sputtering can provide films of high density and good adhesion to substrates, and the development of the magnetron sputtering technique has made its deposition rate feasible for production of high temperature protective coatings. Because magnetron sputtering is a D.C. (direct current) sputtering technique, standard processing cannot be used to deposit non-conducting species like oxides; consequently, a modified method must be used to sputter deposit the composite coating. Efforts have been made on a method specially designed to obtain the composite coating with a magnetron sputtering technique. The present paper reports the work on sputtering coatings and their metallographic and chemical analyses.

Experimental Details

Sputtering Devices

Two magnetron sputtering devices of different types, both made by Beijing Instrument Manufacturing Corporation, were employed. One is a prototype, with a single circular permanent-magnetron target 65mm in diameter, the working power of which is usually less than 0.2Kw. The other one is a commercial type, with double-facing rectangular electro-magnetron targets, 250x120mm, the working power of each is usually less than 5Kw, limited by the cooling capacity of its water-cooling system. The latter type has an electric-resistance furnace under the sputtering chamber to heat the substrate before deposition.

Sputtering Targets and Deposition Process

The base targets on which the oxides were superimposed for alloy-oxide co-sputtering were made by vacuum melting and casting of alloys, namely Co-25Cr-10Ni-4Al-5Ta-0.5Y or Co-30Cr (wt%). The oxides used were Y_2O_3 or SiO_2. The use of SiO_2 was expected to increase the sulfidizing resistance of the coating.

The prototype sputtering device was used to deposit thin films suitable for direct TEM studies. The deposition parameters were as follows:

Primary Vacuum:	6.7×10^{-3} Pa
Working Ar Pressure:	1.1Pa
Working Power:	0.11Kw
Substrate Bias:	Floating

Further experiments were conducted on the commercial-type sputtering device to obtain the coatings of practical, feasible thickness on superalloy substrates. After being heated to 900°C under the primary vacuum in the furnace inside the device, the substrate was raised into the sputtering chamber for sputter deposition. The deposition parameters were as follows:

Primary Vacuum:	5.3×10^{-3} Pa
Working Ar Pressure:	1.1Pa
Working Power:	4.2-4.5x2Kw
Substrate Bias:	0-400v
Substrate Temperature:	560-600°C at the onset of sputtering, cooled down to 440-460°C in about 4 min, and then maintaining that temperature during sputter deposition
Deposition Rate:	0.4-0.5µm/min

Specimen and Analysis

With the prototype device, films of about 50nm thickness were deposited on carbon film pre-adhered to copper grids, while the specimens for x-ray photoelectron spectrum (XPS) analysis were prepared with glass substrates placed beside the grids. The thin foils of the coatings sputtered with the commercial-type device (45-50µm thick) were prepared as follows: sectioned with a layer of the substrate parallel to the coating's plane; electrolytically thinned the sectioned pieces only on their substrate sides; ion beam thinning done on both sides. To analyze the Y contents in the coatings, the X-ray Fluorescent Spectrum (XFS) analysis was carried out, the results being corrected by chemical analysis of the scraps of the sputtered species spalled from the inside walls of the sputtering chamber (0 bias voltage), and an Electron Probe Microanalysis (EPMA) was used to show the distribution of Y along the cross section of the coating.

Results and Discussion

The Existence of the Oxides in the Coating

Both the oxide-bearing or non-oxide bearing films show a microcrystalline structure, as indicated by the TEM image, Fig. 1. As the grains of the films are very fine (<20nm), it is difficult to identify the dispersive oxides possibly distributed in the films. To determine the existence of the oxides, XPS analysis was conducted.

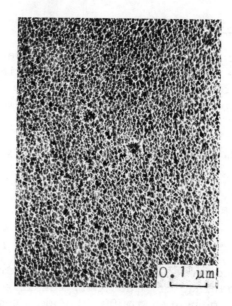

Figure 1 - The TEM image of the sputtered CoCrNiAlTaY-(Y_2O_3) film

Briefly, the XPS analysis directs x-rays onto the specimen so that the inner layer electrons are stimulated; the resulting photoelectronic emission is detected and characterized with respect to kinetic energy and intensity. The kinetic energy of the photoelectrons, E_k, is related to the bonding energy of the electrons, E_b, simply as:

$$E_b = h\nu - E_k + \phi_s \qquad (1)$$

where $h\nu$ is the energy of the directing x-ray and ϕ_s is an instrument parameter. Since the bonding energy of the inner layer electrons is characteristic for every element, the chemistry information of the specimen can be obtained. Furthermore, the XPS has an important feature, the chemical displacement effect (7). In molecules, the density of valence electrons around the atoms is different from that of elemental atoms, and this induces changes in the binding energy of the inner layer electrons; thus, the characteristic energy of the elements in the molecular state will displace correspondingly. By this effect, the chemical state of the elements can be detected.

Fig. 2 is the overall spectrum of the CoCr-(Y_2O_3) film, which shows the existence of Y. Fig. 3 shows the magnified feature of the 3d electron peak of Y and the result of mathematical peak identification; Fig. 4 shows the 2s electron peak of Si in the CoCr-(SiO_2) film. Comparing them with standard spectra, it is determined that the Y and Si in the obtained films exist as Y_2O_3 and SiO_2, respectively. For the thick $CoCr-(Y_2O_3)$ coating obtained with the commercial-type sputtering device, the XPS shows the same results.

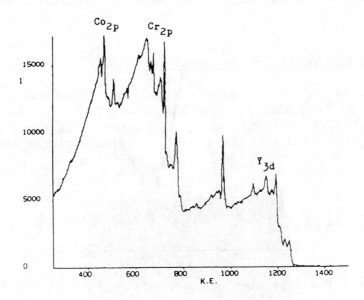

Figure 2 - The overall spectrum of the $CoCr-(Y_2O_3)$ film

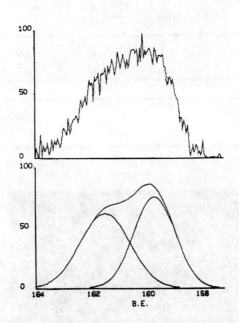

Figure 3 - The magnified Y3d electron peak and its identification

As the XPS cannot give a true quantitative analysis of contents, the XFS analyses were carried out. Table I gives the results obtained, while the characteristic x-ray image obtained by an electron probe microanalysis (Fig. 5) shows that the distribution of yttrium is homogeneous on the cross section of the coating.

The Form of the Oxides

Fig. 6 is a TEM image of the $CoCr-(Y_2O_3)$ coating. The very inhomogeneous contrast inside the grain also has appeared in the coating in which no Y_2O_3 has been introduced. It is, in fact, a reflection of the high density of crystal defects (mainly vacancies) in the high-rate sputter deposited coating. Such a configuration makes the identification of the

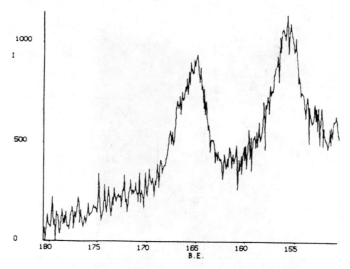

Figure 4 - The Si2s electron peak of the CoCr-(SiO_2) film

Table I. The Analyzed Yttrium Content in the Coatings

Y_2O_3 (% Area) on target	Bias Voltage (V)	Y (w/o) (Chemical Analysis)	I/Im (XFS)	Corrected composition of Y (w/o)
4.7	0	2.07	100	2.07
	-200		83.6	1.73
	-400		30.5	0.78
2.3	0	1.03	40.2	1.03
	-200		33.8	0.86

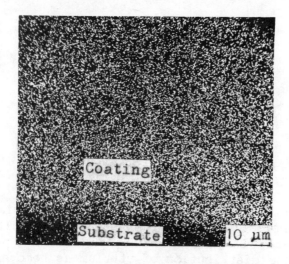

Fig. 5 - The yttrium x-ray image on cross section of the CoCr-(Y_2O_3) coating

oxide particles very difficult. After annealing at 900°C for 13 hours, the grains of the coating without Y_2O_3 grew one or more orders of magnitude and the inhomogeneous contrast disappeared. The grain size and defect density inside the grains of the CoCr-(Y_2O_3) coating changed little after annealing 7 hours longer, as indicated in Fig. 7a, b.

Figure 6 - A TEM image of the as-sputtered CoCr-(Y_2O_3) coating

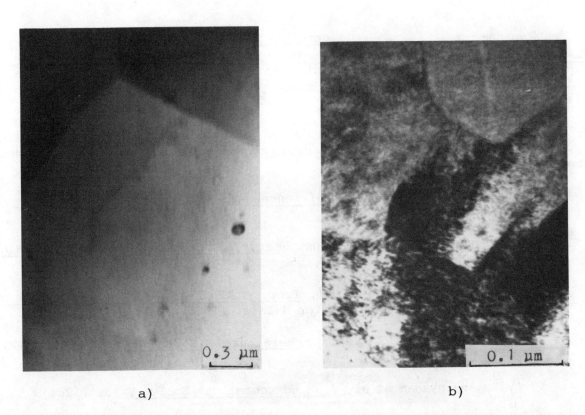

a)   b)

Figure 7 - The TEM image of the annealed coating
     a - CoCr   900°C, 13 hrs annealed
     b - CoCr-(Y_2O_3) 900°C, 20 hrs annealed

After annealing the CoCr-(Y_2O_3) coating at 1200°C for 4 hours, the contrast from the crystal defects in the TEM image was eliminated and there emerged large amounts of particles 5-20nm in size, as indicated in Figure 8. To identify the structure and chemistry of the super fine particles directly is difficult at present, but combined with the XPS analysis and compared with the TEM analysis of the oxide-free CoCr coating, it can be deduced logically that the particles are dispersive yttrium oxide.

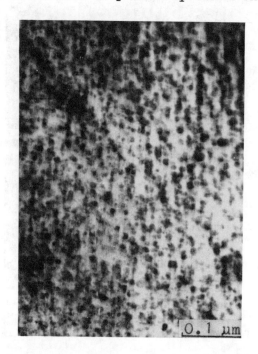

Figure 8 - TEM image of the CoCr-Y_2O_3 coating annealed at 1200°C for 4 hours

Conclusion

1) Substituting the dispersive rare earth oxides for the expensive elemental rare earth additives in coatings is a new way to improve the high temperature protective coating;

2) The co-sputter deposition of alloy-dispersive oxide composite coating by D.C. magnetron sputtering can be realized with the present method.

References

1. D. R. Holmes and A. Rahmel, eds., Materials and Coatings to Resist High Temperature Corrosion, (1978), 55.

2. Fon Jianya, Ye Ruizeng et al, Proc. of the 6th Chinese Conference on Superalloys, Chinese Society for Metals, (1986), Vol. 2.

3. J. K. Tien and F. S. Pettit, Metallurgical Transactions, 3 (1972), 1987.

4. T. A. Ramanarayanan et al, J. Electrochemical Society, 131 (1984).

5. T. A. Ramanarayanan et al, Oxidation of Metals, 22 (1984), 83.

6. As (1) 71.

7. D. Briggs et al, eds., Handbook of X-Ray and Ultraviolet Photoelectron Spectroscopy (Heyden and Son Ltd., 1977)

Subject Index

Age-hardened alloy, 152
Ageing/aging, 667, 670
Aging, 585, 587
Alpid, 516
Aluminum, 152
  effects, 43, 713, 719
Antiphase boundary energy, 566
Atomization, 64
Atomized powders
  inclusions, 143, 144
  production, 142-144
  quality assurance, 145
Auger, 640

Bithermal thermomechanical fatigue, 575-583
Bonding
  brazing, 751
  diffusion, 589, 776
Boron effects, 696

Carbides, 628, 629, 706
  effects, 585, 593
  morphology, 69
Carbon effects, 673, 676, 679, 695, 696, 713, 719, 720
Castability, 308
Casting, 242, 326, 346, 361, 713, 714
  processes, 51
Center segregation, 93
Ceramic filters
  effect on cleanness, 145
Chemistry effects
  Ta, 725-728
  Ti, 725-728
  W, 725-728
Chromium effects, 719, 720
Cleanness
  EB button test, 379, 402, 403
  raft analysis, 381
  effects on properties, 405
Closure, 566
Coarsening, 663, 666, 668, 669
Coating, 136, 747, 759, 816, 856, 866
  corrosion, 142, 747

creep, 765, 806
  (NiCoCrAlY), 575-583
Cobalt effects, 719, 720
Coherent particles, 664
Composites, 174, 186
Composition
  gradient, 452
  profiles, 196, 199
Compression testing, 486, 520
Constitutive laws
  continuum damage mechanics, 686-687
  empirical, 686
  state variables, 686-687
Cooling rate, 173-182, 526
  effect on
    quench crack, 530
    solidification, 443
    yield strength, 528
Crack
  growth, 107, 605, 608, 613, 619, 622
    rates, 16, 157, 618, 621, 623
  initiation, 673-676, 681
  mechanisms, 580-583
  orientation, 605, 607, 608, 611, 613
  propagation, 673-675, 677, 681
Creep, 45, 122, 627, 644, 654, 674, 675, 695, 696, 698, 724, 733, 734, 739-741
  behavior, 45
  data, 684-692
  deformation, 684-686
  ductility, 40
  fatigue,
    interaction, 600
    slow-fast cycling, 599, 600
  fracture, 685
  prediction, 689-691
  primary, 684
  properties, 203, 207, 209
  rupture, 15, 37, 45, 66, 176, 216, 229, 237, 247, 269, 286, 349, 357, 650, 695, 697, 698
  tertiary, 685
Critical resolved shear stress, 281, 316
Cross slip, 316
Crystal structure, 730

Crystallographic deformation, 619, 620, 622
Cyclic hardening, 585

Damage
  accumulation, 566
  tolerance, 4, 13
    creep crack growth rates at 650°C, 9
    fatigue crack growth rates at 650°C, 9
    heat treatment, 6
    Inconel 718, 4, 5
    low cycle fatigue life at 650°C, 9
    microstructural design, 5
    turbine discs, 4
Darcy law, 715
Database
  analysis, 687
  model parameters, 689
Deformation, 566
  behavior, 505, 698, 699
  maps, 505
    microstructure, 510
  mode, 622
  orientation, 698, 699
  stability, 510
Delta
  phase, 43, 54, 57, 62
  precipitation, 35
Dendrite arm spacing, 268, 444
Dendritic solidification, 438
Differential thermal analysis, 695, 717
Diffusion barriers, 198
Directional
  recrystallization, 85, 132
  solidification, 477, 711, 716
    alloys, 816
    Hf
      effects on brazing, 478
      effects on welding, 478
      rich skins, 477
Dislocations, 623, 624, 632
  structure, 220, 277, 349
Dispersion strengthening, 596, 597, 603
  columnar-grained superalloys, 336, 346
DS eutectics, 356
DTA, 724

EDM, 135
Effect of
  Hf and Zr, 336
  HIP on mechanical properties, 467
Engine test evaluation, 243, 363
Elastic
  energy, 663, 664, 665
  interaction, 664, 670, 671
Electro-slag remelt (ESR), 92
Electron beam cold hearth refining, 536
  composition effects, 400
  equipment design, 398, 399
  superalloys, 398
Elevated temperature properties, 203
Environmental effects, 580-583, 656-661
EPMA, 733, 734, 741
Equilibrium solidification, 438
Eta phase, 174-182, 612
Eutectic formation, 703, 706, 707, 708
Extrusion, 536
  ductility, 146, 147
  powders, 146, 147
  temperature effects, 146

Failure mechanisms, 655, 656, 662
Fatigue, 15, 43, 585, 588, 589, 606, 608, 648, 649, 821, 826, 856
  asymmetric, 597, 600
  crack
    initiation, 599, 600
    growth, 66, 593, 605, 607
    propagation, 566
  low cycle, 596, 597, 599
  properties, 107
  Stage I cracks, 597
Field ion microscope-atom probe, 306
Filtering, 377
  efficiency, 382
Fine grain ingot, 535
Flow
  behavior
    Rene 95, 506
  stress, 103
Forgeability, 489
  EB ingots, 401
  trace element effects, 400
Forging, 12, 34, 44, 122, 133, 536
Fractography, 298, 591, 592, 609, 611, 618, 628, 639
Fracture, 14
  surface, 569
  toughness, 300
Freckles, 332

Gamma
  double prime, 35, 43, 54, 57, 62
  grain size control, 827
  matrix, 663, 671
  prime, 43, 54, 57, 62, 75, 102, 162-170,
    217, 233, 237, 249, 290, 300, 306,
    316, 351, 566, 663, 664, 666, 667,
    668, 837, 848
    coarsening, 43
    formers, 65
    size, 69
      distribution, 846-848, 854
      effects, 43
      morphology, 845, 852-854
Gatorize, 102, 541
Grain
  aspect ratio, 79
  boundaries, 629, 631, 638, 640, 641,
    694
    cracking, 345
  size, 15, 102
    banding, 99
    control, 69, 122

Hafnium, 379
  additions to DS castings, 476
  effects, 696
Heat treatment, 34, 43, 64, 104, 122,
  173-182, 237, 247, 337, 349
High cycle fatigue, 242, 270, 359
HIP, 64, 266, 460, 746, 758
  castings, 307
Homogenization, 44, 54, 57
Hot
  corrosion, 76, 674, 711, 865, 866
    resistance, 734, 739
  die forging, 487
  isostatic pressing (HIP), 102, 694
Hydrogen effects, 295

Inclusions, 67
Inelastic strain, 585
Interdiffusion
  coefficients, 194-197
  kinetics, 193, 197, 198, 199
Interfacial energy, 630
Intergranular-sulfide, 796
Intermetallic compound, 174, 184, 203
Internal energy, 154, 155, 156
Investment casting, 714
Fe-Ni-Co matrix, 152

Isocon, 546
Isothermal forging, 546

Laser-Miller plot, 691
Laser drilling, 554
Lattice
  misfit, 663, 664, 734, 739-741
  mismatch, 567
  parameter change, 44
Laves, 54, 57, 62
LCF, 575-583
Life predictions, 856
Long term stability, 38
Low
  coefficient of expansion alloy, 173-182
  cycle fatigue (LCF), 70, 92, 122, 262,
    270, 338, 566, 585, 589
  expansion superalloy
    Incoloy alloy 903, 152
    Incoloy alloy 907, 152
    Incoloy alloy 909, 152
  premier plasma spraying, 496

MCrAlY coatings, 554
Macrostructure, 83
  EB ingots, 401
  forged, 401
Magnesium effects, 621, 627, 632,
  635-642
Magnetron sputtering, 865, 866, 867,
  872
Mechanical
  alloying, 75
  properties, 43, 54, 57, 61, 76, 122, 157
    creep, 635, 636, 641
    fatigue, 635-638, 640, 641
    stress rupture, 635-638, 640, 641
    tensile, 637, 638, 640, 641
Mechanisms
  cavitation, 685
  creep, 684-686
  strain softening, 685
Melt rate, 93
Metal
  matrix composites, 193-201
  powders, 497
Microporosity, 418, 713, 714, 717
Microstructural
  development, 663
  effects, 585, 593, 623

stability, 226, 239, 256
Microstructure, 43, 83, 102, 122, 568, 637, 713
Minor element effects, 693-702
Modeling, 715
Molybdenum effects, 719, 720
Multiple correlation coefficient, 735, 740, 741

NDT, 846
Ni3Al alloying effects, 316
Nickel aluminide, 209, 210
Nimonic alloy
  AP-1
    cleanness, 144, 145
    melt filtering, 145
  PK50
    composition, 400
    electron beam melting, 399, 400
Niobium, 152
  effects, 43
Nitrogen effects, 419
Notched stress rupture, 673-681

Ordered alloys, 43
Oxidation, 75, 746
  /hot corrosion resistance, 176, 240, 359
  resistance, 865, 869
  testing, 136
Oxide
  dispersion strengthening (ODS), 74, 82, 132
  inclusion, 378

Paris region, 617, 623
Partitioning ratios, 735, 736
Peening, 496
PHACOMP, 736
Phase
  analysis, 44, 57
    compositions of phases, 153
    experimental procedures, 152
    fully processed material, 157
    x-ray data of phases, 155
  calculation
    gamma/gamma prime, 733, 734, 735, 737, 738
  composition, 707
  diagram, 725-728, 731
  equilibria, 723-731
  formation, 705

stability, 43
Phases
  carbo-nitrides, 153
  carbo-sulfide, 153
  delta, 153, 155, 160
  epsilon, 154, 155, 160
    double prime, 153, 155
  gamma prime, 153, 159
    double prime, 153, 154, 255
  laves, 154, 155, 157, 159
  silicides, 153
Planar slip, 605
Polycrystal, 606, 613
Polycrystalline alloys, 693, 696, 701
Porosity, 267, 277, 498
Potential drop, 673-681
Powder
  cleanliness, 66
  P/M superalloys, 64, 516
Precipitation, 43
  hardening, 623
  strengthening, 663, 664
Properties, 15
  hot corrosion, 76
  notch stress rupture, 176
  oxidation, 75
  stress rupture, 76
  tensile, 79, 175

Quantitative analysis, 836

Reaction zone, 197, 198
Rafting, 810
Rare earths, 866
Recrystallization, 78
  defects, 603
  directional, 75
  grain structure, 596, 598
Repair, 746, 756
Residual
  niobium, 155, 157, 159
  stress, 527
Rhenium, 379
  containing alloys, 216, 230, 237, 247, 256, 306
Ribbon, 766

Schmid's law, 610
Segregation, 57, 703, 712
Service exposure, 756, 806, 816, 836
Sessile drop test, 384
Sigma phase, 709

Silicon, 152, 153, 154, 155, 160
Single crystals, 216, 226, 236, 246, 266, 276, 286, 296, 306, 326, 575-583, 606, 610, 615, 653-655, 657-662, 693, 697, 856
 orientation, 216, 247, 297, 316
Slip
 bands, 597, 603
 character, 596, 597
 deformation, 618, 622, 624, 698, 699
 dispersal, 598
 mode, 566
 systems, 610, 612
Small angle neutron scattering, 845, 846, 849-854
Solid solution strengthening, 585
Solidification, 54, 62, 440, 703, 705, 713, 714, 719
 rate, 450
Spray forming, 486, 501
Strain
 energy, 610
 rate, 517
Stress
 accelerated grain boundary oxygen embrittlement, 152, 155, 160
 intensity, 618, 621
 relief, 105
 rupture, 133, 626, 653-655, 657-662, 711
Structural
 castings, 306
 characterization, 54, 57
Structure, 451
Superplastic forging, 491
Superplasticity, 67
 APK6, 147

Ta effects, 54, 57, 62
Tag heat treatment, 57
TEM, 725, 727, 729, 868, 870-872
 /SEM 845-848
Tensile properties, 104, 203, 204, 587
Tension compression asymmetry, 316
Tests
 creep properties, 404
 low cycle fatigue, 405
 procedures, 156
 tensile properties, 404
Thermal
 expansion, 152
  coefficient, 174
 fatigue, 674
 stability, 69

Thermomechanical processing, 156
Time-temperature-transformation, 35
 diagram 153
Titanium, 152
 effects, 43, 719, 720
Torsional creep, 644, 648, 652
Transmission electron microscopy, 44
Tubes, 643, 645
Tungsten fiber reinforcing, 186, 194
Turbine
 blades
  forging, 147
  production, 147
  requirements, 142
 discs, 4
  damage tolerance, 4
  FCGR at 650°C, 9
  Inconel 718, 4, 5
  LCF life at 600°C, 9
  microstructural design, 5

Ultrasonic inspection, 547

Vacancy activation energy, 626, 631, 632
Vacuum
 arc remelt (VAR), 92
 induction melting, 419
Vader, 449, 536
VAR, 449
VIM, 449
 melting, 327

Water elutriation, 67
Welding, 749
White spots, 92

X-ray
 fluorescent spectrum, 867-869
 photoelectron spectrum

Yield strengths, 66, 587, 700
 effect of grain size/cooling rate, 532
 modeling, 281
 tensile properties, 205, 240, 247, 269, 275, 298, 337, 352, 359
Yttrium, 379, 867
 effects, 865, 866, 872

Zirconia effects, 673, 676, 679
Zirconium effects, 696

Alloy Index

Alloy 89, 198
Alloy 454, 216, 734
Alloy 617, 643-652
  Inconel 718, 635-642
Alloy 800, 26, 256
AF 115, 485
AF2-1DA, 485
AF2-1DA-6, 122, 127
APK-6, 148
  composition, 146
  creep properties, 145, 148, 149
  fatigue properties, 149
  structure, 143
  tensile properties, 46, 147
Astroloy, 65, 122, 123, 485

B1900, 370

CMSX-2, 216, 226, 246, 275, 291, 306, 724
CMSX-2+RE, 307
CMSX-3, 132, 133, 286, 328, 653, 654, 657-661
CMSX-4, 216, 226, 246

DS MAR-M 246+Hf, 266
DS MAR-M 247, 133
DS MER-M 200+Hf, 336, 345
DZ-3, 345
DZ 38G, 711
DZ 22, 476

FeCrAlY, 196, 198

GH 33, 625
GH 169, 710
GH 220, 625

h Mo, 437
Hastelloy X, 437, 554, 585-594
Haynes alloy 188, 585-594
Haynes alloy 230, 585-594

IC-50, 204
IC-218, 204
IC-221, 204
IN-100, 65, 352, 417, 835

IN-600, 775
IN-625, 437
IN-713, 77, 775
IN-738, 75, 79, 370, 597, 755, 795
IN-738 FC, 76, 765
IN-738 LC, 76, 688-691, 765, 805
IN-792
  composition, 142
  microstructure, 142
IN 901, 525, 605-614
IN 939, 75
IN MA 6000, 74, 75
IN MA 754, 75
IN MA 760, 75
Incipient melting, 349
Incoloy 903, 196, 197, 198
Incoloy 907, 152, 196, 198
Incoloy 909, 152
Incoloy MA 956, 500
Inconel 600, 370
Inconel 713, 370
Inconel 718, 3, 4, 5, 34, 44, 54, 61, 71, 92, 306, 437, 460, 635-642
  CCGR at 650°C, 9
  cast alloy, 54
  casting, 460
  composition, 400
  damage tolerance, 4
  design, 227, 236
  developments, 43
  effects, 43, 693, 694
    aluminum, 43
    niobium, 43
    titanium, 43
  electron beam melting, 399, 400
  FCGR at 650°C, 9
  grain size effect, 9
  heat treatment, 6
  LCF life at 650°C, 9
  microstructural design, 5
  microstructure, 205, 237, 297, 337, 348, 356
  properties, 54
  recycling, 367, 373
  scrap, 367, 371
  serrated grain boundaries, 6

stability, 43
TA 718, 54
turbine discs, 4
IN 792
  composition, 142
  microstructure, 142

K-3, 345
K3H, 476
K5, 476
K5H, 476
K19H, 476

M17F, 711
M36, 711
M38G, 710, 711
M40, 711
MA 754, 595, 596, 597-600
MA 6000, 82, 132, 595, 596, 597-602
MAR-M002, 352, 688-691
MAR-M200, 336, 693-702
MAR-M200 + Zr, 336
MAR-M246, 653, 655, 657-662
MAR-M247, 775
MERL 76, 65, 102, 485, 535
MM 007, 775
MM 200, 476
MMT-143, 741
MXON, 216, 226, 741

NASAIR-100, 226, 328
N18, 65
Ni-270, 795
Ni-20Al-5Cr, 765
NiAl, 663-672
NiCoCrAlY (PWA 276), 575-579, 581, 582
NiCr-Al, 663, 672
Ni-Cu-Si, 663-672
Ni-Si-Al, 663, 672
Ni$_3$Al, 316
Ni$_3$Al+Hf, 316
Ni$_3$Al+Ta, 316
Ni$_3$Al+B, 316
Ni$_3$Al+Zr, 316
NiTaC-13, 356

NiTaC-14B, 356
Nimonic 75, 597
Nimonic 80A, 437, 625
Nimonic 86, 498
Nimonic 263, 498
Nimonic AP-1, 144
Niobium, 193, 194

PWA 1422, 345
PWA 1480, 226, 236, 266, 275, 295, 306, 575-579, 581, 582, 615-624, 693-702, 724
PWA 1480+Re, 307
PWA 1484, 237
Pyromet 901, 24
Pyromet CTX-3, 162, 173-182

Rene, 114
Rene 77, 370
Rene 80, 358, 765
Rene 80H, 815
Rene 95, 65, 485, 505
Rene 150, 357
Rene N-4, 855

SC-53A, 226
SC-83, 741
SRR 99, 605-614, 688-691

TM-321, 741
TMD-5, 741
TMO-2, 741
TMP-3, 741
TMS-1, 226, 741
TMS-12, 226, 328, 740
TMS-26, 328
TUT 92, 231

U-700b, 377
Udimet 720, 13, 14, 449, 845-848

Waspaloy, 196, 198, 370, 546, 618

X 750, 370, 437
X 751, 370

Author Index

Antolovich, S.D., 565
Antony, K.C., 745
Argon, A.S., 285
Arzt, E., 595
Ashby, M.F., 683
Au, P., 3

Bain, K.R., 13
Ballou, O.W., 469
Barbosa, A., 683
Beckman, J.P., 795
Benn, R.C., 73
Bellinger, N., 3
Bernstein, I.M., 275, 295
Bhowal, P.R., 525
Bianchi, P., 845
Biederman, R.R., 505
Blavette, D., 305
Borofka, J.C., 111
Bowman, R., 565
Bridges, P.J., 33
Brooks, J.W., 33
Bruch, C.A., 355

Caless, R.H., 101
Cameron, D.W., 605
Caron, P., 215, 305, 335
Carsughi, F., 845
Castillo, R., 755, 805
Cetel, A.D., 235
Chang, K.-M., 485
Changxu, S., 703
Chao, C.H., 785
Chen, G., 635
Chenggong, L., 475
Chou, C.P., 785
Cockburn, C., 141
Coffey, M.W., 469
Cole, G., 459
Collier, J.P., 43

D'Angelo, D., 845
Darolia, R., 255
DeAntonio, D.A., 161
Delgado, H.E., 515
Doi, M., 663

Dollar, M., 275, 295
Doner, M., 653
Ducrocq, C., 63
Duhl, D.N., 235, 693
Durand-Charre, M., 723
Dyson, B.F., 683

Elder, J.E., 815
Elliott, I.C., 141, 397
Elzey, D.M., 595
Evans, M.D., 91
Ewing, B.A., 131
Ezaki, H., 225

Fanxiu, L., 865
Fiedler, H.C., 485
Field, R.D., 255
Forget, P., 553
Foster, S.M., 245
Frank, R.B., 23
Fritzemeier, L.G., 265
Funamoto, T., 765

Gabb, T.P., 575
Gambone, M.L., 13
Gayda, J., 575
Genereux, P.D., 535
Ghosn, L.J., 615
Gigliotti, M.F., 355
Goward, G.W., 745

Harada, H., 733
Haubert, R.C., 355
Heck, K.A., 151
Heckler, J.A., 653
Heredia, F.E., 315
Heubner, U., 437
Hilborn, M., 407
Hoeppner, D.W., 605
Holderby, M.A., 151
Holmes, J.W., 855
Honnorat, Y., 63
Howson, T.E., 515
Huang, S., 345
Huff, H., 835
Hyzak, J.M., 13, 121

Immarigeon, J-P.A., 3, 755, 805
Inoue, S., 225

Jain, S.K., 131
Jeandin, M., 553
Jing, B., 703
Jizhou, X., 335

Kanagawa, A., 407
Kang, C., 775
Keefe, P.W., 449
Kelly, T., 53, 459
Khan, T., 215, 305, 335
Kissinger, R.D., 111
Klarstrom, D.L., 585
Kobayashi, M., 765
Kodama, H., 765
Köhler, M., 437
Kopp, M.W., 193
Koul, A.K., 3, 755, 805
Kruzynski, G.E., 91

Lahrman, D.F., 255
Lai, G.Y., 585
Lang, Z., 865
Lasalmonie, A., 63
Lechervy, P., 553
Lecomte-Beckers, J., 713
Leying, X., 703
Lian, G., 865
Liburdi, J., 755
Lin, T.L., 345
Loewenkamp, S.A., 53
Loria, E.A., 203
Lowden, P., 755

Ma, P., 625
Macintyre, C.A., 121
Magnani, M., 845
Mahidhara, R.K., 23, 161
Mancuso, S.O., 377
Maurer, G.E., 377, 449
McClintock, F.A., 855
McColvin, G.M., 73
McLean, M., 387, 683
Menzies, R.G., 355
Miner, R.V., 575
Mino, K., 81
Mitchell, A., 407
Miyagawa, O., 825
Miyazaki, T., 663
Moody, N.R., 295

Morimoto, S., 325
Morinaga, M., 225
Morra, J.M., 505
Murata, Y., 225

Nagy, P., 245
Nakagawa, Y.G., 81, 215
Nakao, Y., 775
Nathal, M.V., 183
Nickel, H., 643
Nielsen, T.A., 245
Nishimoto, K., 775
Niyama, E., 325

Ohno, K., 733
Ohta, Y., 215
Okayama, A., 765
Olchini, A., 845

Painter, R.E., 417
Papp, J.F., 367
Patel, S., 397
Patnaik, P.C., 815
Paulonis, D.F., 101, 535
Penkalla, H.J., 643
Petrasek, D.W., 193
Pillhöfer, H., 835
Pollock, T.M., 285
Pope, D.P., 315
Prinz, B., 437

Qu, B., 635
Quested, P.N., 387

Radavich, J.F., 53, 459, 635
Ranke, H., 397
Restall, J.E., 495
Ruizeng, Y., 865
Rustichelli, F., 845

Samuelsson, E., 407
Schubert, F., 643
Sczerzenie, F.E., 377, 449
Selius, A.O., 43
Shah, D.M., 693
Shaw, S.W.K., 141
Shinozaki, K., 775
Shouhua, C., 865
Shunnan, Z., 703
Siereveld, P., 459
Sikka, V.K., 203
Sims, C.T., 173

Sisson, R., 255
Smith, D.F., 151
Smith, J.A., 355
Smith, J.S., 151
Stefanon, M., 845
Stephens, J.R., 183
Stewart, D.P., 545
Stumpp, H., 397
Sun, C., 345
Sundberg, D.V., 121
Suwa, M., 765

Taylor, N.G., 683
Telesman, J., 615
Terashima, H., 81
Thamburaj, R., 3, 815
Thomas, M.C., 13
Tien, J.K., 43, 111, 193
Tuler, F.R., 505

Varela, D., 553
Vasatis, I., 673

Wallace, W., 3
Wallis, R.A., 525
Walston, W.S., 295
Wanner, E.A., 161
Wegman, D.D., 427

Wells, D.A., 151
Widmer, R., 459
Willemin, P., 723
Williams, J.C., 295
Winstone, M.R., 387
Woodford, D.A., 795
Wright, J.V., 495
Wu, D.C., 605

Xie, X., 635
Xu, Z., 635

Yamagata, T., 733
Yamazaki, M., 733
Yaoxiao, Z., 703
Yasuda, K., 765
Yokokawa, T., 733
Yoshiba, M., 825
Yoshinari, A., 325
Young, J.M., 417
Yuan, Y., 625
Yukawa, N., 225
Yunrong, Z., 335, 475
Yuping, W., 335

Zhong, Z., 625
Zhuangqi, H., 703